UNITED STATES
DEPARTMENT OF AGRICULTURE

NATIONAL AGRICULTURAL STATISTICS SERVICE

AGRICULTURAL STATISTICS 2003

UNITED STATES
GOVERNMENT PRINTING OFFICE
WASHINGTON: 2003

For sale by the Superintendent of Documents, U.S. Government Printing Office
Internet: bookstore.gpo.gov Phone: toll free (866) 512-1800; DC area (202) 512-1800
Fax: (202) 512-2250 Mail: Stop SSOP, Washington, DC 20402-0001

ISBN 0-16-051387-1

Agricultural Statistics, 2003

Agricultural Statistics, 2003 was prepared under the direction of FORESTINE CHAPMAN, Agricultural Statistics Board, National Agricultural Statistics Service. ROSE M. PETRONE was responsible for coordination and technical editorial work.

The USDA and NASS invite you to explore their information on the Internet. The USDA Home Page address is **http://www.usda.gov/** and the NASS Home Page address is: **http://www.usda.gov/nass/.**

For information on NASS products you may call the **Agricultural Statistics Hotline, 1–800–727–9540 or send e-mail to nass@nass.usda.gov.**

The cooperation of the many contributors to this publication is gratefully acknowledged. Source notes below each table credit the various Government agencies which collaborated in furnishing information.

CONTENTS

Introduction

Agricultural Statistics is published each year to meet the diverse need for a reliable reference book on agricultural production, supplies, consumption, facilities, costs, and returns. Its tables of annual data cover a wide variety of facts in forms suited to most common use.

Inquiries concerning more current or more detailed data, past and prospective revisions, or the statistical methodology used should be addressed directly to the agency credited with preparing the table. Most of the data were prepared or compiled in the U.S. Department of Agriculture.

The historical series in this volume have been generally limited to data beginning with 1992 or later.

Foreign agricultural trade statistics include Government as well as non-Government shipments of merchandise from the United States and Territories to foreign countries. They do not include U.S. shipments to the U.S. Armed Forces abroad for their own use or shipments between the States and U.S. Territories. The world summaries of production and trade of major farm products are prepared by the U.S. Department of Agriculture from reports of the U.S. Department of Commerce, official statistics of foreign governments, other foreign source materials, reports of U.S. Agricultural Attachés and Foreign Service Officers, and the result of office research.

Statistics presented in many of the tables represent actual counts of the items covered. Most of the statistics relating to foreign trade and to Government programs, such as numbers and amounts of loans made to farmers, and amounts of loans made by the Commodity Credit Corporation, etc., are data of this type. A large number of other tables, however, contain data that are estimates made by the Department of Agriculture.

The estimates for crops, livestock, and poultry made by the U.S. Department of Agriculture are prepared mainly to give timely current State and national totals and averages. They are based on data obtained by sample surveys of farmers and of people who do business with farmers. The survey data are supplemented by information from the Censuses of Agriculture taken every five years and check data from various sources. Being estimates, they are subject to revision as more data become available from commerical or Government sources. Unless otherwise indicated, the totals for the United States shown in the various tables on area, production, numbers, price, value, supplies, and disposition are based on official Department estimates. They exclude States for which no official estimates are compiled.

DEFINITIONS

''Value of production'' as applied to crops in the various tables, is derived by multiplying production by the estimated season average price received by farmers for that portion of the commodity actually sold. In the case of fruits and vegetables, quantities not harvested because of low prices or other economic factors are not included in value of production. The word ''Value'' is used in the inventory tables on livestock and poultry to mean value of the number of head on the inventory date. It is derived by multiplying the number of head by an estimated value per head as of the date.

The word ''Year'' (alone) in a column heading means calendar year unless otherwise indicated. ''Ton'' when used in this book without qualifications means a short ton of 2,000 pounds.

WEIGHTS, MEASURES, AND CONVERSION FACTORS

The following table on weights, measures, and conversion factors covers the most important agricultural products, or the products for which such information is most frequently asked of the U.S. Department of Agriculture. It does not cover all farm products nor all containers for any one product.

The information has been assembled from State schedules of legal weights, various sources within the U.S. Department of Agriculture, and other Government agencies. For most products, particularly fruits and vegetables, there is a considerable variation in weight per unit of volume due to differences in variety or size of commodity, condition and tightness of pack, degree to which the container is heaped, etc. Effort has been made to select the most representative and fairest average for each product. For those commodities which develop considerable shrinkage, the point of origin weight or weight at harvest has been used.

The approximate or average weights as given in this table do not necessarily have official standing as a basis for packing or as grounds for settling disputes. Not all of them are recognized as legal weight. The table was prepared chiefly for use of workers in the U.S. Department of Agriculture who have need of conversion factors in statistical computations.

WEIGHTS, MEASURES, AND CONVERSION FACTORS
(See explanatory text just preceding this table)

WEIGHTS AND MEASURES

Commodity	Unit[1]	Approximate net weight	
		U.S.	Metric
		Pounds	*Kilograms*
Alfalfa seed	Bushel	60	27.2
Apples	.....do	48	21.8
Do	Loose pack	38–42	17.2–19.1
Do	Tray pack	40–45	18.1–20.4
Do	Cell pack	37–41	16.8–18.6
Apricots	Lug (brentwood)[2]	24	10.9
Western	4–basket crate[3]	26	11.8
Artichokes:			
Globe	Ctn, by count and loose pack	20–25	9.1–11.3
Jerusalem	Bushel	50	22.7
Asparagus	Crate (NJ)	30	13.6
Avocados	Lug[4]	12–15	5.4–6.8
Bananas	Fiber folding box[5]	40	18.1
Barley	Bushel	48	21.8
Beans:			
Lima, dry	.....do	56	25.4
Other, dry	.....do	60	27.2
	Sack	100	45.4
Lima unshelled	Bushel	28–32	12.7–14.5
Snap	.....do	28–32	12.7–14.5
Beets:			
Topped	Sack	25	11.3
Bunched	½ crate 2 dz-bchs	36–40	16.3–18.1
Berries frozen pack:			
Without sugar	50–gal. barrel	380	172
3 + 1 pack	.....do	425	193
2 + 1 pack	.....do	450	204
Blackberries	12, ½-pint basket	6	2.7
Bluegrass seed	Bushel	14–30	6.4–13.6
Broccoli	Wirebound crate	20–25	9.1–11.3
Broomcorn (6 bales per ton)	Bale	333	151
Broomcorn seed	Bushel	44–50	20.0–22.7
Brussels sprouts	Ctn, loose pack	25	11.3
Buckwheat	Bushel	48	21.8
Butter	Block	55,68	25,30.9
Cabbage	Open mesh bag	50	22.7
Do	Flat crate (1¾ bu)	50–60	22.7–27.2
Do	Ctn, place pack	53	24.0
Cantaloups	Crate[6]	40	18.1
Carrots	Film plastic Bags, mesh sacks & cartons holding 48 1 lb. film bags	55	24.9
Without tops	Burlap sack	74–80	33.6–36.3
Castor beans	Bushel	41	18.6
Castor oil	Gallon	[7]8	3.6
Cauliflower	W.G.A. crate	50–60	22.7–27.2
Do	Fiberboard box wrapper leaves removed film-wrapped, 2 layers	23–35	10.4–15.9
Celery	Crate[8]	60	27.2
Cherries	Lug (Campbell)[9]	16	7.3
Do	Lug	20	9.1
Clover seed	Bushel	60	27.2
Coffee	Bag	132.3	60
Corn:			
Ear, husked	Bushel	[10]70	31.8
Shelled	......do	56	25.4
Meal	......do	50	22.7
Oil	Gallon	[7]7.7	3.5
Syrup	.....do	11.72	5.3
Sweet	Wirebound crate	50	22.7
Do	Ctn, packed 5 oz. ears	50	22.7
Do	WDB crate, 4½–5 oz. (from FL & NJ)	42	19.1
Cotton	Bale, gross	[11]500	227
Do	Bale, net	[11]480	218
Cottonseed	Bushel	[12]32	14.5
Cottonseed oil	Gallon	[7]7.7	3.5
Cowpeas	Bushel	60	27.2
Cranberries	Barrel	100	45.4
Do	¼–bbl. box[13]	25	11.3
Cream, 40–percent butterfat	Gallon	8.38	3.80
Cucumbers	Bushel	48	21.8
Dewberries	24–qt. crate	36	16.3
Eggplant	Bushel	33	15.0
Eggs, average size	Case, 30 dozen	47.0	21.3
Escarole	Bushel	25	11.3
Figs, fresh	Box single layer[14]	6	2.7
Flaxseed	Bushel	56	25.4
Flour, various	Bag	100	45.4
Do	Ctn or Crate, Bulk	30	13.6
Garlic	Ctn of 12 tubes or 12 film bag pkgs 12 cloves each	10	4.5
Grapefruit:			
Florida and Texas	½–box mesh bag	40	18.1
Florida	1⅗ bu. box	85	38.6
Texas	1⅖ bu. box	80	36.3
California and Arizona	Box[15]	[16]67	30.4
Grapes:			
Eastern	12–qt. basket	20	9.1
Western	Lug	28	12.7
Do	4–basket crate[17]	20	9.1
Hempseed	Bushel	44	20.0
Hickory nuts	.....do	50	22.7
Honey	Gallon	11.84	5.4
Honeydew melons	⅔ Ctn	28–32	12.7–14.5
Hops	Bale, gross	200	90.7

See footnotes on page ix.

WEIGHTS AND MEASURES—Continued

Commodity	Unit [1]	Approximate net weight	
		U.S.	Metric
		Pounds	*Kilograms*
Horseradish roots	Bushel	35	15.9
Do	Sack	50	22.7
Hungarian millet seed	Bushel	48–50	21.8–22.7
Kale	Ctn or crate	25	11.3
Kapok seed	......do	35–40	15.9–18.1
Lard	Tierce	375	170
Lemons:			
California and Arizona	Box [18]	76	34.5
Do	Carton	38	17.2
Lentils	Bushel	60	27.2
Lettuce, iceberg	Iceberg, carton packed 24	43–52	19.5–23.6
Lettuce, hot-house	24-qt. basket	10	4.5
Limes (Florida)	Box	88	39.9
Linseed oil	Gallon	[7]7.7	3.5
Malt	Bushel	34	15.4
Maple syrup	Gallon	11.02	5.0
Meadow fescue seed	Bushel	24	10.9
Milk	Gallon	8.6	3.9
Millet	Bushel	48–60	21.8–27.2
Molasses:			
edible	Gallon	11.74	5.3
inedible	......do	11.74	5.3
Mustard seed	Bushel	58–60	26.3–27.2
Oats	......do	32	14.5
Olives	Lug	25–30	11.3–13.6
Olive oil	Gallon	[7]7.6	3.4
Onions, dry	Sack	50	22.7
Onions, green bunched	Ctn, 24-dz bchs	10–16	4.5–7.3
Oranges:			
Florida	Box	90	40.8
Texas	Box	85	38.5
California and Arizona	Box [15]	75	34.0
Do	Carton	38	17.2
Orchardgrass seed	Bushel	14	6.4
Palm oil	Gallon	[7]7.7	3.5
Parsnips	Bushel	50	22.7
Peaches	......do	48	21.8
Do	2 layer ctn or lug	22	10.0
Do	¾-Bu, Ctn/crate	38	17.2
Peanut oil	Gallon	[7]7.7	3.5
Peanuts, unshelled:			
Virginia type	Bushel	17	7.7
Runners, South-eastern	......do	21	9.5
Spanish:			
Southeastern	......do	25	11.3
Southwestern	......do	25	11.3
Pears:			
California	Bushel	48	21.8
Other	......do	50	22.7
Do	Std box, 4/5 bu	45–48	20.4–21.8
Do	Ctn, Tight-fill pack	36–37	16.3–16.7
Peas:			
Green, unshelled	Bushel	28–30	12.7–13.6
Dry	......do	60	27.2
Peppers, green	......do	25–30	11.3–13.6
Do	1½ bu carton	28	12.7
Perilla seed	Bushel	37–40	16.8–18.1
Pineapples	Carton	40	18.1
Plums and prunes:	Ctn & lugs	28	12.7
Do	½-bu. basket	30	13.6
Popcorn:			
On ear	Bushel	[10]70	31.8
Shelled	......do	56	25.4
Poppy seed	......do	46	20.9
Potatoes	Bushel	60	27.2
Do	Barrel	165	74.8
Do	Box	50	22.7
Do	......do	100	45.4
Quinces	Bushel	48	21.8
Rapeseed	......do	50–60	22.7–27.2
Raspberries	½-pint baskets	6	2.7
Redtop seed	Bushel	50–60	22.7–27.2
Refiners' syrup	Gallon	11.45	5.2
Rice:			
Rough	Bushel	45	20.4
Do	Bag	100	45.4
Do	Barrel	162	73.5
Milled	Pocket or bag	100	45.4
Rosin	Drum, net	520	236
Rutabagas	Bushel	56	25.4
Rye	......do	56	25.4
Sesame seed	......do	46	20.9
Shallots	Crate (4–7 doz. bunches)	20–35	9.1–15.9
Sorgo:			
Seed	Bushel	50	22.7
Syrup	Gallon	11.55	5.2
Sorghum grain [19]	Bushel	56	25.4
Soybeans	......do	60	27.2
Soybean oil	Gallon	[7]7.7	3.5
Spelt	Bushel	40	18.1
Spinach	......do	18–20	8.2–9.1
Strawberries	24-qt. crate	36	16.3
Do	12-pt. crate	9–11	4.1–5.0
Sudangrass seed	Bushel	40	18.1
Sugarcane:			
Syrup (sulfured or un-sulfured)	Gallon	11.45	5.2
Sunflower seed	Bushel	24–32	10.9–14.5
Sweetpotatoes	......do	[20]55	24.9
Do	Crate	50	22.7
Tangerines:			
Florida	Box	95	43.1
Arizona	Box	75	34.0
California	Box	75	34.0

See footnotes on page ix.

WEIGHTS AND MEASURES—Continued

Commodity	Unit [1]	Approximate net weight	
		U.S.	Metric
		Pounds	*Kilograms*
Timothy seed	Bushel	45	20.4
Tobacco:			
Maryland	Hogshead	775	352
Flue-cured	.....do	950	431
Burley	.....do	975	442
Dark air-cured	.....do	1,150	522
Virginia fire-cured	.....do	1,350	612
Kentucky and Tennessee fire-cured	.....do	1,500	680
Cigar-leaf	Case	250–365	113–166
Do	Bale	150–175	68.0–79.4
Tomatoes	Crate	60	27.2
Do	Lug box	32	14.5
Do	2-layer flat	21	9.5
Tomatoes, hot-house	12-qt. basket	20	9.1
Tung oil	Gallon	[7] 7.8	3.5
Turnips:			
Without tops	Mesh sack	50	22.7
Bunched	Crate [6]	70–80	31.8–36.3
Turpentine	Gallon	7.23	3.3
Velvetbeans (hulled)	Bushel	60	27.2
Vetch seed	.....do	60	27.2
Walnuts	Sacks	50	22.7
Water 60° F	Gallon	8.33	3.8
Watermelons	Melons of average or medium size	25	11.3
Wheat	Bushel	60	27.2
Various commodities	Short ton	2,000	907
Do	Long ton	2,240	1,016
Do	Metric ton	2,204.6	1,000

See footnotes on page ix.

To Convert From Avoirdupois Pounds

To	Multiply by
Kilograms	0.45359237
Metric tons	0.00045359237

Conversion Factors

1 Metric ton=2,204.622 pounds
1 Kilogram=2.2046 pounds
1 Acre=0.4047 hectares
1 Hectare=2.47 acres
1 Square mile=640 acres=259 hectares
1 Gallon=3.7853 liters

CONVERSION FACTORS

Commodity	Unit	Approximate equivalent
Apples	1 pound dried	7 pounds fresh; beginning 1943, 8 pounds fresh
Do	1 pound chops	5 pounds fresh
Do	1 case canned [21]	1.4 bushels fresh
Applesauce	......do [21]	1.2 bushels fresh
Apricots	1 pound dried	6 pounds fresh
Barley flour	100 pounds	4.59 bushels barley
Beans, lima	1 pound shelled	2 pounds unshelled
Beans, snap or wax	1 case canned [22]	0.008 ton fresh
Buckwheat flour	100 pounds	3.47 bushels buckwheat
Calves	1 pound live weight	0.611 pound dressed weight (1999 average)
Cattle	......do	0.607 pound dressed weight (1999 average)
Cane syrup	1 gallon	5 pounds sugar
Cherries, tart	1 case canned [21]	0.023 ton fresh
Chickens	1 pound live weight	0.72 pound ready-to-cook weight
Corn, shelled	1 bushel (56 lbs.)	2 bushels (70 pounds) of husked ear corn
Corn, sweet	1 case canned [22]	0.030 ton fresh
Cornmeal:		
Degermed	100 pounds	3.16 bushels corn, beginning 1946
Nondegermed	......do	2 bushels corn, beginning 1946
Cotton	1 pound ginned	3.26 pounds seed cotton, including trash [23]
Cottonseed meal	1 pound	2.10 pounds cottonseed
Cottonseed oil	......do	5.88 pounds cottonseed
Dairy products:		
Butter	......do	21.1 pounds milk
Cheese	......do	10 pounds milk
Condensed milk, whole	......do	2.3 pounds milk
Dry cream	......do	19 pounds milk
Dry milk, whole	......do	7.6 pounds milk
Evaporated milk, whole	......do	2.14 pounds milk
Malted milk	......do	2.6 pounds milk
Nonfat dry milk	......do	11 pounds liquid skim milk
Ice cream [24]	1 gallon	15 pounds milk
Ice cream [24] (eliminating fat from butter and concentrated milk).	......do	12 pounds milk
Eggs	1 case	47 pounds
Eggs, shell	......do	41.2 pounds frozen or liquid whole eggs
Do	......do	10.3 pounds dried whole eggs
Figs	1 pound dried	3 pounds fresh in California; 4 pounds fresh elsewhere
Flaxseed	1 bushel	About 2½ gallons oil
Grapefruit, Florida	1 case canned juice [22]	0.64 box fresh fruit
Hogs	1 pound live weight	0.737 pound dressed weight, excluding lard (1999 average)
Linseed meal	1 pound	1.51 pounds flaxseed
Linseed oil	......do	2.77 pounds flaxseed
Malt	1 bushel (34 lbs.)	1 bushel barley (48 lbs.)
Maple syrup	1 gallon	8 pounds maple sugar
Nuts:		
Almonds, imported	1 pound shelled	3½ pounds unshelled
Almonds, California	......do	2.22 pounds unshelled through 1949; 2 pounds thereafter
Brazil	......do	2 pounds unshelled
Cashews	......do	4.55 pounds unshelled
Chestnuts	......do	1.19 pounds unshelled
Filberts	......do	2.22 pounds unshelled through 1949; 2.5 pounds thereafter
Pecans:		
Seedling	......do	2.78 pounds unshelled
Improved	......do	2.50 pounds unshelled
Pignolias	......do	1.3 pounds unshelled
Pistachios	......do	2 pounds unshelled
Walnuts:		
Black	......do	5.88 pounds unshelled
Persian (English)	......do	2.67 pounds unshelled
Oatmeal	100 pounds	7.6 bushels oats, beginning 1943
Oranges, Florida	1 case canned juice [22]	0.53 box fresh
Peaches, California, freestone	1 pound dried	5⅓ pounds fresh through 1918; 6 pounds fresh for 1919–28; and 6½ pounds fresh from 1929 to date
Peaches, California, clingstone	......do	7½ pounds fresh
Peaches, clingstone	1 case canned [21]	1 bushel fresh
Do	......do	0.0230 ton fresh
Peanuts	1 pound shelled	1½ pounds unshelled
Pears	1 pound dried	6½ pounds fresh
Pears, Bartlett	1 case canned [22]	1.1 bushels fresh
Do	......do	0.026 ton fresh

See footnotes on page ix.

CONVERSION FACTORS—Continued

Commodity	Unit	Approximate equivalent
Peas, green	1 pound shelled	2½ pounds unshelled
Do	1 case canned[22]	0.009 ton fresh (shelled)
Prunes	1 pound dried	2.7 pounds fresh in California; 3 to 4 pounds fresh elsewhere
Raisins	1 pound	4.3 pounds fresh grapes
Rice, milled (excluding brewers)	100 pounds	152 pounds rough or unhulled rice
Rye flour	......do	2.23 bushels rye, beginning 1947
Sheep and lambs	1 pound live weight	0.504 pound dressed weight (1999 average)
Soybean meal	1 pound	1.27 pounds soybeans
Soybean oil	......do	5.49 pounds soybeans
Sugar	1 ton raw	0.9346 ton refined
Tobacco	1 pound farm-sales weight	Various weights of stemmed and unstemmed, according to aging and the type of tobacco. (See circular 435, U.S. Dept. of Agr.)
Tomatoes	1 case canned[22]	0.018 ton fresh
Turkeys	1 pound live weight	0.80 pound ready-to-cook weight
Wheat flour	100 pounds	2.30 bushels wheat[25]
Wool, domestic apparel shorn	1 pound greasy	0.48 pounds scoured
Wool, domestic apparel pulled	......do	0.73 pound scoured

[1] Standard bushel used in the United States contains 2,150.42 cubic inches; the gallon, 231 cubic inches; the cranberry barrel, 5,826 cubic inches; and the standard fruit and vegetable barrel, 7,056 cubic inches. Such large-sized products as apples and potatoes sometimes are sold on the basis of a heaped bushel, which would exceed somewhat the 2,150.42 cubic inches of a bushel basket level full. This also applies to such products as sweetpotatoes, peaches, green beans, green peas, spinach, etc.

[2] Approximate inside dimensions, 4⅝ by 12½ by 16⅛ inches.

[3] Approximate inside dimensions, 4½ by 16 by 16⅛ inches.

[4] Approximate dimensions, 4½ by 13½ by 16⅛ inches.

[5] Approximate inside dimensions, 13 by 12 by 32 inches.

[6] Approximate inside dimensions, 13 by 18 by 21⅝ inches.

[7] This is the weight commonly used in trade practices, the actual weight varying according to temperature conditions.

[8] Approximate inside dimensions, 9¾ by 16 by 20 inches.

[9] Approximate inside dimensions, 4⅛ by 11½ by 14 inches.

[10] The standard weight of 70 pounds is usually recognized as being about 2 measured bushels of corn, husked, on the ear, because it required 70 pounds to yield 1 bushel, or 56 pounds, of shelled corn.

[11] For statistical purposes the bale of cotton is 500 pounds or 480 pounds net weight. Prior to Aug. 1, 1946, the net weight was estimated at 478 pounds. Actual bale weights vary considerably, and the customary average weights of bales of foreign cotton differ from that of the American square bale.

[12] This is the average weight of cottonseed, although the legal weight in some States varies from this figure of 32 pounds.

[13] Approximate inside dimensions, 9¼ by 10½ by 15 inches.

[14] Approximate inside dimensions, 1¾ by 11 by 16⅛ inches.

[15] Approximate inside dimensions, 11½ by 11½ by 24 inches.

[16] Beginning with the 1993-94 season, net weights for California Desert Valley and Arizona grapefruit were increased from 64 to 67 pounds, equal to the California other area net weight, making a 67 pound net weight apply to all of California.

[17] Approximate inside dimensions, 4¾ by 16 by 16⅛ inches.

[18] Approximate inside dimensions, 9⅞ by 13 by 25 inches.6 by 16 by 16⅛ inches.

[19] Includes both sorghum grain (kafir, milo, hegari, etc.) and sweet sorghum varieties.

[20] This average of 55 pounds indicates the usual weight of sweetpotatoes when harvested. Much weight is lost in curing or drying and the net weight when sold in terminal markets may be below 55 pounds.

[21] Case of 24 No. 2½ cans.

[22] Case of 24 No. 303 cans.

[23] Varies widely by method of harvesting.

[24] The milk equivalent of ice cream per gallon is 15 pounds. Reports from plants indicate about 81 percent of the butterfat in ice cream is from milk and cream, the remainder being from butter and concentrated milk. Thus the milk equivalent of the milk and cream in a gallon of ice cream is about 12 pounds.

[25] This is equivalent to 4.51 bushels of wheat per barrel (196 pounds) of flour and has been used in conversions, beginning July 1, 1957. Because of changes in milling processes, the following factors per barrel of flour have been used for earlier periods: 1790–1879, 5 bushels; 1880–1908, 4.75 bushels, 1909–17, 4.7 bushels; 1918 and 1919, 4.5 bushels; 1920, 4.6 bushels; 1921–44, 4.7 bushels; July 1944–Feb. 1946, 4.57 bushels; March 1946–Oct. 1946, average was about 4.31 bushels; and Nov. 1946–June 1957, 4.57 bushels.

CHAPTER I

STATISTICS OF GRAIN AND FEED

This chapter contains tables for wheat, rye, rice, corn, oats, barley, sorghum grain, and feedstuffs. Estimates are given of area, production, disposition, supply and disappearance, prices, value of production, stocks, foreign production and trade, price-support operations, animal units fed, and feed consumed by livestock and poultry.

Table 1-1.—Total grain: Supply and disappearance, United States, 1993–2002[1]

Year[2]	Supply				Disappearance			Ending stocks
	Beginning stocks	Production	Imports	Total	Domestic use	Exports	Total disappearance	
	Million metric tons	*Million metric tons*	*Million metric tons*	*Million metric tons*	*Million metric tons*	*Million metric tons*	*Million metric tons*	*Million metric tons*
1993	79.3	258.8	7.3	345.4	224.2	77.2	301.4	44.0
1994	44.0	355.6	6.3	405.9	246.1	99.2	345.3	60.5
1995	60.5	277.3	5.0	342.9	216.5	100.6	317.1	25.8
1996	25.8	335.5	5.9	367.2	244.5	82.4	326.9	40.3
1997	40.3	336.3	5.9	382.5	245.9	77.5	323.4	59.1
1998	59.1	349.2	6.4	414.6	248.1	88.4	336.5	78.1
1999	78.1	335.1	5.8	419.0	253.0	90.1	343.0	76.0
2000	76.0	342.5	5.7	424.1	257.0	89.3	346.3	77.8
2001[3]	77.8	324.9	6.1	408.8	255.7	85.1	340.8	68.0
2002[4]	68.0	298.6	5.3	371.8	250.5	81.2	331.7	40.2

[1] Aggregate data on corn, sorghum, barley, oats, wheat, rye, and rice. [2] The marketing year for corn and sorghum begins September 1; for oats, barley, wheat, and rye, June 1; and for rice, August 1. [3] Preliminary. [4] Projected as of February 11, 2003, World Agricultural Supply and Demand Estimates. Totals may not add due to independent rounding.

ERS, Market and Trade Economics Division, (202) 694–5296.

Table 1-2.—Wheat: Area, yield, production, and value, United States, 1993–2002

Year	Area		Yield per harvested acre	Production	Marketing year average price per bushel received by farmers[2]	Value of production[2]
	Planted[1]	Harvested				
	1,000 acres	*1,000 acres*	*Bushels*	*1,000 bushels*	*Dollars*	*1,000 dollars*
1993	72,168	62,712	38.2	2,396,440	3.26	7,647,527
1994	70,349	61,770	37.6	2,320,981	3.45	7,968,237
1995	69,031	60,955	35.8	2,182,708	4.55	9,787,766
1996	75,105	62,819	36.3	2,277,388	4.30	9,782,238
1997	70,412	62,840	39.5	2,481,466	3.38	8,286,741
1998	65,821	59,002	43.2	2,547,321	2.65	6,780,623
1999	62,714	53,823	42.7	2,299,010	2.48	5,593,989
2000	62,629	53,133	42.0	2,232,460	2.62	5,782,107
2001	59,597	48,633	40.2	1,957,043	2.78	5,440,217
2002	60,358	45,817	35.3	1,616,441	3.60	5,863,378

[1] Includes area seeded in preceding fall for winter wheat. [2] Includes allowance for loans outstanding and purchases by the Government valued at the average loan and purchase rate, by States, where applicable.

NASS, Crops Branch, (202) 720–2127.

Table 1-3.—Wheat, by type: Area, yield, production, and value, United States, 1993–2002

Year	Area: Planted [1]	Area: Harvested	Yield per harvested acre	Production	Marketing year average price per bushel received by farmers [2]	Value of production [2]
			Winter wheat			
	1,000 acres	*1,000 acres*	*Bushels*	*1,000 bushels*	*Dollars*	*1,000 dollars*
1993	51,587	43,811	40.2	1,760,143	3.03	5,287,607
1994	49,197	41,355	40.2	1,661,943	3.37	5,578,351
1995	48,591	40,987	37.7	1,545,303	4.41	6,720,901
1996	51,445	39,574	37.1	1,469,618	4.33	6,396,217
1997	47,985	41,340	44.6	1,845,528	3.23	5,948,655
1998	46,449	40,126	46.9	1,880,773	2.52	4,740,361
1999	43,331	35,486	47.8	1,696,580	2.29	3,870,955
2000	43,393	35,072	44.7	1,566,023	2.51	3,893,961
2001	41,078	31,295	43.5	1,361,479	2.72	3,684,817
2002	41,735	29,651	38.5	1,142,802	3.45	3,939,217
			Durum wheat			
	1,000 acres	*1,000 acres*	*Bushels*	*1,000 bushels*	*Dollars*	*1,000 dollars*
1993	2,241	2,100	33.6	70,476	4.48	324,049
1994	2,823	2,715	35.6	96,747	4.62	449,041
1995	3,436	3,356	30.5	102,280	5.65	567,541
1996	3,630	3,556	32.6	116,090	4.67	541,993
1997	3,310	3,177	27.6	87,783	4.92	422,497
1998	3,805	3,728	37.0	138,119	3.15	452,860
1999	4,035	3,569	27.8	99,322	2.73	284,677
2000	3,937	3,572	30.7	109,805	2.66	301,356
2001	2,910	2,789	30.0	83,556	3.08	269,391
2002	2,909	2,703	29.4	79,450	4.05	331,183
			Other spring wheat [3]			
	1,000 acres	*1,000 acres*	*Bushels*	*1,000 bushels*	*Dollars*	*1,000 dollars*
1993	18,340	16,801	33.7	565,821	3.58	2,035,871
1994	18,329	17,700	31.8	562,291	3.42	1,940,845
1995	17,004	16,612	32.2	535,125	4.59	2,499,324
1996	20,030	19,689	35.1	691,680	4.20	2,844,028
1997	19,117	18,323	29.9	548,155	3.53	1,915,589
1998	15,567	15,148	34.9	528,469	3.00	1,587,402
1999	15,348	14,768	34.1	503,108	2.88	1,438,357
2000	15,299	14,489	38.4	556,632	2.85	1,586,790
2001	15,609	14,549	35.2	512,008	2.90	1,486,009
2002	15,714	13,463	29.3	394,189	3.90	1,592,978

[1] Seeded in preceding fall for winter wheat. [2] Obtained by weighting State prices by quantity sold. [3] Includes small quantities of Durum wheat grown in other States.

NASS, Crops Branch, (202) 720–2127.

Table 1-4.—Wheat: Stocks on and off farms, United States, 1993–2002

Year beginning September	All wheat							
	On farms				Off farms [1]			
	Sept. 1	Dec. 1	Mar. 1	Jun. 1	Sept. 1	Dec. 1	Mar. 1	Jun. 1
	1,000 bushels	*1,000* bushels	*1,000* bushels	*1,000* bushels	*1,000* bushels	*1,000* bushels	*1,000* bushels	*1,000* bushels
1993	987,000	653,100	363,200	175,255	1,145,607	932,634	664,787	393,229
1994	859,800	570,550	335,250	163,410	1,209,694	920,554	633,949	343,175
1995	743,600	477,000	220,550	74,610	1,137,499	861,267	602,914	301,410
1996	824,500	584,150	320,750	154,560	899,696	634,660	501,069	289,047
1997	794,350	604,000	399,920	224,210	1,281,998	1,015,242	766,644	498,268
1998	885,720	680,200	471,220	277,710	1,499,595	1,215,481	979,191	668,208
1999	888,060	647,400	424,680	226,780	1,556,983	1,236,344	991,841	722,968
2000	808,390	623,420	384,750	197,270	1,544,280	1,182,705	953,648	678,912
2001	696,850	517,890	338,500	216,830	1,458,964	1,105,565	871,268	560,282
2002	580,200	384,800	NA	NA	1,170,787	936,199	NA	NA

Year beginning September	Durum wheat [2]							
	On farms				Off farms [1]			
	Sept. 1	Dec. 1	Mar. 1	Jun. 1	Sept. 1	Dec. 1	Mar. 1	Jun. 1
	1,000 bushels	*1,000* bushels	*1,000* bushels	*1,000* bushels	*1,000* bushels	*1,000* bushels	*1,000* bushels	*1,000* bushels
1993	83,950	50,500	21,645	11,022	20,918	26,398	19,853	16,836
1994	74,220	44,428	26,060	12,240	17,526	21,113	16,801	13,752
1995	72,560	47,220	27,250	12,470	24,655	23,886	21,311	12,931
1996	79,700	66,100	33,100	17,800	22,410	19,541	21,855	12,938
1997	51,000	37,000	22,000	13,380	36,712	30,280	20,473	12,448
1998	88,000	75,300	58,200	37,500	37,908	33,300	30,372	17,302
1999	96,900	74,500	51,700	30,300	39,830	35,449	29,617	19,532
2000	85,700	72,000	44,200	29,100	37,573	32,306	28,616	16,073
2001	63,300	49,600	30,200	20,600	33,841	26,997	21,690	12,390
2002	66,000	49,500	NA	NA	26,854	26,917	NA	NA

[1] Includes stocks at mills, elevators, warehouses, terminals, and processors. [2] Included in all wheat. NA-not available.

NASS, Crops Branch, (202) 720–2127.

Table 1-5.—Wheat: Supply and disappearance, by class, United States, 1998–2002 [1]

Item	Year beginning June				
	1998	1999	2000	2001	2002
	Million bushels	*Million bushels*	*Million bushels*	*Million bushels*	*Million bushels*
All wheat:					
Stocks, June 1	722	946	950	876	777
Production	2,547	2,299	2,232	1,957	1,616
Supply [2]	3,373	3,339	3,272	2,941	2,469
Exports [3]	1,046	1,089	1,062	961	925
Domestic disappearance	1,381	1,301	1,334	1,202	1,126
Stocks, May 31	946	950	876	777	418
Hard red winter:					
Stocks, June 1	307	435	458	411	363
Production	1,179	1,051	846	767	609
Supply [2]	1,487	1,486	1,304	1,179	973
Exports [3]	444	477	393	349	355
Domestic disappearance	608	551	500	467	445
Stocks, May 31	435	458	411	363	173
Soft red winter:					
Stocks, June 1	80	136	133	135	78
Production	443	454	471	400	332
Supply [2]	523	590	604	535	417
Exports [3]	100	175	180	199	110
Domestic disappearance	287	282	289	257	247
Stocks, May 31	136	133	135	78	60
Hard red spring:					
Stocks, June 1	220	233	218	210	230
Production	486	448	502	476	357
Supply [2]	765	741	779	750	612
Exports [3]	243	224	227	216	275
Domestic disappearance	289	298	342	303	247
Stocks, May 31	233	218	210	230	90
Durum:					
Stocks, June 1	26	55	50	45	33
Production	138	99	110	84	79
Supply [2]	197	182	185	163	147
Exports [3]	41	51	56	50	35
Domestic disappearance	101	81	85	80	86
Stocks, May 31	55	50	45	33	27
White:					
Stocks, June 1	90	87	91	75	73
Production	301	247	303	232	239
Supply [2]	401	340	399	315	319
Exports [3]	217	161	206	147	150
Domestic disappearance	97	88	118	95	101
Stocks, May 31	87	91	75	73	68

[1] Data except production are approximations. [2] Total supply includes imports. [3] Imports and exports include flour and products in wheat equivalent.

ERS, Market and Trade Economics Division, (202) 694–5285.

Table 1-6.—Wheat: Area, yield, and production, by States, 2000–2002

State	Area planted[1]			Area harvested			Yield per harvested acre			Production		
	2000	2001	2002	2000	2001	2002	2000	2001	2002	2000	2001	2002
	1,000 acres	*1,000 acres*	*1,000 acres*	*1,000 acres*	*1,000 acres*	*1,000 acres*	*Bushels*	*Bushels*	*Bushels*	*1,000 bushels*	*1,000 bushels*	*1,000 bushels*
AL	140	170	150	90	70	60	54.0	48.0	40.0	4,860	3,360	2,400
AZ	92	94	99	92	93	99	95.4	91.6	95.5	8,775	8,517	9,455
AR	1,180	1,100	960	1,100	970	840	54.0	52.0	46.0	59,400	50,440	38,640
CA	635	615	625	487	461	390	76.0	76.1	80.8	37,000	35,105	31,500
CO	2,548	2,397	2,375	2,396	2,044	1,674	29.8	33.8	23.1	71,370	69,168	38,700
DE	65	60	60	63	57	58	66.0	61.0	70.0	4,158	3,477	4,060
FL	13	10	9	9	9	7	49.0	41.0	43.0	441	369	301
GA	300	300	350	200	200	200	54.0	53.0	41.0	10,800	10,600	8,200
ID	1,370	1,280	1,260	1,300	1,200	1,200	83.4	71.0	73.1	108,450	85,150	87,660
IL	950	750	680	920	720	650	57.0	61.0	49.0	52,440	43,920	31,850
IN	550	400	350	510	380	330	69.0	66.0	53.0	35,190	25,080	17,490
IA	20	25	20	18	18	16	47.0	54.0	50.0	846	972	800
KS	9,800	9,800	9,600	9,400	8,200	8,100	37.0	40.0	33.0	347,800	328,000	267,300
KY	670	550	550	420	360	340	57.0	66.0	53.0	23,940	23,760	18,020
LA	200	175	230	185	160	220	53.0	50.0	40.0	9,805	8,000	8,800
MD	220	190	195	200	175	180	63.0	63.0	66.0	12,600	11,025	11,880
MI	530	570	500	500	560	490	72.0	64.0	67.0	36,000	35,840	32,830
MN	2,022	1,867	2,040	1,971	1,815	1,834	49.0	43.9	33.9	96,526	79,655	62,240
MS	250	250	250	235	225	205	55.0	52.0	44.0	12,925	11,700	9,020
MO	1,050	900	900	950	760	760	52.0	54.0	45.0	49,400	41,040	34,200
MT	5,330	5,360	5,790	4,920	4,215	4,765	27.5	22.9	23.1	135,210	96,570	109,895
NE	1,750	1,750	1,650	1,650	1,600	1,520	36.0	37.0	32.0	59,400	59,200	48,640
NV	18	15	13	15	3	5	98.0	90.0	81.0	1,470	270	405
NJ	40	31	38	35	27	32	57.0	45.0	58.0	1,995	1,215	1,856
NM	470	500	520	175	240	170	24.0	34.0	22.0	4,200	8,160	3,740
NY	150	125	130	140	120	128	53.0	53.0	58.0	7,420	6,360	7,424
NC	720	680	650	550	470	480	50.0	39.0	42.0	27,500	18,330	20,160
ND	10,170	9,450	9,080	9,413	9,080	7,920	33.7	32.2	27.3	316,985	292,400	216,610
OH	1,120	950	860	1,110	900	810	72.0	67.0	62.0	79,920	60,300	50,220
OK	6,100	5,600	6,000	4,200	3,700	3,500	34.0	33.0	28.0	142,800	122,100	98,000
OR	935	910	950	910	855	850	58.8	38.2	40.0	53,540	32,650	34,010
PA	200	170	190	195	160	185	53.0	52.0	54.0	10,335	8,320	9,990
SC	200	220	210	195	210	190	49.0	43.0	37.0	9,555	9,030	7,030
SD	3,020	3,025	3,030	2,878	2,044	1,630	39.7	37.6	25.9	114,268	76,766	42,235
TN	550	500	470	380	340	300	55.0	54.0	46.0	20,900	18,360	13,800
TX	6,000	5,600	6,400	2,200	3,200	2,700	30.0	34.0	29.0	66,000	108,800	78,300
UT	173	160	155	166	141	136	41.3	42.8	36.0	6,850	6,034	4,892
VA	240	200	230	205	170	170	63.0	60.0	63.0	12,915	10,200	10,710
WA	2,475	2,490	2,420	2,420	2,380	2,365	68.1	55.7	54.8	164,880	132,580	129,695
WV	13	12	12	9	8	7	61.0	58.0	48.0	549	464	336
WI	149	178	198	143	167	177	61.0	64.1	60.9	8,730	10,708	10,771
WY	201	168	159	178	126	124	24.2	24.2	19.2	4,312	3,048	2,376
US	62,629	59,597	60,358	53,133	48,633	45,817	42.0	40.2	35.3	2,232,460	1,957,043	1,616,441

[1] Includes area planted preceding fall.

NASS, Crops Branch, (202) 720–2127.

Table 1-7.—Wheat: Supply and disappearance, United States, 1993–2002

Year beginning June	Supply				Disappearance						Ending stocks May 31
					Domestic use				Exports[1]	Total disappearance	
	Beginning stocks	Production	Imports[1]	Total	Food	Seed	Feed[2]	Total			
	Million bushels	*Million bushels*	*Million bushels*	*Million bushels*	*Million bushels*	*Million bushels*	*Million bushels*	*Million bushels*	*Million bushels*	*Million bushels*	*Million bushels*
1993	531	2,396	109	3,036	872	96	272	1,240	1,228	2,467	568
1994	568	2,321	92	2,981	853	89	344	1,287	1,188	2,475	507
1995	507	2,183	68	2,757	883	103	154	1,140	1,241	2,381	376
1996	376	2,277	92	2,746	891	102	308	1,301	1,002	2,302	444
1997	444	2,481	95	3,020	914	92	251	1,257	1,040	2,298	722
1998	722	2,547	103	3,373	909	81	391	1,381	1,046	2,427	946
1999	946	2,299	95	3,339	921	92	288	1,301	1,089	2,390	950
2000	950	2,223	90	3,272	950	80	304	1,334	1,062	2,396	876
2001	876	1,958	108	2,941	926	82	194	1,202	961	2,164	777
2002[3]	777	1,616	75	2,469	940	86	100	1,126	925	2,051	418

[1] Imports and exports include flour and other products expressed in wheat equivalent. [2] Approximates feed and residual use and includes negligible quantities used for distilled spirits. [3] Preliminary. Totals may not add due to independent rounding.

ERS, Market and Trade Economics Division, (202) 694–5296.

Table 1-8.—Wheat, by type: Area, yield, and production, by States, 2000–2002

State	Area planted [1]			Area harvested			Yield per harvested acre			Production		
	2000	2001	2002	2000	2001	2002	2000	2001	2002	2000	2001	2002
	Winter wheat											
	1,000 acres	*1,000 acres*	*1,000 acres*	*1,000 acres*	*1,000 acres*	*1,000 acres*	*Bushels*	*Bushels*	*Bushels*	*1,000 bushels*	*1,000 bushels*	*1,000 bushels*
AL	140	170	150	90	70	60	54.0	48.0	40.0	4,860	3,360	2,400
AZ	7	6	10	7	6	10	100.0	100.0	100.0	700	600	1,000
AR	1,180	1,100	960	1,100	970	840	54.0	52.0	46.0	59,400	50,440	38,640
CA	535	530	530	390	380	300	70.0	70.0	75.0	27,300	26,600	22,500
CO	2,500	2,350	2,350	2,350	2,000	1,650	29.0	33.0	22.0	68,150	66,000	36,300
DE	65	60	60	63	57	58	66.0	61.0	70.0	4,158	3,477	4,060
FL	13	10	9	9	9	7	49.0	41.0	43.0	441	369	301
GA	300	300	350	200	200	200	54.0	53.0	41.0	10,800	10,600	8,200
ID	780	760	730	730	710	690	90.0	73.0	79.0	65,700	51,830	54,510
IL	950	750	680	920	720	650	57.0	61.0	49.0	52,440	43,920	31,850
IN	550	400	350	510	380	330	69.0	66.0	53.0	35,190	25,080	17,490
IA	20	25	20	18	18	16	47.0	54.0	50.0	846	972	800
KS	9,800	9,800	9,600	9,400	8,200	8,100	37.0	40.0	33.0	347,800	328,000	267,300
KY	670	550	550	420	360	340	57.0	66.0	53.0	23,940	23,760	18,020
LA	200	175	230	185	160	220	53.0	50.0	40.0	9,805	8,000	8,800
MD	220	190	195	200	175	180	63.0	63.0	66.0	12,600	11,025	11,880
MI	530	570	500	500	560	490	72.0	64.0	67.0	36,000	35,840	32,830
MN	20	15	35	19	13	30	46.0	29.0	30.0	874	377	900
MS	250	250	250	235	225	205	55.0	52.0	44.0	12,925	11,700	9,020
MO	1,050	900	900	950	760	760	52.0	54.0	45.0	49,400	41,040	34,200
MT	1,500	1,300	1,450	1,350	870	750	33.0	22.0	28.0	44,550	19,140	21,000
NE	1,750	1,750	1,650	1,650	1,600	1,520	36.0	37.0	32.0	59,400	59,200	48,640
NV	10	9	6	9	2	3	100.0	95.0	85.0	900	190	255
NJ	40	31	38	35	27	32	57.0	45.0	58.0	1,995	1,215	1,856
NM	470	500	520	175	240	170	24.0	34.0	22.0	4,200	8,160	3,740
NY	150	125	130	140	120	128	53.0	53.0	58.0	7,420	6,360	7,424
NC	720	680	650	550	470	480	50.0	39.0	42.0	27,500	18,330	20,160
ND	120	150	80	113	80	70	45.0	40.0	38.0	5,085	3,200	2,660
OH	1,120	950	860	1,110	900	810	72.0	67.0	62.0	79,920	60,300	50,220
OK	6,100	5,600	6,000	4,200	3,700	3,500	34.0	33.0	28.0	142,800	122,100	98,000
OR	750	750	800	730	700	710	62.0	40.0	41.0	45,260	28,000	29,110
PA	200	170	190	195	160	185	53.0	52.0	54.0	10,335	8,320	9,990
SC	200	220	210	195	210	190	49.0	43.0	37.0	9,555	9,030	7,030
SD	1,350	1,300	1,300	1,280	370	625	42.0	32.0	29.0	53,760	11,840	18,125
TN	550	500	470	380	340	300	55.0	54.0	46.0	20,900	18,360	13,800
TX	6,000	5,600	6,400	2,200	3,200	2,700	30.0	34.0	29.0	66,000	108,800	78,300
UT	150	140	140	145	125	125	40.0	42.0	35.0	5,800	5,250	4,375
VA	240	200	230	205	170	170	63.0	60.0	63.0	12,915	10,200	10,710
WA	1,850	1,850	1,800	1,800	1,750	1,750	73.0	61.0	59.0	131,400	106,750	103,250
WV	13	12	12	9	8	7	61.0	58.0	48.0	549	464	336
WI	140	170	190	135	160	170	62.0	65.0	62.0	8,370	10,400	10,540
WY	190	160	150	170	120	120	24.0	24.0	19.0	4,080	2,880	2,280
US	43,393	41,078	41,735	35,072	31,295	29,651	44.7	43.5	38.5	1,566,023	1,361,479	1,142,802
	Durum wheat											
AZ	85	88	89	85	87	89	95.0	91.0	95.0	8,075	7,917	8,455
CA	100	85	95	97	81	90	100.0	105.0	100.0	9,700	8,505	9,000
MN	2	2	5	2	2	4	51.0	39.0	35.0	102	78	140
MT	480	510	590	470	495	565	28.0	24.0	23.0	13,160	11,880	12,995
ND	3,250	2,200	2,100	2,900	2,100	1,950	27.0	26.0	25.0	78,300	54,600	48,750
SD	20	25	30	18	24	5	26.0	24.0	22.0	468	576	110
US	3,937	2,910	2,909	3,572	2,789	2,703	30.7	30.0	29.4	109,805	83,556	79,450
	Other spring wheat											
CO	48	47	25	46	44	24	70.0	72.0	100.0	3,220	3,168	2,400
ID	590	520	530	570	490	510	75.0	68.0	65.0	42,750	33,320	33,150
MN	2,000	1,850	2,000	1,950	1,800	1,800	49.0	44.0	34.0	95,550	79,200	61,200
MT	3,350	3,550	3,750	3,100	2,850	3,450	25.0	23.0	22.0	77,500	65,550	75,900
NV	8	6	7	6	1	2	95.0	80.0	75.0	570	80	150
ND	6,800	7,100	6,900	6,400	6,900	5,900	36.5	34.0	28.0	233,600	234,600	165,200
OR	185	160	150	180	155	140	46.0	30.0	35.0	8,280	4,650	4,900
SD	1,650	1,700	1,700	1,580	1,650	1,000	38.0	39.0	24.0	60,040	64,350	24,000
UT	23	20	15	21	16	11	50.0	49.0	47.0	1,050	784	517
WA	625	640	620	620	630	615	54.0	41.0	43.0	33,480	25,830	26,445
WI	9	8	8	8	7	7	45.0	44.0	33.0	360	308	231
WY	11	8	9	8	6	4	29.0	28.0	24.0	232	168	96
US	15,299	15,609	15,714	14,489	14,549	13,463	38.4	35.2	29.3	556,632	512,008	394,189

[1] Includes area planted preceding fall.

NASS, Crops Branch, (202) 720–2127.

Table 1-9.—Wheat: Support operations, United States, 1993–2002

Marketing year beginning June 1	Income support payment rates per bushel [1]	Program price levels per bushel		Put under loan [4]		Acquired by CCC under loan program [5]	Owned by CCC at end of marketing year [6]
		Loan [2]	Target [3]	Quantity	Percentage of production		
	Dollars	*Dollars*	*Dollars*	*Million bushels*	*Percent*	*Million bushels*	*Million bushels*
1993/94	1.03	2.45	4.00	258	10.8	0	150
1994/95	0.61	2.58	4.00	231	10.0	0	142
1995/96	0.00	2.58	4.00	114	5.2	0	118
1996/97	0.874	2.58	NA	194	8.5	0	93
1997/98	0.631	2.58	NA	264	10.6	2	94
1998/99	0.993	2.58	NA	363	14.3	30	128
1999/2000	1.274	2.58	NA	154	6.7	13	104
2000/2001	1.225	2.58	NA	181	8.1	27	97
2001/2002	1.014	2.58	NA	197	10.1	17	99
2002/2003	0.981	2.80	3.86	....	....	....	....

[1] Payment rates for the 1995/96 and prior crop years were calculated according to the provisions of the now defunct deficiency payment/production adjustment program. Payment rates for the 1996/97 through 2001/2002 crops were calculated according to the provisions of the Production Flexibility Contract (PFC) program, following enactment of the Federal Agriculture Improvement and Reform Act of 1996 (1996 Act) and included supplemental PFC payment rates for 1998 through 2001. Payment rates for the 2002/2003 and subsequent crops are calculated according to the provisions of the Direct Payment program, following enactment of the Farm Security and Rural Investment Act of 2002 (2002 Act) and includes a PFC payment rate for 2002 only. Payment rates are rounded to the nearest tenth of a cent. [2] The national average loan rate was also known as the price support rate prior to enactment of the 1996 Act. [3] Between the 1996/97 and 2001/2002 marketing years, target prices were no longer applicable; however, with enactment of the 2002 Act, target prices were reestablished. [4] Represents loans made, purchases, and purchase agreements entered into. Purchases and purchase agreements are no longer authorized for the 1996 and subsequent crops following enactment of the 1996 Act. [5] Acquisition of all loans forfeited during the marketing year including loans made in previous year. [6] Includes 147 million bushels in Food Security Reserve for 1991/92 through 1993/94, 141 million in 1994/95, 118 million in 1995/96, and 93 million each year thereafter. (The Food Security Reserve became the Food Security Commodity Trust in July of 1999 and the Bill Emerson Humanitarian Trust in July of 2002.) NA-not applicable.

FSA, Food Grains Analysis Group, (202) 720–4146.

Table 1-10.—Wheat: Marketing year average price and value, by States, crop of 2000, 2001, and 2002

State	Marketing year average price per bushel			Value of production		
	2000	2001	2002 [1]	2000	2001	2002 [1]
	Dollars	*Dollars*	*Dollars*	*1,000 dollars*	*1,000 dollars*	*1,000 dollars*
AL	2.35	2.55	2.90	11,421	8,568	6,960
AZ	3.45	3.95	4.00	30,279	33,222	37,720
AR	2.35	2.36	2.90	139,590	119,038	112,056
CA	2.82	3.11	3.65	103,834	110,852	114,300
CO	2.70	2.72	3.85	191,991	187,852	148,875
DE	2.10	2.45	3.15	8,732	8,519	12,789
FL	2.25	2.25	2.40	992	830	722
GA	2.45	2.05	2.50	26,460	21,730	20,500
ID	2.63	3.18	3.85	281,124	269,963	339,628
IL	2.09	2.49	3.05	109,600	109,361	97,143
IN	2.11	2.41	3.20	74,251	60,443	55,968
IA	2.15	2.50	2.85	1,819	2,430	2,280
KS	2.65	2.69	3.45	921,670	882,320	922,185
KY	2.26	2.50	3.00	54,104	59,400	54,060
LA	2.40	2.75	3.00	23,532	22,000	26,400
MD	2.10	2.45	3.15	26,460	27,011	37,422
MI	2.11	2.43	3.30	75,960	87,091	108,339
MN	2.89	2.93	4.00	278,473	233,243	248,690
MS	2.54	2.45	2.70	32,830	28,665	24,354
MO	2.28	2.42	3.15	112,632	99,317	107,730
MT	3.02	3.14	4.25	409,007	304,487	466,748
NE	2.61	2.75	3.70	155,034	162,800	179,968
NV	2.55	3.15	3.55	3,773	840	1,445
NJ	2.10	2.30	3.00	4,190	2,795	5,568
NM	2.70	2.75	3.45	11,340	22,440	12,903
NY	1.94	2.64	3.25	14,395	16,790	24,128
NC	2.35	2.40	3.00	64,625	43,992	60,480
ND	2.71	2.79	4.00	857,369	812,938	864,828
OH	2.02	2.46	3.20	161,438	148,338	160,704
OK	2.57	2.74	3.50	366,996	334,554	343,000
OR	2.63	3.27	4.00	140,899	106,718	135,565
PA	2.19	2.68	3.35	22,634	22,298	33,467
SC	2.40	2.30	2.55	22,932	20,769	17,927
SD	2.68	2.78	4.05	305,577	214,059	170,162
TN	2.35	2.50	2.90	49,115	45,900	40,020
TX	2.52	2.78	3.00	166,320	302,464	234,900
UT	3.25	3.30	4.75	22,578	19,912	23,226
VA	2.00	2.29	2.80	25,830	23,358	29,988
WA	2.70	3.23	4.10	443,369	427,838	537,039
WV	1.93	2.46	3.10	1,060	1,141	1,042
WI	1.85	2.40	2.95	16,230	25,668	31,809
WY	2.70	2.70	4.35	11,642	8,263	10,340
US	2.62	2.78	3.60	5,782,107	5,440,217	5,863,378

[1] Preliminary.

NASS, Crops Branch, (202) 720–2127.

Table 1-11.—Wheat: Area, yield, and production in specified countries, 1999/2000–2001/2002 [1]

Continent and country	Area [2]			Yield per hectare			Production		
	1999/2000	2000/2001	2001/2002 [3]	1999/2000	2000/2001	2001/2002 [3]	1999/2000	2000/2001	2001/2002 [3]
	1,000 hectares	*1,000 hectares*	*1,000 hectares*	*Metric tons*	*Metric tons*	*Metric tons*	*1,000 metric tons*	*1,000 metric tons*	*1,000 metric tons*
North America:									
Canada	10,367	10,963	10,585	2.59	2.44	1.94	26,900	26,804	20,568
Mexico	640	690	685	4.77	4.93	4.77	3,050	3,400	3,270
UnitedStates	21,781	21,502	19,681	2.87	2.83	2.71	62,569	60,758	53,262
Total	32,788	33,155	30,951	2.82	2.74	2.49	92,519	90,962	77,100
South America:									
Argentina	6,153	6,408	6,800	2.67	2.53	2.28	16,400	16,230	15,500
Bolivia	120	117	110	0.84	0.89	1.06	101	104	117
Brazil	1,252	1,468	1,725	1.92	1.13	1.88	2,403	1,660	3,250
Chile	392	414	426	3.81	4.30	3.90	1,493	1,780	1,660
Colombia	13	14	14	2.08	2.07	2.07	27	29	29
Ecuador	25	25	20	0.60	0.60	0.60	15	15	12
Paraguay	125	125	125	1.60	1.28	1.60	200	160	200
Peru	130	146	154	1.28	1.28	1.26	167	187	194
Uruguay	197	128	125	1.95	2.53	1.15	384	324	144
Total	8,407	8,845	9,499	2.52	2.32	2.22	21,190	20,489	21,106
Guatemala	3	3	1	1.67	1.67	1.00	5	5	1
Europe:									
Austria	258	291	284	5.40	4.37	5.22	1,392	1,273	1,483
Belgium-Luxembourg	215	235	216	7.60	7.64	8.00	1,634	1,795	1,728
Denmark	638	627	652	7.01	7.48	7.55	4,471	4,693	4,924
Finland	121	149	150	2.70	3.63	3.50	327	541	525
France	5,114	5,269	4,780	7.24	7.13	6.59	37,002	37,560	31,500
Germany	2,601	2,969	2,896	7.54	7.28	7.89	19,615	21,622	22,838
Greece	645	655	635	2.14	2.12	1.92	1,380	1,390	1,217
Ireland	75	80	80	8.15	8.75	8.69	611	700	695
Italy	2,387	2,322	2,289	3.07	3.14	2.76	7,328	7,282	6,308
Netherlands	103	136	123	8.26	8.40	8.05	851	1,143	990
Portugal	226	226	181	1.56	1.57	0.93	352	355	168
Spain	2,442	2,370	2,201	2.01	3.09	2.46	4,900	7,330	5,416
Sweden	276	401	401	6.01	5.99	5.89	1,659	2,400	2,363
United Kingdom	1,847	2,086	1,635	8.05	8.01	7.08	14,870	16,700	11,570
Total	16,948	17,816	16,523	5.69	5.88	5.55	96,392	104,784	91,725
Eastern Europe:									
Albania	120	100	100	2.75	3.00	3.00	330	300	300
Bosnia-Hercegovina	95	114	103	2.72	2.41	2.61	258	275	269
Croatia	1,000	1,050	1,100	2.90	2.76	2.82	2,900	2,900	3,100
Bulgaria	169	235	215	3.30	4.39	3.72	558	1,032	800
Czech Republic	867	970	925	4.65	4.21	4.83	4,028	4,084	4,470
Hungary	734	1,040	1,200	3.59	3.55	4.31	2,636	3,692	5,176
Macedonia (Skopje)	114	115	118	2.89	2.78	1.82	330	320	215
Poland	2,583	2,635	2,627	3.50	3.23	3.53	9,051	8,503	9,283
Romania	1,600	1,942	2,552	3.00	2.37	2.63	4,800	4,600	6,700
Serbia	619	653	700	3.23	2.76	3.29	2,000	1,800	2,300
Slovakia	296	400	450	4.01	3.13	4.22	1,187	1,250	1,900
Slovenia	32	36	40	3.66	3.06	4.50	117	110	180
Total E. Europe	8,229	9,290	10,130	3.43	3.11	3.42	28,195	28,866	34,693
Oth W. Eur.:									
Malta & Gozo	1	1	1	5.00	5.00	5.00	5	5	5
Norway	51	68	63	4.51	4.60	4.00	230	313	252
Switzerland	94	96	91	5.38	6.10	5.71	506	586	520
Total	146	165	155	5.08	5.48	5.01	741	904	777
Fmr. Soviet Union:									
Armenia	90	90	95	2.33	1.11	2.11	210	100	200
Azerbaijan	423	493	600	2.05	2.64	2.67	866	1,300	1,600
Belarus	411	425	425	1.73	2.24	2.35	711	950	1,000
Estonia	68	69	56	1.29	2.13	2.34	88	147	131
Georgia	111	90	111	2.04	1.00	2.70	226	90	300
Kazakhstan	8,730	10,500	10,700	1.28	0.87	1.19	11,200	9,100	12,700
Kyrgyzstan	450	450	480	2.44	2.22	2.60	1,100	1,000	1,250
Latvia	146	158	167	2.41	2.70	2.71	352	427	452
Lithuania	333	371	352	2.62	3.33	2.98	871	1,237	1,050
Moldova	336	370	420	2.34	1.96	2.86	785	725	1,200
Russian Fed.	23,000	23,200	23,800	1.35	1.48	1.97	31,000	34,450	46,900
Tajikistan	326	301	290	1.12	0.94	0.79	366	283	230
Turkmenistan	600	700	700	1.67	1.64	1.71	1,000	1,150	1,200
Ukraine	5,931	5,161	6,900	2.29	1.98	3.09	13,585	10,197	21,300
Uzbekistan	1,400	1,200	1,200	2.64	3.00	2.83	3,700	3,600	3,400
Total	42,355	43,578	46,296	1.56	1.49	2.01	66,060	64,756	92,913

See footnotes at end of table.

Table 1-11.—Wheat: Area, yield, and production in specified countries, 1999/2000–2001/2002 [1]—Continued

Continent and country	Area [2]			Yield per hectare			Production		
	1999/2000	2000/2001	2001/2002 [3]	1999/2000	2000/2001	2001/2002 [3]	1999/2000	2000/2001	2001/2002 [3]
	1,000 hectares	*1,000 hectares*	*1,000 hectares*	*Metric tons*	*Metric tons*	*Metric tons*	*1,000 metric tons*	*1,000 metric tons*	*1,000 metric tons*
Middle East:									
Cyprus	7	6	6	2.00	1.67	1.67	14	10	10
Iran	6,000	6,000	6,000	1.42	1.25	1.25	8,500	7,500	7,500
Iraq	1,300	1,300	1,350	0.62	0.62	0.67	800	800	900
Israel	10	25	60	3.50	3.20	2.25	35	80	135
Jordan	15	23	27	0.80	1.26	0.89	12	29	24
Lebanon	10	15	15	3.00	2.00	2.00	30	30	30
Saudi Arabia	400	400	400	4.50	4.50	4.45	1,800	1,800	1,780
Syria	1,200	1,200	1,500	2.17	2.25	3.00	2,600	2,700	4,500
Turkey	8,650	8,700	8,500	1.91	2.07	1.82	16,500	18,000	15,500
Yemen	110	100	100	1.45	1.50	1.50	160	150	150
Total	17,702	17,769	17,958	1.72	1.75	1.70	30,451	31,099	30,529
Africa:									
Algeria	1,372	827	1,836	1.07	0.92	1.09	1,470	760	2,010
Angola	5	5	5	0.80	0.80	0.80	4	4	4
Chad	4	4	4	1.25	0.75	1.00	5	3	4
Egypt	1,035	1,190	1,002	6.14	5.34	6.12	6,350	6,350	6,130
Ethiopia	1,650	1,675	1,650	1.27	1.46	1.39	2,100	2,450	2,300
Kenya	100	20	145	1.35	5.25	1.59	135	105	230
Lesotho	13	12	12	1.15	0.92	1.25	15	11	15
Libya	290	250	250	0.55	0.50	0.50	160	125	125
Morocco	2,692	2,902	2,701	0.80	0.48	1.23	2,154	1,381	3,316
Mozambique	3	3	3	1.00	1.00	1.00	3	3	3
Nigeria	35	35	35	1.29	1.29	1.43	45	45	50
South Africa, Rep. of	718	934	959	2.40	2.51	2.60	1,725	2,349	2,490
Sudan	140	90	120	1.21	2.39	2.08	170	215	250
Tanzania, United Rep. of	60	60	60	1.37	1.50	1.50	82	90	90
Tunisia	910	910	825	1.53	1.45	1.36	1,393	1,320	1,120
Zaire	10	10	10	1.80	1.00	1.00	18	10	10
Zambia	12	11	12	7.50	6.82	6.67	90	75	80
Zimbabwe	57	46	45	5.68	5.54	5.56	324	255	250
Total	9,106	8,984	9,674	1.78	1.73	1.91	16,243	15,551	18,477
Asia:									
Afghanistan	2,027	2,029	1,779	1.22	0.72	0.90	2,480	1,469	1,597
Bangladesh	805	773	750	2.47	2.16	2.33	1,988	1,673	1,750
Bhutan	13	13	13	1.54	1.54	1.54	20	20	20
Burma	90	90	90	1.11	1.11	1.11	100	100	100
China, Peop. Rep.	28,855	26,650	24,700	3.95	3.74	3.80	113,880	99,640	93,870
India	27,400	27,486	25,068	2.58	2.78	2.74	70,780	76,369	68,763
Japan	169	183	197	3.45	3.76	3.55	583	688	700
Korea, Dem. Rep.	125	90	90	2.00	1.67	1.11	250	150	100
Korea, Rep.	2	1	1	3.00	2.00	3.00	6	2	3
Mongolia	300	300	300	0.67	0.50	0.50	200	150	150
Nepal	640	640	640	1.64	1.64	1.64	1,050	1,050	1,050
Pakistan	8,231	8,463	8,250	2.17	2.49	2.31	17,854	21,079	19,023
Total	68,657	66,718	61,878	3.05	3.03	3.02	209,191	202,390	187,126
Oceania:									
Australia	12,168	13,002	12,500	2.03	1.83	1.92	24,757	23,766	24,000
New Zealand	55	55	55	5.00	5.00	5.00	275	275	275
Total	12,223	13,057	12,555	2.05	1.84	1.93	25,032	24,041	24,275
World Total	216,564	219,380	215,620	2.71	2.66	2.68	586,019	583,847	578,722

[1] Years shown refer to years of harvest. Harvests of Northern Hemisphere countries are combined with those of the Southern Hemisphere which immediately follow; thus the crop harvested in the Northern Hemisphere in 1994 is combined with estimates for the Southern Hemisphere Harvests, which begin late in 1994 and end early in 1995. [2] Harvested area as far as possible. [3] Preliminary.

FAS, Production Estimates and Crop Assessment Division, (202) 720–0888. Prepared or estimated on the basis of official statistics of foreign governments, other foreign source materials, reports of U.S. Agricultural Counselors, Attachés and Foreign Service Officers, results of office research, and related information.

Table 1-12.—Wheat and flour: United States imports,1992–2001

Year beginning June	Wheat grain [1]	Flour (wheat equivalent)	Other products (wheat equivalent) [2]	Total wheat, flour, and other products
	1,000 bushels	*1,000 bushels*	*1,000 bushels*	*1,000 bushels*
1992	56,859	3,122	9,435	69,416
1993	91,287	6,486	11,086	108,859
1994	70,561	8,073	13,313	91,947
1995	47,753	6,687	13,493	67,933
1996	71,727	6,386	14,220	92,333
1997	73,245	6,055	15,623	94,923
1998	79,766	7,423	15,815	103,004
1999	72,408	7,116	14,986	94,510
2000	66,313	8,863	14,649	89,825
2001	82,615	9,903	15,027	107,544

[1] Starting January 1989, Census ceased reporting wheat suitable for milling and unfit for human consumption. [2] Includes macaroni, semolina, and similar products. Beginning in 1988/89 total wheat grain is reported under the suitable for milling column.

ERS, Market and Trade Economics Division, (202) 694–5285.

Table 1-13.—Wheat, flour, and products:[1] International trade, 1999/2000–2001/2002[2]

Country	1999/2000	2000/2001	2001/2002[3]
	1,000 metric tons	*1,000 metric tons*	*1,000 metric tons*
Principal exporters:			
Argentina	11,083	11,396	11,477
Australia	17,124	16,682	16,494
Canada	19,373	17,351	16,758
India	200	2,357	3,234
Kazakhstan	6,514	3,668	3,780
Russia	518	696	4,372
Turkey	1,984	1,601	548
Ukraine	1,952	78	5,486
EU	17,432	15,225	11,494
Eastern Europe	3,401	2,336	4,140
Others	3,746	4,245	6,066
Subtotal	83,327	75,635	83,849
United States	29,449	27,845	26,139
Total	112,776	103,480	109,988
Principal importers:			
Algeria	4,750	5,600	4,500
Bangladesh	1,624	1,293	1,565
Bolivia	458	485	500
Brazil	7,298	7,453	7,111
Chile	732	438	450
China	1,010	195	1,092
Colombia	1,135	1,164	1,200
Cuba	1,119	963	1,100
Ecuador	485	490	500
Egypt	5,872	6,050	7,000
Ethiopia	1,082	888	400
India	1,311	45	33
Indonesia	3,744	4,069	3,677
Iran	7,363	6,245	5,600
Iraq	2,650	3,300	3,000
Israel	1,611	1,251	1,550
Japan	5,960	5,885	5,836
Jordan	741	650	800
Kenya	683	806	624
Korea, North	334	300	300
Korea, South	3,811	3,127	3,979
Libya	1,582	1,400	1,500
Malaysia	1,278	1,265	1,300
Mexico	2,632	3,066	3,200
Morocco	3,094	3,632	3,150
Nigeria	1,282	1,913	2,500
Pakistan	1,766	50	350
Peru	1,215	1,451	1,500
Philippines	2,982	3,050	2,922
Russia	5,083	1,604	550
South Africa	806	438	560
Sri Lanka	834	779	850
Sudan	792	920	900
Taiwan	1,138	1,033	1,000
Thailand	809	941	950
Tunisia	1,186	1,595	1,261
Turkey	1,462	446	1,000
UAE	1,386	1,100	1,100
Uzbekistan	550	550	500
Venezuela	1,386	1,394	1,400
Vietnam	550	650	950
Yemen	2,002	2,117	1,700
EU	4,176	3,159	9,822
O.W. Europe	730	523	760
Eastern Europe	1,620	2,732	1,735
United States	2,503	2,417	2,850
Subtotal	96,617	88,922	95,127
Other Countries	13,203	12,409	13,891
Unaccounted	2,956	2,149	970
Total	112,776	103,480	109,988

[1] Flour and products reported in terms of grain equivalent. [2] Year beginning July 1. [3] Preliminary.

FAS, Grain and Feed Division, (202) 720–6219. Prepared or estimated on the basis of official statistics from foreign governments, other foreign source materials, reports of U.S. Agricultural Counselors, Attachés and Foreign Service Officers, results of office research, and related information.

Table 1-14.—Wheat and flour: [1] United States exports by country of destination, 1999/2000 and 2001/2002

Country of destination	Year [2]		
	1999/2000	2000/2001	2001/2002
	1,000 metric tons	*1,000 metric tons*	*1,000 metric tons*
Wheat:			
Egypt	4,168	4,181	3,547
Japan	3,222	3,018	2,995
Mexico	1,799	1,963	2,166
Nigeria	1,088	1,487	1,882
Philippines	2,113	1,941	1,418
Korea, Republic of	1,580	1,434	1,210
Italy	690	729	1,071
Taiwan	1,018	942	921
Canadian Trans.	171	254	609
Israel	826	670	608
Peru	367	445	561
Yemen	527	644	539
Colombia	702	626	535
Jordan	525	488	456
Venezuela	422	603	442
Sri Lanka	465	465	432
Spain	209	72	395
Thailand	289	353	363
Dominican Republic	278	254	289
China, Peoples Rep.	192	64	221
Others	7,342	6,290	4,622
Total	27,991	26,925	25,283
Wheat flour:			
Yemen	87	117	53
Germany	7	0	51
Tajikistan, Republic	31	67	45
Moldova, Republic of	11	15	37
Yugoslavia	162	66	32
India	51	37	30
Latvia	1	5	28
Bolivia	56	30	27
Mexico	26	19	25
Canada	47	48	23
Albania	20	0	23
Netherlands	10	0	21
Haiti	97	55	19
Russian Federation	18	45	17
Chad	8	9	14
Afghanistan	0	0	14
Armenia, Republic of	7	0	12
Kyrgyzstan, Republic	1	5	12
Kenya	3	3	11
Uzbekistan, Republic	0	0	10
Other	647	323	124
Total	1,290	844	625

[1] Flour reported in terms of grain equivalent. [2] Year beginning Jul 1.

FAS, Grain and Feed Division, (202) 720–6219. Compiled from reports of the U.S. Department of Commerce.

Table 1-15.—Rye: Area, yield, production, disposition, and value, United States, 1993–2002

Year	Area		Yield per harvested acre	Production	Marketing year average price per bushel received by farmers [2]	Value of production [2]
	Planted [1]	Harvested				
	1,000 acres	*1,000 acres*	*Bushels*	*1,000 bushels*	*Dollars*	*1,000 dollars*
1993	1,493	381	27.1	10,340	2.55	27,149
1994	1,613	407	27.9	11,341	2.70	30,520
1995	1,602	385	26.1	10,064	2.90	28,948
1996	1,457	345	25.9	8,936	3.70	33,118
1997	1,400	316	25.7	8,132	3.75	30,120
1998	1,566	418	29.1	12,161	2.49	30,404
1999	1,582	383	28.8	11,038	2.27	25,084
2000	1,329	296	28.3	8,386	2.60	21,830
2001	1,328	255	27.3	6,971	2.93	20,422
2002	1,395	286	24.4	6,985	3.40	23,717

[1] Area planted in preceding fall. [2] Preliminary.

NASS, Crops Branch, (202) 720–2127.

Table 1-16.—Rye: Supply and disappearance, United States, 1993–2002

Year beginning June	Supply				Disappearance							Ending stocks May 31
	Beginning stocks	Production	Imports	Total	Domestic use					Exports	Total disappearance	
					Food	Seed	Industry	Feed [1]	Total			
	1,000 bushels	*1,000 bushels*	*1,000 bushels*	*1,000 bushels*	*1,000 bushels*	*1,000 bushels*	*1,000 bushels*	*1,000 bushels*	*1,000 bushels*	*1,000 bushels*	*1,000 bushels*	*1,000 bushels*
1993	1,555	10,340	4,607	16,502	3,538	3,000	2,000	6,977	15,515	16	15,531	971
1994	971	11,341	4,386	16,698	3,312	3,000	2,000	6,900	15,212	35	15,247	1,451
1995	1,451	10,064	3,760	15,275	3,318	3,000	2,000	6,018	14,336	41	14,377	898
1996	898	8,936	4,327	14,161	3,459	3,000	2,000	4,916	13,375	32	13,407	754
1997	754	8,132	5,562	14,448	3,298	2,000	3,000	5,306	13,604	80	13,684	764
1998	764	12,161	3,322	16,247	3,639	3,000	3,000	4,392	14,031	33	14,064	2,183
1999	2,449	11,038	3,424	16,911	3,300	3,000	3,000	5,736	15,036	286	15,322	1,589
2000	1,589	8,368	3,230	13,187	3,300	3,000	3,000	2,307	11,607	390	11,997	1,190
2001	1,190	6,971	4,945	13,106	3,300	3,000	3,000	3,045	12,345	193	12,538	568
2002 [2]	568	6,985	5,000	12,553	3,300	3,000	3,000	2,435	11,735	250	11,985	568

[1] Residual, approximates total feed use. [2] Preliminary. Totals may not add due to independent rounding.

ERS, Market and Trade Economics Division, (202) 694–5302.

Table 1-17.—Rye: Support operations, United States, 1993–2002

Marketing year beginning June 1	Loan rate per bushel [1]	Put under loan [2]		Acquired by CCC under support program [3]	Owned by CCC at end of marketing year
		Quantity	Percentage of production		
	Dollars	*Million bushels*	*Percent*	*Million bushels*	*Million bushels*
1993/94	1.46	0.1	1.0	0.0	0.0
1994/95	1.61	0.1	0.9	0.0	0.0
1995/96	1.61	0.1	1.0	0.0	0.0
1996/97	(4)	....	....	....	....
1997/98	(4)	....	....	....	....
1998/99	(4)	....	....	....	....
1999/2000	(4)	....	....	....	....
2000/2001	(4)	....	....	....	....
2001/2002	(4)	....	....	....	....
2002/2003	(4)	....	....	....	....

[1] The national average loan rate was also known as the price support rate prior to enactment of the Federal Agriculture Improvement and Reform Act of 1996. [2] Includes loans made, purchases, and purchase agreements entered into. [3] Acquisition of all loans forfeited during the marketing year including loans made in previous years. [4] The Federal Agriculture Improvement and Reform Act of 1996 did not provide authority for rye loans after the 1995/96 marketing year.

FSA, Feed Grains & Oilseeds Analysis Group, (202) 720–2711.

Table 1-18.—Rye: Area, yield, and production, by States, 2000–2002

State	Area planted [1]			Area harvested			Yield per harvested acre			Production		
	2000	2001	2002	2000	2001	2002	2000	2001	2002	2000	2001	2002
	1,000 acres	*1,000 acres*	*1,000 acres*	*1,000 acres*	*1,000 acres*	*1,000 acres*	*Bushels*	*Bushels*	*Bushels*	*1,000 bushels*	*1,000 bushels*	*1,000 bushels*
GA	230	300	260	45	35	45	26.0	25.0	16.0	1,170	875	720
ND	20	13	10	16	10	9	44.0	34.0	38.0	704	340	342
OK	290	250	300	70	50	70	21.0	23.0	19.0	1,470	1,150	1,330
SD	14	10	15	13	10	10	42.0	35.0	27.0	546	350	270
Oth Sts [2]	775	755	810	152	150	152	29.6	28.4	28.4	4,496	4,256	4,323
US	1,329	1,328	1,395	296	255	286	28.3	27.3	24.4	8,386	6,971	6,985

[1] Includes area planted preceding fall. [2] Other States include IL, KS, MI, MN, NE, NY, NC, PA, SC, TX, and WI.
NASS, Crops Branch, (202) 720–2127.

Table 1-19.—Rye: Marketing year average price and value, by States, crop of 2000, 2001, and 2002

State	Marketing year average price per bushel			Value of production		
	2000	2001	2002 [1]	2000	2001	2002 [1]
	Dollars	*Dollars*	*Dollars*	*1,000 dollars*	*1,000 dollars*	*1,000 dollars*
GA	4.90	6.00	6.00	5,733	5,250	4,320
ND	1.31	1.55	2.75	922	527	941
OK	3.40	3.40	4.00	4,998	3,910	5,320
SD	1.47	2.50	3.35	803	875	905
Oth Sts [2]	2.08	2.32	2.85	9,374	9,860	12,231
US	2.60	2.93	3.40	21,830	20,422	23,717

[1] Preliminary. [2] Other States include IL, KS, MI, MN, NE, NY, NC, PA, SC, TX, and WI.
NASS, Crops Branch, (202) 720-2127.

Table 1-20.—Rye: Area, yield, and production in specified countries, 1999/2000–2001/2002 [1]

Continent and country	Area [2]			Yield per hectare			Production		
	1999/ 2000	2000/ 2001	2001/ 2002 [3]	1999/ 2000	2000/ 2001	2001/ 2002 [3]	1999/ 2000	2000/ 2001	2001/ 2002 [3]
	1,000 hectares	*1,000 hectares*	*1,000 hectares*	*Metric tons*	*Metric tons*	*Metric tons*	*1,000 metric tons*	*1,000 metric tons*	*1,000 metric tons*
North America:									
Canada	169	115	123	2.29	2.26	1.85	387	260	228
United States	155	120	103	1.81	1.78	1.72	280	213	177
Total	324	235	226	2.06	2.01	1.79	667	473	405
South America:									
Argentina	92	87	62	1.26	1.44	1.29	116	125	80
Brazil	10	7	7	1.20	1.29	1.29	12	9	9
Chile	5	5	5	2.00	2.00	2.00	10	10	10
Total	107	99	74	1.29	1.45	1.34	138	144	99
Europe:									
Austria	56	53	51	3.89	3.40	4.18	218	180	213
Belgium-Luxembourg	2	2	1	4.50	5.00	5.00	9	10	5
Denmark	51	51	62	4.86	5.14	5.31	248	262	329
Finland	12	44	45	2.00	2.61	1.49	24	115	67
France	41	32	29	4.54	4.56	4.10	186	146	119
Germany	748	843	837	5.79	4.93	6.13	4,329	4,154	5,132
Greece	15	15	15	2.20	2.27	2.20	33	34	33
Italy	7	3	3	2.57	2.67	2.67	18	8	8
Netherlands	3	6	4	4.67	4.83	4.25	14	29	17
Portugal	60	43	43	0.82	0.74	0.79	49	32	34
Spain	125	111	115	1.60	1.89	0.93	200	210	107
Sweden	25	35	35	4.68	5.34	5.20	117	187	182
United Kingdom	8	7	6	5.38	6.29	5.00	43	44	30
Total	1,153	1,245	1,246	4.76	4.35	5.04	5,488	5,411	6,276
Eastern Europe:									
Albania	10	10	10	1.00	1.00	1.00	10	10	10
Bosnia-Hercegovina	4	5	5	2.25	2.40	2.60	9	12	13
Bulgaria	10	10	10	1.50	1.50	1.50	15	15	15
Croatia	2	2	2	3.00	3.00	3.00	6	6	6
Czech Republic	55	44	40	3.64	3.41	3.85	200	150	154
Hungary	40	40	50	2.00	2.00	2.42	80	80	121
Macedonia (Skopje)	7	7	7	1.57	1.57	1.57	11	11	11
Poland	2,243	2,130	2,002	2.31	1.88	2.43	5,181	4,003	4,863
Romania	20	20	20	2.50	2.50	2.50	50	50	50
Serbia	6	6	6	1.67	1.33	1.67	10	8	10
Slovakia	30	32	40	2.33	2.00	3.00	70	64	120
Slovenia	1	1	1	3.00	3.00	3.00	3	3	3
Total	2,428	2,307	2,193	2.32	1.91	2.45	5,645	4,412	5,376
Oth Western Europe:									
Norway	3	3	3	3.67	3.67	3.67	11	11	11
Switzerland	3	3	3	6.67	6.67	6.67	20	20	20
Total	6	6	6	5.17	5.17	5.17	31	31	31
Fmr. Soviet Union:.									
Byelarus	641	780	800	1.45	1.86	2.00	929	1,450	1,600
Estonia	38	29	20	1.03	2.10	2.05	39	61	41
Kazakhstan	20	70	70	1.00	0.71	1.07	20	50	75
Latvia	47	55	56	1.89	2.02	1.91	89	111	107
Lithuania	135	133	111	1.93	2.34	2.11	261	311	234
Russian Fed.	3,400	3,500	3,600	1.41	1.56	1.83	4,800	5,450	6,600
Ukraine	623	637	900	1.48	1.52	2.00	919	966	1,800
Total	4,904	5,204	5,557	1.44	1.61	1.88	7,057	8,399	10,457
Asia:									
Turkey	180	180	180	1.39	1.39	1.39	250	250	250
Africa:									
South Africa	20	20	20	0.15	0.15	0.15	3	3	3
Oceania:									
Australia	35	35	35	0.57	0.57	0.57	20	20	20
World Total	9,157	9,331	9,537	2.11	2.05	2.40	19,299	19,143	22,917

[1] Years shown refer to years of harvest. Harvests of Northern Hemisphere countries are combined with those of the Southern Hemisphere which immediately follow: thus the crop harvested in the Northern Hemisphere in 1994 is combined with estimates for the Southern Hemisphere harvests, which begin late in 1994 and end early in 1995. [2] Harvested area as far as possible. [3] Preliminary.

FAS, Production Estimates and Crop Assessment Division, (202) 720–0888. Prepared or estimated on the basis of official statistics of foreign Governments, other foreign source materials, reports of Agricultural Counselors, Attachés, and Foreign Service Officers, results of office research, and related information.

Table 1-21.—Rice, rough: Area, yield, production, and value, United States, 1993–2002

Year	Area planted	Area harvested	Yield per acre	Production	Marketing year average price per cwt. received by farmers	Value of production
	1,000 acres	*1,000 acres*	*Pounds*	*1,000 cwt.*	*Dollars*	*1,000 dollars*
1993	2,920.0	2,833.0	5,510	156,110	7.98	1,246,875
1994	3,353.0	3,316.0	5,964	197,779	6.78	1,336,570
1995	3,121.0	3,093.0	5,621	173,871	9.15	1,587,236
1996	2,824.0	2,804.0	6,120	171,599	9.96	1,690,270
1997	3,125.0	3,103.0	5,897	182,992	9.70	1,756,136
1998	3,285.0	3,257.0	5,663	184,443	8.89	1,654,157
1999	3,531.0	3,512.0	5,866	206,027	5.93	1,231,207
2000	3,060.0	3,039.0	6,281	190,872	5.61	1,049,961
2001	3,334.0	3,314.0	6,496	215,270	4.25	925,055
2002	3,240.0	3,207.0	6,578	210,960	3.85	840,727

NASS, Crops Branch, (202) 720–2127.

Table 1-22.—Rice, rough: Stocks on and off farms, United States, 1994–2003

Year beginning previous December	On farms			Off farms[1]		
	Dec. 1	Mar. 1	Aug. 1	Dec. 1	Mar. 1	Aug. 1
	1,000 cwt.	*1,000 cwt.*	*1,000 cwt.*	*1,000 cwt.*	*1,000 cwt.*	*1,000 cwt.*
1994	24,164	11,703	517	85,081	67,900	20,390
1995	41,223	23,239	862	100,088	72,686	21,902
1996	32,936	20,520	486	88,395	65,326	19,485
1997	32,719	16,003	428	86,350	64,141	21,365
1998	33,470	21,205	1,136	90,873	66,846	19,855
1999	35,584	22,290	1,560	85,394	57,960	15,066
2000	50,185	27,212	1,141	89,191	63,025	20,829
2001	38,085	18,715	921	95,842	67,305	21,097
2002[2]	52,680	31,725	5,180	101,881	81,783	25,629
2003[2]	53,220	NA	NA	103,850	NA	NA

[1] Stocks at mills and in attached warehouses, in warehouses not attached to mills, and in ports or in transit. [2] Preliminary. NA-not available.

NASS, Crops Branch, (202) 720–2127.

Table 1-23.—Rice, by length of grain: Area, yield, and production, United States, 1993–2002

Year	Area harvested			Yield per acre			Production		
	Long grain	Medium grain	Short grain	Long grain	Medium grain	Short grain	Long grain	Medium grain	Short grain
	1,000 acres	*1,000 acres*	*1,000 acres*	*Pounds*	*Pounds*	*Pounds*	*1,000 cwt.*	*1,000 cwt.*	*1,000 cwt.*
1993	2,028.0	789.0	16.0	5,082	6,575	7,331	103,064	51,873	1,173
1994	2,379.0	925.0	12.0	5,609	6,853	7,867	133,445	63,390	944
1995	2,312.0	769.0	12.0	5,265	6,663	7,500	121,730	51,241	900
1996	1,967.0	822.0	15.0	5,777	6,922	7,127	113,629	56,901	1,069
1997	2,309.0	776.0	18.0	5,391	7,357	7,867	124,485	57,091	1,416
1998	2,568.0	656.0	33.0	5,426	6,616	5,185	139,328	43,404	1,711
1999	2,718.0	742.0	52.0	5,587	6,811	6,969	151,863	50,540	3,624
2000	2,189.0	814.0	36.0	5,882	7,311	7,228	128,756	59,514	2,602
2001	2,697.0	591.0	26.0	6,213	7,801	6,192	167,555	46,105	1,610
2002	2,512.0	668.0	27.0	6,260	7,815	5,615	157,243	52,201	1,516

NASS, Crops Branch, (202) 720–2127.

Table 1-24.—Rice, rough, by length of grain: Stocks in all positions, United States, 1994–2003

Year beginning previous December	Long grain				Medium grain				Short grain			
	Dec. 1	Mar. 1	Aug. 1	Oct. 1[1]	Dec. 1	Mar. 1	Aug. 1	Oct. 1[1]	Dec. 1	Mar. 1	Aug. 1	Oct. 1[1]
	1,000 cwt.	*1,000 cwt.*	*1,000 cwt.*	*1,000 cwt.*	*1,000 cwt.*	*1,000 cwt.*	*1,000 cwt.*	*1,000 cwt.*	*1,000 cwt.*	*1,000 cwt.*	*1,000 cwt.*	*1,000 cwt.*
1994	61,356	44,741	12,543	114	46,524	33,732	7,708	1,906	1,365	1,130	656	449
1995	87,281	57,779	10,188	219	52,676	36,963	11,872	4,186	1,354	1,183	704	522
1996	76,256	52,630	7,418	(3)	43,941	32,264	12,022	4,506	1,134	952	531	(3)
1997	68,687	47,871	10,839	(3)	49,015	31,665	10,723	4,334	1,367	608	231	(3)
1998	78,329	53,329	9,863	(3)	44,477	33,596	10,640	4,368	1,537	1,126	488	(3)
1999	84,346	57,636	10,947	(3)	34,774	21,329	5,037	861	1,858	1,285	642	(3)
2000[2] ..	96,383	62,755	12,511	160	39,601	25,381	8,299	3,287	3,392	2,101	1,160	680
2001[2] ..	82,718	51,428	8,305	116	48,438	32,504	12,841	5,066	2,771	2,088	872	732
2002[2] ..	109,953	83,723	22,743	434	42,525	28,515	8,477	2,691	2,083	1,270	589	363
2003[2] ..	113,897	NA	NA	NA	40,918	NA	NA	NA	2,255	NA	NA	NA

[1] California only. [2] Preliminary. [3] Not published to avoid disclosing individual reports. NA-not available.
NASS, Crops Branch, (202) 720–2127.

Table 1-25.—Rough and milled rice (rough equivalent): Supply and disappearance, United States, 1993–2002[1]

Year beginning August	Supply				Disappearance					Ending stocks July 31
	Beginning stocks	Production	Imports	Total	Food, industrial, & residual[2]	Seed	Total	Exports[3]	Total disappearance	
	Million cwt.	*Million cwt.*	*Million cwt.*	*Million cwt.*	*Million cwt.*	*Million cwt.*	*Million cwt.*	*Million cwt.*	*Million cwt.*	*Million cwt.*
1993 ..	39.4	156.1	6.9	202.5	96.2	4.1	100.3	76.4	176.7	25.8
1994 ..	25.8	197.8	8.1	231.6	97.6	3.9	101.5	98.8	200.3	31.3
1995 ..	31.3	173.9	7.7	212.9	101.1	3.5	104.6	83.2	187.8	25.0
1996 ..	25.0	171.6	10.5	207.2	97.7	3.9	101.6	78.3	179.9	27.2
1997 ..	27.2	183.0	9.3	219.5	99.8	4.1	103.9	87.7	191.6	27.9
1998 ..	27.9	184.4	10.6	223.0	109.6	4.4	114.0	86.8	200.9	22.1
1999 ..	22.1	206.0	10.1	238.2	117.9	4.0	121.9	88.8	210.7	27.5
2000 ..	27.5	190.9	10.9	229.2	134.4	4.1	138.5	83.2	200.7	28.5
2001 ..	28.5	215.3	13.2	256.9	119.9	4.0	123.9	94.1	210.1	39.0
2002[4]	39.0	211.0	12.0	261.9	121.0	4.0	125.0	105.0	200.7	32.0

All data updated as of January 2003. Totals may not add due to independent rounding.
[1] Consolidated supply and disappearance of rough and milled rice. Milled rice data converted to a rough basis using annually derived extraction rates as factors. [2] The residual includes unaccounted losses in drying, processing, and handling. [3] Trade data from Bureau of the Census. [4] Preliminary.
ERS, Market and Trade Economics Division, (202) 694–5292.

Table 1-26.—Rice, by length of grain: Area, yield, and production, by States, 2000–2002

State	Area harvested			Yield per acre			Production		
	2000	2001	2002 [1]	2000	2001	2002 [1]	2000	2001	2002 [1]
	Long grain								
	1,000 acres	*1,000 acres*	*1,000 acres*	*Pounds*	*Pounds*	*Pounds*	*1,000 cwt.*	*1,000 cwt.*	*1,000 cwt.*
AR	1,130.0	1,472.0	1,340.0	6,060	6,330	6,430	68,478	93,178	86,162
CA	9.0	13.0	7.0	7,100	7,700	6,400	639	1,001	448
LA	455.0	538.0	525.0	5,080	5,500	5,500	23,114	29,590	28,875
MS	218.0	253.0	253.0	5,900	6,600	6,400	12,862	16,698	16,192
MO	168.0	206.0	182.0	5,700	6,000	6,050	9,576	12,360	11,011
TX	209.0	215.0	205.0	6,740	6,850	7,100	14,087	14,728	14,555
US	2,189.0	2,697.0	2,512.0	5,882	6,213	6,260	128,756	167,555	157,243
	Medium grain								
	1,000 acres	*1,000 acres*	*1,000 acres*	*Pounds*	*Pounds*	*Pounds*	*1,000 cwt.*	*1,000 cwt.*	*1,000 cwt.*
AR	278.0	148.0	162.0	6,300	6,500	6,500	17,514	9,620	10,530
CA	505.0	433.0	495.0	8,000	8,300	8,300	40,400	35,939	41,085
LA	25.0	8.0	10.0	5,150	5,300	5,250	1,288	424	525
MO	1.0	1.0	....	5,700	5,950	....	57	60	....
TX	5.0	1.0	1.0	5,100	6,200	6,100	255	62	61
US	814.0	591.0	668.0	7,311	7,801	7,815	59,514	46,105	52,201
	Short grain								
	1,000 acres	*1,000 acres*	*1,000 acres*	*Pounds*	*Pounds*	*Pounds*	*1,000 cwt.*	*1,000 cwt.*	*1,000 cwt.*
AR	2.0	1.0	1.0	6,000	6,000	6,000	120	60	60
CA	34.0	25.0	26.0	7,300	6,200	5,600	2,482	1,550	1,456
US	36.0	26.0	27.0	7,228	6,192	5,615	2,602	1,610	1,516

[1] Preliminary.

NASS, Crops Branch, (202) 720–2127.

Table 1-27.—Rice: Area, yield, and production, by States, 2000–2002

State	Area planted			Area harvested			Yield per harvested acre			Production		
	2000	2001	2002 [1]	2000	2001	2002 [1]	2000	2001	2002 [1]	2000	2001	2002 [1]
	1,000 acres	*1,000 acres*	*1,000 acres*	*1,000 acres*	*1,000 acres*	*1,000 acres*	*Pounds*	*Pounds*	*Po unds*	*1,000 cwt.*	*1,000 cwt.*	*1,000 cwt.*
AR	1,420.0	1,631.0	1,516.0	1,410.0	1,621.0	1,503.0	6,110	6,350	6,440	86,112	102,858	96,752
CA	550.0	473.0	533.0	548.0	471.0	528.0	7,940	8,170	8,140	43,521	38,490	42,989
LA	485.0	548.0	540.0	480.0	546.0	535.0	5,080	5,500	5,500	24,402	30,014	29,400
MS	220.0	255.0	255.0	218.0	253.0	253.0	5,900	6,600	6,400	12,862	16,698	16,192
MO	170.0	211.0	190.0	169.0	207.0	182.0	5,700	6,000	6,050	9,633	12,420	11,011
TX	215.0	216.0	206.0	214.0	216.0	206.0	6,700	6,850	7,100	14,342	14,790	14,616
US ...	3,060.0	3,334.0	3,240.0	3,039.0	3,314.0	3,207.0	6,281	6,496	6,578	190,872	215,270	210,960

[1] Preliminary.

NASS, Crops Branch, (202) 720–2127.

Table 1-28.—Rice: Marketing year average price and value, by States, crop of 2000–2002

State	Marketing year average price per cwt.			Value of production		
	2000	2001	2002 [1]	2000	2001	2002 [1]
	Dollars	*Dollars*	*Dollars*	*1,000 dollars*	*1,000 dollars*	*1,000 dollars*
AR	5.60	3.93	3.60	482,227	404,232	348,307
CA	4.99	5.28	5.00	217,170	203,227	214,945
LA	5.82	4.47	3.90	142,020	134,163	114,660
MS	5.68	4.15	3.85	73,056	69,297	62,339
MO	5.40	3.70	3.55	52,018	45,954	39,089
TX	5.82	4.61	4.20	83,470	68,182	61,387
US	5.61	4.25	3.85	1,049,961	925,055	840,727

[1] Preliminary.

NASS, Crops Branch, (202) 720–2127.

Table 1-29.—Rice, milled, by length of grain: Stocks in all positions, United States, 1994–2003

Year beginning previous Dec.	Whole kernels (head rice)											
	Long grain				Medium grain				Short grain			
	Dec. 1	Mar. 1	Aug. 1	Oct. 1 [1]	Dec. 1	Mar. 1	Aug. 1	Oct. 1 [1]	Dec. 1	Mar. 1	Aug. 1	Oct. 1 [1]
1994	2,399	2,753	1,860	43	1,258	1,414	1,168	596	27	26	28	51
1995	1,884	2,394	3,127	51	1,723	2,243	2,333	749	48	53	44	37
1996	3,509	2,258	1,927	17	1,184	870	1,187	780	39	26	50	88
1997	3,358	2,622	2,312	10	1,448	1,079	788	584	56	85	34	66
1998	3,443	3,871	3,228	37	1,248	1,080	728	313	46	35	100	25
1999	2,980	2,361	2,159	21	1,613	1,114	657	301	30	62	131	49
2000	2,732	2,160	2,506	14	770	758	644	540	63	46	30	66
2001	3,624	2,470	2,287	26	1,348	1,164	1,207	342	67	84	87	57
2002	3,796	3,222	2,788	23	986	622	1,032	388	62	110	72	53
2003 [2] ..	4,390	NA	NA	NA	1,674	NA	NA	NA	58	NA	NA	NA

Year beginning previous Dec.	Broken kernels [3]											
	Second heads				Screenings				Brewers			
	Dec. 1	Mar. 1	Aug. 1	Oct. 1 [1]	Dec. 1	Mar. 1	Aug. 1	Oct. 1 [1]	Dec. 1	Mar. 1	Aug. 1	Oct. 1 [1]
1994	632	434	364	164	107	47	82	....	458	1,131	93	21
1995	690	734	579	279	89	31	79	....	411	164	146	33
1996	626	406	329	159	14	6	14	....	337	133	107	25
1997	553	623	467	97	49	20	13	....	187	182	205	19
1998	642	711	524	273	27	27	33	....	276	155	182	14
1999	662	612	588	30	29	92	102	....	102	103	140	12
2000 [2] ..	608	937	652	238	64	85	61	....	348	238	285	46
2001 [2] ..	1,006	1,035	667	403	66	3	72	....	251	228	117	31
2002 [2] ..	825	648	696	246	123	139	133	....	115	72	209	16
2003 [2] ..	1,026	NA	NA	NA	91	NA	NA	....	242	NA	NA	NA

[1] California only. [2] Preliminary. [3] Screenings included in second heads in California. NA-not available.

NASS, Crops Branch, (202) 720–2127.

Table 1-30.—Rice, rough: Support operations, United States, 1993–2002

Marketing year beginning August 1	Income support payment rates per cwt [1]	Program price levels per cwt		Put under loan [4]		Acquired by CCC under loan program [5]	Owned by CCC at end of marketing year
		Loan [2]	Target [3]	Quantity	Percentage of production		
	Dollars	*Dollars*	*Dollars*	*Million cwt*	*Percent*	*Million cwt*	*Million cwt*
1993/94	3.98	6.50	10.71	30.9	19.8	0.0	0.0
1994/95	3.79	6.50	10.71	131.2	66.3	0.0	0.0
1995/96	3.22	6.50	10.71	100.9	58.0	0.0	0.0
1996/97	2.766	6.50	NA	68.9	40.2	0.0	0.0
1997/98	2.710	6.50	NA	67.6	36.9	0.0	0.0
1998/99	4.374	6.50	NA	80.3	43.5	0.0	0.0
1999/2000	5.640	6.50	NA	110.8	53.8	0.0	0.0
2000/2001	5.420	6.50	NA	97.4	51.0	0.2	0.0
2001/2002	4.490	6.50	NA	128.0	60.1	0.1	0.0
2002/2003	4.400	6.50	10.50	....	....	....	....

[1] Payment rates for the 1995/96 & prior crop years were calculated according to the provisions of the now defunct deficiency payment/production adjustment program. Payment rates for the 1996/97 through 2001/2002 crops were calculated according to the provisions of the Production Flexibility Contract (PFC) program, following enactment of the Federal Agriculture Improvement and Reform Act of 1996 (1996 Act) and included supplemental PFC payment rates for 1998 through 2001. Payment rates for the 2002/2003 and subsequent crops are calculated according to the provisions of the Direct Payment program, following enactment of the Farm Security and Rural Investment Act of 2002 (2002 Act) and includes a PFC payment rate for 2002 only. Payment rates are rounded to the nearest tenth of a cent. [2] The national average loan rate was also known as the price support rate prior to enactment of the 1996 Act. [3] Between the 1996/97 and 2001/2002 marketing years, target prices were no longer applicable; however, with enactment of the 2002 Act, target prices were reestablished. [4] Includes loans make and purchase agreements entered. Excludes quantities on which loan deficiency payments were made. [5] Acquisition through loan forfeitures, crop–year basis. NA-not applicable.

FSA, Food Grains Analysis Group, (202) 720-2891.

Table 1-31.—Rice: United States exports (milled basis), by country of destination, 1999–2001 [1]

Country of destination	Year		
	1999	2000	2001
	1,000 metric tons	*1,000 metric tons*	*1,000 metric tons*
Mexico	304	411	367
Haiti	213	219	184
Canada	180	180	183
Honduras	61	80	97
Costa Rica	42	49	42
Jamaica	26	34	19
Dominican Republic	73	55	15
Bahamas	9	9	7
Total Western Hemisphere [2]	1,110	1,210	1,203
European Union:			
United Kingdom	113	127	93
Germany	69	54	59
Netherlands	47	56	47
France	23	25	36
Switzerland	25	28	28
Belgium-Luxembourg	33	21	21
Sweden	5	2	3
Total Europe [2]	480	419	383
Japan	311	347	307
Saudi Arabia	119	164	139
Indonesia	169	79	21
Total Asia, Middle East, and Oceania [2]	829	1,014	753
Ghana	77	65	79
South Africa	84	55	56
Cote D'ivoire	25	17	17
Liberia	10	14	14
Guinea	5	4	5
Total Africa [2]	225	204	201
World total [2]	2,644	2,847	2,540

[1] Year beginning Jan 1. [2] Includes countries not shown.

FAS, Grain and Feed Division, (202) 720–6219.

Table 1-32.—Rice, milled: Area, yield, and production in specified countries, 1999/2000–2001/2002 [1]

Continent and country	Area [2]			Yield per hectare			Production		
	1999/2000	2000/2001	2001/2002 [3]	1999/2000	2000/2001	2001/2002 [3]	1999/2000	2000/2001	2001/2002 [3]
	1,000 hectares	*1,000 hectares*	*1,000 hectares*	*Metric tons*	*Metric tons*	*Metric tons*	*1,000 metric tons*	*1,000 metric tons*	*1,000 metric tons*
North America:									
Mexico	84	79	64	3.20	2.72	2.89	269	215	185
United States	1,421	1,230	1,341	4.58	4.83	4.97	6,502	5,941	6,668
Total	1,505	1,309	1,405	4.50	4.70	4.88	6,771	6,156	6,853
South America:									
Argentina	190	151	120	3.09	3.64	3.42	588	550	410
Bolivia	145	145	145	1.38	1.38	1.38	200	200	200
Brazil	3,655	3,142	3,178	2.13	2.25	2.25	7,768	7,062	7,160
Chile	26	29	30	2.62	2.62	2.67	68	76	80
Colombia	493	460	462	2.83	2.89	2.89	1,394	1,330	1,335
Ecuador	250	180	272	1.76	1.89	1.89	440	340	513
Guyana	147	150	150	2.48	2.43	2.47	365	365	370
Paraguay	26	27	27	2.62	2.63	2.74	68	71	74
Peru	264	304	310	4.73	4.69	4.84	1,250	1,425	1,500
Surinam	55	55	55	1.82	1.82	1.73	100	100	95
Uruguay	183	150	160	4.62	4.67	4.11	846	700	658
Venezuela	150	147	105	3.53	3.40	3.39	530	500	356
Total	5,584	4,940	5,014	2.44	2.57	2.54	13,617	12,719	12,751
Central America:									
Costa Rica	68	65	56	2.69	2.75	2.88	183	179	161
El Salvador	13	6	6	3.08	5.17	4.00	40	31	24
Guatemala	13	13	13	1.92	1.77	1.69	25	23	22
Honduras	3	3	3	2.33	2.33	2.33	7	7	7
Nicaragua	62	93	83	2.19	1.89	1.84	136	176	153
Panama	90	90	90	1.62	1.62	1.62	146	146	146
Total	249	270	251	2.16	2.08	2.04	537	562	513
Carribean:									
Cuba	113	90	100	2.12	1.83	1.69	240	165	169
Dominican Rep.	76	90	99	3.42	3.11	3.13	260	280	310
Haiti	50	52	40	1.20	1.50	1.63	60	78	65
Trinidad	10	10	10	2.00	2.00	2.00	20	20	20
Total	249	242	249	2.33	2.24	2.27	580	543	564
European Union:									
France	17	19	18	3.65	3.37	3.44	62	64	62
Greece	20	20	15	5.45	5.55	6.67	109	111	100
Italy	221	220	218	4.00	3.34	3.38	884	734	737
Portugal	26	24	24	4.08	4.17	4.17	106	100	100
Spain	113	115	113	5.22	4.85	5.50	590	558	621
Total	397	398	388	4.41	3.94	4.18	1,751	1,567	1,620
Eastern Europe:									
Bulgaria	2	2	2	1.50	1.50	1.50	3	3	3
Hungary	5	5	5	2.00	2.00	2.00	10	10	10
Macedonia (Skopje)	10	10	10	1.50	1.50	1.50	15	15	15
Romania	6	6	6	1.17	1.17	1.17	7	7	7
Total	23	23	23	1.52	1.52	1.52	35	35	35
Fmr. Soviet Union:.									
Kazakhstan	70	70	65	1.86	1.86	2.69	130	130	175
Kyrgyzstan	5	5	5	1.00	1.00	1.00	5	5	5
Russian Fed.	173	175	154	1.67	2.18	2.10	289	381	323
Tajikistan	10	13	13	1.10	1.23	1.31	11	16	17
Turkmenistan	30	20	15	0.70	0.90	0.67	21	18	10
Ukraine	22	25	25	2.91	3.60	2.76	64	90	69
Uzbekistan	164	45	27	1.64	2.22	1.56	269	100	42
Total	474	353	304	1.66	2.10	2.11	789	740	641
Middle East:.									
Iran	580	490	480	2.70	2.68	2.71	1,565	1,312	1,300
Iraq	120	100	100	1.00	0.90	0.90	120	90	90
Turkey	80	80	70	2.75	2.88	2.97	220	230	208
Total	780	670	650	2.44	2.44	2.46	1,905	1,632	1,598

See footnotes at end of table.

Table 1-32.—Rice, milled: Area, yield, and production in specified countries, 1999/2000–2001/2002 [1]—Continued

Continent and country	Area [2]			Yield per hectare			Production		
	1999/ 2000	2000/ 2001	2001/ 2002 [3]	1999/ 2000	2000/ 2001	2001/ 2002 [3]	1999/ 2000	2000/ 2001	2001/ 2002 [3]
	1,000 hectares	*1,000 hectares*	*1,000 hectares*	*Metric tons*	*Metric tons*	*Metric tons*	*1,000 metric tons*	*1,000 metric tons*	*1,000 metric tons*
Africa:									
Algeria	1	1	1	1.00	1.00	1.00	1	1	1
Angola	10	10	10	0.90	0.90	0.90	9	9	9
Benin	18	24	25	1.33	0.92	1.28	24	22	32
Burkina	38	40	39	1.63	1.38	1.67	62	55	65
Chad	20	20	20	2.00	2.40	2.55	40	48	51
Cameroon	89	51	90	1.07	0.92	0.97	95	47	87
Cote d'Ivorie	550	580	600	0.99	0.98	0.97	545	570	580
Egypt	654	655	563	5.79	6.05	6.35	3,787	3,965	3,575
Gambia, The	16	18	20	1.19	1.17	1.10	19	21	22
Ghana	105	115	150	1.20	1.30	1.24	126	150	186
Guinea	500	580	580	0.98	0.98	0.89	488	566	514
Guinea-Bissau	68	70	55	0.76	0.97	1.07	52	68	59
Kenya	15	17	17	2.87	1.94	2.12	43	33	36
Liberia	154	150	130	0.77	0.80	0.67	118	120	87
Madagascar	1,227	1,200	1,200	1.38	1.23	1.23	1,690	1,470	1,470
Malawi	46	44	57	1.30	1.48	1.09	60	65	62
Mali	325	350	365	1.48	1.41	1.52	480	492	554
Mauritania	20	25	18	3.35	2.80	2.17	67	70	39
Morocco	8	8	8	5.00	5.00	5.00	40	40	40
Mozambique	180	160	174	0.69	0.63	0.63	125	100	110
Niger	20	22	28	2.05	1.82	1.79	41	40	50
Nigeria	1,660	1,650	1,650	1.20	1.21	1.27	2,000	2,000	2,100
Senegal	96	95	120	1.63	1.47	1.47	156	140	176
Sierra Leone	215	185	285	0.70	0.65	0.65	150	120	186
Somalia	3	3	3	1.33	1.33	1.33	4	4	4
Sudan	9	5	9	0.78	1.00	0.78	7	5	7
Swaziland	2	2	2	1.50	1.50	1.50	3	3	3
Tanzania, United Rep. of	475	500	530	0.93	0.96	0.85	442	479	452
Togo	38	32	45	1.39	1.66	1.22	53	53	55
Zaire	463	447	431	0.45	0.46	0.45	210	204	196
Zambia	15	15	15	0.47	0.47	0.47	7	7	7
Total	7,040	7,074	7,240	1.55	1.55	1.49	10,944	10,967	10,815
Asia:									
Afghanistan	140	130	121	1.34	0.81	1.01	188	105	122
Bangladesh	10,708	10,887	11,000	2.15	2.30	2.32	23,066	25,086	25,500
Brunei	3	3	3	1.33	1.33	1.33	4	4	4
Burma	6,000	6,000	6,200	1.64	1.80	1.68	9,860	10,771	10,440
Cambodia	2,070	1,903	1,980	1.23	1.33	1.30	2,545	2,536	2,583
China Peoples Republic	31,284	29,962	28,812	4.44	4.39	4.31	138,936	131,536	124,306
India	45,160	44,361	44,790	1.99	1.91	2.05	89,700	84,871	91,600
Indonesia	11,650	11,600	11,500	2.87	2.81	2.82	33,445	32,548	32,422
Japan	1,788	1,770	1,706	4.67	4.88	4.83	8,350	8,636	8,242
Korea, Democratic Pe	600	550	550	2.67	2.36	2.45	1,600	1,300	1,350
Korea, Republic of	1,066	1,072	1,083	4.94	4.94	5.09	5,263	5,291	5,515
Laos	718	719	690	1.75	1.84	2.04	1,260	1,320	1,410
Malaysia	660	665	672	1.95	2.12	2.20	1,290	1,410	1,480
Nepal	1,500	1,500	1,500	1.65	1.65	1.67	2,470	2,470	2,500
Pakistan	2,515	2,350	2,115	2.05	2.00	1.84	5,156	4,700	3,882
Philippin	3,995	4,030	4,080	1.95	2.02	2.07	7,772	8,135	8,450
Sri Lanka	720	770	735	2.70	2.52	2.48	1,947	1,940	1,820
Taiwan	353	340	338	3.82	3.95	3.68	1,349	1,342	1,245
Thailand	9,970	9,920	9,990	1.65	1.70	1.65	16,500	16,901	16,500
Vietnam	7,660	7,493	7,471	2.73	2.73	2.82	20,926	20,473	21,036
Total	138,560	136,025	135,336	2.68	2.66	2.66	371,627	361,375	360,407
Oceania:									
Australia	133	186	151	5.92	6.77	6.16	787	1,259	930
World total	154,994	151,490	151,011	2.64	2.62	2.63	409,343	397,555	396,727

[1] Crop year beginning Aug. 1. Crops harvested in the Northern Hemisphere during the latter part of the year are combined with those harvested in Asia principally from November to May, and in the Southern Hemisphere harvested during the first part of the following year. [2] Harvested area as far as possible. [3] Preliminary.

FAS, Production Estimates and Crop Assessment Division, (202) 720–0888. Prepared or estimated on the basis of official statistics of foreign governments, other foreign source materials, reports of U.S. Agricultural Counselors, Attachés and Foreign Service Officers, results of office research, and related information.

Table 1-33.—Rice, milled equivalent: [1] International trade, 2000–2002 [2]

Country	2000	2001	2002 [3]
	1,000 metric tons	*1,000 metric tons*	*1,000 metric tons*
Exports:			
Argentina	332	363	250
Australia	617	618	350
Burma	159	670	1,000
China	2,951	1,847	1,963
Egypt	500	705	400
Guyana	167	175	150
India	1,449	1,936	6,600
Pakistan	2,026	2,417	1,500
Thailand	6,549	7,521	7,245
Uruguay	642	806	526
Vietnam	3,370	3,528	3,245
EU	308	264	350
Others	929	1,051	942
Subtotal	19,999	21,901	24,521
United States	2,847	2,541	3,250
Total	22,846	24,442	27,771
Imports:			
Bangladesh	638	402	300
Brazil	700	673	550
Canada	250	262	235
China	278	267	305
Colombia	60	163	28
Costa Rica	48	41	87
Cote d'Ivoire	450	654	700
Cuba	415	481	538
El Salvador	30	76	90
Ghana	186	211	325
Guinea	275	325	300
Haiti	245	250	300
Honduras	80	96	100
Indonesia	1,500	1,500	3,500
Iran	1,100	765	1,000
Iraq	1,274	959	1,250
Jamaica & Dep	75	75	20
Japan	656	680	650
Korea, North	400	537	654
Korea, South	151	99	120
Malaysia	596	633	600
Mexico	415	388	500
Nigeria	1,250	1,906	1,700
Nicaragua	60	117	110
Peru	86	62	32
Philippines	900	1,175	1,200
Russia	400	247	410
Saudi Arabia	992	1,053	900
Senegal	502	863	900
Singapore	354	444	375
South Africa	523	572	800
Sri Lanka	18	35	80
Syria	150	172	175
Taiwan	3	23	125
Turkey	309	231	325
Uzbekistan	30	142	100
UAE	75	75	80
Yemen	210	202	200
EU	852	923	700
O.W. Europe	50	55	50
Eastern Europe	343	381	409
United States	308	413	400
Subtotal	17,237	18,628	21,223
Other Countries	3,844	4,112	4,401
Unaccounted	1,765	1,702	2,147
World Total	22,846	24,442	27,771

[1] Includes milled, semi-milled, broken, and rough rice in terms of milled equivalent. [2] Year beginning Jan 1. [3] Preliminary.

FAS, Grain and Feed Division, (202) 720–6219. Prepared or estimated on the basis of official statistics frpm foreign governments, other foreign source materials, reports of U.S. Agricultural Counselors, Attachés and Foreign Service Officers, results of office research, and related information.

Table 1-34.—Food grains: Average price, selected markets and grades, 1995–2002[1]

Crop year[2]	Kansas City: Wheat, No. 1 Hard Winter, Ordinary Protein	Kansas City: Wheat, No. 1 Hard Winter, 13% protein	Kansas City: Wheat, No. 2 Soft Red Winter	Minneapolis: Wheat, No. 1 Hard Amber Durum (milling)	Minneapolis: Wheat, No. 1 Dark Northern Spring, 14% protein	Minneapolis: Rye, No. 2, 20 day delivery	Portland Wheat No. 1 Soft White	St. Louis Wheat, No. 2 Soft Red Winter
	Dollars per bushel	*Dollars per bushel*	*Dollars per bushel*	*Dollars per bushel*	*Dollars per bushel*	*Dollars per bushel*	*Dollars per bushel*	*Dollars per bushel*
1995	3.97	4.06	3.76	5.98	4.26	2.94	4.16	3.62
1996	5.49	5.69	5.40	7.04	5.72	3.33	5.27	4.82
1997	4.88	4.92	4.61	5.59	4.97	2.94	4.54	4.10
1998	3.70	3.96	3.38	5.97	4.31	....	3.81	3.43
1999	3.08	3.47	2.68	4.05	3.83	....	2.98	2.41
2000	2.87	3.41	2.53	4.23	3.61	....	3.02	2.39
2001	3.34	3.44	2.79	4.98	....	2.69	3.43	2.61
2002	3.94	3.99	3.50	4.97	4.15	3.57	3.89	3.28

Crop year[2]	Chicago Wheat, No. 2 Soft Red Winter	Denver Wheat, No. 1 Hard Winter	S.W. Louisiana Milled Rice: Medium	S.W. Louisiana Milled Rice: Long	Arkansas Milled Rice: Medium	Arkansas Milled Rice: Long	Texas Milled Rice: Long
	Dollars per bushel	*Dollars per bushel*	*Dollars per cwt.*	*Dollars per cwt.*	*Dollars per cwt.*	*Dollars per cwt.*	*Dollars per cwt.*
1995	3.52	3.57	14.07	14.54	13.16	13.25	13.39
1996	4.82	5.03	19.40	19.46	19.22	19.63	19.62
1997	3.92	4.43	19.91	19.98	18.66	20.42	20.87
1998	3.29	3.33	18.70	18.27	18.34	18.46	18.94
1999	2.46	2.79	18.45	15.31	19.96	15.41	17.02
2000	2.19	2.29	13.10	12.22	13.56	12.24	14.82
2001	2.04	2.89	....	....	....	....	....
2002	2.30	3.53	....	....	....	....	....

[1] Simple average of daily prices. [2] For wheat and rye, crop year begins in June. For rice, crop year begins in August.
AMS, Livestock and Grain Market News branch, (202) 720–6231.

Table 1-35.—Corn: Area, yield, production, and value, United States, 1993–2002

Year	Area planted, all purposes	Corn for grain: Area harvested	Corn for grain: Yield per harvested acre	Corn for grain: Production	Corn for grain: Marketing year average price per bushel	Corn for grain: Value of production	Corn for silage: Area harvested	Corn for silage: Yield per harvested acre	Corn for silage: Production
	1,000 acres	*1,000 acres*	*Bushels*	*1,000 bushels*	*Dollars*	*1,000 dollars*	*1,000 acres*	*Tons*	*1,000 tons*
1993	73,239	62,933	100.7	6,337,730	2.50	16,035,515	6,823	11.9	81,131
1994	78,921	72,514	138.6	10,050,520	2.26	22,874,154	5,717	15.8	90,170
1995	71,479	65,210	113.5	7,400,051	3.24	24,202,234	5,321	14.7	78,181
1996	79,229	72,644	127.1	9,232,557	2.71	25,149,013	5,607	15.4	86,581
1997	79,537	72,671	126.7	9,206,832	2.43	22,351,507	6,054	16.1	97,192
1998	80,165	72,589	134.4	9,758,685	1.94	18,922,084	5,913	16.1	95,479
1999	77,386	70,487	133.8	9,430,612	1.82	17,103,991	6,037	15.8	95,633
2000	79,551	72,440	136.9	9,915,051	1.85	18,499,002	6,082	16.8	102,156
2001	75,752	68,808	138.2	9,506,840	1.97	18,888,389	6,148	16.6	102,077
2002[1]	79,054	69,313	130.0	9,007,659	2.35	21,213,159	7,490	14.0	104,979

[1] Preliminary.
NASS, Crops Branch, (202) 720–2127.

Table 1-36.—Corn: Stocks on and off farms, United States, 1994–2003

Year beginning previous December	On farms: Dec. 1	On farms: Mar. 1	On farms: Jun. 1	On farms: Sep. 1[1]	Off farms[2]: Dec. 1	Off farms[2]: Mar. 1	Off farms[2]: Jun. 1	Off farms[2]: Sep. 1[1]
	1,000 bushels	*1,000 bushels*	*1,000 bushels*	*1,000 bushels*	*1,000 bushels*	*1,000 bushels*	*1,000 bushels*	*1,000 bushels*
1994	3,803,000	2,210,200	1,203,000	395,400	2,133,548	1,785,514	1,156,881	454,743
1995	5,417,500	3,502,000	2,072,000	740,900	2,662,980	2,089,693	1,342,896	816,940
1996	3,960,000	2,000,200	780,100	196,600	2,145,774	1,799,341	937,794	229,342
1997	4,800,000	2,870,000	1,501,000	475,000	2,102,974	1,624,128	995,550	408,161
1998	4,822,000	2,975,000	1,830,000	640,000	2,424,756	1,964,898	1,209,757	667,803
1999	5,320,000	3,570,000	2,257,000	797,000	2,731,846	2,128,428	1,359,225	989,977
2000	5,195,000	3,300,000	2,029,800	793,000	2,844,443	2,301,895	1,556,138	924,549
2001	5,550,000	3,600,000	2,230,800	753,150	2,979,634	2,442,999	1,693,158	1,145,958
2002	5,275,000	3,355,000	2,020,600	586,800	2,989,715	2,440,263	1,576,290	1,009,626
2003[3]	4,800,000	NA	NA	NA	2,833,427	NA	NA	NA

[1] Old crop only. [2] Includes stocks at mills, elevators, warehouses, terminals, and processors. [3] Preliminary. NA-not available.
NASS, Crops Branch, (202) 720–2127.

Table 1-37.—Corn: Area, yield, and production, by States, 2000–2002

State	Area planted for all purposes			Corn for grain								
				Area harvested			Yield per harvested acre			Production		
	2000	2001	2002[1]	2000	2001	2002[1]	2000	2001	2002[1]	2000	2001	2002[1]
	1,000 acres	*1,000 acres*	*1,000 acres*	*1,000 acres*	*1,000 acres*	*1,000 acres*	*Bushels*	*Bushels*	*Bushels*	*1,000 bushels*	*1,000 bushels*	*1,000 bushels*
AL	230	180	200	165	150	180	65.0	107.0	88.0	10,725	16,050	15,840
AZ	56	55	60	33	28	28	196.0	208.0	185.0	6,468	5,824	5,180
AR	180	190	270	175	185	260	130.0	145.0	134.0	22,750	26,825	34,840
CA	540	480	545	205	160	150	170.0	170.0	170.0	34,850	27,200	25,500
CO	1,350	1,220	1,200	1,150	1,070	720	126.0	140.0	156.0	144,900	149,800	112,320
CT	36	32	32	(2)	(2)	(2)	(2)	(2)	(2)	(2)	(2)	(2)
DE	165	170	180	155	162	167	162.0	146.0	83.0	25,110	23,652	13,861
FL	85	65	75	25	26	34	75.0	87.0	96.0	1,875	2,262	3,264
GA	360	265	340	240	220	290	107.0	134.0	115.0	25,680	29,480	33,350
ID	195	175	190	57	45	50	160.0	150.0	160.0	9,120	6,750	8,000
IL	11,200	11,000	11,200	11,050	10,850	11,000	151.0	152.0	136.0	1,668,550	1,649,200	1,496,000
IN	5,700	5,800	5,400	5,550	5,670	5,220	146.0	156.0	121.0	810,300	884,520	631,620
IA	12,300	11,700	12,300	12,000	11,400	11,900	144.0	146.0	165.0	1,728,000	1,664,400	1,963,500
KS	3,450	3,450	3,250	3,170	3,050	2,500	130.0	127.0	116.0	412,100	387,350	290,000
KY	1,330	1,200	1,130	1,230	1,100	1,040	130.0	142.0	102.0	159,900	156,200	106,080
LA	380	315	580	370	307	560	116.0	148.0	122.0	42,920	45,436	68,320
ME	29	28	29	(2)	(2)	(2)	(2)	(2)	(2)	(2)	(2)	(2)
MD	480	490	510	405	410	425	155.0	136.0	76.0	62,775	55,760	32,300
MA	25	22	22	(2)	(2)	(2)	(2)	(2)	(2)	(2)	(2)	(2)
MI	2,200	2,200	2,250	1,950	1,900	2,020	124.0	105.0	115.0	241,800	199,500	232,300
MN	7,200	6,800	7,200	6,650	6,200	6,700	145.0	130.0	157.0	964,250	806,000	1,051,900
MS	390	400	550	365	385	530	100.0	130.0	125.0	36,500	50,050	66,250
MO	2,850	2,700	2,800	2,770	2,600	2,700	143.0	133.0	105.0	396,110	345,800	283,500
MT	60	65	65	16	13	13	140.0	148.0	140.0	2,240	1,924	1,820
NE	8,500	8,100	8,400	8,050	7,750	7,350	126.0	147.0	128.0	1,014,300	1,139,250	940,800
NV	4	3	4	(2)	(2)	(2)	(2)	(2)	(2)	(2)	(2)	(2)
NH	15	15	16	(2)	(2)	(2)	(2)	(2)	(2)	(2)	(2)	(2)
NJ	90	80	90	75	66	70	134.0	112.0	58.0	10,050	7,392	4,060
NM	150	130	135	66	46	49	160.0	180.0	180.0	10,560	8,280	8,820
NY	980	1,030	1,040	450	540	450	98.0	105.0	97.0	44,100	56,700	43,650
NC	730	700	790	640	625	700	116.0	125.0	83.0	74,240	78,125	58,100
ND	1,080	880	1,230	930	705	995	112.0	115.0	115.0	104,160	81,075	114,425
OH	3,550	3,400	3,200	3,300	3,170	2,870	147.0	138.0	88.0	485,100	437,460	252,560
OK	270	250	240	240	210	190	140.0	125.0	130.0	33,600	26,250	24,700
OR	55	45	62	27	18	27	180.0	140.0	115.0	4,860	2,520	3,105
PA	1,550	1,500	1,450	1,080	990	870	127.0	98.0	68.0	137,160	97,020	59,160
RI	2	2	2	(2)	(2)	(2)	(2)	(2)	(2)	(2)	(2)	(2)
SC	310	260	320	280	240	260	65.0	108.0	46.0	18,200	25,920	11,960
SD	4,300	3,800	4,400	3,800	3,400	3,200	112.0	109.0	95.0	425,600	370,600	304,000
TN	650	680	690	580	620	620	114.0	132.0	107.0	66,120	81,840	66,340
TX	2,100	1,600	2,050	1,900	1,420	1,820	124.0	118.0	113.0	235,600	167,560	205,660
UT	64	60	55	18	15	14	144.0	142.0	145.0	2,592	2,130	2,030
VT	90	90	92	(2)	(2)	(2)	(2)	(2)	(2)	(2)	(2)	(2)
VA	470	470	500	330	330	305	146.0	123.0	66.0	48,180	40,590	20,130
WA	155	115	130	100	55	70	185.0	190.0	190.0	18,500	10,450	13,300
WV	55	50	50	35	26	30	130.0	120.0	105.0	4,550	3,120	3,150
WI	3,500	3,400	3,650	2,750	2,600	2,900	132.0	127.0	135.0	363,000	330,200	391,500
WY	90	90	80	58	51	36	132.0	125.0	124.0	7,656	6,375	4,464
US	79,551	75,752	79,054	72,440	68,808	69,313	136.9	138.2	130.0	9,915,051	9,506,840	9,007,659

[1] Preliminary. [2] Not estimated.

NASS, Crops Branch, (202) 720–2127.

Table 1-38.—Corn: Supply and disappearance, United States, 1993–2002

Year beginning September 1	Supply				Disappearance					Ending stocks Aug. 31		
					Domestic use							
	Beginning stocks	Production	Imports	Total	Feed and residual	Food, seed, and industrial	Total	Exports	Total disappearance	Privately held[1]	Government	Total
	Million bushels	*Million bushels*	*Million bushels*	*Million bushels*	*Million bushels*	*Million bushels*	*Million bushels*	*Million bushels*	*Million bushels*	*Million bushels*	*Million bushels*	*Million bushels*
1993	2,113	6,338	21	8,472	4,680	1,613	6,293	1,328	7,621	805	45	850
1994	850	10,051	10	10,910	5,460	1,715	7,175	2,177	9,352	1,516	42	1,558
1995	1,558	7,400	16	8,974	4,692	1,628	6,321	2,228	8,548	396	30	426
1996	426	9,233	13	9,672	5,277	1,714	6,991	1,797	8,789	881	2	883
1997	883	9,207	9	10,099	5,482	1,805	7,287	1,504	8,791	1,304	4	1,308
1998	1,308	9,759	19	11,085	5,471	1,846	7,318	1,984	9,298	1,775	12	1,787
1999	1,787	9,431	15	11,232	5,664	1,913	7,578	1,937	9,515	1,704	14	1,718
2000[2]	1,718	9,915	7	11,639	5,842	1,957	7,799	1,941	9,740	1,891	8	1,899
2001	1,899	9,507	10	11,416	5,877	2,054	7,931	1,889	9,820	1,591	5	1,596
2002[3]	1,596	9,008	15	10,619	5,600	2,265	7,865	1,825	9,690	924	5	929

[1] Includes quantity under loan and farmer–owned reserve. [2] Preliminary. [3] Projected as of February 11, 2003, World Agricultural Supply and Demand Estimates. Totals may not add due to independent rounding.

ERS, Market and Trade Economics Division, (202) 694–5296.

Table 1-39.—Corn: Utilization for silage, by States, 2000–2002

State	Silage								
	Area harvested			Yield per acre			Production		
	2000	2001	2002[1]	2000	2001	2002[1]	2000	2001	2002[1]
	1,000 acres	*1,000 acres*	*1,000 acres*	*Tons*	*Tons*	*Tons*	*1,000 tons*	*1,000 tons*	*1,000 tons*
AL	25	25	15	9.0	13.0	12.0	225	325	180
AZ	22	26	31	25.0	27.0	25.0	550	702	775
AR	4	3	5	12.0	12.0	15.0	48	36	75
CA	330	315	390	26.0	26.0	26.0	8,580	8,190	10,140
CO	110	115	200	22.0	23.0	18.0	2,420	2,645	3,600
CT	33	30	29	19.0	19.0	18.0	627	570	522
DE	9	7	10	22.0	18.0	14.0	198	126	140
FL	48	33	34	16.0	18.0	18.0	768	594	612
GA	45	40	40	15.0	18.0	16.0	675	720	640
ID	135	125	135	25.0	25.0	25.5	3,375	3,125	3,443
IL	115	115	130	16.0	16.0	16.0	1,840	1,840	2,080
IN	130	110	160	19.0	19.0	16.0	2,470	2,090	2,560
IA	250	230	330	17.0	18.5	19.0	4,250	4,255	6,270
KS	180	295	420	14.0	14.0	8.5	2,520	4,130	3,570
KY	95	95	85	18.0	19.0	15.0	1,710	1,805	1,275
LA	7	7	10	10.0	14.0	14.0	70	98	140
ME	26	25	26	17.5	19.0	17.0	455	475	442
MD	70	75	80	20.0	15.0	12.0	1,400	1,125	960
MA	20	19	18	19.5	21.0	17.5	390	399	315
MI	230	280	220	14.0	13.0	15.0	3,220	3,640	3,300
MN	475	500	425	16.0	14.0	18.0	7,600	7,000	7,650
MS	20	13	15	11.0	11.0	14.0	220	143	210
MO	60	70	70	15.0	16.0	13.0	900	1,120	910
MT	42	51	49	21.0	22.0	22.0	882	1,122	1,078
NE	290	275	475	14.0	17.0	9.5	4,060	4,675	4,513
NV	3	3	4	25.0	22.0	18.0	75	66	72
NH	14	14	15	19.5	21.0	18.0	273	294	270
NJ	14	13	18	17.0	16.0	10.0	238	208	180
NM	82	82	82	23.0	23.0	25.0	1,886	1,886	2,050
NY	530	485	580	14.0	16.0	13.0	7,420	7,760	7,540
NC	85	72	70	15.0	19.0	13.0	1,275	1,368	910
ND	140	155	180	11.0	11.0	7.0	1,540	1,705	1,260
OH	180	170	270	16.0	17.0	9.5	2,880	2,890	2,565
OK	25	23	25	17.0	18.0	21.0	425	414	525
OR	27	26	33	23.0	21.0	21.0	621	546	693
PA	460	490	560	17.0	16.0	11.5	7,820	7,840	6,440
RI	2	2	2	18.0	20.0	15.0	36	40	30
SC	15	15	10	8.0	20.0	15.0	120	300	150
SD	450	370	920	11.5	10.5	6.0	5,175	3,885	5,520
TN	65	55	55	17.0	19.0	15.0	1,105	1,045	825
TX	130	130	120	20.0	17.0	18.0	2,600	2,210	2,160
UT	45	44	40	21.0	21.0	20.0	945	924	800
VT	85	85	85	16.5	19.0	16.0	1,403	1,615	1,360
VA	135	135	170	19.0	15.5	11.5	2,565	2,093	1,955
WA	55	60	60	26.0	26.0	26.0	1,430	1,560	1,560
WV	19	23	19	19.0	17.0	16.5	361	391	314
WI	720	780	730	16.5	14.5	16.0	11,880	11,310	11,680
WY	30	37	40	21.0	21.0	18.0	630	777	720
US	6,082	6,148	7,490	16.8	16.6	14.0	102,156	102,077	104,979

[1] Preliminary.

NASS, Crops Branch, (202) 720–2127.

Table 1-40.—Corn for grain: Marketing year average price and value, by States, crop of 2000, 2001, and 2002

State	Marketing year average price per bushel			Value of production		
	2000	2001	2002 [1]	2000	2001	2002 [1]
	Dollars	*Dollars*	*Dollars*	*1,000 dollars*	*1,000 dollars*	*1,000 dollars*
AL	2.16	2.35	2.70	23,166	37,718	42,768
AZ	2.78	2.79	3.10	17,981	16,249	16,058
AR	1.75	2.02	2.40	39,813	54,187	83,616
CA	2.44	2.45	2.80	85,034	66,640	71,400
CO	2.08	2.13	2.55	301,392	319,074	286,416
DE	2.01	2.15	2.85	50,471	50,852	39,504
FL	2.24	2.25	2.60	4,200	5,090	8,486
GA	2.06	2.32	2.65	52,901	68,394	88,378
ID	2.63	2.64	2.90	23,986	17,820	23,200
IL	1.91	2.04	2.40	3,186,931	3,364,368	3,590,400
IN	1.90	1.98	2.45	1,539,570	1,751,350	1,547,469
IA	1.75	1.90	2.25	3,024,000	3,162,360	4,417,875
KS	2.00	2.03	2.50	824,200	786,321	725,000
KY	2.07	2.08	2.55	330,993	324,896	270,504
LA	1.75	2.20	2.40	75,110	99,959	163,968
MD	2.01	2.18	2.85	126,178	121,557	92,055
MI	1.90	1.97	2.30	459,420	393,015	534,290
MN	1.71	1.90	2.15	1,648,868	1,531,400	2,261,585
MS	1.91	1.99	2.35	69,715	99,600	155,688
MO	1.78	1.96	2.45	705,076	677,768	694,575
MT	1.53	1.89	2.45	3,427	3,636	4,459
NE	1.90	1.94	2.40	1,927,170	2,210,145	2,257,920
NJ	1.97	2.09	2.75	19,799	15,449	11,165
NM	2.52	2.40	2.85	26,611	19,872	25,137
NY	2.35	2.51	2.85	103,635	142,317	124,403
NC	2.01	2.36	2.90	149,222	184,375	168,490
ND	1.65	1.87	2.15	171,864	151,610	246,014
OH	1.90	2.00	2.50	921,690	874,920	631,400
OK	2.10	2.10	2.45	70,560	55,125	60,515
OR	2.41	2.54	2.90	11,713	6,401	9,005
PA	2.09	2.33	2.95	286,664	226,057	174,522
SC	2.10	2.20	2.60	38,220	57,024	31,096
SD	1.61	1.75	2.20	685,216	648,550	668,800
TN	1.96	2.06	2.55	129,595	168,590	169,167
TX	2.18	2.29	2.60	513,608	383,712	534,716
UT	2.61	2.85	3.30	6,765	6,071	6,699
VA	2.02	2.14	2.75	97,324	86,863	55,358
WA	2.53	2.56	2.90	46,805	26,752	38,570
WV	2.08	2.29	2.90	9,464	7,145	9,135
WI	1.86	1.97	2.20	675,180	650,494	861,300
WY	2.02	2.30	2.70	15,465	14,663	12,053
US	1.85	1.97	2.35	18,499,002	18,888,389	21,213,159

[1] Preliminary.
NASS, Crops Branch, (202) 720–2127.

Table 1-41.—Corn: Area, yield, and production in specified countries, 1999/2000–2001/2002 [1]

Continent and country	Area [2]			Yield per hectare			Production		
	1999/2000	2000/2001	2001/2002 [3]	1999/2000	2000/2001	2001/2002 [3]	1999/2000	2000/2001	2001/2002 [3]
	1,000 hectares	*1,000 hectares*	*1,000 hectares*	*Metric tons*	*Metric tons*	*Metric tons*	*1,000 metric tons*	*1,000 metric tons*	*1,000 metric tons*
North America:									
Canada	1,141	1,088	1,260	8.03	6.27	6.51	9,161	6,827	8,200
Mexico	7,255	7,145	7,780	2.65	2.51	2.52	19,240	17,920	19,600
United States	28,525	29,316	27,846	8.40	8.59	8.67	239,549	251,854	241,485
Total	36,921	37,549	36,886	7.26	7.37	7.30	267,950	276,601	269,285
Central America:									
Costa Rica	5	5	4	2.00	2.00	2.00	10	10	8
El Salvador	290	260	294	2.17	2.23	1.92	630	581	564
Guatemala	835	835	590	1.44	1.44	1.78	1,200	1,200	1,050
Honduras	372	379	320	1.43	1.43	1.31	533	542	420
Nicaragua	262	327	333	1.14	1.26	1.20	298	411	399
Panama	62	62	62	1.29	1.29	1.29	80	80	80
Total	1,826	1,868	1,603	1.51	1.51	1.57	2,751	2,824	2,521

See footnotes at end of table.

Table 1-41.—Corn: Area, yield, and production in specified countries, 1999/2099–2001/2002 [1]—Continued

Continent and country	Area [2]			Yield per hectare			Production		
	1999/ 2000	2000/ 2001	2001/ 2002 [3]	1999/ 2000	2000/ 2001	2001/ 2002 [3]	1999/ 2000	2000/ 2001	2001/ 2002 [3]
	1,000 hectares	*1,000 hectares*	*1,000 hectares*	*Metric tons*	*Metric tons*	*Metric tons*	*1,000 metric tons*	*1,000 metric tons*	*1,000 metric tons*
South America:									
Argentina	3,100	2,818	2,450	5.55	5.46	5.88	17,200	15,400	14,400
Bolivia	300	175	300	2.27	1.71	2.17	680	300	650
Brazil	12,578	12,972	11,988	2.52	3.20	2.96	31,641	41,536	35,536
Chile	69	82	87	9.45	9.49	9.20	652	778	800
Colombia	520	530	550	1.87	2.04	2.02	973	1,080	1,110
Ecuador	300	196	180	1.33	2.63	2.22	400	516	400
Guyana	5	5	5	1.00	1.00	1.00	5	5	5
Paraguay	332	353	375	1.95	2.50	2.13	647	883	800
Peru	420	510	518	2.56	2.41	2.68	1,074	1,230	1,387
Uruguay	42	62	57	1.55	4.31	3.51	65	267	200
Venezuela	450	450	450	2.89	2.67	2.67	1,300	1,200	1,200
Total	18,116	18,153	16,960	3.02	3.48	3.33	54,637	63,195	56,488
Carribean:									
Cuba	75	75	75	0.93	0.93	0.93	70	70	70
Dominican Rep.	30	30	32	1.23	1.27	1.22	37	38	39
Haiti	350	350	350	0.86	0.86	0.86	300	300	300
Jamaica & Dep.	4	4	4	1.25	0.50	0.50	5	2	2
Trinidad	3	3	3	1.67	1.67	1.67	5	5	5
Total	462	462	464	0.90	0.90	0.90	417	415	416
European Union:									
Austria	152	162	170	9.46	9.44	8.71	1,438	1,530	1,481
Belgium-Luxembourg	28	30	40	8.75	8.07	8.00	245	242	320
France	1,759	1,796	1,889	8.89	9.04	8.52	15,643	16,227	16,100
Germany	371	361	397	8.78	9.21	8.83	3,257	3,324	3,505
Greece	120	135	120	8.00	7.90	6.80	960	1,067	816
Italy	1,028	1,064	1,098	9.74	9.53	9.44	10,017	10,137	10,367
Netherlands	24	27	35	5.50	6.04	5.94	132	163	208
Portugal	164	196	205	5.70	4.45	4.45	935	873	913
Spain	399	425	500	9.47	9.17	10.20	3,777	3,897	5,100
Sweden	1	1	1	0.00	0.00	0.00	0	0	0
Total	4,046	4,197	4,455	9.00	8.93	8.71	36,404	37,460	38,810
Eastern Europe:									
Albania	60	60	60	3.33	3.33	3.33	200	200	200
Bosnia-Hercegovina	214	208	213	4.60	2.28	3.00	984	475	640
Bulgaria	380	380	340	4.21	1.97	2.56	1,600	750	870
Croatia	384	350	380	5.56	5.29	5.53	2,135	1,850	2,100
Czech Republic	39	45	55	6.67	6.76	6.35	260	304	349
Hungary	1,200	1,190	1,300	5.83	4.20	5.85	7,000	5,000	7,600
Macedonia (Skopje)	46	45	45	4.35	2.78	2.22	200	125	100
Poland	104	152	224	5.76	6.07	6.08	599	923	1,362
Romania	3,000	2,860	2,910	3.50	1.68	2.41	10,500	4,800	7,000
Serbia	1,267	1,207	1,300	4.85	2.44	4.77	6,140	2,944	6,200
Slovakia	130	120	140	5.99	3.67	5.14	779	440	720
Slovenia	44	44	45	7.00	7.00	7.44	308	308	335
Total	6,868	6,661	7,012	4.47	2.72	3.92	30,705	18,119	27,476
Oth. Western Europe:									
Switzerland	22	22	24	9.00	9.95	9.25	198	219	222
Fmr. Soviet Union:									
Azerbaijan	29	32	30	3.45	3.13	3.33	100	100	100
Georgia	220	178	200	2.23	1.52	1.50	491	270	300
Kazakhstan	65	75	100	3.08	3.33	3.20	200	250	320
Kyrgyzstan	60	60	75	5.00	5.67	5.87	300	340	440
Moldova	407	420	470	2.21	2.45	2.34	900	1,030	1,100
Russian Fed.	700	800	700	1.57	1.94	1.14	1,100	1,550	800
Tajikistan	12	11	13	2.92	2.27	1.92	35	25	25
Turkmenistan	20	25	25	2.50	2.00	2.00	50	50	50
Ukraine	689	1,279	1,150	2.52	3.01	3.13	1,737	3,848	3,600
Uzbekistan	55	20	15	3.00	2.75	2.00	165	55	30
Total	2,257	2,900	2,778	2.25	2.59	2.44	5,078	7,518	6,765
Africa:									
Algeria	1	1	1	1.00	1.00	1.00	1	1	1
Angola	675	650	650	0.63	0.62	0.62	425	400	400
Benin	575	600	600	1.29	1.10	1.10	740	660	660
Botswana	20	20	30	0.25	0.25	0.20	5	5	6
Burkina	250	250	250	1.70	1.70	2.00	425	425	500
Burundi	115	115	115	1.22	1.22	1.22	140	140	140
Cameroon	325	325	325	2.00	2.00	2.00	650	650	650
Cape Verde	15	15	15	1.67	1.33	1.33	25	20	20
Central African Republic	90	90	90	1.06	1.00	1.00	95	90	90
Chad	130	130	130	0.77	0.58	0.96	100	75	125
Congo (Brazzaville)	30	30	30	0.67	0.67	0.67	20	20	20
Cote d'Ivorie	670	690	690	1.01	0.91	0.91	675	625	625
Egypt	730	705	770	7.78	7.99	8.00	5,678	5,636	6,160

See footnotes at end of table.

Table 1-41.—Corn: Area, yield, and production in specified countries, 1990/2000–2001/2002 [1]—Continued

Continent and country	Area [2]			Yield per hectare			Production		
	1999/2000	2000/2001	2001/2002 [3]	1999/2000	2000/2001	2001/2002 [3]	1999/2000	2000/2001	2001/2002 [3]
	1,000 hectares	*1,000 hectares*	*1,000 hectares*	*Metric tons*	*Metric tons*	*Metric tons*	*1,000 metric tons*	*1,000 metric tons*	*1,000 metric tons*
Africa—Continued									
Ethiopia	1,750	1,850	2,000	1.49	1.68	1.63	2,600	3,100	3,250
Gambia, The	15	15	20	1.33	1.47	1.35	20	22	27
Ghana	700	700	710	1.45	1.45	1.32	1,015	1,015	940
Guinea	90	90	90	1.00	1.00	1.00	90	90	90
Guinea-Bissau	15	15	15	1.67	1.80	1.80	25	27	27
Kenya	1,600	1,570	1,690	1.41	1.38	1.65	2,250	2,160	2,792
Lesotho	130	138	90	0.96	0.42	0.39	125	58	35
Madagascar	200	185	190	0.90	0.86	0.92	180	160	175
Malawi	1,500	1,500	1,490	1.67	1.14	1.03	2,500	1,713	1,540
Mali	200	200	200	2.20	1.13	2.20	440	225	440
Mauritania	3	13	13	5.33	0.77	0.77	16	10	10
Morocco	250	254	250	0.80	0.21	0.80	200	54	200
Mozambique	1,100	1,275	1,320	0.93	0.90	0.94	1,020	1,150	1,240
Nigeria	3,450	3,200	3,400	1.48	1.25	1.47	5,100	4,000	5,000
Rwanda	75	75	75	0.80	0.80	1.07	60	60	80
Senegal	70	100	90	1.00	0.90	1.28	70	90	115
Sierra Leone	10	10	10	1.00	1.00	1.00	10	10	10
Somalia	160	180	200	0.69	0.83	0.50	110	150	100
South Africa, Rep. of	3,814	3,225	3,350	2.77	2.32	2.72	10,563	7,483	9,100
Swaziland	62	62	60	1.21	1.21	1.17	75	75	70
Tanzania, United Rep.	1,760	2,010	2,000	1.39	1.27	1.35	2,450	2,550	2,700
Togo	380	380	380	1.28	1.28	1.28	485	485	485
Tunisia	1	1	1	1.00	1.00	1.00	1	1	1
Uganda	600	630	630	1.17	1.27	1.27	700	800	800
Zaire	1,350	1,350	1,350	0.86	0.81	0.81	1,160	1,100	1,100
Zambia	750	600	600	1.68	1.33	1.03	1,260	800	620
Zimbabwe	1,417	1,223	1,400	1.52	1.20	0.36	2,148	1,467	500
Total	25,078	24,472	25,320	1.74	1.54	1.61	43,652	37,602	40,844
Asia:									
Afghanistan	160	96	80	1.50	1.20	2.00	240	115	160
Bhutan	45	45	45	1.56	1.56	1.56	70	70	70
Burma	203	210	250	1.69	1.71	2.10	344	359	524
Cambodia	45	50	50	1.33	1.30	1.30	60	65	65
China, People's Republic of	25,904	23,056	24,282	4.94	4.60	4.70	128,086	106,000	114,088
India	6,427	6,557	6,870	1.78	1.84	1.97	11,470	12,068	13,510
Indonesia	3,000	3,000	3,000	2.07	1.97	2.00	6,200	5,900	6,000
Japan	1	1	1	1.00	1.00	1.00	1	1	1
Korea, Democratic People's Rep	575	525	520	2.78	2.57	2.88	1,600	1,350	1,500
Korea, Rep. of	20	16	14	3.95	4.00	4.29	79	64	60
Malaysia	27	21	22	2.11	3.10	3.09	57	65	68
Nepal	880	880	880	1.70	1.70	1.70	1,500	1,500	1,500
Pakistan	870	870	875	1.44	1.44	1.43	1,250	1,250	1,250
Philippines	2,549	2,484	2,485	1.75	1.81	1.81	4,449	4,508	4,505
Taiwan	5	14	12	4.80	4.21	4.42	24	59	53
Thailand	1,150	1,210	1,180	3.39	3.88	3.81	3,900	4,700	4,500
Vietnam	690	730	727	2.54	2.75	2.91	1,750	2,005	2,112
Total	42,551	39,765	41,293	3.79	3.52	3.63	161,080	140,079	149,966
Middle East:									
Iran	150	100	100	4.67	4.50	4.50	700	450	450
Iraq	60	60	50	1.92	0.92	1.00	115	55	50
Jordan	1	1	1	1.00	1.00	1.00	1	1	1
Lebanon	2	2	2	1.00	1.00	1.00	2	2	2
Saudi Arabia	3	3	3	1.33	1.33	1.33	4	4	4
Syria	51	50	50	2.94	3.00	3.00	150	150	150
Turkey	525	550	525	3.81	3.82	3.81	2,000	2,000	2,000
Yemen	35	35	35	1.43	1.43	1.43	50	50	50
Total	827	801	766	3.65	3.51	3.53	3,022	2,812	2,707
Oceania:									
Australia	61	75	93	5.98	4.73	5.60	365	355	521
New Zealand	15	15	15	9.67	9.67	9.67	145	145	145
Total	76	90	108	6.71	5.56	6.17	510	500	666
World total	139,050	136,940	137,669	4.36	4.29	4.33	606,404	587,344	596,166

[1] Years shown refer to years of harvest. Harvests of Northern Hemisphere countries are combined with those of the Southern Hemisphere which immediately follow; thus the crop harvested in the Northern Hemisphere in 1994 is combined with estimates for the Southern Hemisphere harvest, which begins late in 1994 and ends early in 1995. [2] Harvested area as far as possible. [3] Preliminary.

FAS, Production Estimates and Crop Assessment Division, (202) 720–0888. Prepared or estimated on the basis of official statistics of foreign governments, other foreign source materials, reports of U.S. Agricultural Counselors, Attachés and Foreign Service Officers, results of office research, and related information.

Table 1-42.—Corn: International trade, 2000–2002[1]

Country	1999/2000	2000/2001	2001/2002[2]
	1,000 metric tons	*1,000 Metric tons*	*1,000 metric tons*
Exports:			
Argentina	8,859	12,229	8,581
Brazil	50	3,741	3,857
Canada	449	127	211
China	9,935	7,276	8,611
Hungary	1,786	730	2,751
Romania	400	50	200
South Africa	836	1,415	1,182
Thailand	75	407	184
Ukraine	55	397	349
EU	210	266	70
Others	1,291	1,590	1,221
Subtotal	23,946	28,228	27,217
United States	49,378	48,192	46,917
Total	73,324	76,420	74,134
Imports:			
Algeria	1,300	1,500	1,537
Brazil	1,789	671	297
Canada	1,115	2,843	4,022
Chile	1,260	1,362	1,278
China	71	89	50
Colombia	2,005	1,857	1,911
Costa Rica	550	513	463
Dominican Republic	1,000	968	1,038
Ecuador	225	149	309
Egypt	4,600	5,268	4,950
Guatemala	500	549	584
Indonesia	1,229	1,280	1,149
Iran	1,249	1,265	1,261
Israel	800	993	1,021
Japan	16,117	16,340	16,395
Jordan	450	454	439
Korea, North	150	688	288
Korea, South	8,694	8,743	8,608
Malaysia	2,296	2,588	2,425
Mexico	4,911	5,928	4,025
Morocco	750	966	850
Peru	862	861	858
Philippines	582	246	263
Poland	250	222	123
Russia	870	150	534
Saudi Arabia	1,500	1,389	1,268
South Africa	350	0	726
Syria	750	794	892
Taiwan	5,023	4,924	4,585
Tunisia	566	776	793
Turkey	1,250	608	1,193
Venezuela	1,300	1,207	515
Zimbabwe	50	50	301
EU	2,296	2,857	2,927
O.W. Europe	276	329	393
United States	229	120	202
Subtotal	67,215	69,547	68,473
Other Countries	5,117	5,784	4,922
Unaccounted	992	1,089	739
World Total	73,324	76,420	74,134

[1] Year beginning Oct 1. [2] Preliminary.

FAS, Grain and Feed Division, (202) 720–6219. Prepared or estimated on the basis of official statistics from foreign governments, other foreign source materials, reports of U.S. Agricultural Counselors, Attachés and Foreign Service Officers, results of office research, and related information.

Table 1-43.—Corn: Support operations, United States, 1993–2002

Marketing year beginning September 1	Income support payment rates per bushel [1]	Program price levels per bushel		Put under loan [4]		Acquired by CCC under loan program [5]	Owned by CCC at end of marketing year
		Loan [2]	Target [3]	Quantity	Percentage of production		
	Dollars	*Dollars*	*Dollars*	*Million bushels*	*Percent*	*Million bushels*	*Million bushels*
1993/94	0.28	1.72	2.75	618	9.8	0	45
1994/95	0.57	1.89	2.75	2,002	19.9	0	42
1995/96	0.00	1.89	2.75	676	9.1	0	30
1996/97	0.251	1.89	NA	970	10.5	0	2
1997/98	0.486	1.89	NA	1,141	12.4	2	4
1998/99	0.564	1.89	NA	1,775	18.2	24	12
1999/2000	0.726	1.89	NA	1,421	15.1	45	15
2000/2001	0.697	1.89	NA	1,394	14.1	31	8
2001/2002	0.576	1.89	NA	1,395	14.7	24	6
2002/2003	0.28	1.98	2.60	....	....	....	....

[1] Payment rates for the 1995/96 & prior crop years were calculated according to the provisions of the deficiency payment/production adjustment program. Payment rates for the 1996/97 through 2001/2002 crops were calculated according to the provisions of the Production Flexibility Contract (PFC) program, following enactment of the Federal Agriculture Improvement and Reform Act of 1996 (1996 Act) and included supplemental PFC payment rates for 1998 through 2001. Payment rates for the 2002/2003 and subsequent crops are calculated according to the provisions of the Direct and counter-cyclical Payment program, following enactment of the Farm Security and Rural Investment Act of 2002 (2002 Act) and includes a PFC payment rate for 2002 only. Payment rates are rounded to the nearest tenth of a cent. [2] The national average loan rate was also known as the price support rate prior to enactment of the 1996 Act. [3] Between the 1996/97 and 2001/2002 marketing years, target prices were no longer applicable; however, with enactment of the 2002 Act, target prices were reestablished. [4] Represents loans made, purchases, and purchase agreements entered into. Purchases and purchase agreements are no longer authorized for the 1996 and subsequent crops following enactment of the 1996 Act. Percentage of production is on a grain basis. [5] Acquisition of all loans forfeited during the marketing year including loans made in previous year. NA-not applicable.

FSA, Feed Grains & Oilseeds Analysis Group, (202) 720–2711.

Table 1-44.—Corn: United States exports, specified by country of destination, 1999/2000–2001/2002 [1]

Country of destination	Year		
	1999/2000	2000/2001	2001/2002
	1,000 metric tons	*1,000 metric tons*	*1,000 metric tons*
Japan	14,939	14,091	14,714
Mexico	4,910	5,928	4,024
Taiwan	4,989	4,894	4,519
Egypt	3,542	4,116	4,325
Canada	1,080	2,797	3,979
Korea, Republic Of	2,822	3,109	1,108
Colombia	1,741	1,656	1,696
Algeria	1,099	1,180	1,343
Dominican Republic	998	968	1,038
Saudi Arabia	1,197	1,003	670
Venezuela	1,146	1,152	502
Israel	748	621	847
Turkey	917	524	747
Syria	583	583	767
Canadian Trans.	958	555	197
Morocco	621	519	539
Tunisia	419	534	625
Guatemala	488	488	579
Costa Rica	481	481	463
El Salvador	368	463	287
Other	5,331	2,527	3,942
Total	49,378	48,192	46,913

[1] Year beginning Oct. 1. Compiled from U.S. Census data.

FAS, Grain and Feed Division, (202) 720–6219.

Table 1-45.—Oats: Area, yield, production, and value, United States, 1993–2002

Year	Area: Planted [1]	Area: Harvested	Yield per harvested acre	Production	Marketing year average price per bushel received by farmers	Value of production
	1,000 acres	*1,000 acres*	*Bushels*	*1,000 bushels*	*Dollars*	*1,000 dollars*
1993	7,937	3,803	54.4	206,731	1.36	290,948
1994	6,637	4,008	57.1	228,844	1.22	299,627
1995	6,225	2,952	54.6	161,094	1.67	278,941
1996	4,668	2,655	57.7	153,245	1.96	313,910
1997	5,068	2,813	59.5	167,246	1.60	273,284
1998	4,892	2,755	60.2	165,981	1.10	199,748
1999	4,673	2,453	59.6	146,193	1.12	175,172
2000	4,477	2,329	64.2	149,545	1.10	175,797
2001	4,403	1,905	61.4	117,024	1.59	195,711
2002 [2]	5,005	2,098	56.8	119,132	1.75	211,849

[1] Relates to the total area of oats sown for all purposes, including oats sown in the preceding fall. [2] Preliminary.
NASS, Crops Branch, (202) 720–2127.

Table 1-46.—Oats: Stocks on and off farms, United States, 1993–2002

Year beginning September	On farms: Sep. 1	On farms: Dec. 1	On farms: Mar. 1	On farms: Jun. 1	Off farms [1]: Sep. 1	Off farms [1]: Dec. 1	Off farms [1]: Mar. 1	Off farms [1]: Jun. 1
1993	161,000	124,200	85,050	53,940	58,004	69,517	61,502	51,583
1994	144,300	113,400	78,400	46,750	75,551	78,664	70,575	53,848
1995	107,200	87,200	57,350	32,600	72,967	65,804	55,268	33,708
1996	93,400	80,650	56,200	33,100	38,459	45,218	39,330	33,576
1997	105,950	83,200	58,800	34,500	48,972	61,051	52,418	39,498
1998	110,300	81,500	61,700	40,700	51,502	61,835	50,850	40,678
1999	97,300	79,800	53,300	36,000	51,151	53,872	48,500	40,031
2000	101,200	86,900	55,800	32,050	49,177	57,237	54,128	40,677
2001	74,800	58,100	40,200	28,650	41,592	56,117	53,158	34,552
2002 [2]	70,500	52,300	NA	NA	41,212	51,294	NA	NA

[1] Inlcudes stocks at mills, elevators, warehouses, terminals, and processors. [2] Preliminary. NA-not available.
NASS, Crops Branch, (202) 720–2127.

Table 1-47.—Oats: Supply and disappearance, United States, 1993–2002

Year beginning June 1	Supply: Beginning stocks	Supply: Production	Supply: Imports	Supply: Total	Disappearance: Domestic use: Feed and residual	Disappearance: Domestic use: Food, seed and industrial	Disappearance: Domestic use: Total	Disappearance: Exports	Disappearance: Total disappearance	Ending stocks May 31: Privately held [1]	Ending stocks May 31: Government	Ending stocks May 31: Total
	Million bushels	*Million bushels*	*Million bushels*	*Million bushels*	*Million bushels*	*Million bushels*	*Million bushels*	*Million bushels*	*Million bushels*	*Million bushels*	*Million bushels*	*Million bushels*
1993	113	207	107	427	230	88	318	3	321	106	0	106
1994	106	229	93	428	243	83	326	1	327	101	0	101
1995	101	161	81	342	195	79	274	2	276	66	0	66
1996	66	153	97	317	172	76	248	3	250	67	0	67
1997	67	167	98	332	185	72	256	2	258	74	0	74
1998	74	166	108	348	196	69	265	2	266	81	0	81
1999	81	146	99	326	180	68	249	2	250	76	0	76
2000	76	150	106	332	189	68	257	2	259	73	0	73
2001 [2]	73	117	96	286	148	72	220	3	223	63	0	63
2002 [3]	63	119	100	282	150	72	222	2	224	58	0	58

[1] Includes quantity under loan and farmer-owned reserve. [2] Preliminary. [3] Projected as of February 11, 2003, World Agricultural Supply and Demand Estimates. Totals may not add due to independent rounding.
ERS, Market and Trade Economics Division, (202) 694–5296.

Table 1-48.—Oats: Support operations, United States, 1993–2002

Marketing Year beginning June 1	Income support payment rates per bushel [1]	Program price levels per bushel		Put under loan [4]		Acquired by CCC under loan program [5]	Owned by CCC at end of marketing year
		Loan [2]	Target [3]	Quantity	Percentage of production		
	Dollars	*Dollars*	*Dollars*	*Million bushels*	*Percent*	*Million bushels*	*Million bushels*
1993/94	0.11	0.88	1.45	1.8	0.9	0.0	0.0
1994/95	0.19	0.97	1.45	1.9	0.8	0.0	0.0
1995/96	0.00	0.97	1.45	0.7	0.4	0.0	0.0
1996/97	0.033	1.03	NA	1.5	1.0	0.0	0.0
1997/98	0.031	1.11	NA	2.4	1.4	0.0	0.0
1998/99	0.047	1.11	NA	4.6	2.8	0.1	0.0
1999/2000	0.060	1.13	NA	1.7	1.2	0.8	0.0
2000/2001	0.058	1.16	NA	1.7	1.1	0.1	0.0
2001/2002	0.047	1.21	NA	1.7	1.5	0.1	0.0
2002/2003	0.024	1.35	1.40	....	....	....	....

[1] Payment rates for the 1995/96 & prior crop years were calculated according to the provisions of the now defunct deficiency payment/production adjustment program. Payment rates for the 1996/97 through 2001/2002 crops were calculated according to the provisions of the Production Flexibility Contract (PFC) program, following enactment of the Federal Agriculture Improvement and Reform Act of 1996 (1996 Act) and included supplemental PFC payment rates for 1998 through 2001. Payment rates for the 2002/2003 and subsequent crops are calculated according to the provisions of the Direct Payment program, following enactment of the Farm Security and Rural Investment Act of 2002 (2002 Act) and includes a PFC payment rate for 2002 only. Payment rates are rounded to the nearest tenth of a cent. [2] The national average loan rate was also known as the price support rate prior to enactment of the 1996 Act. [3] Between the 1996/97 and 2001/2002 marketing years, target prices were no longer applicable; however, with enactment of the 2002 Act, target prices were reestablished. [4] Represents loans made, purchases, and purchase agreements entered into. Purchases and purchase agreements are no longer authorized for the 1996 and subsequent crops following enactment of the 1996 Act. [5] Acquisition of all loans forfeited during the marketing year including loans made in previous year. NA-not applicable.

FSA, Feed Grains & Oilseeds Analysis Group, (202) 720–2711.

Table 1-49.—Oats: Area, yield, and production, by States, 2000–2002

State	Area planted [1]			Area harvested			Yield per harvested acre			Production		
	2000	2001	2002	2000	2001	2002	2000	2001	2002	2000	2001	2002
	1,000 acres	*1,000 acres*	*1,000 acres*	*1,000 acres*	*1,000 acres*	*1,000 acres*	*Bushels*	*Bushels*	*Bushels*	*1,000 bushels*	*1,000 bushels*	*1,000 bushels*
CA	220	260	260	25	15	27	75.0	60.0	80.0	1,875	900	2,160
CO	80	80	65	35	32	8	63.0	60.0	58.0	2,205	1,920	464
GA	70	100	90	35	35	25	72.0	65.0	60.0	2,520	2,275	1,500
ID	80	130	125	20	20	25	70.0	68.0	70.0	1,400	1,360	1,750
IL	75	60	65	55	40	50	73.0	80.0	69.0	4,015	3,200	3,450
IN	40	25	20	25	16	14	78.0	80.0	62.0	1,950	1,280	868
IA	270	240	290	180	130	175	67.0	70.0	76.0	12,060	9,100	13,300
KS	110	100	140	50	40	60	44.0	53.0	52.0	2,200	2,120	3,120
ME	32	33	30	30	31	29	70.0	75.0	90.0	2,100	2,325	2,610
MI	95	70	80	75	55	65	64.0	64.0	64.0	4,800	3,520	4,160
MN	400	300	420	310	210	285	72.0	60.0	56.0	22,320	12,600	15,960
MO	50	40	65	30	20	35	53.0	50.0	48.0	1,590	1,000	1,680
MT	130	130	145	50	60	55	52.0	40.0	49.0	2,600	2,400	2,695
NE	130	155	175	45	60	55	42.0	61.0	43.0	1,890	3,660	2,365
NY	80	95	70	60	80	55	65.0	69.0	66.0	3,900	5,520	3,630
NC	60	60	75	30	30	35	70.0	56.0	57.0	2,100	1,680	1,995
ND	600	575	670	315	240	290	63.0	62.0	44.0	19,845	14,880	12,760
OH	110	100	70	90	85	60	76.0	73.0	62.0	6,840	6,205	3,720
OK	60	55	85	15	10	30	44.0	38.0	37.0	660	380	1,110
OR	50	55	80	25	25	35	98.0	77.0	88.0	2,450	1,925	3,080
PA	175	150	140	145	115	115	57.0	65.0	61.0	8,265	7,475	7,015
SC	60	50	50	35	25	30	60.0	57.0	43.0	2,100	1,425	1,290
SD	350	350	450	220	130	100	61.0	60.0	45.0	13,420	7,800	4,500
TX	600	725	750	100	160	160	43.0	45.0	44.0	4,300	7,200	7,040
UT	50	60	60	7	6	5	70.0	65.0	90.0	490	390	450
WA	35	30	35	15	12	10	75.0	55.0	65.0	1,125	660	650
WI	400	300	430	280	195	250	68.0	64.0	60.0	19,040	12,480	15,000
WY	65	75	70	27	28	15	55.0	48.0	54.0	1,485	1,344	810
US	4,477	4,403	5,005	2,329	1,905	2,098	64.2	61.4	56.8	149,545	117,024	119,132

[1] Relates to the total area of oats sown for all purposes, including oats sown in the preceding fall.

NASS, Crops Branch, (202) 720–2127.

Table 1-50.—Oats: Marketing year average price and value, by States, crop of 2000, 2001, and 2002

State	Marketing year average price per bushel			Value of production		
	2000	2001	2002[1]	2000	2001	2002[1]
	Dollars	*Dollars*	*Dollars*	*1,000 dollars*	*1,000 dollars*	*1,000 dollars*
CA	1.55	2.25	2.70	2,906	2,025	5,832
CO	1.90	2.20	2.85	4,190	4,224	1,322
GA	1.40	1.75	2.10	3,528	3,981	3,150
ID	1.25	1.60	1.70	1,750	2,176	2,975
IL	1.33	1.81	1.80	5,340	5,792	6,210
IN	1.50	1.85	1.80	2,925	2,368	1,562
IA	1.19	1.54	1.70	14,351	14,014	22,610
KS	1.40	1.70	1.85	3,080	3,604	5,772
ME	0.90	1.10	1.45	1,890	2,558	3,785
MI	1.30	1.80	1.75	6,240	6,336	7,280
MN	0.99	1.53	1.65	22,097	19,278	26,334
MO	1.50	1.60	1.85	2,385	1,600	3,108
MT	1.13	1.96	2.00	2,938	4,704	5,390
NE	1.42	1.59	1.95	2,684	5,819	4,612
NY	1.40	1.62	1.65	5,460	8,942	5,990
NC	1.45	1.65	1.80	3,045	2,772	3,591
ND	0.86	1.63	1.65	17,067	24,254	21,054
OH	1.40	1.80	1.75	9,576	11,169	6,510
OK	1.60	1.75	2.00	1,056	665	2,220
OR	1.33	1.96	2.45	3,259	3,773	7,546
PA	1.45	1.72	1.90	11,984	12,857	13,329
SC	1.50	1.75	1.85	3,150	2,494	2,387
SD	1.11	1.67	1.90	14,896	13,026	8,550
TX	1.60	2.20	1.70	6,880	15,840	11,968
UT	1.65	2.25	2.50	809	878	1,125
WA	1.20	1.70	1.70	1,350	1,122	1,105
WI	0.98	1.38	1.65	18,659	17,222	24,750
WY	1.55	1.65	2.20	2,302	2,218	1,782
US	1.10	1.59	1.75	175,797	195,711	211,849

[1] Preliminary.

NASS, Crops Branch, (202) 720–2127.

Table 1-51.—Oats: Area, yield, and production in specified countries, 1999/2000–2001/2002[1]

Continent and country	Area[2]			Yield per hectare			Production		
	1999/2000	2000/2001	2001/2002[3]	1999/2000	2000/2001	2001/2002[3]	1999/2000	2000/2001	2001/2002[3]
	1,000 hectares	*1,000 hectares*	*1,000 hectares*	*Metric tons*	*Metric tons*	*Metric tons*	*1,000 metric tons*	*1,000 metric tons*	*1,000 metric tons*
North America:									
Canada	1,398	1,299	1,238	2.60	2.61	2.17	3,641	3,389	2,691
Mexico	100	100	100	1.00	1.00	1.00	100	100	100
United States	993	943	771	2.14	2.30	2.20	2,122	2,171	1,699
Total	2,491	2,342	2,109	2.35	2.42	2.13	5,863	5,660	4,490
South America:									
Argentina	337	335	381	1.65	1.92	1.78	555	642	678
Brazil	200	249	249	1.25	1.33	1.33	250	330	330
Chile	89	90	93	2.79	3.83	3.01	248	345	280
Colombia	3	3	3	1.33	1.33	1.33	4	4	4
Ecuador	1	1	1	1.00	1.00	1.00	1	1	1
Uruguay	90	90	90	0.33	0.33	0.33	30	30	30
Total	720	768	817	1.51	1.76	1.62	1,088	1,352	1,323
Europe:									
Austria	35	33	31	4.31	3.52	4.13	151	116	128
Belgium	7	7	3	5.00	5.00	5.00	35	35	15
Denmark	26	45	55	5.00	5.18	5.11	130	233	281
Finland	404	400	416	2.45	3.53	3.09	990	1,413	1,287
France	122	103	117	4.52	4.46	4.15	551	459	486
Germany	268	237	233	5.00	4.59	4.94	1,339	1,087	1,151
Greece	45	45	45	2.04	2.27	2.20	92	102	99
Ireland	18	20	19	7.50	6.75	7.37	135	135	140
Italy	132	152	149	3.00	3.00	2.70	396	456	402
Netherlands	3	2	3	4.67	6.50	5.00	14	13	15

See footnotes at end of table.

Table 1-51.—Oats: Area, yield, and production in specified countries, 1999/2000–2001/2002[1]—Continued

Continent and country	Area[2]			Yield per hectare			Production		
	1999/2000	2000/2001	2001/2002[3]	1999/2000	2000/2001	2001/2002[3]	1999/2000	2000/2001	2001/2002[3]
	1,000 hectares	*1,000 hectares*	*1,000 hectares*	*Metric tons*	*Metric tons*	*Metric tons*	*1,000 metric tons*	*1,000 metric tons*	*1,000 metric tons*
Europe, cont.:									
Portugal	100	72	72	0.81	0.76	0.81	81	55	58
Spain	392	427	410	1.40	2.23	1.61	550	952	660
Sweden	306	291	271	3.45	3.96	3.55	1,055	1,151	961
United Kingdom	92	109	112	5.87	5.87	5.49	540	640	615
Total	1,950	1,943	1,936	3.11	3.52	3.25	6,059	6,847	6,298
Eastern Europe:									
Albania	20	20	20	0.75	0.75	0.75	15	15	15
Bosnia-Hercegovina	29	26	26	2.14	2.19	2.08	62	57	54
Bulgaria	35	35	35	1.43	1.43	1.43	50	50	50
Croatia	24	24	24	2.38	2.38	2.38	57	57	57
Czech Republic	55	50	50	3.18	2.72	3.02	175	136	151
Hungary	70	60	60	2.50	1.67	2.50	175	100	150
Macedonia (Skopje)	3	3	3	1.33	1.33	1.33	4	4	4
Poland	572	566	531	2.53	1.89	2.46	1,446	1,070	1,305
Romania	250	225	225	1.50	1.56	1.56	375	350	350
Serbia	67	63	65	1.82	1.52	1.92	122	96	125
Slovakia	23	21	17	2.09	1.19	2.65	48	25	45
Slovenia	2	2	2	3.00	3.00	3.00	6	6	6
Total	1,150	1,095	1,058	2.20	1.80	2.19	2,535	1,966	2,312
Oth. Western Europe:									
Norway	90	94	83	3.96	4.22	4.01	356	397	333
Switzerland	7	7	7	5.00	5.00	5.00	35	35	35
Total	97	101	90	4.03	4.28	4.09	391	432	368
Fmr. Soviet Union:.									
Belarus	293	300	300	1.26	1.73	2.00	368	520	600
Estonia	60	53	48	1.18	2.21	1.94	71	117	93
Georgia	5	5	5	0.60	1.00	1.00	3	5	5
Kazakhstan	140	100	100	1.43	0.80	2.18	200	80	218
Kyrgyzstan	3	3	3	2.00	2.67	2.67	6	8	8
Latvia	47	46	55	1.40	1.74	1.31	66	80	72
Lithuania	51	44	45	1.31	1.89	1.67	67	83	75
Moldova	3	1	0	2.33	1.00	0.00	7	1	0
Russian Fed.	5,350	4,500	4,900	0.82	1.33	1.57	4,400	6,000	7,700
Ukraine	530	481	600	1.43	1.83	1.83	760	881	1,100
Total	6,482	5,533	6,056	0.92	1.41	1.63	5,948	7,775	9,871
Turkey	145	145	145	1.72	1.72	1.72	250	250	250
Africa:.									
Algeria	47	15	50	0.85	0.53	0.88	40	8	44
Morocco	40	40	40	0.75	0.75	0.75	30	30	30
South Africa, Rep. of	700	700	700	0.06	0.06	0.06	45	45	45
Total	787	755	790	0.15	0.11	0.15	115	83	119
Asia:									
China, People's Republic	500	500	500	1.20	1.20	1.20	600	600	600
Japan	1	1	1	2.00	2.00	1.00	2	2	1
Total	501	501	501	1.20	1.20	1.20	602	602	601
Oceania:									
Australia	584	698	678	1.91	1.62	1.80	1,118	1,131	1,222
New Zealand	20	20	20	3.75	3.75	3.75	75	75	75
Total	604	718	698	1.98	1.68	1.86	1,193	1,206	1,297
World total	14,927	13,901	14,200	1.61	1.88	1.90	24,044	26,173	26,929

[1] Years shown refer to years of harvest. Harvests of Northern Hemisphere countries are combined with those of the Southern Hemisphere which immediately follow; thus the crop harvested in the Northern Hemisphere in 1994 is combined with estimates for the Southern Hemisphere harvest, which begins late in 1994 and ends early in 1995. [2] Harvested area as far as possible. [3] Preliminary.

FAS Production Estimates and Crop Assessment Division, (202) 720–0888. Prepared or estimated on the basis of official statistics of foreign governments, other foreign source materials, reports of U.S. Agricultural Counselors, Attachés, and Foreign Service Officers, results of office research, and related information.

Table 1-52.—Barley: Area, yield, production, and value, United States, 1993–2002

Year	Area: Planted[1]	Area: Harvested	Yield per harvested acre	Production	Marketing year average price per bushel received by farmers	Value of production
	1,000 acres	*1,000 acres*	*Bushels*	*1,000 bushels*	*Dollars*	*1,000 dollars*
1993	7,786	6,753	58.9	398,041	1.99	812,889
1994	7,159	6,667	56.2	374,862	2.03	783,709
1995	6,689	6,279	57.2	359,376	2.89	1,028,183
1996	7,094	6,707	58.5	392,433	2.74	1,080,940
1997	6,706	6,198	58.1	359,878	2.38	861,620
1998	6,337	5,864	60.0	352,125	1.98	686,517
1999	5,194	4,734	59.2	280,292	2.13	597,038
2000	5,864	5,213	61.1	318,728	2.11	649,130
2001	4,967	4,289	58.2	249,420	2.22	536,582
2002[2]	5,073	4,135	54.9	226,873	2.70	596,759

[1] Barley sown for all purposes, including barley sown in the preceding fall. [2] Preliminary.
NASS, Crops Branch, (202) 720–2127.

Table 1-53.—Barley: Stocks on and off farms, United States, 1993–2002

Year beginning September	On farms: Sep. 1	On farms: Dec. 1	On farms: Mar. 1	On farms: June 1	Off farms[1]: Sep. 1	Off farms[1]: Dec. 1	Off farms[1]: Mar. 1	Off farms[1]: June 1
	1,000 bushels	*1,000 bushels*	*1,000 bushels*	*1,000 bushels*	*1,000 bushels*	*1,000 bushels*	*1,000 bushels*	*1,000 bushels*
1993	277,600	194,150	116,475	63,870	125,102	139,277	107,807	74,988
1994	220,150	166,300	86,800	37,905	131,740	112,560	106,156	74,688
1995	198,900	121,400	58,825	25,600	113,624	121,110	119,107	73,993
1996	191,700	135,700	82,060	43,715	122,078	110,522	90,840	65,735
1997	195,500	134,500	81,900	43,700	131,347	109,561	98,299	75,533
1998	193,500	149,000	86,900	52,000	132,674	121,535	113,808	89,653
1999	165,100	108,000	59,300	25,400	130,145	119,996	110,917	85,924
2000	151,700	111,500	58,600	28,850	142,341	117,369	103,544	77,409
2001	134,800	92,400	46,000	23,210	110,564	102,587	95,748	69,631
2002[2]	131,600	83,400	NA	NA	92,419	86,601	NA	NA

[1] Includes stocks at mills, elevators, warehouses, terminals, and processors. [2] Preliminary. NA-not available.
NASS, Crops Branch, (202) 720–2127.

Table 1-54.—Barley: Supply and disappearance, United States, 1993–2002

Year beginning June 1	Supply: Beginning stocks	Supply: Production	Supply: Imports	Supply: Total	Disappearance: Domestic use: Feed and residual	Domestic use: Food, seed, and industrial	Domestic use: Total	Exports	Total disappearance	Ending stocks May 31: Privately held[1]	Ending stocks: Government	Ending stocks: Total
	Million bushels	*Million bushels*	*Million bushels*	*Million bushels*	*Million bushels*	*Million bushels*	*Million bushels*	*Million bushels*	*Million bushels*	*Million bushels*	*Million bushels*	*Million bushels*
1993	151	398	71	621	244	172	416	66	482	134	5	139
1994	139	375	66	580	228	173	401	66	467	108	5	113
1995	113	359	41	513	179	172	351	62	413	95	4	100
1996	100	392	37	529	217	172	389	31	419	109	0	109
1997	109	360	40	510	144	172	316	74	390	119	0	119
1998	119	352	30	501	161	170	331	28	360	142	0	142
1999	142	280	28	450	136	172	308	30	338	111	0	111
2000	111	319	29	459	123	172	295	58	353	106	0	106
2001[2]	106	249	24	380	88	172	260	27	287	93	0	93
2002[3]	93	227	20	340	80	172	252	20	272	68	0	68

[1] Includes quantity under loan and farmer–owned reserve. [2] Preliminary. [3] Projected as of February 11, 2003, World Agricultural Supply and Demand Estimates. Totals may not add due to independent rounding.
ERS, Market and Trade Economics Division, (202) 694–5296.

Table 1-55.—Barley: Area, yield, and production, by States, 2000–2002

State	Area planted[1]			Area harvested			Yield per harvested acre			Production		
	2000	2001	2002[2]	2000	2001	2002[2]	2000	2001	2002[2]	2000	2001	2002[2]
	1,000 acres	*1,000 acres*	*1,000 acres*	*1,000 acres*	*1,000 acres*	*1,000 acres*	*Bushels*	*Bushels*	*Bushels*	*1,000 bushels*	*1,000 bushels*	*1,000 bushels*
AZ	40	42	46	36	40	40	114.0	110.0	110.0	4,104	4,400	4,400
CA	130	160	130	95	110	75	68.0	53.0	68.0	6,460	5,830	5,100
CO	110	90	85	105	80	72	115.0	107.0	100.0	12,075	8,560	7,200
DE	30	29	25	28	26	23	81.0	77.0	84.0	2,268	2,002	1,932
ID	750	700	730	730	670	710	76.0	75.0	76.0	55,480	50,250	53,960
KS	8	9	8	7	8	7	35.0	50.0	34.0	245	400	238
KY	9	9	10	8	8	8	75.0	85.0	64.0	600	680	512
ME	25	27	27	24	26	26	70.0	70.0	80.0	1,680	1,820	2,080
MD	55	55	45	50	51	41	82.0	75.0	82.0	4,100	3,825	3,362
MI	20	21	20	19	18	19	60.0	56.0	52.0	1,140	1,008	988
MN	270	160	210	240	145	165	64.0	55.0	39.0	15,360	7,975	6,435
MT	1,250	1,100	1,200	950	720	950	40.0	41.0	42.0	38,000	29,520	39,900
NE	7	5	6	5	4	5	27.0	45.0	43.0	135	180	215
NV	4	4	4	3	1	2	85.0	90.0	97.0	255	90	194
NJ	5	5	4	4	4	3	78.0	54.0	74.0	312	216	222
NY	12	15	11	10	12	10	58.0	51.0	47.0	580	612	470
NC	30	28	31	18	18	20	80.0	67.0	69.0	1,440	1,206	1,380
ND	1,900	1,500	1,600	1,770	1,450	1,240	55.0	55.0	46.0	97,350	79,750	57,040
OH	14	6	6	13	5	5	78.0	76.0	48.0	1,014	380	240
OR	150	110	80	140	100	74	60.0	45.0	50.0	8,400	4,500	3,700
PA	80	70	70	75	60	60	71.0	70.0	74.0	5,325	4,200	4,440
SD	115	90	80	105	78	45	55.0	52.0	41.0	5,775	4,056	1,845
UT	95	85	70	78	65	45	70.0	68.0	64.0	5,460	4,420	2,880
VA	85	70	75	65	50	40	89.0	75.0	77.0	5,785	3,750	3,080
WA	500	430	350	490	420	340	70.0	50.0	54.0	34,300	21,000	18,360
WI	65	47	60	50	35	40	64.0	52.0	45.0	3,200	1,820	1,800
WY	105	100	90	95	85	70	83.0	82.0	70.0	7,885	6,970	4,900
US	5,864	4,967	5,073	5,213	4,289	4,135	61.1	58.2	54.9	318,728	249,420	226,873

[1] Includes area planted in the preceding fall. [2] Preliminary.
NASS, Crops Branch, (202) 720–2127.

Table 1-56.—Barley: Marketing year average price and value, by States, crop of 2000, 2001, and 2002

State	Marketing year average price per bushel			Value of production		
	2000	2001	2002[1]	2000	2001	2002[1]
	Dollars	*Dollars*	*Dollars*	*1,000 dollars*	*1,000 dollars*	*1,000 dollars*
AZ	2.42	2.40	3.00	9,932	10,560	13,200
CA	2.37	2.36	2.45	15,310	13,759	12,495
CO	3.08	2.73	3.10	37,191	23,369	22,320
DE	1.30	1.25	1.40	2,948	2,503	2,705
ID	2.59	2.75	3.05	143,693	138,188	164,578
KS	1.65	1.70	1.80	404	680	428
KY	1.44	1.43	1.50	864	972	768
ME	1.45	1.50	1.50	2,436	2,730	3,120
MD	1.40	1.30	1.30	5,740	4,973	4,371
MI	1.10	1.50	1.60	1,254	1,512	1,581
MN	1.67	1.70	2.20	25,651	13,558	14,157
MT	2.38	2.65	2.90	90,440	78,228	115,710
NE	1.60	1.65	1.80	216	297	387
NV	2.55	2.65	2.95	650	239	572
NJ	1.30	1.25	1.35	406	270	300
NY	1.65	1.60	1.70	957	979	799
NC	1.55	1.50	1.60	2,232	1,809	2,208
ND	1.63	1.71	2.45	158,681	136,373	139,748
OH	1.50	1.50	1.60	1,521	570	384
OR	1.96	2.06	2.40	16,464	9,270	8,880
PA	1.51	1.51	1.60	8,041	6,342	7,104
SD	1.94	2.00	1.85	11,204	8,112	3,413
UT	2.00	2.14	2.35	10,920	9,459	6,768
VA	1.35	1.28	1.40	7,810	4,800	4,312
WA	1.93	1.96	2.60	66,199	41,160	47,736
WI	1.15	1.50	1.55	3,680	2,730	2,790
WY	3.08	3.32	3.25	24,286	23,140	15,925
US	2.11	2.22	2.70	649,130	536,582	596,759

[1] Preliminary.
NASS, Crops Branch, (202) 720–2127.

Table 1-57.—Barley: Area, yield, and production in specified countries, 1999/2000–2001/2002 [1]

Continent and country	Area [2]			Yield per hectare			Production		
	1999/ 2000	2000/ 2001	2001/ 2002 [3]	1999/ 2000	2000/ 2001	2001/ 2002 [3]	1999/ 2000	2000/ 2001	2001/ 2002 [3]
	1,000 hectares	*1,000 hectares*	*1,000 hectares*	*Metric tons*	*Metric tons*	*Metric tons*	*1,000 metric tons*	*1,000 metric tons*	*1,000 metric tons*
North America:									
Canada	4,069	4,551	4,150	3.24	2.89	2.61	13,196	13,172	10,846
Mexico	213	309	315	2.11	2.49	2.43	450	770	767
United States	1,916	2,110	1,736	3.19	3.29	3.13	6,103	6,939	5,430
Total	6,198	6,970	6,201	3.19	3.00	2.75	19,749	20,881	17,043
South America:									
Argentina	180	245	240	2.30	2.76	2.13	414	676	510
Bolivia	85	85	85	0.76	0.76	0.76	65	65	65
Brazil	150	141	141	2.00	2.01	2.01	300	283	283
Chile	17	15	17	3.53	4.33	3.82	60	65	65
Colombia	10	10	10	2.00	2.00	2.00	20	20	20
Ecuador	35	35	35	1.00	1.00	1.00	35	35	35
Peru	130	130	130	1.08	1.15	1.15	140	150	150
Uruguay	55	88	129	2.02	2.43	0.91	111	214	118
Total	662	749	787	1.73	2.01	1.58	1,145	1,508	1,246
Europe:									
Austria	243	224	218	4.71	3.77	4.64	1,145	845	1,012
Belgium-Luxembourg	54	52	47	6.59	6.56	7.00	356	341	329
Denmark	728	741	737	5.05	5.37	5.40	3,675	3,980	3,979
Finland	594	558	560	2.75	3.40	3.20	1,634	1,897	1,792
France	1,534	1,561	1,710	6.22	6.37	5.73	9,539	9,950	9,800
Germany	2,210	2,068	2,112	6.02	5.85	6.39	13,301	12,106	13,495
Greece	165	165	165	2.51	2.50	2.30	414	413	380
Ireland	180	180	185	7.08	6.25	6.70	1,275	1,125	1,240
Italy	353	344	332	3.72	3.67	3.37	1,313	1,262	1,120
Netherlands	58	47	66	6.29	6.13	5.86	365	288	387
Portugal	20	20	20	2.20	2.20	2.20	44	44	44
Spain	3,107	3,307	2,994	2.39	3.41	2.09	7,434	11,283	6,245
Sweden	482	409	392	3.84	4.00	4.17	1,853	1,635	1,633
United Kingdom	1,178	1,128	1,245	5.59	5.75	5.38	6,581	6,490	6,700
Total	10,906	10,804	10,783	4.49	4.78	4.47	48,929	51,659	48,156
Oth. Western Europe:									
Malta	1	1	1	2.00	2.00	2.00	2	2	2
Norway	180	161	173	3.47	3.57	3.54	624	574	612
Switzerland	49	46	44	5.27	6.15	5.70	258	283	251
Total W. Europe	230	208	218	3.84	4.13	3.97	884	859	865
Eastern Europe:									
Albania	10	10	10	3.00	3.00	3.00	30	30	30
Bosnia-Hercegovina	23	23	23	2.43	2.30	2.39	56	53	55
Bulgaria	250	240	260	2.80	2.29	2.88	700	550	750
Croatia	45	45	45	2.78	2.78	2.78	125	125	125
Czech Republic	543	495	500	3.94	3.29	4.04	2,137	1,629	2,018
Hungary	333	325	368	3.12	2.77	3.53	1,038	900	1,300
Macedonia (Skopje)	51	53	50	2.47	2.64	2.20	126	140	110
Poland	1,108	1,096	1,071	3.07	2.54	3.11	3,401	2,783	3,330
Romania	550	213	316	1.85	2.68	3.34	1,015	570	1,055
Serbia	117	111	120	2.56	2.27	2.50	300	252	300
Slovakia	246	199	200	2.94	1.99	3.50	724	397	700
Slovenia	11	11	11	3.00	3.00	3.00	33	33	33
Total E. Europe	3,287	2,821	2,974	2.95	2.65	3.30	9,685	7,462	9,806

See footnotes at end of table.

Table 1-57.—Barley: Area, yield, and production in specified countries, 1999/2000–2001/2002 [1]—Continued

Continent and country	Area [2]			Yield per hectare			Production		
	1999/ 2000	2000/ 2001	2001/ 2002 [3]	1999/ 2000	2000/ 2001	2001/ 2002 [3]	1999/ 2000	2000/ 2001	2001/ 2002 [3]
	1,000 hectares	*1,000 hectares*	*1,000 hectares*	*Metric tons*	*Metric tons*	*Metric tons*	*1,000 metric tons*	*1,000 metric tons*	*1,000 metric tons*
Fmr. Soviet Union:									
Armenia	65	65	85	1.23	0.92	1.41	80	60	120
Azerbaijan	57	109	130	2.63	1.38	2.31	150	150	300
Belarus	811	850	850	1.46	2.00	2.24	1,181	1,700	1,900
Estonia	165	165	136	1.13	2.11	2.09	186	348	284
Georgia	28	20	44	1.82	0.75	2.27	51	15	100
Kazakhstan	1,700	1,625	1,700	1.32	1.03	1.29	2,250	1,675	2,200
Kyrgyzstan	100	100	65	1.80	1.50	2.31	180	150	150
Latvia	148	135	130	1.57	1.93	1.78	233	261	231
Lithuania	421	353	336	1.76	2.44	2.40	742	860	805
Moldova	109	105	100	1.74	1.24	2.35	190	130	235
Russian Fed.	9,850	9,200	10,200	1.08	1.53	1.91	10,600	14,100	19,500
Tajikistan	33	30	30	0.76	0.70	0.67	25	21	20
Turkmenistan	20	20	20	1.00	1.00	1.00	20	20	20
Ukraine	3,475	3,690	3,900	1.85	1.86	2.62	6,425	6,872	10,200
Uzbekistan	75	75	75	1.60	1.60	2.67	120	120	200
Total	17,057	16,542	17,801	1.32	1.60	2.04	22,433	26,482	36,265
Middle East:									
Cyprus	52	45	45	2.17	0.84	2.56	113	38	115
Iran	1,350	1,300	1,300	1.19	1.08	1.15	1,600	1,400	1,500
Iraq	1,100	1,100	1,200	0.55	0.55	0.58	600	600	700
Israel	4	8	5	0.50	0.50	1.00	2	4	5
Jordan	16	16	25	0.81	0.81	0.96	13	13	24
Lebanon	5	5	5	1.20	1.20	1.20	6	6	6
Saudi Arabia	57	17	17	7.02	5.88	5.88	400	100	100
Syria	1,000	250	1,400	0.36	0.52	0.93	360	130	1,300
Turkey	3,550	3,600	3,500	1.86	2.06	1.97	6,600	7,400	6,900
Yemen	50	50	50	1.00	1.00	1.00	50	50	50
Total	7,184	6,391	7,547	1.36	1.52	1.42	9,744	9,741	10,700
Africa:									
Algeria	469	216	515	1.09	0.75	1.11	510	163	574
Egypt	55	55	55	3.09	3.09	3.09	170	170	170
Ethiopia	1,375	1,450	1,325	1.00	1.10	1.09	1,375	1,600	1,450
Kenya	70	70	70	0.86	1.07	1.07	60	75	75
Libya	280	280	280	0.30	0.30	0.30	85	85	85
Morocco	2,070	2,251	2,126	0.71	0.21	0.54	1,474	467	1,155
South Africa, Rep. of	102	72	79	0.89	1.72	1.99	91	124	157
Tunisia	466	260	250	0.88	0.77	0.60	412	200	150
Zimbabwe	3	5	5	5.67	6.40	5.00	17	32	25
Total	4,890	4,659	4,705	0.86	0.63	0.82	4,194	2,916	3,841
Asia:									
Afghanistan	180	124	87	1.20	0.60	1.00	216	74	87
Bangladesh	7	7	7	0.71	0.71	0.71	5	5	5
Bhutan	5	5	5	1.00	1.00	1.00	5	5	5
China, Peoples Rep. of	868	791	770	3.42	3.35	3.29	2,970	2,646	2,535
India	783	725	754	1.88	2.00	1.90	1,470	1,447	1,432
Japan	52	54	61	3.94	3.96	3.38	205	214	206
Korea, Rep. of	100	100	100	4.00	4.00	4.00	400	400	400
Nepal	38	40	40	1.00	1.00	1.00	38	40	40
Pakistan	160	160	160	1.00	1.00	1.03	160	160	165
Total	2,193	2,006	1,984	2.49	2.49	2.46	5,469	4,991	4,875
Oceania:									
Australia	2,596	3,675	3,700	1.94	1.96	2.03	5,032	7,196	7,500
New Zealand	80	80	80	5.00	5.00	5.00	400	400	400
Total	2,676	3,755	3,780	2.03	2.02	2.09	5,432	7,596	7,900
World total	55,283	54,905	56,780	2.31	2.44	2.48	127,664	134,095	140,697

[1] Years shown refer to year of harvest. Harvests of Northern Hemisphere countries are combined with those of the Southern Hemisphere which immediately follow; thus the crop harvested in the Northern Hemisphere in 1994 is combined with estimates of the Southern Hemisphere harvests, which begin late in 1994 and end early in 1995. [2] Harvested area as far as possible. [3] Preliminary.

FAS, Production Estimates and Crop Assessment Division, (202) 720–0888. Prepared or estimated on the basis of official statistics of foreign governments, other foreign source materials, reports of U.S. Agricultural Counselors, Attachés and Foreign Service Officers, results of office research, and related information.

Table 1-58.—Grains and grain products: Total and per capita civilian consumption as food, United States, 1992–2001

Calendar year [1]	Wheat: Total consumed [2]	Wheat: Per capita consumption of food products: Flour [3]	Wheat: Per capita consumption of food products: Cereal	Rye: Total consumed [2]	Rye: Per capita consumption of rye flour	Rice (milled): Total consumed [4]	Rice (milled): Per capita consumption
	Million bushels	*Pounds*	*Pounds*	*Million bushels*	*Pounds*	*Million cwt.*	*Pounds*
1992	817	138	4.7	3.5	0.6	45.4	17.6
1993	853	142	5.0	3.3	0.6	49.6	19.0
1994	871	143	5.2	3.6	0.6	51.4	19.4
1995	858	140	5.4	3.2	0.5	52.2	19.5
1996	896	147	5.4	3.5	0.6	53.2	19.7
1997	902	147	5.4	3.3	0.5	54.3	19.8
1998	909	143	5.4	3.6	0.6	57.0	20.5
1999	921	143	5.3	3.3	0.6	59.3	21.2
2000	950	146	5.1	3.3	0.6	61.0	21.5
2001 [9]	926	141	5.1	3.3	0.6	63.1	21.9

Calendar year [1]	Corn: Total consumed [5]	Corn: Flour and meal	Corn: Hominy and grits	Corn: Syrup	Corn: Dextrose	Corn: Starch	Oats: Total consumed [6]	Oats: Per capita consumption of oat food products	Barley: Total consumed [7]	Barley: Per capita consumption of food products [8]
	Million bushels	*Pounds*	*Pounds*	*Pounds*	*Pounds*	*Pounds*	*Million bushels*	*Pounds*	*Million bushels*	*Pounds*
1992	791	15.3	2.6	66.9	3.6	4.2	77.4	6.6	6.1	0.7
1993	834	15.6	3.1	70.2	3.7	4.4	73	6.3	6.2	0.7
1994	873	15.9	3.6	72.1	3.8	4.5	70	6.0	6.3	0.8
1995	900	16.2	4.0	73.9	4.0	4.7	67	5.7	6.4	0.8
1996	937	16.5	4.5	74.2	4.0	4.9	63	5.4	6.5	0.8
1997	966	16.7	4.9	77.7	3.7	4.8	59	5.1	6.3	0.8
1998	974	17.0	5.3	79.0	3.6	4.8	57	4.9	6.2	0.7
1999	990	17.3	5.8	80.0	3.5	4.7	56.8	4.9	6.1	0.7
2000	976	17.5	6.2	78.5	3.4	4.7	56.7	4.9	6.0	0.7
2001 [9]	986	17.8	6.6	78.3	3.3	4.6	59.0	5.1	5.9	0.7

[1] Data are in marketing year; for corn, September 1-August 31; for oats and barley, June 1-May 31; and rice, August 1-July 31. Wheat, rye, syrup, and sugar are in calendar year. [2] Excludes quantities used in alcoholic beverages. [3] Includes white, whole wheat, and semolina flour. [4] Does not include shipments to U.S. territories. Excludes rice used in alcoholic beverages. Includes imports and rice used in processed foods and pet foods. [5] Includes an allowance for the quantity used as hominy and grits. This series is not adjusted for trade. [6] Oats used in oatmeal, prepared breakfast foods, infant foods, and food products. [7] Malt for food, breakfast food uses, pearl barley, and flour. [8] Malt equivalent of barley food products. [9] Preliminary. Estimates of corn syrup and sugar are unofficial estimates; industry data were not reported after April 1968.

ERS, Market & Trade Economics Division, (202) 694-5290. All figures are estimates based on data from private industry sources, the U.S. Department of Commerce, the Internal Revenue Service, and other Government agencies.

Table 1-59.—Barley: Support operations, United States, 1993–2002

Marketing year beginning June 1	Income support payment rates per bushel [1]	Program price levels per bushel: Loan [2]	Program price levels per bushel: Target [3]	Put under support [4]: Quantity	Put under support [4]: Percentage of production	Acquired by CCC under loan program [5]	Owned by CCC at end of marketing year
	Dollars	*Dollars*	*Dollars*	*Million bushels*	*Percent*	*Million bushels*	*Million bushels*
1993/94	0.67	1.40	2.36	37.7	9.5	0.0	5.2
1994/95	0.52	1.54	2.36	28.2	7.5	0.0	5.0
1995/96	0.00	1.54	2.36	18.0	5.0	0.0	4.2
1996/97	0.332	1.55	NA	28.7	7.3	0.0	0.0
1997/98	0.277	1.57	NA	33.3	9.3	0.0	0.0
1998/99	0.425	1.56	NA	25.9	7.4	1.5	0.3
1999/2000	0.542	1.59	NA	13.6	4.9	3.6	0.1
2000/2001	0.522	1.62	NA	16.0	5.0	1.3	0.1
2001/2002	0.436	1.65	NA	10.6	4.2	0.7	0.0
2002/2003	0.24	1.88	2.21	....	....	....	....

[1] Payment rates for the 1995/96 & prior crop years were calculated according to the provisions of the now defunct deficiency payment/production adjustment program. Payment rates for the 1996/97 through 2001/2002 crops were calculated according to the provisions of the Production Flexibility Contract (PFC) program, following enactment of the Federal Agriculture Improvement and Reform Act of 1996 (1996 Act) and included supplemental PFC payment rates for 1998 through 2001. Payment rates for the 2002/2003 and subsequent crops are calculated according to the provisions of the Direct Payment program, following enactment of the Farm Security and Rural Investment Act of 2002 (2002 Act) and includes a PFC payment rate for 2002 only. Payment rates are rounded to the nearest tenth of a cent. [2] The national average loan rate was also known as the price support rate prior to enactment of the 1996 Act. [3] Between the 1996/97 and 2001/2002 marketing years, target prices were no longer applicable; however, with enactment of the 2002 Act, target prices were reestablished. [4] Represents loans made, purchases, and purchase agreements entered into. Purchases and purchase agreements are no longer authorized for the 1996 and subsequent crops following enactment of the 1996 Act. [5] Acquisition of all loans forfeited during the marketing year including loans made in previous year. NA-not applicable.

FSA, Feed Grains & Oilseeds Analysis Group, (202) 720–2711.

Table 1-60.—Sorghum: Area, yield, production, and value, United States, 1993–2002

Year	Area planted for all purposes [1]	Sorghum for grain [2]					Sorghum for silage		
		Area harvested	Yield per harvested acre	Production	Marketing year average price per Cwt [3]	Value of production [3]	Area harvested	Yield per harvested acre	Production
	1,000 acres	*1,000 acres*	*Bushels*	*1,000 bushels*	*Dollars*	*1,000 dollars*	*1,000 acres*	*Tons*	*1,000 tons*
1993	9,882	8,916	59.9	534,172	4.13	1,234,500	351	11.2	3,914
1994	9,787	8,882	72.7	645,741	3.80	1,317,149	362	11.9	4,316
1995	9,429	8,253	55.6	458,648	5.69	1,389,772	413	10.3	4,242
1996	13,097	11,811	67.3	795,274	4.17	1,986,316	423	11.8	4,976
1997	10,052	9,158	69.2	633,545	3.95	1,408,909	412	13.1	5,385
1998	9,626	7,723	67.3	519,933	2.97	905,468	308	11.4	3,526
1999	9,288	8,544	69.7	595,166	2.80	937,406	320	11.6	3,716
2000	9,195	7,726	60.9	470,526	3.37	847,075	262	10.6	2,773
2001	10,252	8,584	59.9	514,524	3.46	979,794	336	11.1	3,728
2002 [4]	9,580	7,299	50.7	369,758	4.30	883,690	352	9.5	3,360

[1] Grain and sweet sorghum for all uses, including sirup. [2] Includes both grain sorghum for grain, and sweet sorghum for grain or seed. [3] Based on the reported price of grain sorghum. [4] Preliminary.

NASS, Crops Branch, (202) 720–2127.

Table 1-61.—Sorghum grain: Stocks on and off farms, United States, 1994–2003

Year beginning previous Dec.	On farms				Off farms [1]			
	Dec. 1	Mar. 1	Jun. 1	Sep. 1	Dec. 1	Mar. 1	Jun. 1	Sep. 1
	1,000 bushels	*1,000 bushels*	*1,000 bushels*	*1,000 bushels*	*1,000 bushels*	*1,000 bushels*	*1,000 bushels*	*1,000 bushels*
1994	105,950	59,770	32,075	16,155	340,198	216,462	96,035	31,416
1995	126,650	74,620	44,570	17,842	295,809	206,312	114,212	53,772
1996	79,090	33,000	13,955	5,500	222,149	129,641	56,433	12,871
1997	144,590	76,980	38,815	15,487	322,818	197,374	80,329	31,974
1998	99,625	56,760	27,200	13,700	274,244	177,916	68,944	35,203
1999	95,900	60,900	27,400	13,800	239,416	161,536	88,680	51,363
2000	90,300	51,700	27,300	12,200	259,136	173,932	99,606	53,175
2001	74,300	40,100	19,000	8,900	187,681	127,027	57,411	32,851
2002	72,400	38,100	17,300	7,400	241,477	156,007	88,178	53,573
2003	51,600	NA	NA	NA	178,000	NA	NA	NA

[1] Includes stocks at mills, elevators, warehouses, terminals, and processors. NA-not available.

NASS, Crops Branch, (202) 720–2127.

Table 1-62.—Sorghum: Supply and disappearance, United States, 1993–2002

Year beginning September 1	Supply			Disappearance					Ending stocks Aug. 31		
				Domestic use							
	Beginning stocks	Production	Total [1]	Feed and residual	Food, seed and industrial	Total	Exports	Total disappearance	Privately owned [2]	Government	Total
	Million bushels	*Million bushels*	*Million bushels*	*Million bushels*	*Million bushels*	*Million bushels*	*Million bushels*	*Million bushels*	*Million bushels*	*Million bushels*	*Million bushels*
1993	175	534	709	440	20	460	202	662	47	1	48
1994	48	646	693	377	22	399	223	622	71	1	72
1995	72	459	530	295	19	314	198	512	18	0	18
1996	18	795	814	516	45	561	205	766	47	0	47
1997	47	634	681	365	55	420	212	632	49	0	49
1998	49	520	569	262	45	307	197	504	65	0	65
1999	65	595	660	285	55	340	255	595	65	0	65
2000	65	471	536	222	35	258	237	494	41	1	42
2001 [3]	42	515	556	210	45	255	241	495	60	1	51
2002 [4]	61	370	431	160	45	205	200	405	26	0	26

[1] Includes imports. [2] Includes quantity under loan and farmer–owned reserve. [3] Preliminary. [4] Projected as of February 11, 2003, World Agricultural and Supply Demand Estimates. Totals may not add due to independent rounding.

ERS, Market and Trade Economics Division, (202) 694–5296.

Table 1-63.—Sorghum: Area, yield, and production, by States, 2000–2002

State	Area planted for all purposes			Sorghum for grain								
				Area harvested			Yield per harvested acre			Production		
	2000	2001	2002[1]	2000	2001	2002[1]	2000	2001	2002[1]	2000	2001	2002[1]
	1,000 acres	*1,000 acres*	*1,000 acres*	*1,000 acres*	*1,000 acres*	*1,000 acres*	*Bushels*	*Bushels*	*Bushels*	*1,000 bushels*	*1,000 bushels*	*1,000 bushels*
AL	10	12	10	7	7	7	40.0	60.0	45.0	280	420	315
AZ	16	12	15	9	6	6	80.0	80.0	80.0	720	480	480
AR	150	175	240	140	170	230	71.0	86.0	77.0	9,940	14,620	17,710
CA	12	10	15	8	7	11	75.0	90.0	95.0	600	630	1,045
CO	280	310	350	210	220	90	32.0	43.0	20.0	6,720	9,460	1,800
DE	3	2	2	2	1	1	83.0	85.0	48.0	166	85	48
GA	55	50	55	30	25	30	45.0	48.0	43.0	1,350	1,200	1,290
IL	90	80	80	85	77	77	95.0	105.0	83.0	8,075	8,085	6,391
KS	3,500	4,000	3,800	3,200	3,750	3,000	59.0	62.0	45.0	188,800	232,500	135,000
KY	11	11	12	9	10	11	85.0	85.0	75.0	765	850	825
LA	220	230	180	215	210	165	83.0	85.0	81.0	17,845	17,850	13,365
MD	10	9	5	9	8	4	84.0	83.0	47.0	756	664	188
MS	90	90	80	86	87	77	78.0	82.0	81.0	6,708	7,134	6,237
MO	280	230	190	270	220	185	92.0	94.0	85.0	24,840	20,680	15,725
NE	600	550	450	500	425	300	70.0	84.0	50.0	35,000	35,700	15,000
NM	165	170	170	65	140	80	25.0	45.0	35.0	1,625	6,300	2,800
NC	18	15	17	12	11	12	50.0	70.0	45.0	600	770	540
OK	450	500	430	360	420	330	38.0	36.0	45.0	13,680	15,120	14,850
PA	13	11	11	4	4	3	80.0	78.0	48.0	320	312	144
SC	9	8	6	7	6	4	52.0	65.0	35.0	364	390	140
SD	180	240	220	120	150	90	49.0	59.0	34.0	5,880	8,850	3,060
TN	25	30	35	22	27	31	75.0	80.0	80.0	1,650	2,160	2,480
TX	3,000	3,500	3,200	2,350	2,600	2,550	61.0	50.0	51.0	143,350	130,000	130,050
VA	8	7	7	6	3	5	82.0	88.0	55.0	492	264	275
US	9,195	10,252	9,580	7,726	8,584	7,299	60.9	59.9	50.7	470,526	514,524	369,758

[1] Preliminary.

NASS, Crops Branch, (202) 720–2127.

Table 1-64.—Sorghum: Utilization for silage, by States, 2000–2002

State	Silage								
	Area harvested			Yield per acre			Production		
	2000	2001	2002[1]	2000	2001	2002[1]	2000	2001	2002[1]
	1,000 acres	*1,000 acres*	*1,000 acres*	*Tons*	*Tons*	*Tons*	*1,000 tons*	*1,000 tons*	*1,000 tons*
AL	2	1	1	11.0	9.0	9.0	22	9	9
AZ	7	6	9	15.0	19.0	20.0	105	114	180
AR	5	4	2	9.0	9.0	10.0	45	36	20
CA	4	3	4	15.0	20.0	17.0	60	60	68
CO	12	12	15	16.0	20.0	9.0	192	240	135
DE	1	1	1	17.0	18.0	12.0	17	18	12
GA	15	20	20	9.0	10.0	13.0	135	200	260
IL	3	1	2	15.0	9.1	5.0	45	9	10
KS	65	100	100	10.0	12.0	7.0	650	1,200	700
KY	1	....	....	10.0	....	....	10	....	....
LA	1	1	1	11.0	12.0	10.0	11	12	10
MD	1	1	1	15.0	14.0	11.0	15	14	11
MS	3	1	1	9.0	11.0	13.0	27	11	13
MO	3	3	1	7.0	8.0	6.0	21	24	6
NE	20	20	25	11.0	11.0	7.5	220	220	188
NM	5	8	7	18.0	22.0	22.0	90	176	154
NC	5	3	3	11.0	10.0	5.0	55	30	15
OK	17	18	17	9.0	6.0	10.0	153	108	170
PA	7	5	7	11.0	10.0	7.0	77	50	49
SC	2	2	2	7.0	9.0	7.0	14	18	14
SD	20	50	40	9.0	9.5	5.5	180	475	220
TN	2	2	2	8.0	15.0	14.0	16	30	28
TX	60	70	90	10.0	9.0	12.0	600	630	1,080
VA	1	4	1	13.0	11.0	8.0	13	44	8
US	262	336	352	10.6	11.1	9.5	2,773	3,728	3,360

[1] Prelimary.

NASS, Crops Branch, (202) 720–2127.

Table 1-65.—Sorghum grain: Marketing year average price and value, by States, crop of 2000, 2001, and 2002

State	Marketing year average price per cwt			Value of production		
	2000	2001	2002 [1]	2000	2001	2002 [1]
	Dollars	Dollars	Dollars	1,000 dollars	1,000 dollars	1,000 dollars
AL	3.48	3.75	4.20	546	882	741
AZ	4.45	4.00	4.40	1,793	1,075	1,183
AR	2.93	3.56	4.40	16,302	29,146	43,637
CA	4.11	4.38	5.50	1,380	1,545	3,219
CO	3.27	3.27	4.10	12,298	17,323	4,133
DE	3.21	3.48	4.55	299	166	122
GA	3.55	3.30	4.30	2,687	2,218	3,106
IL	3.44	3.61	4.75	15,585	16,345	17,000
KS	3.19	3.24	4.25	337,952	421,848	321,300
KY	3.46	3.42	4.75	1,484	1,628	2,195
LA	2.95	3.75	4.40	29,444	37,485	32,931
MD	3.21	3.48	4.55	1,361	1,294	479
MS	3.07	3.70	3.90	11,538	14,782	13,622
MO	3.14	3.46	4.30	43,718	40,070	37,866
NE	3.28	3.23	4.45	64,400	64,574	37,380
NM	3.93	3.75	4.40	3,575	13,230	6,899
NC	3.57	3.75	4.20	1,200	1,617	1,270
OK	3.10	3.29	4.20	23,803	27,857	34,927
PA	3.20	3.75	4.95	573	655	399
SC	3.49	3.48	3.95	710	760	310
SD	2.70	3.06	4.50	8,879	15,165	7,711
TN	3.23	3.88	4.80	2,987	4,693	6,666
TX	3.28	3.64	4.20	263,764	264,992	305,878
VA	2.90	3.00	4.65	797	444	716
US	3.37	3.46	4.30	847,075	979,794	883,690

[1] Preliminary.

NASS, Crops Branch, (202) 720–2127.

Table 1-66.—Sorghum grain: Support operations, United States, 1993–2002

Marketing year beginning September 1	Income support payment rates per cwt [1]	Program price levels per cwt		Put under support [4]		Acquired by CCC under loan program [5]	Owned by CCC at end of marketing year
		Loan [2]	Target [3]	Quantity	Percentage of production		
	Dollars	Dollars	Dollars	Million cwt.	Percent	Millions cwt.	Million cwt.
1993/94	0.45	2.91	4.66	8.2	2.7	0.0	0.4
1994/95	1.05	3.21	4.66	25.2	7.0	0.0	0.4
1995/96	0.00	3.21	4.66	4.0	1.6	0.0	0.0
1996/97	0.577	3.23	NA	11.4	2.6	0.0	0.0
1997/98	0.971	3.14	NA	9.8	2.8	0.1	0.1
1998/99	1.209	3.11	NA	12.0	4.1	0.6	0.2
1999/2000	1.554	3.11	NA	9.6	2.9	1.0	0.0
2000/2001	1.491	3.05	NA	8.6	3.3	0.4	0.0
2001/2002	1.236	3.05	NA	9.6	3.3	0.1	0.1
2002/2003	0.625	3.54	4.54	....	....	....	....

[1] Payment rates for the 1995/96 & prior crop years were calculated according to the provisions of the now defunct deficiency payment/production adjustment program. Payment rates for the 1996/97 through 2001/2002 crops were calculated according to the provisions of the Production Flexibility Contract (PFC) program, following enactment of the Federal Agriculture Improvement and Reform Act of 1996 (1996 Act) and included supplemental PFC payment rates for 1998 through 2001. Payment rates for the 2002/2003 and subsequent crops are calculated according to the provisions of the Direct Payment program, following enactment of the Farm Security and Rural Investment Act of 2002 (2002 Act) and includes a PFC payment rate for 2002 only. Payment rates are rounded to the nearest tenth of a cent. [2] The national average loan rate was also known as the price support rate prior to enactment of the 1996 Act. [3] Between the 1996/97 and 2001/2002 marketing years, target prices were no longer applicable; however, with enactment of the 2002 Act, target prices were reestablished. [4] Represents loans made, purchases, and purchase agreements entered into. Purchases and purchase agreements are no longer authorized for the 1996 and subsequent crops following enactment of the 1996 Act. [5] Acquisition of all loans forfeited during the marketing year including loans made in previous year. NA-not applicable.

FSA, Feed Grains & Oilseeds Analysis Group, (202) 720–2711.

Table 1-67.—Coarse grains: International trade, 2000–2002[1]

Country	1999/2000	2000/2001	2001/2002[2]
	1,000 metric tons	*1,000 Metric tons*	*1,000 metric tons*
Exports:			
Argentina	9,604	13,016	9,089
Australia	3,836	4,629	4,546
Canada	3,716	3,572	2,442
China	9,954	7,296	8,630
South Africa	873	1,458	1,230
Russia	393	1,032	2,668
Turkey	184	158	501
EU	13,371	8,106	4,790
Others	6,094	9,449	14,186
Subtotal	48,025	48,716	48,082
United States	56,557	55,164	53,452
Total	104,582	103,880	101,534
Imports:			
Algeria	1,957	1,842	2,042
Brazil	2,166	1,014	437
Canada	1,164	2,935	4,178
Chile	1,350	1,388	1,328
China	2,340	2,416	1,969
Colombia	2,112	1,993	2,100
Costa Rica	550	513	463
Dominican Republic	1,000	968	1,038
Ecuador	255	149	336
Egypt	4,600	5,339	5,000
Iran	2,249	2,087	1,711
Israel	1,331	1,402	1,434
Japan	20,414	20,236	19,945
Jordan	1,162	803	938
Korea, North	150	688	288
Korea, South	9,280	8,889	8,834
Libya	475	396	485
Malaysia	2,296	2,588	2,425
Mexico	9,939	11,006	8,874
Morocco	1,497	1,473	1,544
Peru	912	912	933
Poland	756	824	248
Romania	150	463	130
Russia	2,048	563	729
Saudi Arabia	7,400	6,189	7,018
South Africa	491	146	983
Syria	1,325	1,090	1,042
Taiwan	5,231	5,157	4,776
Thailand	451	24	10
Tunisia	805	1,375	1,449
Turkey	1,461	648	1,247
Venezuela	1,300	1,207	515
Zimbabwe	63	66	311
EU	2,699	3,104	4,292
O.W. Europe	1,013	931	924
United States	2,607	2,672	2,347
Subtotal	94,999	93,496	92,323
Other Countries	8,080	8,389	7,666
Unaccounted	1,503	1,995	1,545
World Total	104,582	103,880	101,534

[1] Year beginning Oct 1. [2] Preliminary.

FAS, Grain and Feed Division, (202) 720–6219. Prepared or estimated on the basis of official statistics frpm foreign governments, other foreign source materials, reports of U.S. Agricultural Counselors, Attachés and Foreign Service Officers, results of office research, and related information.

Table 1-68.—Commercial feeds: Disappearance for feed, United States, 1992–2001

Year beginning October	Oilseed cake and meal: Soybean	Oilseed cake and meal: Cotton-seed	Oilseed cake and meal: Linseed	Oilseed cake and meal: Peanut [1]	Oilseed cake and meal: Sun-flower	Oilseed cake and meal: Total	Animal protein: Tankage and meat meal	Animal protein: Fish meal	Animal protein: Dried milk [2]	Animal protein: Total
	1,000 tons	*1,000 tons*	*1,000 tons*	*1,000 tons*	*1,000 tons*	*1,000 tons*	*1,000 tons*	*1,000 tons*	*1,000 tons*	*1,000 tons*
1992	24,229	2,571	102	178	442	27,522	2,372	528	464	3,364
1993	25,273	2,638	109	113	321	28,454	2,446	719	470	3,636
1994	26,545	3,268	101	193	623	30,730	2,551	334	463	3,349
1995	26,610	2,961	129	181	478	30,359	2,536	290	420	3,246
1996	27,320	3,113	149	141	462	31,185	2,783	308	428	3,518
1997	28,895	2,956	185	95	531	32,662	2,501	270	411	3,182
1998	30,657	2,533	169	94	635	34,088	2,620	326	273	3,218
1999	30,345	2,908	185	138	582	34,158	2,342	265	305	2,913
2000	31,643	2,855	193	110	496	35,297	2,398	246	306	2,950
2001 [3]	33,077	3,340	156	143	395	37,111	2,213	258	302	2,773

Year beginning October	Mill products [4]: Wheat millfeeds	Mill products [4]: Gluten feed and meal [5]	Mill products [4]: Rice millfeeds	Mill products [4]: Alfalfa meal	Mill products [4]: Total	Total commercial feeds
	1,000 tons	*1,000 tons*	*1,000 tons*	*1,000 tons*	*1,000 tons*	*1,000 tons*
1992	6,965	595	604	309	8,473	39,359
1993	7,190	912	648	134	8,883	40,973
1994	7,265	139	658	209	8,272	42,351
1995	7,373	881	602	255	9,111	42,715
1996	7,280	1,833	559	248	9,921	44,624
1997	7,030	1,914	547	276	9,767	45,611
1998	7,038	1,349	594	225	9,206	46,513
1999	7,324	1,199	650	NA	9,172	46,243
2000	7,303	1,423	626	NA	9,351	50,549
2001 [3]	6,869	1,475	616	NA	8,960	51,616

[1] Year beginning August 1. [2] Includes dried skim milk, and whey for feed, but does not include any milk products fed on farms. [3] Preliminary. [4] Other mill products that are not listed include screenings, hominy, and oats feed etc., for which no statistics are available. [5] Adjusted for export data. NA-not available.

ERS, Market and Trade Economics Division, (202) 694–5290.

Table 1-69.—High-protein feeds: Quantity for feeding, high-protein animal units, quantity per animal unit, and prices, United States, 1992–2001

Year beginning October	Quantity for feeding[1]						High-protein animal units	Quantity per animal unit	High protein feed prices
	Oilseed meal			Animal protein	Grain protein[3]	Total			
	Soybean meal	Other oilseed meals[2]	Total						
	1,000 tons	*1,000 tons*	*1,000 tons*	*1,000 tons*	*1,000 tons*	*1,000 tons*	*Million units*	*Pounds*	*Index numbers 1992=100*
1992	26,652	3,050	29,702	3,661	354	33,718	126.2	534.474	97
1993	27,800	2,945	30,745	3,991	543	35,279	128.4	549.310	96
1994	29,200	3,875	33,075	3,634	83	36,792	130.7	562.803	81
1995	29,271	3,469	32,740	3,547	524	36,812	132.2	556.942	114
1996	30,052	3,573	33,625	3,869	1,092	38,586	133.1	579.747	136
1997	31,765	3,470	35,254	3,477	1,140	39,871	136.2	585.409	100
1998	33,723	3,154	36,877	3,647	803	41,327	137.0	603.325	70
1999	33,380	3,514	36,893	3,247	714	40,854	137.4	594.522	87
2000	34,807	3,366	38,173	3,291	847	42,312	138.3	612.033	91
2001	36,385	3,733	40,117	3,081	879	44,077	139.8	630.636	89

[1] In terms of 44 percent protein soybean meal equivalent. [2] Includes cottonseed, linseed, peanut meal, and sunflower meal. [3] Beginning 1974, adjusted for exports of corn gluten feed and meal.

ERS, Market and Trade Economics Division (202) 694–5290.

Table 1-70.—Feed concentrates: Fed to livestock and poultry, 1992–2001

Year beginning October	Feed grains				Wheat[2]	Rye[2]	By-product feeds[3]	Total concentrates	Grain consuming animal units	Concentrates fed per grain-consuming animal unit
	Corn[1]	Sorghum[1]	Oats[2] and barley[2]	Total						
	Million tons	*Million tons*	*Million tons*	*Million tons*	*Million tons*	*Million tons*	*Million tons*	*Million tons*	*Millions*	*Tons*
1992	147.1	12.8	8.5	168.3	4.3	0.2	46.6	219.4	82.8	2.65
1993	131.0	12.3	10.4	153.7	10.6	0.2	48.0	212.5	84.0	2.53
1994	152.9	10.6	8.7	172.1	8.2	0.2	50.0	230.0	84.4	2.73
1995	131.4	8.3	7.6	147.2	6.8	0.2	49.4	203.6	85.0	2.40
1996	147.8	14.4	6.9	169.1	8.5	0.1	52.2	230.0	85.3	2.69
1997	153.5	10.2	7.2	170.9	9.7	0.1	53.3	234.0	87.9	2.66
1998	153.1	7.3	6.8	167.3	7.2	0.1	54.1	228.7	88.0	2.60
1999	158.6	8.0	6.6	173.2	10.1	0.1	54.1	237.4	88.9	2.67
2000	163.6	6.2	5.9	175.7	6.9	0.1	58.5	241.2	89.3	2.70
2001	164.5	5.9	4.6	175.1	4.1	0.1	59.5	238.8	89.6	2.67

[1] Marketing year beginning Sept. 1. [2] Marketing year beginning June1. [3] Oilseed meals, animal protein feeds, mill by-products, and mineral supplements.

ERS, Market and Trade Economics Division (202) 694–5290.

Table 1-71.—Feed: Consumed per head and per unit of production, by class of livestock or poultry, with quantity expressed in equivalent feeding value of corn, 1992–2001

Year beginning October	Dairy cattle			Beef cattle				Sheep and lambs	
	Milk cows		Other dairy cattle per head	Cattle on feed per head Jan. 1[1]	Other beef cattle per head	All beef cattle per head	Cattle and calves per 100 pounds produced[2]	Per head	Per 100 pounds produced[3]
	Per head	Per 100 pounds milk produced							
	Pounds	*Pounds*	*Pounds*	*Pounds*	*Pounds*	*Pounds*	*Pounds*	*Pounds*	*Pounds*
1992	12,876	82	6,523	9,613	5,308	5,953	1,330	1,274	1,568
1993	12,668	79	6,488	9,310	5,297	5,896	1,317	1,270	1,530
1994	13,089	80	6,559	9,925	5,319	5,961	1,344	1,278	1,633
1995	12,410	76	6,445	8,932	5,284	5,809	1,292	1,266	1,610
1996	13,031	78	6,549	9,841	5,316	5,992	1,359	1,277	1,631
1997	12,964	76	6,538	9,743	5,313	6,009	1,329	1,276	1,522
1998	12,857	73	6,520	9,586	5,307	5,966	1,295	1,274	1,521
1999	13,017	71	6,547	9,820	5,316	6,058	1,274	1,277	1,548
2000	13,149	73	6,569	10,013	5,322	6,115	1,262	1,279	1,605
2001[4]	13,091	71	6,559	9,928	5,319	6,084	1,262	1,278	1,577

Year beginning October	Poultry								Hogs per 100 pounds produced	Horses and mules 2 years and over per head
	Hens and pullets		Chickens raised		Broilers produced		Turkeys raised			
	Per head Jan. 1	Per 100 eggs	Per head	Per 100 pounds live weight	Per head	Per 100 pounds produced	Per head	Per 100 pounds produced		
	Pounds	*Pounds*	*Pounds*	*Pounds*	*Pounds*	*Pounds*	*Pounds*	*Pounds*	*Pounds*	*Pounds*
1992	126	50	30	822	11.1	256	94	425	603	3,826
1993	122	49	29	792	10.7	243	91	405	594	3,786
1994	131	52	31	843	11.5	257	97	431	625	3,868
1995	117	45	28	827	10.3	226	87	386	545	3,735
1996	130	51	31	976	11.4	242	96	401	619	3,856
1997	128	51	31	981	11.3	238	95	381	640	3,843
1998	126	51	30	993	11.1	233	94	364	595	3,822
1999	129	51	31	898	11.4	231	96	376	590	3,854
2000	132	52	31	874	11.7	235	98	383	597	3,879
2001[4]	131	51	31	911	11.6	235	97	369	595	3,868

[1] Feed consumed by all cattle divided by the number on feed Jan. 1. [2] Feed for all cattle, except milk cows, divided by the net live-weight production of cattle and calves. It includes the growth on dairy heifers and calves as well as all beef cattle. [3] Including wool produced. [4] Preliminary.

ERS, Market and Trade Economics Division, (202) 694–5290.

Table 1-72.—Feed: Consumed by livestock and poultry, by type of feed, with quantity expressed in equivalent feeding value of corn, 1992–2001

Year beginning October	Concentrates	Harvested roughage	Pasture	Total
	Million tons	*Million tons*	*Million*	*Million tons*
1992	230	82	174	487
1993	226	77	183	486
1994	244	81	184	509
1995	218	80	186	484
1996	244	81	179	504
1997	249	81	173	503
1998	244	81	173	498
1999	254	84	166	503
2000	260	86	160	507
2001[1]	259	85	161	504

[1] Preliminary.

ERS, Market and Trade Economics Division, (202) 694–5290.

Table 1-73.—Animal units fed: Grain-consuming, roughage-consuming, and grain-and-roughage-consuming, United States, 1992–2001 [1]

Year beginning October	Grain-consuming [2]	Roughage-consuming [3]	Grain and roughage-consuming [4]
	1,000 units	*1,000 units*	*1,000 units*
1992	82,760	75,311	77,719
1993	84,000	76,450	78,876
1994	84,377	78,040	79,994
1995	85,015	78,128	80,277
1996	85,341	76,420	79,363
1997	87,948	74,861	79,414
1998	88,040	74,526	79,228
1999	88,864	73,281	78,759
2000	89,299	72,481	78,436
2001 [5]	89,590	72,194	78,224

[1] Index series based on average feeding rates for years 1969–71. In calculations for the feeding years 1969 to date, cattle numbers used are the new categories shown in the Livestock and Poultry Inventory, published by NASS, USDA. [2] Livestock and poultry numbers weighted by all concentrates consumed. [3] Livestock and poultry numbers weighted by all roughage (including pasture) consumed. [4] Livestock and poultry numbers weighted by all feed (including pasture) fed to livestock. [5] Preliminary.

ERS, Market and Trade Economics Division, (202) 694–5290.

Table 1-74.—Feed grains: Average price, selected markets and grades, 1994–2002 [1]

Crop year	Kansas City			Minneapolis			
	Corn, No. 2 Yellow	Corn, No. 2 White	Sorghum, No. 2 Yellow	Corn, No. 2 Yellow	Barley, No. 3 or Better malting	Duluth Barley, No. 2 Feed	Oats, No. 2 White
	Dollars per bushel	*Dollars per bushel*	*Dollars per cwt.*	*Dollars per bushel*	*Dollars per bushel*	*Dollars per bushel*	*Dollars per bushel*
1994	2.69	2.78	4.37	2.52	2.48	2.05	1.54
1995	2.47	2.91	4.10	2.27	2.75	2.02	1.49
1996	4.12	4.07	6.66	3.74	3.69	2.67	2.28
1997	2.84	3.09	4.54	2.65	3.18	2.32	2.03
1998	2.49	2.93	4.11	2.39	2.50	1.86	1.70
1999	2.01	2.42	3.29	1.88	2.30	1.86	1.33
2000	1.93	1.94	3.19	1.79	....	1.22	1.28
2001	1.85	....	4.03	....	....	....	....
2002	2.13	2.51	4.27	2.11	2.85	1.70	....

Crop year	Omaha: Corn, No. 2 Yellow	Chicago: Corn, No. 2 Yellow	Texas High Plains: Sorghum, No. 2 Yellow	Memphis		St. Louis: Corn, No. 2 Yellow
				Corn, No. 2 Yellow	Barley, No. 2 Western	
	Dollars per bushel	*Dollars per bushel*	*Dollars per cwt.*	*Dollars per bushel*	*Dollars per bushel*	*Dollars per cwt.*
1994	2.58	2.68	4.95	2.68	6.61	2.67
1995	2.31	2.44	4.75	2.51	5.79	2.51
1996	3.87	3.97	7.30	4.02	7.45	4.06
1997	2.70	2.84	5.02	2.88	7.84	2.90
1998	2.36	2.56	4.72	2.59	6.68	2.62
1999	1.88	2.15	3.79	2.13	....	2.17
2000	1.82	1.97	3.51	2.01	....	2.00
2001	....	1.98	....	2.03	....	....
2002	2.13	2.24	4.27	2.29	....	2.33

[1] Simple average of daily prices.

AMS, Livestock and Grain Market News Branch, (202) 720–6231.

Table 1-75.—Feedstuffs: Average price per ton bulk, in wholesale lots, at leading markets, 1993–2002

Year beginning October	Soybean meal		Cottonseed meal		Linseed meal	Meat meal	Fish meal	Wheat bran	Wheat middlings
	44% protein	49–50% protein	41% protein	41% protein	34% protein	50% protein	60% protein		
	Decatur	Decatur	Kansas City	Memphis	Minneapolis	Kansas City	Gulf Coast	Kansas City	Minneapolis
	Dollars per ton	*Dollars per ton*	*Dollars per ton*	*Dollars per ton*	*Dollars per ton*	*Dollars per ton*	*Dollars per ton*	*Dollars per ton*	*Dollars per ton*
1993	179.80	203.00	180.50	169.60	132.50	204.40	335.40	80.60	68.10
1994	152.50	171.70	129.40	120.80	95.00	168.70	365.00	65.50	53.90
1995	225.10	243.10	208.80	19.70	159.00	231.50	478.70	120.30	108.00
1996	260.40	279.70	207.50	91.40	158.70	271.80	502.60	88.30	80.20
1997	175.00	196.10	162.60	144.00	117.50	180.90	545.60	73.30	61.20
1998	132.00	138.50	132.20	109.60	84.50	138.10	387.50	55.90	60.90
1999	131.99	138.55	130.20	109.60	84.50	138.10	387.80	55.88	49.58
2000	160.03	168.10	146.50	130.70	103.10	166.50	326.40	53.37	45.81
2001	165.21	173.60	165.00	142.70	121.90	166.50	358.20	62.93	50.81
2002	153.82	167.72	160.10	136.20	119.20	166.00	460.00	59.74	58.42

Year beginning October	Wheat shorts or middlings	Wheat millrun	Gluten feed	Hominy feed	Distillers' dried grains	Brewers' dried grains	Alfalfa meal		Blackstrap molasses
			21% protein				Dehydrated, 17% protein	Sun-cured	
	Kansas City	Portland	Illinois Points	Illinois Points	Lawrenceburg	Columbus	Kansas City	Kansas City	New Orleans
	Dollars per ton	*Dollars per ton*	*Dollars per ton*	*Dollars per ton*	*Dollars per ton*	*Dollars per ton*	*Dollars per ton*	*Dollars per ton*	*Dollars per ton*
1993	80.60	88.70	89.30	80.50	123.80	....	124.70	114.2	62.50
1994	66.30	82.00	82.00	83.80	105.40	73.40	118.40	112.7	71.20
1995	120.40	118.70	119.30	132.10	156.10	134.40	130.40	122	77.50
1996	80.30	95.70	90.20	88.90	139.80	134.60	142.70	130.9	60.10
1997	73.30	74.30	67.20	78.90	105.80	107.20	126.80	115	51.50
1998	55.90	54.30	60.40	62.20	85.20	84.20	101.50	92.00	35.50
1999	55.88	54.34	60.41	67.19	85.15	84.19	101.55	91.99	35.48
2000	57.82	60.92	53.64	58.71	79.90	90.94	97.59	86.58	38.72
2001	62.88	63.25	60.55	55.02	80.62	94.00	139.06	130.38	63.16
2002	59.77	75.95	59.63	63.23	80.19	94.00	154.05	134.34	68.63

AMS, Livestock and Grain Market News Branch, (202) 720–6231.

Table 1-76.—Proso millet: Area, yield, production, and value, United States, 1999–2002

Year	Area: Planted	Area: Harvested	Yield per harvested acre	Production	Marketing year average price per bushel received by farmers	Value of production
	1,000 acres	*1,000 acres*	*Bushels*	*1,000 bushels*	*Dollars*	*1,000 dollars*
1999	600	540	33.2	17,910	2.12	38,033
2000	440	370	19.8	7,320	4.79	35,034
2001	650	585	33.2	19,405	2.02	39,109
2002 [1]	450	220	12.5	2,755	7.36	20,266

[1] Preliminary.

NASS, Crops Branch, (202) 720–2127.

Table 1-77.—Proso millet: Area, yield, and production, by States, 2000–2002

State	Area planted			Area harvested		
	2000	2001	2002 [1]	2000	2001	2002 [1]
	1,000 acres	*1,000 acres*	*1,000 acres*	*1,000 acres*	*1,000 acres*	*1,000 acres*
CO	190	240	220	150	230	95
NE	150	190	130	135	180	65
SD	100	220	100	85	175	60
US	440	650	450	370	585	220

State	Yield per acre			Production		
	2000	2001	2002 [1]	2000	2001	2002 [1]
	Bushels	*Bushels*	*Bushels*	*1,000 bushels*	*1,000 bushels*	*1,000 bushels*
CO	19.0	35.0	10.0	2,850	8,050	950
NE	18.0	31.0	13.0	2,430	5,580	845
SD	24.0	33.0	16.0	2,040	5,775	960
US	19.8	33.2	12.5	7,320	19,405	2,755

[1] Preliminary.

NASS, Crops Branch, (202) 720–2127.

Table 1-78.—Proso millet: Marketing year average price and value, by States, crop of 2000, 2001, and 2002

State	Marketing year average price per bushel			Value of production		
	2000	2001	2002 [1]	2000	2001	2002 [1]
	Dollars	*Dollars*	*Dollars*	*1,000 dollars*	*1,000 dollars*	*1,000 dollars*
CO	4.80	2.00	7.40	13,680	16,100	7,030
NE	4.80	1.95	7.20	11,664	10,881	6,084
SD	4.75	2.10	7.45	9,690	12,128	7,152
US	4.79	2.02	7.36	35,034	39,109	20,266

[1] Preliminary.

NASS, Crops Branch, (202) 720–2127.

CHAPTER II

STATISTICS OF COTTON, TOBACCO, SUGAR CROPS, AND HONEY

In addition to tables on cotton, tobacco, sugar, and honey, this chapter includes tables on fibers other than cotton and syrups. Cottonseed data, however, are in the following chapter on oilseeds, fats, and oils.

Table 2-1.—Cotton: Area, yield, production, market year average price, and value, United States, 1993–2002

Year	Area		Yield per harvested acre	Production	Marketing year average price per pound received by farmers	Value of production
	Planted	Harvested				
	1,000 acres	*1,000 acres*	*Pounds*	*1,000 bales* [1]	*Cents*	*1,000 dollars*
1993	13,438.3	12,783.3	606	16,133.6	58.4	4,520,908
1994	13,720.1	13,322.3	708	19,662.0	72.0	6,796,654
1995	16,931.4	16,006.7	537	17,899.8	76.5	6,574,612
1996	14,652.5	12,888.1	705	18,942.0	70.5	6,408,144
1997	13,898.0	13,406.0	673	18,793.0	66.2	5,975,585
1998	13,392.5	10,683.6	625	13,918.2	61.7	4,119,911
1999	14,873.5	13,424.9	607	16,968.0	46.8	3,809,560
2000	15,517.2	13,053.0	632	17,188.3	51.6	4,260,417
2001	15,768.5	13,827.7	705	20,302.8	32.0	3,121,848
2002 [2]	13,962.6	12,413.3	663	17,145.0	43.7	3,593,816

[1] 480-pound net weight bales. [2] Preliminary.
NASS, Crops Branch, (202) 720–2127.

Table 2-2.—Cotton: Area, yield, and production, by type, State, and United States, 2000–2002

State	Area planted			Area harvested			Yield per harvested acre			Production [1]		
	2000	2001	2002 [2]	2000	2001	2002 [2]	2000	2001	2002 [2]	2000	2001	2002 [2]
	1,000 acres	*1,000 acres*	*1,000 acres*	*1,000 acres*	*1,000 acres*	*1,000 acres*	*Pounds*	*Pounds*	*Pounds*	*1,000 bales* [3]	*1,000 bales* [3]	*1,000 bales* [3]
Upland:												
AL	590.0	610.0	590.0	530.0	605.0	540.0	492	730	511	543.0	920.0	575.0
AZ	280.0	295.0	215.0	278.0	290.0	213.0	1,366	1,142	1,262	791.0	690.0	560.0
AR	960.0	1,080.0	960.0	950.0	1,065.0	920.0	720	826	861	1,425.0	1,833.0	1,650.0
CA	775.0	630.0	480.0	770.0	625.0	477.0	1,378	1,359	1,439	2,210.0	1,770.0	1,430.0
FL	130.0	125.0	120.0	106.0	124.0	115.0	480	612	346	106.0	158.0	83.0
GA	1,500.0	1,490.0	1,450.0	1,350.0	1,480.0	1,360.0	591	720	582	1,663.0	2,220.0	1,650.0
KS	40.0	40.5	80.0	37.0	35.5	60.0	288	407	608	22.2	30.1	76.0
LA	710.0	870.0	520.0	695.0	855.0	495.0	629	580	727	911.0	1,034.0	750.0
MS	1,300.0	1,620.0	1,170.0	1,280.0	1,600.0	1,150.0	642	719	826	1,711.0	2,396.0	1,980.0
MO	400.0	405.0	380.0	388.0	400.0	368.0	668	834	796	540.0	695.0	610.0
NM	72.0	68.0	54.0	67.0	65.0	50.0	724	916	960	101.0	124.0	100.0
NC	930.0	970.0	940.0	925.0	965.0	920.0	742	832	412	1,429.0	1,673.0	790.0
OK	280.0	270.0	200.0	145.0	185.0	180.0	503	511	533	152.0	197.0	200.0
SC	300.0	300.0	290.0	290.0	296.0	190.0	627	686	328	379.0	423.0	130.0
TN	570.0	620.0	570.0	565.0	615.0	535.0	603	763	729	710.0	978.0	813.0
TX	6,400.0	6,000.0	5,600.0	4,400.0	4,250.0	4,500.0	430	481	533	3,940.0	4,260.0	5,000.0
VA	110.0	105.0	100.0	108.0	104.0	98.0	738	929	485	166.0	201.3	99.0
US	15,347.0	15,498.5	13,719.0	12,884.0	13,559.5	12,171.0	626	694	651	16,799.2	19,602.4	16,496.0
American-Pima:												
AZ	5.0	7.8	8.0	4.9	7.5	7.9	705	928	972	7.2	14.5	16.0
CA	145.0	240.0	210.0	144.0	239.0	209.0	1,154	1,283	1,332	346.3	639.0	580.0
NM	4.2	5.2	7.1	4.1	5.2	7.1	539	969	946	4.6	10.5	14.0
TX	16.0	17.0	18.5	16.0	16.5	18.3	930	1,059	1,023	31.0	36.4	39.0
US	170.2	270.0	243.6	169.0	268.2	242.3	1,105	1,254	1,286	389.1	700.4	649.0
US, all	15,517.2	15,768.5	13,962.6	13,053.0	13,827.7	12,413.3	632	705	663	17,188.3	20,302.8	17,145.0

[1] Production ginned and to be ginned. [2] Preliminary. [3] 480-pound net weight bales.
NASS, Crops Branch, (202) 720–2127.

Table 2-3.—Cotton: Marketing year average price per pound, and value, by State and United States, crop of 2000, 2001, and 2002

State	Marketing year average price per pound			Value of production		
	2000	2001	2002 [1]	2000	2001	2002 [1]
	Dollars	*Dollars*	*Dollars*	*1,000 dollars*	*1,000 dollars*	*1,000 dollars*
Upland:						
AL	0.528	0.277	0.425	137,618	122,323	117,300
AZ	0.397	0.284	0.489	150,733	94,061	131,443
AR	0.549	0.280	0.420	375,516	246,355	332,640
CA	0.520	0.416	0.538	551,616	353,434	369,283
FL	0.565	0.295	0.380	28,747	22,373	15,139
GA	0.556	0.306	0.446	443,821	326,074	353,232
KS	0.451	0.295	0.380	4,806	4,262	13,862
LA	0.516	0.281	0.411	225,636	139,466	147,960
MS	0.505	0.265	0.399	414,746	304,771	379,210
MO	0.564	0.300	0.430	146,189	100,080	125,904
NM	0.531	0.341	0.520	25,743	20,296	24,960
NC	0.530	0.317	0.420	363,538	254,564	159,264
OK	0.451	0.295	0.380	32,905	27,895	36,480
SC	0.550	0.320	0.400	100,056	64,973	24,960
TN	0.455	0.305	0.410	155,064	143,179	159,998
TX	0.459	0.284	0.383	868,061	580,723	919,200
VA	0.607	0.301	0.408	48,366	29,084	19,388
US	0.498	0.298	0.405	4,073,161	2,833,913	3,330,223
American-Pima:						
AZ	0.967	0.848	0.828	3,342	5,902	6,359
CA	1.010	0.856	0.851	167,886	262,552	236,918
NM	0.938	0.877	0.803	2,071	4,420	5,396
TX	0.938	0.862	0.797	13,957	15,061	14,920
US	1.003	0.856	0.846	187,256	287,935	263,593
US, all	0.516	0.320	0.437	4,260,417	3,121,848	3,593,816

[1] Preliminary.

NASS, Crops Branch, (202) 720–2127.

Table 2-4.—Cotton, American upland: Support operations, United States, 1993–2002

Marketing Year beginning August 1	Income support payment rates per pound [1]	Program price levels per pound		Put under Loan [4]		Acquired by CCC under loan program [5]	Owned by CCC at end of marketing year
		Loan [2]	Target [3]	Quantity	Percentage of production		
	Cents	*Cents*	*Cents*	*1,000 bales* [6]	*Percent*	*1,000 bales* [6]	*1,000 bales* [6]
1993/94	18.600	52.35	72.90	7,721	49.0	3	14
1994/95	4.600	50.00	72.90	4,716	24.4	(7)	(7)
1995/96	0.000	51.92	72.90	3,478	19.8	0	0
1996/97	8.882	51.92	NA	3,340	18.1	0	0
1997/98	7.625	51.92	NA	4,281	23.5	0	0
1998/99	12.237	51.92	NA	4,724	36.8	31	3
1999/2000	15.760	51.92	NA	8,721	54.9	2	1
2000/2001	15.210	51.92	NA	8,837	52.6	6	5
2001/2002	12.660	51.92	NA	13,655	69.7	2	2
2002/2003	12.390	52.00	72.40	....	....	....	....

[1] Payment rates for the 1995/96 & prior crop years were calculated according to the provisions of the now defunct deficiency payment/production adjustment program. Payment rates for the 1996/97 through 2001/2002 crops were calculated according to the provisions of the Production Flexibility Contract (PFC) program, following enactment of the Federal Agriculture Improvement and Reform Act of 1996 (1996 Act) and included supplemental PFC payment rates for 1998 through 2001. Payment rates for the 2002/2003 and subsequent crops are calculated according to the provisions of the Direct Payment program, following enactment of the Farm Security and Rural Investment Act of 2002 (2002 Act) and includes a PFC payment rate for 2002 only. Payment rates are rounded to the nearest thousandth of a cent. [2] The national average loan rate was also known as the price support rate prior to enactment of the 1996 Act. [3] Between the 1996/97 and 2001/2002 marketing years, target prices were no longer applicable; however, with enactment of the 2002 Act, target prices were reestablished. [4] Represents loans made, purchases, and purchase agreements entered into. Purchases and purchase agreements are no longer authorized for the 1996 and subsequent crops following enactment of the 1996 Act. [5] Acquisitions from the crop harvested in the year indicated. [6] Running bales. [7] Less than 500 bales. NA-not applicable.

FSA, Fiber Analysis Group, (202) 720-3392.

Table 2-5.—Cotton: Area, yield, and production in specified countries, 1999/2000–2001/2002 [1]

Continent and country	Area [2]			Yield per hectare			Production		
	1999/ 2000	2000/ 2001	2001/ 2002 [3]	1999/ 2000	2000/ 2001	2001/ 2002 [3]	1999/ 2000	2000/ 2001	2001/ 2002 [3]
	1,000 hectares	*1,000 hectares*	*1,000 hectares*	*Kilograms*	*Kilograms*	*Kilograms*	*1,000 bales* [3]	*1,000 bales* [3]	*1,000 bales* [3]
North America:									
Mexico	152	81	82	958	1,059	1,147	669	394	432
United States	5,433	5,282	5,596	680	708	790	16,968	17,188	20,303
Total	5,585	5,363	5,678	688	714	795	17,637	17,582	20,735
South America:									
Argentina	300	380	165	417	421	396	574	735	300
Bolivia	6	10	10	436	436	501	12	20	23
Brazil	752	853	750	931	1,101	1,021	3,216	4,312	3,518
Colombia	43	46	46	582	611	587	115	129	124
Ecuador	15	15	15	218	290	305	15	20	21
Paraguay	195	300	150	410	327	319	367	450	220
Peru	85	75	75	474	537	534	185	185	184
Venezuela	30	30	30	363	363	363	50	50	50
Total	1,426	1,709	1,241	692	752	779	4,534	5,901	4,440
Central America:									
Costa Rica	1	1	1	218	218	218	1	1	1
El Salvador	1	1	1	218	218	218	1	1	1
Guatemala	2	2	2	327	327	327	3	3	3
Honduras	2	2	2	545	980	436	5	9	4
Nicaragua	4	4	4	544	544	544	10	10	10
Total	10	10	10	436	523	414	20	24	19
Carribean:									
Cuba	4	4	4	109	109	218	2	2	4
Dominican Rep	4	4	4	163	163	163	3	3	3
Haiti	7	7	7	156	156	156	5	5	5
Total	15	15	15	145	145	174	10	10	12
European Union:									
Greece	430	410	410	1,023	1,081	1,111	2,021	2,035	2,093
Italy	1	1	1	218	218	218	1	1	1
Spain	108	90	90	1,222	1,021	1,188	606	422	491
Total	539	501	501	1,062	1,068	1,123	2,628	2,458	2,585
Eastern Europe:									
Albania	1	1	1	218	218	218	1	1	1
Bulgaria	15	15	15	435	435	435	30	30	30
Yugoslavia	1	1	1	218	218	218	1	1	1
Total	17	17	17	410	410	410	32	32	32
Former USSR:									
Azerbaijan, Rep	91	91	82	335	347	385	140	145	145
Kazakhstan, Rep	141	153	184	540	569	769	350	400	650
Kyrgyzstan, Rep	35	34	37	809	832	824	130	130	140
Tajikistan, Rep	248	239	245	395	442	591	450	485	665
Turkmenistan	475	475	515	481	378	359	1,050	825	850
Uzbekistan, Rep	1,500	1,415	1,430	752	677	746	5,180	4,400	4,900
Total	2,490	2,407	2,493	638	578	642	7,300	6,385	7,350
Middle East:									
Cyprus	1	1	1	218	218	218	1	1	1
Iran	210	260	200	674	615	626	650	735	575
Iraq	20	20	20	435	435	348	40	40	32
Israel	15	10	15	1,640	1,633	1,452	113	75	100
Syria	244	270	260	1,294	1,351	1,298	1,450	1,675	1,550
Turkey	719	654	693	1,100	1,198	1,249	3,634	3,600	3,975
Yemen	40	40	40	354	354	354	65	65	65
Total	1,249	1,255	1,229	1,038	1,074	1,116	5,953	6,191	6,298

See footnotes at end of table.

Table 2-5.—Cotton: Area, yield, and production in specified countries, 1999/2000–2001/2002 [1]—Continued

Continent and country	Area			Yield per hectare			Bales		
	1999/ 2000	2000/ 2001	2001/ 2002 [2]	1999/ 2000	2000/ 2001	2001/ 2002 [2]	1999/ 2000	2000/ 2001	2001/ 2002 [2]
	1,000 hectares	*1,000 hectares*	*1,000 hectares*	*Kilograms*	*Kilograms*	*Kilograms*	*1,000 bales* [3]	*1,000 bales* [3]	*1,000 bales* [3]
Africa:									
Angola	10	10	10	436	436	436	20	20	20
Benin	400	380	390	376	344	447	690	600	800
Burkina	300	275	350	367	416	451	505	525	725
Cameroon	180	200	200	416	468	490	344	430	450
Central Afric	70	65	70	156	134	156	50	40	50
Chad	420	400	425	181	163	166	350	300	325
Cote d'Ivoir	290	300	300	563	417	508	750	575	700
Egypt	273	223	315	841	898	996	1,055	920	1,441
Ethiopia	60	60	60	254	254	254	70	70	70
Ghana	41	40	40	372	381	381	70	70	70
Guinea	20	19	19	403	424	424	37	37	37
Kenya	55	37	37	119	118	118	30	20	20
Madagascar	30	30	30	435	435	435	60	60	60
Malawi	50	70	70	174	124	93	40	40	30
Mali	482	275	520	407	380	461	900	480	1,100
Morocco	1	1	1	218	218	218	1	1	1
Mozambique	150	125	155	145	131	155	100	75	110
Niger	5	5	5	218	218	218	5	5	5
Nigeria	280	350	375	179	249	261	230	400	450
Senegal	20	25	32	294	348	442	27	40	65
Somalia	12	12	12	127	127	127	7	7	7
South Africa	75	68	41	351	496	441	121	155	83
Sudan	225	235	150	232	315	399	240	340	275
Tanzania	200	400	450	180	136	121	165	250	250
Togo	160	125	150	354	392	399	260	225	275
Tunisia	2	2	2	1,089	1,089	1,089	10	10	10
Uganda	250	250	250	87	74	78	100	85	90
Zaire	30	30	30	109	109	109	15	15	15
Zambia	150	175	200	160	165	185	110	133	170
Zimbabwe	365	390	400	352	307	191	590	550	350
Total	4,606	4,577	5,089	329	308	345	6,952	6,478	8,054
Asia:									
Afghanistan	60	50	50	363	370	370	100	85	85
Bangladesh	38	16	50	372	463	292	65	34	67
Burma	322	325	320	176	181	184	260	270	270
China, People	3,726	4,058	4,820	1,028	1,089	1,102	17,600	20,300	24,400
India	8,791	8,576	8,851	302	278	303	12,180	10,931	12,300
Indonesia	21	22	22	145	317	544	14	32	55
Korea, Democr	19	19	7	539	527	156	47	46	5
Korea, Republ	1	1	1	218	218	218	1	1	1
Pakistan	2,915	2,928	3,130	642	610	577	8,600	8,200	8,300
Philippines	2	2	5	327	218	305	3	2	7
Sri Lanka	5	5	5	218	218	218	5	5	5
Thailand	14	15	49	451	494	449	29	34	101
Vietnam	23	28	30	331	334	348	35	43	48
Total	15,937	16,045	17,340	532	543	573	38,939	39,983	45,644
Oceania:									
Australia	464	505	420	1,623	1,595	1,659	3,458	3,700	3,200
World total	32,338	32,404	34,033	589	596	629	87,463	88,744	98,369

[1] Harvest season beginning Aug. 1. [2] Preliminary. [3] 480-pound net weight.

FAS, Production Estimates and Crop Assessment Division, (202) 720–0888. Prepared or estimated on the basis of official statistics of foreign governments, other foreign source materials, reports of U.S. Agricultural Counselors, Attachés, and Foreign Service Officers, results of office research, and related information.

Table 2-6.—Cotton: Supply and distribution, United States, 1992–2001

Year beginning August 1	Supply			Distribution				
	Beginning of season total [2]	Ginnings in season [1]	Total supply [2]	Consumption [2]			Exports	Carryover, end of season [2]
				Upland	American Pima	Total		
	1,000 bales	*1,000 bales*	*1,000 bales*	*1,000 bales*	*1,000 bales*	*1,000 bales*	*1,000 bales*	*1,000 bales*
1992	3,569	15,781	19,350	9,789	57	9,846	4,950	4,482
1993	4,482	15,789	20,276	9,951	68	10,019	6,638	3,509
1994	3,509	19,031	22,546	10,653	97	10,750	9,158	2,608
1995	2,608	17,500	20,499	10,089	127	10,216	7,277	2,695
1996	2,695	18,393	21,474	10,601	101	10,702	6,708	3,895
1997	3,985	18,445	22,444	10,793	109	10,902	7,279	4,079
1998	4,079	13,469	17,548	10,067	143	10,210	4,087	3,866
1999	3,866	16,692	20,558	9,711	131	9,842	6,557	4,064
2000	4,064	16,841	20,905	8,430	117	8,547	6,425	5,932
2001 [3]	5,930	19,828	25,758	7,285	99	7,384	10,649	7,284

[1] Ginnings during the 12 months, Aug. 1–July 31. Includes an allowance for "city crop" which consists of rebaled samples and pickings from cotton damaged by fire and weather. [2] May include small volume of foreign growths. [3] Preliminary.

AMS, Cotton Program, (901) 384–3016. Compiled from reports of the Bureau of the Census.

Table 2-7.—Cotton, American Upland: Percentage distribution of fiber strength, United States, 1997–2001

Fiber strength [1]	Year				
	1997	1998	1999	2000	2001
17 and below	*	*	*	*	*
18	*	*	*	*	*
19	*	*	*	*	*
20	*	*	*	*	*
21	*	*	*	0.1	*
22	*	0.1	0.1	0.4	0.2
23	0.1	0.5	0.6	1.5	0.5
24	0.7	2.5	2.2	4.5	1.7
25	2.7	7.0	5.8	10.3	5.0
26	6.9	14.5	11.4	17.5	11.3
27	13.1	21.3	17.1	20.6	18.6
28	19.3	21.0	20.2	16.4	21.4
29	20.0	14.2	17.5	9.8	17.0
30	15.6	7.2	10.8	5.5	10.0
31	10.6	3.5	6.0	3.9	5.5
32	6.3	2.2	3.9	3.6	3.8
33	3.0	2.2	2.5	3.0	2.7
34	1.0	2.1	1.1	1.7	1.4
35	0.3	1.1	0.4	0.7	0.5
36 and above	0.1	0.5	0.2	0.4	0.2
Average	28.9	28.0	28.3	27.6	28.3

[1] Fiber strength expressed in terms of ⅛″ gage (grams per tex). *Less than 0.05 percent.

AMS, Cotton Program, (901) 384–3016.

Table 2-8.—Cotton, American upland: Estimated percentage of the crop forward contracted by growers, by States, 1994–2001

State	Crop of—							
	1994	1995	1996	1997	1998	1999	2000	2001
	Percent	*Percent*	*Percent*	*Percent*	*Percent*	*Percent*	*Percent*	*Percent*
AL	79	42	23	19	21	8	29	19
AZ	5	42	20	11	16	3	2	1
AR	42	69	49	53	53	11	14	5
CA	92	76	28	41	59	15	6	*
FL	....	....	....	12	15	*	3	12
GA	60	56	34	29	37	9	21	8
LA	11	33	17	26	27	8	21	12
MS	38	25	11	18	21	8	11	6
MO	78	89	59	71	94	16	26	11
NM	23	11	17	7	14	9	13	2
NC	70	57	74	37	53	42	23	6
OK	*	5	(1)	(1)	*	....	*	....
SC	55	69	43	49	56	23	30	6
TN	14	25	9	14	29	11	20	10
TX	28	47	18	13	28	5	8	7
US	40	48	28	24	36	10	14	7

[1] Less than 0.5 percent.

AMS, Cotton Program, (901) 384–3016.

Table 2-9.—Cotton, American upland: Carryover and crop, running bales, by grade groupings, United States, 1993–2002

Year beginning August 1	White [1]					Light Spotted				Other Colored [2]	All grades [3]
	21 and higher	31	41	51	61 and 71	22 and higher	32	42	52 and lower		
	1,000 bales	*1,000 bales*	*1,000 bales*	*1,000 bales*	*1,000 bales*	*1,000 bales*	*1,000 bales*	*1,000 bales*	*1,000 bales*	*1,000 bales*	*1,000 bales*
Carryover:											
1993	413	1,307	1,161	300	28	33	181	350	147	176	4,095
1994	602	1,089	1,012	109	3	40	221	299	78	54	3,509
1995	227	634	713	148	6	53	175	403	151	41	2,552
1996	498	764	562	56	1	81	240	303	51	53	2,609
1997	689	1,463	750	43	1	57	423	273	33	57	3,789
1998	570	1,541	871	86	5	160	299	243	116	130	4,021
1999	407	1,094	978	147	3	134	271	471	145	131	3,781
2000	1,274	1,007	981	123	8	68	85	192	42	50	3,830
2001	1,392	1,712	1,464	218	4	63	148	601	150	67	5,819
2002	1,234	2,325	1,976	107	2	99	238	769	54	76	6,700
Crop:											
1992	1,673	5,943	3,713	796	81	132	630	933	490	572	14,964
	New Classification System Beginning With 1993 Crop										
	White Color Grades					Light Spotted Color Grades [4]					
1993	4,458	4,160	2,930	256	7	521	1,102	1,223	182	172	15,010
1994	2,314	5,697	4,138	495	23	783	1,654	2,299	793	316	18,510
1995	3,660	5,453	3,568	232	9	552	1,393	1,505	202	176	16,751
1996	3,219	6,629	3,413	233	16	481	1,602	1,513	339	224	17,669
1997	4,078	5,670	3,571	442	12	739	1,193	1,164	332	375	17,576
1998	1,821	3,893	3,486	251	5	385	849	1,669	419	234	13,012
1999	5,793	4,504	3,215	283	5	359	539	758	168	149	15,773
2000	3,459	5,100	4,454	504	13	292	624	1,408	302	192	16,348
2001	4,950	6,593	3,997	443	8	391	654	1,296	276	431	19,039

[1] "Plus" grades included. [2] Includes all grades of Spotted, Tinged, Stained, Light Gray, Gray, and Below Grade. [3] Carryover as reported by the Bureau of the Census, Crop as reported by AMS, Cotton Program. [4] Includes all Color grades of Spotted, Tinged, Stained, and Below Grade.

AMS, Cotton Program, (901) 384–3016.

Table 2-10.—Cotton, American upland: Carryover and crop, running bales, by staple groupings, United States, 1992–2002

Year beginning August 1	Staple										All staples [1]
	26 and shorter	28	29	30	31	32	33	34	35	36 and longer	
	1,000 bales	*1,000 bales*	*1,000 bales*	*1,000 bales*	*1,000 bales*	*1,000 bales*	*1,000 bales*	*1,000 bales*	*1,000 bales*	*1,000 bales*	*1,000 bales*
Carryover:											
1992	(2)	19	11	56	91	191	229	603	810	1,425	3,437
1993	(2)	(2)	1	10	35	146	285	913	1,284	1,419	4,095
1994	(2)	(2)	4	15	37	197	285	724	952	1,295	3,509
1995	(2)	1	3	14	20	83	135	463	718	1,115	2,552
1996	8	6	3	12	36	143	197	508	771	925	2,609
1997	(2)	(2)	(2)	2	6	30	235	947	1,471	1,099	3,789
1998	(2)	(2)	3	16	62	296	553	802	1,076	1,213	3,782
1999	(2)	1	2	18	81	439	765	902	785	788	3,781
2000	(2)	1	10	46	85	386	651	969	820	862	3,830
2001	1	4	22	88	241	558	1,209	1,385	1,341	970	5,819
2002	....	....	4	9	32	200	705	1,993	2,071	1,686	6,700
Crop:											
1992	(2)	1	8	45	159	405	1,038	2,437	4,428	6,443	14,964
1993	(2)	1	7	45	201	666	1,470	2,509	3,718	6,392	15,010
1994	(2)	2	16	72	250	604	1,271	2,919	5,009	8,366	18,510
1995	(2)	1	7	36	155	489	1,389	3,422	5,208	6,044	16,751
1996	(2)	(2)	2	13	72	277	1,035	3,226	5,668	7,376	17,669
1997	....	(2)	4	27	144	553	1,704	3,306	4,207	7,631	17,576
1998	(2)	2	15	81	307	887	2,282	3,692	3,202	2,544	13,012
1999	1	8	51	177	562	1,553	3,077	4,102	3,415	2,827	15,773
2000	2	20	86	229	558	1,408	2,915	4,196	3,661	3,273	16,348
2001	(2)	1	9	53	256	974	3,084	5,592	4,947	4,123	19,039

[1] Carryover as reported by the Bureau of the Census, Crop as reported by AMS, Cotton Program. [2] Less than 500 bales.

AMS, Cotton Program, (901) 384–3016.

Table 2-11.—Cotton, American Pima: Carryover and crop, running bales, by grade and staple, United States, 1998–2002

Year beginning August 1	Grade					Staple				All grades and staples [1]
	01 and 02	03	04	05	06 and 07	42 and shorter	44	46	48 and longer	
	1,000 bales	*1,000 bales*	*1,000 bales*	*1,000 bales*	*1,000 bales*	*1,000 bales*	*1,000 bales*	*1,000 bales*	*1,000 bales*	*1,000 bales*
Carryover:										
1998	31.8	22.2	2.4	1.9	0.1	0.1	12.1	39.8	6.4	58.4
1999	27.4	45.4	21.6	0.1	....	....	13.5	60.1	20.9	94.5
2000	184.4	34.2	15.0	0.3	....	0.3	70.5	121.4	41.7	233.9
2001	87.5	22.8	2.2	0.1	....	....	28.9	68.8	14.9	112.6
2002	280.5	19.1	2.6	....	0.8	1.1	111.8	163.1	27.0	303.0
Crop:										
1998	193.7	149.5	43.2	8.5	1.5	1.3	41.6	234.3	119.1	396.3
1999	601.8	44.4	3.3	0.3	0.1	1.2	103.9	394.5	150.3	649.9
2000	304.5	58.9	9.4	1.2	0.1	0.1	71.9	255.6	46.5	374.1
2001	618.9	25.1	9.0	8.1	3.6	0.6	160.5	424.1	79.6	664.8

[1] Carryover as reported by the Bureau of the Census; Crop as reported by AMS, Cotton Program.

AMS, Cotton Program, (901) 384–3016.

Table 2-12.—Cotton, upland: Average staple length of upland cotton classed, by States, 1996–2001

State	Average staple length (32ds of an inch)[1]					
	1996	1997	1998	1999	2000	2001
AL	35.3	35.3	34.2	33.4	33.7	34.3
AZ	35.9	35.3	35.5	35.3	35.3	35.3
AR	35.4	36.3	34.8	34.8	34.6	34.8
CA	36.5	36.3	36.8	36.2	36.1	36.1
FL	35.3	35.1	33.9	34.2	34.4	34.4
GA	35.2	34.6	34.1	33.8	34.3	34.1
LA	35.0	35.5	34.4	34.0	33.9	34.2
MS	34.9	36.1	34.6	34.2	34.0	34.3
MO	35.4	36.2	34.9	34.8	34.8	35.4
NM	37.3	36.7	36.2	35.8	36.5	36.3
NC	35.6	34.7	34.1	34.6	34.2	34.6
OK	34.7	34.8	34.5	33.1	34.1	35.4
SC	35.7	35.0	34.3	33.9	34.0	33.8
TN	35.3	36.1	34.4	33.8	33.5	33.9
TX	34.3	33.9	33.2	33.1	32.9	33.8
VA	35.7	35.1	34.5	35.0	34.6	34.9
Other States	(2)	(2)	(2)	(2)	(2)	(2)
US	35.2	35.1	34.3	34.1	34.2	34.5

[1] Average calculated on numerical equivalents of the staple-length designations. For example, 7⁄8-inch = 28, 29⁄32-inch = 29, etc. [2] Not available.

AMS, Cotton Program, (901) 384–3016.

Table 2-13.—Cotton: United States exports, by country of destination, 1999/2000–2001/2002

Country of destination	Year beginning August		
	1999/2000	2000/2001	2001/2002
	1,000 bales[1]	*1,000 bales*[1]	*1,000 bales*[1]
Turkey	787	609	1,595
Mexico	1,500	1,760	1,520
India	97	367	959
Indonesia	573	558	947
Thailand	257	237	692
Taiwan	476	367	692
Korea, Republic of	307	489	578
Pakistan	38	156	428
Hong Kong	316	287	407
Japan	424	355	384
Bangladesh	134	190	335
China, Peoples Republic	147	124	305
Canada	245	322	235
Vietnam	58	71	168
Colombia	117	116	153
Philippines	71	42	126
Peru	59	104	117
Ecuador	66	89	86
El Salvador	62	86	81
Guatemala	87	70	79
Subtotal	5,818	6,401	9,889
Belgium-Luxembourg	27	40	69
Italy	61	52	58
Bahrain	0	14	58
Brazil	282	15	55
Malaysia	47	44	51
Switzerland	15	7	45
Germany	21	29	32
Venezuela	22	34	17
Chile	24	7	17
Dominican Republic	6	42	16
Ireland	0	0	14
Honduras	1	5	7
Sweden	12	14	7
Netherlands	7	4	7
Tunisia	8	8	6
Singapore	2	0	6
Saudi Arabia	0	0	5
Cambodia	1	1	5
Israel	0	0	4
Portugal	2	3	4
Other Countries	45	18	20
Total	6,401	6,739	10,393

[1] 480 pounds net.

FAS, Cotton, Oilseeds, Tobacco, and Seeds Division, (202) 720–9516. Compiled from reports of the U.S. Department of Commerce.

Table 2-14.—Cotton: International trade, 1996/97–2001/2002 [1]

Country	1996/97	1997/98	1998/99	1999/2000	2000/2001	2001/2002
	1,000 bales [2]	*1,000 bales [2]*	*1,000 bales [2]*	*1,000 bales [2]*	*1,000 bales [2]*	*1,000 bales [2]*
Principal exporting countries:						
United States	6,865	7,500	4,298	6,750	6,740	11,000
Uzbekistan, Republic	4,550	4,570	3,812	4,100	3,400	3,400
Australia	2,380	2,712	3,040	3,209	3,904	3,051
Greece	1,200	1,000	964	1,080	1,424	1,000
Syria	700	1,000	850	950	1,050	1,000
Mali	825	825	950	900	575	925
Brazil	0	0	23	12	315	674
Benin	680	645	600	650	625	650
Burkina	360	550	525	520	520	650
Kazakhstan, Republic	200	175	175	270	400	500
Tajikistan, Republic	400	370	425	355	390	475
Turkmenistan	525	500	750	800	675	450
Egypt	211	322	450	425	375	410
Cote d'Ivoire	371	450	550	735	600	400
Cameroon	360	290	300	300	370	350
China, Peoples Republic	10	34	681	1,692	448	342
Chad	360	465	290	330	300	275
Zimbabwe	290	320	320	350	375	250
Sudan	360	325	253	210	159	225
Togo	225	260	300	250	215	225
Argentina	1,332	1,000	1,100	319	441	215
Subtotal	22,204	23,313	20,656	24,207	23,301	26,467
Others	4,638	3,415	3,003	3,086	3,286	2,526
World total	26,842	26,728	23,659	27,293	26,587	28,993

Country	1996/97	1997/98	1998/99	1999/2000	2000/2001	2001/2002
Principal importing countries:						
Turkey	1,355	1,450	1,139	2,400	1,750	2,868
Indonesia	2,147	1,910	2,323	2,076	2,650	2,356
Thailand	1,414	1,236	1,211	1,706	1,584	2,057
Mexico	894	1,371	1,422	1,813	1,865	1,900
Russian Federation	1,000	1,225	850	1,600	1,650	1,850
India	15	145	508	1,600	1,567	1,750
Korea, Republic of	1,504	1,322	1,472	1,524	1,420	1,616
Taiwan	1,300	1,209	1,375	1,438	1,040	1,531
Italy	1,603	1,575	1,315	1,423	1,358	1,350
Japan	1,341	1,341	1,263	1,280	1,138	1,063
Bangladesh	519	597	650	775	900	1,000
Pakistan	279	120	925	475	450	1,000
Germany	737	740	625	640	702	600
Portugal	770	772	714	649	617	550
Hong Kong	606	571	619	517	519	475
Vietnam	165	170	270	330	400	475
France	602	561	516	525	441	460
China, Peoples Republic	3,613	1,834	359	117	241	449
Canada	260	325	329	286	352	400
Czech Republic	284	321	257	262	295	325
Malaysia	363	351	366	462	340	275
Brazil	2,386	1,884	1,334	1,559	603	253
Subtotal	23,157	21,030	19,842	23,457	21,882	24,603
Others	5,739	5,226	5,098	5,026	4,767	4,848
World total	28,896	26,256	24,940	28,483	26,649	29,451

[1] Marketing year beginning Aug. 1. [2] 480-pound net weight.
FAS, Cotton, Oilseeds, Tobacco and Seeds Division, (202) 720–9516. Prepared or estimated on the basis of official statistics of foreign governments, other foreign source materials, reports of U.S. Agricultural Counselors, Attachés, and Foreign Service Officers, results of office research, and related information.

Table 2-15.—Cotton, American upland: High, low, and season average spot prices for the base quality in the designated markets, cents per pound, 1994–2001

Season beginning August 1	Color 41, Leaf 4, Staple 34 [1]		
	Average	High	Low
	Cents	*Cents*	*Cents*
1994	88.14	111.53	65.98
1995	83.03	93.51	75.84
1996	71.59	78.11	68.00
1997	67.79	77.79	59.82
1998	60.12	74.19	47.21
1999	52.36	60.71	45.94
2000	51.56	63.57	35.39
2001	33.10	41.39	25.94

[1] Prices are for mixed lots, net weight, compressed, FOB car/truck.
AMS, Cotton Program, (901) 384–3016.

Table 2-16.—Cotton and cotton linters: United States imports for consumption, by country of origin, 1997/98–2001/2002

Country of origin	Year beginning August				
	1997/98	1998/99	1999/2000	2000/2001	2001/2002
	Bales [1]	*Bales* [1]	*Bales* [1]	*Bales* [1]	*Bales* [1]
Cotton, raw:					
Egypt	106	14,358	43,356	12,345	20,233
Mexico	451	12,039	5,744	451	315
India	602	514	231	1,029	176
Italy	0	0	0	103	2
Guatemala	0	0	0	0	2
China, Peoples Republic	0	74,216	2,835	0	2
Turkey	189	2,033	1,840	1,574	0
United Kingdom	0	6	0	273	0
Barbados	0	72	0	114	0
Spain	0	323	0	6	0
Canada	396	0	9	1	0
Syria	0	33,757	30,224	0	0
Argentina	9,851	21,662	7,403	0	0
Greece	0	232,714	3,090	0	0
Mali	0	4,010	1,746	0	0
Other Countries	1,855	43,035	39	0	0
Total	13,451	438,739	96,518	15,897	20,731
Cotton linters:					
Brazil	0	18,030	8,857	37,563	69,667
Syria	0	34,389	43,518	58,374	24,733
Mexico	40,453	8,771	13,667	2,940	9,885
Turkey	0	46,050	41,124	66,841	3,060
Greece	0	3,181	425	4,119	2,888
Argentina	0	12,572	2,070	3,083	2,267
Israel	0	97	171	621	700
India	0	0	97	810	375
Korea, Republic of	0	0	0	0	27
Latvia	0	1,437	0	25,215	0
Australia	0	0	90	458	0
Germany	0	0	8	219	0
Uzbekistan, Republic of	0	2,212	21,010	0	0
Netherlands	0	0	2,678	0	0
Uruguay	0	0	2,139	0	0
Tanzania	0	0	401	0	0
China, Peoples Republic	0	0	157	0	0
Other Countries	49	5,463	0	0	0
Total	40,502	132,204	136,410	200,241	113,602

[1] 480 pounds net.

FAS, Cotton, Oilseeds, Tobacco and Seeds Division, (202) 720–9516. Compiled from reports of the U.S. Department of Commerce.

Table 2-17.—Cotton, American upland: Percentage distribution of mike readings, by specified groups, United States, 1992–2001

Year beginning August 1	Mike groups						
	26 and below	27 to 29	30 to 32	33 to 34	35 to 49	50 to 52	53 and above
	Percent	*Percent*	*Percent*	*Percent*	*Percent*	*Percent*	*Percent*
1992	0.8	2.4	5.0	5.2	83.2	2.8	0.8
1993	0.1	0.3	1.2	1.8	86.2	8.2	2.2
1994	0.1	0.6	1.8	2.3	90.6	3.8	0.7
1995	0.4	1.2	2.7	3.0	79.6	10.6	2.5
1996	0.4	1.2	3.1	3.5	80.8	8.7	2.5
1997	*	0.4	1.8	2.5	88.4	5.9	0.9
1998	*	0.4	1.2	1.5	83.4	11.1	2.3
1999	0.3	0.5	1.3	1.8	80.2	12.6	3.3
2000	0.1	0.7	2.0	2.8	85.8	7.1	1.3
2001	*	0.2	0.7	1.1	75.9	15.7	6.0

(*) Less than 0.05 percent.

AMS, Cotton Program, (901) 384–3016.

Table 2-18.—Cotton, American upland: Average spot prices for specified grades of staple 34 in the designated markets for mixed lots, net weight, uncompressed in warehouse, cents per pound, 1992–2001

Year beginning August 1	White				Light Spotted			Spotted	
	31	41	51	61	32	42	52	33	43
	Cents	*Cents*	*Cents*	*Cents*	*Cents*	*Cents*	*Cents*	*Cents*	*Cents*
1992	55.12	54.10	50.17	42.07	54.03	51.03	44.90	45.20	42.01

Year beginning August 1[1]	New Classification System Beginning With 1993 Crop								
	White				Light Spotted			Spotted	
	Color 31 Leaf 3	Color 41 Leaf 4	Color 51 Leaf 5	Color 61 Leaf 6	Color 32 Leaf 3	Color 42 Leaf 4	Color 52 Leaf 5	Color 33 Leaf 3	Color 43 Leaf 4
	Cents	*Cents*	*Cents*	*Cents*	*Cents*	*Cents*	*Cents*	*Cents*	*Cents*
1993	66.89	66.12	62.14	55.43	65.88	62.80	56.43	56.77	53.50
1994	89.46	88.14	83.47	77.54	87.65	84.18	78.19	80.61	76.07
1995	84.82	83.03	77.61	73.25	82.49	78.38	72.97	76.31	71.30
1996	72.87	71.59	66.71	63.02	71.12	67.62	62.50	65.93	61.25
1997	69.08	67.79	63.31	60.38	67.08	64.44	60.06	64.47	59.43
1998	61.93	60.12	54.68	51.49	59.43	56.25	51.71	55.47	50.53
1999	53.99	52.36	46.52	42.52	51.50	48.03	43.15	46.50	41.99
2000	52.98	51.56	47.18	43.50	51.36	48.78	45.15	47.81	43.88
2001	34.66	33.10	29.32	26.87	33.26	31.04	28.12	30.42	27.50

[1] Prices are compressed, FOB car/truck.
AMS, Cotton Program, (901) 384–3016.

Table 2-19.—Cotton, American upland: Average spot prices for specified staple lengths of Grade 41 in the designated markets for mixed lots, net weight, uncompressed in warehouse, cents per pound, 1992–2001

Year beginning August 1	Staple							
	28	29	30	31	32	33	34	35
	Cents	*Cents*	*Cents*	*Cents*	*Cents*	*Cents*	*Cents*	*Cents*
1992	46.21	46.21	48.62	50.79	52.45	52.41	54.10	54.76
1993[1]	59.39	59.39	61.38	62.84	64.17	64.16	66.12	66.76
1994[1]	81.51	81.51	83.46	85.01	85.80	86.06	88.14	88.53
1995[1]	76.00	76.00	77.82	79.49	79.99	81.06	83.03	83.58
1996[1]	64.61	64.61	66.38	67.18	67.93	69.81	71.59	72.20
1997[1]	60.97	60.97	62.65	63.48	64.56	66.07	67.79	68.39
1998[1]	47.05	47.05	48.93	51.08	53.75	57.27	60.12	61.05
1999[1]	39.71	39.71	41.15	43.10	44.57	48.02	52.36	54.19
2000[1]	43.14	43.14	44.59	45.90	46.10	48.24	51.56	52.82
2001[1]	29.12	29.12	29.77	30.53	30.01	31.24	33.10	34.31

[1] Staple lengths of Color 41, Leaf 4, compressed, FOB car/truck.
AMS, Cotton Program, (901) 384–3016.

Table 2-20.—Cotton, American upland: Season average spot prices for the base quality, by designated markets, cents per pound, 1996–2001[1]

Market	Color 41, Leaf 4, Staple 34[2]					
	1996	1997	1998	1999	2000	2001
	Cents	*Cents*	*Cents*	*Cents*	*Cents*	*Cents*
Southeast	72.33	68.60	62.06	53.81	52.63	33.02
North Delta	72.11	68.36	61.82	53.34	52.32	33.24
South Delta	72.11	68.36	61.82	53.34	52.32	33.24
East TX–OK	70.29	65.93	57.66	50.49	51.03	32.59
West Texas	69.98	65.88	57.76	50.12	50.71	32.39
Desert SW	69.88	65.63	55.92	48.79	49.47	32.60
SJ Valley	74.47	71.79	63.78	56.67	52.45	34.64
Average	71.59	67.79	60.12	52.36	51.56	33.10

[1] Year beginning August 1. Prices are for mixed lots, net weight, uncompressed in warehouse. [2] Prices are for mixed lots, net weight, compressed, FOB car/truck.
AMS, Cotton Program, (901) 384–3016.

Table 2-21.—Cotton: Supply and distribution, by countries, 2001–2002

Country	Beginning stocks Aug. 1	Production	Imports	Total supply	Consumption[1]	Exports	Ending stocks July 31	Total distribution
	1,000 bales[2]	*1,000 bales*[2]	*1,000 bales*[2]	*1,000 bales*[2]	*1,000 bales*[2]	*1,000 bales*[2]	*1,000 bales*[2]	*1,000 bales*[2]
Importing countries:								
Turkey	882	3,975	2,868	7,725	6,150	125	1,450	7,725
Indonesia	408	55	2,356	2,819	2,300	15	454	2,769
Thailand	390	101	2,057	2,548	1,900	1	622	2,523
Mexico	540	432	1,900	2,872	2,100	80	667	2,847
Russian Federation	222	0	1,850	2,072	1,850	0	222	2,072
India	3,773	12,300	1,750	17,823	13,275	60	4,488	17,823
Korea, Republic of	388	1	1,616	2,005	1,500	4	501	2,005
Taiwan	236	0	1,531	1,767	1,300	4	463	1,767
Italy	152	1	1,350	1,503	1,300	30	163	1,493
Japan	209	0	1,063	1,272	1,050	0	222	1,272
Bangladesh	97	67	1,000	1,164	1,000	0	163	1,163
Pakistan	2,646	8,300	1,000	11,946	8,500	160	3,261	11,921
Germany	120	0	600	720	525	60	135	720
Portugal	148	0	550	698	550	0	148	698
Hong Kong	116	0	475	591	475	30	86	591
Vietnam	88	48	475	611	500	0	111	611
France	75	0	460	535	425	30	75	530
China, Peoples Republic	14,351	24,400	449	39,200	26,000	342	12,858	39,200
Canada	25	0	400	425	375	0	50	425
Czech Republic	48	0	325	373	320	5	48	373
Malaysia	84	0	275	359	290	0	69	359
Brazil	2,968	3,518	253	6,739	3,950	674	2,315	6,939
Belgium-Luxembourg	21	0	250	271	150	100	21	271
Colombia	109	124	250	483	375	0	103	478
Poland	20	0	250	270	250	1	19	270
South Africa, Republic	66	83	250	399	325	5	69	399
Philippines	68	7	224	299	230	0	69	299
Peru	80	184	188	452	360	10	82	452
Morocco	30	1	180	211	180	0	31	211
Austria	37	0	160	197	160	0	37	197
Subtotal	28,397	53,597	26,355	108,349	77,665	1,736	29,002	108,403
Others	14,055	44,752	3,096	61,903	16,906	27,257	17,629	61,792
Total world	42,452	98,349	29,451	170,252	94,571	28,993	46,631	170,195
Exporting countries:								
United States	6,001	20,303	21	26,325	7,721	11,000	7,426	26,147
Uzbekistan, Republic of	743	4,900	0	5,643	1,200	3,400	1,043	5,643
Australia	2,095	3,200	0	5,295	150	3,051	2,259	5,460
Greece	547	2,093	20	2,660	715	1,000	895	2,610
Syria	488	1,550	0	2,038	575	1,000	463	2,038
Mali	137	1,100	0	1,237	15	925	297	1,237
Brazil	2,968	3,518	253	6,739	3,950	674	2,315	6,939
Benin	161	800	0	961	10	650	301	961
Burkina	95	725	0	820	5	650	165	820
Kazakhstan, Republic of	68	650	0	718	40	500	178	718
Tajikistan, Republic of	72	665	0	737	110	475	152	737
Turkmenistan	444	850	0	1,294	375	450	469	1,294
Egypt	359	1,441	130	1,930	600	410	897	1,907
Cote d'Ivoire	88	700	0	788	80	400	308	788
Cameroon	100	450	0	550	45	350	155	550
China, Peoples Republic	14,351	24,400	449	39,200	26,000	342	12,858	39,200
Chad	58	325	0	383	15	275	93	383
Zimbabwe	206	350	0	556	140	250	151	541
Sudan	186	275	0	461	45	225	191	461
Togo	82	275	0	357	15	225	117	357
Argentina	321	300	25	646	275	215	147	637
Paraguay	61	220	0	281	45	190	46	281
Tanzania, United Republic	116	230	0	346	50	175	121	346
Pakistan	2,646	8,300	1,000	11,946	8,500	160	3,261	11,921
Azerbaijan, Republic of	64	145	0	209	7	140	62	209
Turkey	882	3,975	2,868	7,725	6,150	125	1,450	7,725
Spain	152	491	150	793	480	125	188	793
Kyrgyzstan, Republic of	17	140	0	157	15	106	36	157
Belgium-Luxembourg	21	0	250	271	150	100	21	271
Ukraine	64	0	150	214	50	100	64	214
Israel	31	100	5	136	5	100	31	136
Subtotal	33,624	82,471	5,321	121,416	57,533	27,788	36,160	121,481
Others	8,828	15,878	24,130	48,836	37,038	1,205	10,471	48,714
Total world	42,452	98,349	29,451	170,252	94,571	28,993	46,631	170,195

[1] Includes cotton destroyed or unaccounted for. [2] Bales of 480 pounds net weight.

FAS, Cotton, Oilseeds, Tobacco and Seeds Division, (202) 720–9516. Prepared or estimated on the basis of official statistics of foreign governments, other foreign source materials, reports of U.S. Agricultural Counselors, Attaches, and Foreign Service Officers, results of office research, and related information.

Table 2-22.—Sugarbeets: Area, yield, production, marketing year average price per ton received by farmers, value, and sugar production, United States, 1993–2002 [1]

Year	Area		Yield per harvested acre	Production	Price [2]	Value of production	Sugar production	
	Planted	Harvested					Raw value	Refined basis [3]
	1,000 acres	*1,000 acres*	*Tons*	*1,000 tons*	*Dollars*	*1,000 dollars*	*1,000 tons*	*1,000 tons*
1993	1,437.7	1,409.4	18.6	26,249	39.00	1,023,687	4,047	3,792
1994	1,475.8	1,443.0	22.1	31,853	38.80	1,234,470	NA	NA
1995	1,444.6	1,420.1	19.8	28,065	38.10	1,070,663	NA	NA
1996	1,368.4	1,323.3	20.2	26,680	45.40	1,211,001	NA	NA
1997	1,459.3	1,428.3	20.9	29,886	38.80	1,160,029	NA	NA
1998	1,497.8	1,450.7	22.4	32,499	36.40	1,181,494	NA	NA
1999	1,560.6	1,527.3	21.9	33,420	37.20	1,242,895	NA	NA
2000	1,564.2	1,373.0	23.7	32,541	34.20	1,113,030	NA	NA
2001	1,370.5	1,243.4	20.7	25,764	39.80	1,025,306	NA	NA
2002 [4]	1,427.9	1,361.0	20.2	27,550	NA	NA	NA	NA

[1] Relates to year of intended harvest except for overwintered spring planted beets in CA. [2] Prices do not include Government payments under the Sugar Act. [3] Calculated on the basis that 1.07 tons of raw sugar is required to produce one ton of refined sugar. [4] Preliminary. NA-not available.

NASS, Crops Branch, (202) 720–2127.

Table 2-23.—Sugarbeets: Area, yield, and production, by States, 2000–2002 [1]

State	Area planted			Area harvested			Yield per harvested acre			Production		
	2000	2001	2002	2000	2001	2002	2000	2001	2002	2000	2001	2002
	1,000 acres	*1,000 acres*	*1,000 acres*	*1,000 acres*	*1,000 acres*	*1,000 acres*	*Tons*	*Tons*	*Tons*	*1,000 tons*	*1,000 tons*	*1,000 tons*
CA	98.0	46.6	50.0	92.5	44.7	49.0	34.0	35.7	38.0	3,145	1,596	1,862
CO	71.5	41.5	43.9	53.6	36.8	39.5	22.5	22.4	20.1	1,206	824	794
ID	212.0	199.0	212.0	191.0	179.0	210.0	29.3	25.9	24.0	5,596	4,636	5,040
MI	189.0	180.0	180.0	166.0	166.0	178.0	20.5	19.4	18.0	3,403	3,220	3,204
MN	490.0	468.0	505.0	430.0	426.0	476.0	21.5	18.3	18.6	9,245	7,796	8,854
MT	60.7	57.4	58.0	55.2	53.5	55.9	23.9	21.5	19.6	1,319	1,150	1,096
NE	78.2	48.6	57.0	54.8	41.4	42.0	20.3	20.3	18.1	1,112	840	760
ND	258.0	261.0	265.0	232.0	237.0	258.0	22.1	18.1	18.6	5,127	4,290	4,799
OH	1.2	0.8	1.8	0.8	0.6	1.7	21.0	20.0	21.0	17	12	36
OR	16.2	11.9	11.2	13.7	9.7	10.9	30.1	29.9	27.2	412	290	296
WA	28.4	7.2	4.0	27.3	7.1	4.0	29.4	35.6	37.5	803	253	150
WY	61.0	48.5	40.0	56.1	41.6	36.0	20.6	20.6	18.3	1,156	857	659
US	1,564.2	1,370.5	1,427.9	1,373.0	1,243.4	1,361.0	23.7	20.7	20.2	32,541	25,764	27,550

[1] Relates to year of intended harvest except for overwintered spring planted beets in CA.

NASS, Crops Branch, (202) 720–2127.

Table 2-24.—Sugarbeets: Production and value, by States, crop of 2000–2001 [1]

State	Production		Marketing year average price per ton received by farmers		Value of production	
	2000	2001	2000	2001	2000	2001
	1,000 tons	*1,000 tons*	*Dollars*	*Dollars*	*1,000 dollars*	*1,000 dollars*
CA	3,145	1,596	30.40	33.40	95,608	53,306
CO	1,206	824	28.70	34.20	34,612	28,181
ID	5,596	4,636	37.90	40.50	212,088	187,758
MI	3,403	3,220	31.30	34.80	106,514	112,056
MN	9,245	7,796	33.70	40.30	311,557	314,179
MT	1,319	1,150	34.70	38.80	45,769	44,620
NE	1,112	840	29.20	36.90	32,470	30,996
ND	5,127	4,290	37.80	46.10	193,801	197,769
OH	17	12	30.90	35.50	525	426
OR	412	290	37.90	40.50	15,615	11,745
WA	803	253	33.50	40.50	26,901	10,247
WY	1,156	857	32.50	39.70	37,570	34,023
US	32,541	25,764	34.20	39.80	1,113,030	1,025,306

[1] Relates to year of intended harvest except for overwintered spring planted beets in CA.

NASS, Crops Branch, (202) 720–2127.

Table 2-25.—Sugarcane for sugar and seed: Area, yield, production, value, and production of cane sugar and molasses, United States, 1993–2002

Year[1]	Area harvested			Yield of cane per acre			Production		
	For sugar	For seed	Total	For sugar	For seed	For sugar and seed	For sugar	For seed	Total
	1,000 acres	*1,000 acres*	*1,000 acres*	*Tons*	*Tons*	*Tons*	*1,000 tons*	*1,000 tons*	*1,000 tons*
1993	893.3	55.0	948.3	33.2	26.3	32.8	29,623	1,449	31,072
1994	881.7	55.1	936.8	33.3	27.6	33.0	29,404	1,523	30,927
1995	874.7	57.6	932.3	33.3	28.5	33.0	29,137	1,642	30,779
1996	829.5	59.4	888.9	33.4	29.9	33.1	27,687	1,777	29,464
1997	860.3	53.7	914.0	34.9	31.8	34.7	30,003	1,706	31,709
1998	888.3	58.8	947.1	36.9	33.4	36.6	32,743	1,964	34,707
1999	941.4	51.9	993.3	35.7	33.2	35.5	33,577	1,722	35,299
2000	976.7	55.6	1,032.3	35.1	32.8	35.0	34,291	1,823	36,114
2001	970.3	57.5	1,027.8	33.8	31.5	33.7	32,775	1,812	34,587
2002[2]	974.4	51.7	1,026.1	35.1	32.6	35.0	34,245	1,687	35,932

Year[1]	Marketing year average price received by farmers per ton of sugarcane[3]	Value of production		Sugar production			Molasses[7]
		Of cane used for sugar	Of cane used for sugar and seed[4]	Raw value[5]		Refined basis[6]	
				Total	Per ton of cane		
	Dollars	*1,000 dollars*	*1,000 dollars*	*1,000 tons*	*Pounds*	*1,000 tons*	*1,000 gallons*
1993	28.50	845,305	885,459	3,482	235	3,255	194,247
1994	29.20	857,405	900,765	NA	NA	NA	NA
1995	29.50	859,057	906,441	NA	NA	NA	NA
1996	28.30	784,113	833,297	NA	NA	NA	NA
1997	28.10	842,840	890,257	NA	NA	NA	NA
1998	27.30	893,049	944,562	NA	NA	NA	NA
1999	25.60	859,175	901,900	NA	NA	NA	NA
2000	26.10	895,917	941,791	NA	NA	NA	NA
2001	29.00	951,813	NA	NA	NA	NA	NA
2002	NA	NA	NA	NA	NA	NA	NA

[1] In Hawaii, harvest continues throughout the year and production statistics are on a calendar year basis. In other states, harvest is seasonal and the production statistics year relates to the year in which the season begins. [2] Preliminary. [3] Prices do not include Government payments under the Sugar Act. [4] Price per ton of cane for sugar used in evaluating value of production for seed. [5] Raw value is the equivalent in terms of 96° sugar, as defined in the Sugar Act of 1948. [6] Calculated on the basis that 100 pounds of raw sugar is required to produce 93.46 pounds of refined sugar. [7] Excludes edible. NA-not available.

NASS, Crops Branch, (202) 720–2127.

Table 2-26.—Sugarcane for sugar and seed: Production and value, by States, crop of 2000–2001

State	Sugarcane for sugar						Sugar and seed: Value of production[1]	
	Production		Price per ton		Value of production			
	2000	2001	2000	2001	2000	2001	2000	2001
	1,000 tons	*1,000 tons*	*Dollars*	*Dollars*	*1,000 dollars*	*1,000 dollars*	*1,000 dollars*	*1,000 dollars*
FL	17,041	16,338	28.60	31.70	467,610	495,154	487,373	517,915
HI	2,433	1,932	26.30	30.80	62,200	57,842	63,988	59,505
LA	14,851	14,355	22.70	25.70	313,510	342,838	337,118	368,924
TX	1,789	1,962	29.80	28.90	52,597	55,979	53,312	56,702
Total	36,114	34,587	26.10	29.00	895,917	951,813	941,791	1,003,046

[1] Price per ton of cane for sugar used in evaluating value of production for seed.

NASS, Crops Branch, (202) 720–2127.

Table 2-27.—Sugarcane for sugar and seed: Area, yield, and production, by States, 2000–2002

State	Sugarcane for sugar and seed[1]								
	Area harvested			Yield of cane per acre			Cane production		
	2000	2001	2002	2000	2001	2002	2000	2001	2002
	1,000 acres	*1,000 acres*	*1,000 acres*	*Tons*	*Tons*	*Tons*	*1,000 tons*	*1,000 tons*	*1,000 tons*
For sugar:									
FL	436.0	445.0	442.0	37.5	35.1	38.2	16,350	15,620	16,884
HI	30.2	19.3	23.6	78.3	97.3	94.3	2,365	1,878	2,225
LA	465.0	460.0	465.0	29.7	29.0	29.0	13,811	13,340	13,485
TX	45.5	46.0	43.8	38.8	42.1	37.7	1,765	1,937	1,651
US	976.7	970.3	974.4	35.1	33.8	35.1	34,291	32,775	34,245
For seed:									
FL	18.0	20.0	19.0	38.4	35.9	38.0	691	718	722
HI	1.8	1.5	1.5	38.0	36.2	39.1	68	54	59
LA	35.0	35.0	30.0	29.7	29.0	29.0	1,040	1,015	870
TX	0.8	1.0	1.2	30.0	25.0	30.0	24	25	36
US	55.6	57.5	51.7	32.8	31.5	32.6	1,823	1,812	1,687
For sugar and seed:									
FL	454.0	465.0	461.0	37.5	35.1	38.2	17,041	16,338	17,606
HI	32.0	20.8	25.1	76.0	92.9	91.0	2,433	1,932	2,284
LA	500.0	495.0	495.0	29.7	29.0	29.0	14,851	14,355	14,355
TX	46.3	47.0	45.0	38.6	41.7	37.5	1,789	1,962	1,687
US	1,032.3	1,027.8	1,026.1	35.0	33.7	35.0	36,114	34,587	35,932

[1] In Hawaii, harvest continues throughout the year and production statistics are on a calendar year basis. In other states, harvest is seasonal and the production statistics year relates to the year in which the season begins.

NASS, Crops Branch, (202) 720–2127.

Table 2-28.—Sugar, cane (raw value[1]): Refiners' raw stocks, receipts, meltings, continental United States, 1992–2001

Year	Jan. 1 stocks	Receipts[2]	Meltings
	1,000 tons	*1,000 tons*	*1,000 tons*
1992	619	5,730	5,637
1993	507	5,113	5,225
1994	438	5,420	5,425
1995	448	5,220	5,323
1996	334	6,071	6,072
1997	323	5,955	5,954
1998	322	5,636	5,704
1999	332	5,558	5,646
2000	356	5,543	5,575
2001	274	5,362	5,221

[1] Raw value is the equivalent in terms of 96° sugar. [2] Receipts include refiners' total offshore raw sugar receipts in continental U.S. ports, whether entered through the customs or held pending availability of quota and raw cane sugar produced from sugarcane in the continental United States.

FSA, Dairy and Sweeteners Analysis, (202) 720–6733.

Table 2-29.—Sugar (Centrifugal Sugar, raw value): Production in specified countries, 1999/2000–2001/2002 [1]

Continent and country	1999/2000	2000/2001	2001/2002 [2]
	1,000 metric tons	*1,000 metric tons*	*1,000 metric tons*
North America:			
Canada	7,956	7,172	7,534
Mexico	123	80	80
United States	5,220	5,166	4,999
Total	13,299	12,418	12,613
Caribbean:			
Cuba	3,500	3,610	3,200
Dominican Republic	470	460	480
Other	461	468	479
Total	4,431	4,538	4,159
Central America:			
Guatemala	1,632	1,910	1,789
Other	1,868	1,820	1,881
Total	3,500	3,730	3,670
South America:			
Brazil	17,100	20,400	22,750
Colombia	2,225	2,300	2,305
Argentina	1,540	1,600	1,550
Other	2,975	3,146	3,130
Total	23,840	27,446	29,735
Western Europe:			
European Union	18,519	16,230	17,826
Other Western Europe	238	215	215
Total	18,757	16,445	18,041
Eastern Europe:			
Poland	2,188	1,685	2,000
Ukraine	1,687	1,790	1,950
Russian Federation	1,550	1,630	1,700
Other	1,941	2,369	2,378
Total	7,366	7,474	8,028
Africa:			
South Africa, Rep.	2,895	2,542	2,785
Other	5,264	5,223	5,453
Total	8,159	7,765	8,238
Middle East:			
Egypt	1,400	1,450	1,470
Turkey	2,756	1,796	2,336
Other	879	1,207	1,207
Total	5,035	4,453	5,013
Asia and Oceania:			
India	20,480	20,340	18,850
China, Peoples Republic	6,849	7,872	8,410
Thailand	5,107	6,397	6,585
Australia	4,162	4,610	5,027
Pakistan	2,648	3,453	3,778
Indonesia	1,800	1,700	1,800
Philippines	1,805	1,900	2,000
Japan	722	816	800
Other	2,330	2,500	2,023
Total	45,903	49,588	49,273
World total	130,290	133,857	138,770

[1] The following countries are no longer separately identified, Caribbean; Barbados, Haiti, Jamaica, St. Kitts and Nevis, Trinidad and Tobago, Central America; Belize, Costa Rica, El Salvador, Honduras, Nicaragua, Panama, East Europe; Albania, Bulgaria, Croatia, Czech Republic, Former Soviet Union; Belarus, Georgia, Kazakhstan, Kyrgystan, Latvia, Lithuania, Moldova, Middle East; Iran, Iraq, Lebanon, Syria, North Africa; Morocco, Sudan, Tunisia, Oceania; Fiji, Papua New Guinea, Other Africa; Cote de Ivoire, Ethiopia, Kenya, Malawi, Mauritius, Nigeria, Swaziland, Tanzania, Zaire, Zimbabwe, Other Asia; Bangladesh, Bhutan, Burma, Japan, Malaysia, Nepal, Sri Lanka, Taiwan, Vietnam, South America; Bolivia, Chile, Ecuador, Guyana, Paraguay, Peru, Surinam, Uruguay, Venezuela, Other Western Europe; Switzerland. [2] Crop years are on a September/August basis, but include the outturn of sugar from harvests of several Southern Hemisphere countries which begin in September. [3] Preliminary.

FAS, Horticulture and Tropical Products Division, (202) 720–4720. Prepared or estimated on the basis of official statistics of foreign governments, other foreign source materials, reports of U.S. Agricultural Counselors, Attachés, and Foreign Service Officers, results of office research, and related information.

Table 2-30.—Sugar, cane and beet (refined): Stocks, production or receipts, and deliveries, continental United States, 1993–2002

Item and year	Cane sugar refineries	Beet sugar factories	Importers of direct consumption sugar	Mainland cane sugar mills [1]	Total
JAN. 1 STOCKS [2]	*1,000 tons*	*1,000 tons*	*1,000 tons*	*1,000 tons*	*1,000 tons*
1993	178	1,640	0	14	1,832
1994	218	1,696	0	13	1,927
1995	192	1,600	0	13	1,805
1996	195	1,383	0	12	1,590
1997	196	1,520	0	18	1,734
1998	212	1,535	0	22	1,769
1999	255	1,499	0	22	1,776
2000	208	1,554	0	22	1,784
2001	262	1,972	0	19	2,564
2002	288	1,812	0	18	2,115
PRODUCTION OR RECEIPTS					
1992	5,209	4,133	52	15	9,409
1993	4,760	4,249	51	14	9,074
1994	5,531	4,269	78	20	9,898
1995	5,366	4,471	44	14	9,895
1996	6,074	4,149	33	19	10,275
1997	5,968	4,117	27	20	10,132
1998	5,811	4,431	24	23	10,289
1999	5,840	4,767	40	24	10,671
2000	5,681	6,014	36	32	12,093
2001	5,467	4,839	63	26	10,408
DELIVERIES [3]					
1992	5,167	3,661	52	11	8,891
1993	4,673	3,954	51	15	8,693
1994	5,530	4,329	78	12	9,949
1995	5,397	4,645	44	15	10,101
1996	6,074	4,007	33	14	10,128
1997	5,940	4,060	27	16	10,043
1998	5,708	4,410	24	20	10,162
1999	5,777	4,678	40	24	10,519
2000	5,729	5,573	36	15	11,353
2001	5,509	4,961	63	13	10,551

[1] Sugar for human consumption only. [2] Stocks include sugar in bond and in Customs custody and control. [3] Consists of all refined sugar.

FSA, Dairy and Sweeteners Analysis, (202) 720–6733.

Table 2-31.—Sugar, cane and beet: Imports and domestic marketings, by source of supply, continental United States, 1999–2001 [1]

Country or area of supply	1999	2000	2001
	1,000 tons	*1,000 tons*	*1,000 tons*
Argentina	45	44	45
Australia	89	87	87
Belgium-Luxembourg	0	1	2
Belize	12	12	12
Bolivia	9	6	8
Brazil	157	123	152
Canada	10	10	10
Colombia	24	24	25
Congo (Brazzaville)	7	0	7
Costa Rica	16	15	16
Cote d'Ivoire	0	7	7
Dominican Republic	191	185	182
Ecuador	12	12	12
El Salvador	28	27	27
Guatemala	52	51	51
Guyana	13	13	13
Honduras	11	11	11
Hong Kong	0	1	2
India	9	0	0
Jamaica	12	12	12
Madagascar	7	7	6
Malawi	11	0	10
Mauritius	6	6	13
Mexico	24	25	105
Mozambique	14	14	14
Netherlands	1	1	1
Nicaragua	23	22	22
Other Pacific Islands, NE	10	9	9
Panama	31	31	31
Papua New Guinea	7	7	7
Paraguay	7	7	7
Peru	44	22	43
Philippines	145	94	92
South Africa, Republic of	25	24	24
Swaziland	17	17	17
Taiwan	13	13	12
Thailand	15	15	15
Trinidad and Tobago	7	7	7
Uruguay	7	7	7
Zimbabwe	13	13	13
Total foreign countries	1,124	980	1,232
Domestic areas:			
Mainland (beet)	4,678	5,573	4,961
Mainland and Hawaii (cane)	5,759	5,699	5,495
Puerto Rico	43	44	27
Total domestic areas	10,480	11,316	10,483
Grand total	11,604	12,296	11,715

[1] Source: U.S. Census.

FSA, Dairy and Sweeteners Analysis Division, (202) 720–6733 (Domestic). FAS, (202) 720-1061 (Imports).

Table 2-32.—Sugar, cane and beet (raw value): Production, stocks, trade, and supply available for consumption in continental United States includes Puerto Rico, 1993–2002

Year	Production	Visible stocks beginning of period	Receipts from— Foreign sources	Receipts from— Puerto Rico	Commercial exports and shipments	Total deliveries
	1,000 short tons	*1,000 short tons*	*1,000 short tons*	*1,000 short tons*	*1,000 short tons*	*1,000 short tons*
1993	7,829	3,242	2,018	0	358	9,138
1994	7,669	3,512	1,771	0	480	9,321
1995	7,977	3,139	1,759	0	400	9,451
1996	7,268	2,908	2,926	0	331	9,619
1997	7,418	3,195	2,676	0	187	9,755
1998	7,891	3,377	2,148	0	203	9,854
1999	9,083	3,422	1,806	0	203	10,167
2000	8,962	3,855	1,639	0	109	10,091
2001	8,649	4,337	1,641	0	145	10,076
2002 [1]	7,546	4,525	1,543	0	108	9,958

[1] Preliminary

ERS, Specialty Crops Branch, (202) 694–5247.

Table 2-33.—Sugar, centrifugal (raw value): United States exports, by country of destination, 2000–2002 [1]

Country of destination	2000	2001	2002 [2]
	Metric tons	*Metric tons*	*Metric tons*
Angola	17	0	0
Argentina	1,312	593	0
Australia	72	877	3
Bahamas, The	4,648	4,463	2,074
Barbados	1,136	718	685
Belgium-Luxembourg	0	89	0
Belize	7	0	0
Bermuda	2	0	0
Bolivia	0	3	0
Botswana	42	22	46
Brazil	2,462	2,889	186
Canada	13,711	15,094	7,863
Cayman Islands	1,458	321	134
Chile	521	123	3
China, Peoples Republic	6	199	35
Colombia	9,010	262	0
Costa Rica	217	143	0
Denmark	1	0	59
Dominican Republic	6,432	286	270
Ecuador	355	168	0
Egypt	4,605	204	103
El Salvador	231	0	0
France	462	258	188
French West Indies	483	188	0
Germany	1,119	1,160	514
Ghana	2	5	15
Greece	0	57	0
Guatemala	2,287	868	474
Guyana	0	21	0
Haiti	12,209	4	251
Honduras	211	20	0
Hong Kong	21	171	694
Hungary	10	0	0
Iceland	0	0	2
India	0	0	17
Indonesia	0	12	87
Ireland	1	221	0
Israel	0	64	66
Italy	135	45	0
Jamaica & Dep	21,900	30,943	8,136
Japan	165	258	86
Jordan	0	447	141
Korea, Republic of	203	33	1,130
Kuwait	0	6	111
Lebanon	119	7	0
Leeward-Windward Island	353	913	645
Liberia	1,395	0	0
Macau	0	0	14
Malaysia	0	14	0
Marshall Islands	71	0	0
Netherlands	22,481	10,134	4,302
Netherlands Antilles	1,980	2,186	1,491
New Zealand	3,637	2,220	1,056
Nicaragua	227	178	296
Norway	44	12	0
Pakistan	0	0	1
Panama	0	11	0
Peru	6,210	6,800	9
Philippines	17,275	26	0
Qatar	12	98	265
Russian Federation	0	0	3
Saudi Arabia	102	232	209
Singapore	42	52	2
South Africa, Repub	0	21	33
Spain	4	0	2
Suriname	149	0	0
Sweden	54	0	0
Taiwan	0	197	6
Thailand	0	0	4
Trinidad and Tobago	98	86	1
Tunisia	152	0	25
Turkey	8,778	5,064	908
Ukraine	0	0	5
United Arab Emirates	100	158	40
United Kingdom	164	0	0
Uruguay	24	3	15
Venezuela	159	515	601
Western Africa, NEC	537	159	16
Yemen	2,697	4,401	0
Grand total [3]	152,317	94,727	33,325

[1] Data are actual weight x 1.07. [2] October–March. [3] Numbers may not add due to rounding.

FAS, Horticultural and Tropical Products Division, (202) 720–4620. Compiled from reports of the U.S. Department of Commerce.

Table 2-34.—Sugar, centrifugal (raw value): International trade in marketing years 2000/2001–2002/2003

Continent and country	2000/2001		2001/2002		2002/2003[1]	
	Exports	Imports	Exports	Imports	Exports	Imports
	1,000 metric tons	*1,000 metric tons*	*1,000 metric tons*	*1,000 metric tons*	*1,000 metric tons*	*1,000 metric tons*
North America:						
Canada	128	1,443	98	1,396	113	1,420
Mexico	5	1,150	6	1,200	5	1,190
United States	155	43	403	54	180	49
Total	288	2,636	507	2,650	298	2,659
Caribbean:						
Cuba	2,980	0	3,000	0	2,450	0
Dominican Republic	185	26	185	44	185	26
Other	342	280	374	263	342	257
Total	3,507	306	3,559	307	2,977	283
Central America:						
Guatemala	1,190	0	1,310	0	1,310	0
Other	915	0	945	0	837	0
Total	2,105	0	2,255	0	2,147	0
South America:						
Argentina	189	2	135	1	130	1
Brazil	7,700	0	11,600	0	13,100	0
Colombia	965	25	990	7	1,000	1
Other	436	758	475	863	443	701
Total	9,290	785	13,200	871	14,673	703
Western Union:						
European Union	6,607	1,839	4,200	2,018	5,791	2,022
Other	0	392	0	383	0	359
Total	6,607	2,231	4,200	2,401	5,791	2,381
East Europe:						
Poland	437	49	75	70	300	65
Ukraine	8	150	90	250	60	150
Russian Federation	260	5,650	460	4,800	310	4,500
Other	534	2,523	582	2,496	630	2,523
Total	1,239	8,372	1,207	7,616	1,300	7,238
Middle East:						
Egypt	84	946	100	650	100	700
Turkey	826	2	550	1	350	0
Other	812	5,209	902	5,371	1,040	5,583
Total	1,722	6,157	1,552	6,022	1,490	6,283
Other Africa:						
South Africa, Rep	1,580	260	1,235	263	1,350	265
Other	2,044	4,732	1,949	4,577	1,936	4,732
Total	3,624	4,992	3,184	4,840	3,286	4,997
Asia and Oceana:						
Australia	3,056	5	3,447	5	3,864	5
China, Peoples Republic	129	1,083	647	1,392	660	1,420
India	1,360	0	900	30	1,000	50
Indonesia	6	1,591	0	1,500	0	1,600
Japan	10	1,486	10	1,429	10	1,453
Pakistan	0	1,100	0	32	215	0
Philippines	88	215	142	109	142	80
Thailand	3,394	0	4,290	0	4,570	0
Other	1,273	6,757	1333	6,958	1,366	7,222
Total	9,316	12,237	9,436	11,455	10,461	11,830
World total[2]	37,698	37,716	39,100	36,162	42,423	36,374

[1] Preliminary. [2] The following countries are no longer separately identified, Caribbean; Barbados, Haiti, Jamaica, St. Kitts and Nevis, Trinidad and Tobago, Central America; Belize, Costa Rica, El Salvador, Honduras, Nicaragua, Panama, East Europe; Albania, Bulgaria, Other Africa; Cote de Ivoire, Ethiopia, Kenya, Malawi, Mauritius, Nigeria, Swaziland, Tanzania, Zaire, Zimbabwe, Other Asia; Bangladesh, Bhutan, Burma, Japan, Malaysia, Nepal, Sri Lanka, Taiwan, Vietnam, South America; Bolivia, Chile, Ecuador, Guyana, Paraguay.

FAS, Horticultural and Tropical Product Division, (202) 720–4720. Prepared or estimated on the basis of official statistics of foreign governments, other foreign source materials, reports of U.S. Agricultural Counselors, Attaches, and Foreign Service Offices.

Table 2-35.—Sugar (raw and refined): Average price per pound at specified markets, 1993–2002

Year	Cane sugar		Refined beet: Mid-west	Retail price, granulated: United States
	Raw, 96 centrifugal			
	Caribbean ports, f.o.b. and stowed	New York, c.i.f. duty paid		
	Cents	*Cents*	*Cents*	*Cents*
1993	10.03	21.6	25.1	40.5
1994	12.13	22.0	25.1	39.9
1995	13.44	22.9	25.8	39.83
1996	12.24	22.40	29.2	41.79
1997	12.06	21.96	27.09	43.26
1998	9.68	22.06	26.12	42.98
1999	6.54	21.16	26.71	43.27
2000	8.51	19.09	20.80	42.41
2001	9.12	21.11	23.31	43.42
2002	7.88	20.87	25.79	43.42

ERS, Specialty Crops Branch, (202) 694–5247. Compiled from the following sources: (New York) Coffee, Sugar & Cocoa Exchange; and U.S. Department of Labor, Bureau of Labor Statistics.

Table 2-36.—Sugar, centrifugal (raw value): United States imports, by country of origin, 2000–2002 [1] [2]

Country of origin	2000	2001	2002 [3]
	1,000 metric tons	*1,000 metric tons*	*1,000 metric tons*
Argentina	44,320	24,691	65,725
Australia	85,717	95,525	85,931
Belgium-Luxembourg	0	0	38
Belize	14,169	37,413	50,317
Bolivia	6,304	8,329	8,404
Bosnia-Hercegovina	0	0	4
Brazil	152,256	199,594	147,228
Peoples Republic of China	166	66	0
Colombia	61,461	43,274	40
Congo (Brazzaville)	0	7,258	57,692
Costa Rica	32,884	60,969	7,258
Cote d'Ivoire	7,206	7,245	15,809
Dominican Republic	183,086	174,719	0
Ecuador	11,530	11,513	181,869
El Salvador	62,785	46,776	0
Finland	0	0	17
Guatemala	211,819	95,567	1
Guyana	12,560	25,080	95,770
Honduras	5,788	22,961	12,517
India	256	158	0
Iran	0	0	10,529
Jamaica	0	0	1
Republic of Korea	3	0	0
Leeward-Windward Island	0	0	1
Madagascar	7,249	6,469	0
Malawi	10,833	10,531	0
Mauritius	5,813	12,482	0
Mexico	58,845	108,560	0
Mozambique	13,275	13,690	5,885
Netherlands	0	0	10,531
Nicaragua	64,194	22,079	3,810
Other Pacific Islands	9,410	9,361	122,496
Panama	60,559	49,311	13,099
Papua New Guinea	7,204	7,195	0
Paraguay	8,594	7,104	22,114
Peru	22,619	43,103	9,382
Philippines	94,445	92,442	31,678
Saudi Arabia	0	0	7,069
Republic of South Africa	23,976	23,909	7,538
Swaziland	16,705	16,971	41,715
Taiwan	12,383	12,500	75,401
Thailand	14,601	14,577	0
Trinidad and Tobago	7,372	8,963	91
Turkey	8	3	35,566
United Kingdom	0	17	15,783
Uruguay	7,043	7,256	12,211
Zimbabwe	12,766	13,651	0
World total [4]	1,350,201	1,341,311	1,153,520

[1] Imports for consumption. Imports in bonded warehouses (general imports) are tallied when customs as sugar whose content of sucrose, by weight in the dry state, corresponds to a polarimeter reading less than 99.5 degrees. [2] Data are actual weight x 1.035. [3] October–March. [4] Numbers may not add due to rounding.

FAS, Horticultural and Tropical Products Division, (202) 720–4620. Compiled from reports of the U.S. Department of Commerce.

Table 2-37.—Sugar, centrifugal (raw value): Beginning stocks in marketing years 2000/2001–2002/2003 [1]

Country	2000/2001	2001/2002	2002/2003
	1,000 metric tons	*1,000 metric tons*	*1,000 metric tons*
North America:			
Canada	2,013	1,997	1,175
Mexico	208	201	190
United States	1,063	1,548	1,300
Total	3,284	3,746	2,665
Caribbean:			
Cuba	438	238	108
Dominican Republic	26	27	29
Other	122	127	92
Total	586	392	229
Central America:			
Guatemala	96	70	170
Other	368	326	176
Total	464	396	346
South America:			
Argentina	263	146	162
Brazil	710	860	210
Colombia	60	40	42
Other	673	561	630
Total	1,706	1,607	1,044
Western Europe:			
European Union	3,730	3,420	3,278
Other Western Europe	236	305	333
Total	3,966	3,725	3,611
Eastern Europe:			
Poland	189	250	189
Ukraine	277	256	186
Russian Federation	3,000	3,100	2,130
Other	1,361	1,155	1,202
Total	4,827	4,761	3,707
Africa:			
South Africa, Republic of	530	455	450
Other Africa	1,597	1,495	1,378
Total	2,127	1,950	1,828
Middle East:			
Egypt	50	282	247
Turkey	783	865	262
Other	612	560	643
Total	1,445	1,707	1,152
Asia and Oceania:			
India	10,710	11,985	13,000
China, Peoples Republic	1,851	1,004	923
Thailand	608	571	828
Australia	518	634	782
Pakistan	127	425	460
Indonesia	1,330	1,415	1,215
Philippines	330	322	239
Japan	460	365	283
Other	1,525	1,557	1,750
Total	17,459	18,278	19,480
World total	35,864	36,562	34,062

[1] The following countries are no longer separately identified, Caribbean; Barbados, Haiti, Jamaica, St. Kitts and Nevis, Trinidad and Tobago, Central America; Belize, Costa Rica, El Salvador, Honduras, Nicaragua, Panama, East Europe; Albania, Bulgaria, Other Africa; Cote de Ivoire, Ethiopia, Kenya, Malawi, Mauritius, Nigeria, Swaziland, Tanzania, Zaire, Zimbabwe, Other Asia; Bangladesh, Bhutan, Burma, Japan, Malaysia, Nepal, Sri Lanka, Taiwan, Vietnam, South America; Bolivia, Chile, Ecuador, Guyana, Paraguay.

FAS, Horticultural and Tropical Products Division, (202) 720–4720.Prepared or estimated on the basis of official statistics of foreign governments, other foreign source materials, reports of U.S. Agricultural Counselors, Attaches, and Foreign Service Offices.

Table 2-38.—Honey: United States imports for consumption, by country of origin, 1999–2001

Continent and country of origin	1999	2000	2001
	Metric tons	*Metric tons*	*Metric tons*
Central America:			
Guatamala	12	0	5
Total	12	0	5
Eastern Europe:			
Bulgaria	15	0	0
Total	15	0	0
European Union:			
France	92	95	134
Germany	0	100	0
Greece	8,457	40,928	27,299
Italy	24	59	20
Netherlands	0	19	0
Spain	1,750	2,761	7,614
Total	10,322	43,963	35,067
North America:			
Canada	5	5	432
Mexico	17	180	296
Total	22	184	728
Oceania:			
Australia	1,719	352	1,587
New Zealand	35	0	0
Total	1,754	352	1,587
Other Africa:			
South Africa, Rep.	313	1,752	7,429
Swaziland	0	0	7
Total	313	1,752	7,436
Other Asia:			
China, Peoples	549	642	2,553
Hong Kong	32	2	0
Korea, Republic	1	12	25
Philippines	172	0	6
Singapore	10	6	0
Taiwan	1	2	0
Thailand	5	0	817
Total	769	663	3,401
South America:			
Argentina	32	9	67
Brazil	0	21	14
Chile	726	2,392	1,630
Peru	0	1	0
Venezuela	34	1	0
Total	792	2,424	1,711
Grand total	13,999	49,337	49,935

FAS, Horticultural and Tropical Products Division, (202) 720–4620. Compiled from reports of the U.S. Department of Commerce.

Table 2-39.—Beeswax, crude: United States imports for consumption, by country of origin, 1999–2001

Continent and country of origin	1999	2000	2001[1]
	Metric tons	*Metric tons*	*Metric tons*
Carribean:			
Dominican Republic	65	39	35
Total	65	39	35
Eastern Europe:			
Hungary	0	0	13
Total	0	0	13
European Union:			
France	23	26	22
Germany	246	271	170
Greece	0	0	0
Italy	1	0	3
Netherlands	17	3	3
Spain	0	1	0
Sweden	3	0	0
United Kingdom	21	18	33
Total	310	367	231
Former Soviet Union:			
Russian Federation	50	0	0
Total	50	0	0
North America:			
Canada	324	343	298
Mexico	8	18	19
Total	332	361	317
Oceana:			
Australia	77	179	76
New Zealand	20	20	1
Total	97	199	77
Other Africa:			
Cameroon	32	63	0
Djibouti Afars-Issas	0	0	18
Ethiopia	72	134	69
French Ind. Ocean Territory	0	0	13
Guinea	15	0	0
Kenya	21	0	19
Senegal	0	8	0
South Africa, Republic of	1	1	1
Tanzania, United Republic	74	147	55
Total	215	354	174
Other Asia:			
China, Peoples Rep	313	271	224
Japan	4	10	2
Taiwan	2	1	2
TVietnam	0	27	0
Total	319	308	227
Other West Europe:			
Switzerland	0	1	1
Total	0	1	1
South America:			
Argentina	175	339	308
Brazil	0	33	37
Columbia	0	0	6
Total	175	371	350
Grand total	1,563	2,001	1,426

[1] Preliminary.

FAS, Horticultural and Tropical Products Division, (202) 720–4620.

Table 2-40.—Honey: Number of colonies, yield, production, stocks, price and value, United States, 1993–2002 [1] [2]

State	Honey producing colonies	Yield per colony	Production	Stocks Dec 15 [3]	Average price per pound	Value of production
	1,000	*Pounds*	*1,000 pounds*	*1,000 pounds*	*Cents*	*1,000 dollars*
1993 ..	2,875	80.2	230,575	49,046	53.9	124,280
1994 ..	2,783	78.4	218,187	59,877	52.8	115,203
1995 ..	2,655	79.5	211,073	42,313	68.5	144,585
1996 ..	2,581	77.3	199,511	47,206	88.8	177,166
1997 ..	2,631	74.7	196,536	70,696	75.2	147,795
1998 ..	2,633	83.7	220,316	80,808	65.5	147,254
1999 ..	2,688	76.4	205,250	79,375	60.1	126,075
2000 ..	2,620	84.1	220,339	85,328	59.7	132,742
2001 ..	2,506	74.0	185,461	64,556	70.4	132,225
2002 ..	2,524	67.8	171,140	39,047	128.6	221,638

[1] For producers with 5 or more colonies. [2] U.S. price weighted by survey expanded sales. [3] Stocks held by producers.

NASS, Livestock Branch, (202) 720-6351.

Table 2-41.—Honey: Number of colonies, yield, production, stocks, price and value, by State and United States, 2002 [1]

State	Honey producing colonies	Yield per colony	Production	Stocks Dec 15 [2]	Average price per pound [3]	Value of production
	1,000	*Pounds*	*1,000 pounds*	*1,000 pounds*	*Cents*	*1,000 dollars*
AL	12	86	1,032	103	111	1,146
AZ	38	63	2,394	1,197	110	2,633
AR	55	88	4,840	1,065	124	6,002
CA	440	53	23,320	3,498	129	30,083
CO	24	60	1,440	576	122	1,757
FL	220	93	20,460	2,026	112	22,915
GA	50	52	2,600	52	110	2,860
HI	7	136	952	29	110	1,047
ID	100	57	5,700	1,653	135	7,695
IL	9	80	720	130	142	1,022
IN	8	62	496	233	114	565
IA	33	70	2,310	901	137	3,165
KS	17	52	884	407	156	1,379
KY	3	54	162	47	165	267
LA	35	124	4,340	347	114	4,948
ME	11	41	451	266	121	546
MD	3	46	138	21	175	242
MI	72	77	5,544	1,885	134	7,429
MN	117	73	8,541	1,110	142	12,128
MS	18	78	1,404	281	118	1,657
MO	18	53	954	200	137	1,307
MT	134	63	8,442	1,097	135	11,397
NE	43	75	3,225	1,161	145	4,676
NV	9	62	558	61	227	1,267
NJ	10	35	350	74	98	343
NM	11	44	484	252	109	528
NY	60	98	5,880	2,470	117	6,880
NC	14	51	714	86	136	971
ND	320	75	24,000	8,160	142	34,080
OH	16	81	1,296	518	140	1,814
OK	3	53	159	51	160	254
OR	39	49	1,911	803	128	2,446
PA	18	57	1,026	513	130	1,334
SC	5	94	470	14	125	588
SD	225	51	11,475	2,410	140	16,065
TN	8	61	488	137	137	669
TX	114	67	7,638	985	109	8,325
UT	22	59	1,298	208	129	1,674
VT	7	89	623	274	119	741
VA	7	38	266	69	186	495
WA	44	51	2,244	494	125	2,805
WV	9	51	459	151	137	629
WI	70	95	6,650	2,461	130	8,645
WY	38	63	2,394	383	139	3,328
Oth Sts [4]	8	51	408	188	218	891
US [5]	2,524	67.8	171,140	39,047	128.6	221,638

[1] For producers with 5 or more colonies. Colonies which produced honey in more than one State were counted in each State. [2] Stocks held by producers. [3] Price weighted by sales. [4] CT, DE, MA, NH, RI and SC not published separately to avoid disclosing data for individual operations. [5] U.S. total production is the sum of state production. U.S. colonies multiplied by U.S. yield may not exactly equal U.S. production.

NASS, Livestock Branch, (202) 720-3570.

Table 2-42.—Feed and industrial molasses: Mainland production, inshipments and foreign trade, and total market supplies in the United States, 1993–2002

Year	Production in mainland areas				Inshipments from Hawaii
	Mainland cane	Domestic beet	Refiners' blackstrap [1]	Total	
	Metric tons	*Metric tons*	*Metric tons*	*Metric tons*	*Metric tons*
1993	831,661	692,465	113,000	1,637,126	190,371
1994	824,453	1,200,000	114,000	2,138,453	151,172
1995	886,826	1,040,000	114,000	2,040,826	146,000
1996	NA	NA	NA	NA	NA
1997	NA	NA	NA	NA	NA
1998	NA	NA	NA	NA	NA
1999	NA	NA	NA	NA	NA
2000	NA	NA	NA	NA	NA
2001	NA	NA	NA	NA	NA
2002	NA	NA	NA	NA	NA

Year	United States imports	Mainland exports [1]	Total market supplies available
	Total		
	Metric tons	*Metric tons*	*Metric tons*
1993	1,040,858	255,907	2,612,448
1994	1,659,309	277,098	3,671,836
1995	1,048,726	274,868	2,960,684
1996	NA	NA	NA
1997	NA	NA	NA
1998	NA	NA	NA
1999	NA	NA	NA
2000	NA	NA	NA
2001	NA	NA	NA
2002	NA	NA	NA

[1] Does not include exports from Hawaii and Puerto Rico. NA-not available.
AMS, Livestock and Grain Market News Branch, (202) 720–6231.

Table 2-43.—Caloric sweeteners: Per capita food deliveries for domestic food and beverage use, 1992–2001

Year	Refined sugar	Corn sweeteners				Pure honey	Edible syrups	Total caloric sweeteners	U.S. (July 1)
		HFCS	Glucose syrup	Dextrose	Total				
									Millions
1992	64.2	51.8	15.1	3.6	70.5	1.0	0.4	136.1	256.9
1993	63.8	54.5	15.8	3.7	73.9	1.0	0.3	139.1	260.3
1994	64.3	56.2	15.9	3.8	75.9	1.0	0.4	141.5	263.4
1995	64.7	57.6	16.3	4.0	77.9	0.9	0.4	143.8	266.6
1996	65.5	57.8	16.4	4.0	78.2	1.0	0.4	145.0	269.7
1997	65.3	60.4	17.3	3.7	81.5	0.9	0.4	148.1	272.9
1998	65.1	61.9	17.1	3.6	82.7	0.9	0.4	149.1	276.1
1999	66.3	63.7	16.3	3.5	83.5	1.1	0.4	151.3	279.3
2000	65.6	62.7	15.8	3.4	81.9	1.1	0.4	148.9	282.3
2001	64.6	62.6	15.7	3.3	81.6	0.9	0.4	147.4	285.0

Note: Total may not add exactly, due to rounding.
ERS, Market and Trade Economics Division, Specialty Crops Branch, (202) 694–5247.

Table 2-44.—Tobacco: Area, yield, production, and value, United States, 1993–2002

Year	Area harvested	Yield per acre	Production [1]	Marketing year average price per pound received by farmers	Value of production
	Acres	*Pounds*	*1,000 pounds*	*Dollars*	*1,000 dollars*
1993	746,405	2,161	1,613,319	1.754	2,829,161
1994	671,065	2,359	1,582,896	1.758	2,779,056
1995	663,525	1,914	1,269,910	1.820	2,307,168
1996	733,060	2,072	1,518,704	1.882	2,853,739
1997	836,230	2,137	1,787,399	1.802	3,217,176
1998	717,605	2,062	1,479,867	1.828	2,700,795
1999	647,160	1,997	1,292,692	1.828	2,356,304
2000	472,410	2,229	1,052,999	1.910	2,001,775
2001	432,340	2,293	991,552	1.920	1,952,120
2002	430,280	2,068	889,632	1.907	1,726,013

[1] Production figures are on farm-sales-weight basis.
NASS, Crops Branch, (202) 720–2127.

Table 2-45.—Tobacco: Area, yield, and production, by States, 2000–2002

State	Area harvested			Yield per harvested acre			Production		
	2000	2001	2002 [1]	2000	2001	2002 [1]	2000	2001	2002 [1]
	Acres	*Acres*	*Acres*	*Pounds*	*Pounds*	*Pounds*	*1,000 pounds*	*1,000 pounds*	*1,000 pounds*
CT	1,600	2,300	1,850	1,531	1,720	1,702	2,450	3,957	3,148
FL	4,500	4,500	4,600	2,550	2,600	2,600	11,475	11,700	11,960
GA	31,000	26,100	26,500	2,220	2,460	2,100	68,820	64,206	55,650
IN	3,800	4,200	4,000	2,100	2,250	2,000	7,980	9,450	8,000
KY	132,700	115,700	112,200	2,133	2,201	2,018	283,065	254,653	226,430
MD	5,700	2,200	1,700	1,450	1,500	1,400	8,265	3,300	2,380
MA	550	1,140	1,150	836	1,711	1,604	460	1,951	1,845
MO	1,400	1,300	1,300	2,120	2,370	2,350	2,968	3,081	3,055
NC	170,400	161,700	169,300	2,386	2,393	2,111	406,500	386,920	357,350
OH	7,500	6,100	5,500	1,760	1,960	1,720	13,200	11,956	9,460
PA	5,100	3,100	3,400	1,994	1,989	2,004	10,170	6,166	6,815
SC	34,000	32,000	30,500	2,390	2,450	1,950	81,260	78,400	59,475
TN	46,020	39,690	35,900	2,085	2,189	2,021	95,958	86,893	72,540
VA	25,900	29,500	29,570	2,186	2,150	2,238	56,613	63,415	66,180
WV	1,300	1,300	1,300	1,200	1,450	1,500	1,560	1,885	1,950
WI	940	1,510	1,510	2,399	2,397	2,248	2,255	3,619	3,394
US	472,410	432,340	430,280	2,229	2,293	2,068	1,052,999	991,552	889,632

[1] Preliminary.
NASS, Crops Branch, (202) 720–2127.

Table 2-46.—Tobacco: Area, yield, and production in specified countries, 2000–2002

Continent and country	Area harvested			Yield per hectare			Production [2]		
	2000	2001	2002 [1]	2000	2001	2002 [1]	2000	2001	2002 [1]
	Hectares	*Hectares*	*Hectares*	*Metric tons*	*Metric tons*	*Metric tons*	*Metric tons*	*Metric tons*	*Metric tons*
North America:									
Canada	25,091	25,091	25,091	2.59	2.59	2.59	64,864	64,864	64,864
Mexico	21,013	17,851	10,130	1.95	1.90	2.03	40,992	33,842	20,594
United States	202,713	174,992	173,782	2.24	2.57	2.41	453,600	449,745	418,439
Total	313,897	248,817	228,825	2.23	2.25	2.24	699,176	559,456	512,656
South America:									
Argentina	60,000	60,000	65,500	1.68	1.64	1.95	101,000	98,110	127,500
Bolivia	1,060	1,060	1,060	0.92	0.92	0.92	975	980	980
Brazil	329,500	328,000	377,000	1.79	1.65	1.79	589,000	542,400	674,000
Chile	3,508	2,509	2,509	3.00	3.23	3.23	10,521	8,100	8,100
Colombia	22,000	23,000	23,000	1.82	1.83	1.83	40,000	42,000	42,000
Ecuador	1,725	1,774	1,774	0.61	0.62	0.62	1,050	1,100	1,100
Guyana	100	100	100	1.00	1.00	1.00	100	100	100
Paraguay	3,235	4,823	4,823	1.54	1.93	1.93	4,984	9,324	9,324
Peru	2,500	2,500	2,500	5.58	5.08	5.08	13,943	12,707	12,707
Uruguay	830	830	830	3.75	3.75	3.75	3,111	3,111	3,111
Venezuela	5,000	6,000	6,000	1.95	2.04	2.04	9,728	12,222	12,222
Total	455,844	434,844	433,844	1.70	1.68	2.05	774,412	730,154	891,144

See footnotes at end of table.

Table 2-46.—Tobacco: Area, yield, and production in specified countries, 2000–2002—Continued

Continent and country	Area harvested			Yield per hectare			Production [2]		
	2000	2001	2002 [1]	2000	2001	2002 [1]	2000	2001	2002 [1]
	Hectares	*Hectares*	*Hectares*	*Metric tons*	*Metric tons*	*Metric tons*	*Metric tons*	*Metric tons*	*Metric tons*
Central America:									
Costa Rica	117	105	105	1.60	1.85	1.85	187	194	194
El Salvador	580	600	600	2.01	1.95	1.95	1,167	1,167	1,167
Guatemala	8,374	8,793	9,232	2.22	2.22	2.22	18,630	19,562	20,540
Honduras	5,157	5,157	5,157	1.78	1.78	1.78	9,177	9,177	9,177
Nicaragua	1,395	844	844	1.59	1.60	1.60	2,222	1,349	1,349
Panama	1,100	1,160	1,160	1.82	2.04	2.04	2,000	2,367	2,367
Total	17,780	18,517	18,936	2.03	2.04	2.04	36,035	37,763	38,695
Caribbean:									
Cuba	45,327	45,000	45,000	0.84	0.78	0.78	38,000	35,000	35,000
Dominican Rep	23,600	23,600	23,600	1.28	1.28	1.28	30,250	30,250	30,250
Haiti	400	400	400	1.53	1.53	1.53	611	611	611
Jamaica	1,175	1,175	1,175	1.99	1.99	1.99	2,339	2,339	2,339
St. Vincent	70	70	70	1.00	1.00	1.00	70	70	70
Trinidad & Tob.	100	100	100	1.70	1.70	1.70	170	170	170
Total	73,585	83,875	83,875	0.97	0.82	0.82	71,440	68,440	68,440
European Union:									
Austria	111	115	115	2.68	1.74	1.74	297	200	200
Belgium-Lux	320	320	320	3.69	3.69	3.69	1,180	1,180	1,180
France	9,254	8,940	8,800	2.70	2.77	2.75	25,022	24,723	24,200
Germany	4,577	4,623	4,665	2.44	2.35	2.39	11,169	10,864	11,150
Greece	57,000	57,000	57,000	2.22	2.21	2.21	126,700	126,000	126,200
Italy	38,791	39,204	39,400	3.35	3.33	3.31	129,937	130,487	130,400
Portugal	2,132	2,105	2,100	2.90	2.90	2.91	6,193	6,111	6,111
Spain	13,229	13,250	13,110	3.19	3.19	3.20	42,215	42,210	41,900
Total	137,262	135,964	135,285	2.53	2.55	2.55	347,125	346,511	345,406
Western Europe:									
Switzerland	635	635	635	2.09	2.09	2.09	1,325	1,325	1,325
Eastern Europe:									
Albania	7,300	7,300	7,300	1.10	1.10	1.10	8,000	8,000	8,000
Bulgaria	29,920	32,460	37,260	1.08	1.26	1.35	32,455	40,885	50,210
Croatia	5,726	5,850	6,700	2.32	2.30	2.23	13,300	13,455	14,961
Hungary	5,849	5,100	6,000	2.19	2.14	2.02	12,800	10,900	12,140
Macedonia (Skopje)	22,000	22,000	22,000	1.41	1.41	1.41	31,111	31,111	31,111
Poland	14,057	12,600	11,000	2.13	1.76	2.73	30,000	22,200	30,000
Romania	12,000	12,000	12,000	1.23	1.23	1.23	14,750	14,750	14,750
Slovakia	2,000	2,000	2,000	1.75	1.75	1.75	3,500	3,500	3,500
Yugoslavia	6,200	6,200	6,200	1.23	1.23	1.23	7,604	7,604	7,604
Total	109,550	105,893	112,470	1.56	1.46	1.50	170,987	154,389	168,919
FSU-12: [3]									
Armenia	4,304	4,304	4,304	0.92	0.92	0.92	3,966	3,966	3,966
Azerbaijan	6,000	6,000	6,000	1.67	1.75	1.75	10,000	10,500	10,500
Belarus	800	800	800	1.75	1.75	1.75	1,400	1,400	1,400
Georgia	1,801	2,100	2,100	1.14	1.13	1.13	2,061	2,382	2,382
Kazakhstan	4,500	4,500	4,500	1.84	1.84	1.84	8,300	8,300	8,300
Kyrgyzstan	14,465	14,500	14,500	2.66	2.64	2.64	38,458	38,333	38,333
Moldova	23,300	25,000	25,000	1.37	1.33	1.33	31,888	33,333	33,333
Russia	1,950	2,000	2,100	0.82	0.90	0.90	1,600	1,800	1,900
Tajikistan	3,228	3,200	3,200	2.48	2.50	2.50	8,000	8,000	8,000
Turkmenistan	1,600	1,600	1,600	2.24	2.24	2.08	3,578	3,578	3,333
Ukranine	4,300	4,500	4,500	1.03	0.99	0.99	4,444	4,444	4,444
Uzbekistan	6,700	6,700	6,700	3.28	3.28	3.28	22,000	22,000	22,000
Total	65,308	73,458	75,208	1.91	1.84	1.82	124,879	134,860	137,005
North Africa:									
Algeria	5,700	5,700	5,700	1.25	1.26	1.26	7,153	7,200	7,200
Libya	650	650	650	2.56	2.56	2.56	1,667	1,667	1,667
Morocco	4,500	5,948	5,948	1.32	1.21	1.21	5,926	7,194	7,194
Tunisia	3,100	3,100	4,000	1.79	1.79	1.39	5,556	5,556	5,556
Total	13,800	15,398	16,298	1.47	1.40	1.33	20,302	21,617	21,617

See footnotes at end of table.

Table 2-46.—Tobacco: Area, yield, and production in specified countries, 2000–2002—Continued

Continent and country	Area harvested			Yield per hectare			Production [2]		
	2000	2001	2002 [1]	2000	2001	2002 [1]	2000	2001	2002 [1]
	Hectares	*Hectares*	*Hectares*	*Metric tons*	*Metric tons*	*Metric tons*	*Metric tons*	*Metric tons*	*Metric tons*
Other Africa:									
Angola	3,950	3,950	3,950	0.99	0.99	0.99	3,900	3,900	3,900
Benin	917	917	917	0.77	0.77	0.77	702	702	702
Burundi	705	705	705	1.00	1.00	1.00	705	705	705
Cameroon	3,400	3,400	3,400	1.38	1.38	1.38	4,700	4,700	4,700
Central African Rep	750	600	600	0.76	0.83	0.83	570	500	500
Chad	145	145	145	1.38	1.38	1.38	200	200	200
Congo (Brazzaville)	4,000	4,000	4,000	0.45	0.45	0.45	1,800	1,800	1,800
Cote d'Ivoire	20,000	20,000	20,000	0.60	0.60	0.60	12,000	12,000	12,000
Ethiopia	4,500	4,500	4,500	0.74	0.74	0.74	3,333	3,333	3,333
Ghana	3,950	3,950	3,950	0.38	0.38	0.38	1,500	1,500	1,500
Kenya	14,160	14,500	14,500	1.41	1.38	1.38	19,955	20,000	20,000
Liberia	10	10	10	1.00	1.00	0.00	10	10	0
Madagascar	2,110	2,110	2,110	1.05	1.05	1.05	2,222	2,222	2,222
Malawi	100,200	100,200	100,200	1.27	1.27	1.27	127,150	127,150	127,150
Mali	372	824	824	1.33	1.35	1.35	496	1,115	1,115
Mauritius	400	410	410	1.80	1.76	1.76	718	722	722
Mozambique	7,000	7,000	7,000	1.50	1.50	1.50	10,522	10,522	10,522
Niger	1,000	1,000	1,000	0.93	0.93	0.93	930	930	930
Nigeria	10,000	10,000	10,000	2.10	2.10	2.10	21,000	21,000	21,000
Reunion	200	200	2,000	1.00	1.00	0.10	200	200	200
Sierra Leone	540	540	540	1.11	1.11	1.11	600	600	600
South Africa; Rep	15,599	15,500	15,000	1.97	2.19	2.00	30,700	34,000	29,950
Swaziland	200	200	200	1.00	1.00	1.00	200	200	200
Tanzania; United Rep	46,000	35,500	35,500	0.61	0.79	0.79	28,255	27,967	27,967
Togo	400	400	400	0.00	0.00	0.00	2,000	2,000	2,000
Uganda	7,525	7,525	7,525	3.37	3.33	3.33	25,374	25,080	25,080
Zaire	3,700	3,700	3,700	1.11	1.11	1.11	4,110	4,110	4,110
Zambia	2,900	2,900	2,900	1.23	1.23	1.23	3,556	3,556	3,556
Zimbabwe	92,710	77,558	77,558	2.64	2.58	2.58	245,214	200,096	200,096
Total	347,343	320,081	305,080	1.35	1.60	1.54	468,225	512,772	468,904
Other Asia:									
Bangladesh	31,161	31,161	31,161	1.12	1.12	1.12	35,000	35,000	35,000
Burma	36,000	36,000	36,000	1.22	1.22	1.22	44,000	44,000	44,000
Cambodia	9,669	9,685	9,685	0.79	0.80	0.80	7,665	7,748	7,748
China	1,437,200	1,339,600	1,330,408	1.78	1.75	1.75	2,552,000	2,349,627	2,328,978
India	435,600	395,500	425,000	1.52	1.48	1.50	661,600	585,600	635,500
Indonesia	210,000	205,000	205,000	0.88	0.84	0.83	185,121	172,200	170,150
Japan	23,991	23,411	23,720	2.53	2.59	2.53	60,803	60,565	60,000
Korea, North	20,000	20,000	20,000	1.33	1.33	1.33	26,640	26,640	26,640
Korea, South	24,302	23,940	21,200	2.81	2.32	2.47	68,198	55,587	52,350
Laos	6,700	6,700	6,700	5.54	5.54	5.54	37,111	37,111	37,111
Malaysia	16,595	16,804	15,850	0.49	0.56	0.66	8,188	9,395	10,530
Pakistan	53,782	43,319	49,150	1.94	1.91	1.90	104,096	82,854	93,400
Philippines	40,296	36,700	36,700	1.87	1.86	1.86	75,189	68,300	68,300
Sri Lanka	4,480	4,480	4,480	1.34	1.34	1.34	6,000	6,000	6,000
Taiwan	4,230	2,929	2,115	2.77	2.50	2.50	11,702	7,322	5,287
Thailand	43,300	43,800	41,450	1.07	1.21	1.30	46,300	53,100	53,780
Vietnam	24,400	25,000	25,000	1.24	1.28	1.28	30,222	32,000	32,000
Total	2,397,375	2,456,886	2,391,732	1.59	2.00	2.00	3,819,255	3,927,293	3,864,146
Middle East:									
Cyprus	76	80	80	4.92	4.88	4.88	374	390	390
Iran	19,685	20,000	20,000	0.12	1.17	1.17	2,331	23,333	23,333
Iraq	2,400	2,400	2,400	1.04	1.04	1.04	2,500	2,500	2,500
Jordan	3,099	2,900	2,900	0.96	0.78	0.78	2,963	2,255	2,255
Lebanon	9,700	9,700	9,700	1.43	1.55	1.55	13,889	15,000	15,000
Oman	270	270	270	5.35	5.35	5.35	1,444	1,444	1,444
Syria	15,000	15,000	15,000	1.15	1.15	1.15	17,200	17,200	17,200
Turkey	270,475	237,318	207,140	0.93	0.87	0.77	250,495	207,261	160,476
United Arab. Emirates	50	50	50	13.52	13.52	13.52	676	676	676
Yemen	5,947	5,347	5,347	2.17	2.41	2.41	12,903	12,903	12,903
Total	316,936	303,691	259,761	0.99	0.89	0.87	312,525	270,321	226,841
Oceania:									
Australia	2,800	3,185	3,185	2.77	2.44	2.44	7,762	7,762	7,762
New Zealand	600	600	600	2.58	2.58	2.58	1,550	1,550	1,550
Solomon Islands	100	100	100	0.95	0.95	0.95	95	95	95
Total	3,500	3,885	3,885	2.69	2.42	2.42	9,407	9,407	9,407
World Total	4,252,815	4,201,944	4 065,834	1.61	1.61	1.66	6,855,093	6,774,308	6,754,505

[1] Preliminary. [2] Production data in metric tons, on farm-sales-weight basis, which is about 10 percent above dry-weight data normally reported in trade statistics. [3] FSU–12 includes the 12 newly independent States of the former USSR.

FAS,Cotton, Tobacco, and Seeds Division (202) 720-9496. Prepared or estimated on the basis of official statistics of Foreign governments, other foreign source materials, reports of U.S. Agricultural Counselors, Attaches, Foreign Service Officers and results of office research, and related information.

Table 2-47.—Tobacco: Area, yield, production, stocks, supply, disappearance, and price, by types, United States including Puerto Rico, 1993–2002 (farm-sales-weight basis)

Type and crop year	Area	Yield per acre	Production	Stocks [1]	Supply	Disappearance			Average price per pound to growers
						Total	Exports	Domestic	
	Acres	*Pounds*	*1,000 pounds*	*1,000 pounds*	*1,000 pounds*	*1,000 pounds*	*1,000 pounds*	*1,000 pounds*	*Cents*
Total flue-cured, types 11–14: [2]									
1993	400,100	2,217	886,908	1,195,581	2,087,281	792,300	358,938	433,400	168.1
1994	359,500	2,420	869,900	1,294,986	2,101,586	914,800	345,474	569,300	169.8
1995	386,200	1,933	746,616	1,186,786	2,040,968	875,133	344,518	530,615	179.2
1996	422,200	2,151	908,345	1,166,427	2,063,769	947,261	391,200	556,061	183.4
1997	458,300	2,285	1,047,438	1,116,508	2,130,041	876,878	335,900	540,978	172.0
1998	368,800	2,204	812,797	1,253,163	2,067,956	833,676	341,550	492,446	175.5
1999	303,800	2,162	656,752	1,234,280	1,888,172	698,684	261,818	436,866	173.6
2000	250,000	2,396	598,915	1,189,488	1,753,609	717,242	238,025	479,217	179.3
2001	238,100	2,432	579,091	1,036,367	1,580,790	664,912	276,007	388,905	185.8
2002 [3]	246,100	2,137	525,940	915,878	1,480,678	627,678	248,407	379,271	182.5
Total fire-cured, types 21–23:									
1993	16,700	2,449	40,900	68,400	109,300	36,400	17,801	17,700	219.4
1994	18,100	2,669	48,300	72,900	121,200	37,200	15,348	21,100	209.2
1995	16,880	2,322	39,190	83,995	123,185	39,523	14,918	24,640	217.2
1996	16,580	2,668	44,228	83,662	127,890	42,165	0	22,456	224.5
1997	16,550	2,554	42,262	85,725	127,987	40,839	18,235	22,604	225.6
1998	16,840	2,365	39,835	87,148	126,983	37,593	15,727	21,866	222.5
1999	16,420	2,319	38,075	89,390	127,465	36,246	12,979	14,312	226.4
2000	17,540	2,944	51,635	91,219	142,854	44,869	26,292	18,600	216.3
2001	14,620	3,096	45,299	97,962	143,261	38,955	13,492	25,463	214.9
2002 [3]	11,100	2,936	31,150	104,306	136,896	....	....	....	....
Virginia fire-cured, type 21:									
1993	1,200	1,560	1,900	4,400	6,300	3,000	2,300	700	171.9
1994	1,400	1,780	2,400	3,200	5,600	2,100	1,400	700	161.2
1995	1,100	1,400	1,540	3,542	5,082	1,614	800	848	162.5
1996	1,100	1,580	1,738	3,468	5,206	2,812	2,500	312	179.0
1997	1,200	1,640	1,968	2,394	4,362	1,989	554	1,435	212.5
1998	1,500	1,560	2,340	2,373	4,713	2,044	550	1,494	193.6
1999	1,600	1,670	2,672	2,669	5,341	1,897	979	918	181.9
2000	1,300	1,700	2,548	3,444	5,992	1,806	1,000	806	163.7
2001	1,200	1,835	2,202	4,168	6,338	1,567	150	1,417	175.8
2002 [3]	800	1,800	1,440	4,821	6,261	....	....	....	....
Kentucky and Tennessee fire-cured, types 22–23:									
1993	15,450	2,527	39,045	64,015	103,060	33,400	15,501	17,899	219.4
1994	16,730	2,745	45,928	69,660	115,588	35,135	13,948	21,187	209.2
1995	15,780	2,386	37,650	80,453	118,103	37,909	14,118	23,791	217.2
1996	15,480	2,745	42,490	80,194	122,684	39,353	17,209	22,144	224.5
1997	15,350	2,625	40,294	83,331	123,625	38,850	17,681	21,169	225.6
1998	15,340	2,444	37,495	84,775	122,270	35,549	15,177	20,372	222.5
1999	14,970	2,365	35,403	86,721	122,124	34,349	20,955	13,394	229.8
2000	16,240	3,023	49,087	87,775	136,862	43,086	25,292	17,794	216.3
2001	13,420	3,211	43,097	89,766	136,873	37,388	13,342	24,046	217.2
2002 [3]	10,300	3,024	31,150	90,787	130,635	....	....	....	....
Burley, type 31: [2]									
1993	299,700	2,115	633,838	938,900	1,565,500	551,600	152,200	399,400	181.6
1994	266,300	2,300	612,398	1,013,900	1,581,900	623,093	155,000	468,093	184.1
1995	234,200	1,863	436,343	958,807	1,441,307	550,917	164,800	386,117	185.5
1996	268,300	1,940	520,483	890,390	1,406,731	655,740	209,446	446,294	192.2
1997	335,300	1,943	648,633	750,991	1,379,222	547,616	168,395	379,221	188.5
1998	307,100	1,896	582,336	831,606	1,421,903	520,488	168,853	351,635	190.3
1999	300,600	1,829	555,185	901,415	1,452,573	412,531	139,262	273,269	189.9
2000	193,800	1,957	362,788	1,040,042	1,355,481	665,031	142,020	523,012	196.3
2001	167,600	2,033	334,066	690,450	1,034,110	386,229	139,802	246,427	197.5
2002 [3]	159,600	1,904	303,895	647,881	947,881	386,881	132,000	254,881	198.0
Maryland, type 32: [2]									
1993	12,700	1,526	19,380	10,542	29,928	17,317	5,475	11,842	143.0
1994	12,100	1,634	19,770	12,103	31,873	17,666	5,042	12,624	147.0
1995	11,900	1,507	17,935	14,207	32,142	16,162	6,580	9,582	157.0
1996	11,400	1,451	16,030	15,007	31,037	11,549	6,442	6,572	185.6
1997	11,200	1,629	17,700	19,488	37,188	14,645	6,515	8,130	158.5
1998	9,800	1,568	15,370	22,543	37,913	18,855	6,228	12,627	129.1
1999	9,500	1,511	14,350	16,003	30,353	14,353	10,157	4,196	134.5
2000	8,400	1,595	13,395	13,361	26,756	17,071	12,690	4,381	139.0
2001	3,300	1,620	5,346	9,685	15,031	6,719	5,007	1,712	155.0
2002 [3]	3,000	1,595	4,785	8,312	13,097	....	....	....	....

See footnotes at end of table.

Table 2-47.—Tobacco: Area, yield, production, stocks, supply, disappearance, and price, by types, United States including Puerto Rico, 1993–2002 (farm-sales-weight basis)—Continued

Type and crop year	Area	Yield per acre	Production	Stocks[1]	Supply	Disappearance			Average price per pound to growers
						Total	Exports	Domestic	
	Acres	*Pounds*	*1,000 pounds*	*1,000 pounds*	*1,000 pounds*	*1,000 pounds*	*1,000 pounds*	*1,000 pounds*	*Cents*
Total dark air-cured, types 35–37:									
1993	4,890	2,296	11,227	24,386	35,613	10,769	2,186	8,583	171.3
1994	4,690	2,515	11,797	24,844	36,641	([4])	([4])	([4])	168.3
1995	4,180	2,050	8,567	27,280	35,880	11,181	1,266	9,178	176.2
1996	3,850	2,250	8,662	25,472	34,134	10,277	61	10,216	195.0
1997	3,710	2,241	8,315	23,857	32,172	4,260	10	9,250	201.6
1998	4,435	2,206	9,785	22,512	32,297	8,203	42	8,161	195.1
1999	5,100	3,878	11,795	24,094	35,889	9,176	1,433	7,743	2.298
2000	5,580	4,551	16,061	26,713	42,774	9,896	1,022	8,874	2.163
2001	5,070	4,347	14,103	32,878	46,981	8,614	322	8,292	2.149
2002[3]	3,870	4,005	9,625	38,367	47,775	....	....	....	....
One Sucker, Green River type 35-36:									
1993	4,790	2,322	11,123	24,177	35,300	10,570	2,100	8,470	171.5
1994	4,620	2,527	11,673	24,730	36,403	9,123	1,754	7,369	168.5
1995	4,110	2,065	8,488	27,280	35,768	10,344	1,266	9,078	176.2
1996	3,780	2,262	8,550	25,424	33,974	10,141	....	10,141	191.1
1997	3,630	2,258	8,196	23,833	32,029	9,564	10	9,554	201.7
1998	4,335	2,229	9,663	22,465	32,128	8,107	10	8,097	195.5
1999	5,000	2,328	11,640	24,021	35,661	9,036	1,337	7,699	203.9
2000	5,480	2,901	15,896	26,625	42,521	9,824	1,000	8,824	197.1
2001	4,970	2,807	13,949	32,697	46,646	8,391	100	8,291	182.9
2002[3]	3,800	2,505	9,520	38,255	47,775	....	....	....	....
Virginia sun-cured, type 37:									
1993	100	1,040	104	209	313	199	53	146	152.1
1994	70	1,770	124	114	238	131	131	....	147.6
1995	70	1,125	107	186	295	138	138	....	155.5
1996	70	1,600	112	48	160	136	49	27	178.2
1997	80	1,490	119	24	143	96	51	45	190.8
1998	100	1,220	122	47	169	96	38	58	170.9
1999	100	1,550	155	73	228	140	44	96	159.0
2000	100	1,650	165	88	253	72	50	22	180.0
2001	100	1,540	154	181	335	223	1	222	168.6
2002[3]	90	1,500	105	112					
Total continental cigar filler, types 41–44:									
1993	5,800	2,100	12,180	25,472	37,652	15,163	*	15,163	95.0
1994	5,400	2,100	11,340	23,959	35,299	14.832	*	14,832	100.0
1995	4,500	2,050	9,225	20,502	29,702	11,788	*	11,400	145.0
1996	4,600	2,040	9,384	17,939	27,323	14,124	*	12,700	155.0
1997	4,600	2,100	9,660	13,199	22,859	11,000	*	11,000	160.0
1998	4,500	2,100	9,450	12,969	22,419	11,039	*	11,039	130.0
1999	3,200	1,850	5,920	11,380	17,300	7,768	*	7,768	130.0
2000	2,400	2,100	5,040	9,532	14,572	2,453	*	2,453	NA
2001	2,000	2,060	4,120	12,119	16,239	3,968	....	3,968	150.0
2002[3]	2,100	2,100	4,410	12,271	16,681	....	....	....	....
Pennsylvania seedleaf filler, type 41:									
1993	5,800	2,100	12,180	25,462	37,642	12,683	*	13,700	95.0
1994	5,400	2,100	11,340	23,959	35,299	14,832	*	14,893	100
1995	4,500	2,050	9,225	20,467	29,692	11,788	*	11,788	145
1996	4,800	2,140	10,272	17,939	28,211	15,012	*	15,012	155
1997	4,900	2,200	10,780	13,199	23,979	11,010	*	11,010	160
1998	4,500	2,100	9,450	12,969	22,100	10,720	*	10,720	130
1999	3,200	1,850	5,920	11,380	17,300	7,768	*	7,768	130
2000	2,400	2,100	5,040	9,532	14,572	2,453	*	2,453	NA
2001	2,000	2,060	4,120	12,119	16,239	3,968	*	3,968	150
2002[3]	2,100	2,100	4,410	12,271	16,681	....	....	....	....

See footnotes at end of table.

Table 2-47.—Tobacco: Area, yield, production, stocks, supply, disappearance, and price, by types, United States including Puerto Rico, 1993–2002 (farm-sales-weight basis)—Continued

Type and crop year	Area	Yield per acre	Production	Stocks [1]	Supply	Disappearance			Average price per pound to growers
						Total	Exports	Domestic	
	Acres	*Pounds*	*1,000 pounds*	*1,000 pounds*	*1,000 pounds*	*1,000 pounds*	*1,000 pounds*	*1,000 pounds*	*Cents*
Miami Valley (Ohio) filler, types 42–44:									
1993	0	0	0	0	0	....	....	....	....
1994	0	0	0	0	0	....	....	....	NA
1995	0	0	0	0	0	....	....	....	NA
1996	0	0	0	0	0	....	....	....	NA
1997	0	0	0	0	0	....	....	....	NA
1998	....	....	....	....	....	....	....	....	....
1999	....	....	....	....	....	....	....	....	....
2000	....	....	....	....	....	....	....	....	....
Puerto Rican filler, type 46: [5]									
1993	*	NA	*	1,186	1,186	1,090	....	1,090	NA
1994	*	NA	*	96	96	61	61	....	NA
1995	*	*	*	*	*	*	*	*	*
1996	*	*	*	*	*	*	*	*	*
1997	*	*	*	*	*	*	*	*	*
1998	*	*	*	*	*	*	*	*	*
1999	*	*	*	*	*	*	*	*	*
Total cigar binder, types 51–55:									
1993	5,485	1,520	8,337	32,763	41,100	11,576	0	12,052	175.1
1994	3,805	2,017	7,674	29,600	37,300	12,911	*	12,886	181.1
1995	4,000	2,165	8,661	25,931	34,592	11,275	*	11,257	213.8
1996	4,500	1,792	8,063	23,317	31,380	12,398	11,832	12,198	321.1
1997	4,590	2,032	9,327	18,982	28,243	10,655	9,051	10,255	378.4
1998	4,410	1,783	7,863	18,253	26,116	8,335	6,440	1,895	334.7
1999	3,680	1,899	6,987	17,781	24,768	9,321	8,057	1,264	342.7
2000	1,860	1,787	3,325	15,447	18,772	6,735	5,389	1,346	263.3
2001	3,650	2,039	7,441	12,037	19,478	8,954	8,792	162	367.3
2002 [3]	3,610	1,970	7,112	10,524	17,636	....	....	....	....
Connecticut Valley binder, types 51–52:									
1993	885	1,914	1,694	1,918	3,612	1,910	0	2,035	300.0
1994	955	1,893	1,808	1,577	3,385	1,900	0	2,004	306.0
1995	1,240	1,969	2,441	1,381	3,822	2,245	0	2,245	389.0
1996	1,630	1,780	2,901	1,577	4,478	3,298	605	2,932	628.0
1997	2,040	1,783	3,637	3,637	4,817	2,766	605	2,161	741.4
1998	2,360	1,539	3,633	2,051	5,684	2,199	775	1,424	549.9
1999	2,500	1,668	4,169	3,485	7,654	4,888	264	4,624	473.7
2000	900	1,189	1,070	2,766	3,836	1,522	346	1,176	491.6
2001	2,140	1,786	3,822	2,314	6,136	4,308	162	4,146	558.9
2002 [3]	2,100	1,770	3,718	1,828	5,546	....	....	....	....
Wisconsin binder, types 54–55:									
1993	4,600	1,819	6,643	30,845	37,488	9,541	0	9,500	143.2
1994	2,850	1,444	5,866	27,947	33,813	9,263	0	9,263	145.0
1995	2,760	2,254	6,220	24,550	30,770	9,030	0	9,030	145.0
1996	2,870	1,799	5,162	21,740	26,902	9,100	200	8,900	148.6
1997	2,550	2,231	5,690	17,802	23,492	7,290	400	6,890	150.5
1998	2,050	2,063	4,230	16,202	20,432	6,136	1,120	5,016	149.8
1999	1,180	2,388	2,818	14,296	17,114	4,433	1,000	3,433	149.0
2000	960	2,348	2,255	12,681	14,936	5,213	1,000	4,213	155.0
2001	1,510	2,397	3,619	9,723	13,342	4,646	....	4,646	165.0
2002 [3]	1,510	2,248	3,394	8,696	12,090	....	....	....	....
Southern Wisconsin, type 54:									
1993	2,800	1,675	4,690	....	....	....	....	....	....
1994	1,900	2,200	4,180	....	....	....	....	....	145.0
1995	1,900	2,370	4,513	....	....	....	....	....	145.0
1996	1,900	1,900	3,610	....	....	....	....	....	145.0
1997	1,800	2,330	4,194	....	....	....	....	....	148.0
1998	1,500	2,180	3,270	....	....	....	....	....	151.0
1999	890	2,530	2,252	....	....	....	....	....	150.0
2000	710	2,500	1,825	....	....	....	....	....	149.0
2001	1,200	2,535	3,042	....	....	....	....	....	155.0
2002 [3]	1,200	2,350	2,820	....	....	....	....	....	165.0

See footnotes at end of table.

Table 2-47.—Tobacco: Area, yield, production, stocks, supply, disappearance, and price, by types, United States including Puerto Rico, 1993–2002 (farm-sales-weight basis)—Continued

Type and crop year	Area	Yield per acre	Production	Stocks[1]	Supply	Disappearance: Total	Disappearance: Exports	Disappearance: Domestic	Average price per pound to growers
	Acres	*Pounds*	*1,000 pounds*	*1,000 pounds*	*1,000 pounds*	*1,000 pounds*	*1,000 pounds*	*1,000 pounds*	*Cents*
Northern Wisconsin, type 55:									
1993	1,800	1,085	1,953	....	....	....	....	....	139.0
1994	950	1,775	1,686	....	....	....	....	....	148.0
1995	860	1,985	1,707	....	....	....	....	....	145.0
1996	970	1,600	1,552	....	....	....	....	....	150.0
1997	750	1,995	1,496	....	....	....	....	....	151.0
1998	550	1,745	960	....	....	....	....	....	149.0
1999	290	1,952	566	....	....	....	....	....	149.0
2000	230	1,865	430	....	....	....	....	....	155.0
2001	310	1,860	577	....	....	....	....	....	165.0
2002[3]	310	1,850	574	....	....	....	....	....	....
Total cigar wrapper, types 61:									
1993	1,080	1,484	1,577	1,503	3,080	1,100	990	200	990
1994	1,190	1,460	1,666	1,890	3,556	1,610	1,300	200	1,300
1995	1,265	1,582	2,001	2,090	4,091	1,835	1,400	409	NA
1996	1,430	1,473	2,106	2,256	4,362	2,410	2,000	210	[7]
1997	1,680	1,431	2,404	1,952	4,356	2,309	2,100	209	2,520
1998	1,720	1,413	2,431	2,047	4,478	3,202	2,722	480	2,420
1999	1,860	1,951	3,628	1,276	4,904	4,127	3,021	1,106	[7]
2000	1,250	1,472	1,840	777	2,617	1,494	1,300	194	2,530
2001	1,300	1,605	2,086	1,123	3,209	1,422	1,200	222	[7]
2002[3]	900	1,417	1,275	1,787	3,062	....	....	....	....
Total tobacco, types 11–72:[6]									
1993	746,405	2,163	1,614,364	2,300,287	3,914,651	1,435,927	537,533	898,204	175.3
1994	671,065	2,359	1,582,896	2,474,172	4,057,068	1,629,869	524,252	1,131,997	177.4
1995	663,125	1,913	1,268,538	2,319,687	3,588,225	1,362,782	534,034	828,748	182.0
1996	733,060	2,072	1,518,704	2,225,443	3,744,147	1,713,445	631,036	1,082,409	188.2
1997	836,230	2,137	1,787,399	2,030,702	3,818,101	1,567,860	532,211	1,035,649	180.3
1998	717,605	2,062	1,479,867	2,250,241	3,730,108	1,442,223	537,023	905,200	182.8[7]
1999	647,160	1,997	1,292,692	2,300,668	3,593,360	1,194,845	432,876	761,969	182.8[7]
2000	472,410	2,229	1,052,999	2,388,435	3,441,434	1,464,814	414,414	1,050,401	186.9[8]
2001	432,340	2,293	991,552	1,894,081	2,885,633	1,119,773	432,476	687,297	192.0[7]
2002[3]	430,280	2,068	889,632	1,738,398	2,628,030	....	....	....	....

[1] July 1 for flue-cured types 11-14 and cigar types 61 and 62; Oct. 1 for all other types. [2] Flue-cured (type11-14) and Burley (type 31) supply based on actual marketing. Maryland (type 32) based on October 1 stocks. [3] Preliminary. [4] Not available. [5] Puerto Rican crop planted in the fall of the year shown and harvested the following spring. No longer produced. [6] Includes Perique. [7] Does not include cigar wrapper type 61. [8] Does not include cigar filler type 41. NA-not applicable. *Negligible.

ERS, Specialty Crops Branch, (202) 694–5311. Basic export data from the official reports of the Department of Commerce.

Table 2-48.—Tobacco: Stocks owned by dealers and manufacturers, by types, United States, 1997–2001 (farm-sales-weight basis) [1]

Type and year	Jan. 1	Apr. 1	July 1	Oct. 1
	1,000 pounds	*1,000 pounds*	*1,000 pounds*	*1,000 pounds*
Flue-cured, types 11–14:				
1997	1,342,946	1,146,482	988,750	1,281,224
1998	1,479,902	1,278,563	1,123,060	1,420,853
1999	1,529,794	1,311,220	1,189,488	1,451,193
2000	1,416,141	1,202,551	1,036,367	1,218,789
2001	1,077,226	922,184	821,083	1,042,878
Virginia fire-cured, type 21:				
1997	2,163	2,490	2,088	1,861
1998	2,171	2,955	2,024	1,850
1999	3,158	3,698	3,855	3,444
2000	3,593	4,579	4,278	4,186
2001	33,542	4,223	4,020	3,906
Kentucky and Tennessee fire-cured, types 22–23:				
1997	71,517	84,901	79,886	76,997
1998	73,561	89,710	86,187	78,219
1999	85,597	98,039	95,847	87,775
2000	83,955	109,569	105,963	93,776
2001	91,383	106,293	100,652	92,723
Burley, type 31:				
1997	281,518	879,561	770,727	646,096
1998	764,153	754,245	783,324	693,711
1999	1,025,217	1,144,288	1,141,273	1,040,042
2000	1,159,848	1,057,925	786,926	689,459
2001	664,725	639,769	577,785	520,777
Maryland, type 32:				
1997	1,642	17,054	17,371	15,549
1998	16,020	17,128	20,715	18,695
1999	16,003	17,678	17,153	15,142
2000	13,361	36,306	35,339	32,697
2001	8,380	8,604	8,349	7,353
One Sucker and Green River, types 35–36:[2]				
1997	21,651	48,192	22,268	20,307
1998	21,309	23,435	19,433	19,432
1999	25,942	30,274	28,746	26,625
2000	28,222	36,306	35,339	32,697
2001	32,914	36,575	35,505	33,388
Virginia sun-cured, type 37:				
1997	24	2,239	34	17
1998	23	66	37	37
1999	94	103	90	88
2000	90	149	185	181
2001	151	104	100	96
Pennsylvania seedleaf, type 41:				
1997	15,777	20,582	13,765	12,245
1998	11,980	15,525	13,578	12,141
1999	11,358	10,426	9,847	9,532
2000	8,757	11,449	10,314	12,119
2001	9,620	12,603	11,701	10,092
Puerto Rican filler, type 46:[3]				
1997	1	0	0	0
1998	0	0	0	0
1999	0	0	0	0
2000	0	0	0	0
2001	0	0	0	0
Connecticut Valley, types 51–52:				
1997	1,455	1,769	1,437	853
1998	1,916	1,398	1,251	1,493
1999	3,812	2,429	2,481	2,766
2000	2,747	2,356	2,545	2,314
2001	2,209	1,940	1,734	1,459
Wisconsin binder, types 54–55:				
1997	15,495	22,052	18,613	16,997
1998	15,386	19,740	17,188	15,548
1999	13,910	14,742	14,021	12,681
2000	11,384	12,498	10,871	9,723
2001	8,278	10,624	9,478	8,405
Cigar Wrapper, type 61:				
1997	2,184	2,144	1,590	1,157
1998	2,344	1,790	1,698	1,491
1999	1,948	1,544	777	1,827
2000	1,289	1,004	678	1,787
2001	1,169	1,131	770	553

See footnotes at end of table.

Table 2-48.—Tobacco: Stocks owned by dealers and manufacturers, by types, United States, 1997–2001 (farm-sales-weight basis)[1]—Continued

Type and year	Jan. 1	Apr. 1	July 1	Oct. 1
	1,000 pounds	*1,000 pounds*	*1,000 pounds*	*1,000 pounds*
Perique, type 72:				
1997	24	23	22	45
1998	12	16	15	12
1999	27	21	21	24
2000	31	29	21	71
2001	30	20	19	19
Other miscellaneous domestic, type 73:				
1997	2,681	2,398	2,852	1,915
1998	2,182	2,405	2,946	2,207
1999	642	55	54	53
2000	2,198	2,763	2,822	3,098
2001	2,477	2,165	1,753	2,557
Foreign-grown cigar-leaf, types 81–89:				
1997	67,235	75,227	82,116	90,514
1998	87,953	93,279	78,638	92,167
1999	111,158	107,287	106,755	103,773
2000	102,295	100,965	103,126	97,788
2001	74,555	71,520	69,128	72,430
Foreign-grown cigarette and smoking, types 91–99:				
1997	940,670	899,623	1,032,102	1,014,738
1998	1,005,466	968,565	935,595	946,608
1999	959,486	865,615	840,053	836,950
2000	790,892	789,941	753,082	766,382
2001	662,709	654,484	621,554	644,749

[1] Stocks shown have been converted to a farm-sales-weight basis—the equivalent of weight at the time of sale by grower—thereby making these data of leaf-tobacco stocks comparable with data of leaf-tobacco production. [2] One Sucker and Green River combined. [3] Stocks on the island of Puerto Rico are included.

AMS Market Informaiton and Program Analysis Branch, (202) 205–0489.

Table 2-49.—Tobacco products: Consumption, total and per capita (18 years of age and over) in the United States, 1993–2002[1]

Year	Cigarettes			Large cigars[2]			Smoking, chewing, and snuff[3]		All tobacco products[3]	
	Total	Total	Per capita	Total	Total	Per capita	Total	Per capita	Total	Per capita
	Billion	*Million pounds*	*Number*	*Billion*	*Million pounds*[4]	*Number*	*Million pounds*	*Pounds*	*Million pounds*	*Pounds*
1993	485	864	2,539	2.1	34	11	90	0.47	988	5.13
1994	486	816	2,527	2.3	41	12	92	0.47	948	4.93
1995	487	819	2,505	2.6	44	13	91	0.47	953	4.91
1996	487	814	2,482	3.1	52	16	90	0.46	960	4.70
1997	480	805	2,422	3.5	58	18	88	0.45	1,004	4.66
1998	465	781	2,320	3.7	60	18	87	0.43	962	4.49
1999	435	721	2,148	3.8	63	19	87	0.45	876	4.32
2000	430	711	2,092	3.9	63	19	92	0.44	866	4.21
2001	425	696	2,051	3.9	65	19	89	0.43	860	4.15
2002[5]	420	738	2,030	3.8	63	18	89	0.42	902	4.31

[1] Includes consumption by overseas forces. [2] Weighing over 3 pounds per 1,000. [3] Unstemmed-processing weight equivalent. [4] Includes weight of small cigars. [5] Preliminary.

ERS, Market and Trade Economics Division, Specialty Crops Branch, (202) 694–5311. No adjustment made for quantities lost, destroyed, bartered, etc., under war and postwar conditions, but such adjustments probably would be small in relation to totals.

Table 2-50.—Tobacco products: Cigars, cigarettes, chewing and smoking tobacco, and snuff, manufactured in the United States, 1992–2001

Year	Cigars		Cigarettes		Chewing tobacco			
	Large	Small	Large [1]	Small	Firm	Moist	Twist	Looseleaf
	Millions	*Millions*	*Millions*	*Millions*	*1,000 pounds*	*1,000 pounds*	*1,000 pounds*	*1,000 pounds*
1992	1,740.5	1,309.9	0.14	718,667.0	....	5,813	1,176	61,571
1993	1,765.8	1,250.1	0.14	660,724.7	....	5,318	1,137	57,998
1994	1,916.5	1,409.8	0.01	725,006.7	3,307	1,318	1,081	56,780
1995	2,056.8	1,430.4	0.01	743,519.1	2,886	1,247	1,085	57,678
1996	NA	NA	NA	NA	2,905	1,039	1,114	56,012
1997	2,323.6	1,476.1	0.00	765,324.2	2,562	911	977	53,663
1998	2,750.4	1,710.3	0.00	679,746.6	2,359	744	976	49,235
1999	2,938.1	2,316.6	0.00	606,318.5	2,187	633	886	47,177
2000 [2]	2,824.5	2,468.9	0.00	593,173.0	2,048	543	829	45,978
2001 [2]	NA	NA	NA	NA	1,867	475	821	43,872
	Taxable removals and domestic invoices [3]							
1992	2,105.8	1,302.0	0.1	509,494.8	....	5,658	1,169	60,915
1993	2,110.8	1,205.2	0.1	462,858.6	....	4,995	1,122	57,671
1994	2,144.8	1,366.9	00.2	488,606.9	3,095	1,191	1,114	56,431
1995	2,364.6	1,397.0	0.00	489,265.8	3,015	1,147	1,092	57,048
1996	NA	NA	NA	NA	2,797	952	1,104	55,136
1997 [2]	3,031.2	1,587.3	0.00	495,237.5	2,517	806	1,010	52,480
1998 [2]	3,185.1	1,638.0	0.00	457,871.7	2,288	674	968	48,562
1999 [2]	3,348.7	2,195.9	0.00	429,556.4	2,119	581	894	46,916
2000 [2]	3,369.8	2,243.2	0.00	421,597.4	2,049	485	863	45,059
2001 [2]	NA	NA	NA	NA	1,828	429	803	43,532
	Tax-free removals and exports							
1992	78.1	8.1	0	199,240.4	....	119	0	57
1993	68.8	7.8	0.1	193,577.7	....	104	0	42
1994	75.7	12.3	0.2	231,629.4	59	28	0	66
1995	84.6	16.8	0.2	250,765.6	91	33	0	83
1996	NA	NA	NA	NA	73	32	0	100
1997 [2]	115.7	66.3	0.00	310,329.5	65	23	0	94
1998 [2]	134.3	1,323	0.00	212,364.9	48	30	0	73
1999 [2]	121.3	1,152.2	0.00	165,443.8	30	23	0	69
2000 [2]	113.7	228.6	0.00	153,633.8	31	34	0	85
2001 [2]	NA	NA	NA	NA	30	31	0	75

See footnotes at end of table.

Table 2-50.—Tobacco products: Cigars, cigarettes, chewing and smoking tobacco, and snuff, manufactured in the United States, 1992–2001—Continued

Year	Smoking tobacco			Snuff	Total chewing, smoking, and snuff
	Pipe	Granulated	Cigarette cut		
	1,000 pounds	*1,000 pounds*	*1,000 pounds*	*1,000 pounds*	*1,000 pounds*
1992 ..	10,085	84	4,741	57,449	140,919
1993 ..	8,990	37	4,646	59,106	137,232
1994 ..	8,077	50	5,268	59,491	135,372
1995 ..	7,614	93	4,536	60,202	135,026
1996 ..	6,939	90	4,954	61,539	134,592
1997 ..	6,770	72	4,555	64,336	133,846
1998 ..	6,154	60	6,251	65,477	131,256
1999 ..	6,726	61	7,908	66,992	132,570
2000 ..	5,982	50	7,327	69,556	132,313
2001 [2]	5,088	0	7,674	70,893	130,690
	Taxable removals and domestic invoices [3]				
1992 ..	9,491	62	4,527	55,555	137,377
1993 ..	8,526	38	4,731	56,729	133,812
1994 ..	7,380	44	5,128	58,567	132,950
1995 ..	7,032	100	4,686	59,339	133,459
1996 ..	6,469	88	4,790	61,390	132,726
1997 [2]	5,999	70	4,746	62,481	130,109
1998 [2]	5,604	62	6,356	64,051	128,565
1999 [2]	5,701	59	7,688	65,518	129,476
2000 [2]	4,620	50	8,398	68,605	130,129
2001 [2]	4,815	0	10,094	69,661	131,162
	Tax-free removals and exports				
1992 ..	553	0	0	715	1,444
1993 ..	413	0	0	769	1,328
1994 ..	513	0	0	798	1,464
1995 ..	524	0	0	821	1,552
1996 ..	429	0	0	292	926
1997 [2]	532	0	0	862	1,576
1998 [2]	531	0	0	682	1,364
1999 [2]	532	0	0	788	1,442
2000 [2]	546	0	0	742	1,438
2001 [2]	455	0	0	65	1,356

[1] Weighing more than three pounds per thousand. [2] Preliminary. [3] Includes cigars and cigarettes imported or brought into the United States and Puerto Rico. NA-not available.

AMS, Market Information and Program Analysis Branch, (202) 205–0489.

Table 2-51.—Cigarettes and cigars: Total output, domestic consumption, tax-exempt removals, and exports, United States, 1993–2002

Year	Cigarettes				Cigars [3]			
	Total output	Domestic consump-tion [1]	Tax-exempt removals [2]		Total output [4]	Domestic consump-tion [1]	Tax-exempt removals [2]	
			Total	Exports			Total	Exports
	Billion	*Billion*	*Billion*	*Billion*	*Million*	*Million*	*Million*	*Million*
1993	661	485	202	196	1,795	2,138	68	67
1994	726	486	232	220	1,942	2,337	78	75
1995	747	487	251	231	2,058	2,568	85	78
1996	755	487	261	244	2,413	3,054	98	84
1997	720	480	232	217	2,324	3,517	110	136
1998	680	465	213	201	2,751	3,655	112	158
1999	607	435	165	151	2,938	3,845	121	84
2000	595	430	154	148	2,825	3,850	114	113
2001	580	425	140	134	2,993	3,941	48	124
2002 [5]	580	415	140	139	2,900	3,833	225	250

[1] As indicated by taxable removals and imports, and estimated inventory changes. [2] In addition to exports, tax-exempt removals include principally shipments to forces overseas, to United States possessions, and ships' stores. [3] Includes cigarillos but excludes small (approximately cigarette-size) cigars. [4] Includes cigars shipped to mainland United States from Puerto Rico. [5] Preliminary.

ERS, Market and Trade Economics Division, Specialty Crops Branch, (202) 694–5311. Compiled from annual and monthly reports of the Internal Revenue Service, U.S. Treasury Department, and the Commerce Department.

Table 2-52.—Tobacco: Price-support loan operations, United States, 1993–2002 [1]

Year	Flue-cured, types 11–14			Burley, type 31		
	Support price per pound	Placed under loan		Support price per pound	Placed under loan	
		Quantity	Percentage of production		Quantity	Percentage of production
	Cents	*Million pounds*	*Percent*	*Cents*	*Million pounds*	*Percent*
1993	157.7	204.8	23.0	168.3	232.3	37.0
1994	158.3	97.7	12.1	171.4	54.7	9.6
1995	159.7	11.5	1.3	172.5	0.0	0.0
1996	160.1	1.8	0.2	173.7	0.0	0.0
1997	162.1	195.5	19.4	176	124.5	19.8
1998	162.8	82.5	10.1	177.8	73.2	12.4
1999	163.2	136.4	21.2	178.9	230.6	42.0
2000	164.0	27.4	4.4	180.5	19.3	4.8
2001	166.0	15.0	2.6	182.6	12.4	3.5
2002	165.6	....	....	183.5	....	....

[1] Support operations for other kinds of tobacco not shown. Burley and flue-cured usually account for over 95 percent of tobacco loan placements.

FSA, Tobacco and Peanuts Division, (202)–720–5291.

Table 2-53.—Tobacco, unmanufactured: United States imports for consumption, by country of origin, 1999–2001

Type and country of origin	1999	2000	2001[1]
	Metric tons	*Metric tons*	*Metric tons*
Cigarette leaf:			
Brazil	42,393	42,525	44,028
Turkey	35,723	29,303	42,336
Malawi	11,189	16,205	13,142
Argentina	10,441	12,020	11,187
Canada	8,271	6,121	10,318
Italy	5,591	5,520	7,092
Thailand	6,956	5,292	7,027
Greece	6,903	8,243	6,782
Indonesia	4,368	7,421	5,746
Others	50,301	42,744	46,730
Total	182,136	175,394	182,095
Cigar wrapper:			
Ecuador	279	265	424
Dominican Repub	69	242	260
India	0	10	232
Honduras	118	202	217
Others	496	416	214
Total	962	1,135	1,347
Scrap:			
Turkey	4,055	5,997	3,437
Canada	73	915	941
Dominican Repub	763	956	913
Honduras	189	445	478
Greece	113	287	350
Bulgaria	157	78	311
Others	2,791	1,887	951
Total	8,141	10,565	7,381
Stems:			
Brazil	13,680	15,381	16,277
Argentina	3,610	4,865	5,156
Malawi	4,617	59	2,586
Turkey	1,472	1,998	1,285
Zimbabwe	1,423	277	1,284
Others	3,625	5,888	3,270
Total	28,427	28,468	29,858
Grand total[2]	294,249	223,503	221,603

[1] Preliminary. [2] Includes unstemmed and stemmed cigar filler.
FAS, Cotton, Oilseeds, Tobacco and Seeds Division, (202) 720–9496. Compiled from U.S. Bureau of the Census records.

Table 2-54.—Tobacco, unmanufactured: International trade, 1999–2001[1]

Continent and country	1999		2000		2001	
	Exports	Imports	Exports	Imports	Exports	Imports
	Metric tons	*Metric tons*	*Metric tons*	*Metric tons*	*Metric tons*	*Metric tons*
North America:						
Canada	22,377	9,900	22,377	9,900	22,377	9,900
Mexico	10,038	3,350	8,451	9,728	8,451	15,965
United States	179,892	196,601	185,827	254,375	190,000	270,000
Total North America	212,307	209,851	216,655	274,003	220,828	295,865
Central America:						
Belize	0	55	0	55	0	55
Costa Rica	0	100	0	78	0	100
El Salvador	358	808	358	800	358	800
Guatemala	9,671	0	10,105	0	10,610	0
Honduras	4,686	175	4,686	175	4,686	175
Nicaragua	160	400	160	400	160	400
Panama	1,300	153	1,500	154	1,500	154
Total Central America	16,175	1,691	16,809	1,662	17,314	1,684
Carribean:						
Bahamas, The	0	48	0	48	0	48
Barbados	0	150	0	150	0	0
Cuba	5,000	300	5,000	300	5,000	0
Dominican Republic	16,000	270	16,000	270	16,000	270
Grenada	0	20	0	20	0	0
Haiti	0	1,014	0	915	0	915
Jamaica & Dep	200	450	200	450	200	450
Netherland Antilles	0	4	0	4	0	4
St. Lucia	20	15	20	15	20	15
St. Vincent and Gren.	60	6	60	6	60	6
Trinidad and Tobago	0	800	0	800	0	800
Total Carribean	21,280	3,077	21,280	2,978	21,280	2,508

See footnotes at end of table.

Table 2-54.—Tobacco, unmanufactured: International trade, 1999–2001—Continued [1]

Continent and country	1999		2000		2001	
	Exports	Imports	Exports	Imports	Exports	Imports
	Metric tons	*Metric tons*	*Metric tons*	*Metric tons*	*Metric tons*	*Metric tons*
South America:						
Argentina	55,400	4,500	73,700	2,800	75,700	2,000
Bolivia	0	292	0	277	0	270
Brazil	341,500	6,950	435,500	8,950	447,700	8,600
Chile	2,050	3,350	2,050	3,350	2,050	3,350
Colombia	14,000	475	14,000	475	14,000	475
Ecuador	250	800	250	800	250	800
French Guiana	0	10	0	10	0	10
Guyana	0	160	0	160	0	160
Paraguay	3,487	721	7,050	721	7,050	721
Peru	6,136	889	6,150	670	6,150	670
Suriname	0	550	0	550	0	550
Uruguay	228	2,700	228	2,700	228	2,700
Venezuela	0	4,420	0	4,420	0	4,420
Total South America	423,051	25,817	538,928	25,883	553,128	24,726
European Union:						
Austria	50	18,830	120	18,230	120	18,230
Belgium-Luxembourg	30,000	54,080	30,000	54,080	30,000	54,080
Denmark	2,500	16,000	2,500	16,000	2,500	16,000
Finland	207	3,100	207	3,100	0	0
France	19,274	28,594	18,867	31,151	19,000	30,300
Germany	91,710	263,077	102,592	247,066	85,000	257,000
Greece	85,389	19,344	90,000	21,000	100,000	17,000
Ireland	0	5,700	0	5,700	0	5,700
Italy	100,608	34,248	109,524	32,263	105,000	34,500
Netherlands	20,114	88,545	9,285	91,577	5,000	65,000
Portugal	5,100	15,000	4,700	17,000	4,700	17,000
Spain	25,983	58,445	35,911	47,250	27,000	45,100
Sweden	500	12,623	500	12,623	500	12,623
United Kingdom	10,944	108,427	10,510	102,666	10,500	119,000
Total European Union	392,379	726,013	414,716	699,706	389,320	691,533
Western Europe:						
Iceland	0	10	0	10	0	10
Malta & Gozo	150	1,000	150	1,000	150	1,000
Norway	900	8,000	900	8,000	900	8,000
Switzerland	9,000	41,070	9,000	41,070	9,000	41,070
Total Western Europe	10,050	50,080	10,050	50,080	10,050	50,080
Eastern Europe:						
Albania	1,000	32	1,000	32	1,000	32
Bulgaria	18,445	8,283	16,857	7,874	20,500	6,700
Croatia	4,776	4,262	4,500	4,800	5,400	3,400
Czech Republic	250	18,270	250	18,270	250	18,270
Hungary	2,000	8,200	2,000	10,300	1,800	10,000
Macedonia (Skopje)	23,000	0	23,000	0	23,000	0
Poland	4,468	23,873	5,400	30,500	6,000	30,500
Romania	0	13,000	0	13,000	0	13,000
Slovakia	700	2,100	700	2,100	700	2,100
Yugoslavia	500	4,500	500	4,500	500	4,500
Total Eastern Europe	55,139	82,520	54,207	91,376	59,150	88,502
Former Soviet Union:						
Azerbaijan; Republic of	8,000	350	8,300	600	8,300	600
Belarus	0	8,500	0	8,500	0	8,500
Estonia	0	1,800	0	1,800	0	1,800
Georgia; Republic of	1,500	0	1,500	0	1,500	0
Kazakhstan; Republic of	0	2,000	0	2,000	0	2,000
Kyrgyzstan; Republic of	30,513	1,000	31,500	1,000	31,500	1,000
Latvia	0	2,100	0	2,100	0	2,100
Lithuania	0	4,100	0	4,100	0	4,100
Moldova; Republic of	20,900	2,148	22,000	2,500	22,000	2,500
Russian Federation	20	285,000	690	307,500	0	307,500
Tajikistan; Republic of	2,000	0	2,000	0	2,000	0
Turkmenistan	0	2,180	0	2,180	0	2,180
Ukraine	0	45,000	0	45,000	0	45,000
Uzbekistan; Republic of	16,500	4,500	16,500	4,500	16,500	4,500
Total former Soviet Union	79,433	358,678	82,490	381,780	81,800	381,780
Middle East:						
Bahrain	0	175	0	175	0	175
Cyprus	100	1,400	100	1,400	100	1,400
Israel	0	4,900	0	4,900	0	4,900
Jordan	400	2,802	400	2,802	400	2,802
Kuwait	15	65	15	65	0	0
Lebanon	2,000	94	2,000	8	2,000	8
Oman	700	100	700	100	700	100
Saudi Arabia	30	1,600	30	1,600	30	1,600
Syria	5,000	0	5,000	0	5,000	0
Turkey	100,900	58,410	96,450	52,730	90,750	55,750
United Arab Emirates	0	3,333	0	704	0	704
Yemen	2,000	6,000	2,000	6,000	2,000	6,000
Total Middle East	111,145	78,879	106,695	70,484	100,980	73,439

See footnotes at end of table.

Table 2-54.—Tobacco, unmanufactured: International trade, 1999–2001—Continued[1]

Continent and country	1999		2000		2001	
	Exports	Imports	Exports	Imports	Exports	Imports
	Metric tons	*Metric tons*	*Metric tons*	*Metric tons*	*Metric tons*	*Metric tons*
North Africa:						
Algeria	0	18,000	0	18,000	0	18,000
Egypt	0	57,000	0	57,500	0	57,500
Libya	0	2,000	0	2,000	0	2,000
Morocco	0	8,000	0	8,000	0	8,000
Sudan	0	500	0	500	0	500
Tunisia	500	5,500	500	5,500	500	5,500
Total North Africa	500	91,000	500	91,500	500	91,500
Other Africa:						
Angola	0	1,500	0	1,500	0	1,500
Benin	0	475	0	475	0	475
Botswana	0	200	0	200	0	200
Burkina	0	1,100	0	1,100	0	1,100
Cameroon	1,500	2,400	1,500	2,400	1,500	2,400
Cape Verde	0	82	0	82	0	82
Central African Republ	370	350	370	350	370	350
Comoros	0	4	0	4	0	4
Congo (Brazzaville)	0	1,200	0	1,200	0	1,200
Cote d'Ivoire	0	3,050	0	3,000	0	3,000
Djibouti Afars-Issas	0	80	0	80	0	80
Ethiopia	0	400	0	400	0	400
Gabon	0	200	0	200	0	200
Gambia, The	0	370	0	370	0	370
Ghana	0	281	0	281	0	281
Guinea	0	100	0	100	0	100
Guinea-Bissau	0	100	0	100	0	100
Kenya	11,100	100	11,100	100	11,100	100
Liberia	0	315	0	315	0	315
Madagascar	500	1,000	500	1,000	500	1,000
Malawi	101,250	12,000	101,250	12,000	101,250	12,000
Mali	0	200	0	200	0	200
Mauritania	0	800	0	800	0	800
Mauritius	0	70	0	70	0	70
Mozambique	0	100	0	100	0	100
Niger	0	100	0	100	0	100
Nigeria	23	2,300	23	2,300	23	2,300
Reunion	160	200	160	200	160	200
Rwanda	0	600	0	600	0	600
Senegal	0	4,200	0	4,200	0	4,200
Seychelles	0	85	0	85	0	85
Sierra Leone	100	525	100	525	100	525
South Africa, Republic	16,465	14,346	18,450	18,830	15,000	15,800
Tanzania, United Republic	14,000	0	14,000	0	14,000	0
Togo	0	50	0	50	0	50
Uganda	5,000	0	5,000	0	5,000	0
Zaire	0	1,000	0	1,000	0	1,000
Zambia	2,000	20	2,000	20	2,000	20
Zimbabwe	182,072	30	185,000	25	185,000	25
Total Other Africa	334,540	49,933	339,453	54,362	336,003	51,332
Asia:						
Afghanistan	0	6	0	6	0	6
Bangladesh	2,500	1,107	2,500	1,037	2,500	240
Cambodia	0	192	0	0	0	0
China, Peoples Rep	113,005	37,831	139,917	56,686	118,930	59,640
Hong Kong	4,532	17,648	4,532	16,730	4,532	16,730
India	123,185	275	85,500	1,230	105,000	1,130
Indonesia	15,761	33,048	20,882	44,347	22,000	31,400
Japan	322	93,928	784	92,425	800	91,000
Korea, North	1,000	0	1,000	0	1,000	0
Korea, South	3,585	12,311	3,254	13,813	3,000	14,300
Macau	0	60	0	60	0	60
Malaysia	355	19,505	433	28,550	550	21,600
Nepal	0	8,000	0	8,000	0	8,000
Pakistan	6,500	482	5,200	565	6,000	600
Philippines	8,198	25,817	7,700	26,200	7,700	26,200
Singapore	6,200	17,700	6,200	17,700	6,200	17,700
Sri Lanka	300	213	500	1,038	500	1,038
Taiwan	43	7,954	0	6,842	0	8,000
Thailand	26,133	6,516	26,254	5,600	26,270	4,700
Total Asia	311,619	282,593	304,656	320,829	304,982	302,344
Oceania:						
Australia	595	13,426	1,012	14,208	1,012	14,208
Fiji	0	180	0	180	0	0
New Zealand	24	3,342	28	2,605	28	2,605
Papua New Guinea	0	1,000	0	1,000	0	1,000
Solomon Islands	0	150	0	150	0	150
Total Oceania	619	18,098	1,040	18,143	1,040	17,963
World Total	1,968,237	1,978,230	2,107,479	2,082,786	2,096,375	2,073,256

[1] Imports are general imports (actual arrivals).

FAS, Cotton, Oilseeds, Tobacco and Seeds Division, (202) 720–9496, Derived from official statistics of Foreign Governments, or other foreign sources.

Table 2-55.—Tobacco, unmanufactured: United States exports (domestic), by country of destination, total and by types, 1999–2001

Type and country of destination	1999	2000	2001 [1]
	Metric tons	*Metric tons*	*Metric tons*
Total Leaf: [2]			
Germany	32,604	39,039	43,018
Japan	27,347	28,868	23,419
Belgium	8,312	10,538	22,527
Russian Federation	8,103	9,252	12,358
Netherlands	29,242	8,926	9,782
Malaysia	5,408	6,473	8,357
Turkey	12,852	9,795	7,159
Switzerland	7,282	4,293	6,521
Denmark	6,778	6,922	5,537
France	2,522	2,477	5,210
Korea, Republic of	5,741	5,423	4,359
Taiwan	1,993	2,646	3,582
Spain	7,050	4,291	3,578
Thailand	3,104	3,333	3,513
Italy	6,841	7,181	2,822
Dominican Republic	4,936	5,398	2,669
Nigeria	1,614	2,802	2,550
Philippines	2,825	3,367	2,313
Portugal	590	1,885	1,939
Indonesia	1,256	1,483	1,652
Sweden	1,768	1,564	1,636
Australia	1,459	1,644	1,548
Norway	831	900	1,513
Sri Lanka	242	213	1,038
Canada	351	214	833
South Africa, Rep.	911	908	804
United Kingdom	4,055	3,333	704
Honduras	495	980	505
Israel	399	476	447
Egypt	382	547	417
Austria	0	613	408
Pakistan	432	250	365
Kazakhstan, Rep.	0	1	269
China, Peoples Rep	626	109	268
Lebanon	113	94	250
Bangladesh	202	307	237
Singapore	342	183	226
Greece	277	215	216
Romania	0	0	192
Lithuania	0	0	156
Trinidad and Tobag	24	23	151
India	0	91	135
Others	2,663	2,835	1,121
Total	191,972	179,892	186,304
Flue-cured:			
Germany	16,282	18,096	23,175
Japan	16,914	18,916	16,028
Belgium	2,653	6,835	9,568
Netherlands	11,063	3,017	5,032
Switzerland	3,716	2,918	4,599
Malaysia	3,953	3,587	4,576
Korea, Republic of	5,640	5,130	3,461
Turkey	5,879	5,355	2,941
Denmark	3,311	3,362	2,654
Taiwan	1,186	1,571	2,367
Thailand	1,463	1,627	1,818
Spain	144	0	1,480
Indonesia	822	1,040	1,412
Portugal	400	759	1,363
Italy	3,298	3,875	1,270
Norway	549	403	1,191
France	596	113	948
Australia	1,343	1,292	836
Canada	94	33	730
United Kingdom	2,867	2,593	676
Philippines	661	646	615
Sweden	520	489	409
Pakistan	432	250	365
Austria	0	496	313
China, Peoples Rep	550	0	238
Bangladesh	163	276	218
Greece	271	194	216
Russian Federation	252	145	146
Israel	112	83	86
Bulgaria	58	0	76
Singapore	188	176	60
Azerbaijan, Republic	0	0	58
Others	1,457	1,702	316
Total	86,838	84,980	89,241

See footnotes at end of table.

Table 2-55.—Tobacco, unmanufactured: United States exports (domestic), by country of destination, total and by types, 1999–2001—Continued

Type and country of destination	1999	2000	2001 [1]
	Metric tons	*Metric tons*	*Metric tons*
Burley:			
Germany	10,816	9,701	9,840
Belgium	3,684	2,122	8,180
Japan	5,849	6,267	4,948
Malaysia	1,126	2,345	3,286
Turkey	3,460	3,300	2,974
Spain	225	879	1,745
Philippines	2,161	2,651	1,672
Netherlands	13,738	2,142	1,635
Thailand	1,596	1,636	1,619
Denmark	1,487	1,808	1,381
France	558	484	1,020
Italy	1,564	1,107	737
Taiwan	0	6	708
Portugal	123	691	432
Switzerland	1,743	625	415
Sweden	186	231	242
Russian Federation	0	0	108
Israel	89	33	76
Hungary	100	176	73
Indonesia	119	106	68
Australia	80	0	26
China, Peoples Rep	76	100	19
Canada	168	0	14
Norway	24	47	13
Finland	0	9	9
Korea, Republic of	0	6	8
United Kingdom	322	94	6
Greece	0	19	0
Honduras	6	0	0
India	0	3	0
Ireland	0	0	0
Honduras	0	0	0
Others	101	60	0
Total	49,401	36,648	41,254
Dark-fired Kentucky and Tennessee:			
Netherlands	3,562	2,640	1,984
France	672	883	936
Sri Lanka	242	213	915
Nigeria	307	413	778
Italy	308	293	405
Japan	0	169	392
Turkey	410	0	378
Egypt	0	489	372
Korea, Republic of	96	287	326
Sweden	297	539	318
Lebanon	0	0	250
Switzerland	245	439	156
Norway	77	72	152
Belgium	287	82	145
India	0	70	135
Indonesia	18	203	100
Thailand	19	67	76
Australia	0	0	42
Cote d'Ivoire	0	0	50
Germany	0	29	32
Others	129	607	95
Total	6,669	7,495	8,037
VA Fire and Sun-Cured:			
Indonesia	0	0	71
Norway	153	214	11
Cayman Islands	0	0	1
Thailand	0	1	1
United Kingdom	1	7	1
Others	556	25	0
Total	710	247	85
Maryland:			
Germany	579	471	385
Switzerland	190	191	268
Italy	756	200	214
Belgium	245	35	184
Israel	163	146	141
Austria	0	0	18
Argentina	0	0	2
France	127	8	0
Others	169	383	0
Total	2,229	1,434	1,212

See footnotes at end of table.

Table 2-55.—Tobacco, unmanufactured: United States exports (domestic), by country of destination, total and by types, 1999–2001—Continued

Type and country of destination	1999	2000	2001 [1]
	Metric tons	*Metric tons*	*Metric tons*
Blackfat:			
Costa Rica	3	0	0
Spain	0	13	0
Total	3	13	0
Cigar Wrapper:			
Dominican Republic	2,304	1,739	1,426
Nicaragua	9	8	17
Sri Lanka	0	0	17
Honduras	37	8	14
Netherlands	7	10	10
Germany	0	36	4
Belgium	10	12	3
Guatemala	0	0	3
Jamaica	1	1	2
Panama	1	3	2
Chile	14	6	1
Others	89	37	0
Total	2,472	1,860	1,499
Stems Refuse:			
Russian Federation	6,849	7,156	6,468
Germany	4,626	7,629	5,215
Belgium	948	1,375	4,331
France	534	984	2,301
Japan	4,568	3,475	1,502
Denmark	1,870	1,676	1,385
Netherlands	731	753	1,102
Switzerland	1,384	104	829
Australia	20	344	633
Taiwan	557	686	441
Turkey	2,942	1,008	412
Kazakhstan, Republic	0	0	269
Malaysia	184	215	259
Spain	187	6	242
Romania	0	0	192
Singapore	153	0	163
Lithuania	0	0	156
Norway	9	164	145
Portugal	68	434	144
Israel	34	214	138
Peru	48	211	125
Sweden	496	105	119
Others	4,044	4,665	414
Total	30,252	31,204	26,985
Other unmanufactured, Tobacco:			
Spain	5,444	6,273	3,323
Germany	32	273	3,071
Nigeria	761	881	2,389
Russian Federation	227	1,001	1,932
Dominican Republic	1,836	1,459	1,742
South Africa, Republic	931	911	871
Honduras	961	408	624
Malaysia	268	146	325
Italy	189	287	271
Sweden	0	0	203
Canada	311	88	182
Netherlands	554	141	154
Turkey	146	162	130
Belgium	20	486	78
Dominica	0	0	76
Brazil	1	1	76
Philippines	0	0	69
Egypt	0	58	58
Jamaica	158	86	49
Taiwan	540	249	46
Others	981	494	361
Total	13,360	13,404	16,030

[1] Preliminary.

FAS, Cotton, Oilseeds, Tobacco and Seeds Division, (202) 720–9493. Compiled from U.S. Bureau of the Census records.

CHAPTER III

STATISTICS OF OILSEEDS, FATS, AND OILS

This chapter includes information on cottonseed, flaxseed, olive oil, peanuts, soybeans, margarine, and fats and oils. Most butter statistics are included in the chapter on dairy and poultry statistics. Lard data are mostly in the chapter on livestock.

Table 3-1.—Cottonseed: All cotton harvested area and cottonseed production, farm disposition, marketing year average price per ton received by farmers, and value, United States, 1993–2002

Year	Harvested area of all cotton	Cottonseed				
		Production	Farm disposition		Marketing year average price	Value of production
			Sales to oil mills	Other [1]		
	1,000 acres	*1,000 tons*	*1,000 tons*	*1,000 tons*	*Dollars/tons*	*1,000 dollars*
1993	12,783.3	6,343.2	3,927.4	2,415.8	113.00	714,389
1994	13,322.3	7,603.9	5,012.2	2,591.7	101.00	771,315
1995	16,006.7	6,848.7	4,439.1	2,409.6	106.00	731,005
1996	12,888.1	7,143.5	4,363.2	2,780.3	126.00	914,564
1997	13,406.0	6,934.6	4,182.4	2,752.2	121.00	835,371
1998	10,683.6	5,365.4	3,261.1	2,104.3	129.00	687,179
1999	13,424.9	6,353.5	3,340.2	3,013.3	89.00	559,157
2000	13,053.0	6,435.6	3,452.2	2,983.4	105.00	667,800
2001	13,827.7	7,452.2	3,860.9	3,591.3	90.50	667,348
2002 [2]	12,413.3	6,419.3	NA	NA	100.00	638,115

[1] Includes planting seed, feed, exports, inter-farm sales, shrinkage, losses, and other uses. [2] Preliminary. NA-not available.

NASS, Crops Branch, (202) 720–2127.

Table 3-2.—Cottonseed: Production and farm disposition, by State and United States, 2001–2002

State	Production			Farm disposition [1]				Used for planting [3]	
				Sales to oil mills		Other [4]			
	2000	2001	2002 [2]	2000	2001	2000	2001 [2]	2000	2001 [2]
	1,000 tons	*1,000 tons*	*1,000 tons*	*1,000 tons*	*1,000 tons*	*1,000 tons*	*1,000 tons*	*1,000 tons*	
AL	192.0	315.0	201.0	15.0	44.0	177.0	271.0	7.0	6.9
AZ	297.0	261.0	214.0	0.0	3.0	297.0	258.0	2.8	2.4
AR	556.0	708.0	640.0	450.0	568.0	106.0	140.0	10.3	9.2
CA	909.0	849.0	720.0	60.0	20.0	849.0	829.0	7.4	7.1
FL	38.0	53.0	30.0	29.0	38.0	9.0	15.0	1.4	1.2
GA	563.0	764.0	564.0	336.0	434.0	227.0	330.0	18.0	18.0
KS	8.9	11.4	30.0	8.9	7.3	0.0	4.1	0.4	0.7
LA	331.0	380.0	275.0	179.0	170.0	152.0	210.0	7.8	5.9
MS	662.0	877.0	751.0	595.0	820.0	67.0	57.0	16.2	14.0
MO	205.0	268.0	229.0	158.0	183.0	47.0	85.0	4.3	4.3
NM	39.7	47.8	43.3	6.3	15.6	33.4	32.2	0.8	0.6
NC	508.0	559.0	271.0	59.0	79.0	449.0	480.0	8.7	8.6
OK	58.0	80.0	79.0	55.0	66.0	3.0	14.0	3.0	2.9
SC	133.0	137.0	45.0	69.0	89.0	64.0	48.0	2.1	2.0
TN	289.0	351.0	309.0	179.0	296.0	110.0	55.0	4.3	4.1
TX	1,589.0	1,724.0	1,984.0	1,253.0	1,028.0	336.0	696.0	60.2	57.2
VA	57.0	67.0	34.0	0.0	0.0	57.0	67.0	1.0	0.9
US ..	6,435.6	7,452.2	6,419.3	3,452.2	3,860.9	2,983.4	3,591.3	155.7	146.0

[1] 2002 farm disposition not available. [2] Preliminary. [3] Included in 'other' farm disposition. Seed for planting is produced in crop year shown, but used in the following year. [4] Includes planting seed, feed, exports, inter-farm sales, shrinkage, losses, and other uses.

NASS, Crops Branch, (202) 720–2127.

Table 3-3.—Cottonseed: Marketing year average price per ton and value of production, by State and United States, crop of 2000–2002

State	Marketing year average price per ton			Value of production		
	2000	2001	2002 [1]	2000	2001	2002 [1]
	Dollars	*Dollars*	*Dollars*	*1,000 dollars*	*1,000 dollars*	*1,000 dollars*
AL	92.00	76.00	86.00	17,664	23,940	17,286
AZ	124.00	127.00	138.00	36,828	33,147	29,532
AR	96.50	79.00	93.00	53,654	55,932	59,520
CA	139.00	142.00	139.00	126,351	120,558	100,080
FL	100.00	71.50	81.50	3,800	3,790	2,445
GA	88.00	66.50	78.00	49,544	50,806	43,992
KS	90.50	69.50	92.00	805	792	2,760
LA	97.50	76.50	87.50	32,273	29,070	24,063
MS	100.00	76.50	79.00	66,200	67,091	59,329
MO	90.00	85.00	95.00	18,450	22,780	21,755
NM	120.00	112.00	106.00	4,764	5,354	4,590
NC	89.50	79.50	94.00	45,466	44,441	25,474
OK	90.50	69.50	84.00	5,249	5,560	6,636
SC	80.00	69.50	79.50	10,640	9,522	3,578
TN	101.00	85.00	95.50	29,189	29,835	29,510
TX	102.00	92.50	103.00	162,078	159,470	204,352
VA	85.00	78.50	94.50	4,845	5,260	3,213
US	105.00	90.50	100.00	667,800	667,348	638,115

[1] Preliminary.
NASS, Crops Branch, (202) 720–2127.

Table 3-4.—Cottonseed: Crushings, output of products and product prices, United States, 1992–2001

Year beginning August	Quantity crushed	Cottonseed products and prices			
		Oil		Cake and meal	
		Quantity	Price [1]	Quantity	Price [2]
	1,000 tons	*Million pounds*	*Cents per pound*	*1,000 tibs*	*Dollars per ton*
1992	3,629	1,137	30.1	1,533	161.78
1993	3,470	1,119	30.3	1,563	164.30
1994	3,947	1,312	29.2	1,830	112.02
1995	3,882	1,229	26.5	1,748	190.74
1996	3,860	1,216	25.6	1,752	192.00
1997	3,889	1,224	28.8	1,769	145.00
1998	2,719	832	27.3	1,232	110.00
1999	3,064	939	21.5	1,390	127.43
2000	2,753	847	16.0	1,338	143.35
2001	2,791	878	18.0	1,296	136.17

[1] Tanks, f.o.b. Valley Points. [2] 41 percent protein, solvent, Memphis.
ERS, Field Crops Branch, (202) 694–5300. Compiled from annual reports of the U.S. Department of Commerce.

Table 3-5.—Cottonseed oil and cottonseed cake and meal: United States exports by country of destination 1999/2000–2001/2002 (Marketing year October–September)

Continent and country of destination	Cottonseed oil [1]			Cottonseed cake and meal		
	1999/2000	2000/2001	2001/2002 [2]	1999/2000	2000/2001	2001/2002 [2]
	Metric tons	*Metric tons*	*Metric tons*	*Metric tons*	*Metric tons*	*Metric tons*
North and Central America; incl. Caribbean:						
Canada	35,466	43,043	29,949	1,459	316	820
Dominican Republic	0	2	13	0	0	0
El Salvador	1,798	0	3,550	0	0	0
Guatemala	0	0	520	119	0	120
Mexico	3,703	6,038	9,666	84,530	78,326	121,783
Total [3]	42,408	49,936	46,620	86,132	78,689	122,743
South America:						
Venezuela	0	0	0	2	0	0
Total [3]	32	2	40	211	0	76
Europe:						
Austria	0	0	0	14,618	6,612	0
Belgium and Luxembourg	0	0	0	774	552	375
Denmark	0	0	0	0	0	0
France	0	20	157	10	599	0
Germany	0	15	56	1,244	2,284	3,402
Netherlands	510	380	2,129	0	83	167
United Kingdom	7	451	69	80	818	870
Total [3]	1,076	1,133	2,493	20,685	13,649	7,004
Africa:						
Egypt	0	0	0	0	0	7,826
Total [3]	0	0	1,267	0	0	7,826
Asia:						
Hong Kong	0	17	0	0	0	8
Israel	936	0	140	0	0	0
Japan	4,821	5,912	7,219	754	240	300
Korea, Rep. of	652	116	0	2,175	1,600	42
Total [3]	6,681	13,120	8,892	3,132	1,845	634
Oceania	0	0	13	0	185	920
Grand total [4]	50,196	64,192	59,325	110,159	94,368	139,202

[1] Crude and refined (includes shipments under P.L. 480). [2] Preliminary. [3] Includes quantities exported to countries not shown. [4] May not add due to rounding.

FAS, Cotton, Oilseeds, Tobacco and Seeds Division, (202) 720–5710. (Compiled from reports of the U.S. Department of Commerce.)

Table 3-6.—Cottonseed: Area, yield, and production in specified countries, 1998/99–2000/2001 [1]

Continent and country	Area [2]			Yield per hectare			Production		
	1999/ 2000	2000/ 2001	2001/ 2002 [3]	1999/ 2000	2000/ 2001	2001/ 2002 [3]	1999/ 2000	2000/ 2001	2001/ 2002 [3]
	1,000 hectares	*1,000 hectares*	*1,000 hectares*	*Metric tons*	*Metric tons*	*Metric tons*	*1,000 metric tons*	*1,000 metric tons*	*1,000 metric tons*
North and Central America, and the Caribbean:									
Mexico	152	81	82	1.51	1.65	1.79	230	134	147
United States	5,433	5,282	5,596	1.06	1.11	1.21	5,764	5,838	6,761
Total	5,585	5,363	5,678	1.07	1.11	1.22	5,994	5,972	6,908
South America:									
Argentina	300	380	165	0.73	0.75	0.67	220	285	110
Brazil	752	853	750	1.48	1.78	1.75	1,110	1,522	1,310
Paraguay	195	300	150	0.62	0.52	0.53	121	155	80
Total	1,247	1,533	1,065	1.16	1.28	1.41	1,451	1,962	1,500
European Union:									
Greece	430	410	410	1.56	1.57	1.62	670	645	663
Spain	108	90	90	1.52	1.27	1.27	164	114	114
Total	539	501	501	1.55	1.52	1.55	835	760	778
Former Soviet Union:									
Azerbaijan	91	91	82	0.60	0.66	0.71	55	60	58
Kazakhstan	141	153	184	0.99	1.01	1.36	140	155	250
Kyrgyzstan	35	34	37	1.43	1.47	1.62	50	50	60
Tajikistan	248	239	245	0.71	0.79	1.04	175	190	255
Turkmenistan	475	475	515	0.88	0.69	0.66	420	329	339
Uzbekistan	1,500	1,415	1,430	1.43	1.36	1.47	2,150	1,920	2,100
Total	2,490	2,407	2,493	1.20	1.12	1.23	2,990	2,704	3,062
Africa:									
Cameroon	180	200	200	0.90	1.08	1.09	162	215	217
Egypt	273	223	315	1.37	1.37	1.37	375	306	432
Mali	482	275	520	0.75	0.73	0.89	363	200	464
South Africa, Republic of	75	68	41	0.71	0.99	0.85	53	67	35
Sudan	225	235	150	0.55	0.74	0.94	123	174	141
Tanzania, United Republic of	200	400	450	0.36	0.28	0.23	72	110	103
Zimbabwe	365	390	400	0.62	0.44	0.34	225	171	135
Total	1,800	1,791	2,076	0.76	0.69	0.74	1,373	1,243	1,527
Asia and the Middle East:									
China, Peoples Republic of	3,726	4,058	4,820	1.85	1.96	1.98	6,900	7,960	9,560
India	8,791	8,576	8,730	0.59	0.57	0.58	5,170	4,900	5,100
Iran	210	260	200	0.90	0.83	0.84	190	215	168
Pakistan	2,915	2,928	3,130	1.31	1.22	1.11	3,822	3,570	3,484
Syria	244	270	260	2.62	2.74	2.71	639	739	705
Turkey	719	654	693	1.75	1.79	1.83	1,260	1,170	1,267
Total	16,605	16,746	17,833	1.08	1.11	1.14	17,981	18,554	20,284
Oceania:									
Australia	464	505	420	2.00	2.10	2.30	1,047	1,062	968
Total	396	438	553	2.00	2.15	1.79	842	941	990
World Total [4]	31,354	31,429	32,834	1.05	1.07	1.11	32,926	33,526	36,595

[1] Split year includes Northern Hemisphere crop harvested in the late months of the first year shown combined with Southern Hemisphere and certain Northern Hemisphere crops harvested in the early months of the following year. [2] Harvested area. [3] Preliminary. [4] Includes all countries in USDA data base.

FAS, Cotton, Oilseeds, Tobacco and Seeds Division, (202) 720-5710. (Compiled from reports of the U.S. Department of Commerce.)

Table 3-7.—Flaxseed: Area, yield, production, disposition, and value, United States, 1993–2002

Year	Area planted	Area harvested	Yield per harvested acre	Production	Marketing year average price per bushel received by farmers	Value of production
	1,000 acres	*1,000 acres*	*Bushels*	*1,000 bushels*	*Dollars*	*1,000 dollars*
1993	206	191	18.2	3,482	4.25	14,847
1994	178	171	17.1	2,922	4.63	13,590
1995	165	147	15.0	2,212	5.20	11,481
1996	96	92	17.4	1,602	6.37	10,197
1997	151	146	16.6	2,420	5.81	14,046
1998	336	329	20.4	6,708	5.05	33,809
1999	387	381	20.6	7,864	3.79	30,098
2000	536	517	20.8	10,730	3.30	35,569
2001	585	578	19.8	11,455	4.29	49,004
2002 [1]	785	704	17.9	12,569	5.80	72,944

[1] Preliminary.
NASS, Crops Branch, (202) 720–2127.

Table 3-8.—Flaxseed: Supply and disappearance, United States, 1992–2001

Year beginning June	Supply				Disappearance			
	Stocks June 1	Production	Imports	Total	Total used for seed	Exports	Crushings [1]	Total domestic disappearance [2]
	1,000 bushels	*1,000 bushels*	*1,000 bushels*	*1,000 bushels*	*1,000 bushels*	*1,000 bushels*	*1,000 bushels*	*1,000 bushels*
1992	1,556	3,288	6,035	10,879	167	230	8,600	9,104
1993	1,545	3,482	5,118	10,145	144	126	8,650	8,864
1994	1,155	2,922	6,005	10,082	134	72	8,550	8,840
1995	1,170	2,212	7,248	10,630	78	119	9,000	9,281
1996	1,230	1,602	8,390	11,222	122	144	10,000	10,625
1997	453	2,420	9,636	12,509	272	174	10,500	11,154
1998	1,181	6,708	5,992	13,881	313	476	10,600	11,247
1999	2,158	7,864	6,629	16,651	434	215	11,500	14,669
2000	1,767	10,730	2,850	15,347	474	1,015	12,000	13,024
2001	1,308	11,455	1,903	14,666	684	2,386	10,000	11,387

[1] From domestic and imported seed. [2] Total supply minus exports and stocks June 1 of following year.
ERS, Field Crops Branch, (202) 694–5300.

Table 3-9.—Flaxseed: Area, yield, and production, by States, 2000–2002

State	Area planted			Area harvested			Yield per harvested acre			Production		
	2000	2001	2002	2000	2001	2002	2000	2001	2002	2000	2001	2002
	1,000 acres	*1,000 acres*	*1,000 acres*	*1,000 acres*	*1,000 acres*	*1,000 acres*	*Bushels*	*Bushels*	*Bushels*	*1,000 bushels*	*1,000 bushels*	*1,000 bushels*
MN	10	4	6	9	4	5	22.0	13.0	18.0	198	52	90
MT	16	14	17	14	12	15	14.0	15.0	13.0	196	180	195
ND	490	550	750	475	545	680	21.0	20.0	18.0	9,975	10,900	12,240
SD	20	17	12	19	17	4	19.0	19.0	11.0	361	323	44
US	536	585	785	517	578	704	20.8	19.8	17.9	10,730	11,455	12,569

NASS, Crops Branch, (202) 720–2127.

Table 3-10.—Flaxseed: Marketing year average price and value of production, by States, crop of 2000, 2001, and 2002

State	Marketing year average price per bushel			Value of production		
	2000	2001	2002 [1]	2000	2001	2002 [1]
	Dollars	*Dollars*	*Dollars*	*1,000 dollars*	*1,000 dollars*	*1,000 dollars*
MN	3.39	4.40	5.70	671	229	513
MT	4.00	4.35	6.15	784	783	1,199
ND	3.31	4.29	5.80	33,017	46,761	70,992
SD	3.04	3.81	5.45	1,097	1,231	240
US	3.30	4.29	5.80	35,569	49,004	72,944

[1] Preliminary.

NASS, Crops Branch, (202) 720–2127.

Table 3-11.—Flaxseed: Support operations, United States, 1993–2002

Marketin year beginning June 1	Income support payment rates per bushels [1]	Program price levels per bushel		Put under loan [4]		Acquired by CCC under loan program [5]	Owned by CCC at end of marketing year
		Loan [2]	Target [3]	Quantity	Percentage of production		
	Dollars	*Dollars*	*Dollars*	*1,000 bushels*	*Percent*	*1,000 bushels*	*1,000 bushels*
1993/94	....	4.98	....	105.4	3.0	30.4	0
1994/95	....	4.87	....	185.7	6.4	58.9	4
1995/96	....	4.87	....	51.8	2.3	8.9	0
1996/97	....	5.07	....	39.3	2.5	0.0	0
1997/98	....	5.21	....	105.4	4.4	0.0	0
1998/99	....	5.21	....	516.1	7.7	1.8	0
1999/2000	0.1212	5.21	....	430.4	5.5	225.0	5
2000/2001	0.2332	5.21	....	353.6	3.3	151.8	0
2001/2002	....	5.21	....	105.4	0.9	35.7	1
2002/2003	0.4479	3.91	5.49	....	....	....	....

[1] Oilseeds payment rates for 1999/2000 were calculated according to the provisions of the Agriculture, Rural Development, Food and Drug Admisistration, and Related Agencies Appropriations Act, 2000. Rates for 2000/2001 were calculated according to the provisions of the Agricultural Risk Protection Act of 2000, and included supplemental oilseeds payment rates. Payment rates for the 2002/2003 and subsequent crops are calculated according to the provisions of the Direct and counter-cyclical Payment program, following enactment of the Farm Security and Rural Investment Act of 2002 (2002 Act). Payment rates are rounded to the nearest tenth of a cent. [2] The national average loan rate was also known as the price support rate prior to enactment of the Federal Agricultural Improvement and Reform Act of 1996 (1996 Act). [3] Target prices for the 2002/2003 and subsequent crops were enacted with the 2002 Act. [4] Includes loans made. Does not include quantity on which loan deficiency payments were made. [5] Acquisition of all loans forfeited during the marketing year including loans made in previous year.

FSA, Feed Grains & Oilseeds Analysis Group, (202) 720–2711.

Table 3-12.—Flaxseed and linseed oil and meal: Average price Minneapolis, 1992–2001

Year [1]	Average price received by farmers per bushel	Minneapolis	
		Oil, per pound [2]	Meal, per ton [3]
	Dollars	*Cents*	*Dollars*
1992	4.12	31.5	133.60
1993	4.25	31.8	139.55
1994	4.63	33.7	91.96
1995	5.25	36.5	133.60
1996	6.21	36.0	169.75
1997	5.75	37.8	131.40
1998	5.25	37.5	91.63
1999	3.79	37.8	93.77
2000	3.30	36.0	116.23
2001	4.29	36.0	121.29

[1] Year beginning June 1. [2] Raw oil in tank cars. [3] Bulk carlots, 34 percent protein.

ERS, Field Crops Branch, (202) 694–5300.

Table 3-13.—Flaxseed and products: Flaxseed crushed; production, imports, and exports of linseed oil, cake, and meal; and June 1 stocks of oil, United States, 1993–2002

Year beginning June	Total flaxseed crushed	Linseed oil			Linseed cake and meal		
		Stocks June 1	Production	Exports	Production	Imports for consumption	Exports
	1,000 bushels	*Million pounds*	*Million pounds*	*Million pounds*	*1,000 tons*	*1,000 tons*	*1,000 tons*
1993	8,650	54	174	3	160	2	49
1994	8,550	63	172	24	158	5	58
1995	9,000	45	180	26	167	2	35
1996	10,000	50	200	66	185	13	44
1997	10,500	35	205	58	189	15	19
1998	10,600	42	207	63	191	4	26
1999	11,500	48	224	76	207	4	26
2000	12,000	49	234	100	216	4	27
2001	12,000	43	195	90	180	4	28
2002	10,000	45	198	92	183	4	29

ERS, Field Crops Branch, (202) 694–5300.

Table 3-14.—Sunflowerseed, sunflowerseed oil, and sunflowerseed cake and meal: United States exports by country of destination 1999/2000–2001/2002 [1]

Continent and country of destination	Sunflowerseed			Sunflowerseed oil [2]			Sunflowerseed cake and meal		
	1999/2000	2000/2001	2001/2002 [3]	1999/2000	2000/2001	2001/2002 [3]	1999/2000	2000/2001	2001/2002 [3]
	Metric tons	*Metric tons*	*Metric tons*	*Metric tons*	*Metric tons*	*Metric tons*	*Metric tons*	*Metric tons*	*Metric tons*
North and Central America; incl. Caribbean:									
Canada	17,280	16,157	29,172	24,037	22,988	24,327	1,957	1,424	2,165
Mexico	5,343	12,190	6,065	169,578	39,095	17,761	3,921	2,731	2,451
Total [4]	22,813	28,876	35,673	199,541	66,971	44,549	5,939	4,155	4,616
South America:									
Venezuela	25	151	2	25	0	26	0	0	0
Total [4]	357	168	43	699	1,312	212	0	0	0
Europe:									
France	15,662	2,447	2,204	60	0	0	0	0	0
Germany	55,291	59,475	73,165	0	49	4	13	0	0
Netherlands	9,089	16,032	34,905	2,700	57,546	22,914	7,282	0	0
Spain	19,840	25,573	28,097	0	5,812	1,499	0	0	0
Total [4]	149,948	128,540	169,060	6,158	65,209	34,665	14,871	3,933	21,024
Former Soviet Union [5]	708	993	1,644	1,000	849	0	0	0	0
Africa:									
Algeria	17	12,562	160	32,593	62,700	47,897	0	0	0
Egypt	150	100	81	21,827	5,924	12,499	0	0	0
Total [4]	407	13,951	2,065	54,420	68,624	64,894	0	0	0
Asia:									
Japan	3,411	2,955	2,503	6,582	5,457	6,143	0	0	20
Korea, Rep. of	786	1,033	1,393	115	350	309	0	0	0
Turkey	7,624	4,794	7,557	0	12,574	15,697	0	0	0
Total [4]	29,555	26,818	24,800	23,993	44,129	61,114	0	20	20
Oceania	998	1,553	1,513	173	14	4	0	0	0
Grand total [6]	204,786	200,900	234,798	285,982	247,109	205,438	20,810	8,109	25,661

[1] For sunflowerseed, year begins September 1; for sunflowerseed oil cake and meal, year begins October 1. [2] Crude and refined oil. [3] Preliminary. [4] Includes quantities exported to countries not shown. [5] Former Soviet Union; includes all 15 Republics of the USSR. [6] May not add due to rounding.

FAS, Cotton, Oilseeds, Tobacco and Seeds Division, (202) 720–5710. Compiled from reports of the U.S. Department of Commerce.

Table 3-15.—Peanuts: Area, yield, production, disposition, marketing year average price per pound received by farmers, and value, United States, 1993–2002

Year	Area planted	Peanuts for nuts				
		Area harvested	Yield per acre	Production [1]	Marketing year average	Value of production
	1,000 acres	*1,000 acres*	*Pounds*	*1,000 pounds*	*Cents*	*1,000 dollars*
1993	1,733.5	1,689.8	2,008	3,392,415	30.4	1,030,904
1994	1,641.0	1,618.5	2,624	4,247,455	28.9	1,229.012
1995	1,537.5	1,517.0	2,282	3,461,475	29.3	1,013,323
1996	1,401.5	1,380.0	2,653	3,661,205	28.1	1,029,774
1997	1,434.0	1,413.8	2,503	3,539,380	28.3	1,002,703
1998	1,521.0	1,467.0	2,702	3,963,440	28.4	1,125,919
1999	1,534.5	1,436.0	2,667	3,829,490	25.4	971,608
2000	1,536.8	1,336.0	2,444	3,265,505	27.4	896,097
2001	1,541.2	1,411.9	3,029	4,276,704	23.4	1,000,512
2002 [2]	1,358.0	1,296.7	2,561	3,320,490	17.9	594,426

[1] Estimates comprised of quota and non-quota peanuts. [2] Preliminary.
NASS, Crops Branch, (202) 720–2127.

Table 3-16.—Peanuts, farmers' stock: Stocks, production, and quantity milled, United States, 1992–2001

Year beginning August	Stocks Aug. 1 [1]	Production harvested for nuts [1]	Total supply	Milled [1 2]
	1,000 pounds	*1,000 pounds*	*1,000 pounds*	*1,000 pounds*
1992	117,121	4,284,416	4,401,537	4,097,584
1993	61,252	3,392,415	3,453,667	3,147,078
1994	67,428	4,247,455	4,314,883	3,222,198
1995	48,574	3,461,475	3,510,049	2,558,954
1996	66,392	3,661,205	3,727,597	2,919,054
1997	22,714	3,539,380	3,562,094	2,899,138
1998	27,284	3,963,440	3,990,724	3,652,670
1999	158,646	3,829,490	3,988,136	3,703,266
2000	139,210	3,265,505	3,404,715	3,254,950
2001	116,994	4,276,704	4,393,698	3,663,304

[1] Net weight basis. [2] Includes peanuts milled for seed.
NASS, Crops Branch, (202) 720–2127.

Table 3-17.—Peanuts: Crushings, and oil and meal stocks, production, and foreign trade, United States, 1992–2001

Year beginning August	Peanuts crushed (shelled basis)	Peanut oil				Peanut cake and meal	
		Stocks Aug. 1 [1]	Production of crude	Imports	Exports [2]	Stocks Aug. 1 [3]	Production
	1,000 pounds	*1,000 pounds*	*1,000 pounds*	*1,000 pounds*	*1,000 pounds*	*1,000 pounds*	*1,000 pounds*
1992	669,942	15,572	285,904	460	59,338	9,804	377,301
1993	503,674	19,779	212,216	11,421	61,145	11,395	292,093
1994	738,221	4,272	314,189	3,773	96,866	13,405	415,394
1995	751,281	19,763	320,909	4,678	108,146	14,910	420,919
1996	520,413	26,992	220,877	14,445	20,625	7,212	294,590
1997	409,249	22,936	175,853	8,118	13,097	8,291	228,276
1998	345,825	29,297	145,254	72,534	10,516	24,004	192,425
1999	536,164	6,770	228,839	12,835	17,519	2,847	291,491
2000	411,558	10,881	178,523	79,119	13,824	4,721	230,099
2001	521,173	3,812	230,791	38,632	8,386	3,800	296,874

[1] Crude plus refined. [2] Reported as edible peanut oil and crude peanut oil; in this tabulation added without converting. [3] Holding at producing mills only.
NASS, Crops Branch, (202) 720–2127, ERS, and Bureau of the Census.

Table 3-18.—Cleaned peanuts (roasting stock): Supply and disposition, United States, 1992–2001

Year beginning August	Supply				Disposition		
	Stocks Aug. 1	Production	Imports	Total	Exports	Domestic disappearance	
						Total	Per capita
	1,000 pounds	*1,000 pounds*	*1,000 pounds*	*1,000 pounds*	*1,000 pounds*	*1,000 pounds*	*Pounds*
1992	69,018	274,285	0	343,303	136,391	135,024	0.53
1993	71,938	186,232	9	258,179	83,438	141,056	0.54
1994	33,685	296,057	7,308	337,050	102,351	164,079	0.63
1995	70,620	188,954	8,628	268,202	73,937	163,283	0.62
1996	30,981	264,337	6,988	302,306	80,607	167,928	0.63
1997	53,771	229,912	5,907	289,590	75,154	183,868	0.68
1998	30,568	277,552	5,320	313,440	58,864	181,468	0.66
1999	73,108	235,756	5,341	314,205	53,406	200,877	0.72
2000	59,922	228,185	7,622	295,729	41,121	216,236	0.77
2001	38,372	245,783	0	284,155	39,202	179,986	0.63

NASS, Crops Branch, (202) 720–2127, and ERS. Foreign trade from the Bureau of the Census.

Table 3-19.—Shelled peanuts (all grades): Supply, exports, and quantity crushed, United States, 1992–2001

Year beginning August	Supply						Exports	Crushed
	Stocks Aug. 1		Production		Imports	Total		
	Edible	Oil stock	Edible	Oil stock				
	1,000 pounds	*1,000 pounds*	*1,000 pounds*	*1,000 pounds*	*1,000 pounds*	*1,000 pounds*	*1,000 pounds*	*1,000 pounds*
1992	595,022	57,829	2,376,782	533,641	2,000	3,565,274	611,250	669,942
1993	871,207	43,760	1,748,734	425,710	1,420	3,090,831	352,500	503,674
1994	679,639	42,054	1,741,824	511,635	55,385	3,030,537	583,142	738,221
1995	752,814	58,188	1,253,451	491,818	108,303	2,664,574	564,021	751,281
1996	370,431	126,318	1,692,581	305,674	95,041	2,590,045	440,438	520,413
1997	498,954	41,000	1,694,016	290,882	101,792	2,626,644	455,264	409,249
1998	580,370	14,091	2,227,037	310,459	112,643	3,249,600	377,171	345,825
1999	855,572	16,587	2,157,828	448,875	129,714	3,608,576	505,559	536,164
2000	707,554	70,103	1,939,736	337,324	147,201	3,201,918	353,732	411,558
2001	693,209	14,463	2,090,776	485,092	150,276	3,433,816	495,516	521,173

NASS, Crops Branch, (202) 720–2127, and ERS. Foreign trade from the U.S. Bureau of the Census.

Table 3-20.—Peanuts: Shelled (raw basis) by types, used in primary products and apparent disappearance of peanuts (cleaned in shell), United States, 1992–2001

Type, and year beginning August	Shelled uses					Apparent disappearance (cleaned in shell)[2]
	Peanut butter[1]	Snack	Candy	Other	Total	
	1,000 pounds	*1,000 pounds*	*1,000 pounds*	*1,000 pounds*	*1,000 pounds*	*1,000 pounds*
Virginia:						
1992	92,355	124,875	49,223	9,442	275,895	....
1993	63,270	99,381	44,889	15,101	222,641	....
1994	51,354	97,389	26,857	15,316	190,916	....
1995	71,310	93,041	25,176	13,656	203,183	....
1996	64,274	91,882	24,158	12,852	193,166	....
1997	59,228	80,309	28,428	14,135	182,100	....
1998	57,864	99,401	36,178	3,492	196,935	....
1999	73,926	100,384	23,173	3,321	200,804	....
2000	102,050	100,650	19,101	3,271	225,072	....
2001	106,573	97,046	26,640	3,097	233,356	....
Runner:						
1992	674,962	203,732	259,498	14,583	1,152,775	....
1993	635,047	227,286	298,325	18,738	1,179,396	....
1994	644,711	185,377	302,697	19,325	1,152,110	....
1995	634,350	169,142	304,285	15,942	1,123,719	....
1996	634,387	176,851	318,924	19,185	1,149,347	....
1997	676,839	206,718	302,791	20,598	1,206,946	....
1998	670,705	234,486	321,838	17,719	1,244,748	....
1999	690,564	278,440	315,467	15,922	1,300,393	....
2000	643,229	247,739	320,304	15,884	1,227,156	....
2001	702,454	250,079	303,668	13,575	1,269,776	....
Spanish:						
1992	30,593	24,168	19,603	956	75,320	....
1993	28,689	22,200	19,204	2,462	72,555	....
1994	13,758	18,782	20,076	2,213	54,829	....
1995	22,416	14,906	21,202	2,417	60,941	....
1996	28,870	21,369	17,764	1,788	69,791	....
1997	24,163	19,881	19,798	738	64,580	....
1998	16,137	15,919	22,161	920	55,137	....
1999	7,614	15,297	16,313	984	40,208	....
2000	7,960	13,127	16,205	843	38,135	....
2001	9,900	13,791	19,421	612	43,724	....
All types:						
1992	797,910	352,775	328,324	24,981	1,503,990	2,648,303
1993	727,006	348,867	362,418	36,301	1,474,592	2,307,718
1994	709,823	301,548	349,630	36,854	1,397,855	2,164,150
1995	728,076	277,089	350,663	32,015	1,387,843	2,059,522
1996	727,531	290,102	360,846	33,825	1,412,304	1,998,270
1997	760,230	306,908	351,017	35,471	1,453,626	1,930,391
1998	744,706	349,806	380,177	22,131	1,496,820	2,259,798
1999	772,104	394,121	354,953	20,227	1,541,405	2,701,205
2000	753,239	361,516	355,610	19,998	1,490,363	2,347,528
2001	818,927	360,916	349,729	17,284	1,546,856	2,586,177

[1] Excludes peanut butter made by manufacturers for own use in candy. Includes peanut butter used in spreads, sandwiches, and cookies. [2] Apparent disappearance represents stocks beginning of year plus production, minus stocks at end of year.

NASS, Crops Branch, (202) 720–2127, and ERS.

Table 3-21.—Peanuts: Area, yield, and production, by States, 2000–2002

State	Area planted			Peanuts for nuts								
				Area harvested			Yield per harvested acre			Production[2]		
	2000	2001	2002[1]	2000	2001	2002[1]	2000	2001	2002[1]	2000	2001	2002[1]
	1,000 acres	*1,000 acres*	*1,000 acres*	*1,000 acres*	*1,000 acres*	*1,000 acres*	*Pounds*	*Pounds*	*Pounds*	*1,000 pounds*	*1,000 pounds*	*1,000 pounds*
AL	190.0	200.0	190.0	182.0	199.0	185.0	1,490	2,675	2,050	271,180	532,325	379,250
FL	94.0	90.0	96.0	86.0	82.0	86.0	2,485	3,050	2,300	213,710	250,100	197,800
GA	494.0	515.0	510.0	492.0	514.0	505.0	2,700	3,330	2,600	1,328,400	1,711,620	1,313,000
NM	27.3	22.2	18.0	26.0	22.2	18.0	2,115	3,020	3,000	54,990	67,044	54,000
NC	123.0	123.0	101.0	123.0	122.5	100.0	2,750	2,910	2,100	338,250	356,475	210,000
OK	97.0	80.0	60.0	67.0	77.0	57.0	1,800	2,570	2,800	120,600	197,890	159,600
SC	10.5	11.0	10.0	10.0	10.2	8.7	2,950	3,000	2,200	29,500	30,600	19,140
TX	425.0	425.0	315.0	275.0	310.0	280.0	2,540	2,890	3,100	698,500	895,900	868,000
VA	76.0	75.0	58.0	75.0	75.0	57.0	2,805	3,130	2,100	210,375	234,750	119,700
US	1,536.8	1,541.2	1,358.0	1,336.0	1,411.9	1,296.7	2,444	3,029	2,561	3,265,505	4,276,704	3,320,490

[1] Preliminary. [2] Estimates comprised of quota and non-quota peanuts.

NASS, Crops Branch, (202) 720–2127.

Table 3-22.—Peanuts: Marketing year average price, and value of production, by States, crop of 2000, 2001, and 2002

State	Marketing year average price per pound			Value of production		
	2000	2001	2002 [1]	2000	2001	2002 [1]
	Dollars	*Dollars*	*Dollars*	*1,000 dollars*	*1,000 dollars*	*1,000 dollars*
AL	0.266	0.242	0.168	72,134	128,823	63,714
FL	0.300	0.215	0.177	64,113	53,772	35,011
GA	0.286	0.227	0.175	379,922	388,538	229,775
NM	0.320	0.260	0.190	17,597	17,431	10,260
NC	0.273	0.248	0.208	92,342	88,406	43,680
OK	0.293	0.274	0.170	35,336	54,222	27,132
SC	0.254	0.259	0.166	7,493	7,925	3,177
TX	0.246	0.226	0.182	171,831	202,473	157,976
VA	0.263	0.251	0.198	55,329	58,922	23,701
US	0.274	0.234	0.179	896,097	1,000,512	594,426

[1] Preliminary.

NASS, Crops Branch, (202) 720–2127.

Table 3-23.—Peanuts, farmers' stock: Price-support operations, United States, 1992–2002

Marketing year beginning August 1	Income support payment rates per pound [1]	Price support level per pound		Put under support [4] [5]		Owned by CCC at end of marketing year [5]
		Quota [2]	Additional [3]	Quantity	Percentage of production	
	Cents	*Cents*	*Cents*	*Million pounds*	*Percent*	*Million pounds*
1993/94		33.8	6.6	324	9.6	0
1994/95		33.9	6.6	820	19.3	0
1995/96		33.9	6.6	818	23.6	0
1996/97		30.5	6.6	320	8.7	0
1997/98		30.5	6.6	417	11.8	0
1998/99		30.5	6.6	802	20.2	0
1999/2000		30.5	6.6	459	12.0	0
2000/2001		30.5	6.6	450	13.9	0
2001/2002		30.5	6.6	937	21.9	0
		Loan [6]	Target [6]			
2002/2003	1.80	17.75	24.75	....	....	....

[1] Enactment of the Farm Security and Rural Investment Act of 2002 (2002 Act) repealed the peanut quota marketing program; and established payment rates for the 2002/2003 and subsequent crops according to the provisions of the Direct Payment program. [2] Quota peanuts are those peanuts grown within the farm poundage quota. [3] Additional peanuts are those peanuts grown in excess of the quota. [4] Includes loans made and direct purchases. [5] Includes shelled peanuts converted to farmers' stock basis. [6] Loan rates and target prices for the 2002/03 and subsequent crops were enacted with the 2002 Act.

FSA, Tobacco and Peanuts Division, (202) 720–5291.

Table 3-24.—Peanuts: Simple average of monthly f.o.b. price per pound of cleaned and shelled peanuts by approximate crop years, 1992–2001 [1]

Classification	1992	1993	1994	1995	1996	1997	1998	1999	2000	2001
Southeastern area:	*Cents*	*Cents*	*Cents*	*Cents*	*Cents*	*Cents*	*Cents*	*Cents*	*Cents*	*Cents*
Georgia, Alabama, and Florida:										
Shelled:.										
Runner Jumbo	68¾	63⅞	68⅛	67½	60¾	62	59	60¾	62	59½
Runner Medium	64¾	60¼	62½	66	59½	60⅝	56	58¼	59½	57½
Runner U.S. Splits	63¾	59½	62¾	65½	57	57¾	56¼	56¾	58½	56¾
Virginia-North Carolina:										
Clean unshelled Virginias:										
Fancy	47¾	50¾	72¾	57¼	49¾	54⅛	45¾	49⅛	50⅞	51
Shelled Virginias:										
Extra large	68⅛	74¼	77⅛	68½	66½	69⅛	62	66	70½	66½
Medium	64⅝	61¼	75½	67½	60¼	64⅝	59¼	60¾	60⅝	59¼
No. 2 with 70% splits	59⅝	59¼	61¾	60½	55½	56⅜	53¾	53⅞	55⅛	55
Southwestern area:										
Texas and Oklahoma:										
Shelled:										
Spanish No. 1	64⅝	61⅝	62⅞	65½	60¼	60⅝	72	79½	70⅞	62¾
Spanish U.S. Splits	63½	61½	64⅞	66¼	58	58¼	60½	57¼	59	NA
New Mexico:										
Handpicked unshelled Valencias [2]	55½	58	65	61¼	62½	65⅞	63½	NA	NA	NA

[1] Crop year begins about Oct. 1 in the Virginia-North Carolina area; in August, in the Southeastern and the Southwestern States. Prices are for shipment within 6 months. [2] No U.S. grade standards apply to Valencia peanuts. NA-not available.

AMS, Fruit and Vegetable Division, Market News Branch (229) 228–1208.

Table 3-25.—Peanuts:[1] Area, yield, and production in specified countries and the world, 1999–2000/2001–2002

Continent and country	Area[2] 1999/2000	Area[2] 2000/2001	Area[2] 2001/2002[3]	Yield per hectare 1999/2000	Yield per hectare 2000/2001	Yield per hectare 2001/2002[3]	Production 1999/2000	Production 2000/2001	Production 2001/2002[3]
	1,000 hectares	*1,000 hectares*	*1,000 hectares*	*Metric tons*	*Metric tons*	*Metric tons*	*1,000 metric tons*	*1,000 metric tons*	*1,000 metric tons*
North America and Caribbean:									
Mexico	94	78	78	1.38	1.72	1.73	130	134	135
United States	581	541	571	2.99	2.74	3.40	1,737	1,481	1,940
Total[4]	675	619	649	2.77	2.61	3.20	1,867	1,615	2,075
South America:									
Argentina	218	251	218	1.93	1.57	1.51	420	395	330
Brazil	104	102	97	1.65	1.92	2.01	172	196	195
Total[4]	322	353	315	1.84	1.67	1.67	592	591	525
Nicaragua	23	23	18	2.96	2.96	3.56	68	68	64
Spain	1	1	1	1.00	1.00	1.00	1	1	1
Turkey	35	35	35	2.57	2.57	2.57	90	90	90
Africa:									
Benin	125	100	100	0.81	0.81	0.83	101	81	83
Burkina	215	215	215	0.95	0.95	0.95	205	205	205
Cameroon	320	320	320	0.30	0.30	0.31	95	95	100
Central African Republic	100	100	100	1.10	1.16	1.22	110	116	122
Chad	385	385	385	0.97	0.97	0.97	372	372	372
Cote d'Ivoire	145	145	145	0.99	1.00	1.00	144	145	145
Egypt	59	60	60	3.07	3.12	3.12	181	187	187
Gambia,The	99	118	131	1.24	1.17	1.17	123	138	153
Ghana	185	190	200	1.03	1.11	1.10	190	210	220
Guinea	177	177	177	1.03	1.03	1.03	182	182	182
Mali	145	145	145	0.97	0.97	0.97	140	140	140
Morocco	22	24	24	1.91	1.88	1.88	42	45	45
Mozambique	248	248	286	0.46	0.46	0.38	115	115	110
Niger	250	280	280	0.42	0.39	0.39	104	110	110
Nigeria	1,200	1,210	1,220	1.21	1.21	1.22	1,450	1,470	1,490
Senegal	620	1,030	920	1.23	0.97	0.98	764	1,003	903
South Africa, Republic of	83	160	92	1.99	1.16	1.30	165	186	120
Sudan	550	550	550	0.67	0.67	0.67	370	370	370
Tanzania	117	117	117	0.64	0.64	0.64	75	75	75
Togo	65	65	65	0.54	0.54	0.54	35	35	35
Uganda	196	205	205	0.70	0.63	0.63	137	130	130
Zaire	509	491	491	0.78	0.78	0.78	396	382	382
Zambia	135	135	135	0.41	0.41	0.41	55	55	55
Zimbabwe	268	276	270	0.72	0.62	0.22	194	172	60
Total[4]	6,218	6,746	6,633	0.92	0.89	0.87	5,745	6,019	5,794
Asia:									
Bangladesh	34	35	35	1.24	1.14	1.14	42	40	40
Burma	490	530	530	1.15	1.21	1.21	562	640	640
China, Peop.	4,268	4,856	4,990	2.96	2.97	2.89	12,639	14,437	14,415
India	8,000	8,100	8,200	0.69	0.70	0.93	5,500	5,700	7,600
Indonesia	678	650	635	1.50	1.60	1.56	1,020	1,040	990
Japan	11	11	10	2.36	2.45	2.30	26	27	23
Korea, Rep.	8	5	4	1.75	1.80	1.75	14	9	7
Malawi	178	176	175	0.74	0.70	0.71	131	123	125
Malaysia	1	1	1	5.00	5.00	5.00	5	5	5
Pakistan	93	82	82	1.06	1.15	1.15	99	94	94
Philippines	26	27	27	1.00	1.00	1.00	26	27	27
Taiwan	39	39	39	1.64	1.67	1.67	64	65	65
Thailand	86	90	90	1.59	1.50	1.50	137	135	135
Vietnam	248	244	270	1.28	1.45	1.31	318	353	353
Total[4]	14,160	14,846	15,088	1.45	1.53	1.63	20,583	22,695	24,519
Australia	20	21	21	2.00	1.95	1.95	40	41	41
World total[4]	21,454	22,644	22,760	1.35	1.37	1.45	28,986	31,120	33,109

[1] Peanuts in the shell. Split year includes Northern Hemisphere crop harvested in the late months of the first year shown combined with Southern Hemisphere and certain Northern Hemisphere crops harvested in the early months of the following year. [2] Harvested area as far as possible. [3] Preliminary. [4] Regional totals include other countries not shown. World total for all countries in USDA data base.

FAS, Production Estimates and Crop Assessment Division, (202) 720–0888. Prepared or estimated on the basis of official statistics of foreign governments, other foreign source materials, reports of U.S. Agricultural Counselors, Attachés, and Foreign Service Officers, results of office research, and related information.

Table 3-26.—Soybeans: Area, yield, production, and value, United States, 1993–2002

Year	Area planted	Soybeans for beans				
		Area harvested	Yield per acre	Production	Marketing year average price per bushel received by farmers	Value of production
	1,000 acres	*1,000 acres*	*Bushels*	*1,000 bushels*	*Dollars*	*1,000 dollars*
1993	60,085	57,307	32.6	1,869,718	6.40	11,941,449
1994	61,620	60,809	41.4	2,514,869	5.48	13,746,071
1995	62,495	61,544	35.3	2,174,254	6.72	14,599,145
1996	64,195	63,349	37.6	2,380,274	7.35	17,439,971
1997	70,005	69,110	38.9	2,688,750	6.47	17,372,628
1998	72,025	70,441	38.9	2,741,014	4.93	13,493,891
1999	73,730	72,446	36.6	2,653,758	4.63	12,205,352
2000	74,266	72,408	38.1	2,757,810	4.54	12,466,572
2001	74,075	72,975	39.6	2,890,682	4.38	12,605,717
2002	73,758	72,160	37.8	2,729,709	5.40	14,755,470

NASS, Crops Branch, (202) 720–2127.

Table 3-27.—Soybeans: Stocks on and off farms, United States, 1993–2002

Year	On farms				Off farms [1]			
	Dec. 1	Mar. 1	June 1 [2]	Sep. 1	Dec. 1	Mar. 1	June 1 [2]	Sep. 1
	1,000 bushels	*1,000 bushels*	*1,000 bushels*	*1,000 bushels*	*1,000 bushels*	*1,000 bushels*	*1,000 bushels*	*1,000 bushels*
1993	697,400	425,700	195,000	59,080	876,220	595,917	360,260	150,037
1994	985,800	635,300	348,800	105,130	1,116,156	734,898	443,072	229,684
1995	861,500	512,000	234,100	59,523	971,929	678,356	388,701	123,935
1996	935,100	514,000	216,000	43,600	889,984	541,754	283,890	88,233
1997	1,048,000	637,000	318,000	84,300	951,417	565,922	275,654	115,499
1998	1,187,000	815,000	458,000	145,000	999,440	642,338	390,573	203,482
1999	1,150,000	730,000	370,000	112,500	1,032,666	665,986	404,425	177,662
2000	1,217,000	780,000	365,000	83,500	1,022,991	623,908	343,180	164,247
2001	1,240,000	687,000	301,200	62,700	1,035,618	648,987	383,721	145,320
2002 [3]	1,170,000	NA	NA	NA	944,482	NA	NA	NA

[1] Includes stocks at mills, elevators, warehouses, terminals, and processors. [2] Old crop only. [3] Preliminary. NA-not available.

NASS, Crops Branch, (202) 720–2127.

Table 3-28.—Soybeans, soybean meal, and oil: Average price at specified markets, 1992–2001

Year [1]	Soybeans per bushel: No. 1 Yellow Chicago	Soybean oil per pound crude, tanks, f.o.b. Decatur	Soybean meal per short ton: 48 percent protein Decatur
	Dollars	*Cents*	*Dollars*
1992	5.95	21.40	193.75
1993	6.59	27.10	192.86
1994	5.73	27.58	162.55
1995	7.39	24.75	236.00
1996	7.80	22.50	270.90
1997	6.64	25.83	185.28
1998	5.00	19.80	138.55
1999	4.90	15.59	167.70
2000	4.77	14.15	173.60
2001	4.79	16.46	168.00

[1] Year beginning September for soybeans and October for oil and meal.

ERS, Field Crops Branch, (202) 694–5300.

Table 3-29.—Soybeans: Supply and disappearance, United States, 1992–2001

Year beginning September	Supply				
	Stocks by position			Production	Total [1]
	Farm	Terminal market, interior mill, elevator, and warehouse	Total		
	1,000 bushels	*1,000 bushels*	*1,000 bushels*	*1,000 bushels*	*1,000 bushels*
1992	105,000	173,437	278,437	2,190,354	2,470,846
1993	124,970	167,314	292,284	1,869,718	2,168,418
1994	59,080	150,037	209,117	2,514,869	2,729,464
1995	105,130	229,684	334,814	2,174,254	2,513,524
1996	59,523	123,935	183,458	2,380,274	2,572,636
1997	43,600	88,233	131,833	2,688,750	2,825,589
1998	84,300	115,499	199,799	2,741,014	2,944,334
1999	145,000	203,482	348,482	2,653,758	3,006,411
2000	112,500	177,662	290,162	2,757,810	3,051,540
2001 [2]	83,500	164,247	247,747	2,890,682	3,140,749

Year beginning September	Disappearance			
	Crushed [3]	Seed, feed and residual	Exports	Total
	1,000 bushels	*1,000 bushels*	*1,000 bushels*	*1,000 bushels*
1992	1,278,953	129,016	770,595	2,178,564
1993	1,275,648	95,548	588,105	1,959,301
1994	1,405,156	149,282	840,214	2,394,652
1995	1,369,541	111,441	849,084	2,330,066
1996	1,436,961	118,954	885,888	2,440,803
1997	1,596,983	154,476	874,334	2,625,793
1998	1,589,787	201,414	804,651	2,595,852
1999	1,577,650	164,084	973,405	2,715,139
2000	1,639,670	162,671	995,871	2,798,212
2001 [2]	1,699,733	173,000	1,062,556	2,935,289

[1] Includes imports, beginning with 1988. [2] Preliminary. [3] Reported by the U.S. Department of Commerce.
ERS, Field Crops Branch, (202) 694–5300.

Table 3-30.—Soybeans: Support operations, United States, 1993–2002

Marketin year beginning September 1	Income support payment rates per bushels [1]	Program price levels per bushel		Put under loan [4]		Acquired by CCC under loan program [5]	Owned by CCC at end of marketing year
		Loan [2]	Target [3]	Quantity	Percentage of production		
	Dollars	*Dollars*	*Dollars*	*1,000 bushels*	*Percent*	*1,000 bushels*	*1,000 bushels*
1993/94	....	5.02	....	87.1	4.7	0.0	0.0
1994/95	....	4.92	....	375.0	14.9	0.0	0.5
1995/96	....	4.92	....	181.8	8.4	0.0	0.0
1996/97	....	4.97	....	195.9	8.2	0.0	0.0
1997/98	....	5.26	....	266.3	9.9	0.5	0.6
1998/99	....	5.26	....	342.5	12.5	7.7	3.8
1999/2000	0.1409	5.26	....	286.9	10.8	13.7	6.5
2000/2001	0.2633	5.26	....	313.0	11.3	8.7	2.0
2001/2002	....	5.26	....	311.8	10.8	4.9	2.7
2002/2003	0.4400	5.00	5.80	....	....	....	....

[1] Oilseeds producer payment rates for 1999/2000 were calculated according to the provisions of the Agriculture, Rural Development, Food and Drug Admisistration, and Related Agencies Appropriations Act, 2000. Rates for 2000/2001 were calculated according to the provisions of the Agricultural Risk Protection Act of 2000, and included supplemental oilseeds payment rates. Payment rates for the 2002/2003 and subsequent crops are calculated according to the provisions of the Direct Payment program, following enactment of the Farm Security and Rural Investment Act of 2002 (2002 Act). Payment rates are rounded to the nearest tenth of a cent. [2] The national average loan rate was also known as the price support rate prior to enactment of the Federal Agricultural Improvement and Reform Act of 1996 (1996 Act). [3] Target prices for the 2002/2003 and subsequent crops were enacted with the 2002 Act. [4] Includes loans made. Does not include quantity on which loan deficiency payments were made. [5] Acquisition of all loans forfeited during the marketing year including loans made in previous year.
FSA, Feed Grains & Oilseeds Analysis Group, (202) 720–2711.

Table 3-31.—Soybeans: Area, yield, and production, by States, 2000–2002

State	Area planted			Soybeans for beans								
				Area harvested			Yield per harvested acre			Production		
	2000	2001	2002	2000	2001	2002	2000	2001	2002	2000	2001	2002
	1,000 acres	*1,000 acres*	*1,000 acres*	*1,000 acres*	*1,000 acres*	*1,000 acres*	*Bushels*	*Bushels*	*Bushels*	*1,000 bushels*	*1,000 bushels*	*1,000 bushels*
AL	190	140	170	160	135	155	18.0	35.0	24.0	2,880	4,725	3,720
AR	3,350	2,900	2,950	3,150	2,850	2,880	25.5	32.0	33.5	80,325	91,200	96,480
DE	215	205	190	213	201	185	43.0	39.0	25.0	9,159	7,839	4,625
FL	20	10	10	15	9	8	19.0	29.0	31.0	285	261	248
GA	170	165	160	140	155	140	24.0	26.0	21.0	3,360	4,030	2,940
IL	10,500	10,700	10,550	10,450	10,620	10,460	44.0	45.0	43.0	459,800	477,900	449,780
IN	5,500	5,600	5,800	5,480	5,590	5,750	46.0	49.0	41.0	252,080	273,910	235,750
IA	10,700	11,000	10,400	10,680	10,920	10,310	43.5	44.0	48.0	464,580	480,480	494,880
KS	2,950	2,850	2,750	2,500	2,730	2,540	20.0	32.0	23.0	50,000	87,360	58,420
KY	1,180	1,240	1,290	1,160	1,220	1,260	39.0	40.0	32.5	45,240	48,800	40,950
LA	930	640	790	850	610	650	24.0	33.0	32.0	20,400	20,130	20,800
MD	520	520	490	515	515	470	43.0	39.0	23.0	22,145	20,085	10,810
MI	2,050	2,150	2,050	2,030	2,130	2,030	36.0	30.0	38.5	73,080	63,900	78,155
MN	7,300	7,300	7,200	7,150	7,200	7,100	41.0	37.0	43.5	293,150	266,400	308,850
MS	1,700	1,160	1,440	1,580	1,120	1,370	22.0	33.0	32.0	34,760	36,960	43,840
MO	5,150	4,950	5,050	5,000	4,900	5,000	35.0	38.0	34.0	175,000	186,200	170,000
NE	4,650	4,950	4,700	4,575	4,900	4,580	38.0	45.5	38.5	173,850	222,950	176,330
NJ	100	103	100	98	101	97	40.0	31.0	23.0	3,920	3,131	2,231
NY	135	160	140	132	158	138	33.0	33.0	32.0	4,356	5,214	4,416
NC	1,400	1,380	1,360	1,360	1,350	1,280	32.5	32.0	23.5	44,200	43,200	30,080
ND	1,900	2,150	2,670	1,850	2,110	2,630	32.0	33.5	33.0	59,200	70,685	86,790
OH	4,450	4,600	4,750	4,440	4,580	4,710	42.0	41.0	30.0	186,480	187,780	141,300
OK	440	415	270	290	255	250	15.0	19.0	28.0	4,350	4,845	7,000
PA	390	400	365	385	395	350	43.0	35.0	26.0	16,555	13,825	9,100
SC	450	440	435	430	420	415	25.0	21.0	17.0	10,750	8,820	7,055
SD	4,400	4,500	4,250	4,370	4,470	4,090	35.0	32.0	31.0	152,950	143,040	126,790
TN	1,180	1,070	1,160	1,150	1,040	1,120	25.0	34.0	31.0	28,750	35,360	34,720
TX	290	260	230	260	225	215	27.0	26.0	28.0	7,020	5,850	6,020
VA	490	500	480	480	480	440	38.5	35.5	23.0	18,480	17,040	10,120
WV	16	17	18	15	16	17	47.0	42.0	37.0	705	672	629
WI	1,550	1,600	1,540	1,500	1,570	1,520	40.0	37.0	44.0	60,000	58,090	66,880
US	74,266	74,075	73,758	72,408	72,975	72,160	38.1	39.6	37.8	2,757,810	2,890,682	2,729,709

NASS, Crops Branch, (202) 720–2127.

Table 3-32.—Soybeans: Crushings, and oil and meal stocks, production, and foreign trade, United States, 1992–2001

Year beginning October	Soybeans crushed					Soybean oil			Soybean cake and meal		
	Oct.-Dec.	Jan.-Mar.	Apr.-Jun.	Jul.-Sep.	Total	Stocks Oct. 1	Production	Exports	Stocks Oct. 1	Production	Exports
	1,000 bushels	*1,000 bushels*	*1,000 bushels*	*1,000 bushels*	*1,000 bushels*	*Million pounds*	*Million pounds*	*Million pounds*	*1,000 tons*	*1,000 tons*	*1,000 tons*
1992	343,196	331,987	312,342	288,639	1,276,164	2,239	13,778	1,461	230	30,364	6,254
1993	342,257	327,307	305,734	307,371	1,282,668	1,555	13,951	1,531	204	30,514	5,365
1994	370,273	371,891	339,189	325,301	1,406,654	1,103	15,613	2,683	150	33,265	6,715
1995	369,123	349,481	325,929	318,490	1,363,023	1,137	15,240	992	223	32,527	6,004
1996	398,225	392,509	334,449	320,681	1,445,865	2,015	15,752	2,033	212	34,211	6,994
1997	438,067	436,997	375,370	359,693	1,610,077	1,520	18,143	3,079	210	38,176	9,330
1998	430,007	403,987	377,640	388,027	1,599,660	1,382	18,078	2,372	218	37,797	7,122
1999	435,943	395,117	360,423	381,274	1,572,757	1,520	17,825	1,376	330	37,591	7,331
2000	434,530	417,420	391,733	395,327	1,639,010	1,995	18,420	1,401	293	39,385	7,703
2001 [1]	452,756	443,946	414,407	382,738	1,693,847	2,887	18,898	2,500	383	40,346	7,600

[1] Preliminary.

ERS, Field Crops Branch, (202) 694–5300. Data from the U.S. Department of Commerce.

Table 3-33.—Soybeans for beans: Marketing year average price and value of production, by States, crop of 2000, 2001, and 2002

State	Marketing year average price per bushel			Value of production		
	2000	2001	2002[1]	2000	2001	2002[1]
	Dollars	*Dollars*	*Dollars*	*1,000 dollars*	*1,000 dollars*	*1,000 dollars*
AL	4.75	4.60	5.35	13,680	21,735	19,902
AR	4.73	4.37	5.65	379,937	398,544	545,112
DE	4.50	4.25	5.70	41,216	33,316	26,363
FL	4.45	4.20	5.35	1,268	1,096	1,327
GA	4.43	4.35	5.35	14,885	17,531	15,729
IL	4.62	4.55	5.50	2,124,276	2,174,445	2,473,790
IN	4.61	4.42	5.40	1,162,089	1,210,682	1,273,050
IA	4.49	4.35	5.40	2,085,964	2,090,088	2,672,352
KS	4.50	4.16	5.40	225,000	363,418	315,468
KY	4.71	4.47	5.75	213,080	218,136	235,463
LA	4.86	4.39	5.55	99,144	88,371	115,440
MD	4.50	4.20	5.75	99,653	84,357	62,158
MI	4.54	4.47	5.50	331,783	285,633	429,853
MN	4.38	4.32	5.25	1,283,997	1,150,848	1,621,463
MS	4.71	4.39	5.55	163,720	162,254	243,312
MO	4.55	4.32	5.40	796,250	804,384	918,000
NE	4.44	4.19	5.35	771,894	934,161	943,366
NJ	4.35	4.15	5.50	17,052	12,994	12,271
NY	4.55	4.55	5.55	19,820	23,724	24,509
NC	4.51	4.29	5.50	199,342	185,328	165,440
ND	4.23	4.05	5.20	250,416	286,274	451,308
OH	4.63	4.46	5.45	863,402	837,499	770,085
OK	4.25	4.30	5.40	18,488	20,834	37,800
PA	4.37	4.26	5.70	72,345	58,895	51,870
SC	4.50	4.45	5.40	48,375	39,249	38,097
SD	4.27	4.13	5.15	653,097	590,755	652,969
TN	4.69	4.46	5.70	134,838	157,706	197,904
TX	4.40	4.60	5.20	30,888	26,910	31,304
VA	4.35	4.30	5.45	80,388	73,272	55,154
WV	4.66	4.33	5.55	3,285	2,910	3,491
WI	4.45	4.31	5.25	267,000	250,368	351,120
US	4.54	4.38	5.40	12,466,572	12,605,717	14,755,470

[1] Preliminary.

NASS, Crops Branch, (202) 720–2127.

Table 3-34.—Soybeans: Area, yield, and production in specified countries and the world, 1999–2000/2001–2002[1]

Continent and country	Area[2]			Yield per hectare			Production		
	1999/2000	2000/2001	2001/2002[3]	1999/2000	2000/2001	2001/2002[3]	1999/2000	2000/2001	2001/2002[3]
	1,000 hectares	*1,000 hectares*	*1,000 hectares*	*Metric tons*	*Metric tons*	*Metric tons*	*1,000 metric tons*	*1,000 metric tons*	*1,000 metric tons*
North America:									
Canada	1,002	1,061	1,070	2.77	2.55	1.52	2,776	2,703	1,630
Mexico	79	70	77	1.56	1.47	1.56	123	103	120
United States	29,318	29,303	29,532	2.46	2.56	2.66	72,224	75,055	78,672
Total	30,399	30,434	30,679	2.47	2.56	2.62	75,123	77,861	80,422
South America:									
Argentina	8,583	10,400	11,400	2.47	2.67	2.63	21,200	27,800	30,000
Bolivia	606	600	640	2.05	1.72	1.94	1,240	1,030	1,240
Brazil	13,600	13,934	16,350	2.51	2.80	2.66	34,200	39,000	43,500
Colombia	26	23	25	2.00	2.70	2.32	52	62	58
Ecuador	41	58	38	1.61	1.79	1.47	66	104	56
Paraguay	1,150	1,350	1,420	2.52	2.61	2.18	2,900	3,520	3,100
Venezuela	5	4	6	2.00	1.75	1.67	10	7	10
Total	24,011	26,369	29,879	2.49	2.71	2.61	59,668	71,523	77,964

See footnotes at end of table.

Table 3-34.—Soybeans: Area, yield, and production in specified countries and the world, 1999–2000/2001–2002 [1]—Continued

Continent and country	Area [2]			Yield per hectare			Production		
	1999/ 2000	2000/ 2001	2001/ 2002 [3]	1999/ 2000	2000/ 2001	2001/ 2002 [3]	1999/ 2000	2000/ 2001	2001/ 2002 [3]
	1,000 hectares	*1,000 hectares*	*1,000 hectares*	*Metric tons*	*Metric tons*	*Metric tons*	*1,000 metric tons*	*1,000 metric tons*	*1,000 metric tons*
Central America:									
Guatemala	18	18	18	2.83	2.78	2.72	51	50	49
Nicaragua	9	3	2	2.22	2.00	2.00	20	6	4
Total	27	21	20	2.63	2.67	2.65	71	56	53
European Union:									
Austria	19	16	16	2.63	2.06	2.13	50	33	34
France	102	79	119	2.62	2.62	2.45	267	207	291
Germany	3	1	1	2.33	2.00	2.00	7	2	2
Italy	239	246	251	3.41	3.25	3.50	814	800	879
Spain	4	3	4	2.25	2.00	2.00	9	6	8
Total	367	345	391	3.13	3.04	3.10	1,147	1,048	1,214
Eastern Europe:									
Bosnia-Hercegovina	4	4	4	2.25	1.00	2.00	9	4	8
Croatia	46	47	42	2.52	1.38	2.19	116	65	92
Hungary	32	22	25	2.19	1.36	2.20	70	30	55
Romania	80	91	42	2.29	0.77	1.57	183	70	66
Serbia	108	100	110	2.72	1.50	2.18	294	150	240
Total	270	264	223	2.49	1.21	2.07	672	319	461
Other Western Europe:									
Switzerland	2	1	1	4.00	4.00	2.00	8	4	2
Former Soviet Union:									
Russian Fed.	439	340	420	0.76	1.01	0.83	334	342	350
Ukraine	42	61	73	1.07	1.05	1.01	45	64	74
Total	481	401	493	0.79	1.01	0.86	379	406	424
Middle East:									
Iran	90	90	90	1.56	1.56	1.56	140	140	140
Syria	3	3	3	1.00	1.33	1.33	3	4	4
Turkey	28	30	20	2.14	1.33	2.25	60	40	45
Total	121	123	113	1.68	1.50	1.67	203	184	189
Africa:									
Egypt	7	6	5	2.71	2.67	2.00	19	16	10
Morocco	1	1	1	1.00	1.00	1.00	1	1	1
Nigeria	90	125	125	1.00	0.92	0.92	90	115	115
South Africa, Republic of	94	134	124	1.59	1.69	1.63	149	226	202
Uganda	84	106	106	1.20	1.13	1.13	101	120	120
Zambia	13	13	13	2.31	2.31	2.31	30	30	30
Zimbabwe	63	77	60	2.06	2.27	1.42	130	175	85
Total	352	462	434	1.48	1.48	1.30	520	683	563
Asia:									
Burma	102	85	85	0.83	0.92	0.92	85	78	78
China, Peop.	8,000	9,300	9,480	1.79	1.66	1.63	14,290	15,400	15,410
India	5,645	5,800	6,000	0.92	0.91	0.90	5,200	5,250	5,400
Indonesia	1,100	825	679	1.18	1.24	1.22	1,300	1,020	826
Japan	108	123	144	1.73	1.91	1.88	187	235	271
Korea, Dem.	300	310	310	1.13	1.13	1.13	340	350	350
Korea, Rep.	87	86	78	1.33	1.31	1.51	116	113	118
Pakistan	2	2	2	0.50	1.00	1.00	1	2	2
Philippines	1	1	1	1.00	1.00	1.00	1	1	1
Taiwan	3	3	3	2.00	2.00	2.00	6	6	6
Thailand	220	220	210	1.50	1.42	1.39	330	312	292
Vietnam	129	122	130	1.14	1.16	1.15	147	142	150
Total	15,697	16,877	17,122	1.40	1.36	1.34	22,003	22,909	22,904
Oceania:									
Australia	50	50	50	2.20	2.10	2.10	110	105	105
World total	71,777	75,347	79,405	2.23	2.32	2.32	159,904	175,098	184,301

[1] Split year includes Northern Hemisphere crop harvested in the late months of the first year shown combined with Southern Hemisphere and certain Northern Hemisphere crops harvested in the early months of the following year. [2] Harvested area as far as possible. [3] Preliminary.

FAS, Production Estimates and Crop Assessment Division, (202) 720–0888. Prepared or estimated on the basis of official statistics of foreign governments, other foreign source materials, reports of U.S. Agricultural Counselors, Attachés, and Foreign Service Officers, results of office research, and related information.

Table 3-35.—Soybeans, soybean oil, and soybean cake and meal: United States exports by country of destination, 1999/2000–2001/2002 [1]

Continent and country of destination	Soybeans			Soybean oil [2]			Soybean cake and meal		
	1999/2000	2000/2001	2001/2002 [3]	1999/2000	2000/2001	2001/2002 [3]	1999/2000	2000/2001	2001/2002 [3]
	Metric tons	*Metric tons*	*Metric tons*	*Metric tons*	*Metric tons*	*Metric tons*	*Metric tons*	*Metric tons*	*Metric tons*
North and Central America; incl. Caribbean:									
Canada	351,286	342,180	804,904	22,715	54,909	87,116	798,399	934,171	1,112,408
Costa Rica	185,647	218,506	225,915	83	805	113	17	5,084	3
Jamaica & Dep	22	111	17	21,586	16,392	23,919	66,395	68,584	77,002
Mexico	3,418,712	3,803,109	3,881,836	118,079	72,456	161,760	262,650	246,802	379,420
Total [4]	4,051,597	4,483,098	5,104,257	265,937	211,916	436,131	2,013,944	2,128,621	2,551,678
South America:									
Brazil	0	0	281	0	25	83	142	0	5
Columbia	90,070	92,526	193,403	10,225	21,859	8,907	97,622	44,299	44,929
Peru	5,903	0	0	66,686	60,606	37,677	211,315	21,923	105,410
Venezuela	127,045	53,340	80,088	414	577	635	309,844	124,061	66,228
Total [4]	223,821	151,485	305,309	78,419	93,301	48,510	685,345	287,346	363,907
Europe:									
Belgium and Luxembourg	536,945	447,705	456,105	0	0	0	0	498	3,056
Denmark	0	0	83,348	18	0	0	123,844	75,612	63,328
France	170,846	175,856	398,659	23	38	0	7,439	28,240	30,352
Germany	703,485	541,304	478,753	6,480	7,082	19,722	3,637	86	1,621
Greece	89,459	125,632	124,622	509	0	0	32,355	8,452	3,599
Italy	250,255	215,714	633,286	0	0	10	18,834	126,287	40,344
Netherlands	2,313,494	2,627,597	2,397,128	6,822	5,801	5,669	109,462	219,155	106,833
Norway	0	102	0	0	0	0	0	0	390
Spain	1,249,799	1,114,235	1,573,818	9,742	0	4,055	80,610	148,848	78,284
United Kingdom	402,211	347,314	242,667	2,285	0	44	22,522	2,582	1,238
Total [4]	6,037,631	5,952,812	7,060,478	35,517	13,319	30,987	531,010	817,820	492,560
Former Soviet Union: [5]									
Russia	0	0	0	1,771	2,983	216	87,661	118,061	111,502
Ukraine	6,693	19	10,498	0	0	0	0	0	0
Total [4]	39,480	55,719	75,006	2,531	6,904	2,673	92,057	159,117	126,365
Africa:									
Egypt	120,282	124,830	365,724	123	48,245	55,745	171,111	272,512	159,681
Morocco	89,931	101,736	67,618	14,091	0	39,439	60,523	38,087	63,299
Total [4]	229,794	249,141	445,460	71,739	105,023	136,666	517,241	527,334	541,170
Asia:									
China	4,563,338	5,366,949	4,230,417	29,558	5,597	0	23	171	300
India	0	0	0	23,413	54,061	88,528	0	0	0
Israel	519,785	466,034	644,033	789	1,701	42	61,353	54,647	41,591
Japan	3,585,623	3,570,902	3,727,152	286	94	538	270,664	206,377	189,165
Korea, Rep. of	1,406,221	1,188,056	1,245,443	83,872	50,213	86,063	45,126	121,418	38,376
Philippines	358,950	279,622	261,261	30	50	273	863,872	750,789	812,660
Taiwan	1,939,854	1,866,893	2,118,831	11	124	7,500	1,067	7,754	13,850
Total [4]	14,389,029	15,062,347	15,002,148	169,508	205,021	487,765	2,622,411	2,859,778	2,326,091
Oceania:									
Australia	8,841	358	3,771	0	0	71	130,041	147,977	261,912
Total [4]	8,843	358	3,883	0	10	117	189,834	208,106	403,513
Grand Total [6] [7]	26,536,483	27,103,182	28,918,048	623,651	635,493	1,142,850	6,651,842	6,988,122	6,805,285

[1] For soybeans, year begins September 1; for soybean oil and cake and meal, year begins October 1. [2] Crude and refined oil (includes shipments under P.L. 480). [3] Preliminary. [4] Includes quantities exported to countries not shown. [5] Former Soviet Union; includes all 15 Republics of the USSR. [6] May not add due to rounding. [7] Includes quantities transshipped via Canada to unidentified countries.

FAS, Cotton, Oilseeds, Tobacco and Seeds Division, (202) 720–5710. Compiled from reports of the U.S. Department of Commerce.

Table 3-36.—Soybeans: International trade, 1997/98–2001/2002 [1]

Country	1997/98	1998/99	1999/2000	2000/2001	2001/2002
	1,000 metric tons	*1,000 metric tons*	*1,000 metric tons*	*1,000 metric tons*	*1,000 metric tons*
Principal exporting countries:					
United States	23,760	21,898	26,537	27,103	28,918
Brazil	8,750	8,930	11,161	15,470	15,000
Argentina	3,230	3,233	4,131	7,415	6,005
Paraguay	2,390	2,350	2,120	2,550	2,110
Canada	769	876	949	747	450
Netherlands	510	543	841	1,090	1,650
Subtotal	39,409	37,830	45,739	54,375	54,133
Others	1,050	885	944	701	868
World total	40,459	38,715	46,683	55,076	55,001

Country	1997/98	1998/99	1999/2000	2000/2001	2001/2002
Principal importing countries:					
Belgium-Luxembourg	1,279	1,243	1,203	1,429	1,635
China, Peoples Republic	2,940	3,850	10,100	13,245	10,385
Germany	4,000	4,095	3,444	4,567	4,650
Indonesia	826	1,128	1,415	1,261	1,575
Italy	854	715	804	836	820
Japan	4,873	4,807	4,907	4,767	5,023
Korea, Republic of	1,340	1,400	1,606	1,389	1,435
Mexico	3,502	3,766	4,039	4,381	4,510
Netherlands	4,875	5,007	5,264	5,660	6,000
Spain	3,044	3,054	2,741	3,167	3,350
Taiwan	2,387	2,150	2,300	2,330	2,400
Thailand	600	950	1,100	1,290	1,560
United Kingdom	1,045	893	601	865	950
Subtotal	31,565	33,058	39,524	45,187	44,293
Others	8,260	7,497	8,284	9,749	11,699
World total	39,825	40,555	47,808	54,936	55,992

[1] Marketing year beginning Aug. 1.

FAS, Cotton, Oilseeds, Tobacco and Seeds Division, (202) 720–9516. Prepared or estimated on the basis of official statistics of foreign governments, other foreign source materials, reports of U.S. Agricultural Counselors, Attachés, and Foreign Service Officers, results of office research, and related information.

Table 3-37.—Sunflower: Area, yield, production, and value, United States, 1993–2002 [1]

Year	Area planted	Area harvested	Yield per harvested acre	Production	Price per cwt.	Value of production
	1,000 acres	*1,000 acres*	*Pounds*	*1,000 pounds*	*Dollars*	*1,000 dollars*
1993	2,757	2,486	1,035	2,572,063	12.90	326,432
1994	3,567	3,430	1,410	4,835,825	10.70	512,747
1995	3,478	3,368	1,190	4,009,332	11.50	457,573
1996	2,536	2,479	1,436	3,559,343	11.70	414,842
1997	2,888	2,792	1,317	3,676,952	11.60	426,766
1998	3,568	3,492	1,510	5,273,162	10.60	536,971
1999	3,553	3,441	1,262	4,341,862	7.53	339,993
2000	2,840	2,647	1,339	3,544,428	6.89	246,869
2001	2,633	2,555	1,338	3,418,759	9.62	325,950
2002	2,585	2,205	1,133	2,497,236	12.70	317,244

[1] Estimates include all States except AK and HI.
NASS, Crops Branch, (202) 720–2127.

Table 3-38.—Sunflower, Oil Varieties: Area, yield, production, and value, United States, 1993–2002 [1]

Year	Area planted	Area harvested	Yield per harvested acre	Production	Price per cwt.	Value of production
	1,000 acres	*1,000 acres*	*Pounds*	*1,000 pounds*	*Dollars*	*1,000 dollars*
1993	2,297	2,074	1,042	2,160,268	12.30	264,585
1994	3,041	2,943	1,435	4,223,475	10.10	426,859
1995	2,911	2,829	1,201	3,398,437	10.90	371,967
1996	1,967	1,934	1,470	2,843,763	10.80	309,057
1997	2,284	2,212	1,350	2,985,700	11.00	329,858
1998	2,953	2,897	1,549	4,486,360	9.37	423,775
1999	2,757	2,695	1,298	3,497,820	6.33	229,593
2000	2,248	2,116	1,375	2,909,844	5.89	175,306
2001	2,117	2,060	1,361	2,803,704	9.07	254,705
2002	2,125	1,837	1,128	2,072,410	12.40	259,365

[1] Estimates include all States except AK and HI.
NASS, Corps Branch, (202) 720–2127.

Table 3-39.—Sunflower, non-oil varieties: Area, yield, production, and value, United States, 1993–2002 [1]

Year	Area planted	Area harvested	Yield per harvested acre	Production	Price per cwt.	Value of production
	1,000 acres	*1,000 acres*	*Pounds*	*1,000 pounds*	*Dollars*	*1,000 dollars*
1993	460	412	1,000	411,795	15.20	61,847
1994	526	487	1,257	612,350	13.80	85,888
1995	567	539	1,133	610,895	14.10	85,606
1996	569	545	1,313	715,580	13.80	105,785
1997	604	580	1,192	691,252	14.30	96,908
1998	615	595	1,322	786,802	14.60	113,196
1999	796	746	1,131	844,042	13.40	110,400
2000	592	531	1,195	634,584	11.20	71,563
2001	516	495	1,243	615,055	11.60	71,245
2002	460	368	1,154	424,826	13.70	57,879

[1] Estimates include all States except AK and HI.
NASS, Crops Branch, (202) 720–2127.

Table 3-40.—Sunflower: Area, yield, production, and value, by States, 2001–2002

Variety and State	Area planted		Area harvested		Yield per harvested acre	
	2001	2002 [1]	2001	2002 [1]	2001	2002 [1]
	1,000 acres	*1,000 acres*	*1,000 acres*	*1,000 acres*	*Pounds*	*Pounds*
Oil:						
CO	130	95	120	80	1,140	500
KS	300	200	290	155	1,200	900
MN	30	40	28	37	1,400	1,450
NE	52	45	50	34	1,000	500
ND	850	1,150	835	1,105	1,440	1,310
SD	670	535	661	375	1,410	850
TX	35	12	33	11	1,100	800
Other States [2]	50	48	43	40	1,372	1,179
US	2,117	2,125	2,060	1,837	1,361	1,128
Non-oil:						
CO	65	35	62	20	1,150	1,150
KS	35	15	33	13	1,330	930
MN	30	30	28	27	1,250	1,200
NE	30	13	29	11	1,150	700
ND	220	220	215	210	1,260	1,250
SD	45	105	44	55	1,450	1,000
TX	73	30	70	23	1,200	1,000
Other States [2]	18	12	14	9	915	1,015
US	516	460	495	368	1,243	1,154
Total:						
CO	195	130	182	100	1,143	630
KS	335	215	323	168	1,213	902
MN	60	70	56	64	1,325	1,345
NE	82	58	79	45	1,055	549
ND	1,070	1,370	1,050	1,315	1,403	1,300
SD	715	640	705	430	1,412	869
TX	108	42	103	34	1,168	935
Other States [2]	68	60	57	49	1,260	1,149
US	2,633	2,585	2,555	2,205	1,338	1,133

Variety and State	Production		Marketing year average price per cwt.		Value of production	
	2001	2002 [1]	2001	2002 [1]	2001	2002 [1]
	1,000 pounds	*1,000 pounds*	*Dollars*	*Dollars*	*1,000 dollars*	*1,000 dollars*
Oil:						
CO	136,800	40,000	8.10	12.30	11,081	4,920
KS	348,000	139,500	8.28	11.50	28,814	16,043
MN	39,200	53,650	10.30	12.50	4,038	6,706
NE	50,000	17,000	8.10	12.00	4,050	2,040
ND	1,202,400	1,447,550	9.39	12.50	112,905	180,944
SD	932,010	318,750	9.08	13.00	84,627	41,438
TX	36,300	8,800	8.50	13.00	3,086	1,144
Other States [2]	58,994	47,160	10.30	13.00	6,104	6,130
US	2,803,704	2,072,410	9.07	12.40	254,705	259,365
Non-oil:						
CO	71,300	23,000	12.10	13.30	8,627	3,059
KS	43,890	12,090	11.40	14.00	5,003	1,693
MN	35,000	32,400	11.40	13.50	3,990	4,374
NE	33,350	7,700	11.20	13.00	3,735	1,001
ND	270,900	262,500	11.70	13.80	31,695	36,225
SD	63,800	55,000	11.20	12.20	7,146	6,710
TX	84,000	23,000	11.00	15.00	9,240	3,450
Other States [2]	12,815	9,136	14.10	15.00	1,809	1,367
US	615,055	424,826	11.60	13.70	71,245	57,879
Total:						
CO	208,100	63,000	9.50	12.70	19,708	7,979
KS	391,890	151,590	8.95	12.60	33,817	17,736
MN	74,200	86,050	10.90	12.80	8,028	11,080
NE	83,350	24,700	9.20	12.40	7,785	3,041
ND	1,473,300	1,710,050	10.00	12.60	144,600	217,169
SD	995,810	373,750	9.18	13.00	91,773	48,148
TX	120,300	31,800	10.30	14.50	12,326	4,594
Other States [2]	71,809	56,296	11.00	13.30	7,913	7,497
US	3,418,759	2,497,236	9.62	12.70	325,950	317,244

[1] Preliminary. [2] Other States include CA, GA, IL, LA, MI, MO, MT, NM, NY, OH, OK, PA, SC, UT, WA, WI, and WY.
NASS, Crops Branch, (202) 720–2127.

Table 3-41.—Sunflowerseeds: Area, yield and production in specified countries, 1999/2000–2001/2002 [1]

Continent and country	Area [2]			Yield per hectare			Production		
	1999/ 2000	2000/ 2001	2001/ 2002 [3]	1999/ 2000	2000/ 2001	2001/ 2002 [3]	1999/ 2000	2000/ 2001	2001/ 2002 [3]
	1,000 hectares	*1,000 hectares*	*1,000 hectares*	*Metric tons*	*Metric tons*	*Metric tons*	*1,000 metric tons*	*1,000 metric tons*	*1,000 metric tons*
North America:									
Canada	79	69	63	1.54	1.72	1.56	122	119	98
Mexico	1	1	1	1.00	1.00	1.00	1	1	1
United States	1,393	1,071	1,034	1.41	1.50	1.50	1,969	1,608	1,551
Total	1,473	1,141	1,098	1.42	1.51	1.50	2,092	1,728	1,650
South America:									
Argentina	3,477	1,886	1,980	1.73	1.62	1.87	6,000	3,050	3,700
Bolivia	135	178	178	1.11	0.97	0.97	150	173	173
Brazil	58	48	48	1.67	1.54	1.54	97	74	74
Paraguay	50	20	30	1.50	1.50	1.33	75	30	40
Uruguay	50	48	104	0.66	1.19	1.49	33	57	155
Total	3,770	2,180	2,340	1.69	1.55	1.77	6,355	3,384	4,142
European Union:									
Austria	23	23	25	2.70	2.43	2.40	62	56	60
France	817	710	709	2.35	2.55	2.23	1,919	1,813	1,579
Germany	33	26	25	2.55	2.46	2.16	84	64	54
Greece	12	11	10	1.92	1.91	2.00	23	21	20
Italy	257	262	261	2.00	1.83	1.52	515	480	398
Portugal	54	48	50	0.57	0.60	0.58	31	29	29
Spain	840	835	850	0.69	1.02	1.02	579	848	871
Total	2,036	1,915	1,930	1.58	1.73	1.56	3,213	3,311	3,011
Eastern Europe:									
Bulgaria	520	450	398	1.10	0.93	0.98	570	420	392
Croatia	42	26	25	1.71	2.08	1.72	72	54	43
Czech Republic	28	31	30	2.25	2.10	2.17	63	65	65
Hungary	530	300	337	1.50	1.67	1.96	795	500	661
Macedonia (Skopje)	10	10	10	1.40	1.00	1.00	14	10	10
Romania	1,050	822	790	1.24	0.87	0.94	1,300	717	744
Serbia	184	146	170	1.48	1.48	1.65	273	216	280
Slovakia	95	69	62	1.32	1.70	2.19	125	117	136
Total	2,459	1,854	1,822	1.31	1.13	1.28	3,212	2,099	2,331
Frm. USSR (non-Baltics):									
Kazakhstan	213	262	255	0.49	0.40	0.42	104	105	108
Moldova	217	220	220	1.32	1.27	1.27	286	280	280
Russian Federation	5,585	4,629	3,800	0.74	0.85	0.71	4,150	3,915	2,700
Ukraine	2,780	2,842	2,400	0.98	1.22	0.94	2,720	3,457	2,250
Total [4]	8,795	7,953	6,675	0.83	0.98	0.80	7,260	7,757	5,338
Middle East:									
Iran	80	78	78	0.56	0.51	0.51	45	40	40
Israel	8	13	11	1.38	1.23	1.18	11	16	13
Turkey	540	540	510	1.48	1.06	1.02	800	575	520
Total	628	631	599	1.36	1.00	0.96	856	631	573
Africa:									
Egypt	1	1	2	2.00	2.00	2.00	2	2	4
Morocco	75	75	75	0.67	0.27	0.67	50	20	50
South Africa, Republic of	395	520	645	1.38	1.28	1.30	545	664	840
Total	471	596	722	1.27	1.15	1.24	597	686	894
Asia and the Middle East:									
Burma	225	440	440	0.84	0.61	0.61	190	270	270
China,Peoples Republic of	1,140	1,229	1,016	1.58	1.59	1.45	1,800	1,954	1,478
India	2,300	2,200	2,400	0.57	0.57	0.60	1,300	1,250	1,450
Pakistan	144	114	38	1.35	1.32	1.21	195	150	46
Total	3,809	3,983	3,894	0.91	0.91	0.83	3,485	3,624	3,244
Oceania:									
Australia	120	69	65	1.23	1.01	0.97	147	70	63
World total [4]	23,561	20,322	19,145	1.16	1.15	1.11	27,217	23,290	21,246

[1] Split year includes Northern Hemisphere crop harvested in the late months of the first year shown combined with Southern Hemisphere and certain Northern Hemisphere crops harvested in the early months of the following year. [2] Harvested area as far as possible. [3] Preliminary. [4] Regional totals include other countries not shown. World total for all countries in USDA data base.

FAS, Production Estimates and Crop Assessment Division, (202) 720–0888. Prepared or estimated on the basis of official statistics of foreign governments, other foreign source materials, reports of U.S. Agricultural Counselors, Attaches, and Foreign Service Officers, results of office research, and related information.

Table 3-42.—Peppermint oil: Area, yield, production, and value, United States, 1993–2002

Year	Area harvested	Yield per harvested acre	Production	Price per pound	Value of production
	1,000 acres	*Pounds*	*1,000 pounds*	*Dollars*	*1,000 dollars*
1993	98.3	61	6,027	13.30	80,139
1994	108.5	69	7,459	14.60	109,255
1995	135.3	70	9,449	13.80	130,048
1996	132.0	72	9,446	13.60	128,778
1997	135.6	74	9,971	12.90	128,846
1998	124.0	78	9,727	11.90	116,037
1999	106.3	71	7,537	10.70	80,951
2000	90.5	78	7,067	10.80	76,277
2001	78.5	81	6,343	10.90	68,964
2002 [1]	80.2	85	6,818	11.60	79,030

[1] Preliminary.
NASS, Crops Branch (202), 720–2127.

Table 3-43.—Spearmint oil: Area, yield, production, and value, United States, 1993–2002

Year	Area harvested	Yield per harvested acre	Production	Price per pound	Value of production
	1,000 acres	*Pounds*	*1,000 pounds*	*Dollars*	*1,000 dollars*
1993	32.5	84	2,722	12.30	33,488
1994	28.4	78	2,213	12.50	27,592
1995	28.8	79	2,274	12.30	27,858
1996	23.1	94	2,167	12.00	26,094
1997	25.5	96	2,441	11.90	29,128
1998	27.4	109	2,987	11.00	32,731
1999	24.4	101	2,454	9.70	23,925
2000	21.7	101	2,199	9.06	19,919
2001	19.5	105	2,052	9.10	18,645
2002 [1]	18.0	108	1,942	9.10	17,696

[1] Preliminary.
NASS, Crops Branch, (202) 720–2127.

Table 3-44.—Mint oil: Production and value, by States, 2000–2002

State	Production			Price per pound			Value of production		
	2000	2001	2002 [1]	2000	2001	2002 [1]	2000	2001	2002 [1]
	1,000 pounds	*1,000 pounds*	*1,000 pounds*	*Dollars*	*Dollars*	*Dollars*	*1,000 dollars*	*1,000 dollars*	*1,000 dollars*
Peppermint:									
ID	1,425	1,288	1,530	10.30	10.70	11.30	14,678	13,782	17,289
IN	576	490	414	9.50	10.30	10.70	5,472	5,047	4,430
MI [2]	50	50	50	9.20	9.90	9.00	460	495	450
OR	2,541	2,184	2,112	12.70	11.80	13.00	32,271	25,771	27,456
WA	2,160	2,021	2,400	9.30	10.20	10.90	20,088	20,614	26,160
WI	315	310	312	10.50	10.50	10.40	3,308	3,255	3,245
US	7,067	6,343	6,818	10.80	10.90	11.60	76,277	68,964	79,030
Spearmint:									
ID	130	95	88	8.00	8.40	8.40	1,040	798	739
IN	99	96	84	9.70	10.10	9.00	960	970	756
MI	77	85	85	9.20	9.80	9.00	708	833	765
OR	115	132	119	10.50	9.00	8.85	1,208	1,188	1,053
WA	1,616	1,484	1,445	8.80	8.90	9.20	14,221	13,208	13,294
WI	162	160	121	11.00	10.30	9.00	1,782	1,648	1,089
US	2,199	2,052	1,942	9.06	9.10	9.10	19,919	18,645	17,696

[1] Preliminary.
NASS, Crops Branch, (202) 720–2127.

Table 3-45.—Olive oil: World production, 1998–2000 [1] [2]

Continent and country	1998	1999	2000 [3]
	1,000 metric tons	*1,000 metric tons*	*1,000 metric tons*
Europe:			
France	3	4	4
Greece	420	370	380
Italy	742	507	533
Portugal	49	25	34
Spain	650	963	1,196
Total [4]	1,864	1,869	2,147
Middle East:			
Israel	20	20	20
Jordan	8	16	9
Lebanon	8	9	9
Syria	88	170	105
Turkey	70	190	65
Total [4]	194	405	208
Africa:			
Algeria	35	30	45
Morocco	17	17	17
Tunisia	220	135	70
Libya	9	7	9
Total [4]	309	208	174
World total [4]	2,368	2,482	2,529

[1]Marketing year begins November 1. [2]Production excludes residue oil. [3]Preliminary. [4]Includes other countries not listed separately.

FAS, Cotton, Oilseeds, Tobacco and Seeds Division, (202) 720–5710. Prepared or estimated on the basis of official statistics of foreign governments, other foreign source materials, reports of U.S. Agricultural Counselors, AttacheAE1s, and Foreign Service Officers, results of office research, and related information.

Table 3-46.—Margarine, actual weight: Supply and disposition, United States, 1991–2000

Year	Supply			Disposition		
	Production	Stocks, Jan. 1	Total supply	Exports	Domestic disappearance	
					Total	Per capita
	Million pounds	*Million pounds*	*Million pounds*	*Million pounds*	*Million pounds*	*Pounds*
1991	2,698	92	2,791	9	2,691	10.6
1992	2,818	91	2,909	13	2,821	11.0
1993	2,892	75	2,969	15	2,887	11.1
1994	2,623	66	2,693	21	2,610	9.9
1995	2,490	62	2,557	36	2,463	9.3
1996	2,480	58	2,544	29	2,471	9.2
1997	2,367	44	2,417	29	2,344	8.6
1998	2,311	44	2,363	32	2,297	8.3
1999	2,274	35	2,319	36	2,241	8.0
2000	2,398	42	2,453	31	2,353	8.3

ERS, Field Crops Branch. (202) 694–5300. Totals and per capita estimates computed from unrounded numbers.

Table 3-47.—Margarine: Selected reported fats and oils used in manufacture, United States, 1990–2000

Year	Vegetable oils			Animal fats [1]	Total [2]
	Soybean oil	Cottonseed oil	Corn oil		
	Million pounds	*Million pounds*	*Million pounds*	*Million pounds*	*Million pounds*
1990	1,749	NA	208	35	2,102
1991	1,853	25	196	43	2,160
1992	1,926	24	176	37	2,174
1993	2,013	26	161	31	2,239
1994	1,793	NA	NA	42	2,003
1995	1,684	NA	NA	41	1,847
1996	1,694	NA	77	28	1,816
1997	1,650	NA	61	14	1,733
1998	1,606	NA	55	22	1,692
1999	1,574	NA	NA	21	1,664
2000	1,673	NA	56	13	1,755

[1] Lard and edible tallow. [2] Includes small quantities of nuts, coconut, palm, and sunflower oil. NA-not available.

ERS, Field Crops Branch, (202) 694–5300. Compiled from reports of the U.S. Department of Commerce. Totals computed from unrounded numbers.

Table 3-48.—Shortening: Supply and disposition, United States, 1991–2000

Year	Supply			Disposition		
	Factory and warehouse stocks, Jan. 1	Production	Total supply	Exports and shipments	Domestic disappearance	
					Total	Per capita
	Million pounds	*Million pounds*	*Million pounds*	*Million pounds*	*Million pounds*	*Pounds*
1991	116	5,724	5,841	43	5,694	22.4
1992	147	5,719	5,866	43	5,764	22.4
1993	102	6,524	6,626	45	6,532	25.0
1994	94	6,334	6,427	44	6,337	24.0
1995	90	5,975	6,065	45	5,959	22.3
1996	106	5,929	6,035	43	5,954	22.0
1997	81	5,656	5,737	42	5,646	20.6
1998	91	5,724	5,815	54	5,723	20.6
1999	92	5,945	6,037	65	5,951	21.1
2000	86	6,593	6,680	69	6,582	23.1

ERS, Market and Trade Economics Division, Field Crops Branch, (202) 694–5300. Compiled from reports of the Commerce and Agriculture Departments.

Table 3-49.—Shortening: Fats and oils used in manufacture, United States, 1991–2000

Year	Vegetable oils				Animal fats		Total primary and secondary fats and oils [1]
	Cottonseed oil	Soybean oil	Coconut oil	Palm oil	Lard	Edible tallow	
	Million pounds	*Million pounds*	*Million pounds*	*Million pounds*	*Million pounds*	*Million pounds*	*Million pounds*
1991	260	4,152	(2)	(2)	274	460	5,767
1992	241	4,140	(2)	(2)	310	427	5,761
1993	266	4,951	(2)	(2)	296	404	6,544
1994	216	4,929	(2)	(2)	287	405	6,365
1995	212	4,673	(2)	(2)	325	374	6,031
1996	237	4,690	(2)	(2)	284	320	5,935
1997	256	4,517	(2)	(2)	272	312	5,679
1998	200	4,748	(2)	(2)	280	259	5,749
1999	167	5,069	(2)	(2)	241	262	5,968
2000	166	5,714	(2)	(2)	239	284	6,617

[1] Includes small quantities of corn, peanut, safflower, and sunflower oil. [2] Not included to avoid disclosure.

ERS, Market and Trade Economics Division, Field Crops Branch, (202) 694–5300. Compiled from reports of the U.S. Department of Commerce. Totals computed from unrounded numbers.

Table 3-50.—Inedible tallow and grease: Supply and disposition, United States, and price per pound at Chicago, 1991–2000

Year	Supply			Disposition				Price of inedible tallow No. 1 at Chicago, per pound
	Stocks Jan. 1	Production	Total	Exports	Factory consumption			
					Total	Use in soap	Use in feed	
	Million pounds	*Million pounds*	*Million pounds*	*Million pounds*	*Million pounds*	*Million pounds*	*Million pounds*	*Cents*
1991	357	5,759	6,116	1,959	2,949	392	1,748	13.3
1992	349	5,768	6,117	2,285	3,050	334	1,954	14.4
1993	309	6,621	6,930	2,125	3,018	300	1,995	14.9
1994	320	6,712	7,032	2,176	3,190	301	2,102	17.4
1995	348	6,745	7,093	2,683	3,223	264	2,167	19.2
1996	373	6,376	6,749	2,004	3,289	245	2,253	21.7
1997	266	6,249	6,516	1,689	3,399	245	2,401	20.7
1998	339	6,575	6,914	2,300	3,442	228	2,452	17.7
1999	437	7,076	7,513	1,940	3,728	229	2,751	13.00
2000	405	7,149	7,554	1,745	3,662	148	2,756	10.18

ERS, Market and Trade Economics Division, Field Crops Branch, (202) 694–5300.

Table 3-51.—Fats, oils, and oilseeds (fat or oil equivalent): World production, 1998–99/2001–2002

Commodity	World production[1]			
	1998–99	1999–2000	2000–2001	2001–2002[2]
	1,000 metric tons	*1,000 metric tons*	*1,000 metric tons*	*1,000 metric tons*
Edible vegetable oils:				
Cottonseed	3,570	3,567	3,520	3,817
Olive[3]	2,504	2,368	2,482	2,529
Peanut	4,435	4,146	4,288	4,751
Rapeseed	11,807	13,645	13,060	12,216
Soybean	24,650	24,736	26,797	28,823
Sunflower	9,182	9,629	8,425	7,534
Total	56,148	58,091	58,572	59,670
Tropical oils:				
Coconut	2,711	3,335	3,610	3,255
Palm	19,251	21,796	24,331	25,339
Palm kernel	2,426	2,754	3,001	3,110
Total	24,388	27,885	30,942	31,704
Animal fats:				
Butter (fat content)	3,560	3,727	3,933	4,253
Tallow and grease	8,374	8,584	8,312	8,400
Total	11,934	12,311	12,245	12,653
Grand total[4]	92,470	98,287	101,759	104,027

[1] Split year includes Northern Hemisphere crop harvested in the late months of the first year shown combined with Southern Hemisphere and certain Northern Hemisphere crops harvested in the early months of the following year. [2] Preliminary. [3] Excludes olive residue oil. [4] Excludes linseed oil.

FAS, Cotton, Oilseeds, Tobacco and Seeds Division, (202) 720–5710. Prepared or estimated on the basis of official statistics of foreign governments, other foreign source materials, reports of U.S. Agricultural Counselors, Attachés, and Foreign Service Officers, results of office research, and related information.

Table 3-52.—Fats, oils, oilseeds, and oilseed cake and meal: Exports of selected items, United States, 1992–2001

Year beginning January	Lard	Inedible animal tallow, greases, and oils[1]	Oilseeds				
			Cottonseed	Flaxseed	Peanuts unshelled	Peanuts shelled	Soybeans
	Metric tons	*Metric tons*	*Metric tons*	*Metric tons*	*Metric tons*	*Metric tons*	*Metric tons*
1992	61,811	1,362,824	196,408	5,091	63,702	319,677	19,794,760
1993	51,922	1,277,797	126,394	3,421	56,534	216,294	19,213,704
1994	62,122	1,343,096	188,321	3,387	33,244	169,221	18,080,877
1995	56,260	1,616,550	147,600	2,317	44,529	285,530	22,757,454
1996	45,690	1,192,335	83,078	2,890	30,005	208,733	25,565,559
1997	53,729	1,032,246	114,804	5,319	41,245	225,640	26,206,065
1998	59,412	1,363,415	127,554	10,987	27,718	182,324	20,302,984
1999	66,867	1,238,001	117,406	3,167	22,708	184,388	23,141,059
2000	78,930	1,108,129	194,780	27,282	23,963	229,715	26,985,945
2001[2]	46,869	964,642	245,418	60,733	13,843	144,563	28,723,781

Year beginning January	Vegetable oils							Oilseed cake and meal	
	Cocoa butter	Coconut oil	Cottonseed oil[3]	Linseed oil	Margarine	Peanut oil	Soybean oil[3]	Soybean	Other[4]
	Metric tons	*Metric tons*	*Metric tons*	*Metric tons*	*Metric tons*	*Metric tons*	*Metric tons*	*Metric tons*	*Metric tons*
1992	2,821	9,449	123,293	3,942	5,928	64,697	764,845	6,236,001	781,272
1993	4,296	6,363	83,087	3,799	6,978	23,984	705,561	5,500,814	811,855
1994	2,139	8,495	135,639	5,373	9,350	27,027	824,169	4,850,754	555,557
1995	2,972	9,090	137,693	15,422	17,080	47,741	1,037,306	5,858,626	511,125
1996	2,862	3,987	96,014	14,925	13,015	37,300	571,209	5,861,575	195,306
1997	3,886	5,170	110,575	34,691	13,210	8,818	1,015,356	6,992,801	187,401
1998	5,606	3,799	86,902	29,237	14,326	4,183	1,463,885	8,034,793	167,932
1999	5,566	4,276	56,973	28,373	16,090	5,806	845,206	6,566,824	218,166
2000	9,313	5,636	58,484	33,390	14,059	5,515	595,409	6,263,048	160,176
2001[2]	17,506	4,394	64,573	28,575	14,171	6,490	688,336	7,146,543	260,923

[1] Includes edible and inedible tallow, choice white grease, wool grease, yellow grease, edible and inedible oleo and oleo stearine, and animal oils, fats and oils, n.e.c. [2] Preliminary. [3] Includes shipments under PL480. [4] Includes corn meal.

FAS, Cotton, Oilseeds, Tobacco and Seeds Division, (202) 720–5710. Compiled from reports of the U.S. Department of Commerce.

Table 3-53.—Oilseeds, oils, and oilseed cake and meal: Imports of selected items, United States, 1992–2001

Year beginning January	Oilseeds					
	Castor beans	Copra	Flaxseed	Peanuts unshelled	Peanuts shelled [1]	Poppy seed
	Metric tons	*Metric tons*	*Metric tons*	*Metric tons*	*Metric tons*	*Metric tons*
1992	1	291	125,565	48	9,948	4,989
1993	417	396	110,655	0	651	5,162
1994	5	1,515	143,079	2,633	1,296	5,618
1995	0	1,387	186,153	1,628	32,901	6,122
1996	2	1,023	202,313	2,986	38,051	6,243
1997	0	1,170	223,519	3,176	41,272	5,238
1998	0	1,367	171,093	2,686	46,792	5,865
1999	0	605	182,859	2,413	50,225	6,192
2000	0	1,008	122,330	2,423	59,682	5,300
2001 [2]	0	879	49,667	3,458	66,022	4,742

Year beginning January	Oilseeds—continued			Vegetable oils			
	Rapeseed	Sesame seed	Soybeans	Cocoa butter	Castor oil	Coconut oil	Linseed oil
	Metric tons	*Metric tons*	*Metric tons*	*Metric tons*	*Metric tons*	*Metric tons*	*Metric tons*
1992	1,543	34,993	69,519	99,512	34,017	501,466	351
1993	162,093	36,924	97,361	83,399	42,215	443,496	160
1994	409,500	40,516	179,270	54,547	44,109	441,332	427
1995	200,054	39,356	130,141	57,158	41,417	491,151	1,744
1996	261,348	46,563	86,981	68,762	39,938	423,189	2,699
1997	318,249	42,629	258,602	87,689	41,025	589,192	3,102
1998	350,469	47,437	148,780	65,307	48,477	587,062	4,306
1999	210,262	42,214	84,263	80,587	46,671	308,902	5,635
2000	241,585	49,042	110,451	94,629	40,739	477,466	6,102
2001 [2]	237,426	49,072	93,686	80,746	45,395	477,278	4,478

Year beginning January	Vegetable oils—continued						Total oilseed cake and meal
	Olive oil	Palm oil	Palm kernel oil	Peanut oil	Rapeseed oil	Tung oil	
	Metric tons	*Metric tons*	*Metric tons*	*Metric tons*	*Metric tons*	*Metric tons*	*Metric tons*
1992	114,271	102,211	167,512	495	381,456	4,996	605,734
1993	121,248	129,371	119,075	200	395,310	4,270	692,255
1994	126,024	160,827	141,372	5,281	425,533	5,401	783,569
1995	122,270	101,621	121,949	3,165	430,648	4,427	831,047
1996	112,778	125,383	148,719	1,639	522,183	3,943	999,513
1997	163,469	134,519	161,706	6,648	491,083	6,265	1,071,926
1998	164,972	115,871	149,305	30,336	499,574	3,879	1,293,373
1999	162,733	138,864	208,145	9,633	523,219	5,822	1,130,363
2000	203,960	165,606	167,756	19,548	533,485	3,554	1,252,473
2001 [2]	211,893	174,054	150,366	32,789	539,796	11,430	1,074,434

[1] Includes blanched or roasted peanuts. [2] Preliminary.

FAS, Cotton, Oilseeds, Tobacco and Seeds Division, (202) 720–5710. Compiled from reports of the U.S. Department of Commerce.

Table 3-54.—Animal tallow, greases, and oils: [1] United States exports by region and country of destination 1997–2001

Continent and country	1997	1998	1999	2000	2001 [2]
	Metric tons	*Metric tons*	*Metric tons*	*Metric tons*	*Metric tons*
Mexico	272,974	348,265	282,363	284,168	319,082
Turkey	141,149	110,598	102,593	117,073	88,399
European Union	118,009	128,856	92,194	79,547	80,063
Canada	58,966	61,997	68,948	58,792	62,866
Dominican Republic	38,159	36,053	44,929	48,445	57,137
Guatemala	34,494	55,083	65,209	55,973	56,807
Venezuela	61,824	57,655	38,113	52,947	46,969
Colombia	37,989	61,081	58,813	21,869	33,304
Japan	39,494	36,763	42,338	34,764	31,925
Nigeria	4,000	7,298	26,853	30,500	25,539
Korea, Republic	33,372	57,010	107,967	74,371	25,002
El Salvador	32,317	37,850	43,258	40,671	22,988
Russian Federation	1,493	4,591	26,652	16,818	21,319
Honduras	15,251	22,979	24,802	21,802	17,249
China, Peoples Republic	2,344	8,637	37,223	32,193	15,520
Nicaragua	10,925	10,940	7,900	10,445	13,149
Peru	102	10,115	1,888	509	12,290
Haiti	3,069	9,379	12,678	11,426	10,397
Jamaica	9,144	8,090	4,731	4,714	7,320
Costa Rica	177	6,833	7,530	6,158	6,701
Saudi Arabia	6,818	8,068	4,797	8,460	5,794
Egypt	16,951	23,395	22,601	13,294	4,557
Trinidad and Tobago	3,751	4,551	3,814	4,843	4,144
Morocco	12,680	10,470	12,413	10,023	4,000
Panama	6,279	9,866	6,505	5,753	3,433
Others	57,838	212,788	92,733	66,844	17,146
Grand total [3]	1,019,567	1,349,210	1,239,843	1,112,398	993,098

[1] This category includes edible tallow; inedible tallow; choice white grease; pig and poultry fat, yellow grease; sheep or goat, other fat, raw or rendered; lard stearin, lard oil, oleo-oil and tallow oil; other animal fat, and baking and frying fat. [2] Preliminary. [3] May not add due to rounding.

FAS, Dairy, Livestock and Poultry Division, (202) 720–8031. Bureau of the Census, U.S. Department of Commerce.

Table 3-55.—Fats and oils: Use in products for civilian consumption, total and per capita, United States, 1991–2000

Calendar year	Food products[1]													
	Butter (actual weight)		Lard and tallow (direct use)[2]		Margarine (actual weight)		Baking and frying fats (shortening)		Salad and cooking oils		Other edible use		All food products (fat content)	
	Total	Per capita	Total	Per capita	Total	Per capita	Total	Per capita	Total	Per capita	Total	Per capita	Total	Per capita
	Million lbs	*Lbs*	*Million lbs*	*Lbs*	*Million lbs*	*Lbs*	*Million lbs*	*Lbs*	*Million lbs*	*Lbs*	*Million lbs*	*Lbs*	*Million lbs*	*Lbs*
1991	1,100	4.3	796	3.2	2,691	10.6	5,662	22.4	6,743	26.7	321	1.3	16,556	65.5
1992	1,156	4.3	901	3.5	2,821	11.0	5,732	22.4	6,946	27.1	367	1.4	17,127	66.8
1993	1,209	4.6	711	2.8	2,887	11.1	6,495	25.0	6,907	26.6	451	1.7	17,841	68.7
1994	1,255	4.8	1,110	4.2	2,610	9.9	6,305	24.0	6,845	26.0	426	1.6	17,778	67.6
1995	1,187	4.4	963	3.6	2,463	9.3	5,926	22.3	7,057	26.5	434	1.6	17,300	65.0
1996	1,148	4.3	1,059	3.9	2,471	9.2	5,914	22.0	6,924	25.7	361	1.4	17,153	63.7
1997	1,116	4.1	1,102	4.0	2,344	8.6	5,606	20.6	7,652	28.1	297	1.1	17,426	64.0
1998	1,208	4.4	1,409	5.1	2,297	8.3	5,670	20.6	7,532	27.3	365	1.3	17,780	64.5
1999	1,307	4.7	1,545	5.6	2,241	8.0	5,886	21.1	8,030	28.8	431	1.6	18,731	67.2
2000[3] ...	1,277	4.5	1,686	5.9	2,353	8.3	6,513	23.1	9,522	33.7	432	1.6	21,056	74.6

Calendar year	Industrial products											
	Soap		Fatty acids		Animal feeds		Other inedible products		All inedible products[4]		All products[5]	
	Total	Per capita	Total	Per capita	Total	Per capita	Total	Per capita	Total	Per capita	Total	Per capita
	Million pounds	*Pounds*	*Million pounds*	*Pounds*	*Million pounds*	*Pounds*	*Million pounds*	*Pounds*	*Million pounds*	*Pounds*	*Million pounds*	*Pounds*
1991	833	3.3	2,235	8.8	1,974	7.8	286	1.1	5,719	22.6	22,275	88.9
1992	739	2.9	2,041	8.0	2,177	8.5	549	2.1	5,904	23.1	23,031	90.9
1993	749	2.9	1,898	7.3	2,200	8.5	589	2.3	5,846	22.6	23,687	92.4
1994	687	3.0	1,959	8.8	2,340	8.7	654	2.9	6,103	25.2	23,881	92.2
1995	594	2.8	1,964	8.6	2,341	8.8	747	2.5	6,101	23.0	23,401	89.5
1996	469	1.8	1,921	7.2	2,430	9.1	782	2.9	6,018	22.7	23,171	87.8
1997	567	2.1	2,342	8.7	2,646	9.9	557	2.1	6,535	24.4	23,961	89.9
1998	561	2.1	2,187	8.1	2,878	10.6	578	2.1	6,573	24.2	24,353	90.5
1999	565	2.1	2,028	7.4	3,200	11.7	553	2.0	6,733	24.7	25,464	93.8
2000[3] ...	423	1.5	1,906	7.0	3,220	11.8	463	1.7	6,416	23.4	27,472	100.3

[1] Domestic disappearance data are computed by ERS. [2] Includes edible tallow direct use beginning in 1979. [3] Preliminary. [4] Including paint, varnish, resin, plastic, and lubricants. [5] Including only fat content of butter and margarine.

ERS, Market and Trade Economics Division, Field Crops Branch, (202) 694–5300.

Table 3-56.—Fats and oils: Index numbers of wholesale prices, leading markets, United States, 1991–2000

[1982=100]

Year	All fats and oils excluding butter	Seventeen major fats and oils									
		All fats and oils	Classified by origin		Classified by use						
			Animal	Vegetable	Edible			Industrial			
				Domestic origin	Butter	Lard	All edible	Soap fats	Drying oils	All industrial
1991	69.3	53.9	53.6	112.9	NA	104.6	58.0	87.5	99.6	86.5
1992	67.9	51.3	50.8	131.1	NA	102.5	53.8	95.0	89.3	91.9
1993	74.5	54.4	50.2	147.3	NA	116.8	57.1	97.9	94.3	94.9
1994	88.9	61.5	53.0	179.3	NA	128.1	65.7	112.9	93.1	107.4
1995	91.9	115.5	59.6	174.4	NA	151.1	66.1	242.8	92.9	124.0
1996	87.8	655.3	68.1	159.3	NA	162.5	67.1	1,346.9	10.8	129.0
1997	90.2	67.8	74.2	156.8	NA	157.1	67.6	150.2	11.0	141.4
1998	94.3	78.4	88.5	174.3	NA	146.4	84.5	143.6	10.8	135.1
1999	70.1	58.1	67.0	128.0	NA	140.3	60.4	106.6	10.8	102.7
2000	57.7	51.8	63.3	107.68	NA	128.9	55.4	82.9	10.2	81.4

NA-not available.

ERS, Market and Trade Economics Division, Field Crops Branch, (202) 694–5300.

Table 3-57.—Fats and oils: Wholesale price per pound, 1996–2001 [1]

Item and market	1996	1997	1998	1999	2000	2001
	Cents	*Cents*	*Cents*	*Cents*	*Cents*	*Cents*
Castor oil, No. 1, Brazilian, tanks, imported, New York	41.70	41.50	41.50	48.00	47.42	47.92
Coconut oil, crude, tanks, f.o.b. New York	42.62	39.40	31.17	39.89	23.34	24.15
Corn oil, crude, tank cars, f.o.b. Decatur	26.50	24.85	30.33	23.31	20.50	15.75
Cottonseed oil, crude, tank cars, f.o.b. Valley	25.90	26.51	31.03	23.95	20.86	15.41
Linseed oil, raw, tank cars, Minneapolis	37.10	36.25	36.00	36.00	35.83	36.79
Palm oil, U.S. ports, refined	26.60	27.25	31.88	22.86	16.28	15.73
Rapeseed oil, refined (denatured), tanks, New York	58.10	88.00	90.00	80.00	90.00	91.67
Safflower oil, tanks, New York	59.00	59.00	59.00	59.00	59.00	78.75
Soybean oil, crude, tank cars, f.o.b. Decatur	23.80	23.27	25.73	17.72	15.01	14.49
Sunflower oil, crude, Minneapolis	24.66	23.45	24.24	19.09	15.85	17.34
Tallow, inedible, number delivered Chicago	20.10	20.75	17.67	12.99	10.18	11.50
Tung oil, imported, drums, f.o.b. New York	63.30	101.50	100.00	84.83	79.33	61.63

[1] All prices are calendar year basis.

ERS, Market and Trade Economics Division, Field Crops Branch, (202) 694–5300. Compiled from the Chemical Marketing Reporter, the National Provisioner, the Wall Street Journal, and the U.S. Department of Labor.

CHAPTER IV

STATISTICS OF VEGETABLES AND MELONS

This chapter contains statistics on potatoes, sweet potatoes, and commercial vegetables and melons.

For potatoes and sweet potatoes, the estimates of area, production, value, and farm disposition pertain to the total crop and include quantities produced both for sale and for use on farms where grown. Potato statistics are shown on a within-year seasonal grouping of winter, spring, summer, and fall crops, by States. Some States have production in more than one seasonal group.

For processing vegetables, the estimates of area, production, and value for each of 10 crops relate to production used by commercial canners, freezers, and other processors, except dehydrators. These estimates include raw products grown by processors themselves and those grown under contract or purchased on the open market. This production and the actual area harvested are not duplicated in the fresh market estimates for the same commodities. The production of those vegetables used for processing for which regular processing estimates are not made is included in the fresh market estimates. The processed segment of production for asparagus, broccoli, and cauliflower, combined with fresh market production during the year, is published at the end of the season, separately. In 2000, estimates were added for collard greens, kale, mustard greens, turnip greens, okra, chili peppers, pumpkins, radishes, and squash. In 2002, estimates for fresh market lima beans, beets for canning, Brussels sprouts, cabbage for kraut, eggplant, escarole/endive, collard greens, kale, mustard greens, turnip greens, okra, and radishes were discontinued. Additionally, States were removed from the program for certain commodities. For details on the 2002 program changes see the following website: http:/www.usda.gov/nass/events/programchg/vegprogchgs.htm.

Seasonal Groups and Marketing Period

Prospective Area For Harvest

Winter: January, February, March

Spring: April, May, June

Summer: July, August, September

Fall: October, November, December

Annual Acreage, Yield, Production, and Value

The seasonal patterns of harvest do not correspond precisely in all States to the estimating period or periods designated. In some cases, only one seasonal group is shown for a State, but marketing may be active in earlier or later months. Because of the small volume from this earlier or later period, the crop estimate has been placed in the seasonal group where the largest portion is harvested.

In 2002, commercial vegetables for fresh market include 24 principal vegetable and melon crops in the major producing States. These estimates relate to crops which are grown primarily for sale, and they do not include vegetables and melons produced in farm and nonfarm gardens. The bulk of the production of the principal vegetable and melon crops is for consumption in the fresh state. However, quantities used by processors of artichokes, celery, garlic, onions, bell peppers, chile peppers, pumpkins, and squash are included, and separate estimates of commercial processing are not made for these crops. The commercial estimates of the principal crops include local market production from areas near consuming centers as well as production from well recognized commercial areas which specialize in producing supplies for shipment to distant markets.

For fresh market vegetables and melons, value per unit and total value are on a f.o.b. basis. For processed vegetables, value per unit and total value are at processing plant door.

Aggregate data for the years 1999 and 2000 and 2000 and 2001 lack comparability with data from other years because of program changes altering the crops included.

Table 4-1.—Vegetables, commercial: Area, production, and value of principal crops, United States, 1993–2002

Year	Area [1]		
	For fresh market [2]	For processing [3]	Total
	Acres	*Acres*	*Acres*
1993	1,855,000	1,405,190	3,260,190
1994	1,896,400	1,588,450	3,484,850
1995	1,851,860	1,580,400	3,432,260
1996	1,886,780	1,485,020	3,371,800
1997	1,849,730	1,423,000	3,272,730
1998	1,860,070	1,443,510	3,303,580
1999	1,907,120	1,512,750	3,419,870
2000	2,060,390	1,449,230	3,334,700
2001	2,037,660	1,333,610	3,197,530
2002	1,934,200	1,349,020	3,164,220

Year	Production [4]		
	For fresh market [2]	For processing [3]	Total
	Tons	*Tons*	*Tons*
1993	19,526,400	14,914,797	34,441,197
1994	20,808,650	18,316,538	39,125,188
1995	19,856,250	17,543,177	37,399,427
1996	20,600,500	17,547,062	38,147,562
1997	21,828,450	16,229,609	38,058,059
1998	20,934,600	22,535,750	36,410,830
1999	22,484,150	15,476,230	41,547,180
2000	22,557,450	17,026,330	39,583,780
2001	22,368,750	14,990,550	37,359,300
2002	21,859,000	17,116,930	38,975,930

Year	Value [5]		
	For fresh market [2]	For processing [3]	Total
	1,000 dollars	*1,000 dollars*	*1,000 dollars*
1993	7,134,587	1,195,154	8,329,741
1994	6,874,816	1,472,835	8,347,651
1995	7,703,693	1,463,400	9,167,093
1996	6,883,050	1,470,784	8,353,834
1997	8,070,906	1,372,269	9,443,175
1998	8,053,929	1,354,696	9,408,625
1999	7,610,780	1,660,051	9,270,831
2000	8,574,891	1,415,091	9,989,982
2001	8,463,647	1,256,032	9,719,679
2002	8,910,719	1,346,892	10,257,611

[1] Area for fresh market is area for harvest, including any partially harvested or not harvested because of low prices or other economic factors. Area for processing is area harvested. [2] Area, production, and farm value of the following crops for which regular seasonal estimates are prepared in major producing States: Artichokes, asparagus, snap beans, lima beans, broccoli, brussels sprouts, cabbage, cantaloups, carrots, cauliflower, celery, sweet corn, cucumbers, eggplant, escarole/endive, garlic, honeydew melons, head lettuce, leaf lettuce, romaine lettuce, onions, green peppers, spinach, tomatoes, and watermelons. In 2000, collard greens, kale, mustard greens, turnip greens, okra, chile peppers, pumpkins, radishes, and squash were added. In 2002, fresh market lima beans, beets for canning, Brussels sprouts, cabbage for kraut, eggplant, escarole/endive, collard greens, kale, mustard greens, turnip greens, okra, and radishes were discontinued. Additionally in 2002, States were removed from the program for certain commodities. See table footnotes when comparing years. [3] Area, production, and farm value of the following 10 crops in all States: Lima beans, snap beans, beets, cabbage (sauerkraut), carrots, sweet corn, cucumbers (pickles), green peas, spinach, and tomatoes. Production of other vegetables processed included in fresh market series of estimates. [4] Production for fresh market excludes some quantities not marketed because of low prices or other economic factors. [5] Value for all fresh market vegetables. For processing vegetables, value at processing plant door.

NASS, Crops Branch, (202) 720–2127.

Table 4-2.—Vegetables, commercial: Area of principal crops, by States, 2000–2002 [1] [2]

State	For fresh market [3]			For processing [5]			Total		
	2000	2001	2002 [4]	2000	2001	2002 [4]	2000	2001	2002 [4]
	Acres	*Acres*	*Acres*	*Acres*	*Acres*	*Acres*	*Acres*	*Acres*	*Acres*
AL	9,200	7,570	6,400	....	....	....	9,200	7,570	6,400
AZ	122,100	124,800	122,700	....	....	....	122,100	124,800	122,700
AR	5,400	5,430	3,700	7,200	4,300	4,400	12,600	9,730	8,100
CA	860,100	835,600	818,000	311,600	289,700	326,300	1,171,700	1,125,300	1,144,300
CO	30,200	30,200	31,000	....	....	....	30,200	30,200	31,000
CT	4,730	4,970	4,400	....	....	....	4,730	4,970	4,400
DE	7,020	7,330	6,500	49,560	41,700	46,100	56,580	49,030	52,600
FL	199,750	196,300	186,600	....	....	....	199,750	196,300	186,600
GA	139,300	140,550	120,700	5,000	5,600	7,620	144,300	146,150	128,320
HI [6]	1,700	1,670	....	....	....	....	1,700	1,670	....
ID	7,400	7,800	7,900	....	....	....	7,400	7,800	7,900
IL	15,830	15,270	15,680	47,800	41,500	45,900	63,630	56,770	61,580
IN	16,700	17,000	16,400	....	....	....	16,700	17,000	16,400
IA	....	....	....	8,300	8,200	1,400	8,300	8,200	1,400
LA	2,200	1,700	1,800	....	....	....	2,200	1,700	1,800
ME	2,100	2,000	2,000	....	....	....	2,100	2,000	2,000
MD	22,130	22,850	21,700	20,000	11,900	12,500	42,130	34,750	34,200
MA	7,020	7,150	6,700	....	....	....	7,020	7,150	6,700
MI	64,850	62,300	62,600	58,450	50,100	55,100	123,300	112,400	117,700
MN	290	320	250	227,070	204,110	217,600	227,360	204,430	217,850
MS	2,800	2,500	2,300	....	....	....	2,800	2,500	2,300
MO	5,800	5,000	5,100	11,900	15,900	18,200	17,700	20,900	23,300
NV	5,500	4,300	4,400	....	....	....	5,500	4,300	4,400
NH	1,800	1,800	1,500	....	....	....	1,800	1,800	1,500
NJ	34,300	33,700	29,300	10,450	10,500	9,800	44,750	44,200	39,100
NM	26,700	25,400	25,000	....	....	....	26,700	25,400	25,000
NY	79,200	91,520	84,700	77,640	74,150	53,500	156,840	165,670	138,200
NC	63,100	62,460	54,500	....	....	....	63,100	62,460	54,500
OH	30,370	32,840	29,570	8,900	10,650	8,700	39,270	43,490	38,270
OK	6,000	6,000	5,500	....	....	....	6,000	6,000	5,500
OR	30,770	28,660	28,200	92,710	73,860	73,500	123,480	102,520	101,700
PA	32,580	31,180	30,400	11,900	10,150	10,500	44,480	41,330	40,900
RI	900	900	700	....	....	....	900	900	700
SC	21,100	22,200	16,600	....	....	....	21,100	22,200	16,600
TN	15,130	11,950	12,400	4,720	4,720	3,000	19,850	16,670	15,400
TX	105,800	108,100	93,700	21,900	28,500	31,300	127,700	136,600	125,000
UT	2,400	2,100	2,100	....	....	....	2,400	2,100	2,100
VT	900	1,000	900	....	....	....	900	1,000	900
VA	20,320	20,740	18,200	2,430	1,970	1,600	22,750	22,710	19,800
WA	43,300	41,100	40,200	157,300	142,900	142,400	200,600	184,000	182,600
WI	13,600	13,400	13,900	233,800	227,300	210,700	247,400	240,700	224,600
Other States [7]	....	....	....	80,600	75,900	68,900	80,600	75,900	68,900
US	2,060,390	2,037,660	1,934,200	1,449,230	1,333,610	1,349,020	3,509,620	3,371,270	3,283,220

[1] Area for fresh market and for processing is area harvested. [2] Commodity estimates for 2002 and 2001 are comparable. Estimates for 2002 are not comparable due to vegetable estimation program changes. These changes are documented in footnotes found under each individual commodity table. [3] Area of the following crops for which regular seasonal estimates are prepared in major producing States: Artichokes, asparagus, snap beans, lima beans, broccoli, brussels sprouts, cabbage, cantaloups, carrots, cauliflower, celery, sweet corn, cucumbers, eggplant, escarole/endive, garlic, honeydew melons, head lettuce, leaf lettuce, romaine lettuce, onions, green peppers, spinach, tomatoes, and watermelons. In 2000, collard greens, kale, mustard greens, turnip greens, okra, chile peppers, pumpkins, radishes, and squash were added. In 2002, fresh market lima beans, beets for canning, Brussels sprouts, cabbage for kraut, eggplant, escarole/endive, collard greens, kale, mustard greens, turnip greens, okra, and radishes were discontinued. Additionally in 2002, States were removed from the program for certain commodities. See table footnotes when comparing years. [4] Preliminary. [5] Includes Lima beans, snap beans, beets, cabbage (sauerkraut), carrots, sweet corn, cucumbers (pickles), green peas, spinach, and tomatoes. Other vegetables processed (dual purpose) included in fresh market series of estimates. [6] Hawaii removed from the vegetable estimation program for all commodities in 2002. [7] Processing, 2000 - AL, FL, ID, IN, ME, MA, NC, and SC. 2001 - AL, FL, ID, IN, MA, NC, and SC. 2002 - AL, FL, ID, IN, MA, NC, and SC.

NASS, Crops Branch, (202) 720–2127.

Table 4-3.—Vegetables, commercial: Production of principal crops, by States, 2000–2002 [1]

State	For fresh market [2]			For processing [4]			Total		
	2000	2001	2002 [3]	2000	2001	2002 [3]	2000	2001	2002 [3]
	1,000 Cwt	*1,000 Cwt*	*1,000 Cwt*	*Tons*	*Tons*	*Tons*	*Tons*	*Tons*	*Tons*
AL	985	901	879	....	....	....	49,250	45,050	43,950
AZ	36,956	40,245	37,011	....	....	....	1,847,800	2,012,250	1,850,550
AR	671	872	799	28,800	15,480	15,840	62,350	59,080	55,790
CA	229,678	218,412	217,292	10,598,690	8,892,050	11,287,050	22,082,590	19,812,650	22,151,650
CO	9,417	9,523	9,390	....	....	....	470,850	476,150	469,500
CT	298	288	308	....	....	....	14,900	14,400	15,400
DE	1,120	1,610	1,538	161,570	144,980	139,000	217,570	225,480	215,900
FL	47,610	43,913	43,452	....	....	....	2,380,500	2,195,650	2,172,600
GA	25,392	24,311	21,058	18,000	20,860	27,580	1,287,600	1,236,410	1,080,480
HI [5]	461	419	....	....	....	....	23,050	20,950	....
ID	4,810	4,992	5,056	....	....	....	240,500	249,600	252,800
IL	4,370	3,899	3,700	170,060	183,370	162,950	388,560	378,320	347,950
IN	2,899	4,034	3,537	....	....	....	144,950	201,700	176,850
IA	....	....	....	55,780	55,100	8,500	55,780	55,100	8,500
LA	242	187	216	....	....	....	12,100	9,350	10,800
ME	116	110	110	....	....	....	5,800	5,500	5,500
MD	1,803	2,546	2,041	114,130	66,220	67,060	204,280	193,520	169,110
MA	448	497	469	....	....	....	22,400	24,850	23,450
MI	8,493	9,154	9,150	390,580	318,280	362,930	815,230	775,980	820,430
MN	94	109	78	957,840	945,450	1,023,220	962,540	950,900	1,027,120
MS	185	425	357	....	....	....	9,250	21,250	17,850
MO	1,218	1,150	1,326	32,920	44,830	50,080	93,820	102,330	116,380
NV	1,848	1,508	2,035	....	....	....	92,400	75,400	101,750
NH	108	99	75	....	....	....	5,400	4,950	3,750
NJ	5,575	5,494	4,688	77,230	69,660	56,340	355,980	344,360	290,740
NM	5,522	5,393	6,200	....	....	....	276,100	269,650	310,000
NY	17,169	17,878	13,836	389,320	377,290	174,850	1,247,770	1,271,190	866,650
NC	7,994	8,048	7,692	....	....	....	399,700	402,400	384,600
OH	4,500	5,331	5,476	205,440	231,140	178,330	430,440	497,690	452,130
OK [6]	420	750	715	....	....	....	21,000	37,500	35,750
OR	12,752	12,153	12,617	546,420	417,630	412,390	1,184,020	1,025,280	1,043,240
PA	3,385	2,538	2,971	82,530	72,990	75,340	251,780	199,890	223,890
RI	50	68	46	....	....	....	2,500	3,400	2,300
SC	3,505	3,789	2,705	....	....	....	175,250	189,450	135,250
TN	1,605	879	1,647	8,210	9,830	7,200	88,460	53,780	89,550
TX	20,702	23,225	20,555	137,940	191,760	194,330	1,173,040	1,353,010	1,222,080
UT	1,140	956	1,050	....	....	....	57,000	47,800	52,500
VT	41	50	45	....	....	....	2,050	2,500	2,250
VA	2,804	3,372	2,949	8,170	8,120	4,570	148,370	176,720	152,020
WA	10,762	11,248	11,653	1,146,940	1,102,930	1,137,040	1,685,040	1,665,330	1,719,690
WI	2,164	1,941	1,905	1,245,990	1,182,620	1,126,400	1,354,190	1,279,670	1,221,650
Other States [6]	....	....	....	649,770	639,960	605,930	649,770	639,960	605,930
US	479,312	472,317	456,627	17,026,330	14,990,550	17,116,930	40,991,930	38,606,400	39,948,280

[1] Commodity estimates for 2000 and 2001 are comparable. Estimates for 2002 are not comparable due to vegetable estimation program changes. These changes are documented in footnotes found under each individual commodity table. [2] Production of the following crops for which regular seasonal estimates are prepared in major producing States: Artichokes, asparagus, snap beans, lima beans, broccoli, brussels sprouts, cabbage, cantaloups, carrots, cauliflower, celery, sweet corn, cucumbers, eggplant, escarole/endive, garlic, honeydew melons, head lettuce, leaf lettuce, romaine lettuce, onions, green peppers, spinach, tomatoes, and watermelons.In 2002, collard greens, kale, mustard greens, turnip greens, okra, chile peppers, pumpkins, radishes, and squash were added. In 2002, fresh market lima beans, beets for canning, Brussels sprouts, cabbage for kraut, eggplant, escarole/endive, collard greens, kale, mustard greens, turnip greens, okra, and radishes were discontinued. Additionally in 2002, states were removed from the program for certain commodities. See table footnotes when comparing years. [3] Preliminary. [4] Includes Lima beans, snap beans, beets, cabbage (sauerkraut), carrots, sweet corn, cucumbers (pickles), green peas, spinach, and tomatoes. Other vegetables processed (dual purpose) included in fresh market series of estimates. [5] Hawaii removed from the vegetable estimation program for all commodities in 2002. [6] 2000 - AL, FL, ID, IN, ME, MA, NC, and SC. 2001 - AL, FL, ID, IN, NC, and SC. 2002 - AL, FL, ID, IN, NC, and SC.

NASS, Crops Branch, (202) 720–2127.

Table 4-4.—Vegetables, commercial: Value of principal crops, by States, 2000–2002 [1]

State	For fresh market [2]			For processing [4]			Total		
	2000	2001	2002 [3]	2000	2001	2002 [3]	2000	2001	2002 [3]
	1,000 dollars	*1,000 dollars*	*1,000 dollars*	*1,000 dollars*	*1,000 dollars*	*1,000 dollars*	*1,000 dollars*	*1,000 dollars*	*1,000 dollars*
AL	14,087	12,959	12,343	....	....	....	14,087	12,959	12,343
AZ	592,260	755,799	1,215,315	....	....	....	592,260	755,799	1,215,315
AR	11,031	15,264	17,423	5,184	2,477	2,693	16,215	17,741	20,116
CA	4,812,193	4,445,749	4,366,206	655,122	545,351	676,778	5,467,315	4,991,100	5,042,984
CO	105,610	98,303	108,557	....	....	....	105,610	98,303	108,557
CT	8,885	8,599	7,700	....	....	....	8,885	8,599	7,700
DE	11,924	17,176	17,072	26,252	26,431	25,608	38,176	43,607	42,680
FL	1,201,734	1,175,523	1,168,639	....	....	....	1,201,734	1,175,523	1,168,639
GA	404,129	342,769	312,804	4,871	5,300	7,425	409,000	348,069	320,229
HI [5]	15,450	15,525	....	....	....	....	15,450	15,525	....
ID	44,577	34,765	37,404	....	....	....	44,577	34,765	37,404
IL	29,972	28,826	20,999	30,200	29,337	23,399	60,172	58,163	44,398
IN	43,182	59,888	52,981	....	....	....	43,182	59,888	52,981
IA	....	....	....	4,016	4,408	595	4,016	4,408	595
LA	1,670	1,253	1,469	....	....	....	1,670	1,253	1,469
ME	3,828	3,575	3,740	....	....	....	3,828	3,575	3,740
MD	35,207	45,285	42,200	14,988	10,818	11,594	50,195	56,103	53,794
MA	17,760	20,381	14,070	....	....	....	17,760	20,381	14,070
MI	156,650	157,708	157,609	62,590	50,413	51,239	219,240	208,121	208,848
MN	888	1,039	517	103,219	92,352	93,066	104,107	93,391	93,583
MS	1,169	2,338	2,321	....	....	....	1,169	2,338	2,321
MO	4,446	4,888	7,691	7,062	9,334	10,257	11,508	14,222	17,948
NV	29,139	18,966	27,764	....	....	....	29,139	18,966	27,764
NH	4,320	3,960	3,188	....	....	....	4,320	3,960	3,188
NJ	144,971	131,811	120,252	9,458	9,396	7,014	154,429	141,207	127,266
NM	81,667	102,381	107,480	....	....	....	81,667	102,381	107,480
NY	330,153	334,588	291,190	42,553	42,519	25,467	372,706	377,107	316,657
NC	129,529	123,471	117,090	....	....	....	129,529	123,471	117,090
OH	91,480	134,695	143,129	20,587	26,200	22,335	112,067	160,895	165,464
OK	2,940	5,250	5,005	....	....	....	2,940	5,250	5,005
OR	113,491	105,445	109,206	66,652	48,947	48,116	180,143	154,392	157,322
PA	65,857	51,371	60,560	10,079	9,458	9,255	75,936	60,829	69,815
RI	1,650	2,108	1,426	....	....	....	1,650	2,108	1,426
SC	54,592	59,715	42,388	....	....	....	54,592	59,715	42,388
TN	47,275	20,334	54,154	1,353	1,388	954	48,628	21,722	55,108
TX	314,922	390,903	390,477	21,536	32,248	38,367	336,458	423,151	428,844
UT	9,579	6,422	7,355	....	....	....	9,579	6,422	7,355
VT	1,353	1,675	1,553	....	....	....	1,353	1,675	1,553
VA	53,302	63,438	59,890	2,031	1,505	1,097	55,333	64,943	60,987
WA	145,161	142,622	151,964	107,124	93,929	95,047	252,285	236,551	247,011
WI	22,811	20,690	20,773	117,144	108,727	96,312	139,955	129,417	117,085
Other States [6]	....	....	....	103,070	105,494	100,274	103,070	105,494	100,274
US	9,160,844	8,967,457	9,281,904	1,415,091	1,256,032	1,346,892	10,575,935	10,223,489	10,628,796

[1] Commodity estimates for 2000 and 2001 are comparable. Estimates for 2002 are not comparable due to vegetable estimation program changes. These changes are documented in footnotes found under each individual commodity table. [2] Value of the following crops for which regular seasonal estimates are prepared in major producing States: Artichokes, asparagus, snap beans, lima beans, broccoli, brussels sprouts, cabbage, cantaloups, carrots, cauliflower, celery, sweet corn, cucumbers, eggplant, escarole/endive, garlic, honeydew melons, head lettuce, leaf lettuce, romaine lettuce, onions, green peppers, spinach, tomatoes, and watermelons. In 2002, collard greens, kale, mustard greens, turnip greens, okra, chile peppers, pumpkins, radishes, and squash were added. In 2002, fresh market lima beans, beets for canning, Brussels sprouts, cabbage for kraut, eggplant, escarole/endive, collard greens, kale, mustard greens, turnip greens, okra, and radishes were discontinued. Additionally in 2002, states were removed from the program for certain commodities. See table footnotes when comparing years. [3] Preliminary. [4] Includes Lima beans, snap beans, beets, cabbage (sauerkraut), carrots, sweet corn, cucumbers (pickles), green peas, spinach, and tomatoes. Other vegetables processed (dual purpose) included in fresh market series of estimates. [5] Hawaii removed from the vegetable estimation program for all commodities in 2002. [6] 2000 - AL, FL, ID, IN, ME, MA, NC, and SC. 2001 - AL, FL, ID, IN, NC, and SC. 2002 - AL, FL, ID, IN, NC, and SC.

NASS, Crops Branch, (202) 720–2127.

Table 4-5.—Asparagus, commercial crop: Area, yield, production, value per hundredweight and per ton, and total value, United States, 1993–2002

Year	Total crop					For fresh market			For processing		
	Area for harvest	Yield per acre	Produc-tion	Value [1]		Produc-tion	Value [1]		Produc-tion	Value [2]	
				Per cwt.	Total		Per cwt.	Total		Per ton	Total
	Acres	*Cwt.*	*1,000 cwt.*	*Dollars*	*1,000 dollars*	*1,000 cwt.*	*Dollars*	*1,000 dollars*	*Tons*	*Dollars*	*1,000 dollars*
1993	81,150	27	2,203	76.10	167,645	1,252	95.70	119,773	47,540	1,010.00	47,872
1994	76,750	29	2,197	80.90	177,641	1,311	100.00	131,173	44,300	1,050.00	46,468
1995	72,340	28	2,024	87.50	177,170	1,100	113.00	124,171	46,180	1,150.00	52,999
1996	73,560	27	1,989	78.70	156,623	1,114	92.90	103,480	43,780	1,210.00	53,143
1997	74,030	27	2,026	90.10	182,531	1,248	108.00	134,860	38,920	1,220.00	47,571
1998	74,430	27	1,979	101.00	199,482	1,264	124.00	156,734	35,720	1,200.00	42,748
1999	75,890	29	2,176	107.00	233,170	1,455	131.00	190,719	36,070	1,180.00	42,451
2000	77,400	29	2,272	97.40	221,299	1,504	117.00	176,017	38,400	1,180.00	45,282
2001	70,150	30	2,078	110.00	228,925	1,372	140.00	192,346	35,290	1,040.00	36,579
2002 [3] ...	66,000	28	1,868	92.50	172,876	1,267	110.00	139,609	30,050	1,110.00	33,267

[1] Price and value on F.O.B. basis. [2] Price and value at processing plant door. [3] Preliminary.

NASS, Crops Branch, (202) 720–2127.

Table 4-6.—Asparagus, commercial crop: Area, production, and value per hundredweight and per ton, by States, 2000–2002

State	Area harvested [1]			Production			Value per unit		
	2000	2001	2002 [2]	2000	2001	2002 [2]	2000	2001	2002 [2]
	Acres	*Acres*	*Acres*	*1,000 cwt.*	*1,000 cwt.*	*1,000 cwt.*	*Dollars per cwt.*	*Dollars per cwt.*	*Dollars per cwt.*
CA [3]	37,000	35,000	34,000	1,184	1,050	1,020	122.00	156.00	114.00
MI	16,500	14,300	15,000	283	290	219	63.90	43.20	53.40
NJ [4]	1,000	1,100	....	32	26	....	75.00	85.00	....
OR [4]	900	750	....	25	28	....	60.00	53.20	....
WA	22,000	19,000	17,000	748	684	629	73.40	71.50	71.40
US	77,400	70,150	66,000	2,272	2,078	1,868	97.40	110.00	92.50

State	For fresh market						For processing					
	Production			Value per unit			Production			Value per unit		
	2000	2001	2002 [2]	2000	2001	2002 [2]	2000	2001	2002 [2]	2000	2001	2002 [2]
	1,000 cwt.	*1,000 cwt.*	*1,000 cwt.*	*Dollars per cwt.*	*Dollars per cwt.*	*Dollars per cwt.*	*Tons*	*Tons*	*Tons*	*Dollars per ton*	*Dollars per ton*	*Dollars per ton*
CA [3]	1,184	1,050	1,020	122.00	156.00	114.00	....	....	....	....	....	....
MI	41	48	21	69.00	49.00	67.00	12,100	12,100	9,900	1,260.00	840.00	1,040.00
NJ [4]	32	26	....	75.00	85.00	....	....	....	....	....	....	....
OR [4]	7	6	....	60.00	44.60	....	900	1,090	....	1,200.00	1,120.00	....
WA	240	242	226	108.00	98.00	97.00	25,400	22,100	20,150	1,140.00	1,140.00	1,140.00
US	1,504	1,372	1,267	117.00	140.00	110.00	38,400	35,290	30,050	1,180.00	1,040.00	1,110.00

[1] Asparagus for fresh market and for processing is frequently harvested from the same area; therefore it is not practical to make individual area estimates for these segments. [2] Preliminary. [3] Includes a small amount of processing asparagus. [4] Estimate discontinued in 2002.

NASS, Crops Branch, (202) 720–2127.

Table 4-7.—Lima beans for fresh market: Area, production, and value per hundredweight, by States, 2000–2002

State	Area harvested			Production			Value per unit		
	2000	2001	2002[1]	2000	2001	2002[1]	2000	2001	2002[1]
	Acres	*Acres*	*Acres*	*1,000 cwt.*	*1,000 cwt.*	*1,000 cwt.*	*Dollars per cwt.*	*Dollars per cwt.*	*Dollars per cwt.*
AL	800	800	....	23	16	....	41.00	44.00	....
GA	3,100	3,700	....	62	126	....	31.00	20.00	....
NC	1,000	1,100	....	50	33	....	35.00	45.00	....
SC	1,000	1,200	....	20	18	....	40.00	35.00	....
VA	170	290	....	3	9	....	63.00	58.00	....
Total	6,070	7,090	....	158	202	....	35.50	29.00	....

[1] Estimates discontinued in 2002.

NASS, Crops Branch, (202) 720–2127.

Table 4-8.—Lima beans for processing: Area, production, and value per ton, by States, 2000–2002[1]

State	Area harvested			Production			Value per unit		
	2000	2001	2002[2]	2000	2001	2002[2]	2000	2001	2002[2]
	Acres	*Acres*	*Acres*	*Tons*	*Tons*	*Tons*	*Dollars per ton*	*Dollars per ton*	*Dollars per ton*
US	49,750	51,200	48,500	58,890	67,160	63,220	432.00	459.00	430.00

[1] 2000 - CA, DE, IL, MD, NJ, OR, TN, WA, and WI. 2001 - CA, DE, IL, MD, NJ, OR, TN, WA, and WI. 2002 - CA, DE, IL, MD, OR, TN, WA, and WI. NJ estimates discontinued in 2002. [2] Preliminary.

NASS, Crops Branch, (202) 720–2127.

Table 4-9.—Snap beans for fresh market: Area, production, and value per hundredweight, by States, 2000–2002

State	Area harvested			Production			Value per unit		
	2000	2001	2002[1]	2000	2001	2002[1]	2000	2001	2002[1]
	Acres	*Acres*	*Acres*	*1,000 cwt.*	*1,000 cwt.*	*1,000 cwt.*	*Dollars per cwt.*	*Dollars per cwt.*	*Dollars per cwt.*
CA	5,500	5,400	6,500	550	432	520	56.00	64.40	62.30
FL	34,500	33,000	35,000	2,691	2,772	2,940	45.30	52.90	50.60
GA	15,500	16,000	17,500	868	960	753	34.00	22.00	32.00
MD	1,900	2,200	2,100	48	79	42	41.00	30.00	41.00
MI	2,000	3,800	4,300	84	190	194	25.00	35.00	38.00
NJ	3,300	3,300	3,100	165	178	115	36.60	34.50	40.50
NY	7,600	11,400	10,200	517	638	561	61.00	63.00	67.40
NC	7,900	7,600	7,300	474	342	365	29.00	26.00	30.00
SC	1,200	1,300	1,400	31	53	42	39.00	30.00	42.00
TN	9,200	7,200	7,600	331	295	236	26.00	30.00	35.00
VA	4,500	4,700	5,000	135	216	190	25.00	34.00	23.00
US	93,100	95,900	100,000	5,894	6,155	5,958	42.60	45.10	47.40

[1] Preliminary.

NASS, Crops Branch, (202) 720–2127.

Table 4-10.—Snap beans for processing, commercial crop: Area, yield, production, value per ton, and total value, United States, 1993–2002

Year	Area harvested	Yield per acre	Production	Value [1]	
				Per ton	Total
	Acres	*Tons*	*Tons*	*Dollars*	*1,000 dollars*
1993	196,530	3.37	661,600	178.00	117,690
1994	221,900	3.68	816,830	166.00	135,369
1995	216,040	3.27	705,540	173.00	122,379
1996	207,050	3.79	784,920	178.00	139,755
1997	195,080	3.74	729,250	176.00	128,032
1998	198,700	3.68	730,990	172.00	125,373
1999	212,150	3.67	778,430	173.00	134,501
2000	218,380	3.82	833,490	171.00	142,502
2001	195,480	3.56	695,190	161.00	112,137
2002 [2] ...	210,100	3.96	831,260	152.00	126,761

[1] Price and value at processing plant door. [2] Preliminary.
NASS, Crops Branch, (202) 720–2127.

Table 4-11.—Snap beans for processing, commercial crop: Area, production, and value per ton, by States, 2000–2002

State	Area harvested			Production			Value per unit		
	2000	2001	2002 [1]	2000	2001	2002 [1]	2000	2001	2002 [1]
	Acres	*Acres*	*Acres*	*Tons*	*Tons*	*Tons*	*Dollars per ton*	*Dollars per ton*	*Dollars per ton*
IL	15,700	12,200	20,600	46,250	40,960	62,300	243.00	195.00	163.00
IN	6,900	6,500	6,000	20,090	15,390	17,860	159.00	188.00	154.00
MI	24,400	16,000	16,000	91,580	56,000	60,030	160.00	160.00	160.00
NY	26,500	22,300	21,300	89,310	66,110	63,590	193.00	174.00	175.00
OR	22,000	19,300	18,700	133,170	121,510	116,250	188.00	174.00	180.00
PA	8,300	7,900	7,500	27,450	22,690	20,340	216.00	213.00	215.00
WI	66,800	68,500	72,200	261,490	224,860	317,070	133.00	125.00	113.00
Other States [2]	47,780	42,780	47,800	164,150	147,670	173,820	185.00	182.00	183.00
Total	218,380	195,480	210,100	833,490	695,190	831,260	171.00	161.00	152.00

[1] Preliminary. [2] 2000 - AR, CA, DE, FL, GA, MD, MN, MO, NJ, NC, OH, TN, TX, and VA. 2001- AR, CA, DE, FL, GA, MD, MN, MO, NJ, NC, OH, TN, TX, and VA. 2002 - AR, CA, DE, FL, GA, MD, MN, MO, NJ, NC, TX, and VA. OH and TN estimates discontinued in 2002.
NASS, Crops Branch, (202) 720–2127.

Table 4-12.—Beets for canning: Area, production, and value per ton, by States, 2000–2002

State	Area harvested			Production			Value per unit		
	2000	2001	2002 [1]	2000	2001	2002 [1]	2000	2001	2002 [1]
	Acres	*Acres*	*Acres*	*Tons*	*Tons*	*Tons*	*Dollars per ton*	*Dollars per ton*	*Dollars per ton*
NY	2,500	2,300	.....	33,450	35,420	.....	73.00	75.20	.....
WI	3,800	.....	.....	55,670	.....	.....	51.20	.....	.....
Other States [2]	1,170	4,350	.....	24,040	75,760	.....	67.90	61.40	.....
US	7,470	6,650	.....	113,160	111,180	.....	61.20	65.80	.....

[1] Estimates discontinued in 2002. [2] 2000 - MN and OR. 2001 - MN, OR, and WI.
NASS, Crops Branch, (202) 720–2127.

Table 4-13.—Broccoli, commercial crop: Area, yield, production, value per hundredweight and per ton, and total value, United States, 1993–2002 [1]

Year	Total crop: Area for harvest	Total crop: Yield per acre	Total crop: Production	Total crop: Value [2] Per cwt.	Total crop: Value [2] Total	For fresh market: Production	For fresh market: Value [2] Per cwt.	For fresh market: Value [2] Total	For processing: Production	For processing: Value [3] Per ton	For processing: Value [3] Total
	Acres	*Cwt.*	*1,000 cwt.*	*Dollars*	*1,000 dollars*	*1,000 cwt.*	*Dollars*	*1,000 dollars*	*Tons*	*Dollars*	*1,000 dollars*
1993	119,200	101	12,059	25.80	310,675	10,685	26.60	284,165	68,710	386.00	26,510
1994	134,100	117	15,714	26.70	419,571	14,152	27.50	389,659	78,110	383.00	29,912
1995	129,400	122	15,815	28.00	443,304	13,843	29.30	405,286	98,590	386.00	38,018
1996	133,500	118	15,693	26.50	415,695	14,428	27.10	391,194	63,250	387.00	24,501
1997	130,800	129	16,880	28.50	481,459	15,744	29.10	457,423	56,810	423.00	24,036
1998	134,300	129	17,351	29.50	511,681	16,228	30.20	489,912	56,148	388.00	21,769
1999	148,000	147	21,690	23.90	518,019	20,517	24.10	493,814	58,656	413.00	24,205
2000	144,300	145	20,880	30.40	635,691	20,067	31.00	623,043	40,670	311.00	12,648
2001	133,000	145	19,280	26.00	500,987	18,345	26.50	486,214	46,750	316.00	14,773
2002 [4]	126,000	139	17,475	31.50	551,280	16,695	32.10	536,226	39,000	386.00	15,054

[1] Sprouting broccoli only. Does not include broccoli rabe nor heading (cauliflower) broccoli. [2] Price and value on f.o.b. basis. [3] Price and value at processing plant door. [4] Preliminary.

NASS, Crops Branch, (202) 720–2127.

Table 4-14.—Broccoli, commercial crop: Area, production, and value per hundredweight, and per ton, by States, 2000–2002 [1]

State	Area harvested: 2000	Area harvested: 2001	Area harvested: 2002 [2]	Production: 2000	Production: 2001	Production: 2002 [2]	Value per unit: 2000	Value per unit: 2001	Value per unit: 2002 [2]
	Acres	*Acres*	*Acres*	*1,000 cwt.*	*1,000 cwt.*	*1,000 cwt.*	*Dollars per cwt.*	*Dollars per cwt.*	*Dollars per cwt.*
AZ	11,300	11,000	11,000	2,260	2,200	1,375	26.70	28.00	45.70
CA	133,000	122,000	115,000	18,620	17,080	16,100	30.90	25.70	30.30
US	144,300	133,000	126,000	20,880	19,280	17,475	30.40	26.00	31.50

State	For fresh market: Production 2000	Production 2001	Production 2002 [2]	Value per unit 2000	Value per unit 2001	Value per unit 2002 [2]	For processing: Production 2000	Production 2001	Production 2002 [2]	Value per unit 2000	Value per unit 2001	Value per unit 2002 [2]
	1,000 cwt.	*1,000 cwt.*	*1,000 cwt.*	*Dollars per cwt.*	*Dollars per cwt.*	*Dollars per cwt.*	*Tons*	*Tons*	*Tons*	*Dollars per ton*	*Dollars per ton*	*Dollars per ton*
AZ	2,260	2,200	1,375	26.70	28.00	45.70	....	....	....	....	....	....
CA	17,807	16,145	15,320	31.60	26.30	30.90	40,670	46,750	39,000	311.00	316.00	386.00
US	20,067	18,345	16,695	31.00	26.50	32.10	40,670	46,750	39,000	311.00	316.00	386.00

[1] Sprouting broccoli only. Does not include broccoli rabe nor heading (cauliflower) broccoli. [2] Preliminary.

NASS, Crops Branch, (202) 720–2127.

Table 4-15.—Cabbage for fresh market: Area, production, and value per hundredweight, by States, 2000–2002

State	Area harvested			Production			Value per unit		
	2000	2001	2002 [1]	2000	2001	2002 [1 2]	2000	2001	2002 [1]
	Acres	*Acres*	*Acres*	*1,000 cwt.*	*1,000 cwt.*	*1,000 cwt.*	*Dollars per cwt.*	*Dollars per cwt.*	*Dollars per cwt.*
AZ	1,900	2,900	3,000	855	1,160	1,065	12.10	10.80	9.70
CA	13,000	12,000	12,500	4,810	4,560	4,750	13.10	13.80	14.00
CO	2,000	2,200	2,600	940	990	1,144	8.10	8.70	9.50
FL	7,900	7,500	8,000	1,896	2,325	2,560	10.20	11.90	11.30
GA	7,100	8,100	8,000	2,485	2,430	2,360	10.00	8.70	13.00
IL	1,000	790	780	175	126	152	19.70	19.70	13.20
MD [3]	630	650	....	79	78	....	14.00	17.00	....
MI	1,700	1,800	1,800	425	576	540	12.80	14.00	12.00
NJ	1,600	1,500	1,600	640	510	624	13.50	13.20	17.10
NY	12,900	13,800	11,800	5,676	5,520	4,130	15.50	16.80	13.10
NC	9,000	9,100	9,000	1,620	1,911	1,980	9.00	8.80	9.00
OH	1,300	1,400	1,600	384	588	552	10.40	11.40	11.90
PA	1,800	1,900	1,800	306	171	342	17.00	19.00	13.30
TX	10,000	9,300	8,700	4,100	3,627	2,784	12.80	18.20	16.20
VA	1,500	1,500	1,400	525	555	532	6.00	7.00	10.00
WI	4,600	4,300	4,500	920	817	900	8.80	7.20	7.80
Total	77,930	78,740	77,080	25,836	25,944	24,415	12.30	13.30	12.60
HI [3]	560	500	....	154	125	....	18.00	22.00	....
US	78,490	79,240	77,080	25,990	26,069	24,415	12.30	13.30	12.60

[1] Preliminary. [2] Includes some quantities of fall storage in NY harvested but not sold because of shrinkage and loss: 2000, 552,000 cwt; 2001, 563,000 cwt; and 2002, 417,000 cwt. [3] Estimates discontinued in 2002.
NASS, Crops Branch, (202) 720–2127.

Table 4-16.—Cabbage for kraut: Area, production, and value per ton, by States, 2000–2002

State	Area harvested			Production			Value per unit		
	2000	2001	2002 [1]	2000	2001	2002 [1]	2000	2001	2002 [1]
	Acres	*Acres*	*Acres*	*Tons*	*Tons*	*Tons*	*Dollars per ton*	*Dollars per ton*	*Dollars per ton*
NY	2,800	2,600	....	76,100	73,320	....	53.40	51.50	....
WI	3,700	3,600	....	103,490	81,470	....	43.40	47.00	....
Other States [2]	1,010	750	....	28,680	19,570	....	45.60	45.70	....
US	7,510	6,950	....	208,270	174,360	....	47.40	48.70	....

[1] Estimates discontinued in 2002. [2] 2000–2001 - OH and OR.
NASS, Crops Branch, (202) 720–2127.

Table 4-17.—Cantaloups for fresh market: Area, production, and value per hundredweight, by States, 2000–2002

State	Area harvested			Production			Value per unit		
	2000	2001	2002[1]	2000	2001	2002[1]	2000	2001	2002[1]
	Acres	*Acres*	*Acres*	*1,000 cwt.*	*1,000 cwt.*	*1,000 cwt.*	*Dollars per cwt.*	*Dollars per cwt.*	*Dollars per cwt.*
AZ	14,900	14,600	14,800	3,353	3,942	4,440	19.60	17.90	13.20
CA	57,500	54,500	54,900	13,225	13,353	13,451	17.10	18.90	16.60
CO	1,500	1,700	1,600	360	391	400	13.30	15.30	16.40
DE[2]	420	430	....	46	45	....	17.00	20.00	....
GA	5,500	5,300	5,700	908	848	1,254	8.50	12.00	15.00
IN	3,000	2,900	3,000	645	725	540	15.50	16.30	16.80
MD	1,400	1,600	1,600	132	176	176	21.00	20.00	20.00
MI[2]	750	500	....	105	53	....	15.30	21.00	....
PA	1,100	1,200	1,300	143	112	130	16.30	15.60	22.00
SC	1,000	1,500	1,500	100	150	90	13.50	13.30	13.00
TX	10,800	11,200	9,500	1,836	2,800	2,565	23.10	24.90	31.50
VA[2]	800	700	....	112	123	....	14.00	12.00	....
US	98,670	96,130	93,900	20,965	22,718	23,046	17.50	19.00	17.60

[1] Preliminary. [2] Estimates discontinued in 2002.
NASS, Crops Branch, (202) 720–2127.

Table 4-18.—Carrots for fresh market, commercial crop: Area, production, and value per hundredweight, by States, 2000–2002

State	Area harvested			Production			Value per unit		
	2000	2001	2002[1]	2000	2001	2002[1]	2000	2001	2002[1]
	Acres	*Acres*	*Acres*	*1,000 cwt.*	*1,000 cwt.*	*1,000 cwt.*	*Dollars per cwt.*	*Dollars per cwt.*	*Dollars per cwt.*
AZ	1,800	....	....	486	....	....	8.10	....	....
CA	80,500	81,000	77,000	23,345	23,490	22,330	13.30	18.10	20.30
CO	4,100	3,200	2,800	2,255	2,112	1,400	9.60	10.00	10.20
GA	3,200	4,000	4,700	576	1,160	1,410	11.50	12.00	12.00
MI	4,500	4,800	4,400	1,260	1,680	1,452	13.40	13.80	13.00
TX	4,600	3,200	2,900	874	960	580	16.10	25.90	22.00
WA	3,000	....	....	1,200	....	....	16.10	....	....
Other States[2]	1,010	5,560	5,600	602	2,062	1,855	13.60	16.30	19.00
US	102,710	101,760	97,400	30,598	31,464	29,027	13.10	17.20	19.00

[1] Preliminary. [2] 2000 - MN and OR. 2001 - AZ, MN, OR, and WA. 2002 - AZ and WA. MN and OR estimates discontinued in 2002.
NASS, Crops Branch, (202) 720–2127.

Table 4-19.—Carrots for processing, commercial crop: Area, production, and value per ton, by States, 2000–2002

State	Area harvested			Production			Value per unit		
	2000	2001	2002[1]	2000	2001	2002[1]	2000	2001	2002[1]
	Acres	*Acres*	*Acres*	*Tons*	*Tons*	*Tons*	*Dollars per tons*	*Dollars per tons*	*Dollars per tons*
CA	4,900	3,300	2,100	147,000	96,690	60,270	81.40	90.50	108.00
MI	1,250	1,500	1,400	35,000	31,500	32,200	68.80	69.00	67.00
MN	1,300	1,000	900	23,150	23,000	23,400	65.10	97.60	70.80
OR[2]	310	....	....	9,000	....	....	72.20	....	....
TX	2,100	2,300	1,900	33,600	41,960	34,620	48.00	49.30	56.40
WA	5,000	4,500	4,400	160,000	144,000	136,400	66.00	73.00	64.00
WI[3]	4,600	4,200	4,700	98,390	74,890	108,810	68.70	67.10	61.40
Other States[4]	690	1,880	....	12,740	40,200	....	78.10	72.40	....
US	20,150	18,680	15,400	518,880	452,240	395,700	70.30	74.50	70.00

[1] Preliminary. [2] Estimates discontinued in 2002. [3] Includes a small amount of fresh market carrots. [4] 2000 - NJ and OH. 2001 - NJ, OH, and OR. NJ and OH estimates discontinued in 2002.
NASS, Crops Branch, (202) 720–2127.

Table 4-20.—Cauliflower, commercial crop: Area, yield, production, value per hundredweight and per ton, and total value, 1993–2002 [1]

Year	Total crop: Area for harvest	Total crop: Yield per acre	Total crop: Production	Total crop: Value [2] Per cwt.	Total crop: Value [2] Total	For fresh market: Production	For fresh market: Value [2] Per cwt.	For fresh market: Value [2] Total	For processing: Production	For processing: Value [3] Per ton	For processing: Value [3] Total
	Acres	*Cwt.*	*1,000 cwt.*	*Dollars*	*1,000 dollars*	*1,000 cwt.*	*Dollars*	*1,000 dollars*	*Tons*	*Dollars*	*1,000 dollars*
1993	60,600	130	7,884	30.00	236,681	7,011	31.00	217,450	43,660	440.00	19,231
1994	58,800	139	8,190	28.30	231,411	7,345	28.80	211,529	42,290	470.00	19,882
1995	53,350	137	7,315	33.30	243,778	6,484	34.70	225,020	41,550	451.00	18,758
1996	48,200	153	7,354	32.30	237,342	6,801	33.00	224,168	27,640	477.00	13,174
1997	43,500	158	6,889	31.60	217,534	6,323	32.30	203,957	28,300	480.00	13,577
1998	44,200	156	6,897	32.80	226,560	5,468	34.50	188,477	71,450	533.00	38,083
1999	46,400	167	7,742	29.00	224,725	6,958	30.00	208,659	39,186	410.00	16,066
2000	43,660	183	7,970	31.10	247,592	7,200	32.00	230,545	38,480	443.00	17,047
2001	44,050	159	7,018	28.40	199,044	6,230	29.20	181,649	39,410	441.00	17,305
2002 [4]	42,400	152	6,428	32.30	207,629	6,050	32.80	198,401	18,910	488.00	9,228

[1] Includes heading (cauliflower) broccoli. [2] Price and value on f.o.b. basis. [3] Price and value at processing plant door. [4] Preliminary.

NASS, Crops Branch, (202) 720–2127.

Table 4-21.—Cauliflower, commercial crop: Area, production, and value per hundredweight and per ton, by States, 2000–2002 [1]

State	Area harvested: 2000	Area harvested: 2001	Area harvested: 2002 [2]	Production: 2000	Production: 2001	Production: 2002 [2]	Value per unit: 2000	Value per unit: 2001	Value per unit: 2002 [2]
	Acres	*Acres*	*Acres*	*1,000 cwt.*	*1,000 cwt.*	*1,000 cwt.*	*Dollars per cwt.*	*Dollars per cwt.*	*Dollars per cwt.*
AZ	4,200	4,500	4,500	924	900	788	34.90	29.50	45.90
CA	38,500	38,500	37,000	6,930	5,968	5,550	30.50	28.10	30.20
NY	900	800	900	108	112	90	38.00	38.60	44.10
OR [3]	60	250	....	8	38	....	19.10	17.60	....
US	43,660	44,050	42,400	7,970	7,018	6,428	31.10	28.40	32.30

State	For fresh market: Production 2000	Production 2001	Production 2002 [2]	Value per unit 2000	Value per unit 2001	Value per unit 2002 [2]	For processing: Production 2000	Production 2001	Production 2002 [2]	Value per unit 2000	Value per unit 2001	Value per unit 2002 [2]
	1,000 cwt.	*1,000 cwt.*	*1,000 cwt.*	*Dollars per cwt.*	*Dollars per cwt.*	*Dollars per cwt.*	*Tons*	*Tons*	*Tons*	*Dollars per ton*	*Dollars per ton*	*Dollars per ton*
AZ	924	900	788	34.90	29.50	45.90	....	....	....	....	....	....
Other States [4]	6,276	5,330	5,262	31.60	29.10	30.80	38,480	39,410	18,910	443.00	441.00	488.00
US	7,200	6,230	6,050	32.00	29.20	32.80	38,480	39,410	18,910	443.00	441.00	488.00

[1] Includes heading (cauliflower) broccoli. [2] Preliminary. [3] Estimates discontinued in 2002. [4] 2000–2001 - CA, NY, and OR. 2002 - CA and NY. OR estimates discontinued in 2002.

NASS, Crops Branch, (202) 720–2127.

Table 4-22.—Celery, commercial crop: Area, production, and value per hundredweight, by States, 2000–2002 [1]

State	Area harvested			Production			Value per unit		
	2000	2001	2002 [2]	2000	2001	2002 [2]	2000	2001	2002 [2]
	Acres	*Acres*	*Acres*	*1,000 cwt.*	*1,000 cwt.*	*1,000 cwt.*	*Dollars per cwt.*	*Dollars per cwt.*	*Dollars per cwt.*
CA	24,000	25,500	25,000	17,280	17,723	17,750	18.80	14.40	12.80
MI	1,900	1,900	2,100	950	873	987	14.10	14.50	14.60
TX [3]	300	400	....	195	260	....	16.00	15.00	....
US	26,200	27,800	27,100	18,425	18,856	18,737	18.50	14.40	12.90

[1] Mostly for fresh market use, but includes some quantities used for processing. [2] Preliminary. [3] Estimates discontinued in 2002.

NASS, Crops Branch, (202) 720–2127.

Table 4-23.—Celery, commercial crop: Area, yield, production, value per hundredweight, and total value, United States, 1993–2002 [1]

Year	Area for harvest	Yield per acre	Production	Value [2]	
				Per cwt.	Total
	Acres	*Cwt.*	*1,000 cwt.*	*Dollars*	*1,000 dollars*
1993	28,220	645	18,215	14.80	270,082
1994	28,500	660	18,798	11.80	220,892
1995	27,550	658	18,120	16.30	294,971
1996	27,840	683	19,015	10.50	199,877
1997	26,910	673	18,119	14.70	266,321
1998	27,200	662	18,000	12.30	220,623
1999	27,500	681	18,727	12.00	224,702
2000	26,200	703	18,425	18.50	341,391
2001	27,800	678	18,856	14.40	272,391
2002 [3]	27,100	691	18,737	12.90	241,571

[1] Mostly for fresh market use, but includes quantities used for processing. [2] Price and value on f.o.b. basis. [3] Preliminary.

NASS, Crops Branch, (202) 720–2127.

Table 4-24.—Corn, sweet, commercial crop: Area, yield, production, value per hundredweight and per ton, and total value, United States, 1993–2002

Year	For fresh market					For processing				
	Area for harvest	Yield per acre	Production	Value [1]		Area for harvest	Yield per acre	Production	Value [2]	
				Per cwt.	Total				Per ton	Total
	Acres	*Cwt.*	*1,000 cwt.*	*Dollars*	*1,000 dollars*	*Acres*	*Tons*	*Tons*	*Dollars*	*1, 000 dollars*
1993	209,800	90	18,848	17.80	335,716	472,060	5.76	2,721,190	72.40	197,112
1994	225,900	98	22,121	17.20	380,213	516,100	7.23	3,731,040	68.60	256,087
1995	225,200	97	21,792	18.30	397,769	483,910	6.87	3,324,150	75.60	251,156
1996	227,800	102	23,127	16.90	390,737	474,200	6.95	3,296,330	78.50	258,840
1997	236,400	100	23,641	17.70	418,617	465,800	7.18	3,342,330	74.90	250,329
1998	237,400	111	26,311	17.20	452,410	467,300	6.97	3,255,560	73.30	238,748
1999	237,300	109	25,786	17.20	443,276	466,300	7.07	3,297,390	71.10	234,418
2000	246,100	107	26,401	18.20	480,706	459,700	6.86	3,155,540	73.40	231,600
2001	251,600	109	27,383	19.50	534,586	446,450	7.04	3,142,840	72.90	229,204
2002 [3]	248,800	106	26,430	20.10	531,159	419,900	7.35	3,084,290	68.20	210,335

[1] Price and value on f.o.b. basis. [2] Price and value at processing plant door. [3] Preliminary.

NASS, Crops Branch, (202) 720–2127.

Table 4-25.—Corn, sweet, commercial crop: Area, production, and value per hundredweight and per ton, by States, 2000–2002

Utilization and State	Area harvested			Production			Value per unit		
	2000	2001	2002[1]	2000	2001	2002[1]	2000	2001	2002[1]
FOR FRESH MARKET	*Acres*	*Acres*	*Acres*	*1,000 cwt.*	*1,000 cwt.*	*1,000 cwt.*	*Dollars per cwt.*	*Dollars per cwt.*	*Dollars per cwt.*
AL	1,400	1,500	1,700	90	102	116	13.70	17.60	17.00
CA	24,000	22,000	24,000	3,960	3,740	4,080	16.70	22.30	24.00
CO	7,700	7,900	8,500	1,001	1,106	1,020	12.20	7.70	10.20
CT	4,300	4,600	4,400	258	253	308	25.00	24.50	25.00
DE	3,600	3,800	3,800	270	380	323	19.00	17.00	19.00
FL	37,400	37,900	38,900	5,610	5,306	5,641	18.20	22.10	18.00
GA	21,000	25,000	25,000	3,360	3,250	3,125	12.50	14.60	14.00
IL	5,600	5,700	5,600	532	559	560	23.50	26.00	16.80
IN	6,000	6,000	5,100	420	468	337	21.60	22.40	23.00
ME	2,100	2,000	2,000	116	110	110	33.00	32.50	34.00
MD	10,900	10,900	10,500	556	916	525	19.00	18.00	22.00
MA	6,500	6,700	0,700	390	436	469	33.00	33.50	30.00
MI	10,600	9,000	10,000	742	540	800	18.10	22.00	21.00
NH	1,800	1,800	1,500	108	99	75	40.00	40.00	42.50
NJ	9,000	8,900	8,500	945	890	791	18.90	17.60	21.80
NY	27,500	33,400	31,800	2,613	3,841	3,498	21.60	17.80	23.90
NC	8,100	8,200	8,100	810	861	729	15.00	14.80	15.50
OH	14,800	15,400	14,400	1,302	1,309	1,253	21.60	18.40	21.20
OR	5,700	5,100	4,400	941	867	748	11.00	13.90	10.90
PA	18,900	17,100	15,700	964	701	628	23.40	26.50	31.80
RI	900	900	700	50	68	46	33.00	31.00	31.00
TX	4,300	3,900	3,600	387	390	288	20.70	18.00	20.70
VT	900	1,000	900	41	50	45	33.00	33.50	34.50
VA	3,400	3,300	2,900	102	281	104	38.00	30.00	23.00
WA	2,500	2,300	2,600	300	276	338	21.20	22.30	21.10
WI	7,200	7,300	7,500	533	584	473	19.00	19.00	21.00
US	246,100	251,600	248,800	26,401	27,383	26,430	18.20	19.50	20.10
FOR PROCESSING	*Acres*	*Acres*	*Acres*	*Tons*[2]	*Tons*[2]	*Tons*[2]	*Dollars per ton*	*Dollars per ton*	*Dollars per ton*
ID	19,000	....	....	162,900	....	....	56.30	....	....
IL	16,900	17,500	13,800	99,710	116,610	80,040	110.00	126.00	93.00
MN	129,400	130,200	136,900	756,760	794,720	884,630	67.30	68.80	64.60
NY	29,000	29,200	....	154,650	160,600	....	75.00	73.70	....
OR	35,700	29,100	33,000	306,650	226,870	263,580	80.40	80.10	78.30
PA	2,200	650	1,400	12,520	4,880	6,660	85.80	105.00	77.30
WA	98,600	95,100	95,300	845,210	853,560	899,200	75.60	70.40	72.40
WI	92,900	98,800	84,200	576,700	657,640	596,050	65.40	65.80	56.20
Other States[3]	36,000	45,900	55,300	240,440	327,960	354,130	90.00	79.00	73.30
US	459,700	446,450	419,900	3,155,540	3,142,840	3,084,290	73.40	72.90	68.20
Grand total	705,800	698,050	668,700	4,475,590	4,511,990	4,405,790	....	....	....

[1] Preliminary. [2] Tonnage in husk. [3] 2000 - DE, IA, MD, NJ, and TN. 2001 - DE, ID, IA, MD, NJ, and TN. 2002 - DE, ID, IA, MD, NJ, NY, and TN.

NASS, Crops Branch, (202) 720–2127.

Table 4-26.—Cucumbers for fresh market: Area, production, and value per hundredweight, by States, 2000–2002

State	Area harvested			Production			Value per unit		
	2000	2001	2002 [1]	2000	2001	2002 [1]	2000	2001	2002 [1]
	Acres	*Acres*	*Acres*	*1,000 cwt.*	*1,000 cwt.*	*1,000 cwt.*	*Dollars per cwt.*	*Dollars per cwt.*	*Dollars per cwt.*
CA	4,500	4,700	4,300	1,283	1,128	1,032	28.50	30.00	28.60
FL	10,000	7,300	7,500	3,650	2,665	2,850	20.20	22.60	20.80
GA	11,000	14,500	14,600	1,760	2,538	2,920	13.00	12.80	11.30
MD	1,200	1,300	1,200	90	85	84	24.00	23.00	28.00
MI	6,700	5,500	6,000	1,340	1,210	1,140	18.80	20.00	18.00
NJ	2,900	2,700	3,000	537	594	675	19.60	15.90	17.70
NY	3,800	4,300	4,600	798	731	828	25.40	27.50	27.20
NC	6,800	6,900	7,000	850	828	805	14.00	13.00	15.00
SC	1,500	1,900	2,000	90	266	280	27.50	14.20	25.00
TX	1,800	1,800	1,800	360	378	450	24.10	24.20	24.60
VA	3,100	3,500	3,500	192	333	315	21.00	14.00	16.00
US	53,300	54,400	55,500	10,950	10,756	11,379	19.90	19.60	18.80

[1] Preliminary.

NASS, Crops Branch, (202) 720–2127.

Table 4-27.—Cucumbers (for pickles), commercial crop: Area, yield, production, value per ton, total value, and pickle stocks, United States, 1993–2002

Year	For processing					Pickle stocks on hand Dec. 1 [2]
	Area harvested	Yield per acre	Production	Value [1]		
				Per ton	Total	
	Acres	*Tons*	*Tons*	*Dollars*	*1,000 dollars*	*Tons*
1993	109,150	5.38	586,980	215.00	126,255	451,560
1994	116,640	5.43	633,518	219.00	139,044	458,150
1995	117,090	5.22	611,180	222.00	135,933	372,600
1996	105,200	5.36	563,689	248.00	139,985	392,970
1997	103,370	6.00	620,100	234.00	145,371	282,190
1998	102,870	5.77	593,720	237.00	140,553	359,512
1999	105,300	5.97	628,360	238.00	149,839	452,445
2000	104,710	5.86	613,160	269.00	164,956	387,544
2001	108,260	5.37	581,540	291.00	168,958	552,303
2002 [3]	118,900	5.19	617,300	277.00	171,016	300,580

[1] Price and value at processing plant door. [2] Stocks in hands of original salters of both salt and dill pickles, sold and unsold, in tanks and barrels, on Dec. 1 includes stocks of fresh-pack pickles. [3] Preliminary.

NASS, Crops Branch, (202) 720–2127.

Table 4-28.—Cucumbers (for pickles), commercial crop: Area, production, and value per ton, by States, 2000–2002

State	Area harvested			Production			Value per unit		
	2000	2001	2002 [1]	2000	2001	2002 [1]	2000	2001	2002 [1]
	Acres	*Acres*	*Acres*	*Tons*	*Tons*	*Tons*	*Dollars per ton*	*Dollars per ton*	*Dollars per ton*
FL	6,500	6,500	6,500	70,850	70,850	70,850	465.00	451.00	461.00
IN	1,800	1,800	1,600	12,600	11,300	3,950	188.00	179.00	147.00
MI	30,000	29,500	34,500	180,000	125,380	158,700	215.00	246.00	190.00
NC	18,000	17,600	18,000	90,000	86,240	81,000	270.00	280.00	290.00
OH	2,500	2,700	2,400	22,930	29,400	28,700	300.00	370.00	358.00
SC	3,200	3,300	3,700	8,860	16,170	14,130	225.00	200.00	236.00
TX	6,300	11,000	11,000	28,980	55,000	55,000	380.00	346.00	436.00
WI	5,200	5,000	5,600	40,140	30,100	31,750	182.00	195.00	169.00
Oth Sts [2]	31,210	30,860	35,600	158,800	157,100	173,220	248.00	261.00	238.00
US	104,710	108,260	118,900	613,160	581,540	617,300	269.00	291.00	277.00

[1] Preliminary. [2] 2000 - AL, CA, DE, GA, IL, MD, MA, MO, NJ, NY, VA, and WA. 2001 - AL, CA, DE, GA, IL, MD, MA, MO, NJ, NY, VA, and WA. 2002 - AL, CA, DE, GA, MD, MA, MO, and WA. NJ, NY, and VA estimates discontinued in 2002.

NASS, Crops Branch, (202) 720–2127.

Table 4-29.—Eggplant for fresh market: Area, production, and value per hundredweight, by States, 2000–2002

State	Area harvested			Production			Value per unit		
	2000	2001	2002 [1]	2000	2001	2002 [1]	2000	2001	2002 [1]
	Acres	*Acres*	*Acres*	*1,000 cwt.*	*1,000 cwt.*	*1,000 cwt.*	*Dollars per cwt.*	*Dollars per cwt.*	*Dollars per cwt.*
CA	1,600	1,400	....	328	329	....	27.50	24.50	....
CT	130	100	....	17	8	....	35.00	30.00	....
FL	1,800	1,600	....	468	400	....	29.60	33.50	....
GA	1,100	450	....	484	281	....	19.20	18.00	....
MA	50	50	....	6	5	....	35.00	35.00	....
NJ	800	800	....	200	160	....	23.40	18.90	....
NY	520	520	....	94	120	....	39.30	40.70	....
NC	300	310	....	45	37	....	20.00	21.00	....
Total	6,300	5,230	....	1,642	1,340	....	25.70	26.60	....
HI	80	70	....	16	14	....	75.00	70.00	....
US	6,380	5,300	....	1,658	1,354	....	26.20	27.00	....

[1] Estimates discontinued in 2002.
NASS, Crops Branch, (202) 720–2127.

Table 4-30.—Escarole/Endive for fresh market: Area, production, and value per hundredweight, by States, 2000–2002

State	Area harvested			Production			Value per unit		
	2000	2001	2002 [1]	2000	2001	2002 [1]	2000	2001	2002 [1]
	Acres	*Acres*	*Acres*	*1,000 cwt.*	*1,000 cwt.*	*1,000 cwt.*	*Dollars per cwt.*	*Dollars per cwt.*	*Dollars per cwt.*
CA	2,400	2,400	....	384	360	....	36.70	35.50	....
FL	1,300	....	....	286	....	....	23.00	....	....
NJ	800	....	....	152	....	....	27.30	....	....
NY	240	260	....	67	73	....	38.90	36.60	....
OH	430	410	....	47	94	....	31.80	26.00	....
Oth Sts [2]	....	2,100	....	....	406	....	....	23.70	....
US	5,170	5,170	....	936	933	....	30.90	29.50	....

[1] Estimates discontinued in 2002. [2] 2001 - FL and NJ.
NASS, Crops Branch, (202) 720–2127.

Table 4-31.—Garlic for fresh market and processing: Area, production, and value per hundredweight, by States, 2000–2002

State	Area harvested			Production			Value per unit		
	2000	2001	2002 [1]	2000	2001	2002 [1]	2000	2001	2002 [1]
	Acres	*Acres*	*Acres*	*1,000 cwt.*	*1,000 cwt.*	*1,000 cwt.*	*Dollars per cwt.*	*Dollars per cwt.*	*Dollars per cwt.*
CA	29,000	30,000	27,000	4,785	5,100	4,725	29.40	31.50	29.20
NV	2,200	1,400	1,100	330	174	187	17.00	17.00	20.00
OR	3,600	3,800	4,700	466	603	738	19.10	16.00	16.10
US	34,800	35,200	32,800	5,581	5,877	5,650	27.80	29.40	27.20

[1] Preliminary.
NASS, Crops Branch, (202) 720–2127.

Table 4-32.—Honeydew melons, commercial crop: Area, yield, production, value per hundredweight, and total value, United States, 1993–2002

Year	Area for harvest	Yield per acre	Production	Value [1]	
				Per cwt.	Total
	Acres	*Cwt.*	*1,000 cwt.*	*Dollars*	*1,000 dollars*
1993	23,300	163	3,792	18.20	68,888
1994	27,100	174	4,724	16.40	77,473
1995	26,000	167	4,332	20.60	89,193
1996	27,300	174	4,737	17.00	80,405
1997	26,600	182	4,828	18.90	91,040
1998	25,500	192	4,887	21.10	103,291
1999	27,500	193	5,307	21.60	114,727
2000	26,000	193	5,008	19.20	96,181
2001	24,200	189	4,576	21.10	96,663
2002 [2]	24,600	206	5,059	17.90	90,546

[1] Price and value on f.o.b. basis. [2] Preliminary.
NASS, Crops Branch, (202) 720–2127.

Table 4-33.—Honeydew melons, commercial crop: Area, production, and value per hundredweight, by States, 2000–2002

State	Area harvested			Production			Value per unit		
	2000	2001	2002[1]	2000	2001	2002[1]	2000	2001	2002[1]
	Acres	*Acres*	*Acres*	*1,000 cwt.*	*1,000 cwt.*	*1,000 cwt.*	*Dollars per cwt.*	*Dollars per cwt.*	*Dollars per cwt.*
AZ	3,600	2,400	2,500	756	516	775	17.50	19.70	16.40
CA	20,000	20,000	20,400	3,700	3,700	3,774	18.60	19.70	16.80
TX	2,400	1,800	1,700	552	360	510	25.60	37.80	28.30
US	26,000	24,200	24,600	5,008	4,576	5,059	19.20	21.10	17.90

[1] Preliminary.
NASS, Crops Branch, (202) 720–2127.

Table 4-34.—Head lettuce, commercial crop: Area, production, and value per hundredweight, by States, 2000–2002

State	Area harvested			Production			Value per unit		
	2000	2001	2002[1]	2000	2001	2002[1]	2000	2001	2002[1]
	Acres	*Acres*	*Acres*	*1,000 cwt.*	*1,000 cwt.*	*1,000 cwt.*	*Dollars per cwt.*	*Dollars per cwt.*	*Dollars per cwt.*
AZ:									
Western	50,300	51,800	50,000	17,605	20,202	18,000	13.10	16.50	38.70
Other	1,500	1,500	1,700	450	558	391	15.00	15.00	23.30
CA	130,000	128,000	129,000	50,700	48,640	48,375	18.80	18.50	15.20
CO	2,000	2,000	2,700	680	700	810	16.00	12.00	12.20
NJ	1,100	1,000	1,000	193	250	150	35.70	33.90	37.20
US	184,900	184,300	184,400	69,628	70,350	67,726	17.40	17.90	21.50

[1] Preliminary.
NASS, Crops Branch, (202) 720–2127.

Table 4-35.—Head lettuce, commercial crop: Area, yield, production, value per hundredweight, and total value, United States, 1993–2002

Year	Area for harvest	Yield per acre	Production	Value[1]	
				Per cwt.	Total
	Acres	*Cwt.*	*1,000 cwt.*	*Dollars*	*1,000 dollars*
1993	207,810	327	67,811	16.00	1,086,716
1994	220,950	317	70,058	13.30	934,472
1995	197,160	316	62,349	23.50	1,463,348
1996	217,600	285	62,072	14.70	912,586
1997	203,000	339	68,794	17.50	1,201,89[illegible]
1998	198,400	320	63,401	16.10	1,023,427
1999	192,800	365	70,321	13.30	936,252
2000	184,900	377	69,628	17.40	1,208,306
2001	184,300	382	70,350	17.90	1,258,418
2002[2]	184,400	367	67,726	21.50	1,456,472

[1] Price and value on f.o.b. basis. [2] Preliminary.
NASS, Crops Branch, (202) 720–2127.

Table 4-36.—Leaf lettuce for fresh market: Area, production, and value per hundredweight, by States, 2000–2002

State	Area harvested			Production			Value per unit		
	2000	2001	2002 [1]	2000	2001	2002 [1]	2000	2001	2002 [1]
	Acres	*Acres*	*Acres*	*1,000 cwt.*	*1,000 cwt.*	*1,000 cwt.*	*Dollars per cwt.*	*Dollars per cwt.*	*Dollars per cwt.*
AZ	3,800	5,800	5,800	1,064	1,827	1,827	30.70	41.40	61.80
CA	45,000	46,000	49,000	10,575	9,890	10,535	29.60	24.90	28.10
Oth Sts [2]	700	700	....	126	107	....	22.90	22.70	....
US	49,500	52,500	54,800	11,765	11,824	12,362	29.60	27.40	33.10

[1] Preliminary. [2] 2000 - 2001 FL and NJ. FL and NJ estimates discontinued in 2002.
NASS, Crops Branch, (202) 720–2127.

Table 4-37.—Romaine lettuce for fresh market: Area, production, and value per hundredweight, by States, 2000–2002

State	Area harvested			Production			Value per unit		
	2000	2001	2002 [1]	2000	2001	2002 [1]	2000	2001	2002 [1]
	Acres	*Acres*	*Acres*	*1,000 cwt.*	*1,000 cwt.*	*1,000 cwt.*	*Dollars per cwt.*	*Dollars per cwt.*	*Dollars per cwt.*
AZ	11,800	12,300	12,500	4,130	3,813	3,438	19.20	24.90	43.40
CA	36,000	40,500	42,000	11,340	11,340	12,390	20.10	17.50	19.90
FL [2]	950	....	....	185	....	....	17.50	....	....
NJ [2]	700	....	....	175	....	....	23.80	....	....
Oth Sts [3]	....	1,800	....	....	334	....	....	14.60	....
US	49,450	54,600	54,500	15,830	15,487	15,828	19.90	19.30	25.00

[1] Preliminary. [2] Estimates discontinued in 2002. [3] 2001 - FL and NJ.
NASS, Crops Branch, (202) 720–2127.

Table 4-38.—Misc Crops for fresh market and processing: Area, production, and value per hundredweight, by States, 2000–2002

Crop	Area harvested			Production			Value per unit		
	2000	2001	2002	2000	2001	2002	2000	2001	2002
	Acres	*Acres*	*Acres*	*1,000 cwt.*	*1,000 cwt.*	*1,000 cwt.*	*Dollars per cwt.*	*Dollars per cwt.*	*Dollars per cwt.*
Artichokes	8,800	8,000	8,200	1,012	1,000	943	60.30	58.60	70.80
Brussels Sprouts [1]	2,600	2,200	....	468	396	....	34.40	36.50	....

[1] Estimates discontinued in 2002.
NASS, Crops Branch, (202) 720–2127.

Table 4-39.—Onions (fresh market): Foreign trade, United States, 1992–2001 [1]

Year beginning July	Imports	Domestic exports
	1,000 cwt.	*1,000 cwt.*
1992	4,619	3,756
1993	5,729	4,397
1994	4,751	7,975
1995	5,725	5,942
1996	5,875	5,814
1997	5,537	6,589
1998	5,312	5,666
1999	5,005	7,040
2000	5,671	7,964
2001	5,925	6,788

[1] Includes onion sets and pearl onions.
ERS, Specialty Crops Branch, (202) 694–5253. Compiled from reports of the U.S. Department of Commerce.

Table 4-40.—Onions, commercial crop: Area, yield, production, shrinkage and loss, value per hundredweight, and total value, United States, 1993–2002 [1]

Year	Area for harvest	Yield per acre	Production [2]	Shrinkage and loss	Value [3] Per cwt.	Value [3] Total
	Acres	*Cwt.*	*1,000 cwt.*	*1,000 cwt.*	*Dollars*	*1,000 dollars*
1993 ..	158,080	380	60,124	8,343	16.50	856,963
1994 ..	164,650	397	65,313	6,138	10.80	639,397
1995 ..	166,800	392	65,374	7,219	11.10	645,748
1996 ..	166,210	386	64,106	6,678	10.50	604,789
1997 ..	165,910	414	68,769	7,540	12.60	769,974
1998 ..	171,340	393	67,282	6,368	13.80	838,441
1999 ..	173,400	424	73,562	8,823	9.78	632,969
2000 ..	166,170	432	71,721	6,701	11.30	736,369
2001 ..	161,590	419	67,653	6,166	11.40	697,950
2002 [4]	160,220	420	67,365	6,253	11.70	715,979

[1] Mostly for fresh market use, but includes some quantities used for processing. [2] Includes storage crop onions harvested but not sold because of shrinkage and waste. [3] Price and value on f.o.b. basis. [4] Preliminary.

NASS, Crops Branch, (202) 720–2127.

Table 4-41.—Onions, commercial crop: Area, production, shrinkage and loss, and value per hundredweight, by States, 2000–2002 [1]

Season and State	Area harvested			Production			Shrinkage and loss			Value per unit		
	2000	2001	2002 [2]	2000	2001	2002 [2]	2000	2001	2002 [2]	2000	2001	2002 [2]
	Acres	*Acres*	*Acres*	*1,000 cwt.*	*1,000 cwt.*	*1,000 cwt.*	*1,000 cwt.*	*1,000 cwt.*	*1,000 cwt.*	*Dollars per cwt.*	*Dollars per cwt.*	*Dollars per cwt.*
Spring:												
AZ	3,200	3,000	1,500	1,376	1,290	690	....	....	....	5.80	8.00	8.35
CA	7,100	6,200	6,200	3,089	2,666	2,852	....	....	....	10.60	13.50	14.20
GA	12,400	13,500	11,500	3,162	2,565	1,438	....	....	....	26.00	27.50	32.20
TX	13,500	14,200	15,000	4,185	4,615	4,725	....	....	....	17.20	18.50	21.40
Total	36,200	36,900	34,200	11,812	11,136	9,705	....	....	....	16.50	18.20	20.00
Summer:												
Non-storage:												
CA	6,200	7,000	7,200	2,449	3,150	3,384	....	....	....	12.00	13.50	12.90
NV	3,300	2,900	3,300	1,518	1,334	1,848	....	....	....	15.50	12.00	13.00
NM	7,700	7,700	8,000	3,542	3,773	4,400	....	....	....	9.25	14.40	12.50
TX	3,500	2,600	2,800	1,050	1,040	980	....	....	....	23.20	20.20	22.20
WA	800	800	1,100	264	288	396	....	....	....	22.10	19.40	23.90
Non-storage total	21,500	21,000	22,400	8,823	9,585	11,008	....	....	....	13.10	14.50	14.00
Storage: [3]												
CA [4]	36,300	29,800	30,400	16,154	12,069	12,312	250	250	250	7.98	7.26	7.12
CO	11,500	12,000	11,000	4,083	4,140	4,400	700	660	880	13.40	12.40	14.40
ID	7,400	7,800	7,900	4,810	4,992	5,056	1,000	700	900	11.70	8.10	9.00
MI	3,500	3,700	3,900	945	999	897	189	200	180	12.50	12.20	12.50
MN	90	220	250	19	73	78	10	10	10	7.00	7.35	7.60
NY	12,300	12,800	12,300	4,674	4,224	2,583	1,164	451	284	13.50	9.70	12.20
OH	480	470	470	168	169	122	17	15	7	12.00	15.00	13.70
OR (Malheur)	11,600	11,300	12,000	6,960	7,006	7,800	1,640	1,580	1,540	9.88	9.32	9.01
OR (Other)	6,100	5,700	5,400	3,172	2,964	2,862	500	430	370	9.20	9.60	9.86
UT	2,400	2,100	2,100	1,140	956	1,050	110	122	153	9.30	7.70	8.20
WA	15,000	16,000	16,000	8,250	8,800	8,960	1,030	1,672	1,610	8.14	8.20	8.60
WI	1,800	1,800	1,900	711	540	532	91	76	69	7.40	8.00	8.25
Storage total	108,470	103,690	103,620	51,086	46,932	46,652	6,701	6,166	6,253	9.59	8.74	9.12
Total summer	129,970	124,690	126,020	59,909	56,517	57,660	6,701	6,166	6,253	10.20	9.85	10.20
US	166,170	161,590	160,220	71,721	67,653	67,365	6,701	6,166	6,253	11.30	11.40	11.70

[1] Mostly for fresh market use, but includes some quantities used for processing. [2] Preliminary. [3] Includes some quantities of storage crop onions harvested but not sold because of shrinkage and loss. [4] Includes fresh and processed.

NASS, Crops Branch, (202) 720–2127.

Table 4-42.—Peas, green (for processing), commercial crop: Area, yield, production, value per ton, and total value, United States, 1993–2002

Year	Area harvested	Yield per acre	Production	Value [1]	
				Per ton	Total
	Acres	*Tons* [2]	*Tons*	*Dollars*	*1,000 dollars*
1993	228,600	1.54	351,080	251.00	88,004
1994	289,410	1.69	488,580	254.00	124,138
1995	304,000	1.62	492,590	267.00	131,762
1996	249,800	1.67	417,672	285.00	118,910
1997	271,200	1.77	480,000	288.00	138,482
1998	273,900	1.77	483,900	282.00	136,584
1999	271,640	1.70	461,590	275.00	126,925
2000	277,240	1.91	530,050	248.00	131,701
2001	211,140	1.85	390,220	264.00	103,207
2002 [3]	211,300	1.64	347,090	252.00	87,494

[1] Price and value at processing plant door. [2] Shelled basis: 2½ pounds of peas in the shell produce approximately 1 pound of shelled peas. [3] Preliminary.

NASS, Crops Branch, (202) 720–2127.

Table 4-43.—Peas, green (for processing), commercial crop: Area, production, and value per ton, States, 2000–2002 [1]

State	Area harvested			Production			Value per unit		
	2000	2001	2002 [2]	2000	2001	2002 [2]	2000	2001	2002 [2]
	Acres	*Acres*	*Acres*	*Tons*	*Tons*	*Tons*	*Dollars per ton*	*Dollars per ton*	*Dollars per ton*
MN	86,400	67,100	74,600	140,240	95,350	98,370	322.00	337.00	319.00
NY	16,300	17,400	15,400	32,810	39,490	22,220	204.00	312.00	366.00
OR	32,200	22,900	20,200	64,370	38,540	28,910	210.00	178.00	176.00
WA	49,100	38,400	36,800	110,030	86,190	75,000	224.00	211.00	184.00
WI	50,000	36,300	38,400	102,850	68,750	67,230	209.00	236.00	201.00
Other States [3]	43,240	29,040	25,900	79,750	61,900	55,360	253.00	283.00	282.00
US	277,240	211,140	211,300	530,050	390,220	347,090	248.00	264.00	252.00

[1] Shelled basis; 2½ pounds of peas in the shell produce approximately 1 pound of shelled peas. [2] Preliminary. [3] 2000 - CA, DE, ID, IL, ME, MD, NJ, and TN. 2001 - CA, DE, ID, IL, MD, NJ, and TN. 2002 - CA, DE, ID, IL, MD, and NJ. TN estimates discontinued in 2002.

NASS, Crops Branch, (202) 720–2127.

Table 4-44.—Bell peppers for fresh market: Area, production, and value per hundredweight, by States, 2000–2002

State	Area harvested			Production			Value per unit		
	2000	2001	2002 [1]	2000	2001	2002 [1]	2000	2001	2002 [1]
	Acres	*Acres*	*Acres*	*1,000 cwt.*	*1,000 cwt.*	*1,000 cwt.*	*Dollars per cwt.*	*Dollars per cwt.*	*Dollars per cwt.*
CA	21,000	22,000	21,000	6,300	7,150	6,930	27.30	25.50	28.60
FL	18,400	15,700	17,500	6,348	5,416	5,950	38.40	34.40	36.80
GA	5,300	2,100	2,600	901	420	520	26.60	28.00	20.00
MI	2,100	1,400	1,600	462	364	400	22.50	22.00	24.00
NJ	3,600	3,700	3,700	1,080	1,184	962	27.00	23.50	27.20
NY [2]	770	640	....	162	147	....	43.70	40.90	....
NC	6,900	6,300	7,000	828	788	910	21.00	20.50	22.00
OH	1,900	2,000	1,600	390	246	336	27.60	29.40	23.80
PA [2]	380	380	....	42	29	....	18.00	25.30	....
TX	1,300	1,100	800	260	275	160	34.80	40.30	47.00
VA [2]	700	700	....	88	43	....	38.00	30.00	....
US	62,350	56,020	55,800	16,861	16,062	16,168	31.30	28.60	30.80

[1] Preliminary. [2] Estimates discontinued in 2002.

NASS, Crops Branch, (202) 720–2127.

Table 4-45.—Potatoes: Area, yield, production, season average price, and value, United States, 1993–2002

Year	Area planted	Area harvested	Yield per harvested acre	Production	Season average price per cwt. received by farmers [1]	Value of production
	1,000 acres	*1,000 acres*	*Cwt.*	*1,000 cwt.*	*Dollars*	*1,000 dollars*
1993	1,389.9	1,321.2	326	430,349	6.16	2,642,699
1994	1,421.8	1,385.1	339	469,425	5.56	2,593,446
1995	1,400.7	1,376.1	323	445,099	6.75	2,995,711
1996	1,454.7	1,425.9	350	499,254	4.91	2,423,476
1997	1,383.5	1,353.6	345	467,091	5.64	2,622,621
1998	1,416.6	1,387.7	343	475,771	5.56	2,635,279
1999	1,376.8	1,332.4	359	478,216	5.77	2,745,712
2000	1,383.7	1,348.0	381	513,621	5.08	2,591,091
2001	1,247.7	1,222.2	358	437,888	6.99	3,057,573
2002 [2]	1,310.8	1,276.5	363	463,214	6.82	3,151,178

[1] Obtained by weighting State prices by quantity sold. [2] Preliminary.
NASS, Crops Branch, (202) 720–2127.

Table 4-46.—Potatoes: Production, seed used, and disposition, United States, 1992–2001

Year	Production	Total used for seed	Used on farms where produced: For seed, feed, and household use	Used on farms where produced: Shrinkage and loss	Sold
	1,000 cwt.	*1,000 cwt.*	*1,000 cwt.*	*1,000 cwt.*	*1,000 cwt.*
1992	425,367	28,317	5,923	33,807	385,637
1993	430,349	29,081	5,951	30,181	394,217
1994	469,425	29,428	5,904	37,342	426,179
1995	445,099	30,561	5,755	29,630	409,714
1996	499,254	29,138	6,221	41,238	451,795
1997	467,091	29,975	5,475	32,183	429,433
1998	475,771	29,217	5,766	35,454	434,551
1999	478,216	29,709	5,569	35,531	437,116
2000	513,621	27,145	5,288	43,688	464,645
2001 [1]	437,888	29,251	5,387	31,208	401,293

[1] Preliminary.
NASS, Crops Branch, (202) 720–2127.

Table 4-47.—Fall potatoes: Production and total stocks held by growers and local dealers, 15 Major States, 1992–2001

Crop year	Production	Total stocks: Dec. 1	Following year: Jan. 1	Feb. 1	Mar. 1	Apr. 1	May 1	June 1 [1]
	1,000 cwt.	*1,000 cwt.*	*1,000 cwt.*	*1,000 cwt.*	*1,000 cwt.*	*1,000 cwt.*	*1,000 cwt.*	*1,000 cwt.*
1992	368,516	246,820	215,990	184,560	152,795	115,785	75,040	....
1993	376,954	249,410	217,300	185,490	153,435	115,170	72,860	....
1994	410,839	273,290	238,560	202,470	169,570	129,825	87,600	....
1995	394,785	256,710	223,550	189,360	156,020	115,855	75,870	....
1996	443,704	295,100	261,320	226,080	189,210	147,635	103,210	....
1997	413,513	278,830	246,550	212,562	175,870	134,190	92,840	....
1998	423,170	280,910	246,230	209,640	173,650	131,220	87,895	50,270
1999	420,567	275,100	239,910	207,150	169,620	128,410	86,915	47,220
2000	458,827	310,300	275,270	234,260	197,670	153,520	109,160	61,240
2001 [2]	387,179	258,750	224,680	192,190	158,590	119,950	81,200	42,990

[1] Estimates begun in 1998. [2] Preliminary.
NASS, Crops Branch, (202) 720–2127.

Table 4-48.—Potatoes: Area, production, and marketing year price per hundredweight received by farmers, by States, 2000–2002

Season and State	Area harvested			Production			Price for crop of-		
	2000	2001	2002[1]	2000	2001	2002[1]	2000	2001	2002[1]
	1,000 acres	*1,000 acres*	*1,000 acres*	*1,000 cwt.*	*1,000 cwt.*	*1,000 cwt.*	*Dollars*	*Dollars*	*Dollars*
Winter:									
CA	9.0	9.0	9.0	320	310	270	2,880	2,790	2,430
FL	8.0	5.0	6.7	260	265	265	2,080	1,325	1,776
Total	17.0	14.0	15.7	292	294	268	4,960	4,115	4,206
Spring:									
AZ	9.0	8.2	7.8	280	270	270	2,520	2,214	2,106
CA	18.8	15.5	19.0	395	390	405	7,426	6,045	7,695
FL	21.5	25.0	26.3	295	319	300	6,343	7,970	7,883
Hastings	16.5	18.0	19.0	295	330	315	4,868	5,940	5,985
Other	5.0	7.0	7.3	295	290	260	1,475	2,030	1,898
NC	17.0	18.5	21.0	200	190	170	3,400	3,515	3,570
TX	9.3	9.0	12.0	240	230	170	2,232	2,070	2,040
Total	75.6	76.2	86.1	290	286	271	21,921	21,814	23,294
Summer:									
AL	4.1	3.9	3.0	170	160	185	697	624	554
CA	7.5	8.0	8.0	355	355	390	2,663	2,840	3,120
CO	8.3	5.6	6.4	360	360	360	2,988	2,016	2,304
DE	4.7	4.3	3.6	240	270	260	1,128	1,161	936
IL	5.3	5.3	6.4	350	350	310	1,855	1,855	1,984
KS	2.9	2.4	2.9	340	300	340	986	720	986
MD	4.7	4.7	4.7	260	250	250	1,222	1,175	1,175
MO	6.1	5.6	5.4	275	340	240	1,678	1,904	1,296
NJ	2.5	2.5	2.6	285	255	265	713	638	689
NM	3.0	2.2	2.3	350	350	320	1,050	770	736
TX	7.8	8.0	8.3	380	390	400	2,964	3,120	3,320
VA	6.3	6.3	6.3	205	220	220	1,292	1,386	1,386
Total	63.2	58.8	59.9	304	310	309	19,236	18,209	18,486
Fall:									
CA	8.7	3.2	8.9	430	445	500	3,741	1,424	4,450
CO	75.6	67.8	71.5	370	315	390	27,972	21,357	27,885
ID	413.0	348.0	373.0	369	345	358	152,320	120,200	133,385
10 S.W. counties	28.0	23.0	27.0	490	450	455	13,720	10,350	12,285
Other counties	385.0	325.0	346.0	360	338	350	138,600	109,850	121,100
IN	2.8	2.9	2.8	280	320	260	784	928	728
ME	64.0	62.0	64.0	280	265	265	17,920	16,430	16,960
MA	2.5	2.8	2.9	255	265	255	638	742	740
MI	47.5	45.0	45.5	315	310	305	14,963	13,950	13,878
MN	59.0	55.0	55.0	360	335	340	21,240	18,425	18,700
MT	11.3	10.3	10.4	310	320	310	3,503	3,296	3,224
NE	24.7	22.4	21.8	410	375	395	10,127	8,400	8,611
NV	7.0	6.5	7.6	450	360	340	3,150	2,340	2,584
NM	6.8	4.2	4.0	400	340	400	2,720	1,428	1,600
NY	21.3	23.3	22.0	280	255	250	5,964	5,942	5,500
ND	110.0	110.0	102.0	245	240	230	26,950	26,400	23,460
OH	4.2	4.3	4.2	270	255	240	1,134	1,097	1,008
OR	56.5	44.5	49.8	543	466	501	30,683	20,730	24,936
Malheur County	10.5	9.0	8.0	425	410	400	4,463	3,690	3,200
Other counties	46.0	35.5	41.8	570	480	520	26,220	17,040	21,736
PA	13.0	13.5	14.0	270	235	185	3,510	3,173	2,590
RI	0.5	0.5	0.5	275	280	180	138	140	90
SD	2.8	2.7	1.1	290	240	300	812	648	330
UT	1.5	1.3	0.8	290	265	305	435	345	244
WA	175.0	160.0	170.0	600	590	560	105,000	94,400	95,200
WI	84.5	83.0	83.0	400	385	375	33,800	31,955	31,125
Total	1,192.2	1,073.2	1,114.8	392	367	374	467,504	393,750	417,228
US	1,348.0	1,222.2	1,276.5	381	358	363	513,621	437,888	463,214

[1] Preliminary.

NASS, Crops Branch, (202) 720–2127.

Table 4-49.—Fall potatoes: Total stocks held by growers and local dealers, 15 States, crop of 2000 and 2001 [1]

State	Crop of 2000						
	Dec. 1, 2000	Jan. 1, 2001	Feb. 1, 2001	Mar. 1, 2001	Apr. 1, 2001	May 1, 2001	June 1, 2001
	1,000 cwt.	*1,000 cwt.*	*1,000 cwt.*	*1,000 cwt.*	*1,000 cwt.*	*1,000 cwt.*	*1,000 cwt.*
CA	2,700	2,100	1,700	1,300	1,100	800	500
CO	20,300	17,700	15,100	13,000	10,300	7,400	3,500
ID	111,000	101,000	87,500	76,500	61,500	47,000	26,500
ME	14,100	12,500	10,900	8,700	6,600	4,000	1,900
MI	8,700	6,900	5,200	3,400	1,500	700	
MN	14,200	13,000	11,600	10,100	8,100	5,800	3,500
MT	3,350	3,250	3,000	2,800	2,200	750	320
NE	7,500	6,650	4,600				
NY	2,600	1,900	1,400	1,000	400		
ND	18,500	16,200	13,700	11,500	9,500	6,800	4,000
OH	350	170	60				
OR	25,000	23,000	20,000	17,000	13,600	10,000	6,400
PA	2,000	1,500	1,100	830	300		
WA	59,000	52,000	44,500	37,500	29,500	21,500	13,000
WI	21,000	17,400	13,900	10,500	6,900	3,300	1,400
Other				3,540	2,020	1,110	220
15 State total	310,300	275,270	234,260	197,670	153,520	109,160	61,240

State	Crop of 2001 [2]						
	Dec. 1, 2001	Jan. 1, 2002	Feb. 1, 2002	Mar. 1, 2002	Apr. 1, 2002	May 1, 2002	June 1, 2002
	1,000 cwt.	*1,000 cwt.*	*1,000 cwt.*	*1,000 cwt.*	*1,000 cwt.*	*1,000 cwt.*	*1,000 cwt.*
CA	1,200	1,100	800	600	300		
CO	15,600	13,300	11,000	9,400	7,300	5,100	1,700
ID	87,500	78,000	68,000	58,000	45,000	32,500	18,500
ME	12,200	10,800	8,900	7,100	5,300	3,300	1,800
MI	8,200	6,200	4,800	3,200	1,500	400	
MN	12,000	10,800	9,500	7,900	6,100	4,300	2,600
MT	3,000	2,780	2,630	2,460	1,790	450	
NE	4,850	4,050	3,200				
NY	2,400	1,500	1,000	600	300		
ND	19,500	16,900	14,500	12,300	10,400	6,900	4,000
OH	300	150	60				
OR	17,500	15,500	14,000	11,500	8,500	6,500	3,000
PA	1,800	1,400	1,000	600	300		
WA	53,000	46,000	40,000	32,500	25,000	18,000	10,000
WI	19,700	16,200	12,800	10,000	6,700	3,000	1,100
Other				2,430	1,460	750	290
15 State total	258,750	224,680	192,190	158,590	119,950	81,200	42,990

[1] Blank States combined into Other. [2] Preliminary.
NASS, Crops Branch, (202) 720–2127.

Table 4-50.—Frozen French fries: U.S. imports from principal suppliers for marketing years, 1997/98–2001/2002 [1]

Countries	1997/98	1998/99	1999/2000	2000/2001	2001/2002
	Metric tons	*Metric tons*	*Metric tons*	*Metric tons*	*Metric tons*
Canada	336,293	367,365	464,746	493,680	605,912
Other	2,465	137	220	1,387	1,664
Total	338,758	367,502	464,966	495,067	607,576

[1] July through June.
FAS, Horticultural & Tropical Products Division, (202) 720–0897. Compiled from reports of the U.S. Department of Commerce.

Table 4-51.—Potatoes: Utilization, United States, crop years 1994–2001

Item	1994	1995	1996	1997
SALES	*1,000 cwt.*	*1,000 cwt.*	*1,000 cwt.*	*1,000 cwt.*
Table stock	136,106	124,875	131,446	131,670
For processing:				
Chips and shoestring	49,299	47,284	48,305	48,130
Dehydration	41,381	45,065	54,261	48,389
Frozen french fries	136,531	129,029	145,489	131,628
Other frozen products	26,362	27,073	28,972	33,397
Canned potatoes	2,503	3,342	2,785	2,822
Other canned products (hash, stews, soups)	3,006	2,385	2,167	2,675
Starch and flour	2,176	1,668	1,956	1,311
Total	261,258	255,846	283,935	268,352
Other sales:				
Livestock feed	4,147	3,224	[2] 12,073	3,603
Seed	24,668	25,769	24,341	25,808
Total	28,815	28,993	36,414	29,411
Total sales	426,179	409,714	451,795	429,433
NON-SALES				
Seed used on farms where grown	4,760	4,792	4,797	4,167
Household use	1,144	963	1,424	1,308
Shrinkage and loss	37,342	29,630	41,238	32,183
Total non-sales	43,246	35,385	47,459	37,658
Total production	469,425	445,099	499,254	467,091

Item	1998	1999	2000	2001 [1]
SALES	*1,000 cwt.*	*1,000 cwt.*	*1,000 cwt.*	*1,000 cwt.*
Table stock	125,413	134,130	139,146	122,020
For processing:				
Chips and shoestring	51,471	52,916	52,371	54,139
Dehydration	55,522	50,831	54,880	40,745
Frozen french fries	142,932	140,196	146,869	126,711
Other frozen products	24,964	23,593	26,723	23,598
Canned potatoes	2,730	3,311	2,368	2,590
Other canned products (hash, stews, soups)	1,964	2,394	2,709	1,722
Starch and flour	1,585	1,310	1,966	1,158
Total	281,168	274,551	287,886	250,663
Other sales:				
Livestock feed [2]	3,111	3,141	10,096	3,496
Seed	24,859	25,294	23,353	25,114
Diversion (other than feed)	....	....	4,164	....
Total	27,970	28,435	37,613	28,610
Total sales	434,551	437,116	464,645	401,293
NON-SALES				
Seed used on farms where grown	4,358	4,415	3,792	4,137
Household use	1,408	1,154	1,496	1,250
Shrinkage and loss	35,454	35,531	43,688	31,208
Total non-sales	41,220	41,100	48,976	36,595
Total production	475,771	478,216	513,621	437,888

[1] Preliminary. [2] Includes potatoes under Government Diversion.
NASS, Crops Branch, (202) 720–2127.

Table 4-52.—Potatoes: Production, seed used, and disposition, by seasonal groups, crop of 2001 [1]

Season and State	Production	Total used for seed	Used on farms where produced: For seed, feed, and household use	Used on farms where produced: Shrinkage and loss	Sold
	1,000 cwt.	*1,000 cwt.*	*1,000 cwt.*	*1,000 cwt.*	*1,000 cwt.*
Winter:					
CA	2,790	210	60	80	2,650
FL	1,325	190	....	8	1,317
Total	4,115	400	60	88	3,967
Spring:					
AZ	2,214	156	2	14	2,198
CA	6,045	551	90	605	5,350
FL	7,970	676	1	40	7,929
Hastings	5,940	499	1	24	5,915
Other	2,030	177	....	16	2,014
NC	3,515	252	26	109	3,380
TX	2,070	133	19	41	2,010
Total	21,814	1,768	138	809	20,867
Summer:					
AL	624	60	7	30	587
CA	2,840	256	62	76	2,702
CO	2,016	156	5	111	1,900
DE	1,161	56	5	47	1,109
IL	1,855	139	35	42	1,778
KS	720	72	....	34	686
MD	1,175	72	20	34	1,121
MO	1,904	140	1	19	1,884
NJ	638	25	3	22	613
NM	770	50	1	31	738
TX	3,120	132	10	60	3,050
VA	1,386	91	1	48	1,337
Total	18,209	1,249	150	554	17,505
Fall:					
CA	1,424	231	3	128	1,293
CO	21,357	1,790	1,465	1,887	18,005
ID	120,200	8,690	1,230	8,261	110,709
IN	928	61	5	33	890
ME	16,430	1,344	301	849	15,280
MA	742	67	5	29	708
MI	13,950	1,181	245	945	12,760
MN	18,425	1,159	190	1,700	16,535
MT	3,296	225	152	188	2,956
NE	8,400	557	153	720	7,527
NV	2,340	150	....	280	2,060
NM	1,428	80	50	67	1,311
NY	5,942	495	100	502	5,340
ND	26,400	2,340	320	3,960	22,120
OH	1,097	90	3	36	1,058
OR	20,730	1,212	175	1,059	19,496
Malheur County	3,690	204	5	37	3,648
Other counties	17,040	1,008	170	1,022	15,848
PA	3,173	285	59	120	2,994
RI	140	12	....	3	137
SD	648	33	3	32	613
UT	345	17	2	11	332
WA	94,400	4,200	300	7,000	87,100
WI	31,955	1,615	278	1,947	29,730
Total	393,750	25,834	5,039	29,757	358,954
US	437,888	29,251	5,387	31,208	401,293

[1] Preliminary.

NASS, Crops Branch, (202) 720–2127.

Table 4-53.—Potatoes,[1] white: United States exports by country of destination and imports by country of origin, 1998/99–2001/2002

Item and country	Year beginning October			
	1998/99	1999/2000	2000/2001	2001/2002
	Metric tons	*Metric tons*	*Metric tons*	*Metric tons*
Exports				
Canada	204,413	246,828	234,526	296,226
Mexico	38,710	38,589	32,313	4,762
Other	21,738	19,313	26,914	16,975
Total[2]	264,861	304,730	293,753	317,963
Imports				
Certified seed:				
Canada	141,640	138,241	86,988	116,745
Others	5	8	42	70
Total	141,645	138,249	87,030	116,815

[1] Includes seed potatoes. [2] Totals may not add due to rounding.

FAS, Horticultural & Tropical Products Division, (202) 720–0897. Compiled from reports of the U.S. Department of Commerce.

Table 4-54.—Potatoes (fresh): Foreign trade, United States, 1992–2001 [1]

Year beginning July	Imports for consumption	Domestic exports
	1,000 cwt.	*1,000 cwt.*
1992	6,602	5,040
1993	7,018	6,246
1994	5,158	6,273
1995	10,741	5,501
1996	7,129	6,402
1997	10,481	7,172
1998	9,090	5,718
1999	9,094	6,541
2000	5,986	6,695
2001	8,644	6,542

[1] Includes seed.

ERS, Specialty Crops Branch, (202) 694–5253. Compiled from reports of the U.S. Department of Commerce.

Table 4-55.—Spinach for fresh market: Area, production, and value per hundredweight, by States, 2000–2002

State	Area harvested			Production			Value per unit		
	2000	2001	2002[1]	2000	2001	2002[1]	2000	2001	2002[1]
	Acres	*Acres*	*Acres*	*1,000 cwt.*	*1,000 cwt.*	*1,000 cwt.*	*Dollars per cwt.*	*Dollars per cwt.*	*Dollars per cwt.*
AZ	3,400	3,600	4,200	782	360	945	30.90	31.00	30.70
CA	19,000	17,000	20,000	3,420	2,635	3,200	30.70	32.60	35.20
CO	1,400	1,200	1,800	98	84	216	32.00	30.00	27.20
MD	1,300	1,200	1,100	85	78	50	37.00	33.00	30.00
NJ	1,700	1,700	1,600	170	204	157	34.80	33.50	35.70
NY[2]	670	300	....	54	24	....	59.30	28.50	....
TX	2,600	2,500	2,200	221	225	242	37.70	36.20	46.00
Oth Sts[3]	1,150	1,150	....	54	99	....	33.00	25.80	....
US	31,220	28,650	30,900	4,884	3,709	4,810	31.70	32.50	34.50

[1] Preliminary. [2] Estimates discontinued in 2002. [3] 2000 - 2001 - DE and VA. DE and VA estimates discontinued in 2002.

NASS, Crops Branch, (202) 720–2127.

Table 4-56.—Spinach for processing: Area, production, and value per ton, by States, 2000–2002

State	Area harvested			Production			Value per unit		
	2000	2001	2002[1]	2000	2001	2002[1]	2000	2001	2002[1]
	Acres	Acres	Acres	Tons	Tons	Tons	Dollars per ton	Dollars per ton	Dollars per ton
CA	....	7,200	7,300	....	69,120	74,460	....	119.00	119.00
TX	5,100	4,200	4,400	44,280	43,100	38,910	88.20	84.70	77.70
Other States[2]	9,620	2,540	1,420	92,370	14,880	7,880	137.00	190.00	182.00
US	14,720	13,940	13,120	136,650	127,100	121,250	121.00	116.00	110.00

[1] Preliminary. [2] 2000 - CA, NJ, NY, OH, and TN. 2001 - NJ, NY, OH, and TN. 2002 - GA and NJ. NY, OH, and TN estimates discontinued in 2002.

NASS, Crops Branch, (202) 720–2127.

Table 4-57.—Sweet Potatoes: Area, yield, production, season average price per hundredweight received by farmers, and value, United States, 1993–2002

Year	Area harvested	Yield per acre	Production	Price[1]	Value of production
	1,000 acres	Cwt.	1,000 cwt.	Dollars	1,000 dollars
1993	80.0	138	11,027	15.10	166,506
1994	82.7	162	13,380	14.00	187,206
1995	83.1	154	12,821	15.90	203,799
1996	83.7	158	13,216	14.40	190,529
1997	82.1	162	13,327	15.80	211,177
1998	83.8	148	12,382	15.30	189,789
1999	83.1	147	12,234	17.60	214,980
2000	94.9	145	13,794	15.30	210,564
2001	94.4	155	14,637	15.40	224,795
2002[2]	83.3	150	12,498	17.00	213,020

[1] Obtained by weighting State prices by production. [2] Preliminary.

NASS, Crops Branch, (202) 720–2127.

Table 4-58.—Sweet Potatoes: Area, production, and season average price per hundredweight received by farmers, by States, 2000–2002

State	Area harvested			Production			Price for crop of-		
	2000	2001	2002[1]	2000	2001	2002[1]	2000	2001	2002[1]
	1,000 acres	1,000 acres	1,000 acres	1,000 cwt.	1,000 cwt.	1,000 cwt.	Dollars	Dollars	Dollars
AL	3.2	2.9	2.7	464	493	486	15.20	18.20	14.80
CA	10.5	10.0	10.4	2,625	2,300	2,444	24.10	26.90	29.60
GA[2]	0.5	0.4	....	70	40	....	14.00	13.00	....
LA	24.0	22.0	15.0	3,120	3,080	1,875	13.20	13.30	13.50
MS	12.3	16.0	12.3	1,476	2,400	1,968	14.60	13.60	14.60
NJ	1.2	1.2	1.2	120	126	150	21.50	19.60	25.00
NC	37.0	36.0	37.0	5,550	5,580	4,810	12.30	12.00	13.00
SC	0.6	1.6	0.9	51	128	99	14.20	13.00	14.40
TX	5.1	3.8	3.3	230	380	561	16.50	18.70	17.90
VA	0.5	0.5	0.5	88	110	105	13.30	14.80	16.10
US	94.9	94.4	83.3	13,794	14,637	12,498	15.30	15.40	17.00

[1] Preliminary. [2] Estimates discontinued in 2002.

NASS, Crops Branch, (202) 720–2127.

Table 4-59.—Taro: Area, yield, total production, price, and value, Hawaii, 1993–2002[1]

Year	Total area	Yield per acre[1]	Production	Price per pound	Value of production
	Acres	*1,000 pounds*	*1,000 pounds*	*Dollars*	*1,000 dollars*
1993	510	....	6,000	0.460	2,760
1994	490	....	6,100	0.460	2,806
1995	550	....	6,800	0.480	3,264
1996	530	....	5,700	0.490	2,793
1997	450	....	5,500	0.510	2,805
1998	490	....	6,000	0.530	3,180
1999	500	....	6,800	0.530	3,604
2000	470	....	7,000	0.530	3,710
2001	440	....	6,400	0.530	3,392
2002[2]	430	....	6,100	0.540	3,294

[1] Yield not estimated. [2] Preliminary.

NASS, Crops Branch, (202) 720–2127.

Table 4-60.—Tomatoes: Foreign trade, United States,1992–2001

Year beginning July	Imports			Domestic exports				
	Fresh	Canned[1]	Paste	Fresh[2]	Canned[3] whole	Catsup and sauces	Paste[3]	Juice
	1,000 pounds	*1,000 pounds*	*1,000 pounds*	*1,000 pounds*	*1,000 pounds*	*1,000 pounds*	*1,000 pounds*	*1,000 pounds*
1992	824,806	122,593	69,157	351,410	44,953	185,414	147,292	24,468
1993	885,824	131,347	95,708	329,577	48,176	241,118	171,550	33,877
1994	1,456,261	189,042	50,898	321,828	51,704	266,491	198,165	30,802
1995	1,702,019	221,894	33,590	288,021	59,312	265,503	193,215	51,002
1996	1,678,128	230,685	31,818	307,555	82,467	279,845	284,377	24,455
1997	1,794,808	281,251	23,601	332,416	55,165	319,960	298,407	23,050
1998	1,618,343	305,598	167,627	311,056	69,571	329,251	173,369	27,290
1999	1,596,470	189,526	46,790	356,676	85,641	334,125	198,029	29,417
2000	1,885,424	236,709	32,717	398,458	77,988	355,414	215,569	39,324
2001	1,709,492	455,829	40,744	375,659	78,891	367,564	206,499	14,480

[1] Includes all canned tomato imports except paste, and is on a product-weight-basis. [2] Prior to 1989, U.S. exports to Canada were understated due to underreporting of actual shipments. [3] Includes exports for military-civilian feeding abroad.

ERS, Specialty Crops Branch, (202) 694–5253. Compiled from reports of the U.S. Department of Commerce.

Table 4-61.—Tomatoes, commercial crop: Area, yield, production, value per hundredweight and per ton, and total value, United States, 1993–2002

Year	For fresh market					For processing				
	Area harvested	Yield per acre	Production	Value[1] Per cwt.	Value[1] Total	Area harvested	Yield per acre	Production	Value[2] Per ton	Value[2] Total
	Acres	*Cwt.*	*1,000 cwt.*	*Dollars*	*1,000 dollars*	*Acres*	*Tons*	*Tons*	*Dollars*	*1,000 dollars*
1993	134,250	273	36,663	31.60	1,157,996	307,470	31.47	9,676,667	60.10	581,901
1994	135,220	276	37,387	27.40	1,024,563	340,060	33.93	11,539,710	62.10	716,469
1995	131,020	260	34,098	25.20	870,427	344,380	32.77	11,285,007	63.20	713,479
1996	120,640	279	33,634	28.20	947,031	339,140	33.64	11,407,301	62.30	711,043
1997	115,190	285	32,777	31.70	1,040,382	283,390	35.19	9,973,259	60.70	604,905
1998	121,710	268	32,628	35.20	1,149,713	299,960	31.34	9,402,010	65.30	613,954
1999	132,880	276	36,735	25.90	951,046	350,410	36.63	12,836,020	71.10	912,988
2000	123,170	306	37,665	30.80	1,159,590	289,600	37.49	10,858,240	59.80	649,066
2001	124,250	286	35,527	30.40	1,080,166	274,860	33.65	9,248,720	59.20	547,473
2002[3]	124,900	299	37,302	31.40	1,170,922	311,800	37.39	11,656,820	58.60	683,083

[1] Price and value of f.o.b. basis. [2] Price and value at processing plant door. [3] Preliminary.

NASS, Crops Branch, (202) 720–2127.

Table 4-62.—Tomatoes, commercial crop: Area, production, and value per hundredweight and per ton, by States, 2000–2002

Utilization and State	Area harvested			Production			Value per unit		
	2000	2001	2002[1]	2000	2001	2002[1]	2000	2001	2002[1]
FOR FRESH MARKET	*Acres*	*Acres*	*Acres*	*1,000 cwt.*	*1,000 cwt.*	*1,000 cwt.*	*Dollars per cwt.*	*Dollars per cwt.*	*Dollars per cwt.*
AL	1,100	900	1,200	242	212	322	25.90	21.80	23.10
AR	1,500	1,300	1,200	150	299	336	26.00	35.00	44.00
CA	40,000	38,000	38,500	11,600	10,260	11,550	29.80	24.90	25.40
CT[2]	300	....	....	23	....	....	80.00	....	....
FL	39,400	44,500	45,000	15,760	14,908	14,400	31.30	32.40	35.30
GA	3,900	2,600	2,500	1,365	949	750	24.50	25.00	20.00
IN	1,600	1,700	1,600	248	281	248	56.30	62.70	64.10
MD	1,900	1,900	2,000	247	266	268	34.00	38.00	47.00
MA[2]	470	400	....	52	56	....	90.00	100.00	....
MI	2,400	1,800	1,700	408	378	357	44.40	35.00	30.50
NJ	3,600	3,400	3,300	720	714	759	41.70	39.20	36.00
NY	3,000	3,000	2,700	540	480	378	56.80	59.40	63.50
NC	2,400	2,600	2,700	696	832	891	30.00	28.00	27.00
OH	4,500	5,900	6,700	1,125	1,947	2,479	20.00	37.90	36.00
PA	4,000	4,100	4,100	840	537	701	30.00	36.60	23.80
SC	3,400	3,200	3,100	884	1,088	837	23.00	21.60	24.80
TN	3,900	3,000	3,900	1,131	495	1,326	31.00	20.00	33.00
TX	1,400	1,200	900	182	180	180	32.30	36.00	32.00
VA	3,900	3,900	3,800	1,287	1,443	1,520	24.00	22.00	27.00
Oth Sts[3]	....	850	....	....	202	....	....	56.60	....
Total	122,670	124,250	124,900	37,500	35,527	37,302	30.70	30.40	31.40
HI[2]	500	....	....	165	....	....	52.00	....	....
US	123,170	124,250	124,900	37,665	35,527	37,302	30.80	30.40	31.40
FOR PROCESSING	*Acres*	*Acres*	*Acres*	*Tons*	*Tons*	*Tons*	*Dollars per ton*	*Dollars per ton*	*Dollars per ton*
CA	271,000	254,000	291,000	10,286,500	8,640,140	11,056,000	58.60	57.50	57.20
IN	6,600	8,000	8,100	229,020	251,120	256,450	84.10	85.80	86.10
MI	2,800	3,100	3,200	84,000	105,400	112,000	81.00	80.00	83.00
OH	5,400	6,200	6,300	158,710	164,610	149,630	78.70	80.30	80.60
PA	1,400	....	....	42,560	....	....	72.10	....	....
Other States[4]	2,400	3,560	3,200	57,450	87,450	82,740	81.00	85.40	87.60
US	289,600	274,860	311,800	10,858,240	9,248,720	11,656,820	59.80	59.20	58.60
Grand total	412,770	399,110	436,700	12,741,490	11,025,070	13,521,920	....	....	....

[1] Preliminary. [2] Estimates discontinued in 2002. [3] 2001 - CT and HI. [4] 2000 - MD and NJ. 2001 - DE, MD, NJ, PA, and VA. 2002 - MD, NJ, and PA. DE and VA estimates discontinued in 2002.

NASS, Crops Branch, (202) 720–2127.

Table 4-63.—Vegetables and melons, fresh: Total reported domestic rail, truck, and air shipments, 2001

Commodity	Jan.	Feb.	Mar.	Apr.	May	June	July	Aug.	Sep.	Oct.	Nov.	Dec.	Total
	1,000 cwt.	*1,000 cwt.*	*1,000 cwt.*	*1,000 cwt.*	*1,000 cwt.*	*1,000 cwt.*	*1,000 cwt.*	*1,000 cwt.*	*1,000 cwt.*	*1,000 cwt.*	*1,000 cwt.*	*1,000 cwt.*	*1,000 cwt.*
Vegetables:													
Artichokes	30	23	83	73	91	38	42	35	51	63	32	23	584
Asparagus	33	58	315	452	315	100	1	...	...	...	...	...	1,274
Beans	81	148	273	207	364	204	50	113	88	214	285	223	2,250
Broccoli	848	612	846	607	753	605	585	542	508	613	604	544	7,667
Cabbage	1,016	1,151	1,559	1,034	973	627	417	646	692	929	979	1,134	11,157
Carrots	835	762	997	917	933	820	672	690	671	768	809	724	9,598
Cauliflower	524	347	499	308	485	359	327	355	288	403	375	336	4,606
Celery	1,267	985	1,127	960	1,143	1,228	1,361	1,155	1,060	1,307	1,739	1,237	14,569
Chinese cabbage	22	12	6	7	1	...	...	...	...	...	5	22	75
Corn, sweet	115	140	313	1,694	2,245	2,265	557	342	259	291	238	267	8,726
Cucumbers	32	11	134	378	771	586	460	631	560	450	411	334	4,758
Eggplant	29	10	27	46	96	49	11	...	...	7	50	54	379
Endive	15	11	12	8	3	...	...	...	...	...	7	15	71
Escarole	16	15	17	12	5	...	...	...	...	...	7	17	89
Greens	252	249	360	233	114	47	27	35	47	86	256	354	2,060
Lettuce, iceberg	2,968	2,489	3,394	2,966	3,946	3,495	3,142	3,267	2,873	3,554	3,004	2,640	37,738
Lettuce, other	424	366	449	337	358	306	300	307	278	329	337	381	4,172
Lettuce, romaine	940	790	1,069	794	869	701	645	692	610	745	814	955	9,624
Okra	...	...	...	10	18	16	12	14	13	13	4	1	101
Onions, dry	3,616	2,865	2,436	3,279	3,871	3,403	3,732	4,020	3,947	4,150	3,796	3,648	42,763
Onions, green	9	8	11	10	23	65	69	72	55	52	32	20	426
Parsley	22	21	29	24	5	...	...	...	...	...	9	36	146
Peppers, bell	399	413	564	921	1,062	1,114	655	617	690	1,005	1,023	565	9,028
Peppers, other	24	23	34	42	44	9	8	...	...	10	60	48	302
Potatoes, table	10,299	9,075	10,615	10,028	10,187	9,713	8,100	8,883	8,441	9,020	9,346	9,273	112,980
Potatoes, chipper	2,532	2,477	3,175	2,734	4,447	4,836	2,519	1,784	2,675	2,244	2,476	3,682	35,581
Potatoes, seed	396	1,009	3,070	6,156	4,237	334	...	...	...	45	156	453	15,856
Radishes	25	30	34	32	20	...	...	...	...	...	15	30	186
Spinach	98	73	62	41	64	50	61	66	52	57	52	62	738
Squash	57	136	151	179	87	40	28	85	52	40	146	135	1,136
Sweet potatoes	259	262	285	309	213	214	166	191	302	340	691	425	3,657
Tomatoes	1,643	1,456	907	2,472	2,124	2,155	1,611	1,658	1,643	2,251	2,234	2,487	22,641
Tomatoes, cherry	61	65	54	141	180	162	35	13	6	22	109	179	1,027
Tomatoes, plum	177	88	45	263	206	230	203	246	241	264	263	293	2,519
Vegetables, other[1]	3	1	...	...	...	...	...	...	...	...	...	...	4
Total	29,067	26,181	32,952	37,674	40,253	33,771	25,796	26,459	26,102	29,272	30,364	30,597	368,488
Melons:													
Cantaloups	...	...	...	74	2,646	3,610	1,148	498	237	1,473	424	28	10,138
Honeydews	...	...	...	...	498	619	135	45	46	338	76	12	1,769
Mixed and miscellaneous	...	...	...	20	49	...	...	...	...	6	5	...	80
Watermelons	...	...	...	344	5,556	8,770	6,417	4,775	1,031	433	264	...	27,590
Total	...	...	...	438	8,749	12,999	7,700	5,318	1,314	2,250	769	40	39,577
Grand total	29,066	26,174	32,922	38,065	48,986	46,770	33,496	31,777	27,416	31,522	31,133	30,637	408,065

[1] Not otherwise specified.

AMS, Fruit and Vegetable Division, Market News Branch, (202) 720–3343.

Table 4-64.—Vegetables (fresh), melons, potatoes, sweet potatoes: Per capita civilian utilization (farm-weight basis), United States, 1992–2001 [1]

Year	Cabbage	Cucumbers	Tomatoes	Asparagus	Broccoli	Carrots	Head Lettuce	Leaf/romaine
	Pounds	*Pounds*	*Pounds*	*Pounds*	*Pounds*	*Pounds*	*Pounds*	*Pounds*
1992	8.7	4.9	15.5	0.6	3.4	8.3	25.8	4.7
1993	9.3	5.2	16.3	0.6	3.3	10.9	24.4	5.0
1994	9.1	5.4	16.2	0.6	4.4	12.7	25.0	5.7
1995	8.1	5.6	16.9	0.6	4.3	11.2	22.2	5.9
1996	8.3	5.9	17.4	0.6	4.5	12.4	21.6	5.8
1997	9.0	6.4	16.8	0.7	5.0	14.1	23.9	6.6
1998	8.5	6.5	17.6	0.7	5.1	12.7	21.6	6.7
1999 [3]	7.6	6.8	17.8	0.9	6.5	10.9	23.9	7.7
2000	9.1	6.4	17.6	1.0	6.0	10.4	23.4	8.6
2001	9.1	6.4	17.9	0.9	5.8	10.6	24.2	8.5

Year	Snap beans	Garlic	Cauliflower	Celery	Sweet Corn	Onions	Spinach	Bell peppers
	Pounds	*Pounds*	*Pounds*	*Pounds*	*Pounds*	*Pounds*	*Pounds*	*Pounds*
1992	1.4	1.5	1.8	7.3	6.9	16.1	0.8	5.6
1993	1.5	1.8	2.1	7.3	7.0	17.2	0.7	6.1
1994	1.5	1.8	2.0	7.2	8.2	16.9	0.8	6.4
1995	1.6	1.9	1.6	6.9	7.8	17.8	0.7	6.2
1996	1.5	2.3	1.7	7.0	8.3	18.4	0.6	7.1
1997	1.4	2.0	1.8	6.6	8.3	18.8	1.1	6.4
1998	1.7	2.6	1.5	6.5	9.3	18.3	1.0	6.4
1999 [3]	1.9	3.3	1.9	6.6	9.1	18.4	1.0	6.7
2000	2.0	2.2	2.0	6.4	9.2	18.3	1.6	7.0
2001	2.1	2.1	2.0	6.6	9.4	18.1	1.3	6.2

Year	Watermelon	Cantaloupe	Honeydew melons	Others [2]	Total vegetables and melons	Potatoes	Sweet potatoes
	Pounds	*Pounds*	*Pounds*	*Pounds*	*Pounds*	*Pounds*	*Pounds*
1992	14.7	8.5	2.1	5.9	144.5	48.4	4.1
1993	14.2	8.6	1.7	5.9	149.1	50.2	3.7
1994	15.0	8.4	2.0	6.3	155.6	49.7	4.5
1995	15.2	9.0	1.9	5.8	151.2	49.3	4.2
1996	16.6	10.3	2.0	6.0	158.3	50.0	4.3
1997	15.5	10.5	2.2	6.1	163.2	48.5	4.3
1998	14.3	10.7	2.3	6.5	160.5	47.0	3.9
1999 [3]	15.4	11.5	2.5	6.2	166.6	48.0	3.7
2000	13.9	10.8	2.2	15.1	173.2	47.2	4.2
2001	14.9	11.2	2.1	13.9	173.3	46.7	4.3

[1] Fresh vegetable consumption computed for total commercial production for fresh market. Does not include production for home use. Consumption obtained by dividing the total apparent consumption by total July 1 population as reported by the Bureau of the Census. All data for calendar year. [2] Includes artichokes, eggplant, radishes, brussels sprouts, squash, green limas, and escarole/endive. Beginning in 2000, also includes collards, mustard greens, turnip greens, kale, okra, and pumpkins [3] Preliminary.

ERS, Market and Trade Economics Division, Specialty Crops Branch, (202)694–5253.

Table 4-65.—Vegetables, frozen: Commercial pack, United States, 1992–2001

Commodity	1992	1993	1994	1995	1996
	1,000 pounds	*1,000 pounds*	*1,000 pounds*	*1,000 pounds*	*1,000 pounds*
Artichokes	1,142	733	1,364		
Asparagus	16,525	14,522	14,967	13,985	10,364
Beans, butter	3,906	3,642	7,491	7,664	4,577
Beans, green, regular cut	187,377	203,333	239,059	208,246	232,711
Beans, green, French cut	49,704	69,478	71,104	70,144	65,041
Beans, green, wax	6,897	6,091	9,144	7,120	7,592
Beans, green, Italian	20,061	18,639	25,197	17,267	17,711
Beans, whole	42,622	33,811	41,173	53,538	50,517
Beans, baby lima	44,592	66,464	95,695	101,295	85,053
Beans, lima, Fordhook	15,129	13,322	18,942	20,489	22,645
Broccoli	168,518	165,183	151,150	205,471	138,387
Brussels sprouts	42,805	46,950	34,393	25,519	30,154
Carrots	371,232	403,174	412,709	418,816	397,967
Cauliflower	70,317	73,335	66,878	64,207	48,541
Celery	27,161	36,668	34,334	32,279	40,343
Collards	15,981	17,714	20,871	17,647	17,709
Corn, cut	548,749	473,105	686,140	694,110	646,389
Corn-on-cob	375,666	377,657	492,522	414,662	383,840
Kale	4,106	4,010	3,212	3,664	2,671
Mushrooms	26,536	21,535	27,200	26,778	23,219
Mustard greens	7,928	12,311	9,987	10,302	11,294
Okra	61,738	34,593	65,114	64,810	60,847
Onions	85,222	91,576	100,138	99,639	111,021
Peas, blackeye	27,195	25,582	33,427	29,352	22,901
Peas, green	465,375	391,943	499,744	505,350	369,390
Peppers, green and red	57,935	46,331	42,787	54,155	44,858
Potato products	6,798,442	7,252,774	7,934,849	8,364,909	8,419,203
Pumpkin and cooked squash	21,534	24,196	27,356	28,097	20,058
Rhubarb	6,426	10,081	7,525	8,228	6,910
Spinach	184,642	167,147	204,495	185,640	182,999
Squash, summer	58,184	57,366	58,050	54,957	42,387
Sweet potatoes and yams	19,456	16,234	19,506	14,127	14,532
Turnip greens	14,213	17,666	19,834	18,203	19,112
Turnip greens with turnips	18,290	23,776	21,998	17,041	19,568
Miscellaneous vegetables	14,964	15,164	38,575	41,004	36,715
Total	9,880,573	10,236,106	11,536,930	11,898,715	11,607,226

Commodity	1997	1998	1999	2000	2001
	1,000 pounds	*1,000 pounds*	*1,000 pounds*	*1,000 pounds*	*1,000 pounds*
Artichokes					
Asparagus	10,103	9,792	12,439	10,224	9,348
Beans, butter	5,170	7,298	8,515	7,771	7,688
Beans, green, regular cut	231,625	214,871	207,345	235,279	240,514
Beans, green, French cut	75,086	73,613	54,955	52,454	55,503
Beans, green, wax	7,324	8,927	7,132	8,512	8,341
Beans, green, Italian	18,364	18,531	23,537	15,979	15,818
Beans, whole	58,605	71,481	63,714	77,619	74,005
Beans, baby lima	97,598	101,757	83,381	71,637	67,000
Beans, lima, Fordhook	25,872	22,962	14,369	15,443	11,920
Broccoli	109,013	144,361	174,283	138,944	137,705
Brussels sprouts	26,436	33,062	35,264	22,583	19,951
Carrots	409,044	388,094	424,691	412,744	206,085
Cauliflower	44,590	44,607	42,445	40,405	35,550
Celery	32,985	29,001	32,966	48,402	45,683
Collards	22,244	26,915	17,685	26,949	23,420
Corn, cut	736,545	753,429	665,209	666,617	622,984
Corn-on-cob	464,942	435,271	404,831	372,794	343,516
Kale	4,213	2,936	2,319	2,850	2,169
Mustard greens	13,139	10,349	11,152	13,799	12,841
Okra	57,725	85,762	90,004	75,770	70,735
Onions	115,980	122,289	123,414	108,891	107,592
Peas, blackeye	25,363	27,684	16,328	26,652	24,973
Peas, green	468,074	491,377	444,050	451,995	446,389
Peppers, green and red	40,746	42,599	53,876	46,165	45,712
Potato products	8,381,370	8,723,707	8,765,173	7,622,629	7,847,656
Pumpkin and cooked squash	24,570	19,329	17,889	20,346	19,133
Rhubarb	7,505	6,558	6,353	3,276	2,594
Spinach	198,715	190,028	174,149	172,543	173,850
Squash, summer	46,579	42,976	34,842	31,117	28,700
Sweet potatoes and yams	20,992	[1]	[1]	[1]	12,736
Turnip greens	20,490	21,719	32,499	22,933	15,049
Turnip greens with turnips	25,119	19.485	[2]	15,780	0
Mushrooms	22,537	20,415	25,117	36,928	32,332
Miscellaneous vegetables	46,363	42,849	51,341	39,467	55,383
Total	11,895,026	12,253,884	12,121,317	10,915,497	10,822,875

[1] Included with miscellaneous vegetables. [2] Included with Turnip Greens.

ERS, Specialty Crops Branch, (202) 694–5253. Data from American Frozen Food Institute.

Table 4-66.—Vegetables, frozen: Commercial pack, by kinds and regions, United States, 2000

Commodity	East and South [1]	Midwest [2]	West [3]	Northwest [4]	California	Total
	1,000 pounds	*1,000 pounds*	*1,000 pounds*	*1,000 pounds*	*1,000 pounds*	*1,000 pounds*
Asparagus	....	8,603	1,621	....	....	10,224
Beans, butter	7,771	....	....	....	....	7,771
Beans, green, regular cut	71,992	55,104	108,183	....	....	235,279
Beans, green, French cut	21,695	11,949	18,810	....	....	52,454
Beans, wax	....	8,512	....	....	....	8,512
Beans, green, Italian	....	....	15,979	....	....	15,979
Beans, whole	....	18,042	32,475	....	....	50,517
Beans, baby lima	....	28,290	....	....	43,347	71,637
Beans, lima, Fordhook	....	....	15,443	....	....	15,443
Broccoli	....	....	138,994	....	....	138,944
Brussels sprouts	....	....	22,583	....	....	22,583
Carrots	65,192	48,725	298,827	....	....	412,744
Cauliflower	....	....	....	....	40,405	40,405
Celery [7]	....	....	....	....	....	48,402
Collards [7]	....	....	....	....	....	26,949
Corn, cut	91,786	122,829	452,002	....	....	666,617
Corn-on-cob	....	109,918	262,876	....	....	372,794
Kale [7]	....	....	....	....	....	1,643
Mustard greens [7]	....	....	....	....	....	13,799
Okra [7]	....	....	....	....	....	75,770
Onions [7] [8]	....	....	....	....	....	108,891
Peas, blackeye [7]	....	....	....	....	....	26,652
Peas, green	82,124	94,076	275,795	....	....	451,995
Peppers, green and red	17,215	....	28,950	....	....	46,165
Potato products	....	1,340,511 [6]	....	6,271,028	11,090	7,622,629
Pumpkins and cooked squash [5]	....	....	....	....	....	20,346
Rhubarb	....	....	3,276	....	....	3,276
Spinach	90,278	....	....	....	82,265	172,543
Squash, summer [7]	....	....	....	....	....	31,117
Sweet potatoes and yams	....	....	....	....	....	[9]
Turnip greens	....	....	....	....	....	22,933
Turnip greens with turnips	....	....	....	....	....	15,780
Mushrooms	....	....	....	....	....	36,928
Miscellaneous vegetables	....	....	....	....	....	39,467

[1] Alabama, Arkansas, Connecticut, Delaware, Florida, Georgia, Kentucky, Louisiana, Maine, Maryland, Massachusetts, Mississippi, Missouri, New Jersey, New Mexico, New York, North Carolina, Oklahoma, Pennsylvania, Tennessee, Texas, and Virginia. [2] Illinois, Indiana, Michigan, Minnesota, Nebraska, North Dakota, Ohio, South Dakota, and Wisconsin. [3] California, Colorado, Idaho, Montana, Oregon, Washington, and Wyoming. [4] Idaho, Oregon, and Washington. [5] Midwest and West. [6] East, South, and Midwest. [7] Regional packs not reported. [8] Excludes breaded, battered, and extruded onion rings. [9] Included with Miscellaneous Vegetables.

ERS, Specialty Crops Branch, (202) 694–5253. Data from American Frozen Food Institute.

Table 4-67.—Vegetables, canning: Per capita utilization (farm weight), United States, 1992–2001

Year	Cabbage for kraut	Asparagus	Snap beans	Carrots	Green peas
	Pounds	*Pounds*	*Pounds*	*Pounds*	*Pounds*
1992	1.1	0.3	4.0	1.7	2.1
1993	1.4	0.3	4.0	1.0	1.6
1994	1.2	0.3	3.8	1.4	1.4
1995	1.4	0.3	3.5	1.6	1.6
1996	1.0	0.2	3.8	1.7	1.5
1997	1.4	0.2	3.6	1.5	1.5
1998	1.4	0.2	3.8	1.5	1.4
1999 [1]	1.2	0.2	3.7	1.4	1.4
2000	1.4	0.2	3.8	1.1	1.5
2001	1.3	0.2	3.7	0.9	1.4

Year	Tomatoes	Corn	Pickles	Other [2]	Total [3]
	Pounds	*Pounds*	*Pounds*	*Pounds*	*Pounds*
1992	73.4	11.9	4.5	8.2	107.2
1993	75.8	11.1	4.3	7.2	106.7
1994	76.3	10.1	4.8	6.9	106.2
1995	74.6	10.4	5.0	6.1	104.5
1996	73.1	10.4	4.1	6.8	102.6
1997	72.6	9.2	5.2	6.8	102.0
1998	74.1	9.2	4.0	6.9	102.5
1999 [1]	71.2	9.1	4.2	6.5	98.9
2000	70.1	9.0	4.9	7.6	99.6
2001	68.7	9.2	3.8	7.1	96.3

[1] Preliminary. [2] Includes beets, chile peppers (all uses), green lima beans and spinach. [3] Totals may not add due to rounding.

ERS, Specialty Crops Branch, (202) 694–5253.

Table 4-68.—Vegetables, freezing: Per capita utilization (farm weight basis), United States, 1992–2001

Year	Leafy, green, and yellow vegetables				
	Asparagus	Snap beans	Carrots	Peas	Broccoli
	Pounds	*Pounds*	*Pounds*	*Pounds*	*Pounds*
1992	0.1	1.7	2.3	2.0	2.4
1993	0.1	1.7	2.8	1.9	2.3
1994	0.1	1.9	2.8	2.1	2.3
1995	0.1	1.7	2.6	2.1	2.6
1996	0.1	1.9	2.8	1.9	2.5
1997	0.1	1.8	2.6	2.0	2.3
1998	0.1	2.0	2.8	1.9	2.1
1999 [1]	0.1	2.0	2.4	2.0	2.1
2000	0.1	1.8	2.7	2.1	2.4
2001	0.1	1.9	2.4	1.9	2.2

Year	Cauliflower	Sweet Corn	Other [2]	Total vegetables excluding potatoes	Potato products	Grand total
	Pounds	*Pounds*	*Pounds*	*Pounds*	*Pounds*	*Pounds*
1992	0.7	9.0	2.5	20.7	49.9	70.6
1993	0.7	9.7	2.7	21.9	53.5	75.4
1994	0.6	9.1	2.9	21.8	55.7	77.5
1995	0.6	10.4	2.8	22.9	56.2	79.1
1996	0.5	10.4	3.1	23.2	60.2	83.4
1997	0.4	10.1	2.9	22.3	59.4	81.7
1998	0.8	9.8	3.0	22.4	58.2	80.6
1999 [1]	0.5	10.1	3.2	22.4	58.6	81.0
2000	0.5	9.0	3.2	21.8	57.8	79.6
2001	0.6	9.2	3.3	21.6	58.8	80.4

[1] Preliminary. [2] Includes green lima beans, spinach, and miscellaneous freezing vegetables.

ERS, Specialty Crops Branch, (202) 694–5253.

Table 4-69.—Commercially produced vegetables: Per capita utilization, United States, 1992–2001 [1]

Year	Farm weight equivalent					Percentage of annual total			
	Total fresh and processed	Fresh [2]	Processed [3]			Fresh	Processed		
			Total	Canning	Freezing		Total	Canning	Freezing
	Pounds	*Pounds*	*Pounds*	*Pounds*	*Pounds*	*Percent*	*Percent*	*Percent*	*Percent*
1992	272.4	144.5	127.9	107.2	20.7	53.0	47.0	39.4	7.6
1993	277.7	149.1	128.6	106.7	21.9	53.7	46.3	39.4	7.9
1994	283.6	155.6	128.0	106.2	21.8	54.9	45.1	38.4	7.7
1995	278.6	151.2	127.4	104.5	22.9	54.3	45.7	37.5	8.2
1996	284.1	158.3	125.8	102.6	23.2	55.7	44.3	36.1	8.2
1997	287.5	163.2	124.3	102.0	22.3	56.8	43.2	35.5	7.7
1998	285.4	160.5	124.9	102.5	22.4	56.2	43.8	35.9	7.8
1999	287.9	166.6	121.3	98.9	22.4	57.9	42.1	34.4	7.7
2000	294.6	173.2	121.4	99.6	21.8	58.8	41.2	33.8	7.4
2001	291.2	173.3	117.9	96.3	21.6	59.5	40.5	33.1	7.4

[1] Excludes potatoes, sweet potatoes, pulses, dehydrating onions, and mushrooms. [2] See table 4-64 for items included. Includes melons. [3] See table 4-65 for items included.

ERS, Market and Trade Economics Division, Specialty Crops Branch, (202) 694–5253.

Table 4-70.—Frozen Vegetables and potato products: Cold storage holdings, end of month, United States, 2001 and 2002

Month	Asparagus		Green beans, regular cut		Green beans, French cut		Green beans, total	
	2001	2002	2001	2002	2001	2002	2001	2002
	1,000 pounds	*1,000 pounds*	*1,000 pounds*	*1,000 pounds*	*1,000 pounds*	*1,000 pounds*	*1,000 pounds*	*1,000 pounds*
January	10,190	9,581	134,273	117,027	25,951	18,171	160,224	135,198
February	9,757	8,599	115,214	102,337	21,171	14,995	136,385	117,332
March	9,318	7,698	90,844	89,335	20,145	13,304	110,989	102,639
April	8,796	7,357	73,172	64,490	12,658	11,408	85,830	75,898
May	10,403	7,481	57,759	55,541	10,713	8,913	68,472	64,454
June	14,705	10,954	49,355	48,312	8,955	7,398	58,310	55,710
July	14,476	10,884	95,986	99,842	19,708	21,683	115,694	121,525
August	13,881	9,996	184,085	195,536	26,694	25,601	210,779	221,137
September	12,726	7,858	191,945	219,248	29,968	30,399	221,913	249,647
October	13,065	7,122	188,387	204,734	27,755	28,673	216,142	233,407
November	12,238	7,188	163,900	187,561	23,835	24,026	187,735	211,587
December	10,686	6,988	140,311	167,669	20,517	19,726	160,828	187,395

Month	Broccoli spears		Broccoli, chopped & cut		Broccoli, total		Brussels sprouts	
	2000	2001	2000	2001	2000	2001	2000	2001
	1,000 pounds	*1,000 pounds*	*1,000 pounds*	*1,000 pounds*	*1,000 pounds*	*1,000 pounds*	*1,000 pounds*	*1,000 pounds*
January	36,685	41,808	78,986	77,672	115,671	119,480	31,099	18,078
February	38,872	47,431	83,454	78,297	122,326	125,728	29,996	15,144
March	41,030	51,381	78,051	72,073	119,081	123,454	19,677	14,429
April	38,209	59,042	79,203	73,317	117,412	132,359	13,621	13,241
May	36,162	59,382	80,369	70,348	116,531	129,730	12,269	11,369
June	35,926	56,679	89,062	66,486	124,988	123,165	12,244	10,645
July	33,372	54,847	84,445	70,778	117,817	125,625	10,306	10,460
August	29,479	50,436	78,543	79,052	108,022	129,488	8,868	8,131
September	26,303	45,364	78,521	74,247	104,824	119,611	7,945	7,485
October	25,612	45,034	77,645	70,017	103,257	115,051	11,997	11,879
November	27,689	46,383	77,976	69,585	105,665	115,968	15,771	17,935
December	38,035	49,326	76,491	67,727	114,526	117,053	17,595	20,220

Month	Fordhook lima beans		Baby lima beans		Mixed vegetables		Okra	
	2001	2002	2001	2002	2001	2002	2001	2002
	1,000 pounds	*1,000 pounds*	*1,000 pounds*	*1,000 pounds*	*1,000 pounds*	*1,000 pounds*	*1,000 pounds*	*1,000 pounds*
January	12,604	12,236	32,529	45,714	47,504	56,270	39,353	31,978
February	10,315	10,445	29,452	42,887	48,739	55,344	36,470	33,964
March	8,736	9,032	26,555	40,856	41,962	49,273	32,852	28,151
April	8,061	7,964	23,476	37,769	46,115	48,809	30,398	26,756
May	6,809	7,610	20,885	34,265	47,796	45,965	30,558	29,521
June	6,742	7,323	17,169	30,017	47,648	45,248	42,863	35,964
July	5,564	5,682	14,238	26,740	50,301	44,052	48,248	49,258
August	10,033	9,921	21,645	29,417	49,351	50,415	51,271	50,548
September	16,334	14,599	41,627	45,372	50,739	47,766	53,168	56,797
October	16,393	15,064	57,570	52,905	50,852	46,276	51,866	54,395
November	15,334	14,681	53,737	48,481	48,872	40,489	43,600	51,187
December	14,304	13,068	50,358	45,544	51,453	41,396	35,939	46,616

Month	Carrots, diced		Carrots, other		Carrots, total		Cauliflower	
	2001	2002	2001	2002	2001	2002	2001	2002
	1,000 pounds	*1,000 pounds*	*1,000 pounds*	*1,000 pounds*	*1,000 pounds*	*1,000 pounds*	*1,000 pounds*	*1,000 pounds*
January	118,346	81,072	153,117	158,778	271,463	239,850	42,827	35,155
February	110,850	74,028	144,179	139,790	255,029	213,818	38,715	33,288
March	99,715	71,754	118,215	124,741	217,930	196,495	36,682	33,253
April	87,824	66,031	112,844	114,756	200,668	180,787	31,099	31,505
May	75,929	57,583	103,692	99,168	179,621	156,751	28,456	29,011
June	60,047	47,110	101,103	91,705	161,150	138,815	26,866	25,589
July	54,148	42,529	96,230	85,488	150,378	128,017	26,502	24,074
August	46,485	38,824	101,199	85,169	147,684	123,993	21,431	25,186
September	43,832	44,179	117,917	95,017	161,749	139,196	25,404	26,801
October	64,065	71,081	170,102	132,087	234,167	203,168	33,423	33,600
November	89,910	99,574	197,187	144,808	287,097	244,382	34,946	38,733
December	89,361	99,188	186,831	143,368	276,192	242,556	36,443	35,635

See end of table.

Table 4-70.—Frozen Vegetables and potato products: Cold storage holdings, end of month, United States, 2001 and 2002—Continued

Month	Corn, cut		Corn, cob		Corn, total		Onion rings	
	2001	2002	2001	2002	2001	2002	2001	2002
	1,000 pounds	*1,000 pounds*	*1,000 pounds*	*1,000 pounds*	*1,000 pounds*	*1,000 pounds*	*1,000 pounds*	*1,000 pounds*
January	290,817	327,775	223,175	235,144	513,992	562,919	9,615	7,348
February	248,314	290,103	194,204	187,863	442,518	477,966	9,747	7,138
March	194,544	255,128	148,930	173,544	343,474	428,672	7,744	6,803
April	160,573	209,752	128,176	144,328	288,749	354,080	6,429	7,189
May	128,853	170,241	105,266	113,482	234,119	283,723	7,307	7,331
June	97,639	137,242	74,548	85,566	172,187	222,808	7,799	8,619
July	100,301	141,636	75,563	89,916	175,864	231,552	6,801	6,990
August	227,508	289,980	159,099	169,375	386,607	459,355	7,888	7,270
September ..	382,000	451,612	288,534	278,782	670,534	730,394	8,082	7,914
October	409,944	462,317	289,305	301,921	699,249	764,238	7,298	9,420
November ...	379,256	425,682	269,562	288,565	648,818	714,247	8,571	9,241
December ...	351,820	402,164	263,457	262,338	615,277	664,502	6,978	9,636

Month	Onions, other		Blackeye peas		Green peas		Peas & carrots mixed	
	2001	2002	2001	2002	2001	2002	2001	2002
	1,000 pounds	*1,000 pounds*	*1,000 pounds*	*1,000 pounds*	*1,000 pounds*	*1,000 pounds*	*1,000 pounds*	*1,000 pounds*
January	39,722	26,298	3,748	5,161	253,772	195,178	8,213	10,619
February	39,407	27,713	4,279	5,891	226,238	174,097	8,620	10,746
March	36,925	28,674	5,035	6,332	179,626	149,678	8,078	7,581
April	36,503	33,600	4,491	4,980	154,378	133,409	8,695	6,848
May	38,705	33,735	3,769	5,708	131,960	111,980	8,343	6,893
June	38,275	32,143	3,257	5,896	240,088	198,452	8,571	6,746
July	29,993	26,380	3,331	4,519	379,529	320,162	9,144	6,799
August	28,912	29,625	2,913	5,128	338,859	305,107	9,018	7,332
September ..	26,133	28,206	5,491	4,611	320,198	282,648	8,816	7,935
October	27,726	30,376	5,595	7,663	276,547	253,753	9,502	8,762
November ...	28,987	30,069	4,958	8,208	241,553	206,345	8,322	8,365
December ...	29,893	29,844	4,473	6,323	224,715	171,161	8,847	6,597

Month	Spinach		Squash		Southern greens		Other vegetables	
	2001	2002	2001	2002	2001	2002	2001	2002
	1,000 pounds	*1,000 pounds*	*1,000 pounds*	*1,000 pounds*	*1,000 pounds*	*1,000 pounds*	*1,000 pounds*	*1,000 pounds*
January	44,969	66,170	38,577	42,175	37,235	31,158	302,029	316,085
February	46,785	66,518	32,762	40,114	32,819	30,252	305,608	301,858
March	61,186	80,447	27,806	34,785	29,430	27,421	267,798	277,155
April	90,766	103,363	23,608	30,002	28,573	25,500	248,164	258,207
May	98,871	108,686	25,686	31,332	26,667	25,145	236,601	246,237
June	97,278	106,198	31,839	36,878	28,131	26,478	248,246	241,875
July	88,050	98,343	35,755	43,371	28,942	23,477	262,102	263,996
August	77,489	91,412	40,704	49,233	26,421	21,563	298,665	289,438
September ..	67,525	79,605	43,068	49,029	26,733	19,268	329,012	318,368
October	67,221	70,050	45,878	50,243	29,143	21,187	362,264	341,089
November ...	66,631	63,057	45,388	46,891	31,641	26,275	364,365	331,863
December ...	63,368	56,193	43,198	44,617	35,861	22,986	336,555	313,421

Month	Total vegetables		French fries		Other frozen potatoes		Total frozen potatoes	
	2001	2002	2001	2002	2001	2002	2001	2002
	1,000 pounds	*1,000 pounds*	*1,000 pounds*	*1,000 pounds*	*1,000 pounds*	*1,000 pounds*	*1,000 pounds*	*1,000 pounds*
January	2,015,336	1,966,651	982,978	1,047,759	245,612	226,392	1,228,590	1,274,151
February	1,865,967	1,798,842	998,227	1,031,776	256,493	239,726	1,254,720	1,271,502
March	1,590,884	1,652,828	998,851	1,005,122	222,076	266,267	1,220,927	1,271,389
April	1,455,832	1,519,623	1,032,416	964,478	247,987	258,182	1,280,403	1,222,660
May	1,333,828	1,376,927	1,016,976	912,498	253,299	269,837	1,270,275	1,182,335
June	1,388,356	1,369,523	1,083,482	937,933	271,493	285,605	1,354,975	1,223,538
July	1,573,035	1,571,906	1,012,479	848,204	270,109	258,381	1,282,588	1,106,585
August	1,860,441	1,923,695	953,767	797,929	243,701	242,662	1,197,468	1,040,591
September ..	2,202,021	2,243,110	1,071,789	888,831	251,973	252,551	1,323,762	1,141,382
October	2,319,155	2,329,648	1,105,469	994,421	233,072	257,736	1,338,541	1,252,157
November ...	2,254,229	2,235,192	1,067,240	957,161	230,137	257,214	1,297,377	1,214,375
December ...	2,137,489	2,081,751	1,010,098	877,292	229,659	253,920	1,239,757	1,131,212

NASS, Livestock Branch, (202) 720–8784.

Table 4-71.—Watermelon for fresh market: Area, production, and value per hundredweight, by States, 2000–2002

State	Area harvested			Production			Value per unit		
	2000	2001	2002 [1]	2000	2001	2002 [1]	2000	2001	2002 [1]
	Acres	*Acres*	*Acres*	*1,000 cwt.*	*1,000 cwt.*	*1,000 cwt.*	*Dollars per cwt.*	*Dollars per cwt.*	*Dollars per cwt.*
AL	3,900	2,400	3,500	507	432	441	5.60	6.60	6.65
AZ	7,100	6,100	6,100	2,663	2,623	2,562	6.80	10.20	8.30
AR	2,700	3,000	2,500	405	510	463	4.20	4.90	5.70
CA	12,300	11,800	11,800	6,581	6,254	6,018	11.70	10.90	12.70
DE	2,600	2,700	2,700	780	1,161	1,215	7.00	8.00	9.00
FL	27,000	24,000	23,000	8,640	7,440	7,590	5.25	5.70	8.20
GA	24,000	22,000	20,000	4,680	5,830	5,100	4.60	5.00	5.80
IN	6,100	6,400	6,700	1,586	2,560	2,412	6.40	7.80	8.40
LA [2]	2,200	....	1,800	242	....	216	6.90	....	6.80
MD	2,900	3,100	3,200	566	868	896	9.00	8.00	10.00
MS	2,800	2,500	2,300	185	425	357	6.32	5.50	6.50
MO	5,800	5,000	5,100	1,218	1,150	1,326	3.65	4.25	5.80
NC	10,000	9,500	9,600	1,600	1,473	1,632	6.00	5.70	6.50
OK	6,000	6,000	5,500	420	750	715	7.00	7.00	7.00
SC	7,500	7,700	7,500	1,500	1,386	1,313	5.10	5.70	6.00
TX	40,000	45,000	37,000	5,600	7,200	6,660	3.90	4.50	8.50
VA	1,500	1,400	1,600	330	294	288	5.00	7.00	6.00
Total	164,400	158,600	149,900	37,503	40,356	39,204	6.35	6.72	8.38
HI [3]	560	....	....	126	....	....	23.00	....	....
Oth Sts [4]	....	2,220	....	....	292	....	....	12.90	....
US	164,960	160,820	149,900	37,629	40,648	39,204	6.41	6.77	8.38

[1] Preliminary. [2] 2001 data not published to avoid disclosure of individual operations. Data have been included in the Other States total. [3] Estimates discontinued in 2002. [4] 2001 - HI and LA.

NASS, Crops Branch, (202) 720–2127.

CHAPTER V

STATISTICS OF FRUITS, TREE NUTS, AND HORTICULTURAL SPECIALTIES

For most fruits, production is estimated at two levels—total and utilized. Total production is the quantity of fruit harvested plus quantities which would have been acceptable for fresh market or processing but were not harvested or utilized because of economic and other reasons. Utilized production is the amount sold plus the quantities used on farms where grown and quantities held in storage. The difference between total and utilized production is the quantity of marketable fruit not harvested and fruit harvested but not sold or utilized because of economic and other reasons. Production relates to the crop produced on all farms, except for apples and strawberries. In accordance with Congressional enactment, the Department's estimates of apple production since 1938 have related only to commercial production. The estimates for strawberries cover production on area grown primarily for sale. Statistics on utilization of fruit by commercial processors refer to first utilization, not necessarily final utilization. For example, frozen fruit includes fruit which may later be used for preserves.

The price shown for each crop is a marketing year average price for all methods of sales. Prices for most fresh fruit are the average prices producers received at the point of first sale, commonly referred to as the "average price as sold." Since the point of first sale is not the same for all producers, prices for the various methods of sale are weighted by the proportionate quantity sold. For example, if in a given State part of the fruit crop is sold f.o.b. packed by growers, part sold as bulk fruit at the packinghouse door, and some sold retail at roadside stands, the fresh fruit average price as sold is a weighted average of the average price for each method of sale.

The annual estimates are checked and adjusted at the end of each marketing season on the basis of shipment and processing records from transportation agencies, processors, cooperative marketing associations, and other industry organizations. The estimates are reviewed (and revised if necessary) at 5-year intervals, when the Census of Agriculture data become available. The Department's available statistics are limited to the major tree fruits and nuts and to grapes, cranberries, and strawberries, and exclude some States where census data indicate production is of only minor importance.

Table 5-1.—Fruits and planted nuts: Bearing area, United States, 1993–2002

Year	Citrus fruits [1]	Major deciduous fruits [2]	Miscellaneous fruits [3]	Planted nuts [4]	Fruits and planted nuts
	1,000 acres	*1,000 acres*	*1,000 acres*	*1,000 acres*	*1,000 acres*
1993	946.7	1,738.2	290.9	700.5	3,677.0
1994	977.8	1,756.6	288.1	725.6	3,748.4
1995	1,054.0	1,762.8	289.1	718.6	3,825.6
1996	1,104.5	1,796.1	287.8	732.1	3,919.2
1997	1,152.5	1,810.9	292.2	748.6	4,004.2
1998	1,125.3	1,834.3	295.6	769.7	4,024.9
1999	1,111.6	1,870.0	296.9	790.1	4,068.7
2000	1,089.9	1,882.9	299.3	814.0	4,086.0
2001	1,081.1	1,852.6	291.3	844.9	4,069.9
2002	1,053.3	1,853.9	227.9	858.6	3,993.7

[1] Oranges, tangerines, Temples, grapefruit, lemons, limes, tangelos, and K-Early Citrus. Area is for the year of harvest. [2] Commercial apples, apricots, cherries, grapes, nectarines, peaches, pears, plums, and prunes. [3] Avocados, bananas, berries, cranberries, dates, figs, guavas, kiwifruit, olives, papayas, pineapples, and strawberries. [4] Almonds, hazelnuts, macadamia nuts, pistachios, and walnuts.

NASS, Crops Branch, (202) 720–2127.

Table 5-2.—Fruits: Total production in tons, United States, 1993–2002 [1]

Year	Apples, commercial crop [2]	Peaches	Pears	Grapes (fresh basis)	Cherries	Prunes and plums (fresh basis)	Apricots	Figs (fresh basis)	Olives
	1,000 tons	*1,000 tons*	*1,000 tons*	*1,000 tons*	*1,000 tons*	*1,000 tons*	*1,000 tons*	*1,000 tons*	*1,000 tons*
1993	5,343	1,322	948	6,023	339	588	97	61	122
1994	5,751	1,255	1,046	5,873	359	879	153	57	84
1995	5,289	1,145	948	5,922	363	744	61	52	78
1996	5,191	1,052	821	5,554	290	952	79	46	166
1997	5,162	1,312	1,043	7,291	372	926	139	58	104
1998	5,823	1,200	970	5,820	386	559	119	51	90
1999	5,315	1,263	1,016	6,236	344	735	91	47	142
2000	5,292	1,290	967	7,688	351	902	97	56	53
2001	4,714	1,217	1,002	6,570	415	651	83	40	134
2002 [3]	4,296	1,288	912	7,144	212	690	90	49	99

Year	Oranges [4]	Tangerines [4]	Grapefruit [4] [5]	Lemons [4]	K-Early Citrus [4]	Limes [4]	Tangelos [4]	Temples [4]	Strawberries, commercial crop [7]	Pineapples [7]
	1,000 tons	*1,000 tons*	*1,000 tons*	*1,000 tons*	*1,000 tons*	*1,000 tons*	*1,000 tons*	*1,000 tons*	*1,000 tons*	*1,000 tons*
1993	10,992	247	2,791	942	8	44	137	113	723	370
1994	10,329	318	2,661	984	9	9	150	101	824	365
1995	11,432	287	2,912	897	5	10	142	114	804	345
1996	11,426	349	2,718	992	7	14	110	97	813	347
1997	12,692	425	2,885	962	7	14	178	108	814	324
1998	13,670	360	2,593	897	2	19	128	101	820	332
1999	9,824	327	2,513	747	4	22	115	81	916	352
2000	12,997	458	2,763	840	5	26	99	88	951	354
2001	12,221	373	2,462	996	2	11	95	56	826	323
2002 [3]	12,543	420	2,427	828	1	7	97	70	985	320

Year	Avocados [6]	Nectarines	Cranberries	Bananas [7]	Kiwifruit	Dates	Papayas [7]	Berries [8]	Guavas	Total
	1,000 tons	*1,000 tons*	*1,000 tons*	*1,000 tons*	*1,000 tons*	*1,000 tons*	*1,000 tons*	*1,000 tons*	*1,000 tons*	*1,000 tons*
1993	144	205	196	6	49	29	32	149	8	32,028
1994	175	242	234	7	39	23	31	140	9	32,099
1995	190	176	210	7	38	23	25	148	8	32,375
1996	191	247	234	7	32	23	21	119	8	31,906
1997	178	264	275	7	35	21	19	162	8	35,785
1998	159	224	272	11	37	25	20	180	7	34,875
1999	183	274	317	12	27	22	21	201	5	31,152
2000	239	267	286	15	34	17	27	232	8	36,398
2001	221	275	267	14	26	20	28	220	8	33,270
2002 [3]	NA	299	282	10	25	20	23	200	NA	33,336

[1] For some crops in certain years, production includes some quantities unharvested for economic reasons or excess cullage fruit. [2] Estimates of the commercial crop refer to production in orchards of 100 or more bearing-age trees. [3] Preliminary. [4] Year harvest was complete. [5] Excludes economic abandanment in 1996 of 127,500 tons; in 1997 of 255,000 tons; and in 1998 of 255,000 tons. [6] Year of bloom. [7] Utilized production only. [8] Wild Blueberries added in 1998. NA-not available.

NASS, Crops Branch, (202) 720–2127.

Table 5-3.—Apples, commercial crop: [1] Production and season average price per pound, by States, 2000–2002

State	Total production			Utilized production			Price [3] for crop of—		
	2000	2001	2002 [2]	2000	2001	2002 [2]	2000	2001	2002 [2]
	Million pounds	*Million pounds*	*Million pounds*	*Million pounds*	*Million pounds*	*Million pounds*	*Dollars*	*Dollars*	*Dollars*
AZ	95.0	5.4	26.2	94.5	5.3	26.0	0.074	0.066	0.173
AR	7.2	5.5	4.5	3.4	4.3	3.4	0.252	0.250	0.278
CA	570.0	520.0	420.0	510.0	490.0	420.0	0.164	0.148	0.172
CO	30.0	23.0	21.0	27.0	23.0	20.0	0.139	0.208	0.184
CT	20.5	20.5	12.0	20.0	20.0	12.0	0.302	0.322	0.385
GA	14.0	9.0	10.0	13.0	8.9	9.9	0.188	0.230	0.180
ID	140.0	80.0	80.0	140.0	79.0	80.0	0.107	0.141	0.197
IL	42.0	43.6	43.0	36.0	40.6	35.3	0.287	0.237	0.351
IN	45.0	53.0	40.0	43.8	46.0	33.0	0.245	0.185	0.235
IA	7.5	8.8	7.5	7.5	7.7	4.2	0.321	0.333	0.428
KS	3.0	4.0	3.5	1.3	3.0	2.6	0.268	0.283	0.301
KY	6.5	8.3	5.6	5.6	7.0	4.0	0.249	0.290	0.318
ME	39.0	47.0	49.0	35.0	40.0	44.0	0.218	0.290	0.338
MD	33.7	40.8	32.0	33.0	40.7	31.9	0.137	0.155	0.143
MA	50.0	39.0	31.0	43.0	34.0	27.0	0.320	0.324	0.363
MI	800.0	930.0	500.0	795.0	900.0	500.0	0.093	0.094	0.124
MN	22.0	24.0	25.0	17.4	15.5	17.7	0.432	0.475	0.509
MO	38.0	41.0	38.0	34.0	41.0	36.0	0.169	0.171	0.178
NH	34.0	30.0	24.0	32.5	28.5	24.0	0.236	0.250	0.304
NJ	50.0	55.0	35.0	46.0	50.0	35.0	0.134	0.162	0.138
NM	8.0	6.0	2.0	8.0	6.0	1.8	0.254	0.318	0.326
NY	995.0	1,000.0	650.0	935.0	940.0	610.0	0.117	0.119	0.164
NC	190.0	112.0	160.0	145.0	107.0	150.0	0.126	0.149	0.126
OH	103.0	86.0	77.0	96.0	84.0	77.0	0.227	0.236	0.259
OR	167.0	142.0	150.0	162.0	141.0	150.0	0.119	0.121	0.117
PA	475.0	480.0	390.0	475.0	480.0	390.0	0.114	0.096	0.103
RI	2.3	1.8	2.3	2.2	1.4	2.0	0.359	0.383	0.392
SC	20.0	6.0	9.0	16.0	2.0	7.0	0.129	0.187	0.130
TN	9.5	9.0	6.5	8.5	6.5	5.4	0.244	0.235	0.260
UT	49.0	30.0	8.0	43.0	24.0	6.0	0.118	0.172	0.183
VT	41.5	41.0	27.0	38.5	38.0	24.0	0.225	0.241	0.240
VA	320.0	310.0	250.0	314.0	306.0	243.0	0.103	0.106	0.121
WA	6,000.0	5,050.0	5,300.0	6,000.0	5,050.0	5,300.0	0.125	0.178	0.210
WV	85.0	105.0	95.0	83.0	95.0	90.0	0.095	0.085	0.096
WI	71.0	62.0	58.0	58.0	49.0	52.0	0.280	0.293	0.349
US	10,583.7	9,428.7	8,592.1	10,322.2	9,214.4	8,474.2	0.128	0.157	0.191

[1] In orchards of 100 or more bearing-age trees. [2] Preliminary. [3] Fresh fruit prices are equivalent packinghouse-door returns for CA, NY, OR, and WA, and the average price as sold for other States. Quantities processed are priced at the equivalent processing plant door level.

NASS, Crops Branch, (202) 720–2127.

Table 5-4.—Apples: Production and value, United States, 1993–2002

Year	Apples, commercial crop [1]			
	Total production	Utilized production	Marketing year average price [2]	Value
	Million pounds	*Million pounds*	*Cents per pound*	*1,000 dollars*
1993	10,685.1	10,574.3	12.9	1,363,820
1994	11,500.9	11,332.8	12.9	1,467,282
1995	10,578.4	10,383.9	17.0	1,767,001
1996	10,381.9	10,330.0	15.9	1,641,462
1997	10,323.8	10,254.3	15.4	1,575,403
1998	11,646.4	10,762.5	12.2	1,316,172
1999	10,630.7	10,446.5	15.0	1,563,582
2000	10,583.7	10,322.2	12.8	1,320,781
2001	9,428.7	9,214.4	15.7	1,448,348
2002 [3]	8,592.1	8,474.2	19.1	1,622,135

[1] In orchards of 100 or more bearing-age trees. [2] Fresh fruit prices are equivalent returns at packinghouse-door for CA, NY, OR, and WA, and for other States, returns at point of first sale. Processing prices are equivalent returns at processing plant door. [3] Preliminary.

NASS, Crops Branch, (202) 720–2127.

Table 5-5.—Apples, fresh (dessert and cooking): Production in specified countries, 2000/2001–2002/2003 [1]

Continent and country	2000/2001	2001/2002	2002/2003
	1,000 metric tons	*1,000 metric tons*	*1,000 metric tons*
North America:			
Canada	532.2	495.0	510.0
United States	4,837.0	4,367.7	4,050.0
Total	5,369.2	4,862.7	4,560.0
South America:			
Argentina [2]	1,330.8	900.0	NA
Chile [2]	1,000.0	960.0	NA
Total	2,330.8	1,860.0	NA
European Union:			
Belgium and Luxembourg	511.6	343.6	314.0
France	2,300.0	2,055.0	2,140.0
Germany	2,630.8	1,522.4	1,612.0
Greece	315.0	260.0	230.0
Italy	2,267.0	2,220.0	2,370.0
Netherlands	500.0	475.0	355.0
Spain	698.5	884.0	723.0
Sweden	68.0	63.1	54.0
United Kingdom	162.2	169.1	134.2
Total EC	9,453.1	7,992.2	7,932.2
Other Europe:			
Hungary	700.0	605.0	470.0
Russia	1,589.6	1,227.6	1,400.0
Turkey	2,400.0	2,450.0	2,500.0
Total	3,989.6	3,677.6	3,900.0
Total Europe	15,923.5	14,563.4	14,023.7
Africa:			
South Africa, Rep [2]	667.7	690.0	NA
Total	667.7	690.0	NA
Asia:			
China	20,431.2	21,000.0	20,500.0
Japan	799.6	930.7	911.9
Taiwan	7.7	8.1	9.6
Total	21,238.5	21,938.8	21,421.5
Oceania:			
Australia [2]	285.0	295.0	NA
New Zealand [2]	405.0	462.0	NA
Total	690.0	757.0	NA
Total specified countries	47,888.8	46,491.8	NA

[1] May include some cider varieties in countries not reporting separately. [2] It is too early to make reliable forecast for the Southern Hemisphere countries for the 2002/03 season. NA - not available.

FAS, Horticultural and Tropical Products Division, (202) 720–3423. Prepared or estimated on the basis of official statistics of foreign governments, other foreign source materials, reports of U.S. Agricultural Counselors, Attachés, and Foreign Service Officers, results of office research, and related information.

Table 5-6.—Apples, commercial crop: Production and utilization, United States, 1993–2002

Crop of—	Total production	Utilized production	Utilization of quantities sold				
			Fresh [1]	Processed (fresh basis)			
				Canned	Dried	Frozen	Other [2] [3]
	Million pounds	*Million pounds*	*Million pounds*	*Million pounds*	*Million pounds*	*Million pounds*	*Million pounds*
1993	10,685.1	10,574.3	6,124.6	1,334.8	366.4	281.5	2,467.0
1994	11,500.9	11,332.8	6,368.8	1,405.7	414.6	304.4	2,839.3
1995	10,578.4	10,383.9	5,840.2	1,291.0	333.7	304.9	2,614.1
1996	10,381.9	10,330.0	6,206.9	1,294.2	316.6	267.8	2,244.5
1997	10,323.8	10,254.3	5,814.5	1,498.8	267.0	349.0	2,325.0
1998	11,646.4	10,762.5	6,412.5	1,173.8	329.9	266.0	2,580.3
1999	10,630.7	10,446.5	5,994.8	1,318.6	263.2	271.3	2,598.6
2000	10,583.7	10,322.2	6,266.8	1,183.7	248.2	195.9	2,427.6
2001	9,428.7	9,214.4	5,469.8	1,258.4	221.0	248.5	2,016.7
2002 [4]	8,592.1	8,474.2	NA	NA	NA	NA	NA

[1] Includes "Home use." [2] Mostly crushed for vinegar, cider, and juice. For some States, small quantities canned, dried, and frozen are included. [3] Includes the following quantities of juice and cider (million pounds): 1993—2,382.3; 1994—2,707.2; 1995—2,538.2. [4] Preliminary. NA-not available.

NASS, Crops Branch, (202) 720–2127.

Table 5-7.—Apples, commercial crop: Production and utilization, by States, crop of 2001

State	Total production	Utilized production	Utilization					
			Fresh [1]	Processed (fresh basis)				
				Canned	Dried	Frozen	Juice and cider	Other [2]
	Million pounds	*Million pounds*	*Million pounds*	*Million pounds*	*Million pounds*	*Million pounds*	*Million pounds*	*Million pounds*
CA	520.0	490.0	220.0	17.0	....	....	218.0	....
MI	930.0	900.0	270.0	220.0	....	115.0	280.0	15.0
NY	1,000.0	940.0	420.0	320.0	....	30.0	155.0	15.0
OR	142.0	141.0	94.0	....	....	....	24.0	....
PA	480.0	480.0	120.0	226.0	....	....	125.0	....
VA	310.0	306.0	88.0	168.0	....	....	48.0	....
WA	5,050.0	5,050.0	3,700.0	155.0	....	....	910.0	....
WV	105.0	95.0	15.0	58.0	....	....	20.0	....
Other States [3]	891.7	812.4	542.8	94.4	....	103.5	165.7	41.0
US	9,428.7	9,214.4	5,469.8	1,258.4	221.0	248.5	1,945.7	71.0

[1] Includes "Home use." [2] Mostly vinegar, wine, and fresh slices for pie making. [3] AZ, AR, CO, CT, GA, ID, IL, IN, IA, KS, KY, ME, MD, MA, MN, MO, NH, NJ, NM, NC, OH, RI, SC, TN, UT, VT, and WI.

NASS, Crops Branch, (202) 720–2127.

Table 5-8.—Apples, fresh: United States exports by country of destination and imports by country of origin, 1999/2000–2001/2002

Country	Year beginning July		
	1999/2000	2000/2001	2001/2002
	Metric tons	*Metric tons*	*Metric tons*
Exports			
Europe:			
Finland	307	867	797
France	0	25	19
Iceland	1,441	1,139	1,586
Ireland	725	360	401
Netherlands	1,287	597	1,425
Norway	169	361	916
Sweden	525	1,540	908
United Kingdom	23,699	29,462	32,500
Other countries	1,415	2,686	5,429
Total	29,568	37,037	43,981
Latin America:			
Brazil	608	511	152
Colombia	5,238	4,417	2,442
Costa Rica	6,001	8,854	6,546
El Salvador	3,925	4,445	4,469
Guatemala	6,089	8,440	4,294
Nicaragua	895	525	168
Mexico	155,067	223,471	167,173
Honduras	1,956	2,720	3,955
Panama	2,462	4,474	2,719
Ecuador	101	1,266	394
Venezuela	12,443	16,227	10,846
Other countries	642	1,120	645
Total	195,427	276,470	203,803
Caribbean:			
Bahamas	32	106	379
Barbados	348	505	406
Bermuda	10	110	58
Dominican Republic	5,315	8,269	9,366
French West Indies	246	51	3
Haiti	919	136	376
Leeward and Windward Islands	60	93	41
Netherlands Antilles	70	107	113
Trinidad and Tobago	1,737	2,718	4,277
Other countries	8	203	198
Total	8,745	12,298	15,217
Asia:			
Taiwan	84,474	110,483	74,095
Hong Kong	24,590	47,978	40,180
Indonesia	28,411	42,472	41,904
Malaysia	7,425	19,266	22,795
Japan	1,774	789	116
Philippines	10,378	11,346	6,584
Singapore	4,314	5,626	4,429
Thailand	10,668	13,693	12,378
Other countries	5,126	18,021	21,845
Total	177,160	269,674	224,326
Other countries:			
Canada	83,954	95,362	93,126
French Pacific Islands	219	119	118
Kuwait	2,340	2,501	2,602
New Zealand	273	100	107
Saudi Arabia	15,929	17,213	10,867
United Arab Emirates	15,704	24,565	16,920
Other countries	11,406	13,803	8,746
Total	129,825	153,663	132,486
Grand total	540,725	749,142	619,813
Imports			
Canada	46,111	37,875	39,048
Chile	42,269	52,812	61,554
New Zealand	62,617	61,262	57,783
South Africa	11,209	8,122	6,167
Other countries	5,386	3,277	3,559
Total	165,503	163,610	166,539

FAS, Horticultural and Tropical Products Division, (202) 720–6086. Compiled from reports of the U.S. Department of Commerce.

Table 5-9.—Apples:[1] Foreign trade, United States, 1991–2001

Year beginning October	Imports, fresh and dried, in terms of fresh	Domestic exports	
		Fresh	Dried, in terms of fresh[1]
	Metric tons	*Metric tons*	*Metric tons*
1991	143,926	530,073	44,192
1992	139,271	487,817	22,080
1993	130,938	662,906	19,224
1994	171,695	663,054	25,120
1995	196,067	564,953	24,624
1996	197,340	689,749	20,368
1997	173,564	539,081	18,040
1998	171,774	660,251	15,728
1999	195,258	571,860	21,520
2000	180,616	743,644	33,308
2001	193,895	592,428	21,358

[1] Dried converted to terms of fresh apples on following basis; 1 pound dried is equivalent to 8 pounds fresh. No re-exports reported.

ERS, Specialty Crops Branch, (202) 694–5260.

Table 5-10.—Apricots: Production and value, United States, 1993–2002[1]

Year	Total production	Utilized production	Market year average price per ton[2]	Value
	Tons	*Tons*	*Dollars*	*1,000 dollars*
1993	97,350	97,340	398.00	38,694
1994	153,200	140,180	349.00	48,883
1995	60,500	60,500	456.00	27,572
1996	79,300	79,290	444.00	35,171
1997	139,230	129,630	332.00	43,072
1998	118,490	108,080	327.00	35,358
1999	90,500	90,500	391.00	35,377
2000	96,900	87,760	369.00	32,346
2001	82,460	75,430	353.00	26,598
2002[3]	90,140	80,130	354.00	28,326

[1] Production, price, and value for CA, UT, and WA. [2] Fresh fruit prices are equivalent packinghouse-door returns for CA and WA. Quantities processed are priced at the equivalent processing plant door level. [3] Preliminary.

NASS, Crops Branch, (202) 720–2127.

Table 5-11.—Apricots: Production and marketing year average price per ton, by States, 2000–2002

State	Total production			Utilized production			Price[2] for crop of—		
	2000	2001	2002[1]	2000	2001	2002[1]	2000	2001	2002[1]
	Tons	*Tons*	*Tons*	*Tons*	*Tons*	*Tons*	*Dollars*	*Dollars*	*Dollars*
CA	90,000	77,000	85,000	81,000	70,000	75,000	329.00	319.00	306.00
UT	400	260	140	260	230	130	612.00	852.00	708.00
WA	6,500	5,200	5,000	6,500	5,200	5,000	847.00	783.00	1,050.00
Total	96,900	82,460	90,140	87,760	75,430	80,130	369.00	353.00	354.00

[1] Preliminary. [2] Fresh fruit prices are equivalent packinghouse-door returns for CA and WA. Quantities processed are priced at the equivalent processing plant door level.

NASS, Crops Branch, (202) 720–2127.

Table 5-12.—Apricots: Production and utilization, United States,[1] 1993–2002

Crop of—	Total production	Utilized production	Utilization of quantities sold			
			Fresh[2]	Processed[3]		
				Canned[4]	Dried (fresh basis)	Frozen
	Tons	Tons	Tons	Tons	Tons	Tons
1993	97,350	97,340	21,410	45,000	18,000	11,500
1994	153,200	140,180	26,740	71,000	30,500	10,500
1995	60,500	60,500	16,400	19,200	9,600	5,700
1996	79,300	79,290	13,490	20,000	15,000	9,000
1997	139,230	129,630	26,830	46,700	12,000	15,100
1998	118,490	108,080	22,880	40,700	9,000	10,400
1999[5]	90,500	90,500	25,800			
2000	96,900	87,760	26,580	32,000	8,000	10,000
2001	82,460	75,430	18,230	31,000	6,000	9,000
2002[6]	90,140	80,130	18,090	30,500	8,000	10,500

[1] CA, UT, and WA. [2] Includes "Home use." [3] CA only. [4] Includes some quantities frozen or otherwise processed. [5] Breakdown of processed utilization for 1999 unpublished to avoid disclosure of individual operations. [6] Preliminary.

NASS, Crops Branch, (202) 720–2127.

Table 5-13.—Apricots: Production and utilization, by States, crop of 2002 (preliminary)

State	Total production	Utilized production	Utilization			
			Fresh	Processed[1]		
				Canned[2]	Dried (fresh basis)	Frozen
	Tons	Tons	Tons	Tons	Tons	Tons
CA	85,000	75,000	14,000	37,500	8,000	10,500
UT[3]	140	130				
WA[3]	5,000	5,000				
US	90,140	80,130	18,090	37,500	8,000	10,500

[1] CA only. [2] Some quantities used for juice are included in "Canned" to avoid disclosure of individual operations. [3] Missing data not published to avoid disclosure of individual operations, but included in U.S. total.

NASS, Crops Branch, (202) 720–2127.

Table 5-14.—Apricots: Foreign trade, United States, 1992–2001

Year beginning October	Domestic exports				
	Fresh	Canned[1]	Dried[1]	Dried, in fruit salad[2]	Total, in terms of fresh[3]
	Metric tons	Metric tons	Metric tons	Metric tons	Metric tons
1992	5,167	1,269	1,214	428	14,517
1993	7,650	760	1,188	488	16,810
1994	4,031	1,450	1,105	569	13,675
1995	3,492	1,338	1,376	519	14,192
1996	6,604	1,086	1,034	396	14,733
1997	6,980	968	927	428	14,639
1998	9,197	980	1,797	340	20,884
1999	6,204	1,695	1,349	176	15,258
2000	7,663	769	2,251	312	21,389
2001	7,732	1,600	2,004	202	20,219

[1] Net processed weight. [2] Dried apricots are 12⅓ percent of total dried fruit for salad. [3] Dried fruit converted to unprocessed dry weight by dividing by 1.07. Unprocessed dry weight converted to terms of fresh fruit on the basis that 1 pound dried equals 5.5 pounds fresh. Canned apricots converted to terms of fresh on the basis that 1 pound canned equals 0.717 pounds fresh.

ERS, Specialty Crops Branch, (202) 694–5260.

Table 5-15.—Avocados: Foreign trade, United States, 1992–2001

Year beginning October	Imports
	Metric tons
1992	18,466
1993	14,213
1994	18,871
1995	23,118
1996	27,667
1997	39,847
1998	59,637
1999	63,944
2000	73,069
2001	103,339

ERS, Specialty Crops Branch, (202) 694–5260.

Table 5-16.—Avocados: Production, marketing year average price per ton, and value, California and Florida, 1992–93 to 2001–2002

Season	California			Florida			California and Florida [3]		
	Production [1]	Price [2]	Value	Production [1]	Price [2]	Value	Production [1]	Price [2]	Value
	Tons	*Dollars*	*1,000 dollars*	*Tons*	*Dollars*	*1,000 dollars*	*Tons*	*Dollars*	*1,000 dollars*
1992–93	284,000	400	113,600	7,200	583	4,198	291,200	405	117,798
1993–94	139,000	1,810	251,590	4,400	820	3,608	143,400	1,780	255,198
1994–95	155,000	1,480	229,894	20,000	616	12,320	175,000	1,384	242,214
1995–96	171,000	1,370	234,831	19,000	596	11,324	190,000	1,296	246,155
1996–97	167,000	1,560	260,162	23,500	528	12,408	190,500	1,431	272,570
1997–98	154,000	1,710	263,473	24,000	584	14,016	178,000	1,559	277,489
1998–99	136,000	2,400	327,002	23,000	716	16,468	159,000	2,160	343,470
1999–2000	161,000	2,110	339,594	22,000	748	16,456	183,000	1,946	356,050
2000–2001	213,000	1,480	315,842	26,000	584	15,184	239,000	1,361	331,026
2001–2002 [4] ...	198,000	1,750	346,500	23,000	676	15,548	221,000	....	362,048

[1] Production is the quantity sold or utilized. [2] Quantities processed are priced at the equivalent processing plant door level. [3] Season from Nov. 1 to Nov. 30 (following year) for California and June 20 to Feb. 28 for Florida. [4] Preliminary.

NASS, Crops Branch, (202) 720–2127.

Table 5-17.—Bananas: Area, yield, utilized production, marketing year average price, and value, Hawaii, 1993–2002

Year	Area harvested	Yield per acre	Production	Price per pound	Value
	Acres	*1,000 pounds*	*1,000 pounds*	*Cents*	*1,000 dollars*
1993	830	14.1	11,700	38.0	4,446
1994	880	15.6	13,700	37.0	5,069
1995	880	14.8	13,000	40.0	5,200
1996	960	13.5	13,000	40.0	5,200
1997	950	14.4	13,700	38.0	5,206
1998	1,420	14.8	21,000	35.0	7,350
1999	1,420	17.3	24,500	35.0	8,575
2000	1,460	19.9	29,000	36.0	10,440
2001	1,490	18.2	28,000	38.0	10,640
2002 [1]	1,250	15.2	19,000	42.0	7,980

[1] Preliminary.

NASS, Crops Branch, (202) 720–2127.

Table 5-18.—Kiwifruit: Area, yield, utilized production, marketing year average price, and value, California, 1993–2002

Year	Bearing acreage	Yield [1]	Production	Price per ton	Value
	Acres	*Tons*	*Tons*	*Dollars*	*1,000 dollars*
1993 ...	6,900	7.13	44,600	370	16,502
1994 ...	6,500	6.06	37,500	491	18,413
1995 ...	6,100	6.20	33,600	459	15,434
1996 ...	5,700	5.53	28,000	470	13,157
1997 ...	5,300	6.60	31,800	518	16,483
1998 ...	5,300	6.91	33,000	744	24,544
1999 ...	5,300	5.09	24,000	634	15,215
2000 ...	5,300	6.42	30,500	455	13,888
2001 ...	4,900	5.27	23,000	667	15,340
2002 ...	4,500	5.58	22,300	NA	NA

[1] Yield based on total production.

NASS, Crops Branch, (202) 720-2127.

Table 5-19.—Cherries: Foreign trade, United States, 1992–2001

Year beginning October	Imports		Domestic exports	
	Fresh	Dried and preserved	Fresh	Canned
	Metric tons	*Metric tons*	*Metric tons*	*Metric tons*
1992	1,982	2,224	25,742	13,193
1993	1,633	2,568	30,640	14,774
1994	1,376	1,415	30,268	15,168
1995	1,905	1,871	34,702	18,975
1996	1,908	1,884	39,401	15,044
1997	968	2,054	37,591	13,358
1998	2,089	1,750	42,655	14,500
1999	2,815	2,184	43,289	14,969
2000	3,858	2,560	42,880	20,515
2001	6,697	2,942	36,210	19,368

ERS, Specialty Crops Branch, (202) 694–5260.

Table 5-20.—Sweet cherries: Production and value, United States, 1993–2002

Year	Total production	Utilized production	Marketing year average price per ton [1]	Value
	Tons	*Tons*	*Dollars*	*1,000 dollars*
1993	168,350	160,395	1,190.00	190,630
1994	207,100	192,720	1,040.00	200,224
1995	165,300	152,880	1,260.00	193,068
1996	154,100	151,700	1,470.00	223,022
1997	225,770	223,490	1,250.00	278,511
1998	211,400	208,410	1,090.00	226,236
1999	216,120	213,260	1,100.00	234,879
2000	206,500	204,020	1,340.00	274,225
2001	230,380	219,620	1,230.00	270,914
2002 [2]	180,225	176,165	1,550.00	273,694

[1] Fresh fruit prices are equivalent packinghouse-door returns for Western States, and the average price as sold for other States. Quantities processed are priced at the equivalent processing plant door level. [2] Preliminary.
NASS, Crops Branch, (202) 720–2127.

Table 5-21.—Tart cherries: Production and value, United States, 1993–2002

Year	Total production	Utilized production	Marketing year average price per ton [1]	Value
	Tons	*Tons*	*Dollars*	*1,000 dollars*
1993	340.4	273.6	0.121	33,145
1994	304.2	296.3	0.163	48,386
1995	395.6	311.2	0.059	18,456
1996	271.8	260.1	0.161	41,747
1997	292.9	283.3	0.159	44,911
1998	348.1	305.6	0.145	44,186
1999	256.1	254.1	0.218	55,505
2000	288.5	281.4	0.187	52,488
2001	370.1	307.9	0.186	57,150
2002 [2]	62.5	62.2	0.448	27,879

[1] Fresh fruit prices are equivalent packinghouse-door returns for Western States, and the average price as sold for other States. Quantities processed are priced at the equivalent processing plant door level. [2] Preliminary.
NASS, Crops Branch, (202) 720–2127.

Table 5-22.—Sweet cherries: Production and season average price, by States, 2000–2002

State	Total production			Utilized production			Price [2]		
	2000	2001	2002 [1]	2000	2001	2002 [1]	2000	2001	2002 [1]
	Tons	*Tons*	*Tons*	*Tons*	*Tons*	*Tons*	*Dollars per ton*	*Dollars per ton*	*Dollars per ton*
CA	47,000	55,300	55,500	45,700	51,100	53,700	1,570.00	1,560.00	1,740.00
ID	3,000	1,400	1,700	2,970	1,390	1,700	1,590.00	1,300.00	1,490.00
MI	19,600	23,000	2,700	19,600	23,000	2,600	486.00	482.00	855.00
MT	1,100	2,300	2,220	1,050	1,850	2,080	1,490.00	1,130.00	1,820.00
NY	900	1,100	350	900	1,050	350	1,370.00	1,530.00	1,730.00
OR	37,000	40,000	31,000	36,000	34,000	29,000	760.00	842.00	932.00
PA	500	580	355	500	580	355	2,180.00	2,260.00	2,480.00
UT	2,400	700	400	2,300	650	380	1,060.00	791.00	1,540.00
WA	95,000	106,000	86,000	95,000	106,000	86,000	1,630.00	1,360.00	1,660.00
Total 9 States	206,500	230,380	180,225	204,020	219,620	176,165	1,340.00	1,230.00	1,550.00

[1] Preliminary. [2] Fresh fruit prices are equivalent packinghouse-door returns for CA, OR, and WA, and the average price as sold for other States. Quantities processed are priced at the equivalent processing plant door level.
NASS, Crops Branch, (202) 720–2127.

Table 5-23.—Tart cherries: Production and season average price, by States, 2000–2002

State	Total production			Utilized production			Price [2]		
	2000	2001	2002 [1]	2000	2001	2002 [1]	2000	2001	2002 [1]
	Tons	*Tons*	*Tons*	*Tons*	*Tons*	*Tons*	*Dollars per ton*	*Dollars per ton*	*Dollars per ton*
CO	0.9	0.6	0.3	0.8	0.6	0.3	0.265	0.333	0.400
MI	200.0	297.0	15.0	200.0	242.0	15.0	0.182	0.184	0.479
NY	16.6	14.7	12.7	16.6	14.1	12.7	0.213	0.196	0.506
OR	4.4	2.4	3.2	4.2	2.3	3.1	0.210	0.257	0.370
PA	6.1	3.9	3.8	5.6	3.9	3.8	0.149	0.127	0.558
UT	33.0	12.0	3.0	27.0	11.5	2.8	0.220	0.218	0.240
WA	17.5	26.5	20.5	17.5	20.5	20.5	0.143	0.172	0.398
WI	10.0	13.0	4.0	9.7	13.0	4.0	0.228	0.204	0.512
Total 8 States	288.5	370.1	62.5	281.4	307.9	62.2	0.187	0.186	0.448

[1] Preliminary. [2] Fresh fruit prices are equivalent packinghouse-door returns for OR and WA, and the average price as sold for other States. Quantities processed are priced at the equivalent processing plant door level.
NASS, Crops Branch, (202) 720–2127.

Table 5-24.—Sweet cherries: Production and utilization, by States, crop of 2002 (preliminary)

State	Total production	Utilized production	Utilization			
			Fresh [1]	Processed		
				Canned and otherwise processed	Brined	Other [2]
	Tons	Tons	Tons	Tons	Tons	Tons
CA	55,500	53,700	43,300			10,400
MI	2,700	2,600	200	280	1,700	420
OR	31,000	29,000	13,000	2,000	12,000	2,000
WA	86,000	86,000	66,000	3,500	10,000	6,500
Other States [3]	5,025	4,865	3,955		640	270
US	180,225	176,165	126,455	5,780	24,340	19,590

[1] Includes "Home use." [2] Includes California canned utilization and other processed utilizations from all States. [3] ID, MT, NY, PA, and UT.
NASS, Crops Branch, (202) 720–2127.

Table 5-25.—Tart cherries: Production and utilization, by States, crop of 2002 [1]

State	Total production	Utilized production	Utilization			
			Fresh [2]	Processed		
				Canned and otherwise processed [3]	Frozen	Other [3]
	Million pounds	Million pounds	Million pounds	Million pounds	Million pounds	Million pounds
MI	15.0	15.0	0.1	6.5	8.0	0.4
Other States [4]	47.5	47.2	0.7	10.9	20.9	14.7
US	62.5	62.2	0.8	17.4	28.9	15.1

[1] Preliminary. [2] Includes "Home use." [3] Some quantities used for juice, wine, brined, and dried. [4] CO, NY, OR, PA, UT, WA, and WI.
NASS, Crops Branch, (202) 720–2127.

Table 5-26.—Sweet cherries: Production and utilization, United States, [1] 1993–2002

Crop of—	Total production	Utilized production	Utilization of quantities sold		
			Fresh [2]	Processed	
				Other [3]	Brined
	Tons	Tons	Tons	Tons	Tons
1993	168,350	160,395	79,630	23,000	57,765
1994	207,100	192,720	99,270	29,000	64,450
1995	165,300	152,880	64,230	29,290	59,360
1996	154,100	151,700	80,670	22,070	48,960
1997	225,770	223,490	115,440	30,400	77,650
1998	211,400	208,410	108,960	30,200	69,250
1999	216,120	213,260	123,410	26,065	63,785
2000	206,500	204,020	120,760	26,660	56,600
2001	230,380	219,620	145,710	25,730	48,180
2002 [4]	180,225	176,165	126,455	18,570	31,140

[1] CA, ID, MI, MT, NY, OR, PA, UT, and WA. [2] Includes "Home use." [3] Includes canned utilization and other processed utilizations from all States. [4] Preliminary.
NASS, Crops Branch, (202) 720–2177.

Table 5-27.—Tart cherries: Production and utilization, United States,[1] 1993–2002

Crop of—	Total production	Utilized production	Utilization of quantities sold		
			Fresh [2]	Processed	
				Other [3]	Frozen
	Million pounds	Million pounds	Million pounds	Million pounds	Million pounds
1993	340.4	273.6	5.3	113.0	155.3
1994	304.2	296.3	3.5	119.1	173.7
1995	395.6	311.2	2.7	125.7	182.8
1996	271.8	260.1	2.5	87.4	170.2
1997	292.9	283.3	2.6	107.1	173.6
1998	348.1	305.6	2.3	103.5	199.8
1999	256.1	254.1	1.8	114.4	137.9
2000	288.5	281.4	1.8	135.3	144.3
2001	370.1	307.9	1.9	129.2	176.8
2002 [4]	62.5	62.2	0.8	32.5	28.9

[1] CO, MI, NY, OR, PA, UT, WA, and WI. [2] Includes "Home use." [3] Includes canned utilization and other processed utilizations from all states. [4] Preliminary.
NASS, Crops Branch, (202) 720–2177.

Table 5-28.—Citrus fruit: Utilized production and value, United States, for season of 1992–93 to 2001–2002

Season [1]	Production	Marketing year average returns per box [2]	Value	Quantities processed [3]	Production	Marketing year average returns per box [2]	Value	Quantities processed [3]
	Oranges [4]				Grapefruit			
	1,000 boxes	*Dollars*	*1,000 dollars*	*1,000 boxes*	*1,000 boxes* [5]	*Dollars*	*1,000 dollars*	*1,000 boxes*
1992–93	255,760	5.77	1,489,938	193,992	68,375	4.40	301,614	36,710
1993–94	240,450	6.37	1,541,296	182,435	65,100	5.26	341,977	33,231
1994–95	263,605	6.08	1,624,061	207,351	71,050	4.29	307,525	38,387
1995–96	263,890	6.85	1,821,579	207,365	66,200	4.33	290,152	33,582
1996–97	293,020	6.16	1,836,662	228,565	70,100	4.00	284,749	36,665
1997–98	315,525	6.13	1,965,358	247,004	63,150	4.13	268,598	32,460
1998–99	224,580	7.41	1,687,928	192,194	61,200	5.33	334,626	30,027
1999–2000	299,760	5.56	1,666,100	244,582	66,980	6.07	409,716	38,509
2000–2001	280,935	5.88	1,682,790	223,232	59,750	4.69	285,065	32,600
2001–2002 [6]	228,260	6.20	1,834,028	229,133	58,760	4.78	285,546	32,177
	Lemons				Limes (FL)			
1992–93	24,800	9.67	239,872	12,287	1,000	6.55	6,547	270
1993–94	25,900	9.94	257,362	13,449	200	18.46	3,691	50
1994–95	23,600	11.16	263,441	10,287	230	14.47	3,329	40
1995–96	26,100	10.01	261,281	12,533	300	13.05	3,914	60
1996–97	25,300	12.00	303,476	12,206	320	11.93	3,816	65
1997–98	23,600	10.21	240,846	12,250	440	11.90	5,235	110
1998–99	19,650	12.79	251,397	7,523	500	17.83	8,813	90
1999–2000	22,100	13.51	298,677	8,476	600	16.21	9,728	100
2000–2001	26,200	9.06	237,362	12,793	250	17.00	4,249	30
2001–2002 [6]	21,800	15.65	341,174	6,858	150	11.55	1,732	25
	Tangerines [7]				Tangelos (FL)			
1992–93	5,850	14.48	83,222	1,618	3,050	5.29	16,147	1,734
1993–94	7,400	12.57	92,055	2,007	3,350	5.47	18,323	2,253
1994–95	6,700	15.01	100,285	1,756	3,150	4.90	15,435	2,044
1995–96	8,100	13.94	110,573	2,390	2,450	6.16	15,100	1,432
1996–97	9,650	12.47	122,172	3,096	3,950	4.75	18,759	2,918
1997–98	8,200	11.78	96,524	2,642	2,850	4.19	11,950	1,937
1998–99	7,400	15.74	116,632	2,047	2,550	7.17	18,277	1,712
1999–2000	10,350	10.43	108,192	3,640	2,200	5.11	11,232	1,464
2000–2001	8,450	11.26	96,789	2,517	2,100	3.90	8,193	1,358
2001–2002 [6]	9,420	12.98	124,789	2,675	2,150	5.10	10,967	1,454
	Temples (FL)				K-Early Citrus (FL)			
1992–93	2,500	4.98	12,458	1,553	185	7.43	1,375	52
1993–94	2,250	5.64	12,687	1,465	210	4.47	939	50
1994–95	2,550	5.61	14,301	1,764	120	4.48	538	55
1995–96	2,150	6.52	14,024	1,457	160	4.82	771	98
1996–97	2,400	5.23	12,541	1,845	150	3.95	592	93
1997–98	2,250	5.12	11,510	1,684	40	1.13	45	10
1998–99	1,800	7.25	13,050	1,207	80	4.45	356	56
1999–2000	1,950	4.70	9,173	1,510	110	3.24	356	95
2000–2001	1,250	4.23	5,282	907	40	4.68	187	19
2001–2002 [6]	1,550	4.55	7,054	1,140	30	3.83	115	24

[1] See footnote 1, table 5-25. [2] Equivalent packing-house door returns. [3] Includes quantities used for juice, concentrates, grapefruit segments, and other citrus products. In some seasons, includes appreciable quantities of oranges and lemons in CA delivered to processing plants which were not utilized, but for which growers received payment. [4] Includes small quantities of tangerines in TX. Excludes FL Temples. [5] Excludes FL economic abandonment in 1995–96 of 3 million boxes of Colored Seedless; in 1996–97 of 3 million boxes of White Seedless and 3 million boxes of Colored Seedless; in 1997–98 of 5 million boxes of White Seedless and 1 million boxes of Colored Seedless. [6] Preliminary. [7] AZ and CA tangelos and tangors included.

NASS, Crops Branch, (202) 720–2127.

Table 5-29.—Citrus fruit: Utilized production and marketing year average returns per box, by States, 2000–2001 to 2001–2002 [1]

Crop and State	Utilized production		Market year average price [2]	
	2000–2001	2001–2002	2000–2001	2001–2002
ORANGES	*1,000 boxes*	*1,000 boxes*	*Dollars*	*Dollars*
Early, midseason, and Navel varieties: [3]				
AZ	480	270	6.54	12.74
CA	35,500	34,000	9.98	12.89
FL	128,000	128,000	4.56	4.51
TX	2,000	1,530	2.38	4.52
Total early, midseason, and Navel varieties	165,980	163,800	5.54	6.03
Valencia:				
AZ	420	250	5.04	3.42
CA	19,000	22,000	8.43	8.37
FL	95,300	102,000	6.02	6.09
TX	235	210	3.32	5.51
Total Valencia	114,955	124,460	6.36	6.43
All oranges:				
AZ	900	520	5.84	8.26
CA	54,500	56,000	9.44	11.11
FL	223,300	230,000	5.18	5.21
TX	2,235	1,740	2.48	4.64
US, all oranges	280,935	288,260	5.88	6.20
GRAPEFRUIT				
AZ	250	160	5.42	6.73
CA	6,300	6,000	8.77	8.59
FL, all	46,000	46,700	4.55	4.49
Colored seedless	27,300	27,800	4.67	4.61
White seedless	18,700	18,900	4.37	4.32
TX	7,200	5,900	2.68	3.91
US, all grapefruit	59,750	58,760	4.69	4.78
LEMONS				
AZ	3,600	2,800	6.84	14.62
CA	22,600	19,000	9.41	15.80
US, lemons	26,200	21,800	9.06	15.65
LIMES				
FL	250	150	17.00	11.55
TANGELOS				
FL	2,100	2,150	3.90	5.10
TANGERINES				
AZ [4]	650	620	13.67	15.83
CA [5]	2,200	2,200	14.15	17.65
FL	5,600	6,600	10.14	11.54
US, tangerines	8,450	9,420	11.26	12.98
TEMPLES				
FL	1,250	1,550	4.23	4.55
K-EARLY CITRUS				
FL	40	30	4.68	3.83

[1] The crop year begins with the bloom of the first year shown and ends with completion of harvest the following year. [2] Equivalent packinghouse-door returns. [3] Includes small quantities of tangerines in TX. Excludes FL Temples. [4] Net lbs. per box: oranges—AZ and CA, 75; FL, 90; and TX, 85; grapefruit—AZ and CA, 67; FL, 85; TX, 80; lemons—76; tangelos, K-Early Citrus and Temples—90; tangerines—AZ and CA, 75. [5] Includes tangelos and tangors. Estimates discontinued.

NASS, Crops Branch, (202) 720–2127.

Table 5-30.—Citrus fruits: Production in specified countries, 1999–2000 to 2001–2002 [1]

Commodity and country	1999–2000	2000–2001	2001–2002 [2]
	1,000 metric tons	1,000 metric tons	1,000 metric tons
Oranges:			
Argentina	789	913	780
Australia	624	437	591
Brazil	17,136	14,729	17,993
China, Peoples Republic of	3,236	2,907	2,924
Cuba	440	450	340
Cyprus	134	102	98
Egypt	1,637	1,610	1,642
Gaza Strip	105	105	105
Greece	1,040	1,100	800
Israel	327	225	190
Italy	1,750	1,800	1,935
Japan	21	19	19
Mexico	3,385	3,885	3,800
Morocco	845	693	680
South Africa, Republic of	1,119	1,150	1,220
Spain	2,828	2,688	2,807
Turkey	1,100	1,070	1,040
United States [3]	11,875	11,225	11,337
Total	48,391	45,108	48,301
Tangerines:			
Argentina	438	501	416
China; Peoples Republic of	6,472	5,132	5,908
Cuba	5	5	2
Egypt	478	481	483
Greece	85	100	40
Israel	120	86	85
Italy	594	592	636
Japan	1,704	1,398	1,532
Korea; Republic of	635	563	651
Mexico	250	0	0
Morocco	511	260	310
Spain	2,070	1,780	1,655
Turkey	500	560	550
United States [4]	505	421	473
Total	8,192	6,897	7,570
Grapefruit:			
Argentina	189	177	170
Cuba	310	310	175
Cyprus	36	38	37
Gaza Strip	10	10	10
Israel	324	320	297
Italy	12	20	30
Mexico	240	250	225
South Africa, Republic of	186	195	190
Turkey	140	135	140
United States [3]	2,507	2,240	2,210
Total	3,954	3,695	3,484
Lemons:			
Argentina	1,163	1,217	1,200
Australia	30	0	0
Cyprus	28	26	21
Gaza Strip	9	9	9
Greece	125	130	57
Israel	18	16	17
Italy	543	610	537
Japan	2	2	2
Morocco	20	20	20
South Africa, Republic of	118	155	175
Spain	892	960	1,037
Turkey	520	460	500
United States	762	907	855
Total	4,230	4,512	4,430

[1] Split years refer to harvest periods which usually begin in the fall and extend through the following spring. This corresponds roughly with October–June in the Northern Hemisphere and April–December of the second year shown in the Southern Hemisphere. [2] Preliminary. [3] Includes temple oranges. [4] Includes tangelos.

FAS, Horticultural and Tropical Products Division, (202) 720–3423. Prepared or estimated on the basis of official statistics of foreign governments, other foreign source materials, reports of U.S. Agricultural Counselors, Attachés, and Foreign Service Officers, results of office research, and related information. U.S. data are from USDA Agricultural Statistics Board.

Table 5-31.—Concentrated citrus juices: Annual packs, Florida, 1991–2000

Season beginning December	Frozen concentrated juice[1]		
	Orange[2]	Grapefruit[2]	Tangerine
	1,000 gallons	*1,000 gallons*	*1,000 gallons*
1991	145,421	19,458	297
1992	208,662	31,527	329
1993	182,231	26,485	529
1994	216,502	31,344	1,192
1995	202,353	26,930	1,102
1996	241,800	30,032	2,386
1997	253,734	24,223	1,461
1998	158,884	24,512	1,191
1999	207,708	28,642	1,646
2000	235,933	27,481	1,065

[1] Net pack. [2] Frozen orange juice reported in 42.0° Brix; Grapefruit 40.0° Brix. Includes concentrated juice for manufacture.

ERS, Specialty Crops Branch, (202) 694–5260.

Table 5-32.—Citrus juice: United States exports by kind, 1991/92–2000/2001

Year beginning November	Orange juice				Grapefruit juice		
	Not frozen, not concentrated	Not frozen, concentrated	Frozen	Total	Not frozen	Frozen	Total
	1,000 gallons	*1,000 gallons*	*1,000 gallons*	*1,000 gallons*	*1,000 gallons*	*1,000 gallons*	*1,000 gallons*
1991/92	0	18,773	87,890	106,663	7,698	15,927	23,625
1992/93	0	24,254	91,935	116,189	6,987	15,846	22,833
1993/94	224	34,815	69,194	104,233	6,361	10,363	16,724
1994/95	153	41,567	76,759	118,479	7,725	14,415	22,140
1995/96	253	41,522	74,645	116,420	6,404	17,304	23,708
1996/97	540	3,081	558,963	562,584	61,317	17,225	78,542
1997/98	763	3,260	558,856	562,879	47,808	21,916	69,724
1998/99	2,743	3,708	547,032	553,484	80,614	15,077	95,691
1999/2000	4,378	4,293	543,551	552,222	114,508	19,961	134,469
2000/2001	9,461	1,539	442,822	453,824	121,189	20,632	141,821

FAS, Horticultural and Tropical Products Division, (202) 720–3423. Compiled from reports of the U.S. Department of Commerce.

Table 5-33.—Oranges, fresh:[1] United States exports by country of destination, 1998/99–2000/2001

Country of destination	Year beginning November		
	1998/99	1999/2000	2000/2001
	Metric tons	*Metric tons*	*Metric tons*
Europe:			
Belgium and Luxembourg	17	4	45
Denmark	17	0	0
France	46	0	27
Germany	71	47	49
Netherlands	48	889	277
Spain	17	3	0
Sweden	20	0	0
United Kingdom	159	216	239
Total EU	395	1,159	637
Western and Eastern Europe:			
Iceland	97	102	58
Latvia	0	53	0
Norway	0	0	2
Russian Federation	226	109	53
Switzerland	0	394	16
Total	323	823	129
South and Central America:			
Argentina	18	35	0
Brazil	5	16	0
Chile	256	617	438
Colombia	5	34	51
Costa Rica	0	0	64
Dominican Republic	0	0	73
Ecuador	106	87	124
El Salvador	3	2	6
Guatemala	99	56	30
Panama	135	203	99
Peru	36	105	310
Venezuela	19	0	0
Total	682	1,155	1,195
Caribbean:			
Aruba	43	1	0
Bahamas, The	38	25	159
Barbados	23	79	42
Jamaica	0	16	0
Netherlands Antilles	9	0	30
Total	113	121	231
North America:			
Canada	90,605	163,840	159,335
Mexico	6,620	8,426	11,457
Total	97,225	172,266	170,792
Africa and Asia:			
Brunei	0	10	0
Cambodia	0	0	22
China, Peoples Republic of	54	10,842	26,472
Hong Kong	35,646	88,418	86,015
India	49	0	20
Indonesia	313	1,941	1,559
Japan	50,185	106,144	113,684
Kenya	0	0	17
Korea, Republic of	15,579	68,320	80,870
Kuwait	18	20	0
Malaysia	9,390	16,345	26,463
Morocco	0	15	0
Philippines	636	3,070	3,572
Qatar	1	0	0
Saudi Arabia	0	0	18
Singapore	9,418	17,096	18,282
Sri Lanka	70	280	265
Taiwan	17,292	17,293	19,644
Thailand	170	413	253
United Arab Emirates	19	101	152
Vietnam	17	293	601
Total	138,857	330,601	377,909
Oceania:			
Australia	3,993	9,010	9,154
French Pacific Islands	76	53	126
New Zealand	3,359	3,614	5,937
Total	7,428	12,677	15,217
Other countries	91	107	227
World Total	245,115	518,745	566,338

[1] Includes Temple oranges.

FAS, Horticultural and Tropical Products Division, (202) 720–3423. Compiled from reports of the U.S. Department of Commerce.

Table 5-34.—Fresh citrus fruits: Foreign trade, United States, 1992–2001

Year [1]	Oranges		Grapefruit		Lemons		Limes		Tangerines	
	Imports	Domestic exports	Imports	Domestic exports	Imports	Domestic exports	Imports	Domestic exports	Imports	Domestic exports
	Metric tons	*Metric tons*	*Metric tons*	*Metric tons*	*Metric tons*	*Metric tons*	*Metric tons*	*Metric tons*	*Metric tons*	*Metric tons*
1992	10,236	562,598	12,509	444,776	9,834	127,336	112,125	2,770	12,396	9,559
1993	16,321	543,320	14,549	461,579	8,429	124,414	113,936	2,579	10,401	11,914
1994	17,585	580,755	13,676	481,743	12,319	126,121	142,990	3,421	4,254	12,546
1995	23,394	513,629	14,797	497,339	11,042	132,269	130,072	3,363	3,786	17,178
1996	29,654	590,428	12,807	484,403	21,736	120,279	147,064	3,517	3,877	15,285
1997	39,961	642,011	5,171	387,282	22,132	113,282	164,577	3,886	3,712	13,282
1998	101,923	263,199	15,521	428,618	22,991	114,109	151,613	3,921	2,576	11,543
1999	48,885	511,852	5,768	390,958	25,160	106,249	179,394	3,752	5,674	10,983
2000	52,785	570,161	19,409	389,629	34,127	110,373	182,412	3,846	4,117	11,786
2001	56,791	500,019	27,329	396,413	36,356	99,909	179,101	3,658	4,324	12,666

[1] Year beginning October for all commodities.

ERS, Specialty Crops Branch, (202) 694–5260.

Table 5-35.—Dates: Area, yield, total production, marketing year average price per ton, and value, California, 1993–2002

Year	Bearing acreage	Yield per acre	Production	Price per ton	Value
	Acres	*Tons*	*Tons*	*Dollars*	*1,000 dollars*
1993	5,500	5.27	29,000	790	22,910
1994	5,500	4.18	23,000	750	17,250
1995	5,200	4.37	22,700	780	17,706
1996	4,680	4.91	23,000	1,090	25,070
1997	4,800	4.38	21,000	1,100	23,100
1998	4,900	5.08	24,900	1,220	30,378
1999	4,900	4.53	22,200	1,240	27,528
2000	4,800	3.63	17,400	1,230	21,402
2001	4,500	4.38	19,700	1,320	26,004
2002 [1]	4,300	4.53	19,700	1,580	30,810

[1] Preliminary.

NASS, Crops Branch, (202) 720–2127.

Table 5-36.—Dates: Foreign trade, United States, 1992–2001

Year beginning October	Imports
	Metric tons
1992	5,426
1993	5,272
1994	2,774
1995	4,683
1996	2,587
1997	3,208
1998	5,179
1999	5,006
2000	2,997
2001	4,347

ERS, Specialty Crops Branch, (202) 694–5260.

Table 5-37.—Cranberries: Area, yield, production, season average price per barrel, value and quantities processed, United States, 1993–2002 [1]

Year	Area harvested	Yield per acre [2]	Total production [3]	Utilized production	Price [4]	Value	Quantities processed [5]
	Acres	*Barrels* [6]	*Barrels* [6]	*Barrels* [6]	*Dollars*	*1,000 dollars*	*Barrels* [6]
1993	29,400	133.3	3,919,000	3,919,000	50.20	196,820	3,619,000
1994	31,100	150.5	4,682,000	4,682,000	49.30	230,795	4,415,000
1995	32,800	127.8	4,193,000	4,193,000	53.40	223,938	3,858,000
1996	34,000	137.4	4,671,000	4,671,000	65.90	307,827	4,330,000
1997	35,700	154.0	5,497,000	5,497,000	63.70	350,147	5,072,000
1998	36,600	148.7	5,444,000	5,444,000	36.60	199,113	5,157,000
1999	37,500	168.9	6,334,000	6,334,000	17.20	108,773	5,977,000
2000	36,600	156.1	5,712,000	5,579,000	18.10	100,851	5,137,000
2001	34,200	155.8	5,329,000	4,783,000	23.10	110,342	4,357,000
2002 [7]	36,400	155.0	5,642,000	5,635,000	27.40	154,255	5,249,000

[1] Estimates relate to MA, NJ, OR, WA, and WI. [2] Derived from total production. [3] Differences between utilized and total production are quantities unharvested for economic reasons or excess cullage and/or set-aside production under provisions of the Cranberry Marketing Order. [4] Average price of utilized production. Equivalent returns at first delivery point, screened basis of utilized production. [5] Mainly for canning. [6] Barrels of 100 pounds. [7] Preliminary.

NASS, Crops Branch, (202) 720–2127.

Table 5-38.—Cranberries: Area, yield, production, and season average price per barrel, by States, 2000–2002

State	Area harvested			Yield per acre			Total production			Price per barrel[2]		
	2000	2001	2002[1]	2000	2001	2002[1]	2000	2001	2002[1]	2000	2001	2002[1]
	Acres	*Acres*	*Acres*	*Bbl.*[3]	*Bbl.*[3]	*Bbl.*[3]	*Bbl.*[3]	*Bbl.*[3]	*Bbl.*[3]	*Dollars*	*Dollars*	*Dollars*
MA ..	13,900	12,000	13,300	140.5	118.0	109.8	1,953,000	1,416,000	1,460,000	19.00	23.20	28.40
NJ ...	3,700	3,100	3,100	132.2	182.6	138.7	489,000	566,000	430,000	17.90	22.30	25.60
OR ..	2,400	2,400	2,500	165.8	152.1	160.0	398,000	365,000	400,000	16.70	21.90	26.40
WA	1,500	1,600	1,600	120.0	88.8	101.3	180,000	142,000	162,000	22.30	27.00	29.20
WI ...	15,100	15,100	15,900	178.3	188.1	200.6	2,692,000	2,840,000	3,190,000	17.40	23.10	27.20
US ..	36,600	34,200	36,400	156.1	155.8	155.0	5,712,000	5,329,000	5,642,000	18.10	23.10	27.40

[1] Preliminary. [2] Average price of utilized production. Equivalent returns at first delivery point, screened basis of utilized production. [3] Barrels of 100 pounds.

NASS, Crops Branch, (202) 720–2127.

Table 5-39.—Figs: Total production, marketing year average price per ton, and value, California, 1993–2002

Year	Dried (dry basis)				Total		
	Production			Price per ton	Production (fresh basis)[1]	Price per ton	Value
	Total	Standard	Substandard				
	Tons	*1,000 tons*	*Tons*	*Dollars*	*Tons*	*Dollars*	*1,000 dollars*
1993	19,300	17,100	2,200	1,097	60,700	401	24,341
1994	18,200	17,000	1,200	1,198	56,700	419	23,768
1995	16,800	15,200	1,600	895	52,400	314	16,429
1996	14,500	13,100	1,400	774	45,500	283	12,894
1997	18,500	15,900	2,600	699	57,500	265	15,209
1998	16,500	13,300	3,200	594	51,300	226	11,611
1999	15,100	13,800	1,300	681	47,300	268	12,685
2000	17,300	15,400	1,900	672	55,900	272	15,226
2001	12,700	11,400	1,300	....	40,100	370	14,849
2002[2]	15,500	14,000	1,500	....	49,000	345	16,907

[1] Dried figs converted to fresh basis at ratio of 3 pounds fresh to 1 pound dried. [2] Preliminary.

NASS, Crops Branch, (202) 720–2127.

Table 5-40.—Figs, dried: Foreign trade, United States, 1992–2001

Year beginning October	Imports for consumption	Domestic exports
	Metric tons	*Metric tons*
1992	8,239	2,623
1993	10,990	3,035
1994	11,653	5,927
1995	4,834	4,850
1996	3,780	2,027
1997	4,414	1,824
1998	3,425	2,010
1999	3,900	2,763
2000	3,071	2,506
2001	6,807	2,399

ERS, Specialty Crops Branch, (202) 694–5260.

Table 5-41.—Ginger Root: Area, yield, production, marketing year average price, and value, Hawaii, 1992/93–2001/2002

Year	Area harvested	Yield per acre	Total production	Price per pound	Value
	Acres	*1,000 pounds*	*1,000 pounds*	*Cents*	*1,000 dollars*
1992–93 ..	360	27.5	9,900	53.0	5,247
1993–94 ..	150	40.0	6,000	87.0	5,220
1994–95 ..	135	43.0	5,800	87.0	5,046
1995–96 ..	200	47.0	9,400	75.0	7,050
1996–97 ..	275	44.0	12,100	67.0	8,107
1997–98 ..	360	50.0	18,000	40.0	7,200
1998–99 ..	350	46.0	16,100	50.0	8,050
1999–2000	270	50.0	13,500	66.0	8,910
2000–2001	360	50.0	18,000	45.0	8,100
2001–2002	320	45.0	14,400	30.0	4,320

NASS, Crops Branch, (202) 720–2127.

Table 5-42.—Grapes: Production, price, and value, United States, 1993–2002

Year	Grapes			
	Production (fresh basis)		Market year average price per ton [1]	Value
	Total	Utilized		
	1,000 tons	*1,000 tons*	*Dollars*	*1,000 dollars*
1993	6,022,950	6,014,350	333.00	2,005,321
1994	5,873,400	5,869,000	321.00	1,882,591
1995	5,921,900	5,912,350	346.00	2,046,737
1996	5,553,600	5,537,325	429.00	2,376,111
1997	7,290,900	7,287,365	429.00	3,126,433
1998	5,819,950	5,816,405	454.00	2,640,470
1999	6,236,260	6,234,730	469.00	2,926,910
2000	7,687,970	7,687,330	403.00	3,099,127
2001	6,569,600	6,568,400	446.00	2,929,038
2002 [2]	7,144,000	7,141,860	408.00	2,912,742

[1] Fresh fruit prices are equivalent packinghouse-door returns for California and Washington, and the average price as sold for other States. Quantities processed are priced at the equivalent processing plant door level. [2] Preliminary.

NASS, Crops Branch, (202) 720–2127.

Table 5-43.—Grapes: Production and marketing year average price per ton, by States, 2000–2002

State	Total production			Utilized production			Price per ton [1]		
	2000	2001	2002 [2]	2000	2001	2002 [2]	2000	2001	2002 [2]
	Tons	*Tons*	*Tons*	*Tons*	*Tons*	*Tons*	*Dollars*	*Dollars*	*Dollars*
AZ [3]	20,000	15,500	8,400	20,000	15,500	8,400	718.00	576.00	966.00
AR	4,200	2,700	5,600	3,900	2,500	5,400	560.00	541.00	462.00
CA: [3]									
All types	7,059,000	5,979,000	6,500,000	7,059,000	5,979,000	6,500,000	401.00	445.00	407.00
Wine	3,364,000	3,051,000	3,100,000	3,364,000	3,051,000	3,100,000	567.00	597.00	566.00
Table varieties [4]	774,000	713,000	750,000	774,000	713,000	750,000	565.00	610.00	597.00
Raisin [4]	2,921,000	2,215,000	2,650,000	2,921,000	2,215,000	2,650,000	166.00	182.00	167.00
GA	3,500	3,200	2,800	3,200	3,200	2,760	1,230.00	1,050.00	1,060.00
MI	87,200	28,900	42,700	87,200	28,500	42,500	277.00	313.00	312.00
MO	2,950	2,300	2,900	2,950	2,300	2,900	555.00	630.00	615.00
NY	154,000	149,000	156,000	154,000	149,000	155,000	298.00	302.00	280.00
NC	2,300	2,000	2,300	2,300	2,000	2,300	1,160.00	1,270.00	1,280.00
OH	7,700	6,000	5,800	7,700	6,000	5,800	327.00	342.00	381.00
OR	18,600	22,800	22,800	18,600	22,800	22,800	1,400.00	1,480.00	1,460.00
PA	63,000	61,500	53,200	63,000	61,500	53,200	265.00	274.00	239.00
SC [5]	520	....	....	480	....	....	669.00	....	....
TX [6]	....	9,500	4,700	....	9,300	4,400	....	900.00	880.00
VA [6]	....	4,200	4,800	....	3,800	4,400	....	1,260.00	1,490.00
WA:.									
All types	265,000	283,000	332,000	265,000	283,000	332,000	481.00	470.00	405.00
Wine	90,000	100,000	115,000	90,000	100,000	115,000	899.00	897.00	878.00
Juice	175,000	183,000	217,000	175,000	183,000	217,000	266.00	237.00	155.00
US	7,687,970	6,569,600	7,144,000	7,687,330	6,568,400	7,141,860	403.00	446.00	408.00

[1] Fresh fruit prices are equivalent packinghouse-door returns for CA and WA, and the average price as sold for other States. Quantities processed are priced at the equivalent processing plant door level. [2] Preliminary. [3] The 2002 prices for CA and AZ grapes utilized for raisins are not available. Price and value of grapes utilized for raisins in 2002 were derived by using the 2001 comparable prices. [4] Fresh equivalent of dried and not dried. [5] Estimates discontinued in 2001. [6] Estimates began in 2001.

NASS, Crops Branch, (202) 720–2127.

Table 5-44.—Grapes: Production and utilization, United States, 1993–2002

Crop of—	Total production [1]	Utilized production	Utilization of quantities sold				
			Fresh	Processed			
				Canned	Dried (fresh basis)	Crushed for wine	Crushed for juice, etc. [2]
	Tons	*Tons*	*Tons*	*Tons*	*Tons*	*Tons*	*Tons*
1993	6,022,950	6,014,350	800,500	45,000	1,677,200	3,025 ,480	466,170
1994	5,873,400	5,869,300	808,550	38,000	1,906,300	2,695,330	420,820
1995	5,921,900	5,912,350	852,600	35,000	1,552,500	2,973,640	498,610
1996	5,553,600	5,537,325	767,025	36,000	1,329,000	3,042,850	362,450
1997	7,290,900	7,287,365	937,115	44,000	1,806,500	4,034,400	465,350
1998	5,819,950	5,816,405	780,795	36,000	1,331,600	3,314,760	353,250
1999	6,236,260	6,234,730	887,221	35,000	1,459,900	3,350,559	502,050
2000	7,687,970	7,687,330	906,825	32,000	2,194,600	4,129,655	424,250
2001	6,569,600	6,568,400	864,400	29,000	1,736,800	3,568,350	369,850
2002 [3]	7,144,000	7,141,860	937,440	31,000	1,810,050	3,945,570	417,800

[1] Total production includes utilized production plus production not harvested and harvested not sold: 1993—8,600 tons fresh equivalent; 1994—4,100 tons fresh equivalent; 1995—9,550 tons fresh equivalent; 1996—16,275 tons fresh equivalent; 1997—3,535 tons fresh equivalent; 1998—3,545 tons fresh equivalent; 1999—1,530 tons fresh equivalent; 2000—640 tons fresh equivalent; 2001—1,200 tons fresh equivalent; and 2002—2,140 tons fresh equivalent. [2] Mostly juice, but includes some quantities used for jam, jelly, etc. [3] Preliminary.

NASS, Crops Branch, (202) 720–2127.

Table 5-45.—Grapes: Production and utilization, by States, crop of 2002 (preliminary)

State	Total production	Utilized production	Utilization				
			Fresh	Processed			
				Canned	Dried (fresh basis) [1]	Crushed for—	
						Wine	Juice, etc. [2]
	Tons	*Tons*	*Tons*	*Tons*	*Tons*	*Tons*	*Tons*
AZ	8,400	8,400	...	...	2,050	...	...
AR	5,600	5,400	...	...	...	...	...
CA:							
All types	6,500,000	6,500,000	926,000	...	1,808,000	3,735,000	...
Wine	3,100,000	3,100,000	45,000	...	...	3,055,000	...
Table	750,000	750,000	646,000	...	24,000	80,000	...
Raisin	2,650,000	2,650,000	235,000	31,000	1,784,000	600,000	...
GA	2,800	2,760	...	...	...	...	...
MI	42,700	42,500	300	...	...	3,000	39,200
MO	2,900	2,900	240	...	...	2,460	...
NY	156,000	155,000	2,000	...	...	44,000	109,000
NC	2,300	2,300	300	...	...	2,000	...
OH	5,800	5,800	100	...	...	1,200	4,500
OR	22,800	22,800	...	...	...	22,800	...
PA	53,200	53,200	1,000	...	...	6,300	45,900
TX	4,700	4,400	...	...	...	4,400	...
VA	4,800	4,400	...	...	...	4,400	...
WA:.							
All types	332,000	332,000	...	...	...	115,000	217,000
Wine	115,000	115,000	...	...	...	...	...
Juice	217,000	217,000	...	...	...	...	...
Other States	...	...	7,500	...	...	5,010	2,200
US	7,144,000	7,141,860	937,440	31,000	1,810,050	3,945,570	417,800

[1] Equivalent raisins produced (dried basis): 402,313 tons. [2] Mostly juice, but includes some quantities used for jam, jelly, etc.

NASS, Crops Branch, (202) 720–2127.

Table 5-46.—Raisins and currants: United States exports by country of destination, 1999/2000–2001/2002

Country of destination	Year beginning August		
	1999/2000	2000/2001	2001/2002
	Metric tons	*Metric tons*	*Metric tons*
North America:			
Canada	10,721.9	11,792.9	12,382.9
Mexico	1,172.6	629.5	448.6
Total	11,894.5	12,422.4	12,831.5
European Union:			
Sweden	3,526.9	4,805.1	4,388.6
Finland	1,725.0	1,614.2	1,968.9
Denmark	3,176.6	3,852.0	3,179.8
United Kingdom	11,515.6	24,048.3	23,694.2
Netherlands	2,811.8	5,551.2	4,281.5
France	624.7	274.8	455.7
Germany	3,279.8	7,113.3	6,560.0
Spain	499.0	504.6	842.5
Belgium-Lux.	691.3	888.7	1,038.2
Other	209.0	507.5	331.7
Total	28,059.7	49,159.7	46,741.1
Asia:			
Singapore	2,161.6	2,025.3	1,663.5
Malaysia	576.5	889.4	1,795.5
Korea, Republic of	1,122.4	2,303.6	2,716.3
Hong Kong	1,368.1	2,436.4	2,736.4
Taiwan	2,638.6	3,853.6	3,908.0
Japan	22,759.7	22,925.1	23,311.3
China	130.8	263.1	1,034.6
Total	30,757.7	34,696.5	37,165.6
Other countries:			
New Zealand	1,003.2	1,300.0	1,031.4
Norway	2,045.4	2,383.7	2,113.0
Philippines	801.8	1,136.5	1,320.8
Israel	1,008.0	1,011.3	1,344.9
All Others	4,424.5	6,957.8	10,824.1
Total	9,282.9	12,789.3	16,634.2
Grand total	79,994.8	109,067.9	113,372.4

FAS, Horticultural and Tropical Products Division, (202) 720–3423. Compiled from reports of the U.S. Department of Commerce.

Table 5-47.—Grapes and raisins: Foreign trade, United States 1992–2001

Year beginning October	Grapes		Raisins [1]	
	Imports, fresh	Domestic exports, fresh	Imports for consumption	Domestic exports
	Metric tons	*Metric tons*	*Metric tons*	*Metric tons*
1992	325,136	184,779	7,425	121,523
1993	311,027	215,505	6,432	122,628
1994	363,687	204,784	11,134	122,133
1995	341,098	240,097	12,525	118,824
1996	351,567	233,558	11,407	115,016
1997	419,956	217,467	10,918	120,923
1998	387,165	222,317	26,193	104,305
1999	452,182	272,901	16,929	83,832
2000	418,012	303,396	11,641	110,035
2001	501,055	293,615	15,204	116,019

[1] Raisins converted to sweatbox or production basis by multiplying by 1.08.

ERS, Specialty Crops Branch, (202) 694–5260.

Table 5-48.—Guavas: Area, yield, utilized production, marketing year average price, and value, Hawaii, 1993–2002

Year	Area harvested	Yield per acre	Production	Price per pound	Value
	Acres	*1,000 pounds*	*1,000 pounds*	*Cents*	*1,000 dollars*
1993	750	20.5	15,400	13.0	2,002
1994	740	23.9	17,700	13.5	2,390
1995	750	21.9	16,400	13.8	2,263
1996	750	21.7	16,300	13.8	2,249
1997	730	21.8	15,900	12.2	1,940
1998	710	20.6	14,600	12.2	1,781
1999	630	17.0	10,700	12.0	1,284
2000	680	23.4	15,900	12.9	2,051
2001	610	25.1	15,300	14.1	2,157
2002	NA	NA	NA	NA	NA

NA-not available.
NASS, Crops Branch, (202) 720–2127.

Table 5-49.—Nectarines: Production, utilization, and value, California, 1993–2002

Crop of—	Production	Utilization		Marketing year average price per ton[2]	Value
		Fresh[1]	Processed (fresh basis)		
	Tons	*Tons*	*Tons*	*Dollars*	*1,000 dollars*
1993	205,000	201,000	4,000	500.00	102,421
1994	242,000	238,000	4,000	282.00	68,168
1995	176,000	170,000	6,000	534.00	93,990
1996	247,000	239,800	7,200	474.00	116,977
1997	264,000	258,500	5,500	375.00	98,895
1998	224,000	207,600	16,400	471.00	105,466
1999	274,000	256,300	17,700	411.00	112,497
2000	267,000	260,700	6,300	398.00	106,256
2001	275,000	265,400	9,600	464.00	127,642
2002[3]	299,000	299,000	0	383.00	114,517

[1] Includes "Home use." [2] Processing fruit prices are equivalent returns at processing plant door. [3] Preliminary.
NASS, Crops Branch, (202) 720–2127.

Table 5-50.—Olives: Total production, marketing year average price, value, and processed utilization, California, 1993–2002

Year	Production	Marketing year average price per ton	Value	Processed utilization			
				Crushed for oil	Canned	Limited	Undersized
	Tons	*Dollars*	*1,000 dollars*	*Tons*	*Tons*	*Tons*	*Tons*
1993	122,000	467	56,991	5,300	93,000	19,700	3,500
1994	84,000	464	38,994	4,400	66,500	8,400	4,200
1995	77,500	646	50,069	4,000	58,500	9,300	5,200
1996	166,000	617	102,364	7,000	123,000	29,000	6,500
1997	104,000	642	66,801	3,600	82,200	10,200	7,500
1998	90,000	459	41,331	4,100	64,200	12,800	8,400
1999	142,000	387	55,011	5,000	86,000	36,500	14,000
2000	53,000	656	34,743	3,000	41,400	5,100	3,000
2001	134,000	672	90,096	3,000	109,700	15,300	5,500
2002[1]	99,000	592	58,562	2,000	82,800	9,900	3,800

[1] Preliminary.
NASS, Crops Branch, (202) 720–2127.

Table 5-51.—Olives and olive oil: Foreign trade, United States, 1992–2001

Year beginning October	Imports			
	Olives		Olive oil	
	In brine	Dried	Edible	Inedible
	Metric tons	*Metric tons*	*Metric tons*	*Metric tons*
1992	62,224	606	123,234	0
1993	64,626	462	123,934	0
1994	60,554	1,045	127,647	0
1995	65,744	396	113,590	125
1996	73,544	271	148,051	43
1997	81,315	401	161,015	1
1998	85,957	438	170,086	1
1999	87,746	314	189,302	0
2000	99,337	416	212,341	0
2001	101,321	367	217,664	276

ERS, Specialty Crops Branch, (202) 694–5260.

Table 5-52.—Peaches: Production and value, United States, 1993–2002

Year	Total production	Utilized production	Marketing year average price [1]	Value
	Million pounds	*Million pounds*	*Cents per pound*	*1,000 dollars*
1993	2,644.3	2,478.4	15.9	395,241
1994	2,509.0	2,354.0	13.3	313,253
1995	2,289.5	2,179.1	18.4	401,393
1996	2,104.6	2,043.8	19.1	389,894
1997	2,624.6	2,508.4	17.7	444,137
1998	2,400.7	2,325.6	19.2	446,534
1999	2,525.6	2,433.4	19.0	462,836
2000	2,579.9	2,488.8	19.4	481,716
2001	2,433.3	2,334.9	21.1	493,298
2002 [2]	2,575.4	2,472.5	20.5	507,089

[1] Fresh fruit prices are equivalent packinghouse-door returns for CA and WA except equivalent returns for bulk fruit at the first delivery point for CA Clingstone, and the average price as sold for other States. Quantities processed are priced at the equivalent processing plant door level. [2] Preliminary.

NASS, Crops Branch, (202) 720–2127.

Table 5-53.—Peaches: Foreign trade, United States, 1992–2001

Year beginning October	Domestic exports				
	Fresh	Canned	Canned, in fruit salad [2]	Dried, in fruit salad [1 3]	Total, in terms of fresh [4]
	Metric tons	*Metric tons*	*Metric tons*	*Metric tons*	*Metric tons*
1992	64,003	21,393	13,965	728	84,151
1993	83,309	18,171	10,402	831	116,499
1994	68,237	20,915	11,291	968	105,821
1995	74,822	17,777	10,696	884	108,206
1996	103,345	16,873	8,990	675	132,958
1997	79,987	21,580	9,189	728	114,801
1998	98,123	27,170	10,224	579	138,734
1999	113,098	16,875	7,869	299	139,503
2000	129,292	13,008	4,677	532	149,933
2001	127,390	10,911	3,887	345	144,105

[1] Net processed weight. [2] Canned peaches are 40 percent of total canned fruit for salad. [3] Dried peaches are 21 percent of total dried fruit for salad. [4] Dried fruit converted to unprocessed dry weight by dividing by 1.08. Unprocessed dry weight converted to terms of fresh fruit on the basis that 1 pound dried equals 6.0 pounds fresh. Canned peaches converted to terms of fresh on basis that 1 pound canned equals about 1 pound fresh.

ERS, Specialty Crops Branch, (202) 694–5260.

Table 5-54.—Peaches: Production and utilization, United States, 1993–2002

Crop of—	Total production [1]	Utilized production	Utilization of quantities sold				
			Fresh [2]	Processed (fresh basis)			
				Canned	Dried	Frozen	Other [3]
	Million pounds	*Million pounds*	*Million pounds*	*Million pounds*	*Million pounds*	*Million pounds*	*Million pounds*
1993	2,644.3	2,478.4	1,166.0	1,044.6	30.0	178.8	59.0
1994	2,509.0	2,354.0	1,038.9	1,044.5	27.0	157.8	85.8
1995	2,289.5	2,179.1	1,125.0	813.5	28.0	150.9	61.7
1996	2,104.6	2,043.8	769.8	994.3	32.7	183.2	63.8
1997	2,624.6	2,508.4	1,126.8	1,107.8	34.1	201.0	38.7
1998	2,400.7	2,325.6	1,000.5	985.2	25.0	185.8	129.1
1999	2,525.6	2,433.4	1,100.3	996.0	31.4	204.1	101.6
2000	2,579.9	2,488.8	1,161.1	1,026.8	25.2	219.5	56.2
2001	2,433.3	2,334.9	1,154.0	906.7	29.2	200.7	44.3
2002 [4]	2,575.4	2,472.5	1,119.2	1,061.4	28.4	204.3	59.2

[1] Includes harvested not sold and unharvested production for California Clingstone peaches (tons): 1990—None. [2] Includes "Home use." [3] Used for jams, preserves, pickles, wine, brandy, baby food, etc. Includes small quantities frozen for some years. [4] Preliminary.

NASS, Crops Branch, (202) 720–2127.

Table 5-55.—Peaches: Production and season average price per pound, 2000–2002

State	Total production			Utilized production			Price [2] for crop of—		
	2000	2001	2002 [1]	2000	2001	2002 [1]	2000	2001	2002 [1]
	Million pounds	*Million pounds*	*Million pounds*	*Million pounds*	*Million pounds*	*Million pounds*	*Dollars*	*Dollars*	*Dollars*
AL	14.0	23.0	21.0	13.0	22.0	19.0	0.405	0.418	0.440
AR	18.0	12.0	10.0	15.7	10.3	7.8	0.370	0.407	0.428
CA:									
All	1,855.0	1,727.0	1,920.0	1,808.0	1,677.0	1,870.0	0.139	0.147	0.145
Clingstone	1,064.0	952.0	1,124.0	1,017.0	902.0	1,074.0	0.125	0.122	0.124
Freestone	791.0	775.0	796.0	791.0	775.0	796.0	0.157	0.176	0.174
CO	19.0	18.0	19.0	17.0	17.5	18.5	0.470	0.540	0.567
CT	2.0	1.9	1.3	2.0	1.9	1.3	0.650	0.650	0.700
GA	115.0	140.0	110.0	110.0	125.0	102.0	0.379	0.280	0.385
ID	13.0	13.0	13.0	13.0	12.9	13.0	0.387	0.494	0.427
IL	23.0	17.8	17.2	20.0	17.1	16.5	0.412	0.420	0.470
IN	2.6	3.0	3.1	2.5	2.9	3.0	0.414	0.583	0.733
KY	1.1	1.8	1.2	1.1	1.8	1.2	0.346	0.487	0.545
LA	1.2	1.4	1.5	1.1	1.3	1.4	0.771	0.871	0.751
MD	9.0	8.8	7.0	9.0	8.8	7.0	0.398	0.367	0.392
MA	2.1	2.2	2.3	2.1	2.1	2.2	0.700	0.700	0.800
MI	47.5	42.0	14.0	45.5	42.0	14.0	0.249	0.298	0.318
MO	9.5	9.0	13.0	9.5	9.0	13.0	0.350	0.365	0.420
NJ	65.0	75.0	62.0	58.0	70.0	57.0	0.427	0.405	0.442
NY	12.0	12.5	10.0	11.3	12.0	10.0	0.400	0.311	0.238
NC	12.0	3.5	10.0	12.0	3.5	10.0	0.370	0.400	0.350
OH	10.4	11.2	9.4	9.1	11.0	9.1	0.468	0.490	0.496
OK	14.0	12.0	7.0	12.5	10.3	5.8	0.509	0.459	0.577
OR	8.0	6.5	7.9	7.8	6.4	7.4	0.423	0.427	0.409
PA	60.0	75.0	60.0	58.0	74.0	59.0	0.287	0.335	0.335
SC	150.0	100.0	160.0	140.0	85.0	130.0	0.204	0.425	0.329
TN	2.5	3.7	4.0	2.3	3.5	3.7	0.545	0.484	0.470
TX	21.0	30.0	12.0	17.3	26.0	11.4	0.580	0.570	0.600
UT	11.0	9.0	6.5	10.0	8.9	6.5	0.300	0.218	0.312
VA	10.0	8.0	7.0	9.0	7.0	6.7	0.300	0.317	0.410
WA	65.0	55.0	56.0	65.0	55.0	56.0	0.325	0.338	0.378
WV	7.0	11.0	10.0	7.0	10.7	10.0	0.256	0.300	0.310
Total above	2,579.9	2,433.3	2,575.4	2,488.8	2,334.9	2,472.5	0.194	0.211	0.205

[1] Preliminary. [2] Fresh fruit prices are equivalent packinghouse-door returns for CA and WA except equivalent returns for bulk fruit at the first delivery point for CA Clingstone, and the average price as sold for other States. Quantities processed are priced at the equivalent processing plant door level.

NASS, Crops Branch, (202) 720–2127.

Table 5-56.—Peaches: Production and utilization, by States, crop of 2002 (preliminary)

State	Total production	Utilized production [1]	Utilization				
			Fresh [2]	Processed (fresh basis)			
				Canned	Dried	Frozen	Other [3]
	Million pounds	*Million pounds*	*Million pounds*	*Million pounds*	*Million pounds*	*Million pounds*	*Million pounds*
CA, all [3]	1,920.0	1,870.0	....	....	....	....	....
Clingstone [3]	1,124.0	1,074.0	....	1,023.0	....	....	....
Freestone	796.0	796.0	555.7	....	28.4	182.2	....
GA	110.0	102.0	....	....	....	....	....
NJ	62.0	57.0	....	....	....	....	....
PA	60.0	59.0	....	....	....	....	....
SC	160.0	130.0	115.0	....	....	....	....
WA	56.0	56.0	....	....	....	....	....
Other States	107.4	198.5	448.5	38.4	....	22.1	59.2
US	2,575.4	2,472.5	1,119.2	1,061.4	28.4	204.3	59.2

[1] Difference between total and utilized production is harvested not sold and unharvested production. [2] Includes "Home use." [3] Used for jams, preserves, brandy, etc.

NASS, Crops Branch, (202) 720–2127.

Table 5-57.—Peaches, canned: United States exports by country of destination, 1999/2000–2001/2002

Country of destination	Year beginning June		
	1999/2000	2000/2001	2001/2002
	Metric tons	*Metric tons*	*Metric tons*
Caribbean:			
Bahamas	10	130	29
Barbados	0	0	3
Bermuda	0	8	5
Dominican Republic	33	14	16
Haiti	5	0	0
Jamaica & Dep	11	0	0
Leeward-Windward Islands	0	0	3
Netherlands Antilles	45	43	9
Total	104	195	65
Central America:			
Costa Rica	395	463	344
El Salvador	270	234	132
Guatamala	435	237	136
Honduras	47	130	191
Nicaragua	3	0	0
Panama	404	285	177
Total	1,555	1,349	979
European Union:			
Belgium and Luxembourg	159	16	0
Germany	101	0	0
Greece	0	19	8
Italy	0	44	0
Netherlands	464	17	134
Sweden	10	0	0
United Kingdom	43	0	55
Total	776	96	197
Former Soviet Union:			
Russian Federation	0	5	0
Total	0	5	0
Middle East:			
Bahrain	0	0	7
Israel	302	298	36
Kuwait	21	14	0
Qatar	0	12	8
Saudi Arabia	77	240	161
United Arab Emirates	10	32	15
Yemen	3	15	11
Total	414	611	238
North America:			
Canada	7,668	6,308	5,189
Mexico	5,981	1,176	304
Total	13,649	7,484	5,493
Oceania:			
French Pacific Islands	12	16	0
Marshal Islands	0	33	0
Micronesia, Federated States	7	0	0
New Zealand	18	0	0
Total	37	49	0

See footnotes at end of table.

Table 5-57.—Peaches, canned: United States exports by country of destination, 1999/2000–2001/2002—Continued

Country of destination	Year beginning June		
	1999/2000	2000/2001	2001/2002
	Metric tons	*Metric tons*	*Metric tons*
Other Africa:			
Mali	0	0	16
Total	0	0	16
Other Asia:			
China, Republic of	30	18	7
Hong Kong	147	97	6
Indonesia	0	3	6
Japan	2,562	1,301	619
Korea, Republic of	1,112	2,043	956
Malaysia	17	26	4
Philippines	314	612	854
Singapore	202	76	19
Taiwan	411	506	77
Thailand	76	670	999
Total	4,872	5,353	3,547
Other West Europe:			
Iceland	0	0	3
Norway	2	0	1
Switzerland	0	59	0
Total	2	59	4
South America:			
Argentina	35	53	0
Brazil	0	100	0
Chile	0	0	0
Colombia	32	4	0
Ecuador	0	0	0
Peru	33	0	34
Venezuela	14	18	2
Total	115	175	36
South Asia:			
Bangaladesh	16	5	0
Sri Lanka	0	5	0
Total	16	10	0
Grand total	21,539	15,384	10,576

FAS, Horticultural and Tropical Products Division, (202) 720–4620. Compiled from reports of the U.S. Department of Commerce.

Table 5-58.—Pineapples: Total area, utilized production, utilization, marketing year average price, and value, Hawaii, 1993–2002

Year	Total area	Utilized production	Utilization		Price per ton	Value
			Fresh	Processed		
	Acres	*Tons*	*Tons*	*Tons*	*Dollars*	*1,000 dollars*
1993	22,000	370,000	135,000	235,000	216	79,850
1994	22,300	365,000	130,000	235,000	216	78,890
1995	19,900	345,000	125,000	220,000	253	87,360
1996	20,000	347,000	115,000	232,000	276	95,914
1997	19,900	324,000	103,000	221,000	283	91,721
1998	21,000	332,000	111,000	221,000	279	92,776
1999	21,000	352,000	122,000	230,000	288	101,448
2000	20,700	354,000	122,000	232,000	287	101,530
2001	20,100	323,000	110,000	213,000	298	96,337
2002 [1]	19,100	320,000	117,000	203,000	314	100,616

[1] Preliminary.

NASS, Crops Branch, (202) 720–2127.

Table 5-59.—Pears: Production and value, United States 1993–2002

Year	Total production	Utilized production	Marketing year average price [1]	Value
	Tons	*Tons*	*Dollars per ton*	*1,000 dollars*
1993	947,850	946,500	245.00	231,850
1994	1,045,950	1,045,450	223.00	233,071
1995	947,950	947,300	272.00	257,849
1996	820,550	820,250	376.00	308,367
1997	1,042,500	1,041,930	276.00	287,822
1998	970,140	967,795	291.00	281,611
1999	1,015,450	1,013,435	294.00	298,009
2000	967,150	949,170	264.00	250,273
2001	1,001,780	964,280	282.00	271,788
2002 [2] ...	911,450	911,000	326.00	297,410

[1] Fresh fruit prices are equivalent packinghouse-door returns for CA, OR, and WA, and the average price as sold for other States. Quantities processed are priced at the equivalent processing plant door level. [2] Preliminary.

NASS, Crops Branch, (202) 720–2127.

Table 5-60.—Pears: Production and season average price per ton, by States, 2000–2002

Variety and State	Total production			Utilized production			Price [2] for crop of—		
	2000	2001	2002 [1]	2000	2001	2002 [1]	2000	2001	2002 [1]
	Tons	*Tons*	*Tons*	*Tons*	*Tons*	*Tons*	*Dollars*	*Dollars*	*Dollars*
CA, all	312,000	305,000	294,000	297,000	270,000	294,000	228.00	289.00	268.00
Bartlett	282,000	275,000	264,000	267,000	240,000	264,000	205.00	287.00	262.00
Other	30,000	30,000	30,000	30,000	30,000	30,000	437.00	313.00	325.00
CO	3,000	1,900	2,400	2,960	1,900	2,400	375.00	500.00	576.00
CT	1,250	480	500	1,250	480	500	562.00	644.00	858.00
MI	5,200	4,600	1,400	5,200	3,900	1,400	270.00	297.00	318.00
NY	14,500	11,000	10,000	12,900	10,000	9,850	353.00	401.00	374.00
OR, all	220,000	230,000	198,000	219,000	229,500	198,000	303.00	278.00	343.00
Bartlett	60,000	70,000	58,000	59,000	69,500	58,000	297.00	302.00	343.00
Other	160,000	160,000	140,000	160,000	160,000	140,000	305.00	267.00	344.00
PA	4,600	5,500	3,800	4,400	5,200	3,500	509.00	569.00	480.00
UT	600	300	350	460	300	350	533.00	583.00	643.00
WA, all	406,000	443,000	401,000	406,000	443,000	401,000	261.00	272.00	356.00
Bartlett	176,000	201,000	167,000	176,000	201,000	167,000	254.00	219.00	305.00
Other	230,000	242,000	234,000	230,000	242,000	234,000	267.00	316.00	392.00
US	967,150	1,001,780	911,450	949,170	964,280	911,000	264.00	282.00	326.00

[1] Preliminary. [2] Fresh fruit prices are equivalent packinghouse-door returns for CA, OR, and WA, and the average price as sold for other States. Quantities processed are priced at the equivalent processing plant door level.

NASS, Crops Branch, (202) 720–2127.

Table 5-61.—Pears: Foreign trade, United States, 1992–2001

Year beginning October	Imports for consumption, fresh	Domestic exports				
		Fresh [1]	Canned	Dried, in fruit salad [1 2]	Canned, in fruit salad [3]	Total, in terms of fresh fruit [4]
	Metric tons	*Metric tons*	*Metric tons*	*Metric tons*	*Metric tons*	*Metric tons*
1992	64,825	98,818	3,441	578	12,219	118,126
1993	65,283	137,046	3,076	600	9,102	153,010
1994	48,065	127,961	5,039	769	9,879	147,732
1995	57,371	144,426	7,245	702	9,359	165,460
1996	78,520	126,409	3,494	536	7,866	141,152
1997	67,636	156,996	5,228	578	8,041	173,913
1998	87,422	145,725	4,327	459	8,947	161,896
1999	89,828	162,629	4,655	238	6,886	175,672
2000	85,219	158,333	5,887	422	4,093	170,976
2000	79,967	175,337	6,181	273	3,401	186,642

[1] Net processed weight. [2] Dried pears are 16⅔ percent of total dried fruit for salad. [3] Canned pears are 35 percent of total canned fruit for salad. [4] Dried converted to unprocessed dry weight by dividing by 1.03. Unprocessed dry weight converted to terms of fresh on the basis that 1 pound dried equals about 6.5 pounds fresh. Canned converted to terms of fresh on basis that 1 pound of canned equals about 1 pound fresh.

ERS, Specialty Crops Branch, (202) 694–5260.

Table 5-62.—Pears, fresh (dessert and cooking): Production in specified countries, 1999/2000–2001/2002 [1]

Continent and country	2000/2001	2001/2002	2002/2003
	1,000 metric tons	*1,000 metric tons*	*1,000 metric tons*
North America:			
Canada	15.1	17.0	17.5
Mexico	31.3	33.0	33.5
United States	877.4	912.5	860.0
Total	923.7	962.4	911.0
South America:			
Argentina [2]	610.1	550.0	NA
Chile [2]	249.0	232.0	NA
Total	859.1	782.0	NA
European Union:			
Belgium/Luxembourg	183.1	89.7	147.4
France	258.0	247.0	257.0
Germany	65.2	46.8	44.0
Greece	60.0	55.0	24.5
Italy	940.0	840.0	910.0
Netherlands	195.0	70.0	175.0
Spain	595.0	661.0	626.1
Sweden	13.4	12.1	15.7
United Kingdom	34.0	35.0	34.5
Total	2,343.6	2,056.6	2,234.2
Other Europe:			
Russia	190.4	226.0	226.0
Turkey	380.0	360.0	375.0
Total	570.4	586.0	601.0
Total Europe	2,914.0	2,642.6	2,835.2
Africa:			
South Africa	246.3	250.0	NA
Total	246.3	250.0	NA
Asia:			
Japan	423.8	397.0	426.3
China	8,400.0	8,820.0	8,800.0
Total	8,823.8	9,217.0	9,226.3
Oceania:			
Australia	160.0	165.0	NA
New Zealand	23.3	23.8	NA
Total	183.3	188.8	NA
Total selected countries	13,950.3	14,042.8	NA

[1] May include some cider varieties in countries not reporting separately. [2] It is too early to make reliable forecast for the Southern Hemisphere countries for the 2002/03 season. NA-not available.

FAS, Horticultural and Tropical Products Division, (202) 720–3423. Prepared or estimated on the basis of official statistics of foreign governments, other foreign source materials, reports of U.S. Agricultural Counselors, Attachés, and Foreign Service Officers, results of office research, and related information.

Table 5-63.—Pears: Production and utilization, by States, crop of 2002 (preliminary)

State and variety	Total production	Utilized production	Utilization	
			Fresh [1]	Processed [2]
	Tons	*Tons*	*Tons*	*Tons*
CA, all	294,000	294,000	(3)	(3)
Bartlett	264,000	264,000	77,000	187,000
Other	30,000	30,000	(3)	(3)
CO	2,400	2,400	(3)	(3)
CT	500	500	(3)	(3)
MI	1,400	1,400	(3)	(3)
NY	10,000	9,850	(3)	(3)
OR, all	198,000	198,000	(3)	(3)
Bartlett	58,000	58,000	23,000	35,000
Other	140,000	140,000	(3)	(3)
PA	3,800	3,500	(3)	(3)
UT	350	350	(3)	(3)
WA, all	401,000	401,000	47,000	120,000
Bartlett	167,000	167,000	(3)	(3)
Other	234,000	234,000	(3)	(3)
US	911,450	911,000	519,870	391,130

[1] Includes "Home use." [2] Mostly canned, but includes small quantities dried, juiced, and other uses. [3] Data not published to avoid disclosure of individual operations, but included in U.S. totals.

NASS, Crops Branch, (202) 720–2127.

Table 5-64.—Pears: Production and utilization, United States, 1993–2002

Crop of—	Total production	Utilized production	Utilization of quantities sold—Fresh [1]
	Tons	*Tons*	*Tons*
1993	947,850	946,500	507,400
1994	1,045,950	1,045,450	551,000
1995	947,950	947,300	544,260
1996	820,550	820,250	459,550
1997	1,042,500	1,041,930	572,310
1998	970,140	967,795	513,795
1999	1,015,450	1,013,435	536,175
2000	967,150	949,170	547,130
2001	1,001,780	964,280	543,170
2002 [2]	911,450	911,000	519,870

[1] Includes "Home use." [2] Preliminary.

NASS, Crops Branch, (202) 720–2127.

Table 5-65.—Papayas: Area, utilized production, utilization, marketing year average price, and value, Hawaii, 1993–2002

Year	Area harvested	Utilized production	Utilization		Price per pound	Value
			Fresh	Processed		
	Acres	*1,000 pounds*	*1,000 pounds*	*1,000 pounds*	*Cents*	*dollars*
1993	2,555	63,700	58,200	5,500	21.5	13,673
1994	2,200	62,000	56,200	5,800	22.3	13,831
1995	2,435	50,800	41,900	8,900	36.4	18,494
1996	1,835	41,800	37,800	4,000	40.8	17,054
1997	1,985	38,800	35,700	3,100	48.9	18,978
1998	2,120	39,900	35,600	4,300	31.6	12,589
1999	1,940	42,400	39,400	3,000	37.6	15,929
2000	1,650	54,500	50,250	4,250	29.4	16,007
2001	1,950	55,000	52,000	3,000	26.5	14,598
2002 [1]	1,700	45,500	42,500	3,000	25.9	11,778

[1] Preliminary.

NASS, Crops Branch, (202) 720–2127.

Table 5-66.—Plums, California: Production, value, and utilization, 1993–2002

Season	Total production	Utilized production	Marketing year average price per ton [1]	Value
	Tons	*Tons*	*Dollars*	*1,000 dollars*
1993	185,000	185,000	508.00	93,954
1994	247,000	247,000	321.00	79,358
1995	124,000	124,000	950.00	117,849
1996	228,000	228,000	420.00	95,831
1997	246,000	246,000	312.00	76,825
1998	188,000	188,000	529.00	99,388
1999	196,000	196,000	419.00	82,041
2000	197,000	197,000	442.00	87,115
2001	210,000	210,000	306.00	64,362
2002 [2]	200,000	200,000	386.00	77,200

[1] Fresh fruit prices are equivalent returns at point of first sale. Processing fruit prices are equivalent returns at processing plant door. [2] Preliminary.

NASS, Crops Branch, (202) 720–2127.

Table 5-67.—Prunes (dried basis): Production, price and value, California, 1993–2002 [1]

Season	Total production	Utilized production	Marketing year average price per ton [2]	Value
	Tons	*Tons*	*Dollars*	*1,000 dollars*
1993	121,000	121,000	1,120.00	135,520
1994	193,000	193,000	1,090.00	210,370
1995	181,000	181,000	1,040.00	188,240
1996	223,000	223,000	839.00	187,097
1997	214,000	205,000	883.00	181,015
1998	108,000	103,000	764.00	78,692
1999	178,000	165,000	861.00	142,065
2000	219,000	201,000	770.00	154,770
2001	150,000	135,000	723.00	97,605
2002 [3]	158,000	149,000	810.00	120,690

[1] The drying ratio is approximately 3 pounds of fresh fruit to 1 pound of dried fruit. [2] Equivalent returns at the processing plant door. [3] Preliminary.

NASS, Crops Branch, (202) 720–2127.

Table 5-68.—Prunes and plums: [1] Production, value, and utilization, 4-States, 1993–2002

Year	Total production	Utilized production	Marketing year average price per ton	Value	Utilization of quantities sold			
					Fresh [2]	Processed (fresh basis)		
						Dried and other	Canned	Frozen
	Tons	*Tons*	*Dollars*	*1,000 dollars*	*Tons*	*Tons*	*Tons*	*Tons*
1993	28,200	25,930	183.00	4,746	13,790	2,890	8,700	550
1994	38,100	32,100	168.00	5,401	13,550	6,800	10,650	1,100
1995	22,500	21,480	313.00	6,718	12,200	2,730	5,650	900
1996	19,500	18,700	442.00	8,272	10,650	1,900	5,700	450
1997	25,500	23,700	273.00	6,481	10,500	2,800	8,700	1,700
1998	25,600	24,800	311.00	7,707	11,750	4,150	7,250	1,650
1999	22,900	21,620	208.00	4,500	11,150	4,120	5,400	950
2000	23,900	21,950	239.00	5,247	9,400	5,650	5,400	1,500
2001	21,200	20,000	273.00	5,459	11,000	3,250	4,470	1,280
2002 [3]	15,850	14,990	285.00	4,271	6,360	3,930	3,540	1,160

[1] ID, MI, OR, and WA. Mostly prunes; however, estimates include small quantities of plums in all States. [2] Includes "Home use." [3] Preliminary.

NASS, Crops Branch, (202) 720–2127.

Table 5-69.—Prunes and plums (fresh basis): Production and season average price per ton, by States, 2000–2002

State	Total production			Utilized production			Price per ton [2]		
	2000	2001	2002 [1]	2000	2001	2002 [1]	2000	2001	2002 [1]
	Tons	*Tons*	*Tons*	*Tons*	*Tons*	*Tons*	*Dollars*	*Dollars*	*Dollars*
ID	3,500	3,000	2,000	3,350	3,000	1,950	397.00	582.00	536.00
MI	3,600	3,600	250	3,300	3,600	240	261.00	358.00	358.00
OR	10,000	9,000	8,000	8,500	7,800	7,200	192.00	166.00	251.00
WA	6,800	5,600	5,600	6,800	5,600	5,600	209.00	201.00	238.00
Total, 4 States	23,900	21,200	15,850	21,950	20,000	14,990	239.00	273.00	285.00

[1] Preliminary. [2] Fresh fruit prices are equivalent packinghouse-door returns for OR and WA, and the average price as sold for other States. Quantities processed are priced at the equivalent processing plant door level.

NASS, Crops Branch, (202) 720–2127.

Table 5-70.—Prunes and plums: Utilization and marketing year average price per ton, by State, 1996–2002

State and season	Quantity				Price [3]			
	Fresh [1]	Dried and other [2]	Canned	Frozen	Fresh	Dried and other	Canned	Frozen
	Tons	*Tons*	*Tons*	*Tons*	*Dollars*	*Dollars*	*Dollars*	*Dollars*
MI:								
1996	1,250	(4)	900	(4)	360.00	(4)	300.00	(4)
1997	1,500	(4)	2,100	(4)	460.00	(4)	267.00	(4)
1998	1,200	(4)	(4)	(4)	390.00	(4)	(4)	(4)
1999	1,100	(4)	(4)	(4)	440.00	(4)	(4)	(4)
2000	1,250	(4)	(4)	(4)	270.00	(4)	(4)	(4)
2001	1,800	(4)	(4)	(4)	442.00	(4)	(4)	(4)
2002	60	(4)	(4)	(4)	600.00	(4)	(4)	(4)
WA:								
1996	(4)	(4)	(4)	(4)	(4)	(4)	(4)	(4)
1997	(4)	(4)	(4)	(4)	(4)	(4)	(4)	(4)
1998	(4)	(4)	(4)	(4)	(4)	(4)	(4)	(4)
1999	(4)	(4)	(4)	(4)	(4)	(4)	(4)	(4)
2000	(4)	(4)	(4)	(4)	(4)	(4)	(4)	(4)
2001	(4)	(4)	(4)	(4)	(4)	(4)	(4)	(4)
2002	(4)	(4)	(4)	(4)	(4)	(4)	(4)	(4)
Total 4 States: [5]								
1996	10,650	1,900	5,700	450	575.00	175.00	294.00	311.00
1997	10,500	2,800	8,700	1,700	448.00	76.00	149.00	159.00
1998	11,750	4,150	7,250	1,650	476.00	160.00	148.00	230.00
1999	11,150	4,120	5,400	950	232.00	218.00	135.00	297.00
2000	9,400	5,650	5,400	1,500	321.00	145.00	182.00	287.00
2001	11,000	3,250	4,470	1,280	345.00	157.00	190.00	241.00
2002	6,360	5,930	3,540	1,160	396.00	245.00	175.00	147.00

[1] Includes "Home use." [2] Some quantities otherwise processed are included to avoid disclosure of individual operations. [3] Prices for fresh sales are average prices as sold for ID and MI; equivalent packinghouse door returns for OR and WA. Quantities processed are priced at the equivalent processing plant door level. [4] Not published to avoid disclosure of individual operations, but is included in total. [5] ID, MI, OR, and WA.

NASS, Crops Branch, (202) 720–2127.

Table 5-71.—Prunes, dried: United States exports by country of destination, 1999/2000–2001/2002

Country of destination	Year beginning September		
	1999/2000	2000/2001	2001/2002
	Metric tons	*Metric tons*	*Metric tons*
North America:			
Canada	4,753.7	4,508.4	4,861.9
Mexico	998.8	1,788.2	1,141.0
Total	5,752.5	6,296.7	6,002.9
Europe:			
Germany	9,483.9	19,535.3	12,336.7
United Kingdom	5,545.0	6,733.2	5,928.6
Belgium-Luxembourg	1,477.7	6,651.0	3,552.0
Italy	5,774.7	6,040.6	5,869.9
Netherlands	1,713.2	1,781.7	2,755.8
Spain	938.4	1,645.5	894.7
Denmark	1,626.0	1,658.7	1,411.2
Finland	1,154.3	1,648.6	1,391.0
Sweden	1,130.7	1,303.7	1,308.4
Greece	56.9	197.3	403.2
France	178.8	181.5	66.8
Austria	54.2	151.2	0.0
Ireland	110.6	82.2	81.6
Total	29,244.4	47,610.5	36,000.1
Other Western Europe:			
Norway	952.9	1,145.8	1,138.5
Switzerland	61.7	31.1	10.3
Iceland	40.0	29.0	59.5
Cyprus	10.1	16.8	3.8
Malta & Gozo	0.0	0.0	10.0
Total	1,064.8	1,222.6	1,222.2
Middle East:			
Israel	874.8	1,015.1	1,686.7
United Arab Emirate	17.9	54.4	39.0
Kuwait	0.0	36.5	16.8
Saudi Arabia	25.9	31.3	52.4
Jordan	0.0	16.4	0.0
Lebanon	4.4	8.9	52.7
Turkey	36.3	0.0	16.3
Total	959.3	1,162.6	1,863.8
Asia:			
Japan	15,768.4	14,469.5	14,328.1
Hong Kong	2,231.5	2,983.0	4,014.7
Singapore	1,373.0	1,134.4	1,120.9
Taiwan	1,040.2	865.8	1,483.1
Malaysia	895.5	703.6	761.1
China, Peoples Republic	69.7	518.4	178.8
Philippines	340.9	384.7	160.9
Korea, Republic of	28.0	92.5	47.7
Thailand	67.8	88.2	89.7
Indonesia	31.8	42.4	42.4
Brunei	11.6	2.0	4.3
Vietnam	276.1	0.0	2.9
Total	22,134.5	21,284.5	22,234.7
Oceana:			
Australia	1,320.1	1,074.8	1,570.6
New Zealand	979.9	869.9	842.5
French Pacific Islands	6.1	0.6	0.0
Total	2,306.1	1,945.3	2,413.1
Other countries	3,100.0	2,768.0	2,731.3
Grand total	64,561.6	82,290.2	72,468.1

FAS, Horticultural and Tropical Products Division, (202) 720–3423. Compiled from reports of the U.S. Department of Commerce.

Table 5-72.—Prunes: Foreign trade, United States, 1992–2001

Year beginning October	Imports				Domestic exports			
	Fresh prunes and plums	Other-wise pre-pared or pre-served	Dried prunes [1]	Total, in terms of fresh [2]	Fresh prunes and plums	Dried prunes [1]	Dried, in fruit salad [1 3]	Total, in terms of fresh [2]
	Metric tons	*Metric tons*	*Metric tons*	*Metric tons*	*Metric tons*	*Metric tons*	*Metric tons*	*Metric tons*
1992	23,639	837	2,052	29,719	56,952	84,755	1,492	280,862
1993	21,741	921	4,588	34,480	69,913	57,936	1,702	224,742
1994	23,294	881	375	25,060	40,429	60,247	1,983	201,988
1995	20,218	826	250	21,610	67,193	62,548	1,811	234,279
1996	22,923	746	450	24,762	71,906	67,535	1,382	250,825
1997	19,509	811	213	20,791	48,855	70,831	1,491	236,614
1998	27,735	735	713	30,247	60,078	68,450	1,185	240,861
1999	22,893	778	510	24,916	61,354	66,304	613	235,081
2000	33,401	792	431	35,232	62,926	83,746	1,089	283,171
2001	32,459	811	969	35,704	62,788	69,653	706	245,451

[1] Net processed weight. [2] Exports and imports of dried prunes converted to unprocessed dry weight by dividing by 1.04. Unprocessed dry weight converted to terms of fresh fruit on the basis that 1 pound dried equals 2.7 pounds fresh. "Other-wise prepared or preserved" converted to terms of fresh fruit on the basis that 1 pound equals 0.899 pound fresh.
[3] Dried prunes in salad estimated at 43 percent of total dried fruit for salad.

ERS, Specialty Crops Branch, (202) 694–5260.

Table 5-73.—Strawberries, commercial crop: Production and value per hundredweight, by States, 2000–2002

Utilization, season, and State	Production			Value per unit		
	2000	2001	2002 [1]	2000	2001	2002 [1]
FOR FRESH MARKET [2]	*1,000 cwt.*	*1,000 cwt.*	*1,000 cwt.*	*Dollars per cwt.*	*Dollars per cwt.*	*Dollars per cwt.*
Winter:						
FL	2,205	1,690	1,760	76.00	99.00	87.20
Total	2,205	1,690	1,760	76.00	99.00	87.20
Spring:						
CA	11,518	10,338	12,642	61.40	70.60	67.40
MI	66	44	36	81.00	100.00	98.00
NJ	16	18	....	94.00	72.20	....
NY	65	60	63	105.00	118.00	140.00
NC	231	196	225	75.00	85.00	85.00
OH	44	46	39	108.00	105.00	124.00
OR	18	22	23	97.00	102.00	106.00
PA	65	86	73	108.00	110.00	137.00
VA	54	41	....	105.00	124.00	....
WA	20	24	18	99.00	108.00	107.00
WI	44	46	43	104.00	110.00	116.00
Total	12,141	10,921	13,162	62.90	72.20	69.00
US	14,346	12,611	14,922	64.90	75.80	71.10
PROCESSING						
Spring:						
CA	4,214	3,390	4,315	22.10	28.50	32.10
MI	17	6	6	47.00	47.00	46.00
OR	335	380	315	47.00	34.00	45.00
WA	109	136	144	44.00	32.00	44.00
US	4,675	3,912	4,780	24.50	29.20	33.30

[1] Preliminary. [2] Mostly for fresh market, but includes some quantities used for processing in States for which processing estimates are not prepared.

NASS, Crops Branch, (202) 720–2127.

Table 5-74.—Strawberries, commercial crop: Area, yield, production, value per hundred weight, and total value, United States, 1993–2002

Year	Fresh market and processing					Fresh market[3]			Processing		
	Area for harvest	Yield per acre	Production[1]	Value[2] Per cwt	Value[2] Total	Production	Value[2] Per cwt	Value[2] Total	Production	Value[2] Per cwt	Value[2] Total
	Acres	*Cwt.*	*1,000 cwt.*	*Dollars per cwt.*	*1,000 dollars*	*1,000 cwt.*	*Dollars per cwt.*	*1,000 dollars*	*1,000 cwt.*	*Dollars per cwt.*	*1,000 dollars*
1993	51,230	282	14,465	46.30	669,937	10,108	54.00	546,033	4,357	28.40	123,904
1994	48,830	338	16,486	50.70	836,142	11,477	60.20	690,414	5,009	29.10	145,728
1995	48,080	333	16,020	50.70	811,634	11,456	60.40	692,249	4,564	26.20	119,385
1996	47,670	341	16,259	47.30	768,943	12,126	56.50	684,661	4,133	20.40	84,282
1997	44,260	368	16,278	55.50	903,350	12,018	65.60	787,974	4,260	27.10	115,376
1998	45,230	363	16,397	61.10	1,001,854	11,338	74.10	840,403	5,059	31.90	161,451
1999	46,760	392	18,326	61.80	1,133,412	13,064	73.50	959,650	5,262	33.00	173,762
2000	47,650	399	19,021	55.00	1,045,998	14,326	64.90	931,529	4,675	24.50	114,469
2001	46,000	359	16,523	64.80	1,070,052	12,611	75.80	955,883	3,912	29.20	114,169
2002[4]	47,600	414	19,702	61.90	1,220,504	14,922	71.10	1,061,205	4,780	33.30	159,299

[1] Excludes the following quantities not harvested and not marked because of economic conditions (1,000 hundredweight): 13 in 1990, 7 in 1991, 8 in 1992, 6 in 1993, and 50 in 1995. [2] Fresh market price and value on f.o.b. basis. Processing price and value at processing plant door. [3] Mostly for fresh market, but includes some quantities used for processing in States for which processing estimates are not prepared. [4] Preliminary.

NASS, Crops Branch, (202) 720–2127.

Table 5-75.—Strawberries, commercial crop: Area harvested, production, value per hundred weight, by States, 2000–2002[1]

Season and State	Area harvested			Production			Value per unit		
	2000	2001	2002[2]	2000	2001	2002[2]	2000	2001	2002[2]
	Acres	*Acres*	*Acres*	*1,000 cwt.*	*1,000 cwt.*	*1,000 cwt.*	*Dollars per cwt.*	*Dollars per cwt.*	*Dollars per cwt.*
Winter:									
FL	6,300	6,500	6,900	2,205	1,690	1,760	76.00	99.00	87.20
Total	6,300	6,500	6,900	2,205	1,690	1,760	76.00	99.00	87.20
Spring:									
CA	27,600	26,400	28,500	15,732	13,728	16,957	50.90	60.20	58.40
MI	1,200	900	900	83	50	42	74.00	93.60	90.60
NJ	450	400	....	16	18	....	94.00	72.20	....
NY	1,600	1,600	1,400	65	60	63	105.00	118.00	140.00
NC	1,700	1,700	1,800	231	196	225	75.00	85.00	85.00
OH	1,000	1,000	1,100	44	46	39	108.00	105.00	124.00
OR	3,500	3,100	3,000	353	402	338	49.50	37.70	49.20
PA	1,300	1,300	1,300	65	86	73	108.00	110.00	137.00
VA	500	500	....	54	41	....	105.00	124.00	....
WA	1,500	1,600	1,800	129	160	162	52.50	43.40	51.00
WI	1,000	1,000	900	44	46	43	104.00	110.00	116.00
Total	41,350	39,500	40,700	16,816	14,833	17,942	52.20	60.90	59.50
US	47,650	46,000	47,600	19,021	16,523	19,702	55.00	64.80	61.90

[1] Includes quantities used for fresh market and processing. [2] Preliminary.

NASS, Crops Branch, (202) 720–2127.

Table 5-76.—Fruits, noncitrus: Production, utilization, and value, United States, 1993–2003[1]

Year	Utilized production	Fresh[2]	Processed: Canned	Processed: Dried	Processed: Juice	Processed: Frozen	Processed: Wine	Processed: Other	Value of utilized production
	1,000 tons	*1,000 tons*	*1,000 tons*	*1,000 tons*	*1,000 tons*	*1,000 tons*	*1,000 tons*	*1,000 tons*	*1,000 dollars*
1993	16,554	6,391	2,042	2,339	1,749	627	3,029	181	6,130,121
1994	17,339	6,710	2,090	2,816	1,886	665	2,711	228	6,268,533
1995	16,348	6,285	1,753	2,400	1,857	647	2,992	205	6,815,962
1996	16,103	6,313	1,873	2,275	1,582	604	3,043	180	7,265,788
1997	18,363	6,643	2,130	2,660	1,675	699	4,035	247	8,158,095
1998	16,560	6,512	1,847	1,911	1,786	710	3,315	207	7,270,414
1999	17,330	6,674	1,986	2,150	1,887	719	3,351	248	8,070,535
2000	18,845	7,002	1,815	3,023	1,713	692	4,130	193	7,888,365
2001	16,731	6,480	1,860	2,289	1,462	669	3,569	164	7,901,997
2002[3]	16,791	NA	NA	NA	NA	NA	NA	NA	8,284,559

[1] Includes the following crops: Apples, apricots, avocados, bananas, berries, cherries, cranberries, dates, figs, grapes, guavas, kiwifruit, nectarines, olives, papayas, peaches, pears, pineapples, plums, prunes, and strawberries. [2] Includes "Home Use," local and roadside sales. [3] Preliminary. NA-not available.

NASS, Crops Branch, (202) 720–2127.

Table 5-77.—Fruits, fresh: Total reported domestic rail, truck, and air shipments, 2001

Commodity	Jan.	Feb.	Mar.	Apr.	May	Jun.	Jul.	Aug.	Sep.	Oct.	Nov.	Dec.	Total
	1,000 cwt.	*1,000 cwt.*	*1,000 cwt.*	*1,000 cwt.*	*1,000 cwt.*	*1,000 cwt.*	*1,000 cwt.*	*1,000 cwt.*	*1,000 cwt.*	*1,000 cwt.*	*1,000 cwt.*	*1,000 cwt.*	*1,000 cwt.*
Citrus:													
Grapefruit	2,165	2,440	2,871	1,651	599	122	8	12	125	1,337	1,765	2,064	15,159
Lemons	98	80	114	73	89	59	36	45	29	34	38	38	733
Limes	4	3	4	4	8	22	16	17	13	6	4	1	102
Oranges	916	762	839	755	754	418	103	117	136	788	1,215	1,646	8,449
Tangelos	97	18	1	....	....	....	....	....	....	2	90	220	428
Temples	123	87	40	....	....	....	....	....	....	....	....	....	250
Total	3,403	3,390	3,869	2,483	1,450	621	163	191	303	2,167	3,112	3,969	25,121
Noncitrus:													
Apples	4,118	4,423	6,107	3,988	3,519	3,306	2,254	2,430	3,895	3,731	3,849	5,191	46,811
Apricots	....	....	....	....	130	147	60	34	2	....	....	....	373
Avocados	311	342	524	476	476	634	529	580	399	290	307	145	5,013
Blueberries	....	....	....	10	50	158	303	101	25	11	4	....	662
Cherries	....	....	....	1	606	1,209	971	1,107	....	....	....	....	3,894
Cranberries	....	....	....	....	....	....	....	....	5	21	45	8	79
Fruits, other[1]	25	17	20	15	3	....	13	63	81	90	52	35	414
Grapes	61	1	....	57	179	1,356	1,741	3,051	2,750	2,292	1,586	573	13,647
Kiwifruit	37	33	22	4	1	....	....	....	2	15	27	64	205
Nectarines	....	....	....	2	501	1,315	1,706	1,217	363	1	....	....	5,105
Papaya	40	32	37	34	38	39	32	35	32	40	36	24	419
Peaches	....	....	....	24	670	1,668	1,784	1,897	959	186	5	....	7,193
Pears	910	1,061	1,453	720	576	299	315	567	873	1,285	1,419	1,344	10,822
Persimmons	1	....	....	....	....	....	....	....	....	14	20	8	43
Pineapples	70	68	98	103	125	95	100	111	75	93	65	70	1,073
Plums	....	....	....	....	143	728	1,355	839	658	41	4	....	3,768
Pomegrantes	....	....	....	....	....	....	....	3	27	65	59	19	173
Prunes	....	....	....	....	....	....	....	21	32	....	....	....	53
Strawberries	322	557	1,054	1,700	2,593	1,680	1,154	848	502	461	166	271	11,308
Total	5,895	6,534	9,315	7,134	9,610	12,634	12,317	12,904	10,680	8,636	7,644	7,752	111,055
Grand total	9,298	9,924	13,184	9,617	11,060	13,255	12,480	13,095	10,983	10,803	9,856	11,721	136,176

[1] Not otherwise specified.

AMS, Fruit and Vegetable Division, Market News Branch, (202) 720–3343.

Table 5-78.—Fruits, dried: Production (dry basis), California, 1993–2002

Year	Apples	Apricots	Dates	Figs [1]	Peaches [2]	Pears [3]	Prunes	Grapes [4]	Total
	Tons	*Tons*	*Tons*	*Tons*	*Tons*	*Tons*	*Tons*	*Tons*	*Tons*
1993	2,290	2,730	29,000	19,300	1,850	1,500	121,000	388,300	565,970
1994	NA	4,800	23,000	18,200	1,450	1,300	193,000	422,000	663,750
1995	NA	1,480	22,700	16,800	1,550	800	181,000	309,200	533,530
1996	NA	2,170	23,000	14,500	1,700	700	223,000	313,900	578,970
1997	NA	1,740	21,000	18,500	1,894	800	205,000	432,800	681,734
1998	NA	1,250	24,900	16,500	1,551	1,100	103,000	281,200	429,501
1999	NA	NA	22,200	15,100	1,849	1,010	165,000	348,300	553,459
2000	NA	1,120	17,400	17,300	1,350	600	201,000	493,700	732,470
2001	NA	820	19,700	12,700	1,450	500	135,000	417,100	587,270
2002	NA	1,120	19,500	15,500	1,525	310	149,000	401,800	588,755

[1] Standard and substandard. [2] Freestone only. [3] Bartlett only. [4] Raisin and table type. NA-not available.

ERS, Specialty Crops Branch, (202) 694–5260.

Table 5-79.—Raisins: Commercial production in specified countries, 1999/2000–2001/2002

Commodity/country	1999/2000	2000/2001	2001/2002 [1]
	Metric tons	*Metric tons*	*Metric tons*
Australia	26,667	13,676	31,000
Chile	45,000	41,500	42,500
Greece	22,500	28,000	28,500
Mexico	12,000	13,000	13,500
South Africa	38,142	34,843	42,355
Turkey	195,000	285,000	220,000
United States	310,653	439,531	353,802
Total	649,962	855,550	731,657

[1] Preliminary.

FAS, Horticultural and Tropical Products Division, (202) 720–3423. Prepared or estimated on the basis of official statistics of foreign governments, other foreign source materials, reports of U.S. Agricultural Counselors, Attaches, and Foreign Service Officers, results of office research, and related information.

Table 5-80.—Fruits, frozen: Commercial pack, United States, 1992–2001

Commodity	1992	1993	1994	1995	1996	1997	1998	1999	2000	2001
	1,000 pounds	*1,000 pounds*	*1,000 pounds*	*1,000 pounds*	*1,000 pounds*	*1,000 pounds*	*1,000 pounds*	*1,000 pounds*	*1,000 pounds*	*1,000 pounds*
Apples	115,525	93,016	91,709	113,278	114,332	119,180	124,866	111,944	141,820	146,145
Apricots	21,237	17,901	18,192	10,046	17,759	24,267	20,929	18,492	22,786	30,638
Cherries, sweet	20,095	16,037	18,480	27,032	14,945	24,515	21,628	13,640	15,901	13,101
Peaches	99,010	94,381	102,073	89,743	109,598	124,220	110,491	123,942	148,083	131,694
Plums and prunes	1,229	870	1,181	1,042	1,736	789	1,518	986	1,331	1,380
Purees, noncitrus	52,315	63,928	63,122	71,862	78,457	85,333	100,239	85,535	74,663	58,924
Berries:										
Blackberries	31,498	23,118	24,226	26,823	20,404	26,272	24,734	23,895	26,857	22,884
Blueberries	123,738	139,059	110,017	118,064	90,085	122,767	90,850	96,567	102,185	98,369
Boysenberries	6,917	4,437	5,105	3,459	5,288	4,983	3,338	4,703	3,597	3,537
Loganberries	1,634	89	59	(1)	(1)	(1)	(1)	(1)	(1)	(1)
Raspberries	32,145	25,479	29,900	40,109	25,118	27,504	23,851	23,324	23,902	21,736
Strawberries	268,494	365,663	369,035	371,138	330,139	328,150	373,824	419,768	439,749	422,371
Miscellaneous fruits and berries	154,957	136,813	173,099	140,140	136,526	110,644	107,716	101,907	135,066	54,799
Total	928,794	980,791	1,006,198	1,012,736	944,387	998,624	1,003,984	1,024,703	1,135,940	1,005,578

[1] Included in miscellaneous.

ERS, Specialty Crops Branch, (202) 694-5260. Data from American Frozen Food Institute.

Table 5-81.—Fruits: Per capita consumption, United States, 1992–2001 [1]

Year	Fruits used fresh: Citrus fruit [2] Per capita	Fruits used fresh: Noncitrus fruits [3] Per capita	Canned fruits [4] Per capita
	Pounds	*Pounds*	*Pounds*
1992	24.2	73.4	19.7
1993	25.8	73.3	17.8
1994	24.7	74.8	18.1
1995	23.8	72.6	14.7
1996	24.6	72.7	16.1
1997	26.5	74.7	17.6
1998	26.6	75.0	15.1
1999	20.3	79.9	16.7
2000	23.4	77.1	15.3
2001 [8]	24.3	73.3	15.6

Year	Juice [5] Per capita	Frozen fruit [6] Per capita	Dried fruits [7] Per capita
	Gallons	*Pounds*	*Pounds*
1992	8.6	3.5	2.7
1993	8.5	3.4	3.0
1994	8.8	3.4	2.9
1995	8.8	3.8	2.7
1996	8.8	3.6	2.8
1997	8.8	3.3	2.7
1998	8.6	3.7	2.8
1999	9.5	3.4	2.5
2000	8.6	3.9	2.5
2001 [8]	8.4	4.0	2.4

[1] Fresh citrus fruits, canned fruit, and fruit juices are on a crop-year basis. Dried fruits are on a pack-year basis. The per capita consumption was obtained by dividing the total consumption by total population. [2] Oranges and temples, tangerines and tangelos, lemons, limes, and grapefruit. [3] Apples, apricots, avocados, bananas, cherries, cranberries, grapes, kiwifruit, mangoes, peaches and nectarines, pears, pineapples, papayas, plums and prunes, and strawberries. [4] Apples, apricots, cherries, olives, peaches, pears, pineapples, and plums and prunes. [5] Orange, grapefruit, lemon, lime, apple, grape, pineapple, prune, and cranberry. [6] Blackberries, blueberries, raspberries, strawberries, other berries, apples, apricots, cherries, and peaches. [7] Apples, apricots, dates, figs, peaches, pears, prunes, and raisins. Dried data in terms of processed weight. [8] Preliminary.

ERS, Specialty Crops Branch, (202) 694–5260.

Table 5-82.—All tree nuts: Supply and utilization, United States, 1992/93–2001/02

Market year[1]	Beginning stocks	Marketable production [2]	Imports	Total supply	Exports	Ending stocks	Domestic consumption: Total	Domestic consumption: Per capita Pounds
	—Million pounds (shelled)—							
1992/93	262.5	860.3	228.1	1,350.8	544.0	237.0	569.9	2.21
1993/94	237.0	947.1	214.5	1,398.7	508.4	279.7	610.6	2.33
1994/95	279.7	1,061.5	218.9	1,560.2	629.2	334.1	596.9	2.25
1995/96	334.1	770.1	204.0	1,308.2	545.7	251.9	510.6	1.91
1996/97	251.9	816.9	226.0	1,294.8	607.2	156.4	523.5	1.93
1997/98	156.4	1,214.4	234.3	1,605.2	682.1	348.7	572.2	2.09
1998/99	348.7	851.8	242.0	1,439.4	612.8	223.9	602.8	2.17
1999/2000	223.9	1,295.7	283.4	1,803.0	730.7	365.6	706.7	2.52
2000/2001	365.6	1,110.1	280.4	1,756.0	851.1	271.8	633.0	2.23
2001/2002 [3]	271.8	1,329.9	309.8	1,910.6	787.8	285.2	837.6	2.91

[1] Marketing season begins July 1 for almonds, hazelnuts, macadamias, pecans, and other nuts; August 1 for walnuts; and September 1 for pistachios. [2] Utilized production (NASS data) minus inedibles and noncommercial useage. [3] Preliminary.

ERS, Specialty Crops Branch, (202) 694–5260.

Table 5-83.—Tree nuts: Commercial production in specified countries, 1999/2000–2001/2002

Commodity and country	1999/2000	2000/2001	2001/2002 [1]
	1,000 metric tons	*1,000 metric tons*	*1,000 metric tons*
Almonds (shelled basis):			
Greece	17.0	15.5	13.0
Italy	17.0	10.0	18.0
Morocco [2]	NA	NA	NA
Spain	66.0	53.0	57.0
Turkey	14.0	15.5	14.0
United States	377.8	323.4	376.0
Total	491.8	417.4	478.0
Hazelnuts (in-shell basis):			
Italy	110.0	83.0	130.0
Spain	25.0	16.0	26.0
Turkey	610.0	490.0	725
United States	36.3	20.4	44.8
Total	781.3	609.4	925.8
Walnuts:			
Chile	10.0	11.8	12.4
China	274.0	310.0	279
France	29.0	27.8	27.8
India	28.0	31.0	29
Italy	18.0	16.0	13
Turkey	70.0	69.0	68
United States	256.7	216.8	276.7
Total	685.7	682.4	705.9

[1] Preliminary. [2] Due to changes in FAS' reporting schedule, effective 8/99, the FAS office in Rabat will no longer be submitting PSD forecasts.

FAS, Horticultural and Tropical Products Division, (202) 720–3423. Prepared or estimated on the basis of official statistics of foreign governments, other foreign source materials, reports of U.S. Agricultural Counselors, Attachés and Foreign Service Officers, results of office research, and related information.

Table 5-84.—Almonds (shelled basis): Bearing acreage, yield, production, price, and value, California, 1993–2002 [1]

Year	Bearing Acreage	Yield per acre	Production	Price per pound	Value
	Acres	*Pounds*	*1,000 pounds*	*Dollars*	*1,000 dollars*
1993	413,000	1,190	490,000	1.94	930,618
1994	433,000	1,700	735,000	1.34	965,202
1995	418,000	885	370,000	2.48	880,896
1996	428,000	1,190	510,000	2.08	1,018,368
1997	442,000	1,720	759,000	1.56	1,160,640
1998	460,000	1,130	520,000	1.41	703,590
1999	480,000	1,740	833,000	0.86	687,742
2000	500,000	1,410	703,000	0.97	666,487
2001	525,000	1,580	830,000	0.91	740,012
2002	530,000	2,000	1,060,000	1.01	1,049,188

[1] Price and value are based on edible portion of the crop only. Included in production are inedible quantities of no value as follows (million pounds): 1993-10.3; 1994-14.7; 1995-14.8; 1996-20.4; 1997-15.0; 1998-21.0; 1999-33.2; 2000-15.9; 2001-16.8; 2001-21.2.

NASS, Crops Branch, (202) 720–2127.

Table 5-85.—Almonds (shelled basis [1]): Foreign trade, United States, 1992–2001

Year beginning October	Imports	Domestic exports
	Metric tons	*Metric tons*
1992	27	114,530
1993	47	125,014
1994	66	213,657
1995	68	279,338
1996	25	162,629
1997	28	182,157
1998	29	184,500
1999	39	197,271
2000	173	225,550
2001	319	261,319

[1] Imports of unshelled nuts converted to shelled basis at ratio of 1.67 to 1. Exports of unshelled nuts converted to shelled basis at ratio of 1.67 to 1.0.

ERS, Market and Trade Economics Division, Specialty Crops Branch, (202) 694–5260.

Table 5-86.—Hazelnuts (in-shell basis): Bearing acreage, yield, production, price, and value, Oregon, Washington, and United States, 1993–2002 [1]

Year	Bearing Acreage	Yield per acre	Production	Price per ton	Value
	Acres	Tons	Tons	Dollars	*1,000 dollars*
			Oregon		
1993	26,700	1.52	40,700	633	25,763
1994	27,200	0.77	21,000	834	17,514
1995	27,600	1.40	38,700	913	35,333
1996	28,200	0.66	18,750	859	16,106
1997	28,600	1.63	46,650	899	41,938
1998	29,100	0.53	15,400	964	14,846
1999	28,800	1.38	39,700	890	35,333
2000	28,300	0.79	22,300	890	19,847
2001	28,100	1.76	49,500	701	34,700
2002	27,800	0.65	18,000	1,000	18,009
			Washington		
1993	330	0.91	300	683	205
1994	350	0.57	200	900	180
1995	380	0.79	300	935	281
1996	400	0.63	250	940	235
1997	400	0.88	350	940	329
1998	430	0.23	100	960	96
1999	400	0.75	300	900	270
2000	350	0.57	200	960	192
2001 [2]	....	....	....	....	....
2002	....	....	....	....	....
			United States		
1993	27,030	1.52	41,000	633	25,968
1994	27,550	0.77	21,200	835	17,694
1995	27,980	1.39	39,000	913	35,614
1996	28,600	0.66	19,000	860	16,341
1997	29,000	1.62	47,000	899	42,267
1998	29,530	0.52	15,500	964	14,942
1999	29,200	1.37	40,000	890	35,603
2000	28,650	0.79	22,500	891	20,039
2001	28,100	1.76	49,500	701	34,700
2002	27,800	0.65	18,000	1,000	18,009

[1] Preliminary. [2] WA discontinued.
NASS, Crops Branch, (202) 720–2127.

Table 5-87.—Hazelnuts (shelled basis [1]): Foreign trade, United States, 1992–2001

Year beginning October	Imports	Domestic exports
	Metric tons	*Metric tons*
1992	3,962	2,676
1993	2,937	3,738
1994	6,927	4,117
1995	3,525	6,849
1996	3,718	5,825
1997	4,265	10,119
1998	5,484	3,999
1999	5,425	6,563
2000	5,129	5,705
2001	6,736	11,110

[1] Imports of unshelled nuts converted to shelled basis at ratio of 2.22 to 1. Exports of unshelled nuts converted to shelled basis at ratio of 2.50 to 1.
ERS, Specialty Crops Branch, (202) 694–5260.

Table 5-88.—Macadamia nuts (in-shell basis): Bearing acreage, yield, production, price, and value, Hawaii, 1993–2002

Year	Bearing Acreage	Yield per acre	Production	Price per pound	Value
	Acres	Pounds	*1,000 pounds*	*Cents*	*1,000 dollars*
1993	18,500	2,620	48,500	68.0	32,980
1994	18,500	2,840	52,500	69.0	36,225
1995	19,300	2,640	51,000	74.0	37,740
1996	19,200	2,940	56,500	78.0	44,070
1997	19,200	3,020	58,000	75.0	43,500
1998	19,200	2,990	57,500	65.0	37,375
1999	18,900	2,990	56,500	67.0	37,855
2000	17,700	2,820	50,000	59.0	29,500
2001	17,800	3,150	56,000	59.0	33,040
2002 [1]	17,800	2,920	52,000	55.0	28,600

[1] Preliminary.
NASS, Crops Branch, (202) 720–2127.

Table 5-89.—Pecans (in-shell basis): Production, price per pound, and value, United States, 1993–2002

Year	Improved varieties			Native and seedling			All pecans		
	Production	Price	Value	Production	Price	Value	Production	Price	Value
	1,000 pounds	*Cents*	*1,000 dollars*	*1,000 pounds*	*Cents*	*1,000 dollars*	*1,000 pounds*	*Cents*	*1,000 dollars*
1993	237,100	62.9	149,189	109,200	39.6	43,270	365,000	58.6	213,862
1994	118,900	115.0	136,945	59,600	76.4	45,531	199,000	104.0	207,345
1995	174,800	112.0	195,216	76,800	72.5	55,678	267,500	101.0	271,377
1996 [1]	165,125	68.9	113,749	44,375	46.4	20,606	209,500	64.1	134,355
1997	202,900	93.3	189,226	132,100	53.0	69,994	335,000	77.4	259,220
1998	112,000	135.0	150,908	34,400	77.2	26,544	146,400	121.0	177,452
1999	219,400	101.0	222,647	186,700	57.7	107,751	406,100	81.4	330,398
2000	160,550	126.0	201,575	49,300	75.4	37,193	209,850	114.0	238,768
2001	246,550	66.2	163,204	91,950	41.2	37,897	338,500	59.4	201,101
2002 [2]	139,700	106.0	147,851	38,700	55.4	21,448	178,400	94.9	169,299

[1] MO and TN discontinued. [2] Preliminary.

NASS, Crops Branch, (202) 720–2127.

Table 5-90.—Pecans (in-shell basis): Production and marketing year average price per pound, by States, 2000–2002

Item and State	Production			Price for crop of—		
	2000	2001	2002 [1]	2000	2001	2002 [1]
IMPROVED VARIETIES [2]	*1,000 pounds*	*1,000 pounds*	*1,000 pounds*	*Dollars*	*Dollars*	*Dollars*
AL	10,000	10,000	6,500	0.900	0.580	0.580
AZ	14,500	21,000	16,000	1.310	0.460	1.220
AR	650	1,950	1,200	1.000	0.600	0.600
CA	3,400	3,700	3,600	1.520	0.810	1.200
FL	1,200	1,200	900	1.050	0.510	0.870
GA	65,000	85,000	42,000	1.230	0.660	1.020
LA	3,500	3,500	2,000	1.250	0.600	0.800
MS	2,500	3,000	2,200	0.970	0.650	0.800
NM	35,000	60,000	36,000	1.370	0.640	1.120
NC	1,400	2,700	1,500	1.150	0.650	1.000
OK	200	2,000	2,000	1.300	0.530	0.600
SC	1,200	2,500	800	1.110	0.640	0.960
TX	22,000	50,000	25,000	1.300	0.800	1.150
US	160,550	246,550	139,700	1.260	0.662	1.060
NATIVE AND SEEDLING						
AL	5,000	5,000	500	0.600	0.360	0.540
AR	250	650	500	0.750	0.400	0.500
FL	2,100	2,100	600	0.600	0.420	0.500
GA	15,000	25,000	3,000	0.980	0.450	0.671
KS	550	2,200	1,700	0.830	0.400	0.750
LA	14,500	10,500	4,000	0.600	0.350	0.500
MS	1,000	1,500	800	0.520	0.450	0.500
NC	200	500	400	0.800	0.450	0.650
OK	2,300	18,000	12,000	0.800	0.420	0.500
SC	400	1,500	200	0.920	0.460	0.650
TX	8,000	25,000	15,000	0.750	0.400	0.570
US	49,300	91,950	38,700	0.754	0.412	0.554
ALL PECANS						
AL	15,000	15,000	7,000	0.800	0.507	0.577
AZ	14,500	21,000	16,000	1.310	0.460	1.220
AR	900	2,600	1,700	0.931	0.550	0.571
CA	3,400	3,700	3,600	1.520	0.810	1.200
FL	3,300	3,300	1,500	0.764	0.453	0.722
GA	80,000	110,000	45,000	1.180	0.612	1.000
KS	550	2,200	1,700	0.831	0.400	0.750
LA	18,000	14,000	6,000	0.726	0.413	0.600
MS	3,500	4,500	3,000	0.841	0.583	0.720
NM	35,000	60,000	36,000	1.370	0.640	1.120
NC	1,600	3,200	1,900	1.110	0.619	0.926
OK	2,500	20,000	14,000	0.840	0.431	0.514
SC	1,600	4,000	1,000	1.060	0.573	0.898
TX	30,000	75,000	40,000	1.150	0.667	0.933
US	209,850	338,500	178,400	1.140	0.594	0.949

[1] Preliminary. [2] Budded, grafted or topworked varieties.

NASS, Crops Branch, (202) 720–2127.

Table 5-91.—Pecans (shelled basis[1]): Foreign trade, United States, 1992–2001

Year beginning October	Imports	Domestic exports
	Metric tons	*Metric tons*
1992	12,864	7,067
1993	9,192	7,616
1994	13,571	6,119
1995	12,782	7,917
1996	10,743	8,819
1997	14,576	9,797
1998	16,071	7,887
1999	12,152	9,237
2000	12,902	8,963
2001	14,323	11,115

[1]Imports of unshelled nuts converted to shelled basis at ratio of 2.50 to 1. Exports of unshelled nuts converted to shelled basis at ratio of 2.50 to 1.

ERS, Specialty Crops Branch, (202) 694–5260.

Table 5-92.—Pistachios (in-shell basis): Bearing acreage, yield, production, price, and value, California, 1993–2002

Year	Bearing Acreage	Yield per acre	Production	Price per ton	Value
	Acres	Pounds	1,000 pounds	*Dollars*	*1,000 dollars*
1993	57,000	2,670	152,000	1.07	162,640
1994	57,500	2,240	129,000	0.921	118,809
1995	60,300	2,450	148,000	1.09	161,320
1996	64,300	1,630	105,000	1.16	121,800
1997	65,400	2,750	180,000	1.13	203,400
1998	68,000	2,760	188,000	1.03	193,640
1999	71,000	1,730	123,000	1.33	163,590
2000	74,600	3,260	243,000	1.01	245,430
2001	78,000	2,060	161,000	1.01	162,610
2002	83,000	3,610	300,000	1.11	333,000

NASS, Crops Branch, (202) 720–2127.

Table 5-93.—Walnuts (English): Bearing acreage, yield, production, price, and value, California, 1993–2002

Year	Bearing Acreage	Yield per acre	Production	Price per ton	Value
	Acres	*Tons*	*Tons*	*Dollars*	*1,000 dollars*
1993	185,000	1.41	260,000	1,390	361,400
1994	189,000	1.23	232,000	1,030	238,960
1995	193,000	1.21	234,000	1,400	327,600
1996	192,000	1.08	208,000	1,580	328,640
1997	193,000	1.39	269,000	1,430	384,670
1998	193,000	1.18	227,000	1,050	238,350
1999	191,000	1.48	283,000	886	250,738
2000	193,000	1.24	239,000	1,240	296,360
2001	196,000	1.56	305,000	1,120	341,600
2002[1]	200,000	1.41	282,000	NA	NA

[1] Preliminary. NA-not available.

NASS, Crops Branch, (202) 720–2127.

Table 5-94.—Walnuts (shelled basis[1]): Foreign trade, United States, 1992–2001

Year beginning October	Imports	Domestic exports
	Metric tons	*Metric tons*
1992	3,835	30,172
1993	336	38,396
1994	319	42,080
1995	1,050	45,997
1996	2,662	46,811
1997	97	39,341
1998	111	41,000
1999	76	41,428
2000	523	41,919
2001	49	47,099

[1]Imports of unshelled nuts converted to shelled basis at ratio of 2.50 to 1. Exports of unshelled nuts converted to shelled basis at ratio of 2.50 to 1.

ERS, Specialty Crops Branch, (202) 694–5260.

Table 5-95.—Cacao beans: United States imports by country of origin, 1999–2001

Continent and country	1999	2000	2001[1]
	Metric tons	*Metric tons*	*Metric tons*
North America:			
Canada	22	56	47
Costa Rica	39	39	19
Dominican Republic	14,631	18,090	27,974
El Salvador	0	0	1
Haiti	1,544	1,560	1,731
Jamaica	0	0	13
Mexico	3,466	200	896
Nicaragua	0	0	2
Panama	116	38	57
St. Lucia	24	16	16
Trinidad and Tobago	87	64	50
Total	19,929	20,063	30,806
South America:			
Argentina	1,001	0	0
Brazil	45	340	20
Colombia	25	67	38
Ecuador	32,396	13,505	20,325
Peru	60	0	0
Venezuela	280	285	225
Total	33,807	14,197	20,608
Africa:			
Cameroon	0	0	20
Cote d'Ivoire	193,151	242,637	191,720
Ghana	23,810	50,633	35,234
Madagascar	102	63	0
Nigeria	2,449	10	1,198
South Africa, Republic of	250	0	0
Total	219,762	293,343	228,172
Asia and Oceania:			
China, Peoples Republic of	0	100	0
Fiji	1	0	0
India	750	0	500
Indonesia	162,217	130,563	140,676
New Zealand (exc. Cook; Niue; & Tokela	25	0	0
Papua New Guinea	24,641	9,235	13,326
Singapore	1,350	1,560	0
Solomon Islands	50	0	0
Western Samoa	0	1	0
Total	189,034	141,472	154,502
Europe:			
France	0	1,802	0
Moldova Republic of	0	13	0
Netherlands	4,562	19	13
Spain	3	5	4
Switzerland	151	0	0
Total	4,716	1,826	17
Grand total	467,250	470,900	434,105

[1] Preliminary.

FAS, Horticultural and Tropical Products Division, (202) 720–3423. Compiled from reports of the U.S. Department of Commerce.

Table 5-96.—Coffee: United States imports by country of origin, 1999–2001 [1]

Continent and country	1999	2000	2001 [2]
	1,000 bags	*1,000 bags*	*1,000 bags*
North America: [3]			
Mexico	3,182	3,610	1,929
Guatemala	2,148	2,377	2,037
El Salvador	550	1,209	506
Honduras	370	806	513
Nicaragua	181	485	455
Costa Rica	684	705	883
Panama	129	106	99
Other countries	43	34	30
Total	7,287	9,332	6,452
South America:			
Colombia	3,359	3,147	3,217
Venezuela	372	32	2
Ecuador	419	164	225
Peru	762	857	692
Brazil	4,658	2,570	2,809
Other countries	15	10	12
Total	9,585	6,632	6,957
Africa:			
Cote d'Ivoire	48	238	86
Burundi	51	74	32
Rwanda	15	24	60
Ethiopia	77	91	80
Uganda	197	153	137
Kenya	68	91	98
Tanzania, United Republic of	13	20	21
Madagascar	36	42	40
Other countries	46	49	79
Total	551	782	633
Asia and Oceania:			
India	164	398	179
Indonesia	724	692	887
Papua New Guinea	144	96	178
Thailand	184	448	416
Vietnam	1,251	2,489	2,903
Other countries	21	16	27
Total	2,488	4,139	4,590
Europe:.			
France	153	80	89
Germany	466	477	542
Other countries	32	30	52
Total	651	587	683
Grand total	20,561	21,622	19,316

[1] Green coffee only; does not include roasted or soluble. Bags of 60 kilograms each (132.276 lbs.). [2] Preliminary. [3] Includes Central America.

FAS, Horticultural and Tropical Products Division, (202) 720–3423. Compiled from reports of the U.S. Department of Commerce.

Table 5-97.—Coffee: Area, yield, total production, marketing year average price, and value, Hawaii, 1993–94 to 2002–03

Season	Area harvested	Yield per harvested acre	Production [1]	Price per pound	Value
	Acres	*1,000 pounds*	*1,000 pounds*	*Dollars*	*1,000 dollars*
1993–94	4,200	0.7	2,900	2.25	6,525
1994–95	4,400	1.0	4,300	2.80	12,040
1995–96	5,500	1.0	5,400	3.00	16,200
1996–97	5,400	1.2	6,400	3.25	20,800
1997–98	5,800	1.6	9,400	3.00	28,200
1998–99	6,100	1.6	9,500	2.60	24,700
1999–2000	6,400	1.6	10,000	2.10	21,000
2000–2001	6,800	1.3	8,700	2.65	23,055
2001–2002	6,300	1.3	8,000	2.45	19,600
2002–2003 [2]	6,200	1.4	8,500	2.30	19,550

[1] Parchment basis. [2] Preliminary.

NASS, Crops Branch, (202) 720–2127.

Table 5-98.—Coffee: Exports from principal producing countries, 1999–2001 [1]

Continent and country of origin	1999	2000	2001 [2]
	1,000 bags	*1,000 bags*	*1,000 bags*
North America:			
Costa Rica	2,195	1,964	2,018
Cuba	77	113	109
Dominican Republic	161	155	95
El Salvador	1,890	2,536	1,533
Guatemala	4,681	4,852	4,110
Haiti	72	72	86
Honduras	1,987	2,879	2,392
Jamaica	24	29	28
Mexico	4,358	5,304	3,408
Nicaragua	984	1,345	1,365
Panama	118	72	57
Trinidad and Tobago	3	2	0
Total	16,550	19,323	15,201
South America:			
Bolivia	124	106	70
Brazil	23,139	18,016	23,172
Colombia	9,996	9,175	9,944
Ecuador	988	696	756
Paraguay	9	10	11
Peru	2,407	2,362	2,663
Venezuela	452	25	49
Total	37,115	30,390	36,665
Africa:			
Angola	50	22	14
Burundi	403	444	301
Cameroon	1,154	1,205	1,125
Central African Republic	195	203	94
Congo, Republic of	0	0	0
Equatorial Guinea	1	0	0
Ethiopia	1,818	1,982	1,376
Gabon	3	1	2
Ghana	88	31	32
Guinea	56	36	76
Cote d'Ivoire	2,406	6,110	4,174
Kenya	1,114	1,328	1,080
Madagascar	464	216	295
Malawi	54	61	64
Nigeria	6	2	5
Rwanda	314	270	300
Sierra Leone	30	53	2
Tanzania	634	756	866
Togo	321	283	193
Uganda	3,841	2,513	3,060
Zaire	373	330	169
Zambia	51	63	112
Zimbabwe	127	106	99
Total	13,503	16,015	13,439
Asia and Oceania:			
India	3,613	4,441	3,769
Indonesia	5,065	5,194	4,992
Papua New Guinea	1,320	1,043	1,095
Philippines	9	5	3
Sri Lanka	14	1	3
Thailand	493	969	1,122
Vietnam	7,742	11,619	13,946
Total	18,256	23,272	24,930
Grand total	85,424	89,000	90,235

[1] Mostly green coffee, some countries include soluble or roasted. Bags of 60 kilograms each (132.276 lbs.). [2] Preliminary.

FAS, Horticultural and Tropical Products Division, (202) 720–3423. International Coffee Organization.

Table 5-99.—Tea: United States imports by country of origin, 1999–2001

Continent and country	1999	2000	2001 [1]
	Metric tons	*Metric tons*	*Metric tons*
North America:			
Canada	220	517	553
Mexico	70	52	8
Total	290	569	561
Central and South America: [2]			
Argentina	36,035	35,250	39,909
Brazil	1,568	1,456	1,571
Ecuador	626	512	615
Other countries	121	146	86
Total	38,350	37,364	42,181
Europe:			
Germany	2,271	2,834	3,436
Netherlands	169	121	115
Turkey	955	917	1,178
Ukraine	0	0	33
United Kingdom	898	1,154	1,317
Other countries	1,135	1,114	1,903
Total	5,428	6,140	7,982
Asia and Oceania:			
China, Peoples Republic of	15,898	14,199	15,651
Hong Kong	155	174	123
India	3,957	4,962	5,447
Indonesia	9,505	7,139	6,423
Japan	328	360	388
Papua New Guinea	892	669	1,157
Sri Lanka	4,646	4,183	3,262
Taiwan	551	509	552
Vietnam	1,036	1,764	1,571
Other countries	241	367	219
Total	37,209	34,326	34,793
Africa:			
Kenya	4,454	4,420	4,747
Malawi	5,343	4,754	5,878
Mozambique	84	0	0
Rwanda	0	26	0
South Africa, Republic of	274	423	309
Tanzania, United Republic of	1,246	165	59
Other countries	189	104	145
Total	11,590	9,892	11,138
Grand total	92,865	88,291	96,658

[1] Preliminary. [2] Includes Mexico.

FAS, Horticultural and Tropical Products Division, (202) 720–3423. Compiled from reports of the U.S. Department of Commerce.

Table 5-100.—Agaricus mushrooms: Area, volume of sales, marketing year average price, and value of sales, United States, 1992/93–2001/2002 [1]

Year	Area in production	Volume of sales	Price per pound	Value of sales		
				Total	Fresh market	Processing
	1,000 sq. ft.	*1,000 pounds*	*Cents*	*1,000 dollars*	*1,000 dollars*	*1,000 dollars*
1992–93	141,909	776,357	86.2	669,398	521,566	147,832
1993–94	135,703	750,799	91.6	687,673	532,863	154,810
1994–95	139,617	782,340	93.5	731,173	560,127	171,046
1995–96	135,320	777,870	93.5	727,578	588,126	139,452
1996–97	136,461	776,677	94.0	730,296	605,728	124,568
1997–98	145,094	808,678	95.7	773,617	670,168	103,449
1998–99	150,017	847,760	97.7	828,098	712,000	116,098
1999–2000	151,487	854,394	97.0	828,551	715,943	112,608
2000–2001	143,873	846,209	97.6	825,500	736,543	88,957
2001–2002	146,263	837,866	104.0	874,536	799,406	75,130

[1] Marketing year begins July 1 and ends June 30 the following year.

NASS, Crops Branch, (202) 720–2127.

Table 5-101.—Mushrooms, canned: United States imports by country of origin, 1997–2001

Country	1997 [1]	1998 [1]	1999 [1]	2000 [1]	2001 [1]
	Metric tons	*Metric tons*	*Metric tons*	*Metric tons*	*Metric tons*
Canada	348	1,054	2,786	1,576	1,559
Chile	2,463	2,955	0	0	0
China, Peoples Republic of	30,459	21,793	145	3,740	8,774
Columbia	450	299	927	1,770	1,601
France	23	11	2,275	3,174	2,490
Hong Kong	1,758	724	0	19	12
India	4,513	5,697	12,639	15,621	13,377
Indonesia	14,420	12,095	13,198	13,174	10,168
Malaysia	43	68	240	665	839
Mexico	2,139	3,267	2,164	2,986	2,285
Netherlands	600	2,029	8,193	12,858	8,173
Spain	172	929	1,356	1,006	1,332
Taiwan	995	1,216	2,358	4,754	2,490
Thailand	329	148	161	90	98
Other	136	151	231	645	666
Total	58,848	52,436	46,673	62,078	53,864

[1] Includes straw mushrooms.

FAS, Horticultural and Tropical Products Division, (202) 720–6086. Compiled from reports of the U.S. Department of Commerce.

Table 5-102.—Specialty mushrooms: Number of growers, total production, volume of sales, price per pound, and value of sales, July 1–June 30, 1999–2000/2001–2002 [1]

Year and variety	Growers [2]	All sales			
		Total production [3]	Volume of sales [4]	Price per pound [5]	Value of sales
	Number	*1,000 pounds*	*1,000 pounds*	*Dollars*	*1,000 dollars*
1999–2000					
Shiitake	153	8,635	8,173	3.29	26,893
Oyster	68	3,573	3,346	2.16	7,218
Other	21	1,326	1,211	3.87	4,689
US [6]	180	13,534	12,730	3.05	38,800
2000–2001					
Shiitake	120	9,391	8,939	3.17	28,314
Oyster	54	3,817	3,629	2.13	7,745
Other	19	1,397	1,316	4.69	6,178
US [6]	159	14,605	13,884	3.04	42,237
2001–2002					
Shiitake	123	8,263	7,893	2.94	23,196
Oyster	51	4,265	4,028	2.00	8,050
Other	23	1,541	1,424	4.02	5,727
US [6]	152	14,069	13,345	2.77	36,973

[1] Specialty mushroom estimates represent growers who have at least 200 natural wood logs in production or commercial indoor growing areas. [2] Growers counted only once for US total if growing more than one specialty type mushroom. Growers growing Agaricus and Specialty are included. [3] Total production includes all fresh market and processing sales plus amount harvested but not sold (shrinkage, cullage, dumped, etc.). [4] Virtually all specialty mushroom sales are for fresh market. [5] Prices for mushrooms are the average prices producers receive at the point of first sale, commonly referred to as the average price as sold. For example, if in a given State, part of the fresh mushrooms are sold F.O.B. packed by growers, part are sold bulk to brokers or repackers, and some are sold retail at roadside stands, the mushroom average price as sold is a weighted average of the average price for each method of sale. [6] 1999-00: AR, CA, CT, FL, GA, HI, ID, IL, IA, KS, KY, ME, MD, MA, MI, MN, MO, MT, NH, NY, NC, OH, OK, OR, PA, SC, VT, VA, WA, WV, and WI. 2000-01: AR, CA, CT, DE, FL, GA, HI, ID, IL, IA, KS, KY, ME, MD, MA, MI, MN, MO, MT, NH, NY, NC, OH, OK, OR, PA, SC, TX, VT, VA, WA, WV, and WI. 2001-02: AR, CA, CT, DE, FL, GA, HI, IA, KY, ME, MD, MA, MI, MN, MO, MT, NH, NY, NC, OH, OK, OR, PA, SC, VT, VA, WA, WV, and WI.

NASS, Crops Branch, (202) 720–2127.

Table 5-103.—Cut flowers: Sales and wholesale value for operations with $100,000+ sales, Surveyed States, 1992–2001

Year	Quantity sold	Wholesale price	Value of sales at wholesale [1]	Quantity sold	Wholesale price	Value of sales at wholesale [1]
	Standard carnations			Miniature carnations		
	1,000 blooms	*Cents*	*1,000 dollars*	*1,000 bunches*	*Dollars*	*1,000 dollars*
1992	213,599	14.4	30,777	9,706	1.51	14,703
1993	194,876	14.0	27,207	8,974	1.35	12,107
1994	116,232	15.6	18,141	7,174	1.40	10,065
1995	112,067	15.3	17,199	9,152	1.21	11,060
1996	92,160	14.5	13,345	5,562	1.32	7,330
1997	74,368	15.8	11,739	4,907	1.52	7,441
1998	63,171	15.7	9,891	4,211	1.43	6,036
1999	38,456	18.7	7,201	2,631	1.46	3,829
2000	40,206	16.0	6,430	(2)	(2)	(2)
2001	29,894	15.3	4,571	(2)	(2)	(2)
	Standard chrysanthemums			Pompon chrysanthemums		
	1,000 blooms	*Cents*	*1,000 dollars*	*1,000 bunches*	*Dollars*	*1,000 dollars*
1992	19,775	39.3	7,769	15,436	1.16	17,971
1993	19,730	43.3	8,541	11,025	1.42	15,695
1994	15,303	47.9	7,325	12,106	1.35	16,345
1995	14,877	53.8	7,997	13,542	1.26	17,079
1996	15,184	55.6	8,438	12,003	1.26	15,145
1997	11,643	47.7	5,550	12,562	1.30	16,341
1998	14,265	51.6	7,362	12,427	1.35	16,828
1999	12,786	55.4	7,083	16,315	0.93	15,181
2000	(2)	(2)	(2)	13,159	1.31	17,214
2001	(2)	(2)	(2)	12,655	1.31	16,578
	Hybrid Tea roses			Sweetheart roses		
	1,000 blooms	*Cents*	*1,000 dollars*	*1,000 blooms*	*Cents*	*1,000 dollars*
1992	458,974	33.9	155,738	74,588	25.2	18,775
1993	429,020	33.8	144,823	78,270	25.0	19,545
1994	404,938	33.1	134,034	64,551	23.7	15,276
1995	352,154	32.5	114,594	45,350	24.5	11,133
1996	309,663	34.2	105,823	43,251	26.6	11,523
1997	343,879	32.0	109,914	44,706	28.2	12,601
1998	268,302	33.6	90,174	48,806	25.0	12,191
1999	223,562	34.3	76,709	39,377	26.1	10,294
2000	(2)	(2)	(2)	(2)	(2)	(2)
2001	(2)	(2)	(2)	(2)	(2)	(2)
	Gladioli			Other cut flowers		
	1,000 spikes	*Cents*	*1,000 dollars*			*1,000 dollars*
1992	187,643	20.0	37,494			175,228
1993	165,683	20.3	33,675			162,318
1994	190,914	20.7	39,520			201,591
1995	149,132	24.2	36,110			208,458
1996	144,941	22.9	33,260			217,836
1997	154,650	22.5	34,861			273,122
1998	129,297	25.6	33,138			235,975
1999	100,457	25.4	25,535			285,792
2000	127,109	22.3	28,339			133,648
2001	109,456	22.1	24,183			135,143

See footnotes at end of table.

Table 5-103.—Cut flowers: Sales and wholesale value for operations with $100,000+ sales, Surveyed States, 1992–2001—Continued

Year	Quantity sold	Wholesale price	Value of sales at wholesale [1]	Quantity sold	Wholesale price	Value of sales at wholesale [1]
	Alstromeria			Delphinium & Larkspur		
	1,000 blooms	*Cents*	*1,000 dollars*	*1,000 bunches*	*Dollars*	*1,000 dollars*
1992						
1993						
1994						
1995						
1996						
1997						
1998						
1999						
2000	25,056	23.7	5,939	43,840	25.0	10,955
2001	21,085	24.7	5,206	49,468	21.8	10,807
	Gerbera Daisy			Iris		
	1,000 blooms	*Cents*	*1,000 dollars*	*1,000 bunches*	*Dollars*	*1,000 dollars*
1992						
1993						
1994						
1995						
1996						
1997						
1998						
1999						
2000	67,520	30.9	20,886	79,012	25.8	20,395
2001	69,351	28.9	20,045	85,174	23.6	20,117
	All Lilies			Lisianthus		
	1,000 blooms	*Cents*	*1,000 dollars*	*1,000 blooms*	*Cents*	*1,000 dollars*
1992						
1993						
1994						
1995						
1996						
1997						
1998						
1999						
2000	81,301	68.8	55,975	15,150	45.5	6,891
2001	86,511	66.4	57,452	20,534	49.2	10,101
	All orchids			All roses		
	1,000 spikes	*Cents*	*1,000 dollars*			*1,000 dollars*
1992						
1993						
1994						
1995						
1996						
1997						
1998						
1999						
2000	11,719	68.9	8,071	185,975	37.3	69,294
2001	12,090	73.5	8,883	187,585	36.1	67,656
	Snapdragons			Tulips		
	1,000 blooms	*Cents*	*1,000 dollars*	*1,000 blooms*	*Cents*	*1,000 dollars*
1992						
1993						
1994						
1995						
1996						
1997						
1998						
1999						
2000	63,711	30.1	19,166	77,039	34.7	26,760
2001	60,367	28.6	17,249	75,443	34.8	26,265

[1] Equivalent wholesale value of all sales. [2] This data series discontinued.

NASS, Crops Branch, (202) 720–2127.

Table 5-104.—Cut Greens: Sales and wholesale value for operations with $100,000+ sales, Surveyed States, 1992–2001

Year	Quantity sold	Wholesale price	Value of sales at wholesale [1]	Quantity sold	Wholesale price	Value of sales at wholesale [1]
	Leatherleaf Ferns			Other cut cultivated greens		
	1,000 bunches	*Dollars*	*1,000 dollars*	*1,000 bunches*	*Dollars*	*1,000 dollars*
1992	78,880	0.85	66,718			44,806
1993	85,239	0.85	72,142			43,837
1994	78,912	0.86	67,544			51,703
1995	74,968	0.85	63,485			49,639
1996	77,982	0.87	67,993			50,192
1997	78,707	0.82	64,373			51,811
1998	72,981	0.83	60,498			57,191
1999	71,928	0.90	64,547			62,128
2000	75,611	0.88	66,245			59,923
2001	62,272	0.87	54,283			56,783

[1] Equivalent wholesale value of all sales.
NASS, Crops Branch, (202) 720–2127.

Table 5-105.—Potted flowering and foliage plants: Sales and wholesale value for operations with $100,000+ sales, Surveyed States, 1992–2001

Year	Quantity sold		Wholesale Price		Value of sales at wholesale [1]
	Less than 5 inches	5 inches or more	Less than 5 inches	5 inches or more	
	African violets				
	1,000 pots	*1,000 pots*	*Dollars*	*Dollars*	*1,000 dollars*
1992	22,625	436	1.13	3.50	27,232
1993	22,492	237	1.07	3.46	24,777
1994	23,963	721	1.06	2.77	27,423
1995	21,366	707	1.08	2.74	25,086
1996	20,683	269	1.06	3.79	23,029
1997	19,736	338	1.09	3.23	22,584
1998	19,277	216	1.14	3.20	22,672
1999	20,678	226	1.15	3.18	24,455
2000	16,043	257	1.12	3.57	18,909
2001	15,196	226	1.17	3.69	18,564
	Florist chrysanthemums				
	1,000 pots	*1,000 pots*	*Dollars*	*Dollars*	*1,000 dollars*
1992	8,448	24,823	1.37	3.54	99,254
1993	8,995	23,129	1.37	3.60	95,564
1994	9,300	21,987	1.44	3.65	93,649
1995	10,358	21,642	1.47	3.56	92,219
1996	8,905	19,825	1.42	3.64	84,726
1997	7,779	19,328	1.49	3.68	82,632
1998	7,207	17,727	1.53	3.52	73,408
1999	6,895	17,218	1.51	3.51	70,923
2000	8,439	19,936	1.50	3.47	81,869
2001	6,344	19,490	1.56	3.46	77,262
	Potted florist roses				
	1,000 pots	*1,000 pots*	*Dollars*	*Dollars*	*1,000 dollars*
1992					
1993					
1994					
1995					
1996					
1997					
1998					
1999					
2000	8,784	2,844	1.77	4.19	27,499
2001	7,242	3,199	1.93	3.89	26,412
	Potted spring flowering bulbs				
	1,000 pots	*1,000 pots*	*Dollars*	*Dollars*	*1,000 dollars*
1992					
1993					
1994					
1995					
1996					
1997					
1998					
1999					
2000	6,408	8,775	1.59	3.33	39,392
2001	6,850	11,039	1.48	3.25	45,959

See footnotes at end of table.

Table 5-105.—Potted flowering and foliage plants: Sales and wholesale value for operations with $100,000+ sales, Surveyed States, 1992–2001—Continued

Year	Quantity sold: Less than 5 inches	Quantity sold: 5 inches or more	Wholesale Price: Less than 5 inches	Wholesale Price: 5 inches or more	Value of sales at wholesale[1]
	Cyclamen[2]				
	1,000 pots	*1,000 pots*	*Dollars*	*Dollars*	*1,000 dollars*
1992	....	....	....	....	....
1993	....	....	....	....	....
1994	2,035	2,441	1.89	3.80	13,122
1995	2,568	2,537	2.01	3.93	15,145
1996	3,087	2,556	2.12	4.02	16,824
1997	3,129	2,767	2.17	4.02	17,902
1998	2,649	2,608	2.25	3.95	16,259
1999	2,683	3,125	2.23	3.68	17,491
2000	(3)	(3)	(3)	(3)	(3)
2001	(3)	(3)	(3)	(3)	(3)
	Florist azaleas				
	1,000 pots	*1,000 pots*	*Dollars*	*Dollars*	*1,000 dollars*
1992	4,149	10,544	1.45	5.23	61,115
1993	3,154	10,989	1.66	4.92	59,241
1994	6,172	9,426	1.53	4.58	52,662
1995	6,371	9,673	1.65	4.88	57,770
1996	3,251	8,567	1.77	4.45	43,843
1997	3,350	9,584	1.82	3.77	42,186
1998	2,714	7,224	2.04	4.72	39,635
1999	2,718	7,207	1.95	5.26	43,185
2000	4,880	10,032	2.08	5.14	61,719
2001	3,936	10,027	2.55	5.28	63,029
	Kalanchoes[2]				
	1,000 pots	*1,000 pots*	*Dollars*	*Dollars*	*1,000 dollars*
1992	....	....	....	....	....
1993	....	....	....	....	....
1994	3,031	2,192	1.34	3.24	11,170
1995	3,279	1,982	1.47	3.53	11,832
1996	3,410	2,692	1.40	3.35	13,800
1997	3,615	2,863	1.44	3.51	15,251
1998	3,944	2,993	1.51	3.66	16,907
1999	3,930	2,853	1.50	3.74	16,550
2000	(3)	(3)	(3)	(3)	(3)
2001	(3)	(3)	(3)	(3)	(3)
	Easter lilies				
	1,000 pots	*1,000 pots*	*Dollars*	*Dollars*	*1,000 dollars*
1992	38	8,951	1.97	3.87	34,729
1993	....	8,640	....	3.96	34,180
1994	....	9,081	....	3.97	36,021
1995	37	9,193	2.97	4.00	36,925
1996	182	9,399	2.88	3.99	38,012
1997	177	12,175	2.72	3.76	46,278
1998	16	9,414	2.79	3.95	37,273
1999	13	9,096	2.31	3.95	35,990
2000	141	9,367	3.22	4.09	37,246
2001	214	9,143	3.07	4.04	38,521

See footnotes at end of table.

Table 5-105.—Potted flowering and foliage plants: Sales and wholesale value for operations with $100,000+ sales, Surveyed States, 1992–2001—Continued

Year	Quantity sold		Wholesale Price		Value of sales at wholesale[1]
	Less than 5 inches	5 inches or more	Less than 5 inches	5 inches or more	
	Other lilies[3]				
	1,000 pots	*1,000 pots*	*Dollars*	*Dollars*	*1,000 dollars*
1992	60	1,318	2.18	3.81	5,153
1993	62	1,479	2.34	3.87	5,870
1994	66	1,677	2.29	3.85	6,601
1995	286	1,971	1.72	3.76	7,895
1996					
1997					
1998					
1999					
2000					
2001					
	Poinsettias				
	1,000 pots	*1,000 pots*	*Dollars*	*Dollars*	*1,000 dollars*
1992	12,496	43,058	1.83	3.96	193,484
1993	11,349	45,155	1.90	3.94	199,464
1994	11,686	46,209	1.76	4.05	207,605
1995	12,407	45,770	1.90	4.10	211,437
1996	12,819	46,873	1.92	4.07	215,248
1997	13,446	48,999	1.85	4.14	227,729
1998	14,915	46,993	1.87	4.18	224,441
1999	12,839	48,848	1.86	4.15	226,816
2000	15,457	50,931	1.81	4.28	246,263
2001	14,881	52,494	1.80	4.37	256,211
	Other flowering[4]				
	1,000 pots	*1,000 pots*	*Dollars*	*Dollars*	*1,000 dollars*
1992	67,932	49,395	2.00	4.10	338,608
1993	50,375	55,815	1.48	3.40	264,250
1994	37,205	39,250	1.77	3.78	214,237
1995	45,951	37,580	1.61	3.97	222,798
1996	52,440	39,266	1.30	3.41	201,865
1997	47,538	37,930	1.32	3.58	198,451
1998	53,219	40,238	1.71	3.74	241,357
1999	53,252	44,032	1.60	3.61	244,030
2000	33,585	35,766	1.73	3.90	197,684
2001	32,769	35,372	1.88	4.10	206,427
	Potted Orchids[5]				
	1,000 pots	*1,000 pots*	*Dollars*	*Dollars*	*1,000 dollars*
1992					
1993					
1994					
1995					
1996	5,212	3,380	4.00	7.74	46,993
1997	4,134	5,447	5.93	8.32	69,856
1998	3,510	5,096	6.00	8.60	64,885
1999	4,805	5,695	5.51	9.29	79,398
2000	4,782	4,912	6.58	11.72	89,018
2001	6,784	5,335	5.86	11.20	99,514
	Foliage				
	1,000 pots	*1,000 pots*	*Dollars*	*Dollars*	*1,000 dollars*
1992					350,915
1993					350,581
1994					400,232
1995					413,566
1996					432,976
1997					421,398
1998					436,243
1999					432,532
2000					(3)
2001					(3)

[1] Equivalent wholesale value of all sales except for potted foliage which is value of sales less cost of plant material purchased from other growers for growing on. [2] Estimates began in 1994. [3] This data series discontinued. [4] 1994-2000 data not comparable with 1991-93 which includes data for cyclamen and kalanchoes. Potted orchids were included in other flowering prior to 1996. [5] Potted orchids reported separately for first time in 1996. Previously orchids were included in other flowering category.

NASS, Crops Branch, (202) 720–2127.

Table 5-106.—Flowering & foliage hanging baskets: Sales and wholesale value for operations with $100,000+ sales, Surveyed States, 1992–2001

Year	Quantity sold	Wholesale price	Value of sales at wholesale [1]	Quantity sold	Wholesale price	Value of sales at wholesale [1]
	Geraniums [2]			Impatiens [2]		
	1,000 baskets	*Dollars*	*1,000 dollars*	*1,000 baskets*	*Dollars*	*1,000 dollars*
1992						
1993						
1994	3,274	6.36	20,833	3,018	5.28	15,922
1995	4,010	6.34	25,407	3,668	5.22	19,163
1996	4,691	6.43	30,142	4,004	5.22	20,891
1997	5,066	6.38	32,333	4,339	5.05	21,912
1998	4,734	6.56	31,058	4,140	4.96	20,536
1999	4,961	6.55	32,502	4,274	5.04	21,559
2000	(3)	(3)	(3)	4,054	5.15	20,859
2001	(3)	(3)	(3)	3,510	5.42	19,036
	New Guinea Impatiens [2]			Petunias [2]		
	1,000 baskets	*Dollars*	*1,000 dollars*	*1,000 baskets*	*Dollars*	*1,000 dollars*
1992						
1993						
1994	3,264	6.06	19,777	852	5.42	4,614
1995	3,812	6.04	23,014	1,345	5.36	7,205
1996	4,210	6.32	26,604	1,664	5.40	8,984
1997	4,674	6.25	29,198	2,013	5.40	10,876
1998	4,522	6.34	28,668	2,203	5.37	11,822
1999	4,911	6.35	31,196	2,823	5.59	15,774
2000	4,635	6.39	29,615	2,941	5.30	15,595
2001	4,661	6.34	29,540	3,170	5.84	18,524
	Other flowering hanging baskets [4]			Foliage		
	1,000 baskets	*Dollars*	*1,000 dollars*	*1,000 baskets*	*Dollars*	*1,000 dollars*
1992	22,498	5.44	122,365	20,007	3.80	76,094
1993	25,588	5.44	139,224	17,196	3.86	66,468
1994	16,480	5.48	90,381	24,095	3.70	89,074
1995	15,857	5.64	89,420	22,725	3.76	85,403
1996	15,595	5.76	89,874	19,736	3.85	75,971
1997	18,093	5.70	103,183	20,066	3.92	78,566
1998	19,404	5.95	115,437	15,993	4.14	66,258
1999	19,210	6.19	118,900	20,630	3.85	79,467
2000	14,760	6.01	88,656	20,983	4.20	88,113
2001	15,291	6.28	96,089	22,964	3.93	90,231
	Begonias [5]			Marigolds [5]		
	1,000 baskets	*Dollars*	*1,000 dollars*	*1,000 baskets*	*Dollars*	*1,000 dollars*
1992						
1993						
1994						
1995						
1996						
1997						
1998						
1999						
2000	2,855	5.51	15,733	59	7.71	455
2001	3,295	5.72	18,847	84	7.74	650

See footnotes at end of table.

Table 5-106.—Flowering & foliage hanging baskets: Sales and wholesale value for operations with $100,000+ sales, Surveyed States, 1992–2001—Continued

Year	Quantity sold	Wholesale price	Value of sales at wholesale [1]	Quantity sold	Wholesale price	Value of sales at wholesale [1]
	Geraniums from seed [5]			Geraniums from cuttings [5]		
	1,000 baskets	*Dollars*	*1,000 dollars*	*1,000 baskets*	*Dollars*	*1,000 dollars*
1992						
1993						
1994						
1995						
1996						
1997						
1998						
1999						
2000	684	5.83	3,991	4,146	7.00	29,024
2001	629	5.96	3,748	4,177	6.88	28,749
	Pansy/Viola [5]					
	1,000 baskets	*Dollars*	*1,000 dollars*	*1,000 baskets*	*Dollars*	*1,000 dollars*
1992						
1993						
1994						
1995						
1996						
1997						
1998						
1999						
2000	303	6.38	1,932			
2001	455	5.99	2,725			

[1] Equivalent wholesale value of all sales. [2] Estimates began in 1994. [3] This data series discontinued. [4] 1994–2000 data are not comparable with 1991–93 which includes geraniums, impatiens, New Guinea impatiens, and petunias. [5] Estimates began in 2000.

NASS, Crops Branch, (202) 720–2127.

Table 5-107.—Bedding plant flats: Sales and wholesale value for operations with $100,000+ sales, Surveyed States, 1992–2001

Year	Quantity sold	Wholesale price	Value of sales at wholesale [1]	Quantity sold	Wholesale price	Value of sales at wholesale [1]
	Geraniums			Impatiens [2]		
	1,000 flats	*Dollars*	*1,000 dollars*	*1,000 flats*	*Dollars*	*1,000 dollars*
1992	3,148	9.62	30,291			
1993	3,254	9.81	31,915			
1994	4,097	9.13	37,402	12,940	6.49	83,916
1995	3,978	9.01	35,859	15,501	6.90	107,014
1996	4,462	9.55	42,598	15,722	6.96	109,475
1997	3,940	9.46	37,281	18,147	6.97	126,394
1998	3,143	8.97	28,189	16,715	6.71	112,105
1999	3,205	9.77	31,325	16,331	7.04	114,939
2000	(5)	(5)	(5)	15,380	7.70	118,381
2001	(5)	(5)	(5)	15,011	7.83	117,476
	New Guinea Impatiens [2]			Petunias [2]		
	1,000 flats	*Dollars*	*1,000 dollars*	*1,000 flats*	*Dollars*	*1,000 dollars*
1992						
1993						
1994	890	8.85	7,873	8,841	6.82	60,252
1995	1,055	9.10	9,601	10,754	7.21	77,487
1996	877	8.98	7,873	10,761	7.10	76,359
1997	861	9.86	8,487	11,818	7.36	87,014
1998	589	10.66	6,277	11,783	7.23	85,189
1999	1,342	9.24	12,403	11,645	7.46	86,848
2000	657	9.71	6,381	12,093	7.90	95,488
2001	543	10.54	5,723	11,866	8.12	96,408
	Other Flowering and foliar type bedding plant flats [3]			Vegetable type bedding plant flats [4]		
	1,000 flats	*Dollars*	*1,000 dollars*	*1,000 flats*	*Dollars*	*1,000 dollars*
1992	67,941	6.73	457,588	11,218	6.98	78,224
1993	74,334	6.92	514,095	9,592	6.96	66,759
1994	58,245	7.11	414,304	9,230	6.97	64,373
1995	55,652	7.09	394,419	10,327	7.23	74,676
1996	57,654	7.20	414,932	10,666	7.46	79,578
1997	70,958	7.46	529,375	12,624	7.82	98,755
1998	63,123	7.58	478,431	12,044	7.70	92,723
1999	69,832	8.02	560,067	11,976	8.12	97,288
2000	47,709	8.04	383,686	8,604	7.97	68,604
2001	43,701	7.79	340,489	8,653	8.32	72,021
	Begonias [6]			Marigolds [6]		
	1,000 flats	*Dollars*	*1,000 dollars*	*1,000 flats*	*Dollars*	*1,000 dollars*
1992						
1993						
1994						
1995						
1996						
1997						
1998						
1999						
2000	6,645	7.83	52,004	5,443	7.75	42,169
2001	7,616	7.93	60,374	6,364	8.04	51,152
	Geraniums from seed [6]			Geraniums from vegetable cuttings [6]		
	1,000 flats	*Dollars*	*1,000 dollars*	*1,000 flats*	*Dollars*	*1,000 dollars*
1992						
1993						
1994						
1995						
1996						
1997						
1998						
1999						
2000	861	9.33	8,035	1,574	9.47	14,906
2001	843	9.65	81.35	1,255	9.75	12,240
	Pansy/Viola [6]					
	1,000 flats	*Dollars*	*1,000 dollars*	*1,000 flats*	*Dollars*	*1,000 dollars*
1992						
1993						
1994						
1995						
1996						
1997						
1998						
1999						
2000	10,153	8.23	83,521			
2001	12,816	8.05	103,179			

[1] Equivalent wholesale value of all sales. [2] Estimates began in 1994. [3] 1994–2000 data not comparable with 1991–93 which includes impatiens, New Guinea impatiens, and petunias. [4] Does not include vegetable transplants grown for use in commercial vegetable production. [5] This data series discontinued. [6] Estimates began in 2000.

NASS, Crops Branch, (202) 720–2127.

Table 5-108.—Potted flowering and foliar type bedding plants: Sales and wholesale value for operations with $100,000+ sales, Surveyed States, 1992–2001

Year	Quantity sold		Wholesale Price		Value of sales at wholesale [1]
	Less than 5 inches	5 inches or more	Less than 5 inches	5 inches or more	
	Hardy/Garden mums				
	1,000 pots	*1,000 pots*	*Dollars*	*Dollars*	*1,000 dollars*
1992	16,413	20,712	0.84	1.84	55,666
1993	13,347	25,694	0.81	1.75	55,745
1994	13,574	25,473	0.83	1.83	57,923
1995	13,761	29,545	0.80	1.83	65,001
1996	13,001	32,588	0.82	1.79	69,078
1997	12,686	36,067	0.89	1.86	78,316
1998	14,965	42,505	0.89	1.89	93,787
1999	14,606	46,755	0.89	1.94	103,564
2000	17,813	48,534	1.11	1.78	106,385
2001	15,381	45,313	1.15	1.89	103,240
	Geraniums (cuttings)				
	1,000 pots	*1,000 pots*	*Dollars*	*Dollars*	*1,000 dollars*
1992	42,206	11,168	1.26	2.46	80,678
1993	41,584	11,265	1.29	2.62	83,038
1994	45,086	10,351	1.29	2.71	86,061
1995	49,726	15,902	1.24	2.64	103,569
1996	40,770	15,287	1.30	2.57	92,457
1997	40,620	15,190	1.41	2.61	97,123
1998	47,855	17,374	1.37	2.57	110,425
1999	45,221	17,099	1.39	2.76	110,291
2000	44,004	18,423	1.41	2.62	110,223
2001	42,303	18,300	1.46	2.75	111,870
	Geraniums (seed)				
	1,000 pots	*1,000 pots*	*Dollars*	*Dollars*	*1,000 dollars*
1992	45,542	1,882	0.80	2.44	41,126
1993	47,211	2,079	0.77	2.08	40,758
1994	48,612	2,107	0.80	2.24	43,551
1995	46,913	2,377	0.80	1.84	42,035
1996	42,339	2,184	0.82	1.91	38,841
1997	44,198	3,781	0.80	1.90	42,675
1998	42,564	2,173	0.80	1.82	37,889
1999	42,550	2,149	0.79	1.87	37,837
2000	46,834	1,295	0.83	2.08	41,756
2001	43,162	1,037	0.82	2.40	37,986
	Impatiens [2]				
	1,000 pots	*1,000 pots*	*Dollars*	*Dollars*	*1,000 dollars*
1992	....	....	....	....	....
1993	....	....	....	....	....
1994	21,820	1,721	0.57	1.45	14,874
1995	21,916	2,334	0.64	1.57	17,585
1996	20,098	2,513	0.64	1.54	16,828
1997	20,791	3,211	0.63	1.53	18,102
1998	21,033	4,137	0.65	1.57	20,078
1999	24,730	4,926	0.64	1.41	22,837
2000	23,903	5,791	0.67	1.46	24,473
2001	25,767	5,332	0.70	1.60	26,598

See footnotes at end of table.

Table 5-108.—Potted flowering and foliar type bedding plants: Sales and wholesale value for operations with $100,000+ sales, Surveyed States, 1992–2001—Continued

Year	Quantity sold		Wholesale Price		Value of sales at wholesale [1]
	Less than 5 inches	5 inches or more	Less than 5 inches	5 inches or more	
	New Guinea Impatiens [2]				
	1,000 pots	*1,000 pots*	*Dollars*	*Dollars*	*1,000 dollars*
1992	....	....	....	....	....
1993	....	....	....	....	....
1994	7,161	1,979	....	....	....
1995	8,452	3,072	1.20	2.44	13,428
1996	10,499	3,732	1.24	2.45	18,003
1997	11,960	4,427	1.20	2.35	21,364
1998	14,851	4,078	1.31	2.38	26,222
1999	15,860	5,090	1.33	2.47	29,789
2000	18,148	5,653	1.33	2.48	33,802
2001	16,374	6,670	1.35	2.60	39,223
			1.39	2.37	38,530
	Petunias [2]				
	1,000 pots	*1,000 pots*	*Dollars*	*Dollars*	*1,000 dollars*
1992	....	....	....	....	....
1993	....	....	....	....	....
1994	11,191	959	....	....	....
1995	10,656	1,129	0.65	1.45	8,679
1996	10,520	1,813	0.66	1.78	9,030
1997	9,061	1,985	0.66	1.58	9,797
1998	10,504	2,726	0.68	1.60	9,321
1999	11,491	3,115	0.70	1.72	12,042
2000	13,340	3,784	0.75	1.75	14,072
2001	13,854	4,447	0.83	1.73	17,580
			0.89	2.19	22,124
	Other flowering/foliar type [3]				
	1,000 pots	*1,000 pots*	*Dollars*	*Dollars*	*1,000 dollars*
1992	240,921	38,322	0.71	1.86	241,175
1993	210,780	37,230	0.72	1.95	224,947
1994	171,536	46,786	0.73	2.06	221,258
1995	159,627	54,664	0.78	1.79	221,883
1996	134,216	83,260	0.82	1.69	250,656
1997	170,526	78,583	0.94	2.58	363,277
1998	232,674	111,416	0.88	2.82	520,239
1999	236,079	93,870	0.91	2.67	466,422
2000	159,357	53,236	0.96	2.35	277,692
2001	157,485	59,680	0.96	2.33	290,217
	Vegetable type [4]				
	1,000 pots	*1,000 pots*	*Dollars*	*Dollars*	*1,000 dollars*
1992	17,537	2,488	0.63	1.60	15,058
1993	13,761	2,311	0.73	1.50	13,530
1994	14,803	2,950	0.70	1.48	14,666
1995	16,185	2,910	0.74	1.56	16,596
1996	19,127	4,827	0.74	1.59	21,802
1997	23,545	6,101	0.76	1.50	27,115
1998	29,293	7,956	0.80	1.82	37,926
1999	26,454	5,721	0.79	1.87	31,513
2000	25,430	4,452	0.86	1.80	29,768
2001	23,197	4,223	0.89	1.91	28,697

See footnotes at end of table.

Table 5-108.—Potted flowering and foliar type bedding plants: Sales and wholesale value for operations with $100,000+ sales, Surveyed States, 1992–2001—Continued

Year	Quantity sold		Wholesale Price		Value of sales at wholesale [1]
	Less than 5 inches	5 inches or more	Less than 5 inches	5 inches or more	
			Begonias [5]		
	1,000 pots	*1,000 pots*	*Dollars*	*Dollars*	*1,000 dollars*
1992					
1993					
1994					
1995					
1996					
1997					
1998					
1999					
2000	12,559	2,321	0.89	1.81	15,427
2001	13,938	4,059	0.90	2.18	21,362
			Marigolds [5]		
	1,000 pots	*1,000 pots*	*Dollars*	*Dollars*	*1,000 dollars*
1992					
1993					
1994					
1995					
1996					
1997					
1998					
1999					
2000	4,994	1,111	0.78	1.33	5,368
2001	4,967	1,499	0.72	1.44	5,740
			Pansy/viola [5]		
	1,000 pots	*1,000 pots*	*Dollars*	*Dollars*	*1,000 dollars*
1992					
1993					
1994					
1995					
1996					
1997					
1998					
1999					
2000	19,985	3,932	0.74	1.55	20,882
2001	17,403	4,804	0.78	1.52	20,855

See footnotes at end of table.

Table 5-108.—Potted flowering and foliar type bedding plants: Sales and wholesale value for operations with $100,000+ sales, Surveyed States, 1992–2001—Continued

Year	Quantity sold			Wholesale Price			Value of sales at wholesale[1]
	Less than 1 gallon	1-2 gallons	2 gallons or more	Less than 1 gallon	1-2 gallons	2 gallons or more	
	Potted Hosta[5]						
	1,000 pots	*1,000 pots*	*1,000 pots*	*Dollars*	*Dollars*	*Dollars*	*1,000 dollars*
1992	....	....	....	....	....	....	....
1993	....	....	....	....	....	....	....
1994	....	....	....	....	....	....	....
1995	....	....	....	....	....	....	....
1996	....	....	....	....	....	....	....
1997	....	....	....	....	....	....	....
1998	....	....	....	....	....	....	....
1999	....	....	....	....	....	....	....
2000	3,858	7,269	296	2.02	3.67	8.28	35,874
2001	2,774	7,511	1,149	2.49	3.51	4.98	39,021
	Other Potted Herbaceous Perennials[5]						
	1,000 pots	*1,000 pots*	*1,000 pots*	*Dollars*	*Dollars*	*Dollars*	*1,000 dollars*
1992	....	....	....	....	....	....	....
1993	....	....	....	....	....	....	....
1994	....	....	....	....	....	....	....
1995	....	....	....	....	....	....	....
1996	....	....	....	....	....	....	....
1997	....	....	....	....	....	....	....
1998	....	....	....	....	....	....	....
1999	....	....	....	....	....	....	....
2000	66,995	56,181	3,904	1.47	3.04	5.84	291,734
2001	79,224	71,817	5,027	1.47	2.78	5.92	345,483

[1] Equivalent wholesale value of all sales. [2] Estimates began in 1994. [3] 1994–2000 data not comparable with 1991–93 which includes data for impatiens, New Guinea impatiens, and petunias. [4] Does not include vegetable transplants grown for use in commercial vegetable production. [5] Estimates began in 2000.

NASS, Crops Branch, (202) 720–2127.

Table 5-109.—Floriculture: Growing area by type of cover, all operations with $10,000+ sales, 36 Surveyed States, 2000–2001

State	Glass greenhouses		Fiberglass and other rigid greenhouses		Film plastic (single/multi) greenhouses	
	2000	2001	2000	2001	2000	2001
	1,000 square feet	*1,000 square feet*	*1,000 square feet*	*1,000 square feet*	*1,000 square feet*	*1,000 square feet*
AL	527	636	274	426	8,216	8,338
AZ	…	…	1,145	1,078	1,093	1,100
AR	24	9	23	16	2,988	2,112
CA	16,461	17,261	33,596	32,266	56,859	56,667
CO	517	471	7,492	7,053	2,939	2,707
CT	1,553	1,844	830	1,615	6,007	5,705
FL	4,265	4,826	10,745	13,640	43,276	38,855
GA	620	435	328	489	7,492	6,731
HI	69	61	3,080	2,769	2,450	2,392
IL	2,282	2,752	2,639	2,784	8,147	8,763
IN	1,109	1,374	972	1,420	4,913	5,157
IA	587	559	1,693	1,567	4,297	4,029
KS	108	117	455	497	4,129	4,127
KY	368	469	1,018	998	3,545	3,486
LA	125	107	178	295	2,896	2,525
MD	807	1,078	582	530	3,265	3,781
MA	1,894	1,877	1,073	1,261	6,704	6,250
MI	4,441	4,694	4,096	3,876	32,665	32,443
MN	501	526	2,016	2,301	7,694	7,945
MS	163	165	210	61	2,313	3,167
MO	740	676	459	561	5,228	5,111
NJ	3,502	3,899	484	848	14,670	14,730
NM	1,458	1,455	168	105	2,454	2,490
NY	5,555	6,283	1,354	1,461	19,520	15,958
NC	2,518	2,889	227	272	14,222	14,220
OH	8,403	8,960	2,038	1,698	18,823	18,163
OK	65	55	384	311	2,096	1,859
OR	1,531	1,525	2,585	2,414	9,462	10,094
PA	4,128	4,078	2,672	2,997	16,010	16,368
SC	921	905	407	140	3,712	3,809
TN	782	753	185	201	5,679	5,548
TX	1,058	1,801	9,406	8,877	22,289	28,053
UT	36	70	952	1,066	3,794	3,359
VA	1,043	1,070	960	592	6,105	7,320
WA	2,093	1,973	799	565	5,991	6,108
WI	1,686	1,503	1,118	1,085	6,603	6,715
36 States	71,940	77,156	96,643	98,135	368,546	366,185

See end of table.

Table 5-109.—Floriculture: Growing area by type of cover, all operations with $10,000+ sales, 36 Surveyed States, 2000–2001—Continued

State	Shade and temporary cover		Total covered area		Open ground	
	2000	2001	2000	2001	2000	2001
	1,000 square feet	*1,000 square feet*	*1,000 square feet*	*1,000 square feet*	*Acres*	*Acres*
AL	222	269	9,239	9,669	201	85
AZ	653	688	2,891	2,866	294	291
AR	49	29	3,084	2,166	79	54
CA	21,444	22,903	128,360	129,097	9,696	8,653
CO	267	164	11,215	10,395	82	59
CT	404	409	8,794	9,573	398	525
FL	330,470	323,335	388,756	380,656	9,960	8,680
GA	384	459	8,824	8,114	155	165
HI	21,074	20,487	26,673	25,709	1,315	1,386
IL	577	1,005	13,645	15,304	440	509
IN	104	140	7,098	8,091	206	197
IA	149	159	6,726	6,314	90	85
KS	119	142	4,811	4,883	71	71
KY	92	102	5,023	5,055	79	53
LA	288	292	3,487	3,219	84	85
MD	172	71	4,826	5,460	219	208
MA	112	133	9,783	9,521	421	384
MI	1,106	1,141	42,308	42,154	3,299	3,235
MN	140	403	10,351	11,175	186	84
MS	77	116	2,763	3,509	44	89
MO	136	131	6,563	6,479	143	457
NJ	260	366	18,916	19,843	2,301	2,435
NM	93	92	4,173	4,142	9	13
NY	527	667	26,956	24,369	914	1,243
NC	476	515	17,443	17,896	384	417
OH	502	610	29,766	29,431	409	447
OK	96	70	2,641	2,295	31	31
OR	1,571	1,668	15,149	15,701	1,807	1,553
PA	419	557	23,229	24,000	575	493
SC	1,238	1,751	6,278	6,605	250	505
TN	230	94	6,876	6,596	227	191
TX	8,693	9,162	41,446	47,893	993	6,232
UT	220	161	5,002	4,656	46	30
VA	477	730	8,585	9,712	215	558
WA	189	257	9,072	8,903	1,208	1,369
WI	455	532	9,862	9,835	172	240
36 States	393,485	389,810	930,614	931,286	37,002	41,110

NASS, Crops Branch, (202) 720-2127.

Table 5-110.—Floriculture: Total operations of $10,000+ sales and expanded wholesale value, 36 Surveyed States, 2000–2001

State	Total operations		Expanded wholesale value [1]	
	2000	2001	2000	2001
	Number	*Number*	*1,000 dollars*	*1,000 dollars*
AL	191	171	58,555	59,134
AZ	31	31	27,706	27,085
AR	70	57	14,633	16,998
CA	973	895	951,532	1,015,739
CO	149	134	87,375	86,238
CT	297	284	72,125	80,124
FL	1,192	1,075	739,807	764,989
GA	235	197	77,381	69,142
HI	434	413	66,564	69,500
IL	317	303	110,291	117,520
IN	285	289	48,878	54,940
IA	139	131	50,754	53,290
KS	136	137	34,306	33,415
KY	184	167	30,707	31,436
LA	116	108	20,922	21,042
MD	151	148	58,131	65,616
MA	471	446	79,546	78,163
MI	747	708	273,517	270,121
MN	249	237	77,602	79,625
MS	53	71	14,543	15,482
MO	206	209	46,426	47,071
NJ	398	376	134,256	136,952
NM	43	41	27,886	32,093
NY	796	731	179,883	172,383
NC	353	335	136,177	161,416
OH	703	650	189,250	189,863
OK	105	100	14,027	17,993
OR	223	257	94,183	89,099
PA	777	745	170,754	180,604
SC	117	105	85,889	90,297
TN	202	204	52,373	45,781
TX	382	333	267,580	277,583
UT	76	82	34,973	35,556
VA	232	224	79,456	79,454
WA	210	197	103,432	98,378
WI	381	374	65,165	74,534
36 States	11,624	10,965	4,576,585	4,738,656

[1] Wholesale value of sales as reported by growers with $100,000 or more in sales of floriculture crops plus a calculated wholesale value of sales for growers with sales below $100,000. The value of sales for growers below the $100,000 level was estimated by multiplying the number of growers in each size group by the mid-point of each dollar value range.

NASS, Crops Branch, (202) 720-2127.

Table 5-111.—Floriculture Crops: Wholesale value of sales by category for operations with $100,000+ sales, 36 Surveyed States, 2000–2001

State	Total cut flowers		Total potted flowering plants		Total foliage for indoor or patio use	
	2000	2001	2000	2001	2000	2001
	1,000 dollars	*1,000 dollars*	*1,000 dollars*	*1,000 dollars*	*1,000 dollars*	*1,000 dollars*
AL	...	...	8,673	10,004	2,475	2,044
AZ	...	...	2,427	2,431	2,385	2,517
AR	...	...	2,059	1,844	176	287
CA	288,441	292,115	172,002	205,177	104,856	112,487
CO	8,156	6,440	10,378	10,056	779	546
CT	983	970	8,863	11,936	786	1,356
FL	28,119	23,168	86,049	87,629	346,946	361,150
GA	...	...	13,717	9,092	1,721	1,812
HI	16,436	16,221	17,505	18,992	16,153	18,381
IL	821	1,026	22,787	26,422	1,993	3,114
IN	...	...	6,287	5,917	2,514	2,542
IA	...	...	11,645	11,114	2,146	2,288
KS	...	...	6,857	7,082	2,942	290
KY	...	...	4,463	4,992	1,205	1,329
LA	...	...	2,615	2,541	3,293	2,688
MD	...	...	7,271	8,600	467	371
MA	3,527	3,076	9,246	9,662	3,644	3,450
MI	7,624	8,119	32,363	29,554	3,601	3,531
MN	6,442	5,184	11,221	11,949	1,737	1,795
MS	...	...	4,139	3,456	439	576
MO	...	...	13,614	10,907	3,783	3,819
NJ	7,128	8,533	28,562	28,772	3,822	5,552
NM	...	...	16,496	17,762	2,269	4,732
NY	6,136	4,506	37,424	40,165	3,701	2,466
NC	2,775	2,796	34,332	36,272	8,373	8,668
OH	3,220	2,879	27,352	26,124	4,537	4,883
OK	...	...	2,637	2,667	949	905
OR	10,195	10,698	29,087	28,409	370	1,297
PA	4,454	4,586	39,782	41,880	2,652	3,383
SC	...	...	9,705	8,014	2,170	1,122
TN	...	...	8,352	8,922	963	899
TX	616	677	56,847	52,061	20,132	16,543
UT	...	...	11,040	8,380	2,282	4,165
VA	...	...	23,554	23,732	1,029	926
WA	19,624	18,857	9,200	8,587	2,425	1,776
WI	1,623	1,915	11,048	10,795	477	1,634
Oth Sts	13,643	12,490	...	...	...	...
36 States	429,963	424,256	799,599	831,899	560,192	585,324

See end of table.

Table 5-111.—Floriculture Crops: Wholesale value of sales by category for operations with $100,000+ sales, 36 Surveyed States, 2000–2001 [1]—Continued

State	Total bedding/garden plants [2]		Total cut cultivated greens		Propagative materials		Total reported wholesale value of floriculture crops	
	2000	2001	2000	2001	2000	2001	2000	2001
	1,000 dollars	*1,000 dollars*	*1,000 dollars*	*1,000 dollars*	*1,000 dollars*	*1,000 dollars*	*1,000 dollars*	*1,000 dollars*
AL	41,185	40,938					52,333	52,986
AZ	19,974	19,442					24,786	24,390
AR	9,496	12,650					11,731	14,781
CA	287,842	303,533	18,008	16,641	60,118	68,506	931,267	998,459
CO	52,626	48,207					71,939	65,249
CT	52,677	56,939				1,048	63,309	72,249
FL	103,813	101,597	100,607	87,430	50,873	82,520	716,407	743,494
GA	50,678	49,154			5,666	4,148	71,782	64,206
HI	2,403	2,613	255	158	2,592	2,260	55,344	58,625
IL	72,854	75,481			3,301	2,777	101,756	108,820
IN	25,573	30,678			1,999	3,084	36,373	42,221
IA	30,199	32,839				2,791	43,990	49,032
KS	20,082	21,622			92	226	29,973	29,220
KY	18,575	19,379				260	24,243	25,960
LA	10,190	11,202					16,098	16,431
MD	43,800	49,547			1,939		53,477	58,518
MA	45,675	44,873			2,092	2,100	64,184	63,161
MI	188,648	189,381			22,717	21,900	254,953	252,485
MN	46,019	48,353					65,419	67,281
MS	8,431	9,376					13,009	13,408
MO	21,232	25,041					38,629	39,767
NJ	75,748	75,018					115,260	117,875
NM	4,513	5,183					23,278	27,677
NY	97,565	96,818			11,852	5,993	156,678	149,948
NC	77,520	101,409			2,687	2,821	125,687	151,966
OH	128,699	131,759			4,693	4,928	168,501	170,573
OK	6,678	11,298					10,264	14,870
OR	39,069	34,157	5,828	4,162	3,199		87,748	78,723
PA	82,410	75,447			15,386	30,872	144,684	156,168
SC	62,382	69,254					74,257	78,390
TN	32,972	26,822					42,287	36,643
TX	171,629	187,159			6,644	10,837	255,868	267,277
UT	17,220	18,011					30,542	30,556
VA	48,071	47,336					72,654	71,994
WA	61,755	58,070			4,353	5,163	97,357	92,453
WI	37,217	46,242				2,952	50,365	63,538
Oth Sts			1,470	2,675	42,435	50,797	57,548	65,962
36 States	2,095,420	2,176,828	126,168	111,066	242,638	305,983	4,253,980	4,435,356

[1] Missing data were included in "Other States" to avoid disclosure of individual operations. [2] Includes Annual Bedding Plants and Herbaceous Perennials.

NASS, Crops Branch, (202) 720-2127.

Table 5-112.—Fruit and orange juice: Cold storage holdings, end of month, United States, 2001 and 2002

Month	Fresh					
	Apples, regular storage		Apples, CA storage		Apples, total	
	2001	2002	2001	2002	2001	2002
	1,000 pounds	*1,000 pounds*	*1,000 pounds*	*1,000 pounds*	*1,000 pounds*	*1,000 pounds*
January	517,877	482,122	3,584,094	3,146,464	4,101,971	3,628,586
February	309,538	324,764	3,098,080	2,632,823	3,407,618	2,957,587
March	233,516	187,258	2,369,296	2,033,890	2,602,812	2,221,148
April	101,138	108,012	1,789,432	1,441,646	1,890,570	1,549,658
May	41,493	51,877	1,288,666	990,961	1,330,159	1,042,838
June	13,574	14,489	884,663	629,212	898,237	643,701
July	28,154	11,957	458,481	304,357	486,635	316,314
August	52,876	7,563	90,503	82,220	143,379	89,783
September	959,272	926,419	1,847,115	1,897,480	2,806,387	2,823,899
October	1,781,866	1,400,871	3,782,496	3,847,680	5,564,362	5,248,551
November	1,141,115	1,027,615	3,834,202	3,952,142	4,975,317	4,979,757
December	759,622	597,986	3,595,582	3,718,089	4,355,204	4,316,075

Month	Fresh					
	Pears, Bartlett		Pears, other		Pears, total	
	2001	2002	2001	2002	2001	2002
	1,000 pounds	*1,000 pounds*	*1,000 pounds*	*1,000 pounds*	*1,000 pounds*	*1,000 pounds*
January	996	648	249,055	237,987	250,051	238,635
February	1,186	916	179,523	187,141	180,709	188,057
March	....	26	112,537	135,727	112,537	135,753
April	....	35	55,357	80,190	55,357	80,225
May	....	35	17,587	43,433	17,587	43,468
June	....	35	....	12,634	....	12,669
July	18,115	29,685	330	100	18,445	29,785
August	89,999	92,178	3,417	26,834	93,416	119,012
September	138,669	98,546	415,005	348,174	553,674	446,720
October	73,688	61,084	443,436	433,730	517,124	494,814
November	28,749	25,500	383,482	359,021	412,231	384,521
December	12,152	8,291	309,726	286,634	321,878	294,925

Month	Frozen							
	Apples		Apricots		Blackberries, IQF		Blackberries, pails & tubs	
	2001	2002	2001	2002	2001	2002	2001	2002
	1,000 pounds	*1,000 pounds*	*1,000 pounds*	*1,000 pounds*	*1,000 pounds*	*1,000 pounds*	*1,000 pounds*	*1,000 pounds*
January	65,964	59,738	7,316	5,983	14,813	19,496	1,841	3,233
February	70,725	65,459	5,597	4,343	13,105	13,207	1,780	2,792
March	70,555	70,656	4,540	3,815	11,938	12,342	1,446	2,478
April	69,609	71,745	3,500	2,662	11,078	9,071	1,367	2,267
May	66,926	67,963	3,103	2,112	9,353	7,170	1,103	1,044
June	64,348	58,610	13,880	5,731	9,029	5,785	1,973	589
July	55,314	49,709	15,139	9,195	20,046	20,425	4,567	2,381
August	45,301	41,174	12,956	9,031	20,880	20,150	4,994	2,345
September	36,202	31,486	11,316	8,536	19,545	18,436	4,792	2,580
October	38,421	32,321	9,661	5,632	17,654	16,638	4,289	2,205
November	41,455	37,649	8,686	5,260	15,336	15,727	3,736	1,904
December	50,967	46,099	7,763	5,054	12,885	14,131	3,146	1,816

Month	Frozen							
	Blackberries, barrels		Blackberries, concentrate		Blackberries, total		Blueberries	
	2001	2002	2001	2002	2001	2002	2001	2002
	1,000 pounds	*1,000 pounds*	*1,000 pounds*	*1,000 pounds*	*1,000 pounds*	*1,000 pounds*	*1,000 pounds*	*1,000 pounds*
January	3,334	6,087	2,174	963	22,162	29,779	77,392	91,259
February	3,098	4,578	2,114	1,215	20,097	21,792	68,791	84,492
March	2,606	3,981	1,876	1,129	17,866	19,930	63,218	75,051
April	2,343	3,870	1,575	937	16,363	16,145	53,932	66,635
May	1,553	3,279	1,680	710	13,689	12,203	46,978	56,245
June	1,647	2,674	1,548	535	14,197	9,583	44,407	51,746
July	9,005	11,445	1,425	505	35,043	34,756	62,904	66,429
August	9,408	10,952	1,500	614	36,782	34,061	128,534	120,995
September	8,836	11,102	1,878	576	35,051	32,694	134,814	119,604
October	7,714	8,270	1,456	772	31,113	27,885	120,955	111,364
November	6,955	8,045	1,189	485	27,216	26,161	111,814	100,274
December	5,026	6,364	1,114	485	22,171	22,796	100,526	88,705

See end of table.

Table 5-112.—Fruit and orange juice: Cold storage holdings, end of month, United States, 2001 and 2002—Continued

Month	Boysenberries		Cherries, Tart (RSP)		Cherries, Sweet		Grapes	
	2001	2002	2001	2002	2001	2002	2001	2002
	1,000 pounds	*1,000 pounds*	*1,000 pounds*	*1,000 pounds*	*1,000 pounds*	*1,000 pounds*	*1,000 pounds*	*1,000 pounds*
January	4,290	2,810	107,783	117,143	12,113	12,424	4,942	4,689
February	3,846	2,448	98,810	115,834	11,729	12,100	4,418	4,124
March	3,387	2,335	88,595	106,151	10,706	10,640	3,895	3,805
April	2,904	1,856	78,721	96,170	9,181	9,018	3,152	2,996
May	2,751	1,691	66,095	86,138	9,264	8,647	4,217	3,187
June	3,655	2,270	56,927	75,917	9,436	8,728	3,298	2,757
July	5,786	5,655	158,160	81,794	20,129	13,529	2,543	2,423
August	4,646	4,846	174,165	78,729	19,467	12,582	2,611	2,366
September	3,846	4,128	155,033	74,498	18,064	11,603	4,113	4,989
October	3,566	3,568	144,013	66,942	16,332	9,888	5,916	7,915
November	3,409	3,356	129,620	59,721	13,918	8,019	5,463	7,605
December	3,152	3,171	127,215	54,724	13,844	7,934	4,936	7,133

Month	Peaches		Raspberries, Black		Red Raspberries, IQF		Red Raspberries, pails & tubs	
	2001	2002	2001	2002	2001	2002	2001	2002
	1,000 pounds	*1,000 pounds*	*1,000 pounds*	*1,000 pounds*	*1,000 pounds*	*1,000 pounds*	*1,000 pounds*	*1,000 pounds*
January	65,640	57,791	1,433	2,762	14,568	15,648	13,916	8,352
February	57,442	51,198	1,088	2,486	12,758	13,881	12,753	7,867
March	48,511	43,959	1,174	2,257	12,185	11,944	11,113	7,181
April	35,842	36,187	1,002	2,188	11,332	10,063	10,139	5,737
May	29,547	28,116	907	2,020	10,047	8,970	8,049	4,845
June	20,133	30,694	901	1,961	10,747	7,480	6,478	4,126
July	28,858	26,069	4,587	4,195	25,627	26,321	16,324	16,514
August	66,996	59,708	4,363	4,119	24,272	27,054	14,736	15,661
September	83,669	91,213	4,030	3,644	22,093	25,314	11,942	13,991
October	77,452	89,133	3,347	2,824	21,513	22,878	10,887	12,538
November	70,764	82,842	3,015	2,585	19,193	19,998	10,489	12,016
December	65,242	72,564	2,840	2,612	17,611	17,502	9,135	10,810

Month	Red Raspberries, barrels		Red Raspberries, concentrate		Red Raspberries, total		Strawberries, IQF & Poly	
	2001	2002	2001	2002	2001	2002	2001	2002
	1,000 pounds	*1,000 pounds*	*1,000 pounds*	*1,000 pounds*	*1,000 pounds*	*1,000 pounds*	*1,000 pounds*	*1,000 pounds*
January	14,924	14,793	1,570	1,040	44,978	39,833	....	124,193
February	11,937	13,174	1,351	1,298	38,799	36,220	....	106,388
March	8,399	11,595	1,129	1,129	32,826	31,849	....	98,112
April	7,174	7,977	1,126	966	29,771	24,743	....	104,480
May	5,733	6,855	935	760	24,764	21,430	....	162,933
June	4,092	5,571	895	606	22,212	17,783	....	271,619
July	31,172	34,212	1,870	538	74,993	77,585	....	296,726
August	30,921	36,445	1,333	614	71,262	79,774	....	282,019
September	24,441	31,196	1,260	843	59,736	71,344	....	256,249
October	20,833	26,185	1,224	677	54,457	62,278	....	221,757
November	18,606	23,792	1,610	635	49,898	56,441	....	197,396
December	14,173	21,807	1,243	541	42,162	50,660	....	176,116

Month	Strawberries, pails & tubs		Strawberries, barrels & drums		Strawberries, juice stock		Strawberries, total	
	2001	2002	2001	2002	2001	2002	2001	2002
	1,000 pounds	*1,000 pounds*	*1,000 pounds*	*1,000 pounds*	*1,000 pounds*	*1,000 pounds*		
January	....	56,372	....	11,686	....	8,639	262,335	200,890
February	....	48,062	....	9,769	....	6,321	242,074	170,540
March	....	33,872	....	11,348	....	6,685	209,761	150,017
April	....	32,216	....	11,272	....	5,635	179,841	153,603
May	....	36,085	....	13,594	....	14,355	241,792	226,967
June	....	71,375	....	25,411	....	21,493	369,331	389,898
July	....	82,233	....	28,117	....	25,371	376,661	432,447
August	....	74,213	....	26,873	....	22,844	346,661	405,949
September	....	69,910	....	23,241	....	20,230	312,681	369,630
October	....	62,358	....	22,039	....	18,237	299,598	324,391
November	....	58,426	....	21,283	....	15,784	266,872	292,889
December	....	57,056	....	20,778	....	12,426	243,716	266,376

See end of table.

Table 5-112.—Fruit and orange juice: Cold storage holdings, end of month, United States, 2001 and 2002—Continued

Month	Other fruit		Total frozen fruit		Orange juice	
	2001	2002	2001	2002	2001	2002
	1,000 pounds	*1,000 pounds*	*1,000 pounds*	*1,000 pounds*	*1,000 pounds*	*1,000 pounds*
January	792,306	388,146	1,470,902	1,015,458	1,610,816	1,721,270
February	746,425	375,838	1,371,509	948,720	1,825,110	1,770,894
March	716,384	337,088	1,272,652	858,986	1,735,462	1,794,375
April	636,114	296,360	1,121,215	781,681	1,872,235	1,886,008
May	489,333	265,570	1,000,354	783,663	2,061,832	1,982,817
June	422,796	238,311	1,046,360	895,480	2,035,592	1,934,000
July	340,817	210,944	1,183,584	1,016,044	1,913,162	1,870,861
August	224,238	191,481	1,141,818	1,048,210	1,691,076	1,680,896
September	239,926	190,911	1,101,916	1,017,263	1,537,665	1,543,617
October	392,232	433,794	1,199,774	1,180,359	1,398,873	1,409,629
November	420,083	435,283	1,155,569	1,120,259	1,406,673	1,471,232
December	419,219	418,911	1,106,334	1,048,138	1,571,735	1,673,647

NASS, Livestock Branch, (202) 720–8784.

Table 5-113.—Nuts: Cold storage holdings, end of month, United States, 2001 and 2002

Month	Peanuts					
	Shelled		In-shell		Total	
	2001	2002	2001	2002	2001	2002
	1,000 pounds	*1,000 pounds*	*1,000 pounds*	*1,000 pounds*	*1,000 pounds*	*1,000 pounds*
January	416,094	485,040	15,047	16,492	431,141	501,532
February	466,017	541,345	17,054	15,645	483,071	556,990
March	496,615	562,626	23,747	17,361	520,362	579,987
April	516,285	578,187	20,063	20,143	536,348	598,330
May	558,875	560,447	22,247	26,301	581,122	586,748
June	618,220	531,862	23,366	34,401	641,586	566,263
July	533,333	458,404	16,373	30,661	549,706	489,065
August	415,123	330,681	8,755	14,589	423,878	345,270
September	338,495	214,618	5,016	9,873	343,511	224,491
October	386,056	261,649	9,604	14,046	395,660	275,695
November	421,106	260,687	24,347	13,589	445,453	274,276
December	448,628	275,060	16,268	13,878	464,896	288,938

Month	Pecans					
	Shelled		In-shell		Total	
	2001	2002	2001	2002	2001	2002
	1,000 pounds	*1,000 pounds*	*1,000 pounds*	*1,000 pounds*	*1,000 pounds*	*1,000 pounds*
January	42,257	33,545	169,830	176,352	212,087	209,897
February	43,857	42,295	170,910	195,471	214,767	237,766
March	45,672	45,795	150,389	181,333	196,061	227,128
April	52,558	47,588	121,101	165,641	173,659	213,229
May	48,592	49,857	103,263	135,285	151,855	185,142
June	51,972	52,955	77,265	120,484	129,237	173,439
July	50,550	53,939	61,021	104,570	111,571	158,509
August	47,423	54,726	34,883	83,033	82,306	137,759
September	42,021	53,461	15,815	52,844	57,836	106,305
October	34,294	40,670	18,212	33,745	52,506	74,415
November	29,336	35,114	32,886	42,369	62,222	77,483
December	28,310	35,135	106,868	76,241	135,178	111,376

Month	Other nuts					
	Shelled		In-shell		Total	
	2001	2002	2001	2002	2001	2002
	1,000 pounds	*1,000 pounds*	*1,000 pounds*	*1,000 pounds*	*1,000 pounds*	*1,000 pounds*
January	194,574	126,974	20,900	14,170	215,474	141,144
February	188,058	135,704	9,382	14,185	197,440	149,889
March	178,425	128,923	16,054	11,447	194,479	140,370
April	159,139	129,075	13,935	7,949	173,074	137,024
May	156,954	78,194	12,771	5,822	169,725	84,016
June	127,590	115,377	10,764	5,473	138,354	120,850
July	79,700	89,831	8,614	5,027	88,314	94,858
August	69,318	63,572	8,007	4,467	77,325	68,039
September	79,549	69,564	9,475	10,419	89,024	79,983
October	86,095	106,313	19,601	13,833	105,696	120,146
November	104,924	107,247	18,136	14,468	123,060	121,715
December	132,103	115,127	13,062	12,812	145,165	127,939

NASS, Livestock Branch, (202) 720-8784.

CHAPTER VI

STATISTICS OF HAY, SEEDS, AND MINOR FIELD CROPS

Chapter VI deals with hay, pasture, seeds, and various minor field crops.

Table 6-1.—Hay, all: Area, yield, production, and value, United States, 1993–2002

Year	Area harvested	Yield per acre	Production	Marketing year average price per ton received by farmers	Value of production
	1,000 acres	*Tons*	*1,000 tons*	*Dollars*	*1,000 dollars*
1993	59,689	2.46	146,699	84.70	10,946,576
1994	58,815	2.55	150,136	86.70	11,113,274
1995	59,764	2.58	154,239	82.20	11,035,838
1996	61,169	2.45	149,779	95.80	12,726,992
1997	61,084	2.50	152,536	100.00	13,249,825
1998	60,076	2.53	151,780	84.60	11,606,734
1999	63,220	2.53	159,707	76.90	11,014,373
2000	59,854	2.54	151,921	84.60	11,416,651
2001	63,521	2.47	156,764	96.50	12,602,534
2002[1]	64,497	2.34	150,962	94.00	12,432,729

[1] Preliminary.

NASS, Crops Branch, (202) 720–2127.

Table 6-2.—Hay, all: Stocks on farms, United States, 1993–2002

Crop year	Dec. 1	May 1[1]
	1,000 tons	*1,000 tons*
1993	100,953	22,096
1994	105,296	20,775
1995	109,438	20,739
1996	105,179	17,424
1997	103,044	21,827
1998	112,066	24,817
1999	108,922	28,817
2000	105,582	21,106
2001	110,510	22,494
2002[2]	103,756	NA

[1] Following year. [2] Preliminary. NA-not available.

NASS, Crops Branch, (202) 720–2127.

Table 6-3.—Hay, alfalfa and alfalfa mixtures: Area, yield, and production, by States, 2000–2002

State	Area harvested			Yield per harvested acre			Production		
	2000	2001	2002 [1]	2000	2001	2002 [1]	2000	2001	2002 [1]
	1,000 acres	*1,000 acres*	*1,000 acres*	*Tons*	*Tons*	*Tons*	*1,000 tons*	*1,000 tons*	*1,000 tons*
AZ	205	215	230	8.30	8.00	8.10	1,702	1,720	1,863
AR	20	20	25	2.50	3.10	3.40	50	62	85
CA	1,020	1,010	1,140	7.00	7.20	7.10	7,140	7,272	8,094
CO	900	950	780	3.70	3.80	2.90	3,330	3,610	2,262
CT	12	8	9	2.20	2.30	2.80	26	18	25
DE	8	8	7	5.00	3.40	3.60	40	27	25
ID	1,130	1,120	1,250	4.20	3.90	4.00	4,746	4,368	5,000
IL	500	500	450	3.80	3.90	3.60	1,900	1,950	1,620
IN	430	330	280	4.10	4.00	3.30	1,763	1,320	924
IA	1,250	1,250	1,250	3.90	3.70	3.90	4,875	4,625	4,875
KS	900	900	950	4.10	4.60	3.70	3,690	4,140	3,515
KY	250	250	300	3.90	3.70	3.00	975	925	900
ME	12	10	8	2.20	2.20	2.60	26	22	21
MD	65	65	60	4.40	3.10	2.60	286	202	156
MA	16	18	18	2.30	2.30	2.80	37	41	50
MI	1,000	900	900	3.70	3.60	3.50	3,700	3,240	3,150
MN	1,550	1,450	1,600	3.60	3.50	3.30	5,580	5,075	5,280
MO	470	450	460	3.10	3.05	3.00	1,457	1,373	1,380
MT	1,200	1,450	1,400	2.10	2.10	2.10	2,520	3,045	2,940
NE	1,350	1,450	1,350	3.10	3.55	3.00	4,185	5,148	4,050
NV	265	265	275	4.60	4.50	4.30	1,219	1,193	1,183
NH	8	7	7	2.00	2.00	2.10	16	14	15
NJ	30	30	25	3.00	3.40	3.00	90	102	75
NM	290	270	260	5.20	5.00	5.60	1,508	1,350	1,456
NY	420	560	570	2.40	2.80	2.30	1,008	1,568	1,311
NC	20	20	20	2.70	3.00	1.80	54	60	36
ND	1,350	1,600	1,450	2.40	2.10	1.30	3,240	3,360	1,885
OH	570	570	590	4.00	3.50	3.00	2,280	1,995	1,770
OK	330	350	340	3.30	2.70	3.50	1,089	945	1,190
OR	390	460	475	4.20	4.30	4.30	1,638	1,978	2,043
PA	650	670	680	3.10	2.50	2.60	2,015	1,675	1,768
RI	1	1	1	2.50	2.20	2.20	3	2	2
SD	2,650	3,000	2,400	2.05	2.20	1.40	5,433	6,600	3,360
TN	35	35	30	3.70	3.90	3.80	130	137	114
TX	120	130	130	4.00	4.90	5.00	480	637	650
UT	550	550	560	4.00	4.00	3.60	2,200	2,200	2,016
VT	50	40	45	2.00	2.00	2.00	100	80	90
VA	120	110	120	4.00	3.10	2.50	480	341	300
WA	470	470	490	5.00	4.80	5.00	2,350	2,256	2,450
WV	50	50	50	3.20	2.50	2.50	160	125	125
WI	1,800	1,700	1,650	3.00	2.50	2.80	5,400	4,250	4,620
WY	620	580	500	2.30	2.20	2.30	1,426	1,276	1,150
US	23,077	23,822	23,135	3.48	3.37	3.19	80,347	80,327	73,824

[1] Preliminary.

NASS, Crops Branch, (202) 720–2127.

Table 6-4.—Hay, all other: Area, yield, and production, by States, 2000–2002

State	Area harvested			Yield per harvested acre			Production		
	2000	2001	2002[1]	2000	2001	2002[1]	2000	2001	2002[1]
	1,000 acres	*1,000 acres*	*1,000 acres*	*Tons*	*Tons*	*Tons*	*1,000 tons*	*1,000 tons*	*1,000 tons*
AL	720	920	800	1.80	2.60	2.20	1,296	2,392	1,760
AZ	42	43	45	4.00	3.30	3.80	168	142	171
AR	1,230	1,300	1,350	2.30	2.10	2.60	2,829	2,730	3,510
CA	510	530	500	2.80	3.10	3.00	1,428	1,643	1,500
CO	500	650	570	1.50	1.80	1.30	750	1,170	741
CT	53	55	50	2.10	1.80	1.80	111	99	90
DE	9	9	8	2.50	2.40	2.30	23	22	18
FL	270	270	280	2.50	2.80	2.80	675	756	784
GA	650	650	650	2.40	3.00	2.60	1,560	1,950	1,690
ID	260	300	320	2.10	1.90	1.90	546	570	608
IL	350	300	350	2.20	2.40	2.10	770	720	735
IN	320	280	320	2.70	2.60	2.10	864	728	672
IA	450	400	350	2.50	2.35	2.20	1,125	940	770
KS	1,900	2,400	2,300	1.50	1.60	1.50	2,850	3,840	3,450
KY	2,200	2,100	2,100	2.40	2.20	2.20	5,280	4,620	4,620
LA	350	450	450	1.90	2.80	2.50	665	1,260	1,125
ME	120	120	125	1.80	1.50	1.70	216	180	213
MD	170	160	160	2.50	2.00	2.20	425	320	352
MA	80	80	75	2.00	1.80	2.00	160	144	150
MI	300	250	250	2.10	2.20	2.20	630	550	550
MN	700	700	700	1.80	1.60	1.90	1,260	1,120	1,330
MS	800	780	750	1.60	2.50	2.50	1,280	1,950	1,875
MO	3,250	3,600	3,800	1.60	1.80	1.70	5,200	6,480	6,460
MT	800	1,000	1,200	1.30	1.40	1.40	1,040	1,400	1,680
NE	1,700	1,800	1,900	1.10	1.35	1.00	1,870	2,430	1,900
NV	225	230	210	1.70	1.70	1.60	383	391	336
NH	50	50	48	1.70	1.70	1.50	85	85	72
NJ	100	90	90	1.70	1.70	1.50	170	153	135
NM	90	110	120	1.80	2.20	1.90	162	242	228
NY	1,100	1,100	1,150	1.90	1.80	2.10	2,090	1,980	2,415
NC	690	690	730	2.60	2.20	1.50	1,794	1,518	1,095
ND	1,100	1,100	1,850	1.70	1.55	1.10	1,870	1,705	2,035
OH	830	950	900	2.70	2.40	2.20	2,241	2,280	1,980
OK	2,100	2,200	2,400	1.70	1.40	1.60	3,570	3,080	3,840
OR	690	565	620	2.00	1.90	2.20	1,380	1,074	1,364
PA	1,150	980	1,120	2.10	1.80	1.60	2,415	1,764	1,792
RI	8	7	6	2.10	1.70	2.20	17	12	13
SC	300	320	330	2.40	2.00	1.90	720	640	627
SD	1,400	1,700	1,600	1.40	1.50	0.90	1,960	2,550	1,440
TN	2,000	2,100	2,000	2.30	2.20	2.20	4,600	4,620	4,400
TX	4,000	5,100	5,500	2.10	2.00	2.40	8,400	10,200	13,200
UT	150	160	150	2.00	2.10	1.80	300	336	270
VT	180	200	195	1.70	1.60	2.00	306	320	390
VA	1,200	1,200	1,250	2.30	2.00	1.40	2,760	2,400	1,750
WA	310	320	320	2.90	2.60	2.80	899	832	896
WV	550	530	520	2.10	1.80	1.80	1,155	954	936
WI	300	300	400	2.00	1.80	1.80	600	540	720
WY	520	550	450	1.30	1.10	1.00	676	605	450
US	36,777	39,699	41,362	1.95	1.93	1.86	71,574	76,437	77,138

[1] Preliminary.

NASS, Crops Branch, (202) 720–2127.

Table 6-5.—Hay, all: Area, yield, and production, by States, 2000–2002

State	Area harvested			Yield per harvested acre			Production		
	2000	2001	2002 [1]	2000	2001	2002 [1]	2000	2001	2002 [1]
	1,000 acres	*1,000 acres*	*1,000 acres*	*Tons*	*Tons*	*Tons*	*1,000 tons*	*1,000 tons*	*1,000 tons*
AL	720	920	800	1.80	2.60	2.20	1,296	2,392	1,760
AZ	247	258	275	7.57	7.22	7.40	1,870	1,862	2,034
AR	1,250	1,320	1,375	2.30	2.12	2.61	2,879	2,792	3,595
CA	1,530	1,540	1,640	5.60	5.79	5.85	8,568	8,915	9,594
CO	1,400	1,600	1,350	2.91	2.99	2.22	4,080	4,780	3,003
CT	65	63	59	2.11	1.86	1.95	137	117	115
DE	17	17	15	3.71	2.88	2.87	63	49	43
FL	270	270	280	2.50	2.80	2.80	675	756	784
GA	650	650	650	2.40	3.00	2.60	1,560	1,950	1,690
ID	1,390	1,420	1,570	3.81	3.48	3.57	5,292	4,938	5,608
IL	850	800	800	3.14	3.34	2.94	2,670	2,670	2,355
IN	750	610	600	3.50	3.36	2.66	2,627	2,048	1,596
IA	1,700	1,650	1,600	3.53	3.37	3.53	6,000	5,565	5,645
KS	2,800	3,300	3,250	2.34	2.42	2.14	6,540	7,980	6,965
KY	2,450	2,350	2,400	2.55	2.36	2.30	6,255	5,545	5,520
LA	350	450	450	1.90	2.80	2.50	665	1,260	1,125
ME	132	130	133	1.83	1.55	1.76	242	202	234
MD	235	225	220	3.03	2.32	2.31	711	522	508
MA	96	98	93	2.05	1.89	2.15	197	185	200
MI	1,300	1,150	1,150	3.33	3.30	3.22	4,330	3,790	3,700
MN	2,250	2,150	2,300	3.04	2.88	2.87	6,840	6,195	6,610
MS	800	780	750	1.60	2.50	2.50	1,280	1,950	1,875
MO	3,720	4,050	4,260	1.79	1.94	1.84	6,657	7,853	7,840
MT	2,000	2,450	2,600	1.78	1.81	1.78	3,560	4,445	4,620
NE	3,050	3,250	3,250	1.99	2.33	1.83	6,055	7,578	5,950
NV	490	495	485	3.27	3.20	3.13	1,602	1,584	1,519
NH	58	57	55	1.74	1.74	1.58	101	99	87
NJ	130	120	115	2.00	2.13	1.83	260	255	210
NM	380	380	380	4.39	4.19	4.43	1,670	1,592	1,684
NY	1,520	1,660	1,720	2.04	2.14	2.17	3,098	3,548	3,726
NC	710	710	750	2.60	2.22	1.51	1,848	1,578	1,131
ND	2,450	2,700	3,300	2.09	1.88	1.19	5,110	5,065	3,920
OH	1,400	1,520	1,490	3.23	2.81	2.52	4,521	4,275	3,750
OK	2,430	2,550	2,740	1.92	1.58	1.84	4,659	4,025	5,030
OR	1,080	1,025	1,095	2.79	2.98	3.11	3,018	3,052	3,407
PA	1,800	1,650	1,800	2.46	2.08	1.98	4,430	3,439	3,560
RI	9	8	7	2.22	1.75	2.14	20	14	15
SC	300	320	330	2.40	2.00	1.90	720	640	627
SD	4,050	4,700	4,000	1.83	1.95	1.20	7,393	9,150	4,800
TN	2,035	2,135	2,030	2.32	2.23	2.22	4,730	4,757	4,514
TX	4,120	5,230	5,630	2.16	2.07	2.46	8,880	10,837	13,850
UT	700	710	710	3.57	3.57	3.22	2,500	2,536	2,286
VT	230	240	240	1.77	1.67	2.00	406	400	480
VA	1,320	1,310	1,370	2.45	2.09	1.50	3,240	2,741	2,050
WA	780	790	810	4.17	3.91	4.13	3,249	3,088	3,346
WV	600	580	570	2.19	1.86	1.86	1,315	1,079	1,061
WI	2,100	2,000	2,050	2.86	2.40	2.60	6,000	4,790	5,340
WY	1,140	1,130	950	1.84	1.66	1.68	2,102	1,881	1,600
US	59,854	63,521	64,497	2.54	2.47	2.34	151,921	156,764	150,962

[1] Preliminary.

NASS, Crops Branch, (202) 720–2127.

Table 6-6.—Hay, all: Marketing year average price and value of production, by States, crop of 2000, 2001, and 2002

State	Marketing year average price per ton, baled			Value of production		
	2000	2001	2002	2000	2001	2002
	Dollars	*Dollars*	*Dollars*	*1,000 dollars*	*1,000 dollars*	*1,000 dollars*
AL	56.00	55.00	55.00	72,576	131,560	96,800
AZ	94.00	99.00	100.00	174,688	183,273	202,289
AR	58.50	62.50	61.10	164,374	171,488	198,835
CA	90.50	115.00	95.50	768,978	1,018,989	913,509
CO	85.50	101.00	117.00	347,880	480,440	366,795
CT	142.00	147.00	142.00	19,386	17,244	16,275
DE	112.00	136.00	145.00	7,244	6,731	6,302
FL	82.00	96.00	97.00	55,350	72,576	76,048
GA	59.00	58.00	57.00	92,040	113,100	96,330
ID	94.50	116.00	96.50	491,547	565,014	534,688
IL	83.00	87.50	89.50	216,095	231,645	204,833
IN	86.00	90.00	123.00	216,628	176,088	175,308
IA	82.00	89.50	84.50	473,438	480,095	464,295
KS	77.00	89.00	93.50	468,465	641,130	594,563
KY	81.50	71.00	71.00	475,875	359,998	360,270
LA	49.00	51.00	45.00	32,585	64,260	50,625
ME	106.00	108.00	109.00	25,732	21,778	25,497
MD	113.00	134.00	143.00	79,934	70,250	72,640
MA	143.00	146.00	146.00	27,668	26,965	29,100
MI	62.50	70.50	80.00	272,040	266,740	291,850
MN	59.00	58.00	68.00	422,730	366,468	465,140
MS	45.00	44.00	40.00	57,600	85,800	75,000
MO	70.50	75.00	67.00	480,585	551,911	500,640
MT	85.50	94.00	84.00	299,100	412,598	379,050
NE	69.50	70.50	85.50	416,103	519,957	502,325
NV	93.50	114.00	103.00	150,484	180,165	159,145
NH	128.00	131.00	137.00	12,951	12,992	11,877
NJ	108.00	111.00	112.00	27,890	28,713	23,430
NM	122.00	124.00	139.00	203,516	195,994	233,200
NY	103.00	104.00	106.00	315,367	353,902	373,980
NC	71.00	72.00	81.50	131,520	113,460	92,100
ND	42.50	45.50	64.50	207,010	228,528	231,665
OH	90.00	87.00	107.00	387,618	365,370	383,430
OK	73.50	76.50	82.50	284,498	280,595	346,750
OR	94.50	112.00	106.00	278,772	333,626	357,729
PA	109.00	117.00	128.00	491,360	406,320	473,096
RI	143.00	145.00	141.00	2,850	2,034	2,115
SC	80.00	85.00	70.00	57,600	54,400	43,890
SD	63.50	65.50	76.00	453,191	574,725	353,040
TN	51.00	54.00	58.00	240,090	257,091	263,796
TX	76.00	75.00	79.50	607,080	743,891	930,400
UT	78.50	95.00	95.50	190,500	232,552	212,085
VT	109.00	111.00	112.00	44,412	44,480	53,550
VA	85.50	84.00	95.50	276,840	230,015	197,300
WA	107.00	120.00	112.00	355,261	375,328	381,080
WV	62.50	66.50	73.00	82,510	71,875	77,334
WI	41.50	55.00	65.50	281,400	276,495	358,980
WY	84.50	109.00	110.00	175,290	203,885	173,750
US	84.60	96.50	94.00	11,416,651	12,602,534	12,432,729

NASS, Crops Branch, (202) 720–2127.

Table 6-7.—Hay: Area and production, by kinds, United States, 1993–2002

Year	Area harvested			Production		
	Alfalfa	All other hay	All hay	Alfalfa	All other hay	All hay
	1,000 acres	*1,000 acres*	*1,000 acres*	*1,000 acres*	*1,000 acres*	*1,000 acres*
1993	24,673	35,016	59,689	80,115	66,584	146,699
1994	24,138	34,677	58,815	81,130	69,006	150,136
1995	24,404	35,360	59,764	84,138	70,101	154,239
1996	24,206	36,963	61,169	79,139	70,640	149,779
1997	23,551	37,533	61,084	78,535	74,001	152,536
1998	23,672	36,404	60,076	82,310	69,470	151,780
1999	24,055	39,165	63,220	84,385	75,322	159,707
2000	23,077	36,777	59,854	80,347	71,574	151,921
2001	23,822	39,699	63,521	80,327	76,437	156,764
2002 [1]	23,135	41,362	64,497	73,824	77,138	150,962

[1] Preliminary.

NASS, Crops Branch, (202) 720–2127.

Table 6-8.—Hay: Supply and disappearance, prices, and number of animal units fed annually, United States, 1993–2002 [1]

Year beginning May	Farm carryover May 1	Production	Total supply	Disappearance	Roughage-consuming animal units	Supply per animal unit	Disappearance per animal unit	Price received per ton
	Million tons	*Million tons*	*Million tons*	*Million tons*	*Million units*	*Tons*	*Tons*	*Dollars*
1993	21.0	146.7	167.7	145.7	76.5	2.19	1.91	84.70
1994	22.1	150.1	172.2	151.4	78.0	2.21	1.94	86.70
1995	20.8	154.2	175.0	154.2	78.1	2.24	1.97	82.20
1996	20.8	149.8	170.5	152.8	76.4	2.23	2.00	95.80
1997	17.4	152.5	170.0	148.1	74.9	2.27	1.98	100.00
1998	21.8	151.8	173.6	148.8	74.5	2.33	2.00	84.60
1999	24.8	159.7	184.5	155.7	73.3	2.52	2.12	76.90
2000	28.8	151.9	181.0	159.6	72.5	2.49	2.20	84.60
2001	21.1	156.8	177.8	155.4	72.2	2.46	2.15	96.50
2002 [2]	22.5	151.0	179.2	NA	72.2	2.40	NA	NA

[1] Excludes trade. [2] Preliminary. NA-not available.

ERS, Market and Trade Economics Division, (202) 694-5296.

Table 6-9.—Field seeds: Average retail price paid by farmers for seed, Apr. 15, United States, 1993–2002

Kind of seed	1993	1994	1995	1996	1997	1998	1999	2000	2001	2002
	Price per 100 pounds									
	Dollars	*Dollars*	*Dollars*	*Dollars*	*Dollars*	*Dollars*	*Dollars*	*Dollars*	*Dollars*	*Dollars*
Alfalfa, uncertified varieties	152.00	161.00	168.00	185.00	185.00	205.00	184.00	165.00	158.00	280.00
Alfalfa, certified varieties	269.00	266.00	274.00	277.00	282.00	288.00	287.00	277.00	278.00	157.00
Clover, ladino	324.00	321.00	320.00	318.00	307.00	308.00	298.00	285.00	285.00	280.00
Clover, red	148.00	148.00	134.00	172.00	184.00	194.00	178.00	143.00	132.00	130.00
Lespedeza, Korean	132.00	84.50	66.00	99.00	90.00	89.00	76.15	77.50	160.00	98.00
Lespedeza, Striate, Kobe	119.00	82.40	95.60	125.00	112.00	108.00	96.00	90.00	180.00	104.00
Lespedeza, Sericea	92.10	169.00	178.00	291.00	220.00	290.00	294.00	310.00	330.00	300.00
Timothy	80.60	76.00	71.00	76.00	73.00	71.20	78.80	115.00	105.00	90.00
Orchardgrass	122.00	161.00	148.00	141.00	119.00	116.00	107.00	108.00	135.00	143.00
Blue Grass, Kentucky:										
Public and common	239.00	211.00	159.00	172.00	153.00	152.00	129.00	158.00	140.00	155.00
Proprietary, including Merion	283.00	274.00	228.00	243.00	224.00	216.00	204.00	214.00	220.00	225.00
Ryegrass, annual	56.70	64.60	67.60	58.80	57.90	65.30	64.20	60.50	55.50	58.00
Tall fescue	87.00	71.50	88.10	109.00	148.00	101	99.50	91.00	114.00	106.00
Sudangrass	45.20	47.90	51.80	51.90	51.40	53.70	52.20	53.00	53.00	56.00
Potatoes	8.60	10.20	7.90	10.30	7.60	9.10	8.50	10.45	8.50	10.90
Peanuts	77.30	86.90	79.50	82.00	81.75	83.60	80.90	81.70	82.60	82.10
Sunflower	300.00	297.00	297.00	313.00	355.00	380.00	400.00	395.00	407.00	407.00
Cottonseed, all	62.70	63.50	68.20	73.00	74.90	79.30	82.40	128.00	154.00	213.00
Biotech [1]	....	....	....	....	....	....	....	....	217.00	271.00
Non-biotech	....	....	....	....	....	....	....	....	87.00	94.00
Grain sorghum, hybrid	74.50	82.10	78.70	84.00	92.00	96.00	97.60	93.00	93.00	96.00
	Price per bushel									
	Dollars	*Dollars*	*Dollars*	*Dollars*	*Dollars*	*Dollars*	*Dollars*	*Dollars*	*Dollars*	*Dollars*
Corn, hybrid, all [2]	72.70	73.40	77.10	77.70	83.50	86.90	88.10	87.50	92.20	92.00
Biotech [1]	....	....	....	....	....	....	....	....	110.00	113.00
Non-biotec	....	....	....	....	....	....	....	....	85.30	85.80
Wheat (spring)	5.98	7.37	7.12	8.10	7.30	6.85	6.10	6.10	6.20	6.50
Wheat (winter)	7.73	7.90	7.80	8.50	10.00	8.25	7.35	7.05	7.20	7.70
Oats (spring)	4.39	4.69	4.51	5.19	5.32	5.02	4.60	4.50	4.70	5.35
Rice	15.40	22.00	15.10	17.50	19.00	19.50	19.10	17.25	15.70	14.90
Barley (spring)	5.00	5.18	5.37	6.49	6.13	6.04	5.80	5.80	5.80	5.80
Soybeans for seed, all	12.40	13.60	13.40	14.80	16.10	17.15	17.00	17.10	20.70	22.50
Biotech [1]	....	....	....	....	....	....	....	....	23.90	27.00
Non-biotec	....	....	....	....	....	....	....	....	17.90	15.00
Flaxseed	7.37	7.74	8.00	8.14	9.31	10.00	8.50	7.90	7.60	7.60

[1] Biotech varities are made to be resistant to herbicides, insects, or both. A technology fee is included within the price. [2] Price per 80,000 kernels.

NASS, Environmental, Economics, and Demographics Branch, (202) 720–6146.

Table 6-10—Beans, dry edible (cleaned basis): Production, by classes, United States, 2000–2002 [1]

Class	2000	2001	2002 [2]
	1,000 cwt.	*1,000 cwt.*	*1,000 cwt.*
Navy (pea beans)	4,771	2,311	5,304
Great northern	2,489	2,108	1,545
Small white	64	39	65
Pinto	10,670	8,721	12,880
Red kidney, light	1,352	776	1,199
Red kidney, dark	1,014	727	1,075
Pink	320	326	607
Small red	313	172	605
Cranberry	449	147	359
Black	1,336	783	3,114
Large lima (CA)	437	326	334
Baby lima (CA)	542	235	501
Blackeye	382	553	543
Garbanzo	1,308	1,612	838
Other	962	747	1,005
Total	26,409	19,583	29,974

[1] Excludes beans grown for garden seed. [2] Preliminary.
NASS, Crops Branch, (202) 720–2127.

Table 6-11.—Beans, dry edible: Area, yield, and production, by States, 2000–2002 [1]

State	Area planted			Area harvested			Yield per acre (cleaned basis)			Production (cleaned basis)		
	2000	2001	2002 [2]	2000	2001	2002 [2]	2000	2001	2002 [2]	2000	2001	2002 [2]
	1,000 acres	*1,000 acres*	*1,000 acres*	*1,000 acres*	*1,000 acres*	*1,000 acres*	*Pounds*	*Pounds*	*Pounds*	*1,000 cwt.*	*1,000 cwt.*	*1,000 cwt.*
CA	115.0	88.0	92.0	112.0	85.0	89.0	1,840	1,760	1,980	2,059	1,496	1,762
CO	120.0	115.0	92.0	110.0	105.0	70.0	1,800	1,700	2,170	1,980	1,785	1,519
ID	90.0	75.0	95.0	88.0	73.0	93.0	1,950	1,950	2,050	1,716	1,424	1,907
KS	18.0	15.0	18.0	16.0	14.0	14.5	1,810	1,850	1,100	289	259	160
MI	285.0	215.0	270.0	275.0	130.0	265.0	1,500	600	1,850	4,125	780	4,903
MN	165.0	115.0	170.0	150.0	105.0	150.0	1,600	1,500	1,650	2,400	1,575	2,475
MT	40.5	43.5	26.9	34.8	28.5	23.0	1,400	1,320	1,570	486	376	361
NE	165.0	160.0	185.0	156.0	148.0	165.0	2,070	2,150	2,100	3,230	3,185	3,465
NM [3]	....	15.0	8.0	....	15.0	8.0	....	2,000	1,800	....	300	144
NY	25.0	23.0	25.0	24.5	22.3	24.5	1,460	870	1,360	358	194	333
ND	610.0	440.0	790.0	525.0	400.0	690.0	1,450	1,550	1,540	7,613	6,200	10,626
OR	12.0	10.0	9.8	11.7	9.5	9.1	1,800	1,810	1,730	211	172	157
SD	11.0	18.0	21.0	10.8	17.0	16.0	2,090	1,590	1,630	226	270	261
TX	20.0	30.0	37.5	16.6	26.4	32.5	950	1,320	970	158	348	315
UT	5.4	6.1	1.8	3.0	5.7	0.3	330	300	1,670	10	17	5
WA	32.0	34.0	41.0	32.0	34.0	41.0	2,000	1,700	2,000	640	578	820
WI	8.3	6.3	7.1	8.1	6.1	7.0	1,800	1,800	1,960	146	110	137
WY	36.0	27.0	32.0	34.0	24.0	29.0	2,240	2,140	2,150	762	514	624
US	1,758.2	1,435.9	1,922.1	1,607.5	1,248.5	1,726.9	1,643	1,569	1,736	26,409	19,583	29,974

[1] Excludes beans grown for garden seed. [2] Preliminary. [3] Estimates discontinued in 2000, reinstated in 2001.
NASS, Crops Branch, (202) 720–2127.

Table 6-12.—Beans, dry edible: Area, yield, production, price, and value, United States, 1993–2002 [1]

Year	Area planted	Area harvested	Yield per acre [2]	Production [2]	Marketing year average price per 100 pounds received by farmers	Value of production
	1,000 acres	*1,000 acres*	*Pounds*	*1,000 cwt.*	*Dollars*	*1,000 dollars*
1993	1,867.9	1,618.0	1,351	21,862	24.60	538,210
1994	2,011.8	1,831.2	1,581	28,950	22.50	631,080
1995	2,066.3	1,896.3	1,618	30,689	20.80	633,620
1996	1,839.0	1,750.7	1,594	27,912	23.50	652,240
1997	1,869.8	1,758.8	1,670	29,370	19.30	576,658
1998	2,014.1	1,917.7	1,586	30,418	19.00	567,243
1999	2,023.0	1,877.0	1,763	33,085	16.40	547,636
2000	1,758.2	1,607.5	1,643	26,409	15.50	413,986
2001	1,435.9	1,248.5	1,569	19,583	22.10	426,475
2002 [3]	1,922.1	1,726.9	1,736	29,974	17.00	519,609

[1] Excludes beans grown for garden seed. [2] Cleaned basis. [3] Preliminary.
NASS, Crops Branch, (202) 720–2127.

Table 6-13.—Beans, dry edible (cleaned basis): Marketing year average price and value of production, by States, crop of 2000, 2001, and 2002 [1]

State	Marketing year average price per cwt.			Value of production		
	2000	2001	2002 [2]	2000	2001	2002 [2]
	Dollars	*Dollars*	*Dollars*	*1,000 dollars*	*1,000 dollars*	*1,000 dollars*
CA	26.80	31.20	32.00	55,181	46,675	56,384
CO	15.60	23.60	18.20	30,888	42,126	27,646
ID	17.30	21.20	19.50	29,687	30,189	37,187
KS	13.90	18.00	16.00	4,017	4,662	2,560
MI	13.70	24.60	14.50	56,513	19,188	71,094
MN	15.90	24.30	16.80	38,160	38,273	41,580
MT	16.30	17.30	15.20	7,922	6,505	5,487
NE	15.80	18.50	19.20	51,034	58,923	66,528
NM [3]	....	19.00	15.00	....	5,700	2,160
NY	18.80	24.70	21.90	6,730	4,792	7,293
ND	12.60	20.40	14.80	95,924	126,480	157,265
OR	17.90	20.50	18.40	3,777	3,526	2,889
SD	14.50	18.70	17.40	3,277	5,049	4,541
TX	19.00	20.00	20.00	3,002	6,960	6,300
UT	20.60	27.00	20.00	206	459	100
WA	18.30	20.80	20.20	11,712	12,022	16,564
WI	21.60	28.40	17.70	3,154	3,124	2,425
WY	16.80	23.00	18.60	12,802	11,822	11,606
US	15.50	22.10	17.00	413,986	426,475	519,609

[1] Excludes beans grown for garden seed. [2] Preliminary. [3] Estimates discontinued in 2000, reinstated in 2001.
NASS, Crops Branch, (202) 720–2127.

Table 6-14.—Beans, dry edible: Season average wholesale price per 100 pounds, selected markets, 1992–2001

Year beginning September	F.o.b. California points			F.o.b. Northern Colorado points: Pinto	F.o.b. Western Nebraska points: Great northern	F.o.b. Southern Idaho points: Small red	F.o.b. Michigan points:		
	Baby lima	Large lima	Blackeye				Pea bean (Navy)	Black	Light red kidney
	Dollars	*Dollars*	*Dollars*	*Dollars*	*Dollars*	*Dollars*	*Dollars*	*Dollars*	*Dollars*
1992	22.70	27.48	25.80	23.81	21.25	30.02	21.15	24.25	34.82
1993	34.65	41.87	42.72	33.28	33.15	29.33	23.73	27.95	30.57
1994	33.81	44.67	40.18	21.32	37.18	28.19	31.63	32.60	28.77
1995	39.66	49.97	29.36	25.74	38.55	28.87	23.88	22.86	29.02
1996	45.58	57.09	32.57	27.56	26.61	39.31	23.11	27.08	37.76
1997	27.51	39.81	31.00	26.48	27.07	28.16	20.82	33.19	28.44
1998	41.27	46.80	37.68	20.89	25.84	27.54	26.32	28.56	32.40
1999	28.13	35.48	23.40	19.75	24.25	21.83	19.19	18.37	26.79
2000	25.68	34.05	25.91	21.00	23.22	24.32	16.40	18.38	25.47
2001 [1]	33.91	41.65	29.94	31.38	23.52	33.43	26.00	37.29	34.15

[1] Preliminary.
ERS, Specialty Crops Branch, (202) 694–5253. Compiled from the Bean Market Summary, Agricultural Marketing Service, U.S. Department of Agriculture, Greeley, Colorado.

Table 6-15.—Beans, dry edible: United States exports by class and quantity, 1992/93–2001/2002

Year[1]	Navy or pea	Great northern	Other white	Pinto	Red kidney	Other[2]	Total
	Metric tons	*Metric tons*	*Metric tons*	*Metric tons*	*Metric tons*	*Metric tons*	*Me tric tons*
1992/93	81,688	17,554	6,974	56,711	35,190	101,103	299,220
1993/94	68,274	16,368	1,481	64,235	33,388	93,478	277,224
1994/95	95,426	36,484	7,438	144,344	38,166	95,316	417,174
1995/96	86,464	28,291	1,126	82,062	32,380	98,920	329,244
1996/97	95,279	40,773	551	74,960	35,335	107,092	353,992
1997/98	107,669	50,201	1,338	96,256	33,605	150,203	439,271
1998/99	90,680	42,011	1,561	94,991	27,680	117,212	374,135
1999/2000	67,222	38,204	1,388	64,337	28,662	127,581	327,394
2000/2001	89,997	50,742	1,565	93,037	31,939	103,492	370,771
2001/2002	63,069	48,563	882	71,310	20,066	76,428	280,317

[1] Marketing year beginning September. [2] Includes other colored, black, blackeye, and limas.

FAS, Grain and Feed Division, (202) 720–6219. Compiled from reports of the U.S. Department of Commerce.

Table 6-16.—Beans, dry edible: United States exports to specified countries, 1999/2000–2001/2002 [1] [2]

Country	1999/2000	2000/2001	2001/2002
	1,000 metric tons	*1,000 metric tons*	*1,000 metric tons*
Mexico	75	90	60
United Kingdom	35	59	40
Algeria	9	18	17
Dominican Republic	7	10	16
Haiti	25	21	16
Japan	17	17	14
Canada	20	22	14
Angola	16	12	9
France	11	14	8
Spain	5	5	6
Guatemala	2	6	6
Greece	4	6	6
Italy	18	15	6
Mozambique	2	0	5
Albania	1	0	4
Australia	5	4	4
Korea, Republic of	2	3	3
New Zealand	2	4	3
Bulgaria	0	0	3
Yugoslavia	15	2	3
Other	55	63	37
Total	327	371	280

[1] Marketing year beginning September. [2] Excluding seed bean exports. Compiled from U.S. Census data.

FAS, Grain and Feed Division, (202) 720–6219.

Table 6-17.—Peas, dry field: United States exports to specified countries, 1999/2000–2001/2002 [1] [2] [3]

Country	1999/2000	2000/2001	2001/2002
	Metric tons	*Metric tons*	*Metric tons*
Canada	8	10	31
Philippines	7	14	19
Spain	7	12	9
India	5	13	9
Korea, Republic of	1	2	4
Netherlands	1	0	3
Uganda	2	0	3
Taiwan	2	2	2
Kenya	2	3	2
Djibouti Afars-Issas	0	0	2
South Africa, Republic	0	0	2
Algeria	1	4	2
Bolivia	1	0	2
Peru	5	4	2
Angola	3	0	1
Mexico	3	3	1
Russian Federation	2	9	1
Japan	2	2	1
Pakistan	2	2	1
Italy	2	3	1
Other	34	20	12
Total	89	104	110

[1] Marketing year beginning September. [2] Excluding seed pea exports. [3] Regional totals include countries not listed.
FAS, Grain and Feed Division, (202) 720–6219.

Table 6-18.—Hops: Area, yield, production, value, and Sept. 1 stocks, United States, 1993–2002

Year	Area harvested	Yield per acre	Production	Marketing year average price per pound received by farmers	Value of production	Stocks Sept. 1
	1,000 acres	*Pounds*	*1,000 pounds*	*Cents*	*1,000 dollars*	*1,000 pounds*
1993	43.1	1,767	76,144	176.0	133,965	58,060
1994	42.4	1,758	74,560	181.0	134,701	63,000
1995	43.2	1,826	78,852	171.0	135,087	55,900
1996	44.2	1,698	74,971	165.0	123,530	58,700
1997	43.3	1,729	74,872	160.0	119,840	62,000
1998	36.6	1,625	59,548	169.0	100,728	55,000
1999	34.3	1,881	64,456	169.0	109,099	54,000
2000	36.1	1,871	67,577	187.0	126,217	48,000
2001	35.9	1,861	66,832	185.0	123,843	54,000
2002 [1]	29.3	1,990	58,337	194.0	113,413	65,000

[1] Preliminary.
NASS, Crops Branch, (202) 720–2127.

Table 6-19.—Hops: Area, yield, and production, by States, 2000–2002

State	Area harvested			Yield per acre			Production		
	2000	2001	2002 [1]	2000	2001	2002 [1]	2000	2001	2002 [1]
	1,000 acres	*1,000 acres*	*1,000 acres*	*Pounds*	*Pounds*	*Pounds*	*1,000 pounds*	*1,000 pounds*	*1,000 pounds*
ID	3,321	3,469	3,399	1,484	1,329	1,624	4,929.8	4,609.3	5,519.6
OR	5,819	6,103	5,577	1,785	1,875	1,692	10,387.0	11,443.2	9,438.0
WA	26,980	26,339	20,333	1,937	1,928	2,133	52,260.0	50,779.6	43,379.0
US	36,120	35,911	29,309	1,871	1,861	1,990	67,576.8	66,832.1	58,336.6

[1] Preliminary.
NASS, Crops Branch, (202) 720–2127.

Table 6-20.—Hops: Marketing year average price and value of production, by States, crop of 2000, 2001, and 2002

State	Marketing year average price per pound			Value of production		
	2000	2001	2002 [1]	2000	2001	2002 [1]
	Dollars	*Dollars*	*Dollars*	*1,000 dollars*	*1,000 dollars*	*1,000 dollars*
ID	1.78	1.59	1.58	8,775	7,329	8,721
OR	2.20	2.15	2.13	22,851	24,603	20,103
WA	1.81	1.81	1.95	94,591	91,911	84,589
Total	1.87	1.85	1.94	126,217	123,843	113,413

[1] Preliminary.
NASS, Crops Branch, (202) 720–2127.

Table 6-21.—Hops: United States exports by country of destination and imports by country of origin, 1999/2000–2001/2002

Item and country	Year beginning September		
	1999/2000	2000/2001	2001/2002
	Metric tons	*Metric tons*	*Metric tons*
Exports			
North America:			
Canada	1,326	1,109	1,147
Mexico	974	898	894
Total	2,300	2,007	2,041
European Union:			
Germany	799	1,171	1,580
Belgium-Luxembourg	310	870	1,224
United Kingdom	1,002	628	550
Netherlands	130	243	320
Finland	75	91	23
Other Countries	385	351	406
Total	2,701	3,354	4,103
Asia:			
Japan	546	385	518
Philippines	134	146	122
Thailand	126	343	313
Hong Kong	94	267	325
Other Countries	105	402	517
Total	1,004	1,543	1,795
Latin America:			
Brazil	865	1,442	1,598
Colombia	268	285	363
Argentina	183	196	225
Dominican Republic	216	216	274
Ecuador	24	97	57
Peru	90	61	30
Other Countries	522	292	499
Total	2,168	2,589	3,046
Other countries:			
Russian Federation	141	140	70
Australia	126	104	164
Other Countries	339	147	635
Total	606	391	869
Grand total	8,779	9,884	11,854
Imports			
Eastern Europe:			
Bulgaria	0	2	0
Czech Republic	79	302	37
Poland	0	9	0
Yugoslavia	10	0	5
Total	89	313	42
European Union:			
Belgium-Luxemborg	20	0	13
France	636	805	1,050
German	3,807	3,374	2,149
United Kingdom	129	219	120
Total	4,592	4,398	3,332
Oceania:			
Australia	130	100	441
New Zealand	139	282	276
Total	269	382	717
Other Africa:			
Ethiopia	0	4	0
Total	0	4	0
Grand total	4,950	5,097	4,091

FAS, Horticultural and Tropical Products Division, (202) 720–3423.

CHAPTER VII

STATISTICS OF CATTLE, HOGS, AND SHEEP

This chapter contains information about most kinds of farm livestock and livestock products, with the exception of dairy and poultry. The information relates to inventories, production, disposition, prices, and income for farm animals, and to livestock slaughter (including horse slaughter), meat production, and market statistics for meat animals.

Table 7-1.—All cattle and calves: Number and value, United States, Jan. 1, 1994–2003

Year	Total number	Value	
		Per head	Total
	Thousands	*Dollars*	*1,000 dollars*
1994	100,974	659	66,512,550
1995	102,785	615	63,185,288
1996	103,548	503	52,055,705
1997	101,656	525	53,383,392
1998	99,744	603	60,193,070
1999	99,115	594	58,833,650
2000	98,198	683	67,099,440
2001	97,277	725	70,495,030
2002	96,704	747	72,283,865
2003 [1]	96,106	726	69,785,870

[1] Preliminary.

NASS, Livestock Branch, (202) 720–3570.

Table 7-2.—All cattle and calves: Number, by classes, United States, Jan. 1, 1994–2003

Year	All cattle and calves [1]	Cows and heifers that have calved		500 pounds and over					Calves under 500 pounds
				Heifers			Steers	Bulls	
		Beef cows	Milk cows	Beef cow replace-ments	Milk cow replace-ments	Other			
	Thou-sands	*Thou-sands*	*Thou-sands*	*Thou-sands*	*Thou-sands*	*Thou-sands*	*Thou-sands*	*Thou-sands*	*Thou-sands*
1994	100,974	34,603	9,507	6,364	4,125	9,104	17,086	2,312	17,873
1995	102,785	35,190	9,482	6,452	4,121	9,302	17,513	2,385	18,341
1996	103,548	35,319	9,420	6,189	4,090	9,948	17,815	2,384	18,384
1997	101,656	34,458	9,318	6,042	4,058	10,212	17,392	2,350	17,826
1998	99,744	33,885	9,199	5,764	3,986	10,051	17,189	2,270	17,401
1999	99,115	33,745	9,133	5,535	4,069	10,170	16,891	2,281	17,290
2000	98,198	33,569	9,190	5,503	4,000	10,147	16,682	2,293	16,815
2001	97,277	33,397	9,183	5,588	4,057	10,131	16,441	2,274	16,206
2002	96,704	33,118	9,112	5,561	4,060	10,057	16,790	2,244	15,763
2003 [2]	96,106	32,947	9,152	5,608	4,104	9,890	16,590	2,253	15,563

[1] Totals may not add due to rounding. [2] Preliminary.

NASS, Livestock Branch, (202) 720–3570.

Table 7-3.—All cattle and calves: Number and value, by States, Jan. 1, 2002–2003

State	Number		Value			
			Value per head		Total value	
	2002	2003 [1]	2002	2003 [1]	2002	2003 [1]
	Thousands	*Thousands*	*Dollars*	*Dollars*	*1,000 dollars*	*1,000 dollars*
AL	1,370	1,440	590	570	808,300	820,800
AK	11.5	11.5	820	820	9,430	9,430
AZ	840	820	770	760	646,800	623,200
AR	1,820	1,850	620	620	1,128,400	1,147,000
CA	5,200	5,250	1,060	930	5,512,000	4,882,500
CO	3,050	2,650	700	770	2,135,000	2,040,500
CT	61	56	980	1,010	59,780	56,560
DE	26	25	1,010	1,000	26,260	25,000
FL	1,780	1,750	700	670	1,246,000	1,172,500
GA	1,240	1,290	640	600	793,600	774,000
HI	152	150	570	560	86,640	84,000
ID	1,990	2,000	910	880	1,810,900	1,760,000
IL	1,430	1,360	700	710	1,001,000	965,600
IN	880	860	790	720	695,200	619,200
IA	3,550	3,600	720	730	2,556,000	2,628,000
KS	6,600	6,350	640	670	4,224,000	4,254,500
KY	2,300	2,430	660	630	1,518,000	1,530,900
LA	850	860	660	640	561,000	550,400
ME	97	93	980	980	95,060	91,140
MD	245	240	940	980	230,300	235,200
MA	51	49	1,000	1,020	51,000	49,980
MI	990	990	930	880	920,700	871,200
MN	2,550	2,470	800	770	2,040,000	1,901,900
MS	1,100	1,070	590	570	649,000	609,900
MO	4,350	4,500	680	660	2,958,000	2,970,000
MT	2,450	2,400	840	820	2,058,000	1,968,000
NE	6,400	6,200	690	720	4,416,000	4,464,000
NV	500	510	760	750	380,000	382,500
NH	41	40	1,060	1,060	43,460	42,400
NJ	44	46	1,020	1,010	44,880	46,460
NM	1,580	1,590	800	810	1,264,000	1,287,900
NY	1,390	1,450	1,160	960	1,612,400	1,392,000
NC	950	920	620	600	589,000	552,000
ND	1,970	1,880	820	840	1,615,400	1,579,200
OH	1,250	1,220	870	810	1,087,500	988,200
OK	5,200	5,400	650	650	3,380,000	3,510,000
OR	1,400	1,360	760	760	1,064,000	1,033,600
PA	1,640	1,630	1,070	930	1,754,800	1,515,900
RI	5.5	5.5	910	900	5,005	4,950
SC	430	435	630	590	270,900	256,650
SD	3,950	3,700	790	790	3,120,500	2,923,000
TN	2,200	2,270	600	590	1,320,000	1,339,300
TX	13,600	14,000	610	600	8,296,000	8,400,000
UT	920	880	770	760	708,400	668,800
VT	285	280	1,190	1,070	339,150	299,600
VA	1,650	1,630	680	620	1,122,000	1,010,600
WA	1,130	1,100	940	870	1,062,200	957,000
WV	415	405	620	600	257,300	243,000
WI	3,300	3,300	1,080	990	3,564,000	3,267,000
WY	1,470	1,290	780	760	1,146,600	980,400
US	96,704.0	96,106.0	747	726	72,283,865	69,785,870

[1] Preliminary.

NASS, Livestock Branch, (202) 720–3570.

Table 7-4.—Cattle and calves, Jan. 1: Number, by sex and weight classes, by States, 2002 and 2003

State	Cows and heifers that have calved				Heifers, 500 pounds and over					
	Beef cows		Milk cows		Beef cow replacements		Milk cow replacements		Other	
	2002	2003 [1]	2002	2003 [1]	2002	2003 [1]	2002	2003 [1]	2002	2003 [1]
	Thousands	*Thousands*	*Thousands*	*Thousands*	*Thousands*	*Thousands*	*Thousands*	*Thousands*	*Thousands*	*Thousands*
AL	750	761	20	19	102	118	7	7	41	40
AK	4.3	4.3	1.2	1.3	0.7	0.7	0.5	0.6	0.4	0.3
AZ	185	165	140	150	36	34	33	35	17	20
AR	927	949	33	31	165	180	15	10	65	75
CA	760	740	1,620	1,680	135	130	770	790	170	170
CO	797	702	93	98	120	102	40	38	750	650
CT	8	6	24	23	2.0	1.0	11.5	11.0	1.5	1.5
DE	4	4	9	9	0.9	0.6	3.5	3.3	1.6	1.1
FL	958	953	152	147	140	130	40	40	25	25
GA	594	625	86	85	87	90	29	29	30	36
HI	78	79	7	6	10	12	3	3	7	6
ID	493	490	377	390	85	90	175	190	200	195
IL	445	426	115	114	65	60	50	50	190	160
IN	226	230	154	145	42	41	66	55	62	56
IA	985	992	205	208	140	130	120	100	590	600
KS	1,505	1,525	95	105	230	250	40	50	1,650	1,620
KY	1,075	1,120	125	120	170	175	40	45	100	130
LA	466	471	54	49	81	86	11	10	17	17
ME	10	10	38	36	4.0	4.5	20.0	19.0	1.0	1.5
MD	41	41	81	80	14	11	33	32	9	9
MA	6	5	21	20	2.0	1.0	9.0	10.0	1.0	1.0
MI	73	89	297	301	30	35	135	135	45	42
MN	410	400	500	480	100	100	295	295	180	180
MS	576	557	34	33	98	95	15	15	27	25
MO	2,060	2,116	140	134	280	305	60	55	290	320
MT	1,451	1,402	19	18	370	380	13	11	177	179
NE	1,932	1,934	68	66	285	280	25	20	1,420	1,390
NV	240	245	25	25	44	46	10	11	38	35
NH	4	4	18	17	1.0	1.5	8.0	8.0	0.5	0.5
NJ	8	9	13	13	3	3	6	6	2	2
NM	500	470	290	310	100	90	60	70	110	110
NY	75	80	675	680	20	25	300	325	40	40
NC	434	417	66	63	73	73	30	28	20	19
ND	1,008	973	42	37	155	155	15	13	260	230
OH	280	260	260	260	70	70	110	110	70	70
OK	1,933	2,042	87	88	330	380	20	30	480	500
OR	605	593	105	117	155	115	55	60	115	110
PA	152	150	588	590	40	45	285	280	45	50
RI	1.4	1.6	1.4	1.4	0.4	0.3	0.7	0.6	0.1	0.1
SC	210	210	20	20	38	38	8	8	13	15
SD	1,792	1,686	98	94	300	290	30	30	540	470
TN	1,060	1,106	90	84	175	200	40	40	75	75
TX	5,440	5,489	310	311	750	760	100	110	1,650	1,710
UT	357	339	93	91	75	75	44	45	71	70
VT	12	10	154	153	4	4	58	59	4	4
VA	690	682	120	118	120	120	60	48	70	67
WA	253	248	247	247	48	55	105	105	127	115
WV	199	195	16	15	40	35	5	7	25	23
WI	230	235	1,280	1,265	60	65	650	650	90	75
WY	815	706	5	4	165	120	1	1	144	149
US	33,117.7	32,946.9	9,111.6	9,151.7	5,561.0	5,607.6	4,060.2	4,103.5	10,057.1	9,890.0

See footnote at end of table.

Table 7-4.—Cattle and calves, Jan. 1: Number, by sex and weight classes, by States, 2002 and 2003—Continued

State	Steers, 500 pounds and over		Bulls, 500 pounds and over		Calves under 500 pounds	
	2002	2003[1]	2002	2003[1]	2002	2003[1]
	Thousands	*Thousands*	*Thousands*	*Thousands*	*Thousands*	*Thousands*
AL	60	75	50	55	340	365
AK	0.8	0.8	1.2	1.1	2.4	2.4
AZ	250	250	22	21	157	145
AR	145	130	50	55	420	420
CA	640	615	65	65	1,040	1,060
CO	1,010	870	45	40	195	150
CT	2.0	2.5	1.0	1.0	11.0	10.0
DE	2.0	1.9	0.4	0.4	4.6	4.7
FL	25	25	60	60	380	370
GA	40	50	39	40	335	335
HI	8	7	4	5	35	32
ID	360	360	40	35	260	250
IL	285	280	25	25	255	245
IN	120	128	20	20	190	185
IA	960	1,040	70	70	480	460
KS	2,280	1,970	85	90	715	740
KY	215	220	75	80	500	540
LA	24	24	32	33	165	170
ME	4.0	3.5	1.5	1.5	18.5	17.0
MD	18	23	4	4	45	40
MA	2.0	2.5	1.0	1.0	9.0	8.5
MI	195	195	17	18	198	175
MN	480	470	35	35	550	510
MS	55	56	40	39	255	250
MO	480	485	120	125	920	960
MT	245	240	100	100	75	70
NE	2,120	2,000	100	100	450	410
NV	50	52	15	16	78	80
NH	1.5	1.5	0.5	0.5	7.5	7.0
NJ	3	4	1	1	8	8
NM	200	230	50	50	270	260
NY	37	37	18	18	225	245
NC	43	48	34	32	250	240
ND	315	300	57	56	118	116
OH	160	180	30	30	270	240
OK	1,160	1,230	130	135	1,060	995
OR	160	150	40	40	165	175
PA	175	170	25	25	330	320
RI	0.5	0.5	0.2	0.2	0.8	0.8
SC	13	15	17	17	111	112
SD	750	730	95	90	345	310
TN	118	140	72	75	570	550
TX	2,540	2,750	370	370	2,440	2,500
UT	126	125	24	22	130	113
VT	3.0	3.0	3.0	3.0	47.0	44.0
VA	175	190	40	40	375	365
WA	189	177	24	23	137	130
WV	45	43	15	14	70	73
WI	330	320	30	30	630	660
WY	170	170	50	45	120	95
US	16,789.8	16,590.2	2,243.8	2,252.7	15,762.8	15,563.4

[1] Preliminary.

NASS, Livestock Branch, (202) 720–3570.

Table 7-5.—Cattle and buffalo: Number in specified countries, 1999–2001[1]

Country	1999	2000	2001[2]
	Thousands	*Thousands*	*Thousands*
India	306,967	312,572	313,774
Brazil	143,893	146,272	150,382
China, Peoples Republic	124,354	126,983	128,663
United States	99,115	98,198	97,277
European Union	81,636	82,235	81,337
Argentina	49,437	49,832	50,167
Australia	26,688	27,588	28,800
Russian Federation	28,600	27,000	25,500
Colombia	20,621	21,700	22,676
Mexico	24,859	23,715	22,551
Canada	13,211	13,201	13,608
South Africa, Republic	13,465	13,195	13,460
Venezuela	12,900	13,100	13,400
Turkey	11,600	11,500	11,350
Uruguay	10,402	10,557	10,423
Ukraine	11,722	10,627	9,424
New Zealand	8,870	8,960	9,020
Egypt	6,120	6,275	6,300
Poland	6,455	6,093	5,722
Philippines	5,442	5,502	5,472
Japan	4,656	4,588	4,530
Romania	3,086	2,932	2,870
Nicaragua	2,283	2,250	2,280
Korea, Republic of	2,922	2,487	2,134
Dominican Republic	1,960	1,923	1,918
Others	17,963	15,720	7,954
Total	1,039,227	1,045,005	1,040,992

[1] Various dates of enumeration are used by the countries reporting animal numbers. Data presented in this table approximate Jan. 1 as closely as possible. [2] Preliminary.

FAS, Dairy Livestock and Poultry Division, (202) 720–8031. Prepared or estimated on the basis of official statistics of foreign governments, other foreign source materials, reports of U.S. Agricultural Counselors, Attachés, and Foreign Service Officers, results of office research, and related information.

Table 7-6.—Cows and calf crop: Cows and heifers that have calved, Jan. 1, 2002–2003, and calves born, by States, 2001 and 2002

State	Cows and heifers that have calved, Jan. 1		Calves born	
	2002	2003[1]	2001	2002[1]
	Thousands	*Thousands*	*Thousands*	*Thousands*
AL	770	780	680	690
AK	5.5	5.6	3.8	3.7
AZ	325	315	270	265
AR	960	980	820	850
CA	2,380	2,420	1,990	2,040
CO	890	800	840	820
CT	32	29	27	24
DE	13	13	10	10
FL	1,110	1,100	940	930
GA	680	710	580	600
HI	85	85	66	63
ID	870	880	860	860
IL	560	540	510	490
IN	380	375	345	340
IA	1,190	1,200	1,120	1,110
KS	1,600	1,630	1,480	1,510
KY	1,200	1,240	1,080	1,110
LA	520	520	405	415
ME	48	46	44	42
MD	122	121	103	102
MA	27	25	22	19
MI	370	390	335	340
MN	910	880	890	880
MS	610	590	540	500
MO	2,200	2,250	2,060	2,090
MT	1,470	1,420	1,550	1,490
NE	2,000	2,000	1,820	1,820
NV	265	270	225	220
NH	22	21	19	19
NJ	21	22	18	18
NM	790	780	610	600
NY	750	760	620	640
NC	500	480	450	445
ND	1,050	1,010	1,000	1,000
OH	540	520	480	470
OK	2,020	2,130	1,890	1,930
OR	710	710	660	660
PA	740	740	660	660
RI	2.8	3.0	2.6	2.6
SC	230	230	185	185
SD	1,890	1,780	1,850	1,840
TN	1,150	1,190	1,050	1,060
TX	5,750	5,800	5,050	5,000
UT	450	430	400	390
VT	166	163	145	140
VA	810	800	720	730
WA	500	495	460	450
WV	215	210	195	190
WI	1,510	1,500	1,360	1,350
WY	820	710	840	780
US	42,229.3	42,098.6	38,280.4	38,193.3

[1] Preliminary.

NASS, Livestock Branch, (202) 720–3570.

Table 7-7.—Cattle and calves: All cattle on feed, United States, Jan. 1, 1995–2003[1]

Year	Number
	Thousands
1995	12,420
1996	12,958
1997	13,181
1998	13,608
1999	13,218.5
2000	14,003
2001	14,201.4
2002	13,860.3
2003[2]	12,916

[1] Cattle and calves on feed are animals for slaughter market being fed a ration of grain or other concentrates and are expected to produce a carcass that will grade select or better. [2] Preliminary.

NASS, Livestock Branch, (202) 720–3570.

Table 7-8.—Cattle and calves: Number on feed, 1,000+ capacity feedlots, by States, Jan. 1, 2002–2003[1]

State	2002	2003[2]
	1,000 Head	*1,000 Head*
AZ	305	289
CA	480	490
CO	1,180	1,020
ID	330	295
IA	355	355
KS	2,480	2,200
NE	2,230	2,140
NM	108	119
OK	360	330
SD	195	205
TX	2,880	2,630
WA	247	185
Other States	415	335
US	11,565	10,593

[1] Cattle and calves on feed are animals for slaughter market being fed a ration of grain or other concentrates and are expected to produce a carcass that will grade Select or better. [2] Preliminary.

NASS, Livestock Branch, (202) 720–3570.

Table 7-9.—Cattle: Average price per 100 pounds, by grades, at Nebraska Direct and South St. Paul, 1993–2002

Year	Steers[1]		Heifers[2]		Sioux Falls, SD		South St. Paul	
					Cows[3]			Cows[3]
	Choice	Select	Choice	Select	Commercial	Breaking utility	Commercial	Utility Boning
	Dollars	*Dollars*	*Dollars*	*Dollars*	*Dollars*	*Dollars*	*Dollars*	*Dollars*
1993	76.23	74.09	76.01	73.77	56.50	52.18	....	50.62
1994	67.60	66.33	67.93	66.14	48.22	43.68	....	45.95
1995	65.64	63.94	65.46	63.69	39.08	38.04	40.24	39.23
1996	74.50	61.83	64.18	61.22	35.24	33.64	37.69	35.22
1997	65.92	63.85	65.66	63.36	36.91	35.64	42.7	38.62
1998	60.07	56.17	59.23	55.17	43.22	39.23	40.15	37.02
1999 65-80%	65.64	....	65.68	....	45.04	40.29	43.52	38.55
2000 65-80%	69.52	....	69.55	....	49.25	44.51	49.26	41.77
2001 65-80%	67.68	....	67.81	....	52.35	46.67	50.35	47.91
2002 65-80%	66.39	....	67.39	....	44.99	40.97	45.16	42.50

[1] 1,100 to 1,300 pound weight range; weighted average of price range. [2] 1,000 to 1,200 pound weight range; simple average of price range. [3] All weights; simple average of price range.

AMS, Livestock and Grain Market News, (202) 720–7316.

Table 7-10.—Cattle and calves: Production, disposition, cash receipts, and gross income, United States, 1992–2001

Year	Calf crop [1]	Death loss		Marketings [2]		Cattle shipped in for feeding and breeding [3]	Farm slaughter
		Cattle	Calves	Cattle	Calves		Cattle and calves
	1,000 head	1,000 head	1,000 head	1,000 head	1,000 head	1,000 head	1,000 head
1992	38,933	1,677	2,689	45,632	9,565	22,492	242
1993	39,369	1,702	2,921	46,238	9,627	23,149	233
1994	40,105	1,589	2,681	46,499	9,571	22,277	229
1995	40,264	1,645	2,739	48,741	9,656	23,507	227
1996	39,823	1,761	2,811	48,722	10,295	22,098	225
1997	38,961	1,847	2,829	49,647	10,154	23,828	223
1998	38,812	1,668	2,542	47,227	9,729	21,939	214
1999	38,796	1,659	2,455	48,373	9,855	22,842	213
2000	38,631	1,711	2,386	48,702	10,002	23,452	203
2001 [4]	38,280	1,723	2,488	46,663	9,605	21,820	194

Year	Quantity produced (live weight) [5]	Value of production	Cash receipts from sales of cattle, calves, beef and veal [6]	Value of cattle and calves slaughtered for home consumption	Gross income [7]	Average price per 100 pounds received by farmers	
						Cattle	Calves
	1,000 pounds	1,000 dollars	1,000 dollars	1,000 dollars	1,000 dollars	Dollars	Dollars
1992	40,252,994	28,632,524	37,272,292	408,169	37,680,461	71.30	89.00
1993	41,139,670	28,848,468	39,486,980	389,997	39,876,977	72.60	91.20
1994	41,572,867	26,533,580	36,253,055	350,864	36,603,919	66.70	87.20
1995	42,533,734	24,699,740	34,044,038	304,751	34,348,789	61.80	73.10
1996	40,883,614	22,034,934	30,976,861	274,011	31,250,872	58.70	58.40
1997	41,110,640	24,941,882	35,999,620	322,053	36,321,673	63.10	78.90
1998	41,620,414	24,153,116	33,415,404	304,406	33,719,810	59.60	78.80
1999	42,468,398	26,051,477	36,529,972	330,862	36,860,834	63.40	87.70
2000	42,841,607	28,391,942	40,711,031	366,744	41,077,775	68.60	104.00
2001 [4]	42,365,499	29,267,264	40,439,864	362,803	40,802,667	71.30	106.00

[1] Calves born during the year. [2] Includes custom slaughter for use on farms where produced and State outshipments, but excludes interfarm sales within the State. [3] Includes cattle shipped in from other States and from central markets, but excludes cattle for immediate slaughter. [4] Preliminary. [5] Adjustments made for changes in inventory and for inshipments. [6] Receipts from marketings and sale of farm slaughter. [7] Cash receipts from sales of cattle, calves, beef, and veal plus value of cattle and calves slaughtered for home consumption.

NASS, Livestock Branch, (202) 720–3570.

Table 7-11.—Cattle: Weighted average weight and price per 100 pounds, Texas-Oklahoma, Kansas, Colorado, Nebraska, Iowa-So. Minnesota Feedlots, 1995–2002 [1]

Year	Steers SE/CH 65-80%			Steers SE/CH 35-65%		
	Price	Average Weight	Number of Head	Price	Average Weight	Number of Head
	Dollars	Pounds		Dollars	Pounds	
1995	66.47	1,249	560,746	66.29	1,188	3,877,665
1996	64.76	1,249	473,484	64.82	1,200	3,711,625
1997	66.07	1,258	468,729	66.03	1,192	3,621,244
1998	61.05	1,282	408,859	61.79	1,211	3,135,109
1999	65.66	1,287	586,293	65.63	1,249	3,057,112
2000	69.82	1,294	584,809	70.15	1,253	2,631,692
2001	NA	NA	NA	NA	NA	NA
2002	66.74	1,327	270,924	67.40	1,263	1,965,036

Year	Heifers SE/CH 65-80%			Heifers SE/CH 35-65%		
	Price	Average Weight	Number of Head	Price	Average Weight	Number of Head
	Dollars	Pounds		Dollars	Pounds	
1995	65.91	1,133	439,383	....	....	....
1996	64.85	1,135	407,843	....	....	....
1997	66.49	1,132	405,819	65.92	1,084	2,674,322
1998	61.31	1,175	278,275	61.83	1,102	2,369,994
1999	65.75	1,182	493,893	65.81	1,135	2,410,684
2000	69.88	1,190	492,881	70.21	1,131	2,353,612
2001	68.22	1,208	377,415	69.01	1,134	2,065,438
2002	67.53	1,229	324,078	67.42	1,142	1,692,785

[1] Sales FOB feedlots and delivered. Estimated net weights after 3-4 % shrink. NA-not available.

AMS, Livestock and Grain Market News, (202) 720–7316.

Table 7-12.—Cattle and calves: Receipts at selected public stockyards, 1993–2002 [1]

Year	Oklahoma City	Greeley	Amarillo	South St. Joseph	South St. Paul	All others reporting	Total markets reporting [2 3]
	Cattle						
	Thousands	*Thousands*	*Thousands*	*Thousands*	*Thousands*	*Thousands*	*Thousands*
1993	538	....	....	120	212	4,409	4,894
1994	575	....	....	127	188	4,169	4,573
1995	563	....	....	121	179	4,068	4,860
1996	554	....	....	118	173	4,236	4,949
1997	526	71	136	121	180	4,214	4,953
1998	546	41	117	102	155	3,058	3,761
1999	516	48	105	110	157	4,598	5,018
2000	497	53	95	101	145	4,454	4,847
2001	512	49	89	113	135	2,698	3,593
2002	516	43	144	30	143	2,293	3,169
	Calves						
	Thousands	*Thousands*	*Thousands*	*Thousands*	*Thousands*	*Thousands*	*Thousands*
1993	....	....	....	....	48	15	63
1994	....	....	....	....	36	12	48
1995	....	....	....	....	4	89	92
1996	....	....	....	....	4	90	93
1997	....	....	....	....	3	92	96
1998	....	....	....	....	3	89	92
1999	....	....	....	....	2	92	94
2000	....	....	....	....	1	89	90
2001	....	....	....	....	1	86	87
2002	....	3	....	....	1	106	113

[1] Total rail and truck receipts unloaded at public stockyards. Saleable receipts 1978 on. [2] Rounded totals of the complete figures. [3] The number of stockyards varies from 23 to 46.

AMS, Livestock & Grain Market News, (202) 720–7316. Compiled from reports received from stockyard companies.

Table 7-13.—Cattle and calves: Number slaughtered, United States, 1993–2002

Year	Cattle slaughter					Calf slaughter				
	Commercial			Farm	Total	Commercial			Farm	Total
	Federally inspected	Other	Total [1]			Federally inspected	Other	Total [1]		
	Thousands	*Thousands*	*Thousands*	*Thousands*	*Thousands*	*Thousands*	*Thousands*	*Thousands*	*Thousands*	*Thousands*
1993	32,593	731	33,324	180	33,504	1,159	35	1,195	47	1,242
1994	33,482	714	34,196	180	34,376	1,237	31	1,268	47	1,315
1995	34,879	760	35,639	178	35,817	1,393	38	1,430	47	1,477
1996	35,721	862	36,583	177	36,760	1,714	55	1,768	47	1,815
1997	35,567	751	36,318	174	36,492	1,534	41	1,575	44	1,619
1998	34,787	678	35,465	172	35,637	1,422	36	1,458	43	1,501
1999	35,486	664	36,150	170	36,320	1,251	31	1,282	40	1,322
2000	35,631	615	36,246	170	36,416	1,089	43	1,132	40	1,172
2001	34,771	599	35,370	160	35,530	981	26	1,007	40	1,047
2002	35,120	614	35,735	153	35,888	1,019	26	1,045	37	1,082

[1] Totals are based on unrounded numbers.

NASS, Iowa Agricultural Statistics Service, (515) 284–4340.

Table 7-14.—Cattle and calves: Number slaughtered commercially, total and average live weight, by States, 2002 [1]

State	Cattle			Calves		
	Number slaughtered	Total live weight	Average live weight	Number slaughtered	Total live weight	Average live weight
	Thousands	*1,000 pounds*	*Pounds*	*Thousands*	*1,000 pounds*	*Pounds*
AL	7.0	5,922	848			
AK	0.8	822	1,064			
AZ	576.8	760,219	1,318			
AR	15.1	12,828	848			
CA	1,293.5	1,667,314	1,289	121.4	17,585	145
CO	2,596.1	3,309,943	1,275			
DE-MD	37.4	47,030	1,257	1.5	501	332
FL	97.0	107,054	1,103			
GA	362.9	445,472	1,228			
HI	11.6	12,145	1,044			
ID	827.8	1,023,273	1,236			
IL	942.3	1,164,745	1,236	103.8	45,612	439
IN	39.4	41,692	1,059	29.2	7,958	272
IA	686.7	870,451	1,268			
KS	7,403.6	9,112,096	1,231			
KY	16.6	14,401	867			
LA	16.9	14,637	866	10.0	4,974	497
MI	455.9	633,312	1,389	41.0	18,093	441
MN	771.9	1,057,677	1,370			
MS						
MO	90.4	112,600	1,245	1.3	483	375
MT	20.2	23,318	1,152			
NE	7,861.9	10,143,727	1,290			
NV	1.0	1,056	1,015			
N ENG [2]	20.4	22,121	1,085	27.2	3,339	123
NJ	27.5	31,971	1,162	116.3	52,533	452
NM	20.4	18,475	907			
NY	51.6	56,808	1,102	169.2	16,707	99
NC	200.3	212,031	1,059	1.7	651	392
ND	14.6	16,593	1,136			
OH	138.4	156,964	1,134	22.5	9,082	404
OK	40.6	40,964	1,009	1.3	556	435
OR	16.2	19,519	1,205			
PA	1,040.9	1,328,715	1,277	190.3	64,874	341
SC						
SD	183.5	233,655	1,273			
TN	15.0	13,127	877			
TX	6,471.6	7,675,296	1,186	20.6	8,863	430
UT	647.1	797,399	1,232			
VA	12.3	12,073	979	0.3	84	296
WA	852.5	1,061,005	1,245			
WV	11.0	10,267	936			
WI	1,665.3	2,209,482	1,327	148.3	65,564	442
WY	7.0	8,212	1,176			
US [3]	35,734.6	44,687,867	1,251	1,045.4	325,678	312

[1] Includes slaughter in federally inspected and other slaughter plants; excludes animals slaughtered on farms. Average live weight is based on unrounded numbers. Totals may not add due to rounding. [2] CT, ME, MA, NH, RI, and VT. [3] States with no data printed are still included in the U.S. total. Data are not printed to avoid disclosing individual operations.

NASS, Iowa Agricultural Statistics Service, (515) 284–4340.

Table 7-15.—Cattle and calves: Number slaughtered under Federal inspection, and average live weight, 1993–2002

Year	Cattle		Calves	
	Number slaughtered	Average live weight	Number slaughtered	Average live weight
	Thousands	*Pounds*	*Thousands*	*Pounds*
1993	32,593	1,164	1,159	387
1994	33,482	1,193	1,237	383
1995	34,879	1,187	1,393	371
1996	35,721	1,173	1,714	340
1997	35,567	1,177	1,534	335
1998	34,787	1,207	1,422	282
1999	35,486	1,212	1,251	288
2000	35,631	1,222	1,089	311
2001	34,771	1,224	981	318
2002	35,120	1,253	1,019	310

NASS, Iowa Agricultural Statistics Service, (515) 284–4340.

Table 7-16.—Cattle and calves: Production, disposition, cash receipts, and gross income, by States, 2001 (preliminary)

State	Marketings [1]		Cattle shipped in for feeding and breeding	Farm slaughter of cattle and calves [2]	Quantity produced (live weight) [3]	Value of production	Cash receipts from sales of cattle, calves, beef, and veal [4]	Value of cattle and calves slaughtered for home consumption	Gross income [5]
	Cattle	Calves							
	1,000 head	*1,000 head*	*1,000 head*	*1,000 head*	*1,000 pounds*	*1,000 dollars*	*1,000 dollars*	*1,000 dollars*	*1,000 dollars*
AL	462.0	170.0	18.0	2.0	493,218	362,106	362,816	2,774	365,590
AK	1.5	0.2	0.4	0.3	2,461	1,923	1,326	171	1,497
AZ	712.0	141.0	625.0	1.0	641,866	403,925	683,361	2,516	685,877
AR	654.0	199.0	145.0	5.0	558,414	391,838	432,272	5,697	437,969
CA	2,054.0	387.0	750.0	14.0	1,969,826	924,113	1,351,500	9,224	1,360,724
CO	2,905.0	100.0	2,180.0	2.0	2,108,813	1,462,694	2,589,204	8,065	2,597,269
CT	14.9	13.4	3.0	1.0	15,315	8,742	10,052	1,132	11,184
DE	7.7	3.5	2.0	0.3	7,391	5,183	6,626	478	7,104
FL	261.0	687.0	60.0	2.0	447,032	336,727	360,515	1,333	361,848
GA	359.0	317.0	125.0	3.0	442,152	278,815	347,677	4,398	352,075
HI	16.6	42.0	....	0.4	36,300	20,175	18,007	388	18,395
ID	1,124.0	151.0	553.0	2.0	1,042,690	700,819	914,863	6,445	921,308
IL	550.0	109.0	175.0	7.0	586,292	441,212	527,954	14,226	542,180
IN	282.0	117.0	93.0	4.0	278,481	185,361	215,717	10,118	225,835
IA	2,014.0	102.0	1,080.0	4.0	1,615,780	1,131,966	1,824,202	7,365	1,831,567
KS	5,559.0	1.5	4,195.0	4.5	3,936,837	2,459,824	4,915,465	9,918	4,925,383
KY	550.0	473.0	100.0	6.0	622,480	483,880	499,782	9,793	509,575
LA	139.0	240.0	2.0	2.0	229,265	174,618	177,947	938	178,885
ME	23.0	20.0	5.0	1.0	25,743	15,664	16,992	1,262	18,254
MD	60.0	24.0	5.0	2.0	90,570	65,853	61,165	2,146	63,311
MA	8.0	11.0	3.0	1.0	12,167	7,076	6,023	1,132	7,155
MI	266.0	36.0	50.0	4.0	353,634	204,736	227,930	7,467	235,397
MN	957.0	137.0	360.0	6.0	1,106,915	726,006	890,842	15,134	905,976
MS	254.0	197.0	9.0	2.0	296,553	210,163	196,774	6,143	202,917
MO	790.0	1,000.0	40.0	5.0	1,186,415	997,115	926,287	29,642	955,929
MT	1,385.0	350.0	175.0	5.0	956,075	795,126	977,613	9,258	986,871
NE	5,603.0	85.0	3,850.0	2.0	4,335,261	2,711,105	5,066,780	8,848	5,075,628
NV	200.0	91.0	67.0	2.0	176,140	146,911	192,470	2,412	194,882
NH	9.3	10.0	2.0	0.9	10,626	6,457	7,063	892	7,955
NJ	9.5	12.5	2.5	0.5	11,447	7,070	10,328	657	10,985
NM	1,155.0	250.0	864.0	3.0	658,825	421,838	807,053	4,449	811,502
NY	170.0	403.0	45.0	2.0	234,174	131,176	151,540	2,681	154,221
NC	243.0	186.0	19.0	2.0	314,118	225,840	227,629	10,827	238,456
ND	525.0	532.0	113.0	3.0	550,129	463,229	524,924	5,824	530,748
OH	380.0	73.0	46.0	6.0	427,488	305,018	312,377	21,516	333,893
OK	2,210.0	340.0	1,100.0	10.0	1,824,699	1,474,754	1,865,073	17,500	1,882,573
OR	385.0	190.0	40.0	12.0	579,525	422,986	407,837	11,450	419,287
PA	448.0	243.0	120.0	10.0	538,450	368,500	418,208	17,252	435,460
RI	1.5	1.5	0.3	0.1	1,284	745	1,058	113	1,171
SC	162.0	25.0	4.0	2.0	172,647	122,440	128,916	4,060	132,976
SD	1,780.0	502.0	514.0	2.0	1,553,999	1,240,268	1,566,898	10,748	1,577,646
TN	583.0	386.0	62.0	3.0	557,697	396,475	409,572	5,511	415,083
TX	7,935.0	120.0	3,420.0	15.0	7,734,268	5,462,435	6,812,228	17,556	6,829,784
UT	380.0	90.0	126.0	4.0	397,185	314,868	374,459	7,170	381,629
VT	63.0	88.0	15.0	2.0	72,111	47,006	58,546	1,749	60,295
VA	477.0	174.0	14.0	5.0	518,291	352,165	349,753	11,021	360,774
WA	670.0	14.0	235.0	10.0	643,794	492,641	654,241	7,300	661,541
WV	123.0	72.0	35.0	3.0	134,407	82,925	85,738	3,742	89,480
WI	806.0	458.0	76.0	7.0	1,203,564	729,605	707,049	15,837	722,886
WY	936.0	230.0	297.0	1.0	652,685	545,147	757,212	6,525	763,737
US	46,663.0	9,604.6	21,820.2	194.0	42,365,499	29,267,264	40,439,864	362,803	40,802,667

[1] Includes custom slaughter for use on farms where produced and State outshipments, but excludes interfarm sales within the State. [2] Excludes custom slaughter for farmers at commercial establishments. [3] Adjustments made for changes in inventory and for inshipments. [4] Includes receipts from marketings and sales of farm-slaughter. [5] Includes cash receipts from sales of cattle, calves, beef, and veal plus value of cattle and calves slaughtered for home consumption.

NASS, Livestock Branch, (202) 720–3570.

Table 7-17.—Cattle: Number slaughtered under Federal inspection and percentage distribution, by classes, 1993–2002 [1]

Year	Number						Percentage of total					
	Steers	Heifers	Cows: Dairy cows	Cows: Other cows	Cows: Total cows	Bulls and stags	Steers	Heifers	Cows: Dairy cows	Cows: Other cows	Cows: Total cows	Bulls and stags
	Thousands	*Thousands*	*Thousands*	*Thousands*	*Thousands*	*Thousands*	*Percent*	*Percent*	*Percent*	*Percent*	*Percent*	*Percent*
1993 ..	16,840	9,152	2,995	2,959	5,954	648	51.7	28.1	9.2	9.1	18.3	2.0
1994 ..	17,651	9,389	2,858	2,955	5,813	629	52.7	28.0	8.5	8.8	17.3	1.9
1995 ..	17,887	10,175	2,862	3,281	6,143	674	51.3	29.2	8.2	9.4	17.6	1.9
1996 ..	17,400	10,502	3,037	4,068	7,105	715	48.7	29.4	8.5	11.4	19.9	2.0
1997 ..	17,172	11,287	2,926	3,498	6,424	683	48.3	31.7	8.2	9.8	18.1	1.9
1998 ..	17,101	11,228	2,620	3,245	5,865	593	49.2	32.3	7.5	9.3	16.9	1.7
1999 ..	17,608	11,648	2,573	3,030	5,603	627	49.6	32.8	7.3	8.5	15.8	1.8
2000 ..	17,758	11,835	2,631	2,796	5,427	612	49.8	33.2	7.4	7.8	15.2	1.7
2001 ..	17,097	11,379	2,582	3,092	5,674	620	49.2	32.7	7.4	8.9	16.3	1.8
2002 ..	17,523	11,342	2,607	3,051	5,658	598	49.9	32.3	7.4	8.7	16.1	1.7

[1] Totals and percentages based on unrounded data and may not equal sum of classes due to rounding.
NASS, Iowa Agricultural Statistics Service, (515) 284–4340.

Table 7-18.—Cattle and calves: Number of operations, 2001–2002, and inventory, Jan 1, 2002–2003, by States and United States [1]

State	Operations with cattle: 2001	Operations with cattle: 2002 [2]	January 1 cattle inventory: 2002	January 1 cattle inventory: 2003 [2]
	Number	*Number*	*1,000 head*	*1,000 head*
AL	26,000	26,000	1,370	1,440
AK	130	120	11.5	11.5
AZ	3,000	2,900	840	820
AR	30,000	30,000	1,820	1,850
CA	21,000	20,000	5,200	5,250
CO	14,600	14,300	3,050	2,650
CT	1,200	1,200	61	56
DE	480	450	26	25
FL	19,500	19,000	1,780	1,750
GA	22,000	21,000	1,240	1,290
HI	800	800	152	150
ID	11,500	11,500	1,990	2,000
IL	24,000	23,000	1,430	1,360
IN	20,000	19,000	880	860
IA	35,000	34,000	3,550	3,600
KS	35,000	34,000	6,600	6,350
KY	47,000	46,000	2,300	2,430
LA	15,000	15,000	850	860
ME	1,800	1,800	97	93
MD	4,500	4,400	245	240
MA	1,200	1,200	51	49
MI	15,500	15,000	990	990
MN	29,000	29,000	2,550	2,470
MS	23,000	23,000	1,100	1,070
MO	67,000	68,000	4,350	4,500
MT	13,300	12,500	2,450	2,400
NE	26,000	25,000	6,400	6,200
NV	1,700	1,700	500	510
NH	900	900	41	40
NJ	1,700	1,600	44	46
NM	8,500	8,500	1,580	1,590
NY	16,000	15,000	1,390	1,450
NC	25,000	24,000	950	920
ND	13,300	13,000	1,970	1,880
OH	27,000	26,000	1,250	1,220
OK	59,000	60,000	5,200	5,400
OR	16,200	16,200	1,400	1,360
PA	28,000	28,000	1,640	1,630
RI	200	200	5.5	5.5
SC	11,000	11,000	430	435
SD	19,500	19,000	3,950	3,700
TN	52,000	51,000	2,200	2,270
TX	151,000	151,000	13,600	14,000
UT	8,000	7,800	920	880
VT	3,000	2,900	285	280
VA	26,000	25,000	1,650	1,630
WA	15,000	14,500	1,130	1,100
WV	13,000	13,000	415	405
WI	40,000	38,000	3,300	3,300
WY	6,400	6,200	1,470	1,290
US	1,049,910	1,032,670	96,704.0	96,106.0

[1] An operation is any place having one or more head of cattle on hand at any time during the year. [2] Preliminary.
NASS, Livestock Branch, (202) 720–3570.

Table 7-19.—Cattle and calves: Average dressed weight under Federal inspection, 1993–2002

Year	Cattle					Calves
	All cattle	Steers	Heifers	Cows	Bulls	
	Pounds	*Pounds*	*Pounds*	*Pounds*	*Pounds*	*Pounds*
1993	694	747	685	542	859	228
1994	717	776	707	536	854	227
1995	711	769	706	534	857	218
1996	702	766	705	524	842	211
1997	706	764	703	539	851	208
1998	730	789	724	554	865	174
1999	736	793	727	560	881	176
2000	745	798	733	579	892	192
2001	744	798	734	584	893	196
2002	765	823	753	590	912	190

NASS, Iowa Agricultural Statistics service (515) 284–4340.

Table 7-20.—Cattle and calves: Number of operations by size group, selected States, and United States, 2001–2002[1]

State	Operations having									
	1–49 head		50–99 head		100–499 head		500–999 head		1000+ head	
	2001	2002[2]	2001	2002[2]	2001	2002[2]	2001	2002[2]	2001	2002[2]
	Number									
AL	18,000	18,000	4,400	4,400	3,300	3,300	250	240	50	60
AZ	1,750	1,650	370	380	610	610	120	120	150	140
AR	19,500	19,500	5,800	5,800	4,400	4,400	240	230	60	70
CA	13,700	12,900	1,900	1,800	3,100	3,000	1,100	1,100	1,200	1,200
CO	7,800	7,700	2,160	2,000	3,500	3,500	680	650	460	450
FL	14,500	14,200	2,300	2,200	2,050	1,950	350	350	300	300
GA	15,000	14,200	3,890	3,800	2,800	2,700	250	250	60	50
ID	6,600	6,600	1,640	1,600	2,500	2,500	440	460	320	340
IL	15,400	14,700	4,500	4,400	3,880	3,680	180	170	40	50
IN	15,100	14,400	2,900	2,700	1,890	1,780	80	90	30	30
IA	15,800	15,600	7,800	7,300	10,300	10,100	800	700	300	300
KS	16,400	16,000	7,100	6,900	9,700	9,400	1,150	1,050	650	650
KY	33,000	32,000	8,200	8,200	5,550	5,550	200	200	50	50
LA	10,800	11,000	2,000	1,800	2,000	2,000	160	160	40	40
MD	3,100	3,000	670	630	700	730	20	30	10	10
MI	10,800	10,400	2,100	2,040	2,350	2,300	180	180	70	80
MN	14,700	14,700	6,800	6,800	7,000	7,000	380	380	120	120
MS	16,000	15,900	3,800	3,800	3,000	3,120	150	130	50	50
MO	41,000	42,000	14,600	14,500	10,800	10,800	460	580	140	120
MT	4,800	4,600	2,350	2,000	4,800	4,700	1,000	850	350	350
NE	9,100	8,700	5,400	5,100	9,000	8,700	1,400	1,500	1,100	1,000
NM	5,000	5,000	1,100	1,100	1,700	1,750	350	320	350	330
NY	8,900	8,000	2,900	2,800	3,880	3,860	230	240	90	100
NC	19,900	18,800	3,100	3,200	1,900	1,900	80	80	20	20
ND	3,800	3,600	3,150	2,950	5,700	5,800	550	550	100	100
OH	20,500	19,500	3,200	3,200	3,120	3,100	150	160	30	40
OK	36,000	37,000	10,700	10,600	10,800	10,900	950	950	550	550
OR	12,400	12,400	1,400	1,400	1,740	1,740	420	410	240	250
PA	16,300	16,500	7,100	6,900	4,420	4,430	150	140	30	30
SD	5,000	4,700	3,400	3,400	9,200	9,000	1,350	1,400	550	500
TN	39,500	38,000	7,800	7,800	4,500	5,000	170	165	30	35
TX	103,000	103,000	22,000	22,000	22,300	22,300	2,400	2,300	1,300	1,400
UT	4,600	4,400	1,200	1,300	1,800	1,700	270	270	130	130
VT	1,400	1,300	650	700	880	830	50	50	20	20
VA	17,000	16,000	4,700	4,700	4,000	4,000	250	250	50	50
WA	11,700	11,400	1,290	1,200	1,600	1,500	270	260	140	140
WI	17,000	16,100	10,200	9,400	12,100	11,800	570	550	130	150
WY	1,900	2,400	1,100	650	2,500	2,300	600	550	300	300
Oth Sts[3]	27,000	27,000	3,400	3,300	3,210	3,200	320	300	180	170
US	653,750	642,850	179,070	174,750	188,580	186,930	18,720	18,365	9,790	9,775

[1] An operation is any place having one or more cattle on hand during the year. [2] Preliminary. [3] Individual State estimates not available for the 12 other States.

NASS, Livestock Branch, (202) 720–3570.

Table 7-21.—Cattle and calves: Percent of inventory by size group, selected States, and United States, 2001–2002 [1]

State	Inventory on operations having									
	1–49 head		50–99 head		100–499 head		500–999 head		1,000+ head	
	2001	2002 [2]	2001	2002 [2]	2001	2002 [2]	2001	2002 [2]	2001	2002 [2]
	Percent									
AL	25.0	26.0	21.0	21.0	39.0	38.0	10.5	9.5	4.5	5.5
AZ	2.0	2.5	2.3	2.5	14.0	14.0	8.7	9.0	73.0	72.0
AR	22.0	21.0	22.0	22.0	42.0	42.0	7.5	7.5	6.5	7.5
CA	3.2	2.9	2.8	2.6	15.0	14.5	15.0	15.0	64.0	65.0
CO	3.8	4.0	4.7	4.5	23.0	24.0	14.5	14.5	54.0	53.0
FL	11.0	11.5	8.8	8.5	24.0	23.0	13.2	13.0	43.0	44.0
GA	20.0	20.0	21.0	21.0	40.0	40.0	11.5	12.0	7.5	7.0
ID	5.3	5.0	5.7	5.5	26.0	25.0	15.0	15.5	48.0	49.0
IL	20.0	19.5	21.0	21.0	46.0	46.0	8.2	7.7	4.8	5.8
IN	30.0	29.0	22.0	21.0	36.0	36.0	5.8	6.8	6.2	7.2
IA	7.5	8.0	14.5	14.0	51.0	52.0	13.0	12.0	14.0	14.0
KS	5.3	5.2	7.2	7.0	29.5	29.0	11.0	10.8	47.0	48.0
KY	28.0	28.0	23.0	23.0	40.0	40.0	5.2	5.5	3.8	3.5
LA	23.0	24.0	16.0	14.0	43.0	43.0	11.0	11.5	7.0	7.5
MD	16.0	15.0	17.0	16.0	55.0	55.0	5.7	8.5	6.3	5.5
MI	16.0	16.0	15.0	14.0	45.0	43.0	12.0	12.0	12.0	15.0
MN	13.0	13.0	19.0	19.0	50.0	50.0	10.0	10.0	8.0	8.0
MS	22.0	22.0	22.0	22.0	40.0	41.0	8.0	7.0	8.0	8.0
MO	19.0	19.0	23.0	22.0	46.0	46.0	7.0	8.5	5.0	4.5
MT	2.5	2.2	6.5	5.8	40.0	42.0	26.0	23.0	25.0	27.0
NE	2.4	2.5	5.6	5.5	28.0	29.0	14.0	16.0	50.0	47.0
NM	5.2	5.0	4.8	4.5	22.0	21.0	15.0	13.5	53.0	56.0
NY	11.0	10.0	16.0	15.0	51.0	50.0	11.0	12.0	11.0	13.0
NC	30.0	29.0	22.3	23.2	38.0	38.0	5.9	6.0	3.8	3.8
ND	4.6	4.3	10.0	9.7	61.0	61.0	17.0	17.0	7.4	8.0
OH	26.0	25.0	17.5	17.3	45.0	45.0	7.5	7.9	4.0	4.8
OK	14.0	14.0	14.0	13.0	37.0	38.0	12.0	12.0	23.0	23.0
OR	12.0	12.0	7.0	6.5	25.0	26.0	20.0	19.5	36.0	36.0
PA	17.0	17.0	29.0	28.0	45.0	46.0	5.8	5.6	3.2	3.4
SD	2.7	2.5	5.8	6.0	50.0	50.0	21.0	23.0	20.5	18.5
TN	33.0	32.0	25.0	23.0	35.0	38.0	5.1	4.8	1.9	2.2
TX	12.0	13.0	11.0	11.0	30.0	29.0	11.0	11.0	36.0	36.0
UT	8.0	7.5	9.0	9.5	41.0	41.0	19.0	19.0	23.0	23.0
VT	7.5	6.5	16.0	17.5	54.0	52.0	11.5	12.0	11.0	12.0
VA	20.0	20.0	19.0	19.0	46.0	46.0	10.0	10.0	5.0	5.0
WA	9.0	10.0	7.0	7.0	28.0	27.0	16.0	15.0	40.0	41.0
WI	9.0	9.0	19.0	19.0	56.0	55.0	10.5	11.0	5.5	6.0
WY	1.8	3.1	4.2	2.9	32.0	32.0	26.0	24.0	36.0	38.0
Oth Sts [3]	21.0	21.0	13.0	13.0	32.0	33.0	12.0	11.0	22.0	22.0
US	11.5	11.7	12.4	12.2	36.2	36.2	12.4	12.4	27.5	27.5

[1] Percents reflect average distributions of various probability surveys conducted during the year but are based primarily on end-of-year surveys. [2] Preliminary. [3] Individual State estimates not available for the 12 other States.

NASS, Livestock Branch, (202) 720–3570.

Table 7-22.—Beef cows: Number of operations, 2001–2002, and inventory, January 1, 2002–2003, by States and United States [1]

State	Operations with beef cows [2]		January 1 beef cow inventory	
	2001	2002 [3]	2002	2003 [3]
	Number	*Number*	*1,000 head*	*1,000 head*
AL	25,000	25,000	750	761
AK	90	80	4.3	4.3
AZ	2,100	2,000	185	165
AR	27,000	27,000	927	949
CA	13,500	12,500	760	740
CO	11,000	10,500	797	702
CT	800	800	8	6
DE	240	230	4	4
FL	16,500	16,500	958	953
GA	21,000	20,000	594	625
HI	700	700	78	79
ID	7,600	7,600	493	490
IL	16,500	15,800	445	426
IN	13,000	12,000	226	230
IA	26,000	26,000	985	992
KS	29,000	28,000	1,505	1,525
KY	39,000	39,000	1,075	1,120
LA	13,200	13,000	466	471
ME	1,000	1,000	10	10
MD	2,800	2,700	41	41
MA	700	700	6	5
MI	7,000	7,000	73	89
MN	15,500	15,500	410	400
MS	21,000	21,000	576	557
MO	57,000	58,000	2,060	2,116
MT	11,900	11,400	1,451	1,402
NE	22,000	22,000	1,932	1,934
NV	1,400	1,400	240	245
NH	530	530	4	4
NJ	950	900	8	9
NM	6,900	6,800	500	470
NY	6,200	6,000	75	80
NC	22,000	21,000	434	417
ND	12,000	11,800	1,008	973
OH	17,000	16,000	280	260
OK	49,000	50,000	1,933	2,042
OR	12,700	12,600	605	593
PA	11,100	11,200	152	150
RI	140	140	1.4	1.6
SC	10,000	10,000	210	210
SD	17,000	16,800	1,792	1,686
TN	45,000	44,000	1,060	1,106
TX	133,000	133,000	5,440	5,489
UT	5,700	5,600	357	339
VT	1,000	1,000	12	10
VA	22,000	22,000	690	682
WA	10,400	10,100	253	248
WV	11,000	11,000	199	195
WI	12,000	12,000	230	235
WY	5,500	5,200	815	706
US	813,650	805,080	33,117.7	32,946.9

[1] An operation is any place having one or more beef cows on hand at any time during the year. [2] Included in operations with cattle. [3] Preliminary.

NASS, Livestock Branch, (202) 720–3570.

Table 7-23.—Beef cows: Number of operations by size group, selected States and United States, 2001–2002 [1]

State	Operations having							
	1–49 head		50–99 head		100–499 head		500+ head	
	2001	2002 [2]	2001	2002 [2]	2001	2002 [2]	2001	2002 [2]
	Number							
AL	20,000	20,000	3,300	3,250	1,650	1,700	50	50
AZ	1,400	1,300	210	220	410	400	80	80
AR	22,000	22,000	3,300	3,200	1,630	1,730	70	70
CA	10,500	9,700	990	810	1,700	1,700	310	290
CO	6,700	6,300	1,740	1,700	2,300	2,250	260	250
FL	13,200	13,300	1,600	1,600	1,400	1,320	300	280
GA	17,500	16,700	2,200	2,000	1,270	1,270	30	30
ID	5,200	5,200	950	930	1,300	1,300	150	170
IL	14,200	13,500	1,690	1,650	600	640	10	10
IN	12,100	11,200	700	600	200	200	....	....
IA	18,900	19,000	4,800	4,600	2,250	2,350	50	50
KS	19,200	18,600	5,400	5,200	4,230	4,020	170	180
KY	33,500	33,400	3,700	4,000	1,770	1,570	30	30
LA	10,900	10,600	1,200	1,300	1,060	1,060	40	40
MN	13,300	13,300	1,600	1,500	585	680	15	20
MS	17,100	17,100	2,800	2,800	1,060	1,060	40	40
MO	45,000	46,000	8,300	8,200	3,600	3,700	100	100
MT	5,000	4,800	2,300	2,130	4,100	4,000	500	470
NE	12,200	12,200	4,300	4,200	4,930	5,070	570	530
NM	4,500	4,500	970	950	1,200	1,150	230	200
NC	19,800	18,800	1,630	1,600	560	590	10	10
ND	5,300	5,000	3,220	3,220	3,400	3,500	80	80
OH	16,100	15,000	700	800	190	190	10	10
OK	37,500	38,500	7,300	7,400	4,000	3,900	200	200
OR	10,200	10,100	1,200	1,100	1,100	1,180	200	220
PA	10,600	10,600	380	470	120	130	....	....
SD	7,000	6,800	3,690	3,670	5,900	5,900	410	430
TN	40,000	39,000	3,600	3,600	1,385	1,380	15	20
TX	104,000	104,000	17,000	17,000	11,100	11,100	900	900
UT	3,700	3,600	950	950	960	960	90	90
VA	18,100	18,000	2,700	2,800	1,160	1,170	40	30
WA	9,000	8,900	750	600	590	550	60	50
WY	2,200	2,300	1,150	900	1,800	1,700	350	300
Oth Sts [3]	52,000	51,000	3,000	3,700	1,380	1,320	170	160
US	637,900	630,300	99,320	98,650	70,890	70,740	5,540	5,390

[1] An operation is any place having one or more beef cows on hand at any time during the year. Missing data combined with other size groups. [2] Preliminary. [3] Individual State estimates are not available for 17 other States.

NASS, Livestock Branch, (202) 720–3570.

Table 7-24.—Beef cows: Percent of inventory by size group, selected States, and United States, 2001–2002[1]

State	Inventory on operations having							
	1–49 head		50–99 head		100–499 head		500+ head	
	2001	2002[2]	2001	2002[2]	2001	2002[2]	2001	2002[2]
	Percent							
AL	39.0	41.0	25.5	24.5	31.0	30.0	4.5	4.5
AZ	7.5	8.5	6.5	7.5	37.0	37.0	49.0	47.0
AR	42.0	42.0	23.5	22.0	29.0	30.0	5.5	6.0
CA	12.5	12.2	8.5	6.8	45.0	45.0	34.0	36.0
CO	12.0	13.0	14.0	15.0	50.0	49.0	24.0	23.0
FL	16.0	17.0	11.0	11.0	27.0	27.0	46.0	45.0
GA	41.0	42.0	23.0	22.0	33.0	33.0	3.0	3.0
ID	14.0	14.0	13.0	12.0	48.0	47.0	25.0	27.0
IL	55.0	52.0	25.0	25.0	18.5	21.0	1.5	2.0
IN	71.0	71.0	18.0	17.0	11.0	12.0	....	....
IA	35.0	34.0	31.0	30.0	31.0	33.0	3.0	3.0
KS	24.5	24.5	23.0	23.0	44.0	44.0	8.5	8.5
KY	52.0	52.0	22.2	24.1	24.0	22.0	1.8	1.9
LA	36.0	35.0	17.0	18.0	40.0	40.0	7.0	7.0
MN	49.0	49.0	26.0	24.0	22.2	24.0	2.8	3.0
MS	40.0	38.0	26.0	26.0	29.0	30.0	5.0	6.0
MO	43.0	42.0	26.0	26.0	27.0	28.0	4.0	4.0
MT	6.0	6.0	10.0	10.0	56.0	57.0	28.0	27.0
NE	12.0	12.0	15.0	15.0	47.0	49.0	26.0	24.0
NM	10.0	11.0	12.0	12.0	42.0	43.0	36.0	34.0
NC	55.0	55.0	23.0	23.0	20.0	20.0	2.0	2.0
ND	12.5	11.0	21.0	21.0	61.0	62.0	5.5	6.0
OH	72.0	69.0	15.0	18.0	10.5	10.5	2.5	2.5
OK	33.0	34.0	24.0	24.0	35.0	34.0	8.0	8.0
OR	18.0	18.0	11.0	10.0	39.0	39.0	32.0	33.0
PA	71.0	67.0	16.0	19.0	13.0	14.0	....	....
SD	9.0	8.5	14.0	13.5	61.0	61.0	16.0	17.0
TN	55.0	56.0	23.0	22.6	20.9	20.0	1.1	1.4
TX	29.0	29.0	20.0	20.0	35.0	37.0	16.0	14.0
UT	14.0	13.0	16.0	16.0	48.0	49.0	22.0	22.0
VA	43.0	43.0	26.0	26.0	27.0	28.0	4.0	3.0
WA	26.0	30.0	18.0	15.0	39.0	38.0	17.0	17.0
WY	5.5	6.0	9.5	8.0	46.0	48.0	39.0	38.0
Other States[3]	46.0	46.0	16.0	20.0	21.0	19.0	17.0	15.0
US	28.9	29.0	19.2	19.2	37.0	37.5	14.9	14.3

[1] Percents reflect average distributions of various probability surveys conducted during the year but are based primarily on end-of-year surveys. Missing data combined with other size groups. [2] Preliminary. [3] Individual State estimates are not available for 17 other States.

NASS, Livestock Branch, (202) 720–3570.

Table 7-25.—Hogs and pigs: Number and value, United States, Dec. 1, 1993–2002

Year	Number	Value	
		Per head	Total
	Thousands	*Dollars*	*1,000 dollars*
1993	57,940	75.00	4,339,509
1994	59,738	53.00	3,178,123
1995	58,201	71.00	4,115,118
1996	56,124	94.00	5,280,742
1997	61,158	82.00	4,985,532
1998	62,206	44.00	2,765,847
1999	59,342	72.00	4,254,293
2000	59,138	77.00	4,542,493
2001	59,804	77.00	4,589,912
2002 [1]	58,943	71.00	4,159,643

[1] Preliminary.

NASS, Livestock Branch, (202) 720–3570.

Table 7-26.—Hogs and pigs: Number and value, by States, Dec. 1, 2001–2002

State	Operations		Number		Value			
					Value per head		Total value	
	2001	2002 [1]	2001	2002 [1]	2001	2002 [1]	2001	2002 [1]
	Number	*Number*	*Thousands*	*Thousands*	*Dollars*	*Dollars*	*1,000 dollars*	*1,000 dollars*
AL	300	300	195	165	76	70	14,820	11,550
AK	50	40	1.0	1.2	160	170	160	204
AZ	200	180	133	143	83	76	11,039	10,868
AR	1,100	1,000	580	305	63	68	36,540	20,740
CA	850	750	110	135	100	92	11,000	12,420
CO	400	390	800	790	92	74	73,600	58,460
CT	200	220	3.5	3.8	100	92	350	350
DE	130	130	26.0	22.0	74	68	1,924	1,496
FL	1,500	1,400	35.0	35.0	80	74	2,800	2,590
GA	900	1,000	315	345	68	63	21,420	21,735
HI	230	230	27.0	24.0	140	140	3,780	3,360
ID	400	400	24.0	22.0	78	72	1,872	1,584
IL	5,300	4,600	4,250	4,050	79	71	335,750	287,550
IN	4,000	3,400	3,200	3,150	78	74	249,600	233,100
IA	10,500	10,000	15,400	15,300	83	75	1,278,200	1,147,500
KS	1,500	1,500	1,570	1,530	73	65	114,610	99,450
KY	1,300	1,200	405	370	63	54	25,515	19,980
LA	600	580	26.0	20.0	83	76	2,158	1,520
ME	400	400	6.5	6.0	83	76	540	456
MD	520	510	52.0	45.0	76	70	3,952	3,150
MA	400	400	18.0	16.5	83	76	1,494	1,254
MI	2,500	2,300	960	860	84	80	80,640	68,800
MN	6,500	6,200	5,800	5,900	90	85	522,000	501,500
MS	1,600	1,500	285	275	82	75	23,370	20,625
MO	3,100	2,800	3,000	2,950	68	61	204,000	179,950
MT	550	500	170	185	80	74	13,600	13,690
NE	3,400	2,900	2,900	2,900	84	75	243,600	217,500
NV	100	100	7.0	6.5	100	92	700	598
NH	250	270	3.5	3.2	91	84	319	269
NJ	400	300	13.0	15.0	91	84	1,183	1,260
NM	400	400	3.0	3.0	83	76	249	228
NY	1,400	1,300	75.0	86.0	76	70	5,700	6,020
NC	3,400	3,200	9,800	9,600	62	58	607,600	556,800
ND	700	600	154	154	80	74	12,320	11,396
OH	4,900	4,500	1,430	1,440	79	76	112,970	109,440
OK	2,700	2,600	2,480	2,490	65	59	161,200	146,910
OR	1,000	1,000	29.0	31.0	83	76	2,407	2,356
PA	2,900	2,900	1,060	1,080	69	67	73,140	72,360
RI	50	50	2.5	2.9	80	74	200	215
SC	700	600	320	300	64	58	20,480	17,400
SD	1,600	1,500	1,290	1,290	83	77	107,070	99,330
TN	1,500	1,300	225	220	66	61	14,850	13,420
TX	4,000	3,800	900	930	71	62	63,900	57,660
UT	500	400	610	670	83	76	50,630	50,920
VT	250	300	2.5	2.5	100	92	250	230
VA	1,100	1,100	415	400	60	59	24,900	23,600
WA	750	750	24.0	24.0	91	84	2,184	2,016
WV	1,000	1,100	11.0	11.0	80	74	880	814
WI	2,700	2,300	540	520	70	68	37,800	35,360
WY	150	150	117	115	91	84	10,647	9,660
US	80,880	75,350	59,804	58,943	77	71	4,589,912	4,159,643

[1] Preliminary. Totals may not add due to rounding

NASS, Livestock Branch, (202) 720–3570.

Table 7-27.—Sows farrowing and pig crop: Number, United States, 1993–2002

Year	Sows farrowing		Pig crop		
	Dec.-May	June-Nov.	Dec.-May	June-Nov.	Total
	Thousands	*Thousands*	*Thousands*	*Thousands*	*Thousands*
1993	6,039	5,977	49,110	48,216	97,326
1994	6,257	6,139	51,217	50,261	101,478
1995	6,046	5,843	50,077	48,739	98,816
1996	5,665	5,449	47,888	46,571	94,459
1997	5,595	5,884	48,394	51,190	99,584
1998	6,014	6,047	52,469	52,535	105,004
1999	5,877	5,764	51,519	50,835	102,354
2000	5,683	5,727	50,087	50,660	100,747
2001	5,619	5,767	49,472	51,031	100,503
2002 [1]	5,779	5,650	50,752	50,006	100,758

[1] Preliminary.
NASS, Livestock Branch, (202) 720–3570.

Table 7-28.—Hogs and pigs: Number for breeding and market, United States, 1993–2002

Year	All hogs and pigs	Kept for breeding	Market hogs by weight groups				
			Under 60 pounds	60 to 119 pounds	120 to 179 pounds	180 pounds and over	Total
	June 1						
	Thousands	*Thousands*	*Thousands*	*Thousands*	*Thousands*	*Thousands*	*Thousands*
1993	58,795	7,260	20,739	12,898	9,877	8,022	51,536
1994	60,847	7,532	22,024	13,092	10,007	8,193	53,315
1995	59,329	7,117	21,042	12,841	9,780	8,549	52,211
1996	56,038	6,682	19,645	12,196	9,757	7,759	49,356
1997	57,366	6,789	19,988	12,574	10,002	8,013	50,577
1998	62,213	6,958	21,482	13,711	11,084	8,978	55,254
1999	60,896	6,515	20,532	13,501	11,076	9,272	54,380
2000	59,117	6,234	19,907	13,247	10,710	9,019	52,884
2001	58,603	6,186	19,923	12,992	10,536	8,967	52,417
2002 [1]	60,288	6,209	20,186	13,588	10,884	9,420	54,078
	Dec. 1						
	Thousands	*Thousands*	*Thousands*	*Thousands*	*Thousands*	*Thousands*	*Thousands*
1993	57,940	7,166	19,171	12,627	10,268	8,709	50,774
1994	59,738	6,998	19,477	13,007	10,927	9,329	52,739
1995	58,201	6,770	18,916	12,755	10,704	9,057	51,431
1996	56,124	6,578	18,503	12,193	10,209	8,641	49,546
1997	61,158	6,957	20,237	13,319	11,188	9,457	54,200
1998	62,206	6,682	20,140	13,631	11,585	10,168	55,523
1999	59,342	6,234	19,241	13,107	11,073	9,687	53,109
2000	59,138	6,270	19,421	12,933	11,846	9,669	52,868
2001	59,804	6,209	19,908	12,924	10,744	10,018	53,594
2002 [1]	58,943	6,012	19,461	12,919	10,736	9,815	52,931

[1] Preliminary.
NASS, Livestock Branch, (202) 720–3570.

Table 7-29.—Hogs: Number in specified countries, 1999–2001 [1]

Country	1999	2000	2001 [2]
	Thousands	*Thousands*	*Thousands*
China, Peoples Republic	422,563	430,198	446,815
European Union	123,708	124,319	123,261
United States	62,206	59,342	59,138
Brazil	31,427	31,860	32,440
Poland	19,275	18,224	16,988
Russian Federation	16,400	16,100	15,780
Canada	12,429	12,904	13,576
Philippines	10,398	10,764	11,715
Mexico	10,860	10,781	10,649
Japan	9,879	9,805	9,785
Ukraine	10,083	10,073	7,652
Taiwan	6,539	7,243	7,495
Korea, Republic of	6,700	7,000	7,350
Hungary	5,450	5,335	4,834
Romania	7,194	5,700	4,797
Czech Republic	3,802	3,688	3,594
Australia	2,626	2,511	2,604
Bulgaria	1,721	1,512	1,143
Kazakhstan, Republic of	750	NA	NA
Switzerland	1,466	NA	NA
Thailand	4,810	4,802	NA
Total	770,286	772,161	779,616

[1] Various dates of enumeration are used by the countries reporting animal numbers. Data presented in this table approximate Jan. 1 as closely as possible. [2] Preliminary. NA-not available.

FAS, Dairy, Livestock and Poultry Division, (202) 720–8031. Prepared or estimated on the basis of official statistics of foreign governments, other foreign source materials, reports of U.S. Agricultural Counselors, Attachés, and Foreign Service Officers, results of office research, and related information.

Table 7-30.—Hogs: Number slaughtered, United States, 1993–2002

Year	Commercial			Farm	Total
	Federally inspected	Other	Total [1]		
	Thousands	*Thousands*	*Thousands*	*Thousands*	*Thousands*
1993	90,933	2,135	93,068	229	93,296
1994	93,435	2,261	95,696	208	95,905
1995	94,203	2,123	96,325	210	96,535
1996	90,534	1,860	92,394	175	92,569
1997	90,228	1,733	91,960	165	92,125
1998	99,285	1,745	101,029	165	101,194
1999	99,739	1,806	101,544	150	101,694
2000	96,436	1,540	97,976	130	98,106
2001	96,528	1,434	97,962	120	98,082
2002	98,915	1,348	100,263	115	100,378

[1] Totals are based on unrounded numbers.

NASS, Iowa Agricultural Statistics Service, (515) 284–4340.

Table 7-31.—Sows farrowing and pig crop: Number by States, 2001 and 2002

State	Sows farrowing							
	Dec.–Feb.		Mar.–May		June–Aug.		Sept.–Nov.	
	2001	2002 [1]	2001	2002 [1]	2001	2002 [1]	2001	2002 [1]
	Thousands	*Thousands*	*Thousands*	*Thousands*	*Thousands*	*Thousands*	*Thousands*	*Thousands*
AR	53	52	54	52	57	52	51	44
CO	76	79	79	76	79	74	82	75
IL	195	210	210	225	210	205	220	210
IN	155	155	155	155	150	150	155	150
IA	430	420	450	470	440	420	460	430
KS	73	79	77	82	78	80	81	79
MI	46	50	50	50	52	54	46	42
MN	260	260	270	270	265	260	265	270
MO	170	170	185	180	180	170	180	175
NE	140	170	160	175	155	160	165	170
NC	530	540	540	550	550	555	530	530
OH	73	74	73	75	75	75	74	73
OK	165	170	165	165	175	170	170	165
PA	51	53	52	54	54	53	55	53
SD	62	70	66	70	66	68	71	61
TX	37	43	40	45	44	50	41	46
WI	34	28	33	29	36	25	30	28
Oth Sts [2]	198	213	211	220	212	212	213	216
US	2,748	2,836	2,870	2,943	2,878	2,833	2,889	2,817

State	Pig crop							
	Dec.–Feb.		Mar.–May		June–Aug.		Sept.–Nov.	
	2001	2002 [1]	2001	2002 [1]	2001	2002	2001	2002 [1]
	Thousands	*Thousands*	*Thousands*	*Thousands*	*Thousands*	*Thousands*	*Thousands*	*Thousands*
AR	472	478	486	499	530	494	485	431
CO	646	664	656	646	668	618	697	645
IL	1,658	1,796	1,859	1,935	1,817	1,784	1,892	1,817
IN	1,364	1,349	1,380	1,364	1,335	1,305	1,349	1,313
IA	3,612	3,612	3,938	4,019	3,784	3,654	4,002	3,655
KS	624	664	666	718	679	704	693	687
MI	403	450	445	443	473	489	421	382
MN	2,327	2,288	2,444	2,417	2,385	2,301	2,385	2,403
MO	1,488	1,505	1,647	1,611	1,602	1,513	1,584	1,540
NE	1,232	1,496	1,408	1,575	1,349	1,440	1,436	1,539
NC	4,691	4,833	4,860	4,950	5,005	5,023	4,770	4,770
OH	621	662	635	671	668	668	651	646
OK	1,526	1,488	1,543	1,460	1,645	1,530	1,547	1,502
PA	449	466	463	470	475	466	484	477
SD	546	623	587	616	584	612	635	549
TX	329	387	352	401	383	445	365	407
WI	281	237	284	248	293	208	254	244
Oth Sts [2]	1,694	1,796	1,856	1,916	1,864	1,860	1,842	1,885
US	23,963	24,794	25,509	25,959	25,539	25,114	25,492	24,892

[1] Preliminary. Totals may not add due to rounding. [2] Individual State estimates not available for the 33 other States.

NASS, Livestock Branch, (202) 720–3570.

Table 7-32.—Hogs: Production, disposition, cash receipts, and gross income, United States, 1992–2001

Year	Marketings [1]	Shipped in for feeding and breeding	Farm slaughter [2]	Quantity produced (live weight) [3]	Value of production [4]	Cash receipts from sales of hogs, pork, and lard [5]	Value of hogs slaughtered for home consumption	Gross income [6]	Average price per 100 pounds received by farmers
	1,000 head	*1,000 head*	*1,000 head*	*1,000 pounds*	*1,000 dollars*	*1,000 dollars*	*1,000 dollars*	*1,000 dollars*	*Dollars*
1992	98,589	4,871	272	23,946,691	9,854,258	10,016,503	60,713	10,077,216	41.60
1993	98,390	5,471	219	23,776,854	10,662,125	10,948,792	46,256	10,995,048	45.20
1994	101,121	6,202	193	24,447,634	9,693,542	9,898,265	43,899	9,942,164	39.90
1995	103,007	7,557	188	24,426,543	9,829,498	10,254,866	41,849	10,296,715	40.50
1996	101,468	10,036	176	23,080,309	11,902,326	12,565,187	49,413	12,614,600	51.90
1997	104,301	14,935	161	23,979,220	12,551,845	13,053,680	48,320	13,102,000	52.90
1998	117,240	19,371	163	25,714,706	8,673,713	9,444,082	34,374	9,478,456	34.40
1999	121,137	22,636	141	25,790,623	7,766,241	8,621,759	28,381	8,650,140	30.30
2000	118,541	24,525	125	25,730,208	10,817,518	11,803,046	34,720	11,837,766	42.30
2001 [7]	119,558	26,920	119	25,936,846	11,442,171	12,455,790	35,711	12,491,501	44.30

[1] Includes custom slaughter for use on farms where produced and State outshipments, but excludes interfarm sales within the State. [2] Excludes custom slaughtered for farmers at commercial establishments. [3] Adjustments made for changes in inventory and for inshipments. [4] Includes allowance for higher average price of State inshipments and outshipments of feeder pigs. [5] Receipts from marketings and sale of farm slaughter includes allowance for higher average price of State outshipments of feeder pigs. [6] Cash receipts from sale of hogs, pork, and lard plus value of hogs slaughtered for home consumption. [7] Preliminary.

NASS Livestock Branch, (202) 720–3570.

Table 7-33.—Hogs: Receipts at selected public stockyards and direct receipts at interior markets, 1992–2001 [1]

Year	Receipts at selected public stockyards: South St. Joseph	South St. Paul	All others reporting	Total markets reporting [2] [3]	Direct receipts in interior Iowa and Southern Minnesota [4]
	Thousands	*Thousands*	*Thousands*	*Thousands*	*Thousands*
1992	439	479	1,704	4,037	25,961
1993	411	443	1,565	2,936	26,536
1994	451	406	1,368	3,018	28,669
1995	371	363	1,132	2,517	28,545
1996	259	330	937	1,972	27,199
1997	192	258	734	1,479	28,624
1998	158	265	565	988	34,082
1999	107	237	1,225	1,569	33,281
2000	59	203	998	1,260	36,504
2001	....	....	....	....	....

[1] Total rail and truck receipts. Saleable receipts 1978 on. [2] Rounded total of complete figures. [3] The number of stockyards reporting varies from 41 to 68. [4] Covers receipts at 14 packing plants and 30 concentration yards.
Prior to 1994 includes numbers from the following areas: Kansas City, National Stockyards and Fort Worth.

AMS, Livestock & Grain Market News, (202) 720–7316. Compiled from reports received from stockyard companies.

Table 7-34.—Hogs and corn: Hog-corn price ratio and average price received by farmers for corn, United States, 1992–2001

Year	Hog-corn price ratio [1]	Price of corn per bushel [2]
		Dollars
1992	18.5	2.29
1993	20.5	2.22
1994	16.4	2.41
1995	16.4	2.56
1996	15.4	3.55
1997	20.0	2.60
1998	14.7	2.20
1999	17.4	1.89
2000	23.3	1.86
2001	23.4	1.89

[1] Number of bushels of corn equal in value to buy 100 pounds of live hogs at local markets, based on average prices received by farmers for hogs and corn. Annual average is a simple average of monthly ratios for the calendar year. [2] Annual average is a simple average of entire month prices for the calendar year.

NASS, Environmental, Economics, and Demographics Branch, (202) 720–6146.

Table 7-35.—Hogs: Production, disposition, cash receipts, and gross income, by States, 2001 (preliminary)

State	Marketings [1]	Shipped in for feeding and breeding	Farm slaughter [2]	Quantity produced (live weight) [3]	Value of production [4]	Cash receipts from sales of hogs, pork, and lard [5]	Value of hogs slaughtered for home consumption	Gross income [6]
	1,000 head	*1,000 head*	*1,000 head*	*1,000 pounds*	*1,000 dollars*	*1,000 dollars*	*1,000 dollars*	*1,000 dollars*
AL	390.0	185.0	1.0	65,514	28,247	31,221	460	31,681
AK	1.2	....	0.5	328	200	146	49	195
AZ	124.0	20.0	1.0	44,487	19,241	14,453	229	14,682
AR	2,111.0	125.0	1.0	286,249	130,041	142,253	450	142,703
CA	288.0	45.0	6.0	54,107	23,872	27,863	925	28,788
CO	2,686.0	56.0	1.0	518,176	251,208	263,659	242	263,901
CT	8.3	0.1	0.1	2,026	831	813	41	854
DE	63.7	0.5	0.2	18,286	7,162	7,203	48	7,251
FL	106.0	8.0	1.0	18,241	6,875	7,365	71	7,436
GA	939.0	48.0	2.0	178,104	80,995	88,182	462	88,644
HI	31.0	....	1.0	6,055	5,074	4,546	147	4,693
ID	48.0	10.0	1.0	13,248	6,061	6,505	230	6,735
IL	8,339.0	1,610.0	3.0	1,818,327	852,946	920,218	922	921,140
IN	6,586.0	1,290.0	1.0	1,475,608	616,115	693,444	747	694,191
IA	26,253.0	12,900.0	11.0	6,407,914	2,751,508	3,131,682	4,352	3,136,034
KS	2,863.0	480.0	1.0	712,593	304,215	322,124	543	322,667
KY	885.0	117.0	5.0	206,470	88,270	94,321	1,861	96,182
LA	46.0	2.0	1.0	9,610	3,495	3,526	77	3,603
ME	13.3	0.5	0.2	3,849	1,606	1,308	315	1,623
MD	74.4	6.0	0.4	16,030	6,198	6,833	96	6,929
MA	30.6	5.5	0.3	7,259	2,805	2,969	183	3,152
MI	1,919.0	280.0	4.0	488,320	199,602	211,337	1,695	213,032
MN	13,009.0	4,050.0	4.0	2,783,244	1,235,194	1,416,925	1,929	1,418,854
MS	615.0	102.0	2.0	144,199	58,630	63,850	426	64,276
MO	7,350.0	1,590.0	2.0	1,177,011	513,137	586,436	1,283	587,719
MT	344.0	28.0	2.0	74,288	31,358	35,689	445	36,134
NE	6,012.0	750.0	1.0	1,450,111	665,255	711,323	1,179	712,502
NV	9.3	3.0	0.3	1,885	791	849	28	877
NH	7.8	1.4	0.2	1,701	679	727	60	787
NJ	12.6	1.1	1.5	1,140	399	410	122	532
NM	5.8	3.4	1.0	1,167	333	423	165	588
NY	136.0	4.0	1.0	28,498	10,850	10,777	384	11,161
NC	17,652.0	158.0	16.0	3,665,347	1,718,662	1,709,794	2,108	1,711,902
ND	520.0	28.0	2.0	92,042	44,657	47,917	489	48,406
OH	3,111.0	600.0	8.0	770,745	332,964	350,125	2,882	353,007
OK	6,656.0	990.0	2.0	1,163,253	487,079	520,257	547	520,804
OR	55.0	....	1.0	13,705	6,482	6,393	326	6,719
PA	1,875.5	117.0	8.5	444,495	178,027	179,860	1,208	181,068
RI	5.5	....	0.1	1,247	513	503	20	523
SC	483.0	42.0	2.0	98,454	42,810	41,033	915	41,948
SD	3,177.0	910.0	3.0	611,515	275,247	320,303	1,341	321,644
TN	511.0	100.0	3.0	94,515	40,850	45,929	739	46,668
TX	1,210.0	111.0	6.0	258,425	100,752	102,455	1,707	104,162
UT	926.0	1.0	1.0	225,290	107,820	106,338	230	106,568
VT	6.0	0.1	0.2	1,429	586	535	71	606
VA	472.0	15.0	2.0	120,956	49,938	49,760	1,295	51,055
WA	50.5	3.1	2.0	11,430	5,121	4,970	527	5,497
WV	10.1	2.0	0.6	2,523	973	889	124	1,013
WI	1,230.0	104.0	2.0	276,986	116,516	126,921	517	127,438
WY	300.0	17.0	2.0	70,444	29,981	32,428	499	32,927
US	119,557.6	26,919.7	119.1	25,936,846	11,442,171	12,455,790	35,711	12,491,501

[1] Includes custom slaughter for use on farms where produced and State outshipments, but excludes interfarm sales within the State. [2] Excludes custom slaughter for farmers at commercial establishments. [3] Adjustments made for changes in inventory and for inshipments. [4] Includes allowance for higher average price of State inshipments and outshipments of feeder pigs. [5] Receipts from marketings and sale of farm-slaughter. [6] Cash receipts from sales of hogs, pork, and lard plus value of hogs slaughtered for home consumption.

NASS, Livestock Branch, (202) 720–3570.

Table 7-36.—Hogs: Number slaughtered commercially, total and average live weight, by States, 2002[1]

State	Number slaughtered	Total live weight	Average live weight
	Thousands	*1,000 pounds*	*Pounds*
AL	136.1	72,072	529
AK	....	....	....
AZ	3.7	890	240
AR	....	....	....
CA	2,428.7	564,027	232
CO	15.8	3,739	236
DE and MD	19.2	5,207	271
FL	105.4	19,573	186
GA	155.2	38,899	251
HI	32.1	6,789	212
ID	....	....	....
IL	9,215.7	2,494,817	271
IN	6,953.0	1,796,449	258
IA	28,831.7	7,639,312	265
KS	286.0	95,889	335
KY	2,237.5	588,903	263
LA	27.2	5,634	207
MI	119.6	43,517	364
MN	8,842.9	2,311,812	261
MS	....	....	....
MO	1,970.9	....	....
MT	15.0	3,540	236
NE	6,944.7	1,827,457	263
NV	....	....	....
N ENG[2]	26.7	6,914	259
NJ	106.1	10,032	95
NM	2.8	599	212
NY	36.1	7,034	195
NC	10,223.5	2,596,812	254
ND	171.4	39,191	229
OH	1,198.4	315,511	263
OK	4,546.3	1,274,148	280
OR	196.2	50,594	258
PA	2,628.4	648,983	247
SC	....	....	....
SD	4,306.4	1,122,733	261
TN	686.0	330,309	482
TX	429.5	121,127	282
UT	59.9	10,949	183
VA	4,124.3	1,064,738	258
WA	20.4	5,173	253
WV	9.6	2,456	256
WI	462.5	201,411	436
WY	4.7	1,176	253
US[3]	100,262.6	26,561,446	265

[1] Includes slaughter in federally inspected and other slaughter plants; excludes animals slaughtered on farms. Average live weight is based on unrounded numbers. Totals may not add due to rounding. [2] CT, ME, MA, NH, RI, and VT. [3] States with no data printed are still included in US total. Data are not printed to avoid disclosing individual operations.

NASS, Iowa Agricultural Statistics Service, (515) 284–4340.

Table 7-37.—Hogs: Number slaughtered, average dressed and live weights, Federally inspected, 1993–2002[1]

Year	Federally inspected											
	Barrows and gilts			Sows			Stags and boars			Total		
	Head	Percent of total	Avg. dressed weight	Head	Percent of total	Avg. dressed weight	Head	Percent of total	Avg. dressed weight	Head	Avg. dressed weight	Avg. live weight
	1,000		*Pounds*	*1,000*		*Pounds*	*1,000*		*Pounds*	*1,000*	*Pounds*	*Pounds*
1993	86,362	95.0	179	3,709	4.1	291	862	0.9	221	90,933	184	254
1994	88,615	94.8	180	3,917	4.2	294	904	1.0	219	93,435	185	256
1995	89,662	95.2	181	3,676	3.9	292	864	0.9	221	94,203	186	257
1996	86,443	95.5	181	3,425	3.8	290	667	0.7	222	90,534	186	254
1997	86,587	96.0	185	3,064	3.4	291	577	0.6	220	90,228	189	257
1998	95,354	96.0	185	3,437	3.5	295	494	0.5	228	99,285	189	257
1999	96,000	96.3	187	3,335	3.3	296	404	0.4	232	99,739	191	259
2000	93,115	96.6	191	3,005	3.1	309	316	0.3	226	96,436	194	262
2001	93,201	96.6	193	3,009	3.1	316	318	0.3	226	96,528	197	265
2002	95,459	96.5	193	3,185	3.2	317	271	0.3	235	98,915	197	265

[1] All weights calculated using unrounded totals. Totals and percentages based on unrounded data and may not equal sum of classes due to rounding.

NASS, Iowa Agricultural Statistics Service, (515) 284–4340.

Table 7-38.—Hogs and pigs: Number of operations and inventory by size groups, selected States, regions, and United States, 2001–2002 [1]

State	Operations having											
	1–99 head		100–499 head		500–999 head		1,000–1,999 head		2,000–4,999 head		5,000+ head	
	2001	2002	2001	2002	2001	2002	2001	2002	2001	2002	2001	2002
	Number											
AR	720	650	90	80	80	80	90	90	90	75	30	25
CO	317	315	35	33	15	13	12	10	4	4	17	15
IL	2,000	1,500	1,400	1,200	670	700	670	630	450	430	110	140
IN	1,500	1,400	1,100	720	520	430	435	430	350	320	95	100
IA	1,500	1,300	3,200	3,000	2,300	2,000	1,800	1,850	1,200	1,350	500	500
KS	820	820	320	350	160	110	80	100	80	75	40	45
MI	1,700	1,500	430	450	90	90	110	100	130	115	40	45
MN	1,700	1,450	2,100	2,100	1,100	1,000	700	700	650	700	250	250
MO	1,700	1,600	700	500	330	300	110	100	180	210	80	90
NE	1,100	800	1,000	900	630	570	380	350	200	190	90	90
NC	1,600	1,460	140	120	100	90	200	190	720	700	640	640
OH	2,900	2,800	1,200	900	360	320	275	270	140	185	25	25
OK	2,400	2,300	120	110	40	50	50	50	50	50	40	40
PA	2,000	2,000	400	390	200	190	110	120	155	165	35	35
SD	600	550	500	450	220	220	130	130	90	90	60	60
TX	3,895	3,680	60	76	16	16	7	5	5	5	17	18
WI	1,900	1,500	510	500	120	130	110	110	50	50	10	10
Oth Sts [2]	17,460	17,100	2,100	1,600	250	180	200	200	250	250	120	130
US	45,812	42,725	15,405	13,479	7,201	6,489	5,469	5,435	4,794	4,964	2,199	2,258

State	Inventory on operations having											
	1–99 head		100–499 head		500–999 head		1,000–1,999 head		2,000–4,999 head		5,000+ head	
	2001	2002	2001	2002	2001	2002	2001	2002	2001	2002	2001	2002
	Percent											
AR	2.0	2.0	3.0	3.0	8.0	9.0	20.0	21.0	39.0	39.0	28.0	26.0
CO	0.6	0.6	0.9	0.8	1.1	1.0	1.8	1.7	1.9	1.9	93.7	94.0
IL	1.0	0.7	7.0	6.3	10.0	11.0	20.0	19.0	33.0	30.0	29.0	33.0
IN	1.0	1.0	7.0	5.5	10.0	9.0	17.0	17.5	29.0	29.0	36.0	38.0
IA	0.2	0.2	5.5	5.5	9.3	8.0	17.0	17.3	23.0	26.0	45.0	43.0
KS	1.0	1.0	4.0	4.0	6.0	4.0	6.5	9.0	14.0	14.0	68.5	68.0
MI	1.0	1.0	6.5	7.5	6.0	6.5	12.0	12.0	38.5	34.0	36.0	39.0
MN	1.0	1.0	6.0	6.0	10.0	9.0	16.0	15.0	24.0	25.0	43.0	44.0
MO	1.0	1.0	5.0	4.0	6.0	6.0	4.0	4.0	17.0	20.0	67.0	65.0
NE	1.0	0.8	8.0	8.0	13.5	13.0	15.5	15.2	21.0	21.0	41.0	42.0
NC	0.2	0.2	0.3	0.2	0.7	0.6	2.9	2.5	20.9	20.0	75.0	76.5
OH	4.0	4.0	19.0	14.0	16.0	15.0	25.0	26.0	26.0	31.0	10.0	10.0
OK	1.5	1.0	1.0	1.0	1.5	1.5	2.5	2.5	6.5	5.0	87.0	89.0
PA	3.0	3.0	9.0	9.0	12.0	11.0	15.0	16.0	40.0	40.0	21.0	21.0
SD	1.5	1.0	10.0	9.0	11.0	11.0	11.5	11.0	21.0	21.0	45.0	47.0
TX	4.0	2.7	2.0	2.5	1.0	1.1	1.0	0.7	2.0	2.0	90.0	91.0
WI	7.0	6.0	20.0	19.0	14.0	15.0	25.0	25.0	22.0	22.0	12.0	13.0
Oth Sts [2]	1.5	1.5	6.5	4.5	3.5	2.5	5.5	5.5	14.0	14.0	69.0	72.0
US	1.0	1.0	5.5	5.0	7.5	7.0	12.0	12.0	22.0	22.0	52.0	53.0

[1] An operation is any place having one or more head of hogs and pigs on hand at any time during the year. Percents reflect average distributions based primarily on mid-year surveys. [2] Individual State estimates not available for the 33 other States.

NASS, Livestock Branch, (202) 720–3570.

Table 7-39.—Lard, including rendered pork fat: Stocks Jan. 1, production, trade, and disappearance, United States, 1992–2001

Year	Supply						Disposition			
	Production				Stocks Jan. 1[1]	Total[2]	Exports	Domestic disappearance		
	Federally inspected	Other commercial	Farm	Total				Total	Direct use	
									Total	Per civilian
	Million pounds	Million pounds	Million pounds	Million pounds	Million pounds	Million pounds	Million pounds	Million pounds	Million pounds	Pounds
1992	....	....	....	1,025	37	1,065	136	906	291	1.1
1993	....	....	....	1,005	23	1,031	114	879	299	1.2
1994	....	....	....	1,034	38	1,074	137	896	471	1.8
1995	....	....	....	1,040	41	1,082	124	920	430	1.6
1996	....	....	....	998	38	1,038	101	918	468	1.7
1997	....	....	....	993	19	1,013	90	901	518	1.9
1998	....	....	....	1,091	22	1,116	131	956	541	2.0
1999	....	....	....	1,097	28	1,127	147	953	547	2.0
2000	....	....	....	1,058	27	1,087	174	895	549	1.9
2001	....	....	....	1,058	18	1,080	103	963	661	2.3

[1] Factory and warehouse stocks as reported by the Bureau of the Census. [2] Includes imports, which were less than 500,000 pounds.

ERS, Field Crops Branch, (202) 694–5300. Compiled from reports of the U.S. Department of Commerce and the U.S. Department of Agriculture. Totals and per capita estimates computed from unrounded numbers.

Table 7-40.—Lard, including rendered pork fat: United States exports by country of destination, 1999–2001

Continent and country	1999	2000	2001[1]
	Metric tons	Metric tons	Metric tons
Mexico	31,984	45,688	29,248
Canada	11,928	9,270	9,270
Taiwan	8,688	6,352	6,084
Belize	1,122	811	861
Honduras	38	19	472
Korea, Republic of	463	1,113	300
Trinidad and Tobago	0	0	95
Japan	73	134	76
Bahamas, the	6	64	58
Hong Kong	5,857	10,593	55
Haiti	39	114	53
Saudi Arabia	0	20	40
Micronesia, Federation	0	0	39
South Africa, Republic	20	4	32
Netherlands Antilles	194	35	30
Jamaica	2,329	830	24
Panama	205	1,065	22
Guatemala	58	18	20
China, Peoples Republic	1,136	0	20
Colombia	23	16	17
Lebanon	0	15	17
Marshal Islands	0	8	9
Chile	0	18	8
Kuwait	0	0	8
European Union	0	14	6
Others	2,705	2,731	3
Grand total[2]	66,867	78,930	46,869

[1] Final. [2] May not add due to rounding.

FAS, Dairy, Livestock and Poultry Division, (202) 720–8031. Compiled from reports of the U.S. Department of Commerce.

Table 7-41.—Sheep and lambs: Number and value, United States, Jan. 1, 1994–2003

Year	Number	Value	
		Per head	Total
	Thousands	*Dollars*	*1,000 dollars*
1994 [1]	9,836	69.90	681,384
1995 [1]	8,989	74.70	663,449
1996 [1]	8,465	86.50	732,197
1997 [1]	8,024	96.00	761,650
1998 [1]	7,825	102.00	797,826
1999 [1]	7,215	88.00	637,634
2000 [1]	7,032	95.00	668,750
2001 [1]	6,965	100.00	694,495
2002 [1]	6,685	94.00	618,123
2003 [1 2]	6,350	104.00	657,663

[1] Beginning in 1994 includes new crop lambs. [2] Preliminary.

NASS, Livestock Branch, (202) 720–3570.

Table 7-42.—Sheep and lambs: Number, by classes, United States, Jan. 1, 1994–2003

Year	All sheep and lambs	Sheep and lambs on feed	Breeding sheep				
			Total [4]	Lambs		1 year and over	
				Ewes	Rams	Ewes	Rams
	Thousands	*Thousands*	*Thousands*	*Thousands*	*Thousands*	*Thousands*	*Thousands*
1994 [1]	9,836	1,839	7,236	1,098	(5)	5,844	294
1995 [1]	8,989	(3)	6,518	857	(5)	5,404	257
1996 [1]	8,465	(3)	6,226	858	(5)	5,134	234
1997 [1]	8,024	(3)	5,919	787	(5)	4,912	220
1998 [1]	7,825	(3)	5,611	839	(5)	4,570	203
1999 [1]	7,215	(3)	5,299	774	(5)	4,322	203
2000 [1]	7,032	(3)	5,164	730	(5)	4,229	206
2001 [1]	6,965	(3)	4,967	675	(5)	4,091	201
2002 [1]	6,685	(3)	4,913	734	(5)	3,980	200
2003 [1 2]	6,350	(3)	4,682	702	(5)	3,789	192

[1] Includes new crop lambs. [2] Preliminary. [3] Discontinued. [4] Categories may not add to total due to rounding. [5] Combined with ewes.

NASS, Livestock Branch, (202) 720–3570.

Table 7-43.—Sheep: Number in specified countries, 1998–2000 [1]

Country	1998	1999	2000 [2]
	Thousand head	*Thousand head*	*Thousand head*
Argentina	13,953	13,800	14,100
Australia	117,091	117,191	118,321
Bulgaria	3,822	3,900	3,910
China, Peoples Republic	268,143	271,130	280,420
Egypt	6,300	6,253	6,171
France	10,640	10,567	10,500
Germany	2,298	2,290	2,243
Greece	8,741	8,430	8,430
India	180,130	180,885	181,440
Ireland	5,624	5,393	5,350
Italy	12,135	12,200	12,200
Kazakhstan, Republic of	9,556	9,000	8,700
Mexico	14,714	15,076	15,438
New Zealand	46,150	45,800	46,000
Poland	422	370	370
Portugal	4,316	4,210	4,210
Romania	9,167	8,700	8,500
Russian Federation	18,213	15,698	15,700
Saudi Arabia	12,312	12,400	12,483
South Africa, Republic of	34,910	35,000	35,220
Spain	24,199	23,700	23,600
Turkey	37,300	34,400	31,000
Ukraine	2,113	1,843	1,700
United Kingdom	31,080	30,800	30,600
United States	7,235	7,215	7,026
Uruguay	16,746	15,500	14,500
Total	897,310	891,751	898,132

[1] Various dates of enumeration are used by the countries reporting animal numbers. Data presented in this table approximate Jan. 1 as closely as possible. [2] Preliminary.

FAS, Dairy, Livestock and Poultry Division, (202) 720–1350. Prepared or estimated on the basis of official statistics of foreign governments, other foreign source materials, reports of U.S. Agricultural Counselors, Attachés, and Foreign Service Officers, results of office research, and related information.

Table 7-44.—Breeding sheep: Number, by classes, by States, Jan. 1, 2002 and 2003

State	Lambs		One year and over			
	Breeding		Ewes		Rams	
	2002	2003[1]	2002	2003[1]	2002	2003[1]
	1,000 head	*1,000 head*	*1,000 head*	*1,000 head*	*1,000 head*	*1,000 head*
AZ	6.0	6.0	57.0	56.0	3.0	3.0
CA	45.0	50.0	310.0	295.0	15.0	10.0
CO	29.0	24.0	170.0	155.0	6.0	6.0
ID	35.0	36.0	184.0	184.0	5.0	5.0
IL	8.0	11.0	44.0	42.0	4.0	3.0
IN	6.5	8.5	39.0	34.0	2.5	2.5
IA	26.0	28.0	142.0	135.0	7.0	7.0
KS	9.0	6.5	53.0	47.0	3.0	2.5
MI	10.0	11.0	40.0	41.0	2.0	2.0
MN	15.0	15.0	95.0	85.0	5.0	5.0
MO	10.0	9.0	46.0	48.0	3.0	3.0
MT	60.0	49.0	232.0	213.0	8.0	8.0
NE	12.0	11.0	58.0	56.0	3.0	3.0
NV	14.0	15.0	69.0	65.0	2.0	2.0
N ENG[2]	8.0	7.0	32.0	32.0	2.0	3.0
NM	27.0	26.0	150.0	140.0	8.0	9.0
NY	9.0	10.0	40.0	40.0	3.0	3.0
ND	17.0	10.0	85.0	77.0	3.0	3.0
OH	20.0	22.0	93.0	97.0	7.0	6.0
OK	6.0	7.0	36.0	41.0	3.0	3.0
OR	29.0	25.0	134.0	125.0	8.0	7.0
PA	12.0	11.0	55.0	55.0	6.0	5.0
SD	46.0	40.0	265.0	265.0	9.0	10.0
TX	130.0	130.0	720.0	680.0	40.0	40.0
UT	36.0	31.0	275.0	250.0	9.0	9.0
VA	7.0	6.5	37.0	38.0	3.0	3.5
WA	8.0	6.0	36.0	34.0	3.0	2.0
WV	6.0	5.0	24.0	23.0	1.0	1.0
WI	10.0	11.0	52.0	52.0	3.0	3.0
WY	59.0	55.0	320.0	295.0	11.0	10.0
Other States[3]	18.0	19.0	87.0	89.0	12.0	12.0
US	733.5	701.5	3,980.0	3,789.0	199.5	191.5

[1] Preliminary. [2] N ENG includes CT, ME, MA, NH, RI, and VT. [3] AL, AK, AR, DE, FL, GA, HI, KY, LA, MD, MS, NJ, NC, SC, and TN.

NASS, Livestock Branch, (202) 720–3570.

Table 7-45.—Sheep and lambs: Average price per 100 pounds at San Angelo, 1993–2002[1]

Year	Sheep			Slaughter lambs			
				Shorn		Spring	
	Good	Utility	Cull	Prime	Choice	Prime	Choice
1993	36.95	36.41	27.94	64.92	64.92	61.62	61.62
1994	40.19	36.06	27.81	66.15	66.15	....	....
1995	33.97	35.61	29.07	75.71	75.71	77.91	77.91
1996	34.50	35.38	23.96	85.68	85.68	82.00	82.00
1997	49.24	46.53	32.80	89.26	89.26	89.20	89.20
1998	40.11	39.01	26.53	71.79	71.79	74.37	74.37
1999	41.71	41.21	27.72	76.37	65.92	76.39	66.42
2000	45.37	42.53	29.84	80.36	80.36	80.10	80.10
2001	44.14	45.11	30.49	70.05	70.05	69.78	69.78
2002	38.04	39.26	24.51	71.69	71.69	72.09	72.09

[1] Simple average of monthly bulk-of-sales prices from data of the livestock reporting service. 1995 to present price reflects wooled lamb as well as the weight range of 110-130.

AMS, Livestock & Grain Market News, (202) 720–7316.

Table 7-46.—Sheep and lambs: Number of breeding sheep and market, by States, Jan. 1, 2002–2003

State	Breeding sheep and lambs		Market sheep and lambs	
	2002	2003[1]	2002	2003[1]
	1,000 head	*1,000 head*	*1,000 head*	*1,000 head*
AZ	66.0	65.0	68.0	66.0
CA	370.0	355.0	430.0	435.0
CO	205.0	185.0	165.0	185.0
ID	224.0	225.0	36.0	35.0
IL	56.0	56.0	14.0	12.0
IN	48.0	45.0	9.0	5.0
IA	175.0	170.0	75.0	80.0
KS	65.0	56.0	35.0	34.0
MI	52.0	54.0	20.0	21.0
MN	115.0	105.0	45.0	40.0
MO	59.0	60.0	11.0	7.0
MT	300.0	270.0	35.0	30.0
NE	73.0	70.0	28.0	19.0
NV	85.0	82.0	15.0	13.0
N ENG[2]	42.0	42.0	5.0	5.0
NM	185.0	175.0	45.0	40.0
NY	52.0	53.0	8.0	12.0
ND	105.0	90.0	40.0	35.0
OH	120.0	125.0	20.0	25.0
OK	45.0	51.0	15.0	14.0
OR	171.0	157.0	114.0	78.0
PA	73.0	71.0	13.0	12.0
SD	320.0	315.0	80.0	65.0
TX	890.0	850.0	240.0	200.0
UT	320.0	290.0	45.0	30.0
VA	47.0	48.0	12.0	14.0
WA	47.0	42.0	9.0	9.0
WV	31.0	29.0	6.0	5.0
WI	65.0	66.0	15.0	14.0
WY	390.0	360.0	90.0	100.0
Other States[3]	117.0	120.0	29.0	28.0
US	4,913.0	4,682.0	1,772.0	1,668.0

[1] Preliminary. [2] N ENG includes CT, ME, MA, NH, RI, and VT. [3] AL, AK, AR, DE, FL, GA, HI, KY, LA, MD, MS, NJ, NC, SC, and TN.

NASS, Livestock Branch, (202) 720–3570.

Table 7-47.—Lamb crop: Per 100 ewes 1+, number and percent of previous year, by State, and United States, 2001–2002

State	Breeding ewes 1 year & older, Jan. 1		Lambs per 100 ewes 1+, Jan. 1		Lamb crop [1]		
	2001	2002 [2]	2001	2002 [2]	2001	2002 [2]	2002 as % of 2001
	1,000 head	*1,000 head*	*Percent*	*Percent*	*1,000 head*	*1,000 head*	*Percent*
AZ	55.0	57.0	73	70	40.0	40.0	100
CA	320.0	310.0	91	92	290.0	285.0	98
CO	165.0	170.0	115	118	190.0	200.0	105
ID	195.0	184.0	131	130	255.0	240.0	94
IL	48.0	44.0	135	134	65.0	59.0	91
IN	45.0	39.0	122	121	55.0	47.0	85
IA	144.0	142.0	153	151	220.0	215.0	98
KS	58.0	53.0	124	117	72.0	62.0	86
MI	40.0	40.0	125	133	50.0	53.0	106
MN	90.0	95.0	172	153	155.0	145.0	94
MO	47.0	46.0	117	130	55.0	60.0	109
MT	265.0	232.0	128	136	340.0	315.0	93
NE	77.0	58.0	112	138	86.0	80.0	93
NV	68.0	69.0	93	90	63.0	62.0	98
N ENG [3]	34.0	32.0	115	119	39.0	38.0	97
NM	165.0	150.0	85	83	140.0	125.0	89
NY	38.0	40.0	129	128	49.0	51.0	104
ND	89.0	85.0	129	118	115.0	100.0	87
OH	86.0	93.0	145	145	125.0	135.0	108
OK	33.0	36.0	121	128	40.0	46.0	115
OR	120.0	134.0	133	112	160.0	150.0	94
PA	54.0	55.0	131	129	71.0	71.0	100
SD	265.0	265.0	115	123	305.0	325.0	107
TX	710.0	720.0	80	75	570.0	540.0	95
UT	300.0	275.0	102	100	305.0	275.0	90
VA	37.0	37.0	135	149	50.0	55.0	110
WA	35.0	36.0	166	156	58.0	56.0	97
WV	24.0	24.0	133	138	32.0	33.0	103
WI	53.0	52.0	136	140	72.0	73.0	101
WY	340.0	320.0	101	106	345.0	340.0	99
Other States [4]	91.0	87.0	91	97	83.0	84.0	101
US	4,091.0	3,980.0	110	110	4,495.0	4,360.0	97

[1] Lamb crop defined as lambs born in the Eastern States and lambs docked or branded in the Western States. [2] Preliminary. [3] N ENG includes CT, ME, MA, NH, RI, and VT. [4] AL, AK, AR, DE, FL, GA, HI, KY, LA, MD, MS, NJ, NC, SC, and TN.

NASS, Livestock Branch, (202) 720–3570.

Table 7-48.—Sheep and lambs: Production, disposition, cash receipts, and gross income, United States, 1992–2001

Year	Lamb crop [1]	Marketings [2]		Inshipments	Farm slaughter	Quantity produced (live weight) [3]
		Sheep	Lambs			
	1,000 head	*1,000 head*	*1,000 head*	*1,000 head*	*1,000 head*	*1,000 pounds*
1992	7,225	1,923	7,007	2,389	87	745,970
1993	6,377	1,911	6,636	2,399	75	739,329
1994	5,968	1,465	6,469	2,149	82	634,647
1995	5,643	1,052	6,286	2,141	76	602,011
1996	5,361	938	6,069	2,196	71	572,344
1997	5,356	1,015	5,676	2,055	71	602,535
1998	5,007	992	5,505	1,744	73	554,918
1999	4,733	790	5,208	1,891	67	533,610
2000	4,622	788	4,827	1,763	69	508,935
2001 [4]	4,495	713	4,797	1,543	66	498,763

Year	Value of production	Cash receipts for sales of sheep, lambs, and mutton and lamb [5]	Value of sheep and lambs slaughtered for home consumption	Gross income [6]	Average price per 100 pounds received by farmers	
					Sheep	Lambs
	1,000 dollars	*1,000 dollars*	*1,000 dollars*	*1,000 dollars*	*Dollars*	*Dollars*
1992	394,409	507,458	8,230	515,688	25.80	59.50
1993	413,756	533,930	8,753	542,683	28.60	64.40
1994	364,477	510,289	9,147	519,786	30.90	65.60
1995	414,366	566,240	10,387	576,627	28.00	78.20
1996	440,686	612,345	10,690	623,035	29.90	88.20
1997	489,564	635,451	11,363	646,814	37.90	90.30
1998	354,437	484,515	8,518	493,033	30.60	72.30
1999	348,742	469,066	8,456	477,522	31.10	74.50
2000	361,231	470,176	9,429	479,605	34.30	79.80
2001 [4]	299,998	397,585	8,206	405,791	34.60	66.90

[1] Lamb crop defined as lambs born in the native States, plus lambs docked or branded in the Western States. [2] Includes custom slaughter for use on farms where produced and state outshipments, but excludes interfarm sales within the State. [3] Adjustments made for changes in inventory and for inshipments. [4] Preliminary. [5] Receipts from marketings and sale of farm-slaughtered meat. [6] Cash receipts from sales of sheep, lambs, and mutton and lamb plus value of sheep and lambs slaughtered for home consumption.

NASS, Livestock Branch, (202) 720–3570.

Table 7-49.—Sheep and lambs: Receipts at selected public stockyards, 1993–2002 [1]

Year	Sioux Falls	South St. Joseph	South St. Paul	All others reporting	Total markets reporting [2] [3]
	Thousands	*Thousands*	*Thousands*	*Thousands*	*Thousands*
1993	....	88	72	1,187	1,377
1994	....	80	64	1,232	1,469
1995	....	7	51	776	846
1996	....	8	47	857	921
1997	100	8	47	998	1,060
1998	73	5	50	811	938
1999	77	5	54	962	1,097
2000	61	3	55	935	1,054
2001	48	4	56	913	1,021
2002	48	4	63	832	947

[1] Total rail and truck receipts unloaded at public stockyards. Saleable receipts only, 1978. [2] Rounded totals of complete figures. [3] The number of stockyards reporting varies from 41 to 68.

AMS, Livestock & Grain Market News, (202) 720–7316. Compiled from reports received from stockyard companies.

Table 7-50.—Sheep and lambs: Production, disposition, cash receipts, and gross income, by States, 2001 (preliminary)

State	Marketings [1]		Shipped in	Slaughter on farms [2]	Quantity produced (live weight) [3]	Value of production	Cash receipts from sales of sheep, lambs, and mutton and lamb [4]	Value of sheep and lambs slaughtered for home consumption	Gross income [5]
	Sheep	Lambs							
	1,000 head	*1,000 head*	*1,000 head*	*1,000 head*	*1,000 pounds*	*1,000 dollars*	*1,000 dollars*	*1,000 dollars*	*1,000 dollars*
AZ	14.0	88.0	81.0	13.0	4,040	2,330	6,601	858	7,459
CA	47.0	553.0	305.0	5.0	51,640	33,349	52,974	712	53,686
CO	40.0	810.0	654.0	1.0	75,775	40,507	81,400	78	81,478
ID	43.0	220.0	16.0	1.0	26,145	13,117	14,708	119	14,827
IL	8.0	51.0	3.0	1.0	4,016	2,688	2,954	188	3,142
IN	17.9	39.6	6.0	2.5	4,486	2,162	2,872	274	3,146
IA	20.0	252.0	74.0	1.0	30,636	18,222	23,470	264	23,734
KS	11.5	68.0	14.0	1.0	8,216	5,066	5,862	188	6,050
MI	6.5	31.5	1.5	2.0	4,245	2,833	2,441	403	2,844
MN	8.0	152.0	28.0	1.0	16,808	9,787	11,503	145	11,648
MO	9.5	41.8	3.0	0.2	4,745	3,034	3,252	78	3,330
MT	79.0	251.0	8.0	1.0	31,681	16,947	18,487	146	18,633
NE	51.0	91.6	57.0	0.4	10,384	5,371	8,603	54	8,657
NV	7.0	45.0	18.0	1.0	3,130	2,045	2,681	115	2,796
N ENG [6]	10.0	27.1	2.0	0.9	2,950	2,922	2,905	338	3,243
NM	33.0	114.0	9.0	7.0	11,029	6,926	7,874	707	8,581
NY	3.0	36.0	2.0	1.0	3,802	3,585	3,591	158	3,749
ND	18.0	71.5	10.0	0.5	9,095	4,983	5,504	73	5,577
OH	13.5	111.0	19.0	1.5	12,687	6,718	9,254	196	9,450
OK	2.5	29.0	7.5	1.0	3,225	2,093	2,028	185	2,213
OR	6.0	130.0	40.0	3.0	13,080	8,043	7,512	573	8,085
PA	6.3	49.7	2.2	1.7	5,811	4,125	3,758	200	3,958
SD	24.0	283.5	39.0	0.5	27,565	18,526	22,051	102	22,153
TX	59.0	496.0	77.0	2.0	48,084	33,679	40,175	175	40,350
UT	51.0	241.0	7.0	5.0	25,350	14,345	15,194	472	15,666
VA	2.5	39.0	1.0	1.0	4,154	3,154	3,202	166	3,368
WA	3.0	49.0	3.0	1.0	4,953	3,080	2,901	236	3,137
WV	3.0	20.8	2.0	0.2	2,508	1,604	1,549	37	1,586
WI	8.0	55.0	2.0	1.0	8,313	4,689	4,673	174	4,847
WY	86.0	296.0	32.0	2.0	33,840	20,471	23,072	185	23,257
Oth Sts [7]	22.0	54.0	20.0	6.0	6,370	3,597	4,534	607	5,141
US	713.2	4,797.1	1,543.2	66.4	498,763	299,998	397,585	8,206	405,791

[1] Includes custom slaughter for use on farms where produced and State outshipments, but excludes interfarm sales within the State. [2] Excludes custom slaughter for farmers at commercial establishments. [3] Adjustments made for changes in inventory and for inshipments. [4] Receipts from marketings and sale of farm-slaughter. [5] Cash receipts from sales of sheep, lambs, and mutton and lamb plus value of sheep and lambs slaughtered for home consumption. [6] N ENG includes CT, ME, MA, NH, RI, and VT. [7] AL, AK, AR, DE, FL, GA, HI, KY, LA, MD, MS, NJ, NC, SC, and TN.

NASS, Livestock Branch, (202) 720–3570.

Table 7-51.—Sheep and lambs: Number slaughtered commercially, total and average live weight, by States, 2002 [1]

State	Number slaughtered	Total live weight	Average live weight
	Thousands	*1,000 pounds*	*Pounds*
AL			
AK			
AZ	0.9	96	104
AR			
CA			
CO	1,148.1	168,222	147
DE and MD	40.0	4,414	110
FL	3.8	285	75
GA	7.1	738	103
HI			
ID	4.8	565	118
IL	205.0	27,594	135
IN	36.3	3,921	108
IA	472.9	67,587	143
KS	1.9	224	118
KY	9.9	1,319	133
LA	2.2	203	94
MI	197.1	23,911	121
MN	3.2	364	115
MS			
MO	6.6	691	105
MT	3.8	465	122
NE	1.4	159	116
NV	1.5	190	126
N ENG [2]	26.6	2,766	104
NJ	114.0	10,894	96
NM	32.1	3,577	112
NY	37.7	3,298	87
NC	0.6	51	91
ND	1.3	156	123
OH	18.0	1,924	107
OK	1.3	153	122
OR	11.7	1,479	126
PA	70.2	7,259	103
SC			
SD	14.6	1,674	114
TN	5.9	451	76
TX	336.8	42,698	127
UT	32.4	4,017	124
VA	9.6	953	99
WA	9.4	1,081	115
WV			
WI	13.5	1,793	133
WY	1.6	218	135
US [3]	3,286.3	437,727	133

[1] Includes slaughter in federally inspected and in other slaughter plants; exludes animals slaughtered on farms. Average live weight is based on unrounded numbers. Totals may not add due to rounding. [2] CT, ME, MA, NH, RI, and VT. [3] States with no data prined are still included in US total. Data are not printed to avoid disclosing individual operations.

NASS, Iowa Agricultural Statistics Service, (515) 284–4340.

Table 7-52.—Sheep and lambs: Number slaughtered, United States, 1993–2002

Year	Commercial			Farm	Total
	Federally inspected	Other	Total [1]		
	Thousands	*Thousands*	*Thousands*	*Thousands*	*Thousands*
1993	5,002	181	5,182	77	5,259
1994	4,755	183	4,938	76	5,014
1995	4,388	172	4,560	69	4,629
1996	4,032	152	4,184	65	4,249
1997	3,771	137	3,907	62	3,969
1998	3,670	133	3,804	57	3,861
1999	3,556	145	3,701	65	3,766
2000	3,308	152	3,460	67	3,527
2001	3,065	157	3,222	68	3,290
2002	3,092	194	3,286	65	3,351

[1] Totals are based on unrounded numbers.

NASS, Iowa Agricultural Statistics Service, (515) 284–4340.

Table 7-53.—Sheep and lambs: Number slaughtered, average dressed and live weights, percentage distribution, by class, Federally inspected, 1993–2002 [1]

Year	Federally inspected								
	Lambs and yearlings			Mature sheep			Total		
	Head	Pct. of total	Avg. dressed weight	Head	Pct. of total	Avg. dressed weight	Head	Avg. dressed weight	Avg. live weight
	1,000		*Pounds*	*1,000*		*Pounds*	*1,000*	*Pounds*	
1993	4,708	94.1	65	293	5.9	57	5,002	64	129
1994	4,463	93.9	63	292	6.1	57	4,756	63	126
1995	4,114	93.8	63	274	6.2	58	4,388	63	127
1996	3,763	93.3	65	270	6.7	58	4,032	64	129
1997	3,558	94.3	67	213	5.7	60	3,771	67	134
1998	3,460	94.3	66	211	5.7	61	3,670	66	133
1999	3,369	94.7	67	188	5.3	59	3,556	67	134
2000	3,141	95.0	68	167	5.0	63	3,308	68	137
2001	2,921	95.3	71	144	4.7	62	3,065	70	142
2002	2,944	95.2	68	148	4.8	63	3,092	68	135

[1] All percents and weights calculated using unrounded totals.

NASS, Iowa Agricultural Statistics Service, (515) 284–4340.

Table 7-54.—Sheep and lambs: Operations with sheep, 2001–2002

State	Operations with sheep	
	2001	2002 [1]
	Number	*Number*
AZ	270	270
CA	2,900	2,800
CO	2,000	1,900
ID	1,000	1,000
IL	2,300	2,200
IN	2,100	2,000
IA	4,700	4,600
KS	1,500	1,400
MI	1,800	1,900
MN	2,400	2,300
MO	1,700	1,600
MT	1,800	1,700
NE	1,600	1,500
NV	300	300
N ENG [2]	2,000	2,000
NM	850	800
NY	1,500	1,500
ND	1,100	1,000
OH	3,600	3,500
OK	1,400	1,500
OR	3,000	3,100
PA	2,500	2,600
SD	2,300	2,300
TX	6,800	6,800
UT	1,500	1,400
VA	1,400	1,500
WA	1,200	1,200
WV	1,000	1,100
WI	2,200	2,300
WY	900	800
Other States [3]	5,500	5,300
US	65,120	64,170

[1] Preliminary. [2] New England includes CT, ME, MA, NH, RI, and VT. [3] Other States include AL, AK, AR, DE, FL, GA, HI, KY, LA, MD, MS, NJ, NC, SC, and TN.

NASS, Livestock Branch, (202) 720–3570.

Table 7-55.—Breeding Sheep: Survey percent by size groups, United States, 2002–2003 [1]

Item	1–99 head		100–499 head		500–4,999 head		5,000+ head	
	2002	2003 [2]	2002	2003 [2]	2002	2003 [2]	2002	2003 [2]
	Percent	*Percent*	*Percent*	*Percent*	*Percent*	*Percent*	*Percent*	*Percent*
Operations	91.1	91.8	7.3	6.7	1.5	1.4	0.1	0.1
Inventory	30.1	29.9	23.5	23.8	32.4	33.1	14.0	13.2

[1] Percents reflect distributions of end-of-year survey. [2] Preliminary.

NASS, Livestock Branch, (202) 720–3570.

Table 7-56.—Wool: Number of sheep shorn, weight per fleece, production, average price per pound received by farmers, value of production, exports, imports, total new supply of apparel wool, and imports of carpet wool, United States, 1992–2001

Year	Sheep and lambs shorn [1]	Weight per fleece	Shorn wool production	Price per pound [2]	Value of production [3]
	Thousands	*Pounds*	*1,000 pounds*	*Cents*	*1,000 dollars*
1992	10,521	7.88	82,943	74.0	60,162
1993	9,976	7.77	77,535	51.0	39,077
1994	8,877	7.73	68,577	78.0	52,377
1995	8,138	7.80	63,513	1.04	64,277
1996	7,215	7.78	56,669	0.70	39,270
1997	6,960	7.70	53,578	0.84	44,909
1998	6,428	7.66	49,255	0.60	29,415
1999	6,158	7.57	46,572	0.38	17,860
2000	6,135	7.56	46,446	0.33	15,377
2001 [4]	5,689	7.55	43,016	0.36	15,311

Year	Shorn wool production	Raw wool supply (clean)				
		Domestic production [5]	Exports [6]	Imports for consumption		Total new supply [9]
				48's and Finer [7]	Not Finer than 46's [8]	
	1,000 pounds	*1,000 pounds*	*1,000 pounds*	*1,000 pounds*	*1,000 pounds*	*1,000 pounds*
1992	82,943	43,794	3,413	65,457	23,802	129,640
1993	77,535	40,938	2,529	76,001	21,876	138,286
1994	68,577	36,209	2,863	64,889	24,645	122,880
1995	63,513	33,535	6,042	63,781	25,039	116,313
1996	56,159	29,921	5,715	54,073	21,296	99,575
1997	53,578	28,289	4,732	51,484	24,962	100,003
1998	49,255	26,007	1,721	45,805	24,702	94,793
1999	46,549	24,575	3,694	21,264	21,810	63,955
2000	46,446	24,413	6,629	23,902	21,099	62,785
2001 [4]	43,016	22,712	6,154	15,843	19,727	52,128

[1] Includes sheep shorn at commercial feeding yards. [2] Price computed by weighting State average prices for all wool sold during the year by sales of shorn wool. [3] Production by States multiplied by annual average price. [4] Preliminary. [5] Conversion factor from grease basis to clean basis are as follows: Shorn wool production—52.8 percent (Stat. Bull. 616) from 1987-1997. [6] Includes carpet wool exports. [7] Prior to 1989, known as dutiable imports. [8] Prior to 1989, known as duty-free imports. In 1994 includes 24,645,306 pounds of imported raw wool not finer than 46's and 2,182,576 pounds of miscellaneous imported raw wool. [9] Production minus exports plus imports; stocks not taken into consideration.

ERS, Field Crops Branch, (202) 694–5300 and NASS. Imports and exports from reports of the U.S. Department of Commerce.

Table 7-57.—Wool: Price-support operations, United States, 1993–2002

Marketing year	Shorn wool price per pound		Payment rate		Marketings covered by payments [2]		Amount of payments [3]		
	Support	Season average received by producers	Average for shorn wool per pound	Unshorn lambs per cwt. [1]	Shorn wool	Unshorn lambs	Shorn wool	Unshorn lambs	Total
	Cents	*Cents*	*Cents*	*Cents*	*Million pounds*	*Million pounds*	*Million dollars*	*Million dollars*	*Million dollars*
1993	204	51.0	153.0	612	73.3	554	107.8	26.1	133.9
1994	209	78.0	131.0	524	64.5	478	60.4	15.2	76.6
1995	212	104.0	108.0	432	54.3	467	27.3	7.5	34.8
1996	([4])	70.0	....	....	....	....	....	....	....
1997	([4])	84.0	....	....	....	....	....	....	....
1998	([4])	60.0	....	....	....	....	....	....	....
1999	([4])	38.0	([5]) 20	([5]) 80	43.5	....	8.7	....	8.7
2000	([4])	33.0	([6]) 40	([6])	43.9	....	33.6	....	33.6
2001	([4])	36.0	....	....	....	....	....	....	....
2002	100	53.0	([7])	....	([7])	....	....	....	....

[1] For unshorn lambs sold. [2] Unadjusted for weight of unshorn lambs purchased. [3] Payments for wool marketed during the year shown are made after Mar. 31 of the following year, and include deductions for the American Sheep Industry Association. Figures for 1994 and 1995 reflect 20- and 50-percent reductions in payments, respectively, as required by Public Law 103-130 whcih also terminated price support for wool as of December 31, 1995. [4] Public Law 103-130 terminated price support for wool as of December 31, 1995. [5] Market Loss Assistance Payments for 1999-crop wool authorized by Public Law 106-224 on June 21, 2000. [6] Market Loss Assistance Payments authorized by Public Law 106-387 for marketing year 2000 production. Additional payments authorized by Public Law 107-25. [7] Nonrecourse Marketing Assistance Loan Program authorized by Public Law 107-171, enacted May 13, 2002.

FSA, Fibers Analysis Group, (202) 720–3392.

Table 7-58.—Wool: Mill consumption, by grades, on the woolen and worsted systems, scoured basis, United States, 1992–2001 [1] [2] [3]

Item	1992	1993	1994	1995	1996	1997	1998	1999	2000	2001
Apparel wool:										
Woolen system:	*Mil. lb.*	*Mil. lb.*	*Mil. lb.*	*Mil. lb.*	*Mil. lb.*	*Mil. lb.*	*Mil. lb.*	*Mil. lb.*	*Mil. lb.*	*Mil. lb.*
60's and finer	33.9	40.9	36.0	30.2	42.1	49.0	31.3	18.4	18.5	16.1
Coarser than 60's	25.6	26.6	26.0	27.1	27.6	21.3	15.1	10.8	13.4	9.8
Total	59.5	67.5	62.0	57.3	69.7	70.3	46.4	29.2	31.9	25.9
Worsted system:										
60's and finer	58.5	58.8	59.6	55.0	46.1	48.2	42.2	27.4	21.7	NA
Coarser than 60's	18.1	15.1	17.0	17.0	13.8	11.9	9.8	7.0	8.4	NA
Total	76.6	73.9	76.6	72.0	59.8	60.1	52.0	34.4	30.1	27.1
Total apparel:										
60's and finer	92.4	99.7	95.6	85.2	88.2	97.2	75.5	45.8	NA	NA
Coarser than 60's	43.7	41.7	43.0	44.1	41.3	33.2	24.9	17.7	NA	NA
Total	136.1	141.4	138.6	129.3	129.5	130.4	98.4	63.5	63.0	53.0
Carpet wool	14.7	15.4	14.7	12.7	12.3	13.6	16.3	13.9	15.2	13.3
Grand total mill	150.8	156.8	153.3	142.0	141.8	144.0	114.7	77.5	77.2	66.3

[1] Scoured wool, plus greasy wool converted to a scoured basis, using assumed average yields. Includes both pulled and shorn, foreign and domestic wool. Wool was considered as consumed (1) on the woolen system when laid in mixes and (2) on the worsted system as the sum of top and noil production. [2] Domestic, duty-paid, and duty-free foreign. [3] Excludes wool consumed on the cotton system and in the manufacture of felt, hat bodies, and other miscellaneous products.

ERS, Field Crops Branch, (202) 694–5300. Compiled from reports of the U.S. Department of Commerce.

Table 7-59.—Wool: United States imports (for consumption), clean content, by grades, 1992–2001 [1] [2]

Grade	1992	1993	1994	1995	1996	1997	1998	1999	2000	2001
48's and finer:	*Mil. lb.*	*Mil. lb.*	*Mil. lb.*	*Mil. lb.*	*Mil. lb.*	*Mil. lb.*	*Mil. lb.*	*Mil. lb.*	*Mil. lb.*	*Mil. lb.*
Finer than 58's [3]	58.1	70.7	59.4	56.9	50.8	48.7	42.8	19.9	22.2	14.2
48's–58's [4]	7.4	5.3	5.5	6.9	3.3	2.8	3.0	1.4	1.7	1.6
Total	65.5	76.0	64.9	63.8	54.1	51.5	45.8	21.3	23.9	15.8
Not Finer than 46's:										
Wool for special use [5]	0.8	13.8	13.8	8.6	3.9	3.0	2.2	2.0	2.2	2.4
Not finer than 40's [6]	7.9	1.4	1.9	3.2	3.2	9.0	7.8	6.3	6.5	6.5
Finer than 40's–44's [7]	6.8	4.7	6.6	9.4	10.1	9.3	10.0	8.1	5.7	6.7
46's [8]	8.3	2.0	2.3	3.9	4.1	3.7	4.7	5.4	5.6	4.1
Total	23.8	21.9	24.6	25.0	21.3	24.9	24.7	21.8	20.0	19.7
Miscellaneous [9] ...	....	2.4	2.2	38.8	0	0	0	0	0	0
Grand total	89.3	100.3	91.7	88.8	75.4	76.4	70.5	43.1	45.0	35.6

[1] Natural fiber grown by sheep or lambs. [2] Beginning 1989 the following Harmonized Tariff Schedule numbers are in the above 7 wool import groups: 5101.19.606060, 5101.19.6060, 5101.21.4000, 5101.21.4000, 5101.29.4060, 0.5(5101.30.4000). [4] 5101.11.6030, 5101.19.6030, 5101.21.4030, 5101.29.4030, 0.5(5101.30.4000). [5] 5101.11.1000, 5101.19.1000, 5101.21.1000, 5101.29.1000. [6] 5101.11.2000, 5101.19.2000, 5101.21.1500, 5101.29.1500, 5101.30.1000. [7] 5101.11.4000, 5101.19.4000, 5101.21.3000, 5101.29.3000, 5101.30.1500. [8] 5101.11.5000, 5101.19.5000, 5101.21.3500, 5101.29.3500, 5101.30.3000. [9] 5101.21.6000, 5101.29.6000, 5101.30.6000. They include wool not carded or combed but processed beyond the scoured or carbonized condition, e.g. dyed. This wool is not identified by use or grade. In 1989 this quantity was 48,074 pounds, 1990 was 32,979 pounds, 1991 was 47,245 pounds, and 1992 was 25,728 pounds.

ERS, Field Crops Branch, (202) 694–5300. Compiled from reports of the U.S. Department of Commerce.

Table 7-60.—Wool: United States imports (for consumption), clean content, by country of origin, 1992–2001 [1]

Country of origin	1992	1993	1994	1995	1996	1997	1998	1999	2000	2001 [2]
48's and finer:	*Mil. lb.*	*Mil. lb.*	*Mil. lb.*	*Mil. lb.*	*Mil. lb.*	*Mil. lb.*	*Mil. lb.*	*Mil. lb.*	*Mil. lb.*	*Mil. lb.*
Argentina	0.1	0.1	0.1	0.1	0.1	0.5	0.6	0.1	0.1	....
Australia	55.0	64.8	53.0	51.6	47.1	44.2	38.8	17.6	20.2	12.7
Canada	1.1	1.2	0.8	1.1	1.2	1.0	1.1	0.7	1.0	0.8
Chile	0.4	0.3	0.6	0.7	0.2	0.1	0.2	....	....	....
New Zealand	3.2	1.7	1.8	2.6	1.9	1.6	1.5	1.3	1.3	1.0
South Africa	1.0	1.1	0.8	0.9	0.7	1.2	1.9	1.1	1.0	1.0
United Kingdom	0.3	0.3	1.5	1.5	0.3	0.2	0.1	....	....	....
Uruguay	3.2	5.9	5.2	2.5	1.6	2.1	0.7	0.2	0.1	0.1
Other	1.2	0.6	1.1	2.8	2.0	1.1	0.9	0.2	0.2	0.2
Total	65.5	76.0	64.9	63.8	54.1	51.5	45.8	21.2	23.9	15.8
Not finer than 46's:										
Argentina	0.7	0.6	0.6	0.1	0.2	0.1	0.5	0.4	0.5	....
Australia	0.3	0.3	0.1	0.3	0.4	0.1	0.2	0.1	0.1	0.4
Canada	0.3	0.1	0.1	0.6	0.1	0.6	0.2	0.2	0.2	0.4
New Zealand	17.9	15.6	18.3	17.0	15.6	18.9	18.5	16.8	15.0	14.9
Uruguay	0.1	0.1	0.3	....	....	....	....	....	....	....
South Africa	....	0.1	0.1	0.0	0.0	0.0	....	....	....	....
United Kingdom	4.2	4.7	4.8	5.7	3.5	3.7	3.6	2.9	3.2	2.6
Other	0.3	0.4	0.3	....	1.2	0.0	0.0	0.0	1.1	0.4
Total	23.8	21.9	24.6	25.0	21.0	24.3	23.1	20.6	20.1	18.7
Grand total	89.3	100.3	91.7	88.8	75.4	75.8	68.9	41.8	44.0	34.5

[1] Wool not advanced in any manner or by any process of manufacture beyond washed, scoured, or carbonized condition. [2] The country of origin for the 2.2 million pounds of miscellaneous imported raw wool was: New Zealand, 1.6 million; and Australia, 0.6 million.

ERS, Field Crops Branch, (202) 694–5300. Compiled from reports of the U.S. Department of Commerce.

Table 7-61.—Wool: Average price per pound, clean basis, delivered to United States mills, 1993–2002 [1]

Year	Territory [2]		Australian 64's good topmaking (in bond, American yield)
	64's (20.60–22.04 microns)	Avg. 58's–56's (24.95–27.84 microns)	
	Cents	*Cents*	*Cents*
1993	137	102	194
1994	212	126	311
1995	249	170	311
1996	193	137	264
1997	238	153	300
1998	162	113	202
1999	110	70	176
2000	108	61	179
2001	121	72	179
2002	190	130	280

[1] Beginning January 1976 the unit designation terminology for wool prices changed to microns. For example 64's (20.60–22.04 microns) formerly was fine good French combing and staple. Two designations 56's (26.40–27.84 microns) and 58's (24.95–26.39 microns) have been averaged in the price data shown here and together were formerly the category fleece ⅜ blood good French combing and staple. [2] Wool grown in the range areas of California, Oregon, Washington, Texas, the intermountain States (including Arizona and New Mexico), and parts of the Dakotas, Kansas, Nebraska, and Oklahoma. These wools vary considerably in shrinkage and color.

ERS, Field Crops Branch, (202) 694–5300 and AMS.

Table 7-62.—Wool: Number of sheep shorn, weight per fleece, production, by States, 2001–2002

State	Sheep and lambs shorn		Weight per fleece		Shorn wool production	
	2001	2002[1]	2001	2002[1]	2001	2002[1]
	1,000 head	*1,000 head*	*Pounds*	*Pounds*	*1,000 pounds*	*1,000 pounds*
AZ	120.0	118.0	6.2	6.0	740	710
CA	510.0	520.0	7.4	7.0	3,750	3,650
CO	430.0	460.0	7.2	6.7	3,080	3,070
ID	230.0	217.0	9.3	9.7	2,140	2,115
IL	70.0	65.0	6.6	6.6	460	430
IN	44.0	42.0	6.6	6.7	290	280
IA	245.0	220.0	5.8	5.9	1,420	1,300
KS	81.0	73.0	7.1	7.2	575	525
MI	77.0	76.0	6.2	6.1	480	460
MN	165.0	150.0	6.6	6.8	1,085	1,020
MO	56.0	59.0	7.0	6.7	390	395
MT	318.0	279.0	9.4	9.6	2,978	2,691
NE	85.0	80.0	7.3	7.6	620	610
NV	71.0	65.0	8.7	9.1	620	590
N ENG[2]	40.0	40.0	7.4	7.1	295	283
NM	235.0	215.0	8.0	8.1	1,880	1,750
NY	47.0	47.0	6.8	6.9	320	325
ND	101.0	99.0	8.9	8.3	900	825
OH	140.0	155.0	6.9	7.1	960	1,100
OK	52.0	55.0	6.0	6.4	310	350
OR	240.0	200.0	6.3	6.3	1,510	1,250
PA	64.0	63.0	6.9	6.8	440	430
SD	360.0	375.0	8.2	8.3	2,965	3,105
TX	870.0	790.0	6.9	7.0	6,003	5,530
UT	295.0	280.0	9.5	9.5	2,800	2,650
VA	43.0	40.0	6.5	6.4	280	255
WA	48.0	44.0	8.2	8.1	395	357
WV	30.0	28.0	6.0	6.2	180	174
WI	70.0	67.0	7.1	7.2	500	480
WY	440.0	420.0	9.0	8.9	3,950	3,750
Oth States[3]	112.0	111.0	6.3	6.2	700	690
US	5,689.0	5,453.0	7.6	7.5	43,016	41,150

[1] Preliminary. [2] N ENG includes CT, ME, MA, NH, RI, and VT. [3] AL, AK, AR, DE, FL, GA, HI, KY, LA, MD, MS, NJ, NC, SC, and TN.

NASS, Livestock Branch, (202) 720–3570.

Table 7-63.—Wool: Price and value of production, by States, 2001–2002

State	Price per pound		Value of production [1]	
	2001	2002 [2]	2001	2002 [2]
	Dollars	*Dollars*	*Dollars*	*1,000 dollars*
AZ	0.30	0.32	222	227
CA	0.35	0.45	1,313	1,643
CO	0.33	0.57	1,016	1,750
ID	0.22	0.67	471	1,417
IL	0.09	0.23	41	99
IN	0.15	0.16	44	45
IA	0.12	0.18	170	234
KS	0.20	0.30	115	158
MI	0.12	0.14	58	64
MN	0.11	0.20	119	204
MO	0.09	0.14	35	55
MT	0.45	0.76	1,340	2,045
NE	0.20	0.30	124	183
NV	0.44	0.85	273	502
N ENG [3]	0.40	0.40	118	113
NM	0.70	0.80	1,316	1,400
NY	0.17	0.20	54	65
ND	0.30	0.60	270	495
OH	0.14	0.15	134	165
OK	0.20	0.25	62	88
OR	0.27	0.38	408	475
PA	0.25	0.25	110	108
SD	0.31	0.45	919	1,397
TX	0.52	0.68	3,122	3,760
UT	0.29	0.60	812	1,590
VA	0.20	0.26	56	66
WA	0.45	0.47	178	168
WV	0.21	0.22	38	38
WI	0.10	0.15	50	72
WY	0.51	0.73	2,015	2,738
Oth States [4]	0.44	0.62	308	428
US	0.36	0.53	15,311	21,792

[1] Production multiplied by marketing year average price. [2] Preliminary. [3] N ENG includes CT, ME, MA, NH, RI, and VT. [4] AL, AK, AR, DE, FL, GA, HI, KY, LA, MD, MS, NJ, NC, SC, and TN.

NASS, Livestock Branch, (202) 720–3570.

Table 7-64.—Mohair: Price-support operations, United States, 1993–2002

Marketing year begining January 1	Price per pound		Average payment rate per pound	Marketings covered by payments	Amount of payments [2]
	Loans [1]	Season average received by producers			
	Cents	*Cents*	*Cents*	*Million pounds*	*Million dollars*
1993	473.8	82.0	391.8	17.4	67.9
1994	473.9	256.0	217.9	13.1	21.8
1995	465.7	184.0	281.7	12.9	18.0
1996	(3)	192.0			
1997	(3)	225.0			
1998	(3)	249.0			
1999	200.0	346.0	(4) 40.0	(4) 3.9	(4) 1.6
2000	200.0	404.0	(5) 40.0	(5) 2.7	(5) 1.1
2001	(3)	213.0			
2002	420.0	158.0	(6)	(6)	(6)

[1] The national average loan rate was also known as the price-support rate prior to enactment of the Farm Security and Riral Investment Act of 2002. [2] Payments for mohair marketed during the year shown are made after Mar. 31 of the following year and includes deductions for the Mohair Council of America. Figures for 1994 and 1995 relect 25- and 50-percent reductions in payments, respectively, as required by Public Law 103-130 which also terminated price support for mohair as of December 31, 1995. [3] No program. [4] Recourse loan program authorized by P.L. 105-277 for fiscal year 1999. No payments were involved. Market Loss Assistance Payments for 1999 crop mohair authorized by P.L. 106-224 on June 21, 2000. [5] Recourse loan program authorized by P.L. 106-78 for fiscal year 2000. No payments were involved. Market Loss Assistance Payments authorized by P.L. 106-387 for marketing year 2000 production. Additional payments authorized by P.L. 107-25. [6] Nonrecourse Marketing Assistance Loan Program authorized by Public Law 107-171, enacted May 12, 2002.

FSA, Fibers Analysis Group, (202) 720–3392.

Table 7-65.—Goats and mohair: Number of goats clipped, mohair production, average price per pound received by farmers, and value of production, Texas, 1993–2002

Year	Goats clipped [1]	Average clip per goat	Mohair production	Price per pound	Value of production [2]
	Thousands	*Pounds*	*1,000 pounds*	*Dollars*	*1,000 dollars*
1993	1,900	7.1	13,490	0.83	11,197
1994	1,600	7.3	11,680	2.62	30,602
1995	1,470	7.7	11,319	1.85	20,940
1996	1,070	7.0	7,490	1.95	14,606
1997	840	7.6	6,384	2.28	14,556
1998	620	7.5	4,650	2.59	12,044
1999	375	6.8	2,550	3.68	9,384
2000	345	6.8	2,346	4.30	10,088
2001	260	6.6	1,716	2.20	3,775
2002	240	7.9	1,896	1.60	3,034

[1] The number clipped is the sum of goats and kids clipped in the spring and kids clipped in the fall. [2] Production multiplied by marketing year average price per pound.

NASS, Livestock Branch, (202) 720–3570.

Table 7-66.—Red meat: Production, by class of slaughter, United States, 1993–2002

Year	Commercial			Farm	Total	Commercial			Farm	Total
	Federally inspected	Other	Total [1]			Federally inspected	Other	Total [1]		
	Beef					Pork, excluding lard				
	Million pounds	*Million pounds*	*Million pounds*	*Million pounds*	*Million pounds*	*Million pounds*	*Million pounds*	*Million pounds*	*Million pounds*	*Million pounds*
1993	22,528	414	22,942	107	23,049	16,684	345	17,029	58	17,087
1994	23,875	403	24,278	108	24,386	17,293	366	17,659	38	17,697
1995	24,691	426	25,117	107	25,224	17,467	343	17,810	38	17,848
1996	24,948	473	25,421	106	25,527	16,780	304	17,084	32	17,116
1997	24,964	420	25,384	106	25,490	16,962	283	17,245	30	17,275
1998	25,264	389	25,653	107	25,760	18,692	289	18,981	30	19,011
1999	25,998	387	26,385	107	26,492	18,977	301	19,278	28	19,306
2000	26,405	371	26,776	111	26,887	18,672	257	18,929	24	18,953
2001	25,743	365	26,108	105	26,213	18,899	240	19,139	22	19,161
2002	26,714	377	27,091	102	27,193	19,437	227	19,664	21	19,685
	Veal					Lamb and Mutton				
	Million pounds	*Million pounds*	*Million pounds*	*Million pounds*	*Million pounds*	*Million pounds*	*Million pounds*	*Million pounds*	*Million pounds*	*Million pounds*
1993	260	8	268	18	286	320	9	329	8	337
1994	276	7	283	10	293	297	9	306	4	310
1995	298	9	307	11	318	275	9	284	4	288
1996	355	13	368	11	379	258	7	265	4	269
1997	314	9	323	10	333	250	7	257	3	260
1998	243	8	251	10	261	242	7	249	3	252
1999	217	7	224	9	233	236	7	243	4	247
2000	205	10	215	10	225	224	8	232	4	236
2001	188	6	194	10	204	216	8	224	4	228
2002	190	6	196	9	205	209	9	218	4	222
	All meat, excluding lard									
	Million pounds	*Million pounds*	*Million pounds*	*Million pounds*	*Million pounds*					
1993	39,793	775	40,568	191	40,759	....	....	....	....	....
1994	41,741	785	42,526	160	42,686	....	....	....	....	....
1995	42,731	787	43,518	160	43,678	....	....	....	....	....
1996	42,340	798	43,138	153	43,291	....	....	....	....	....
1997	42,491	718	43,209	149	43,358	....	....	....	....	....
1998	44,441	692	45,133	150	45,283	....	....	....	....	....
1999	45,428	702	46,130	148	46,278	....	....	....	....	....
2000	45,506	645	46,151	149	46,300	....	....	....	....	....
2001	45,045	619	45,664	141	45,805	....	....	....	....	....
2002	46,549	620	47,169	137	47,305	....	....	....	....	....

[1] Totals are based on unrounded data.

NASS, Iowa Agricultural Statistics Service, (515) 284–4340.

Table 7-67.—Meat: Production by types in specified countries, 2000 and 2001 [1]

Country	Beef and veal		Pork [3]		Total production	
	2000	2001 [2]	2000	2001 [2]	2000	2001 [2]
	1,000 metric tons	*1,000 metric tons*	*1,000 metric tons*	*1,000 metric tons*	*1,000 metric tons*	*1,000 metric tons*
Argentina	2,880	2,680	NA	NA	2,880	2,680
Australia	1,988	2,052	365	379	2,353	2,431
Brazil	6,520	6,895	2,010	2,230	8,530	9,125
Bulgaria	95	62	224	121	319	183
Canada	1,246	1,250	1,638	1,729	2,884	2,979
China, Peoples Republic	5,328	5,488	40,314	41,845	45,642	47,333
Colombia	662	681	NA	NA	662	681
Costa Rica	84	83	NA	NA	84	83
Czech Republic	206	195	590	580	796	775
Dominican Republic	46	46	NA	NA	46	46
Egypt	442	437	NA	NA	442	437
El Salvador	21	21	NA	NA	21	21
European Union	7,462	6,896	17,585	17,419	25,047	24,315
Guatemala	45	45	NA	NA	45	45
Honduras	20	21	NA	NA	20	21
Hong Kong	17	15	161	165	178	180
Hungary	NA	NA	490	460	490	460
India	1,700	1,770	NA	NA	1,700	1,770
Japan	530	458	1,269	1,245	1,799	1,703
Kazakhstan, Republic	320	NA	NA	NA	320	0
Korea, Republic of	278	221	1,004	1,077	1,282	1,298
Mexico	1,900	1,925	1,035	1,065	2,935	2,990
New Zealand	580	590	NA	NA	580	590
Nicaragua	55	60	NA	NA	55	60
Philippines	220	215	1,008	1,064	1,228	1,279
Poland	325	305	1,620	1,547	1,945	1,852
Romania	173	200	360	406	533	606
Russian Federation	1,840	1,760	1,500	1,560	3,340	3,320
Singapore	0	NA	19	19	19	19
South Africa, Republic	630	665	NA	NA	630	665
Taiwan	5	5	921	962	926	967
Turkey	625	640	NA	NA	625	640
Ukraine	754	646	676	591	1,430	1,237
United States	12,196	11,890	8,597	8,691	20,793	20,581
Uruguay	440	317	NA	NA	440	317
Venezuela	350	355	NA	NA	350	355
Total meat	49,983	48,889	81,386	83,155	131,369	132,044

[1] Carcass weight basis: excludes offals, rabbit, and poultry meat. [2] Preliminary. [3] Includes edible pork fat, but excludes lard and inedible greases (except United States). NA-not available.

FAS, Dairy, Livestock and Poultry Division, (202) 720–8031. Prepared or estimated on the basis of official statistics of foreign governments, other foreign source materials, reports of U.S. Agricultural counselors, Attachés, and Foreign Service Officers, results of office research, and related information.

Table 7-68.—Meat: United States exports by type of product, 1992–2001

Year	Beef and veal			Lamb and mutton, fresh or frozen	Pork			Variety meats, fresh, chilled, or frozen	Other meats	Total [1]
	Fresh and chilled	Frozen	Prepared and preserved		Fresh and chilled	Frozen	Prepared and preserved			
	Metric tons	*Metric tons*	*Metric tons*	*Metric tons*	*Metric tons*	*Metric tons*	*Metric tons*	*Metric tons*	*Metric tons*	*Metric tons*
1992	194,164	242,371	12,061	3,278	54,228	62,354	23,717	326,869	225,010	1,144,052
1993	181,282	229,353	14,477	3,608	59,479	69,562	19,230	356,232	234,899	1,168,121
1994	242,391	275,067	13,419	3,766	80,609	68,805	27,960	394,740	301,112	1,407,868
1995	262,381	319,416	13,651	2,511	100,235	127,835	35,766	465,919	353,207	1,680,921
1996	273,276	324,329	14,577	2,478	101,975	166,058	38,481	495,075	434,759	1,851,007
1997	316,534	359,460	15,227	2,545	134,684	151,121	38,301	469,233	435,258	1,922,362
1998	346,403	352,050	17,966	2,528	147,006	209,134	43,789	495,495	423,980	2,038,352
1999	370,184	414,458	19,323	2,219	160,910	225,492	47,898	522,994	455,561	2,219,038
2000	395,588	417,538	21,791	2,184	208,055	185,241	44,836	600,241	503,942	2,379,414
2001 [2]	393,105	362,972	23,932	2,770	227,807	247,461	52,788	682,261	513,969	2,507,064

[1] May not add due to rounding. [2] Preliminary.

FAS, Dairy, Livestock and Poultry Division, (202) 720–8031. Compiled from reports of the U.S. Department of Commerce.

Table 7-69.—Meat: United States exports and imports into the United States, carcass weight equivalent, 1994–2003 [1]

Year	Exports				Imports			
	Beef and veal	Lamb and mutton	Pork [2]	All meat	Beef & veal	Lamb and mutton	Pork [2]	All meat
	Million pounds	*Million pounds*	*Million pounds*	*Million pounds*	*Million pounds*	*Million pounds*	*Million pounds*	*Million pounds*
1994	1,611	9	549	2,169	2,369	49	743	3,161
1995	1,821	6	787	2,614	2,104	64	664	2,832
1996	1,877	6	970	2,853	2,073	72	620	2,764
1997	2,136	5	1,044	3,185	2,343	83	634	3,061
1998	2,171	6	1,230	3,407	2,643	112	705	3,461
1999	2,410	5	1,278	3,693	2,873	112	827	3,812
2000	2,468	6	1,287	3,761	3,031	130	967	4,128
2001	2,269	7	1,559	3,835	3,163	146	951	4,260
2002 [3]	2,447	7	1,614	4,068	3,218	160	1,071	4,449
2003 [4]	2,565	5	1,645	4,215	3,305	172	1,080	4,557

[1] Carcass weight equivalent of all meat, including the meat content of minor meats and of mixed products. Includes shipments to U.S. Territories are included in domestic consumption. [2] The pork series has been revised to a dressed weight equivalent rather than "Pork, excluding lard." [3] Preliminary. [4] Forecast.

ERS, Market and Trade Economics Division, Animal Products Branch, (202) 694–5180. Data on imports and commercial exports are computed from records of the U.S. Department of Commerce, those on exports by the U.S. Department of Agriculture are separately estimated from deliveries and stocks.

Table 7-70.—Meat: United States imports, products weight basis, by country of origin, 2001

Country of origin	Beef and veal			Lamb, mutton, and goat, except canned	Pork			Variety meats, fresh, chilled and frozen	Other livestock meats n.s.e.	Total
	Fresh	Frozen	Other prepared or preserved		Fresh and chilled	Frozen	Other prepared or preserved			
	Metric tons	*Metric tons*	*Metric tons*	*Metric tons*	*Metric tons*	*Metric tons*	*Metric tons*	*Metric tons*	*Metric tons*	*Metric tons*
Canada	332,471	18,347	5,473	82	240,139	44,508	50,111	46,670	6,126	743,926
Australia	19,812	363,904	460	49,112	0	14	0	10,535	135	443,971
New Zealand	3,832	207,970	615	17,275	0	0	0	2,847	1,013	233,552
European Union	0	71	2	0	130	40,163	24,648	469	5,327	70,809
Brazil	0	0	40,846	0	0	0	0	0	238	41,084
Argentina	1,413	3,135	20,671	0	0	0	0	33	42	25,293
Uruguay	1,668	7,996	2,939	273	0	0	0	1,573	134	14,583
Nicaragua	3,791	8,842	0	0	0	0	0	228	0	12,861
Costa Rica	3,577	6,582	3	0	0	0	0	175	0	10,337
Poland	0	0	0	0	0	0	6,927	0	367	7,294
Mexico	1,965	1,448	2,654	0	7	2	13	6	87	6,180
Hungary	0	0	0	0	0	0	1,998	0	0	1,998
China	0	0	0	0	0	0	2	0	1,170	1,172
Honduras	0	602	0	0	0	0	0	0	2	604
Other	0	0	52	42	0	0	25	6	2,083	2,208
Total [1]	368,529	618,897	73,713	66,785	240,275	84,687	83,724	62,541	16,723	1,615,873

[1] May not add due to rounding. n.s.e. not specified elsewhere.

FAS, Dairy, Livestock and Poultry Division, (202) 720–8031. Compiled from reports of the U.S. Department of Commerce.

Table 7-71.—Meat: United States imports by type of product, 1992–2001

Year	Beef and Veal			Lamb, mutton, and goat, except canned	Pork			Variety meats, fresh or frozen	Other livestock meats n.s.e.	Total [1]
	Fresh	Frozen	Other prepared or preserved		Fresh	Frozen	Other prepared or preserved			
	Metric tons	*Metric tons*	*Metric tons*	*Metric tons*	*Metric tons*	*Metric tons*	*Metric t ons*	*Metric tons*	*Metric tons*	*Metric tons*
1992	137,277	591,645	74,643	23,853	129,935	55,737	79,075	20,095	12,859	1,125,118
1993	157,471	561,906	73,880	24,468	126,831	80,821	94,238	25,400	13,371	1,158,385
1994	179,121	535,328	74,704	23,276	130,648	78,378	94,013	27,511	13,012	1,155,990
1995	175,540	466,378	65,399	29,844	133,101	61,286	79,155	26,081	12,539	1,049,324
1996	227,874	412,805	66,719	33,009	125,220	58,336	72,650	32,579	13,744	1,042,934
1997	262,985	469,949	63,181	37,848	126,061	65,000	72,903	44,317	14,215	1,156,457
1998	295,820	527,063	68,884	51,630	146,965	70,227	76,230	47,031	13,058	1,296,907
1999	337,899	542,524	82,669	50,209	188,556	77,638	84,207	51,640	13,625	1,428,966
2000	336,117	608,737	73,750	59,968	229,395	91,446	92,672	57,388	14,281	1,563,753
2001 [2]	368,529	618,897	73,713	66,785	240,275	84,687	83,724	62,541	16,723	1,615,873

[1] May not add due to rounding. [2] Preliminary. n.s.e. not specified elsewhere.

FAS, Dairy, Livestock and Poultry Division, (202) 720–8031. Compiled from reports of the U.S. Department of Commerce.

Table 7-72.—Meat: International trade, selected countries, 1999–2001 [1]

Continent and country	1999		2000		2001 [2]	
	Exports	Imports	Exports	Imports	Exports	Imports
	1,000 metric tons	*1,000 metric tons*	*1,000 metric tons*	*1,000 metric tons*	*1,000 metric tons*	*1,000 metric tons*
Argentina	359	17	357	17	168	13
Australia	1,307	33	1,387	50	1,461	43
Brazil	573	55	655	72	1,085	44
Bulgaria	3	26	0	19	0	37
Canada	1,046	319	1,181	331	1,301	390
China, Peoples Republic	132	56	127	66	199	78
Colombia	2	3	2	1	5	1
Costa Rica	17	0	17	1	12	2
Czech Republic	12	28	8	27	24	23
Dominican Republic	0	5	0	1	0	3
Egypt	0	218	0	236	0	105
El Salvador	0	12	0	14	0	16
European Union	2,244	405	2,115	502	1,807	468
Guatemala	0	5	0	5	0	6
Honduras	0	1	3	1	3	0
Hong Kong	1	284	1	319	1	332
Hungary	131	19	143	35	118	39
India	222	0	365	0	370	0
Japan	0	1,878	0	2,011	1	2,023
Kazakhstan, Republic	17	13	10	3	NA	NA
Korea, Republic of	113	398	30	498	42	369
Mexico	61	548	70	696	71	720
New Zealand	442	7	485	12	500	7
Nicaragua	27	0	31	0	28	0
Philippines	0	122	0	140	0	117
Poland	270	55	188	48	138	26
Romania	5	32	2	43	0	57
Russian Federation	5	1,670	10	998	9	1,213
Saudi Arabia	0	66	NA	NA	NA	NA
Singapore	3	44	3	58	2	37
South Africa, Rep.	13	41	14	37	17	23
Switzerland	6	18	NA	NA	NA	NA
Taiwan	0	180	0	137	0	92
Turkey	0	0	0	0	0	6
Ukraine	216	5	206	2	126	6
United States	1,673	1,679	1,703	1,814	1,737	1,866
Uruguay	189	0	236	0	145	0
Venezuela	0	2	0	2	0	4
Total [3]	9,089	8,244	9,349	8,196	9,370	8,160

[1] Carcass weight equivalent of beef and veal, pork, and lamb, mutton and goat meat. Excludes fat, offals, and live animals. [2] Preliminary. [3] Totals may not add due to rounding.

FAS, Dairy, Livestock and Poultry Division, (202) 720–8031. Prepared or estimated on the basis of official statistics of foreign governments, other foreign source materials, reports of U.S. Agricultural counselors, Attachés, and Foreign Service Officers, results of office research, and related information.

Table 7-73.—Meats and lard: Production and consumption, United States, 1994–2003 [1]

Year	Beef			Veal			Lamb and mutton		
	Production	Consumption		Production	Consumption		Production	Consumption	
		Total	Per capita		Total	Per capita		Total	Per capita
	Million pounds	*Million pounds*	*Pounds*	*Million pounds*	*Million pounds*	*Pounds*	*Million pounds*	*Million pounds*	*Pounds*
1994	24,386	25,125	96.4	293	290	1.1	308	366	1.3
1995	25,222	25,534	97.1	319	326	1.2	285	359	1.3
1996	25,525	25,861	96.0	378	378	1.4	268	333	1.1
1997	25,490	25,611	93.9	334	333	1.2	260	332	1.1
1998	25,760	26,305	95.4	262	265	1.0	251	360	1.2
1999	26,493	26,938	96.5	235	235	0.8	248	358	1.1
2000	26,888	27,337	96.9	225	225	0.8	234	354	1.1
2001	26,212	27,026	94.9	205	204	0.7	227	368	1.1
2002 [2]	27,194	27,840	96.8	206	205	0.7	222	383	1.2
2003 [3]	26,055	27,140	93.5	195	197	0.7	211	373	1.1

Year	Pork			All meats			Lard		
	Production	Consumption		Production	Consumption		Production	Consumption	
		Total	Per capita		Total	Per capita		Total	Per capita
	Million pounds	*Million pounds*	*Pounds*	*Million pounds*	*Million pounds*	*Pounds*	*Million pounds*	*Million pounds*	*Pounds*
1994	17,696	17,811	68.3	42,683	43,577	167.1	1,034	599	2.3
1995	17,849	17,768	67.5	43,677	43,967	167.1	1,040	584	2.23
1996	17,117	16,797	62.3	43,288	43,369	161.1	998	606	2.3
1997	17,274	16,823	61.7	43,358	43,099	158.0	NA	NA	NA
1998	19,011	18,309	66.4	45,283	45,238	164.0	NA	NA	NA
1999	19,308	18,953	67.9	46,284	46,483	166.6	NA	NA	NA
2000	18,952	18,644	66.1	46,299	46,560	165.0	NA	NA	NA
2001	19,160	18,492	64.9	45,804	46,089	161.8	NA	NA	NA
2002	19,688	18,670	66.5	47,310	47,558	165.4	NA	NA	NA
2003 [3]	19,477	19,130	65.1	45,938	46,614	160.6	NA	NA	NA

[1] Carcass weight equivalent or dressed weight. Beginning 1977, pork production was no longer reported as "pork, excluding lard." This series has been revised to reflect pork production in prior years on a dressed weight basis that is comparable with the method used to report beef, veal, and lamb and mutton. Edible offals are excluded. Shipments to the U.S. territories are included in domestic consumption. [2] Preliminary. [3] Forecast. NA-not available.

ERS, Animal Products, (202) 694–5180.

Table 7-74.—Hides and skins: United States imports by country of origin, 1997–2001

Country of origin	1997	1998	1999	2000	2001 [1]
	1,000 pieces	*1,000 pieces*	*1,000 pieces*	*1,000 pieces*	*1,000 pieces*
Cattle and buffalo hides: [2]					
Canada	1,405	1,808	1,742	1,876	1,615
Mexico	48	72	144	48	71
Jamaica	5	6	4	25	11
Australia	0	1	1	0	7
New Zealan	0	0	0	0	7
European Union	19	4	5	11	5
Others	19	19	11	13	5
Total	1,496	1,909	1,906	1,972	1,721
Calf and kip:					
Canada	673	238	222	57	141
Mexico	0	11	1	0	12
Australia	33	4	0	0	10
New Zealan	0	0	0	0	3
Brazil	18	1	4	0	2
European Union	10	11	10	32	1
Others	15	4	3	19	2
Total	750	270	239	108	171
Goat and kid:					
Mali	0	3	64	59	50
China; Peo	0	0	0	10	50
Pakistan	2	5	57	20	16
Peru	31	27	0	2	13
Canada	19	18	10	15	10
Others	7	6	5	2	15
Total	60	58	135	109	155
Sheep and lambs: [3]					
New Zealand	572	465	702	586	443
Canada	337	204	36	97	265
European Union	414	310	199	151	144
Saudi Arab	0	0	36	48	93
Mali	3	30	74	56	51
Australia	54	56	3	19	28
Nigeria	0	0	19	0	22
South Africa	4	18	4	0	18
Ethiopia	493	156	0	0	0
Eritrea	0	63	96	12	12
Others	164	87	13	24	12
Total	2,040	1,389	1,183	993	1,088

[1] Final. [2] Excludes pieces of cattle hides reported in units of weight. [3] Includes cabretta and hair sheepskins; excludes pieces of sheep and lamb skins reported in pounds.

FAS, Dairy, Livestock, and Poultry Division, (202) 720–8031. Compiled from reports of the U.S. Department of Commerce.

Table 7-75.—Hides, packer: Average price per hundred pounds, Central U.S., 1993–2002

Year	Steers					Heifers		
	Heavy native	Light native	Heavy Texas	Butt branded	Colorado branded	Heavy [1] native	Light [1] native	Branded [2]
	Dollars	*Dollars*	*Dollars*	*Dollars*	*Dollars*	*Dollars*	*Dollars*	*Dollars*
1993	79.51	....	57.63	69.83	65.52	82.16	....	74.09
1994	87.40	....	66.07	81.14	77.66	94.99	....	86.00
1995	88.05	....	66.27	82.32	77.98	93.69	....	80.76
1996	87.62	....	63.76	79.75	73.72	92.15	....	82.27
1997	87.66	....	64.60	80.03	77.00	90.99	....	82.39
1998	76.39	....	49.65	62.14	56.54	75.45	....	63.12
1999	72.36	....	....	64.28	60.83	73.80	....	67.25
2000	80.17	....	....	73.67	71.24	83.41	....	77.54
2001	85.84	....	....	79.79	75.90	85.52	....	85.44
2002	82.25	....	....	75.97	71.07	85.73	....	78.75

[1] Heifers. [2] Cows.

AMS, Livestock & Grain Market News, (202) 720–7316.

Table 7-76.—Hides and skins: United States exports by country of destination, 1997–2001

Country of destination	1997	1998	1999	2000	2001[1]
	1,000 pieces	*1,000 pieces*	*1,000 pieces*	*1,000 pieces*	*1,000 pieces*
Cattle and buffalo hides:[2]					
Korea, Republic of	7,468	4,923	5,996	7,048	7,540
China, Peoples Rep	1,578	2,165	1,816	3,558	5,342
Taiwan	2,863	2,703	2,859	2,714	2,736
Mexico	2,463	2,842	2,722	2,187	1,646
Hong Kong	320	349	328	553	1,374
Japan	1,797	1,414	1,253	1,430	1,316
European Union	767	1,752	1,151	1,522	1,299
Thailand	325	334	345	451	862
Canada	1,246	1,092	828	871	716
Brazil	23	89	23	38	76
Others	136	184	248	214	348
Total	18,987	17,848	17,569	20,585	23,254
Sheep and lamb skins:[3]					
Sheep:					
Turkey	1,158	720	1,024	1,450	1,609
European Union	479	315	166	193	215
Poland	184	25	86	122	189
Mexico	49	236	358	288	98
Guatemala	0	0	0	0	70
Hong Kong	60	5	16	112	67
Canada	229	96	69	86	42
Japan	149	7	125	48	37
Korea, Republic of	168	185	20	26	35
China, Peoples Rep	67	16	139	66	18
Others	240	97	30	13	26
Total	2,782	1,702	2,034	2,404	2,404
Calf and kip skins:[4]					
European Union	1,517	1,387	1,004	486	798
Korea, Republic of	227	83	91	976	424
Japan	521	390	282	172	136
Brazil	11	0	3	12	125
Mexico	165	282	248	243	117
Hong Kong	58	44	26	45	99
China, Peoples Rep	167	145	92	58	69
India	1	0	1	23	55
Israel	30	38	64	18	38
Canada	12	44	37	59	30
Others	350	326	218	100	100
Total	3,060	2,738	2,065	2,192	1,990

[1] Final. [2] Excludes pieces of cattle hides reported in units of pounds. [3] Schedule B numbers 211.6000 and 211.7000 have been dropped from hides and skins count due to Schedule B changes implemented January 1, 1978. [4] Under the new harmonized system kip skin exports are included with calf skin exports.

FAS, Dairy, Livestock, and Poultry Division, (202) 720–8031. Compiled from reports of the U.S. Department of Commerce.

Table 7-77.—Hides and skins: United States imports and exports, 1992–2001

Year	Imports				Exports		
	Calf and kip	Cattle and buffalo[1]	Goat and kid	Sheep and lamb[2]	Calf and kip	Cattle and buffalo[3]	Sheep and lamb[2]
	1,000 pieces	*1,000 pieces*	*1,000 pieces*	*1,000 pieces*	*1,000 pieces*	*1,000 pieces*	*1,000 pieces*
1992	335	1,473	189	4,507	2,904	19,127	4,828
1993	310	1,653	108	3,646	2,612	18,207	3,615
1994	104	1,726	50	3,250	4,491	17,983	3,686
1995	337	1,759	57	3,228	5,207	20,048	3,528
1996	678	1,700	171	2,229	3,347	20,296	2,998
1997	750	1,496	60	2,040	3,060	18,987	2,782
1998	270	1,909	58	1,389	2,738	17,854	1,702
1999	239	1,906	135	1,183	2,065	17,575	2,034
2000	108	1,972	109	993	2,192	20,610	2,404
2001[4]	171	1,721	155	1,088	1,990	23,252	2,404

[1] Excludes pieces of cattle hides reported in units of weight. [2] Includes cabretta and hair sheepskins. [3] Schedule B numbers 211.6000 and 211.7000 have been dropped from hides and skins count due to Schedule B changes implemented January 1, 1978. [4] Final.

FAS, Dairy, Livestock, and Poultry Division, (202) 720–8031. Compiled from reports of the U.S. Department of Commerce.

Table 7-78.—Mink: Farms, pelts produced and value of mink pelts, United States, 1992–2001

Year	Mink farms	Pelts produced	Average marketing price	Value of mink pelts
	Number	*Thousand*	*Dollars*	*Million dollars*
1992	571	2,900	23.80	69.0
1993	523	2,620	34.10	89.3
1994	484	2,623	33.00	86.6
1995	478	2,803	53.10	148.8
1996	449	2,783	35.30	98.2
1997	452	2,993	33.10	99.1
1998	438	2,938	24.80	72.9
1999	398	2,813	33.70	94.8
2000	350	2,666	34.00	90.6
2001 [1]	324	2,565	33.50	85.9

[1] Preliminary.

NASS, Livestock Branch, (202) 720–3570.

Table 7-79.—Mink pelts: Number produced by color class, major States, and United States, 2001

State	Standard	Ranch wild	Demibuff [1]	Pastel	Pale brown	Sapphire	Gunmetal	Mahogany
	Number	*Number*	*Number*	*Number*	*Number*	*Number*	*Number*	*Number*
ID	41,000	19,100	...	...	...	...	22,100	46,600
IL	41,100	...	...	...	...	...	...	3,300
IA	55,000	...	...	...	...	...	10,300	9,300
MI	...	...	...	...	...	...	...	...
MN	49,900	26,800	60,700	2,500	...	2,000	56,100	79,300
NY	...	...	...	...	...	...	...	...
OH	29,500	...	...	...	...	5,600	...	18,400
OR	79,400	...	...	...	...	16,900	145,100	...
PA	...	...	...	...	...	16,800	19,700	...
SD	...	...	...	...	...	...	...	...
UT	205,000	...	60,000	...	...	15,500	40,000	233,500
WA	35,500	...	...	...	...	...	63,100	...
WI	327,000	...	...	10,600	...	25,600	108,500	123,000
Other States [2]	84,000	54,800	8,300	28,700	1,900	25,900	7,100	66,600
US	947,400	100,700	129,000	41,800	1,900	108,300	472,000	580,000

State	Pearl	Lavender Hope	Pink	Violet type	White	Miscellaneous and unclassified	Total pelts
	Number	*Number*	*Number*	*Number*	*Number*	*Number*	*Number*
ID	...	...	...	...	...	...	151,200
IL	2,700	...	...	...	4,000	...	51,600
IA	...	...	...	...	...	...	105,900
MI	...	...	...	...	...	...	54,000
MN	...	...	...	...	...	...	286,500
NY	...	...	...	...	...	...	5,500
OH	...	...	...	...	...	...	56,500
OR	...	...	...	...	...	...	251,000
PA	...	...	...	...	...	...	74,900
SD	...	...	...	...	...	...	78,100
UT	...	...	...	...	...	...	610,000
WA	...	...	...	...	...	...	113,100
WI	...	...	...	...	52,600	...	672,000
Other States [2]	71,200	3,500	800	21,400	17,100	10,900	55,000
US	73,900	3,500	800	21,400	73,700	10,900	2,565,300

[1] Includes crosses of Dark brown, Violet, Pastel, Standard, Pearl or others. [2] "Other States" include some pelts from the above listed States which were not published to avoid disclosing individual operations. Published color classes may not add to the State total.

NASS, Livestock Branch, (202) 720–3570.

Table 7-80.—Livestock: Number of animals slaughtered under Federal inspection and number of whole carcasses condemned, 1993–2002

Year	Cattle		Calves		Sheep and lambs	
	Total head	Condemned [1]	Total head	Condemned [1]	Total head	Condemned [1]
	1,000	*1,000*	*1,000*	*1,000*	*1,000*	*1,000*
1993	32,441	159.9	1,202	21.0	5,101	14.3
1994	33,121	164.5	1,187	19.9	4,796	13.0
1995	34,640	174.1	1,367	24.7	4,494	11.6
1996	35,714	181.5	1,641	28.8	4,086	11.0
1997	35,576	176.2	1,581	27.1	3,835	8.8
1998	34,911	157.1	1,466	28.7	3,685	7.3
1999	35,299	155.3	1,296	26.6	3,551	6.5
2000	35,735	188.9	1,125	22.4	3,420	5.8
2001	37,641	198.2	1,333	25.2	3,463	5.6
2002	31,404	165.9	1,034	19.5	2,922	5.4

Year	Goats		Hogs		Horses	
	Total head	Condemned [1]	Total head	Condemned [1]	Total head	Condemned [1]
	1,000	*1,000*	*1,000*	*1,000*	*1,000*	*1,000*
1993	283	2.2	91,476	186.4	186	0.7
1994	352	2.5	91,772	212.1	114	0.5
1995	325	1.9	95,188	248.8	113	0.5
1996	384	2.3	91,849	320.7	104	0.6
1997	378	1.9	88,889	308.9	90	0.4
1998	432	1.5	96,899	395.9	77	0.5
1999	483	2.1	100,537	460.8	67	0.4
2000	542	1.2	97,389	410.8	51	0.3
2001	592	1.1	96,600	437.4	62	0.2
2002	553	1.0	89,855	353.6	43	0.2

[1] Condemnations include ante-mortem and post-mortem inspection. Condemnations are for the fiscal year ending September 30. Data reported by Food Safety and Inspection Service, USDA.

NASS, Iowa Agricultural Statistics Service, (515) 284–4340.

Table 7-81.—Livestock: Number and value, United States, Jan. 1, 2001–2003

Class of livestock and poultry	Number			Value					
				Per head [2]			Total		
	2001	2002	2003 [1]	2001	2002	2003 [1]	2001	2002	2003 [1]
	Thousands	*Thousands*	*Thousands*	*Dollars*	*Dollars*	*Dollars*	*1,000 dollars*	*1,000 dollars*	*1,000 dollars*
Cattle	97,277	96,704	96,106	725.00	747.00	726.00	70,495,030	72,283,865	69,785,870
Hogs [3]	59,138	59,804	58,943	77.00	77.00	71.00	4,542,493	4,589,912	4,159,643
Sheep and lambs	6,965	6,685	6,350	100.00	94.00	104.00	694,495	618,123	657,663
Angora goats [4]	359	302	305	54.20	51.70	55.70	19,440	15,610	17,000
Total [5]	....	....	....	....	....	....	75,751,458	77,507,510	74,620,176
Chickens [3]	435,056	441,509	438,948	2.44	2.42	2.38	1,060,964	1,068,851	1,045,222
Total [6]	....	....	....	....	....	....	76,812,422	78,576,361	75,665,398

[1] Preliminary. [2] Based on reporters' estimates of average price per head in their localities. [3] Dec. 1 of preceding year. [4] AZ, NM, and TX only. [5] Cattle, hogs, sheep, and goats. [6] Includes all cattle, hogs, sheep, goats, and chickens (excluding broilers).

NASS, Livestock Branch, (202) 720–3570.

Table 7-82.—Livestock: Average price per 100 pounds received by farmers, by States, 2000 and 2001

State	Cows [1]		Steers and heifers		Beef cattle [2]		Calves	
	2000	2001	2000	2001	2000	2001	2000	2001
	Dollars	*Dollars*	*Dollars*	*Dollars*	*Dollars*	*Dollars*	*Dollars*	*Dollars*
AL	36.70	39.50	83.10	81.80	70.00	70.30	102.00	100.00
AK	60.00	65.00	84.00	90.00	75.60	79.00	100.00	100.00
AZ	36.20	40.60	69.10	72.20	68.60	71.90	102.00	101.00
AR	37.00	40.20	79.40	82.10	66.00	68.30	101.00	101.00
CA	35.90	38.90	69.80	74.10	50.00	53.80	93.00	94.10
CO	39.20	43.10	71.70	74.70	69.70	72.70	105.00	102.00
CT	42.00	40.00	58.00	55.00	57.00	55.00	65.00	65.00
DE	43.00	46.50	70.30	74.00	67.90	71.40	95.00	90.60
FL	38.40	42.20	83.00	84.00	49.40	52.30	101.00	98.60
GA	37.90	40.90	79.70	79.40	52.20	54.20	98.20	95.70
HI	27.40	28.70	49.90	50.90	38.20	41.30	61.00	71.00
ID	36.10	38.70	72.00	73.90	63.10	65.10	91.60	99.00
IL	38.40	41.40	68.30	73.20	67.70	72.60	115.00	110.00
IN	39.00	44.40	65.70	73.40	57.10	63.70	95.00	90.80
IA	37.50	42.40	70.00	73.10	69.10	72.80	102.00	100.00
KS	37.60	42.20	72.70	74.80	71.70	73.80	107.00	107.00
KY	37.80	40.70	82.70	81.60	71.90	71.70	96.10	96.20
LA	36.60	40.50	83.20	82.20	51.60	53.80	98.50	97.90
ME	50.00	45.00	70.00	65.00	65.00	60.00	55.00	60.00
MD	43.00	46.50	70.30	74.00	67.90	71.40	95.00	90.60
MA	40.00	40.00	57.00	55.00	56.00	55.00	60.00	65.00
MI	38.10	41.70	63.60	66.10	56.00	58.80	102.00	109.00
MN	39.70	42.80	66.80	71.00	62.20	65.90	94.80	95.20
MS	37.90	41.40	77.20	77.70	61.90	63.70	94.20	95.40
MO	38.00	40.80	87.50	87.30	76.00	76.20	102.00	99.50
MT	39.50	41.90	88.30	89.60	78.30	80.50	103.00	99.20
NE	38.40	40.20	70.00	73.60	69.10	72.60	106.00	108.00
NV	37.90	38.50	88.60	88.00	75.90	77.00	112.00	109.00
NH	41.00	40.00	68.00	65.00	62.00	60.00	60.00	65.00
NJ	37.00	41.00	57.00	59.00	42.00	45.00	93.00	105.00
NM	40.70	44.00	85.00	86.10	67.80	69.20	101.00	102.00
NY	36.20	38.40	58.70	64.90	38.00	40.20	104.00	112.00
NC	39.00	42.60	84.50	81.00	68.00	66.80	96.20	91.50
ND	39.70	40.80	83.30	85.70	72.20	77.50	101.00	95.50
OH	38.00	39.80	70.30	73.40	67.10	69.90	94.70	96.60
OK	38.00	40.90	84.80	87.60	79.90	79.20	102.00	102.00
OR	37.70	40.60	84.30	83.10	70.50	70.40	93.00	93.60
PA	38.60	42.20	67.40	71.10	59.90	63.60	109.00	115.00
RI	47.00	45.00	63.00	60.00	57.00	55.00	60.00	65.00
SC	41.60	44.80	77.50	77.30	68.80	69.20	101.00	98.00
SD	38.50	41.40	81.80	83.10	75.50	76.50	108.00	106.00
TN	37.20	39.60	83.80	81.70	65.20	65.00	94.80	94.30
TX	36.00	39.80	74.00	76.60	70.70	73.50	107.00	107.00
UT	38.60	40.80	73.80	79.30	71.30	76.60	98.90	104.00
VT	40.00	40.00	68.00	70.00	63.00	65.00	60.00	65.00
VA	35.90	37.80	81.60	78.80	66.20	65.60	100.00	91.50
WA	38.60	41.40	85.80	84.60	80.60	78.20	96.80	95.70
WV	34.80	37.00	77.10	76.10	59.10	60.20	87.50	83.90
WI	38.80	41.30	68.70	71.70	51.40	54.10	127.00	132.00
WY	39.70	41.50	87.50	88.80	81.50	81.20	107.00	105.00
US	37.90	40.90	72.30	75.10	68.60	71.30	104.00	106.00

See footnotes at end of table.

Table 7-82.—Livestock: Average price per 100 pounds received by farmers, by States, 2000 and 2001—Continued

State	Hogs [3]		Lambs		Sheep	
	2000	2001	2000	2001	2000	2001
	Dollars	*Dollars*	*Dollars*	*Dollars*	*Dollars*	*Dollars*
AL	39.50	41.60	....	....	....	....
AK	58.40	61.00	....	....	....	....
AZ	46.00	47.00	75.00	70.00	38.00	37.80
AR	39.10	40.70	....	....	....	....
CA	44.10	44.90	79.10	67.30	31.50	30.20
CO	44.90	47.10	78.30	65.30	32.60	31.90
CT	40.00	41.00	....	....	....	....
DE	37.30	38.90	....	....	....	....
FL	35.40	37.90	....	....	....	....
GA	40.50	42.40	....	....	....	....
HI	79.30	83.80	....	....	....	....
ID	45.00	48.00	77.00	54.20	32.40	32.00
IL	40.40	43.90	72.00	68.00	30.00	30.00
IN	40.10	42.90	73.80	62.70	35.40	35.40
IA	44.50	46.30	73.80	64.40	34.00	33.20
KS	40.60	41.90	79.50	64.50	29.30	29.90
KY	40.20	42.00	....	....	....	....
LA	35.10	36.50	....	....	....	....
ME	40.00	41.00	....	....	....	....
MD	37.30	38.90	....	....	....	....
MA	40.00	41.00	....	....	....	....
MI	40.70	41.70	75.00	70.00	31.00	31.00
MN	44.00	45.50	75.40	62.00	27.40	24.90
MS	40.00	41.80	....	....	....	....
MO	38.60	41.30	79.00	70.00	35.50	35.00
MT	39.80	43.60	84.50	59.80	31.40	33.90
NE	44.30	46.30	76.20	61.20	34.60	34.50
NV	39.20	41.60	72.40	67.70	27.90	30.50
NH [4]	40.00	41.00	110.00	115.00	40.00	40.00
NJ	35.00	35.60	....	....	....	....
NM	37.60	39.20	79.70	67.30	41.00	35.10
NY	34.10	38.70	91.70	104.00	36.60	38.50
NC	42.40	44.70	....	....	....	....
ND	43.10	45.40	76.90	62.90	30.90	29.00
OH	41.30	43.70	78.40	72.30	32.50	32.80
OK	38.40	40.20	75.00	74.00	35.00	33.00
OR	47.00	47.30	79.40	68.20	27.20	28.80
PA	38.80	40.40	88.80	78.70	35.50	34.20
RI	40.00	41.00	....	....	....	....
SC	40.60	42.50	....	....	....	....
SD	43.40	45.40	89.20	73.60	33.30	33.20
TN	41.00	43.00	....	....	....	....
TX	36.60	39.10	81.20	74.80	41.10	42.70
UT	45.90	47.90	82.90	61.00	28.20	27.10
VT	40.00	41.00	....	....	....	....
VA	40.20	41.40	84.00	80.60	30.20	38.50
WA	43.00	44.10	78.00	65.00	29.70	31.30
WV	38.00	38.90	86.20	69.10	31.30	30.50
WI	37.60	39.10	73.00	61.70	30.80	28.50
WY	41.00	43.60	82.70	70.80	31.70	30.00
Other States [5]	....	....	79.40	70.00	34.20	34.00
US	42.30	44.30	79.80	66.90	34.30	34.60

[1] Includes cull dairy cows sold for slaughter, but not cows for dairy herd replacement. [2] Weighted average of prices for cows, and for steers and heifers. [3] December of preceding year through November. [4] For lambs and sheep, CT, ME, MA, NH, RI and VT are included in NH. [5] AL, AK, AR, DE, FL, GA, HI, KY, LA, MD, MS, NJ, NC, SC and TN.

NASS, Livestock Branch, (202) 720–3570.

Table 7-83.—Frozen meat: Cold storage holdings, end of month, United States, 2001 and 2002

Month	Boneless beef		Beef cuts		Total beef	
	2001	2002	2001	2002	2001	2002
	1,000 pounds	*1,000 pounds*	*1,000 pounds*	*1,000 pounds*	*1,000 pounds*	*1,000 pounds*
January	345,366	397,326	65,520	58,125	410,886	455,451
February	307,150	387,401	53,062	51,627	360,212	439,028
March	280,364	359,552	52,234	50,940	332,598	410,492
April	270,552	350,736	44,792	55,002	315,344	405,738
May	277,920	351,227	47,229	50,533	325,149	401,760
June	297,570	345,231	43,216	51,638	340,786	396,869
July	306,053	360,127	45,307	56,407	351,360	416,534
August	320,850	406,336	52,363	55,421	373,213	461,757
September	330,521	439,605	52,245	55,302	382,766	494,907
October	342,586	463,814	52,532	61,353	395,118	525,167
November	371,177	453,370	56,373	59,197	427,550	512,567
December	402,184	463,057	58,553	61,584	460,737	524,641

Month	Picnics		Bellies		Butts	
	2001	2002	2001	2002	2001	2002
	1,000 pounds	*1,000 pounds*	*1,000 pounds*	*1,000 pounds*	*1,000 pounds*	*1,000 pounds*
January	11,253	11,778	50,145	50,849	11,556	17,612
February	11,403	12,617	47,154	57,569	13,247	17,270
March	10,695	14,297	45,440	60,721	13,532	16,339
April	9,901	15,927	43,878	63,293	13,180	16,037
May	10,318	14,621	46,029	62,269	9,972	14,280
June	10,788	14,304	39,552	51,019	6,082	11,896
July	12,669	13,563	24,996	29,925	4,897	11,188
August	11,088	11,314	12,754	14,250	6,765	9,033
September	12,028	11,958	8,960	9,452	8,520	11,211
October	15,304	16,777	28,216	10,354	12,409	9,858
November	13,675	11,879	36,297	18,059	16,017	14,304
December	13,953	8,132	44,301	28,254	17,279	16,587

Month	Hams					
	Bone-in		Boneless		Total	
	2001	2002	2001	2002	2001	2002
	1,000 pounds	*1,000 pounds*	*1,000 pounds*	*1,000 pounds*	*1,000 pounds*	*1,000 pounds*
January	27,214	28,619	41,784	40,238	68,998	68,857
February	32,755	33,298	41,735	41,172	74,490	74,470
March	24,144	30,143	39,660	42,900	63,804	73,043
April	29,300	45,736	39,410	54,104	68,710	99,840
May	33,977	56,063	46,787	72,135	80,764	128,198
June	36,399	67,086	47,598	63,304	83,997	130,390
July	44,237	76,508	42,714	59,246	86,951	135,754
August	52,261	86,684	38,054	53,515	90,315	140,199
September	60,007	99,406	37,396	56,223	97,403	155,629
October	60,439	91,935	41,042	55,892	101,481	147,827
November	32,252	45,504	33,730	42,181	65,982	87,685
December	19,153	22,777	34,056	36,836	53,209	59,613

Month	Loins					
	Bone-in		Boneless		Total	
	2001	2002	2001	2002	2001	2002
	1,000 pounds	*1,000 pounds*	*1,000 pounds*	*1,000 pounds*	*1,000 pounds*	*1,000 pounds*
January	23,174	21,297	17,906	20,545	41,080	41,842
February	15,883	18,368	17,543	21,850	33,426	40,218
March	14,479	18,322	16,022	21,741	30,501	40,063
April	14,805	19,147	18,396	20,587	33,201	39,734
May	14,363	16,751	13,303	15,583	27,666	32,334
June	12,017	15,535	13,684	14,227	25,701	29,762
July	9,627	12,723	11,893	13,253	21,520	25,976
August	7,964	11,308	13,270	13,293	21,234	24,601
September	9,909	13,185	13,464	12,803	23,373	25,988
October	10,517	13,715	15,711	15,526	26,228	29,241
November	19,232	17,055	18,232	18,372	37,464	35,427
December	24,277	22,557	21,357	19,605	45,634	42,162

See end of table.

Table 7-83.—Frozen meat: Cold storage holdings, end of month, United States, 2001 and 2002—Continued

Month	Ribs		Trimmings		Other frozen pork	
	2001	2002	2001	2002	2001	2002
	1,000 pounds	*1,000 pounds*	*1,000 pounds*	*1,000 pounds*	*1,000 pounds*	*1,000 pounds*
January	62,526	86,302	60,195	59,583	84,850	84,928
February	69,266	83,356	61,659	64,150	81,893	75,735
March	63,912	86,324	55,187	69,371	78,412	83,424
April	56,154	84,410	57,074	75,602	81,774	81,463
May	45,307	67,755	54,346	67,069	75,514	66,637
June	32,925	53,007	42,216	53,681	66,304	61,438
July	29,759	52,211	38,667	48,681	54,964	55,714
August	26,201	51,522	32,509	49,459	58,984	59,749
September	34,411	52,362	37,158	45,136	63,945	63,273
October	50,313	57,985	38,474	49,447	68,425	73,826
November	63,689	77,728	45,578	50,737	77,317	75,805
December	77,590	93,296	54,928	54,510	87,371	77,908

Month	Variety meats		Unclassified pork		Total pork	
	2001	2002	2001	2002	2001	2002
	1,000 pounds	*1,000 pounds*	*1,000 pounds*	*1,000 pounds*	*1,000 pounds*	*1,000 pounds*
January	50,580	49,500	30,199	32,631	471,382	503,882
February	50,478	50,422	25,273	35,122	468,289	510,929
March	49,914	51,371	20,903	36,500	432,300	531,453
April	48,995	52,815	19,737	38,598	432,604	567,719
May	50,091	51,600	21,467	43,239	421,474	548,002
June	45,610	51,589	20,934	40,731	374,109	497,817
July	49,394	58,852	15,730	40,320	339,547	472,184
August	54,804	58,416	17,913	45,834	332,567	464,377
September	63,591	54,991	17,557	50,227	366,946	480,227
October	66,737	41,040	22,977	53,447	430,564	489,802
November	53,776	42,849	22,937	49,469	432,732	463,942
December	40,162	37,926	30,532	50,113	464,959	468,501

Month	Veal		Lamb & mutton		Canned hams	
	2001	2002	2001	2002	2001	2002
	1,000 pounds	*1,000 pounds*	*1,000 pounds*	*1,000 pounds*	*1,000 pounds*	*1,000 pounds*
January	5,013	5,121	13,833	13,110	5,808	4,905
February	3,984	4,873	13,141	11,269	6,105	4,342
March	3,816	3,981	13,729	10,528	5,429	5,053
April	3,716	3,504	13,551	13,172	6,141	5,944
May	3,608	3,736	14,586	12,938	7,062	6,154
June	3,500	3,329	15,443	13,553	7,620	6,182
July	3,038	3,658	15,744	14,215	7,212	5,776
August	3,740	3,720	15,266	14,458	6,914	6,129
September	4,545	4,376	13,979	11,961	6,226	6,045
October	5,344	5,217	13,238	12,004	4,343	6,507
November	5,733	7,511	11,336	9,255	3,517	4,051
December	6,053	6,955	11,905	7,124	3,053	3,446

Month	Other canned meat		Total red meat	
	2001	2002	2001	2002
	1,000 pounds	*1,000 pounds*	*1,000 pounds*	*1,000 pounds*
January	900	105	907,822	982,574
February	901	275	852,632	970,716
March	66	395	787,938	961,902
April	49	524	771,405	996,601
May	618	360	772,497	972,950
June	715	287	742,173	918,037
July	536	239	717,437	912,606
August	856	240	732,556	950,681
September	801	332	775,263	997,848
October	411	170	849,018	1,038,867
November	80	362	880,948	997,688
December	98	804	946,805	1,011,471

NASS, Livestock Branch, (202) 720–8784.

CHAPTER VIII

DAIRY AND POULTRY STATISTICS

Dairy statistics in this chapter include series relating to many phases of production, movement, prices, stocks, and consumption of milk and its products. Two series of number of milk cows on farms are included in this publication. One series is an inventory number of a specific classification estimated as one of the major groups making up the total cattle population on January 1. The other series identified as ''milk cows'' is an annual average number of milk cows during the year (excluding any not yet fresh) and is used in estimating milk production.

In comparing the several series of milk prices, it is important to note that prices received by farmers for all whole milk sold are for milk or milkfat content as actually sold, while certain prices paid by dealers for milk for fluid purposes or for specified manufacturing purposes may be quoted on a 3.5 percent butterfat basis, or for some types of manufacturing milk on the test of the milk used for that particular purpose.

Poultry and poultry products statistics include inventory numbers of chickens by classes; the production, disposition, cash receipts, and gross income from chickens and eggs; poultry and egg receipts at principal markets; commercial broiler production; turkey production, disposition, and gross income; poultry and eggs under Federal inspection; and the National Poultry Improvement Plan. Estimates relating to inventories, production, and income exclude poultry and eggs produced on places not classified as farms.

Table 8-1.—Milk cows and heifers: Number that have calved and heifers 500 pounds and over kept for milk cow replacements, United States, Jan. 1, 1994–2003

Year	Milk cows and heifers that have calved	Heifers 500 pounds and over kept for milk cow replacements
	Thousands	*Thousands*
1994	9,507	4,125
1995	9,482	4,121
1996	9,420	4,090
1997	9,318	4,058
1998	9,199	3,986
1999	9,133	4,069
2000	9,190	4,000
2001	9,183	4,057
2002	9,112	4,060
2003[1]	9,152	4,104

[1] Preliminary.

NASS, Livestock Branch, (202) 720–3570.

Table 8-2.—Milk cows and heifers: Number that have calved and heifers 500 pounds and over kept for milk cow replacements, by States, Jan. 1, 2002 and 2003

State	Milk cows and heifers that have calved		Heifers 500 pounds and over kept for milk cow replacements	
	2002	2003 [1]	2002	2003 [1]
	Thousands	*Thousands*	*Thousands*	*Thousands*
AL	20	19	7	7
AK	1.2	1.3	0.5	0.6
AZ	140	150	33	35
AR	33	31	15	10
CA	1,620	1,680	770	790
CO	93	98	40	38
CT	24	23	11.5	11.0
DE	9	9	3.5	3.3
FL	152	147	40	40
GA	86	85	29	29
HI	7	6	3	3
ID	377	390	175	190
IL	115	114	50	50
IN	154	145	66	55
IA	205	208	120	100
KS	95	105	40	50
KY	125	120	40	45
LA	54	49	11	10
ME	38	36	20.0	19.0
MD	81	80	33	32
MA	21	20	9.0	10.0
MI	297	301	135	135
MN	500	480	295	295
MS	34	33	15	15
MO	140	134	60	55
MT	19	18	13	11
NE	68	66	25	20
NV	25	25	10	11
NH	18	17	8.0	8.0
NJ	13	13	6	6
NM	290	310	60	70
NY	675	680	300	325
NC	66	63	30	28
ND	42	37	15	13
OH	260	260	110	110
OK	87	88	20	30
OR	105	117	55	60
PA	588	590	285	280
RI	1.4	1.4	0.7	0.6
SC	20	20	8	8
SD	98	94	30	30
TN	90	84	40	40
TX	310	311	100	110
UT	93	91	44	45
VT	154	153	58	59
VA	120	118	60	48
WA	247	247	105	105
WV	16	15	5	7
WI	1,280	1,265	650	650
WY	5	4	1	1
US	9,111.6	9,151.7	4,060.2	4,103.5

[1] Preliminary.

NASS, Livestock Branch, (202) 720–3570.

Table 8-3.—Milk-feed price ratios: All milk-price; dairy feed, 16%; Milk-feed price ratios and value per 100 pounds of grain and concentrate rations fed to milk cows, United States, annual 1992–2001

Year	All milk price cwt.	16% dairy feed price cwt [1]	Milk-feed price ratio [2]
	Dollars	*Dollars*	*Pounds*
1992	13.09	8.90	3.01
1993	12.80	9.00	2.80
1994	12.97	9.20	2.65
1995	12.74	8.70	2.59
1996	14.75	11.25	2.44
1997	13.36	10.75	2.38
1998	15.46	9.70	3.34
1999	14.38	9.00	3.59
2000	12.40	8.75	3.05
2001	15.05	9.20	3.38

[1] Commercially prepared 16%dairy ration: Annual average prior to 1995, April price 1995-current. [2] Annual ratios based on average of monthly ratios. Pounds of 16 % mixed dairy feed equal in value to one pound of whole milk. Effective January 1995, prices of commercial prepared feeds are based on current U.S. prices received for corn (51 lbs), soybeans (8 lbs), and alfalfa hay (41 lbs).

NASS, Environmental, Economics, and Demographics Branch, (202) 720–6146.

Table 8-4.—Milk cows: Number of operations, 2001–2002, and inventory, Jan. 1, 2002–2003, by selected States and United States [1]

State	Operations with milk cows		January 1 milk cow inventory	
	2001	2002 [2]	2002	2003 [2]
	Number	*Number*	*1,000 head*	*1,000 head*
AL	250	200	20	19
AK	30	30	1.2	1.3
AZ	250	250	140	150
AR	900	800	33	31
CA	2,500	2,500	1,620	1,680
CO	800	780	93	98
CT	310	290	24	23
DE	120	120	9	9
FL	510	500	152	147
GA	720	660	86	85
HI	50	50	7	6
ID	1,000	950	377	390
IL	1,900	1,800	115	114
IN	2,900	2,600	154	145
IA	3,500	3,200	205	208
KS	1,200	1,200	95	105
KY	2,900	2,500	125	120
LA	610	590	54	49
ME	600	550	38	36
MD	950	900	81	80
MA	350	330	21	20
MI	3,300	3,200	297	301
MN	7,800	7,200	500	480
MS	480	450	34	33
MO	3,700	3,500	140	134
MT	650	600	19	18
NE	1,100	1,000	68	66
NV	150	150	25	25
NH	260	250	18	17
NJ	250	250	13	13
NM	500	500	290	310
NY	7,200	7,100	675	680
NC	900	870	66	63
ND	850	760	42	37
OH	5,200	5,100	260	260
OK	1,700	1,600	87	88
OR	820	800	105	117
PA	10,300	9,900	588	590
RI	30	30	1.4	1.4
SC	240	220	20	20
SD	1,300	1,200	98	94
TN	1,500	1,400	90	84
TX	2,100	1,900	310	311
UT	760	700	93	91
VT	1,600	1,500	154	153
VA	1,500	1,500	120	118
WA	1,000	950	247	247
WV	600	500	16	15
WI	19,100	17,800	1,280	1,265
WY	270	260	5	4
US	97,510	91,990	9,111.6	9,151.7

[1] An operation is any place having one or more milk cows on hand at any time during the year. [2] Preliminary.

NASS, Livestock Branch, (202) 720–3570.

Table 8-5.—Milk cows: Number of operations by size group, selected States, and United States, 2001–2002 [1]

State	Operations having											
	1–29 Head		30–49 Head		50–99 Head		100–199 Head		200-499 Head		500+ Head	
	2001	2002 [2]	2001	2002 [2]	2001	2002 [2]	2001	2002 [2]	2001	2002 [2]	2001	2002 [2]
	Number	*Number*	*Number*	*Number*	*Number*	*Number*	*Number*	*Number*	*Number*	*Number*	*Number*	*Number*
AZ	120	115	....	....	10	10	10	10	20	20	90	95
CA	300	280	70	70	80	90	250	260	700	700	1,100	1,100
CO	580	575	10	10	30	30	55	40	75	75	50	50
FL	290	290	5	5	20	20	35	30	60	55	100	100
GA	320	290	20	10	90	90	170	160	90	80	30	30
ID	230	210	80	80	210	190	150	140	150	150	180	180
IL	480	480	360	350	700	620	280	270	70	65	10	15
IN	1,500	1,200	470	490	570	580	280	250	55	55	25	25
IA	820	715	890	800	1,200	1,050	480	510	95	100	15	25
KY	1,300	1,000	520	470	760	700	270	270	45	55	5	5
MD	160	160	150	150	370	310	210	210	50	60	10	10
MI	1,050	1,050	550	500	800	750	620	590	215	240	65	70
MN	1,100	900	2,700	2,400	3,000	2,900	700	700	250	250	50	50
MO	1,600	1,600	700	600	850	800	500	450	45	40	5	10
NM	335	325	....	....	5	5	5	5	15	15	140	150
NY	1,300	1,200	1,200	1,250	2,800	2,800	1,300	1,200	470	510	130	140
NC	460	460	40	30	150	140	160	155	75	70	15	15
OH	2,400	2,300	850	900	1,250	1,200	560	540	120	140	20	20
OK	1,050	1,000	120	120	260	230	210	190	50	50	10	10
OR	400	370	30	30	80	85	170	170	100	100	40	45
PA	1,800	1,600	3,700	3,500	3,600	3,600	900	900	270	260	30	40
SD	410	370	280	250	370	345	180	170	45	45	15	20
TN	630	550	140	130	390	390	250	250	80	70	10	10
TX	950	900	100	90	250	200	320	290	320	260	160	160
UT	270	240	35	40	140	110	170	160	110	110	35	40
VT	130	110	270	270	780	710	270	260	120	120	30	30
VA	490	500	140	150	420	400	340	350	100	90	10	10
WA	300	290	40	30	110	110	190	180	220	200	140	140
WI	3,150	2,600	5,300	4,800	7,800	7,500	2,000	2,000	680	700	170	200
Oth Sts	4,200	4,000	1,100	1,000	2,100	1,900	1,300	1,150	500	470	100	110
US	28,125	25,680	19,870	18,525	29,195	27,865	12,335	11,860	5,195	5,155	2,790	2,905

[1] An operation is any place having one or more head of milk cows on hand at any time during the year. [2] Preliminary.
NASS, Livestock Branch, (202) 720–3570.

Table 8-6.—Milk cows: Percent of inventory by size group, selected States, and United States, 2001–2002 [1]

State	Inventory on operations having											
	1–29 head		30–49 head		50–99 head		100–199 head		200–499 head		500+ head	
	2001	2002 [2]	2001	2002 [2]	2001	2002 [2]	2001	2002 [2]	2001	2002 [2]	2001	2002 [2]
	Percent	*Percent*	*Percent*	*Percent*	*Percent*	*Percent*	*Percent*	*Percent*	*Percent*	*Percent*	*Percent*	*Percent*
AZ	0.3	0.3	....	....	0.4	0.4	0.8	0.8	4.5	3.5	94.0	95.0
CA	0.1	0.1	0.2	0.2	0.3	0.4	2.4	2.3	15.0	14.0	82.0	83.0
CO	1.1	1.3	0.4	0.4	2.0	2.0	7.5	5.0	26.7	25.8	62.3	65.5
FL	0.4	0.4	0.1	0.1	1.0	1.0	3.0	3.0	12.5	12.5	83.0	83.0
GA	1.0	1.0	1.0	0.5	8.0	7.5	28.0	26.0	27.0	26.0	35.0	39.0
ID	0.4	0.4	0.9	0.8	4.1	3.7	5.6	5.1	13.0	12.0	76.0	78.0
IL	3.0	3.0	10.0	10.0	38.0	34.0	28.0	30.0	16.0	15.5	5.0	7.5
IN	9.0	7.0	12.0	11.0	25.0	25.0	24.0	22.0	10.0	10.0	20.0	25.0
IA	5.0	3.5	13.0	12.0	36.0	32.0	28.0	30.0	12.0	13.0	6.0	9.5
KY	8.0	6.0	15.0	14.5	39.0	39.0	27.0	27.0	8.5	11.0	2.5	2.5
MD	1.5	2.0	6.5	7.0	29.0	24.0	34.0	35.0	16.0	19.0	13.0	13.0
MI	5.0	4.0	7.0	6.0	18.0	17.0	27.0	26.0	21.0	23.0	22.0	24.0
MN	4.0	3.6	20.0	19.0	38.0	39.0	16.0	16.0	14.0	14.0	8.0	8.4
MO	5.0	4.5	17.0	16.0	31.0	31.0	36.0	35.0	7.5	7.0	3.5	6.5
NM	0.2	0.2	....	....	0.1	0.1	0.3	0.3	1.4	1.4	98.0	98.0
NY	2.0	2.0	8.0	7.5	29.0	28.0	27.0	24.5	18.0	20.0	16.0	18.0
NC	1.9	2.2	2.1	1.8	17.0	16.0	32.0	33.0	30.0	29.0	17.0	18.0
OH	9.0	9.0	13.0	13.0	31.0	31.0	28.0	26.0	13.0	15.0	6.0	6.0
OK	4.0	3.0	5.0	5.0	19.0	18.0	29.0	27.0	15.0	17.0	28.0	30.0
OR	0.7	0.9	1.3	1.1	7.0	6.0	21.0	19.0	32.0	28.0	38.0	45.0
PA	4.0	3.0	24.0	23.0	37.0	38.0	20.0	20.0	11.0	11.0	4.0	5.0
SD	4.0	4.0	11.0	10.0	26.0	25.0	25.0	24.0	14.0	14.0	20.0	23.0
TN	2.5	2.5	5.0	5.0	29.0	29.0	35.0	36.0	22.0	21.0	6.5	6.5
TX	1.0	1.0	1.0	1.0	5.0	4.5	14.0	13.5	31.0	27.0	48.0	53.0
UT	1.0	1.0	1.0	1.5	11.0	8.5	24.0	23.0	33.0	31.0	30.0	35.0
VT	1.0	1.0	7.0	7.0	31.0	29.0	23.0	22.0	22.0	23.0	16.0	18.0
VA	3.0	3.5	4.0	4.5	26.0	25.0	38.0	40.0	23.0	21.0	6.0	6.0
WA	0.3	0.2	0.7	0.5	3.0	3.3	11.0	11.0	28.0	26.0	57.0	59.0
WI	3.7	3.5	14.0	13.0	38.0	38.0	19.0	19.0	15.0	14.5	10.3	12.0
Oth Sts	3.0	3.0	6.0	6.0	24.0	23.0	27.0	26.0	22.0	22.0	18.0	20.0
US	2.6	2.3	8.1	7.5	20.9	20.1	17.4	16.7	16.4	16.1	34.6	37.3

[1] Percents reflect average distribution of various probability surveys conducted during the year but are based primarily on beginning-of-year and mid-year surveys. [2] Preliminary.
NASS, Livestock Branch, (202) 720–3570.

Table 8-7.—Milk Production: Percent of production by size groups, selected States, and United States, 2001–2002 [1]

State	Production on operations having											
	1–29 Head		30–49 Head		50–99 Head		100–199 Head		200-499 Head		500+ Head	
	2001	2002 [2]	2001	2002 [2]	2001	2002 [2]	2001	2002 [2]	2001	2002 [2]	2001	2002 [2]
	Percent	*Percent*	*Percent*	*Percent*	*Percent*	*Percent*	*Percent*	*Percent*	*Percent*	*Percent*	*Percent*	*Percent*
AZ	0.1	0.1	....	....	0.2	0.2	0.5	0.5	3.7	3.2	95.5	96.0
CA	0.1	0.1	0.1	0.1	0.2	0.3	2.6	2.5	19.0	17.0	78.0	80.0
CO	1.0	1.0	0.5	0.5	1.5	2.0	6.5	4.5	26.5	25.0	64.0	67.0
FL	0.2	0.2	0.1	0.1	0.6	0.5	2.6	1.7	11.5	12.5	85.0	85.0
GA	0.5	0.5	0.5	0.5	6.0	6.0	27.0	27.0	27.0	26.0	39.0	40.0
ID	0.2	0.2	0.6	0.6	3.1	2.7	4.6	4.0	11.5	11.0	80.0	81.5
IL	2.5	2.5	8.0	9.0	38.0	32.0	28.0	30.5	17.0	17.0	6.5	9.0
IN	7.0	6.0	10.0	9.0	24.0	23.0	25.0	21.0	11.0	11.0	23.0	30.0
IA	4.0	3.0	11.0	10.0	35.0	31.0	29.0	31.0	14.0	14.0	7.0	11.0
KY	5.0	5.0	14.0	13.0	39.0	38.0	30.0	29.0	10.0	12.0	2.0	3.0
MD	1.0	1.0	5.0	6.0	28.0	23.0	35.0	35.0	17.0	20.0	14.0	15.0
MI	3.0	2.5	5.5	4.5	15.5	15.0	27.0	25.0	23.0	25.0	26.0	28.0
MN	3.0	2.7	18.0	17.0	37.0	38.0	17.0	17.0	16.0	16.0	9.0	9.3
MO	3.0	2.0	15.0	15.0	30.0	30.0	38.0	38.0	9.0	7.0	5.0	8.0
NM	0.1	0.1	....	....	0.1	0.1	0.4	0.4	1.4	1.4	98.0	98.0
NY	1.5	1.5	6.5	5.5	26.0	25.0	27.0	25.0	20.0	22.0	19.0	21.0
NC	0.4	0.5	1.6	1.5	17.0	15.0	33.0	32.0	30.0	30.0	18.0	21.0
OH	5.0	5.5	12.0	11.5	30.0	30.0	30.0	28.0	15.0	17.0	8.0	8.0
OK	2.0	2.0	4.0	4.0	17.0	15.0	27.0	25.0	15.0	17.0	35.0	37.0
OR	0.5	0.5	1.5	1.0	6.0	5.5	18.0	18.0	33.0	28.0	41.0	47.0
PA	3.0	2.0	22.0	21.0	37.0	38.0	21.0	21.0	12.0	12.0	5.0	6.0
SD	3.0	3.0	9.0	9.0	22.0	23.0	25.0	24.0	17.0	16.0	24.0	25.0
TN	2.0	1.5	4.5	4.0	28.0	28.0	35.0	37.0	23.0	22.0	7.5	7.5
TX	0.3	0.3	0.7	0.7	4.0	4.0	14.0	12.0	29.0	25.0	52.0	58.0
UT	0.7	0.7	0.8	1.3	9.5	7.0	23.0	21.0	34.0	32.0	32.0	38.0
VT	0.5	0.5	5.5	5.5	28.0	27.0	23.0	22.0	23.0	25.0	20.0	20.0
VA	2.0	2.0	3.0	4.0	24.0	23.0	39.0	41.0	25.0	23.0	7.0	7.0
WA	0.1	0.2	0.4	0.3	2.5	2.5	10.0	9.0	28.0	26.0	59.0	62.0
WI	2.7	2.5	13.3	12.0	36.0	35.0	19.0	19.0	17.0	17.5	12.0	14.0
Oth Sts	1.5	1.5	5.0	5.0	22.0	21.0	27.0	25.0	23.5	23.5	21.0	24.0
US	1.6	1.5	6.7	6.1	18.3	17.4	16.5	15.6	17.9	17.5	39.0	41.9

[1] Percents reflect average distributions of various probability surveys conducted during the year but are based primarily on beginning-of-year and mid-year surveys. [2] Preliminary.

NASS, Livestock Branch, (202) 720–3570.

Table 8-8.—Milk cows: Number of operations, percent of inventory and percent of milk production by size group, United States, 2001–2002 [1]

Head	Operations		Percent of inventory		Percent of production	
	2001	2002 [2]	2001	2002 [2]	2001	2002 [2]
	Number	*Number*	*Percent*	*Percent*	*Percent*	*Percent*
1-29	28,125	25,680	2.6	2.3	1.6	1.5
30-49	19,870	18,525	8.1	7.5	6.7	6.1
50-99	29,195	27,865	20.9	20.1	18.3	17.4
100-199	12,335	11,860	17.4	16.7	16.5	15.6
200-499	5,195	5,155	16.4	16.1	17.9	17.5
500-999	1,700	1,750	12.0	12.5	13.5	13.9
1,000-1,999	770	775	11.5	11.5	13.2	13.0
2,000+	320	380	11.1	13.3	12.3	15.0
Total	97,510	91,990	100.0	100.0	100.0	100.0

[1] An operation is any place having one or more head of milk cows on hand at any time during the year. Percents reflect average distributions of various probability surveys conducted during the year but are based primarily on beginning-of-year and mid-year surveys. [2] Preliminary.

NASS, Livestock Branch, (202) 720–3570.

Table 8-9.—Official Dairy Herd Improvement test plans: Numbers of herds and cows and milk, fat, and protein production, United States, 1992–2001

Year	Herds	Cows	Cows per herd	Average production			Cows with protein information	Average protein production [1]	Average protein production [1]
				Milk	Fat	Fat			
	Number	*Number*	*Number*	*Pounds*	*Percent*	*Pounds*	*Percent*	*Percent*	*Pounds*
1992 ...	30,216	3,137,741	103.8	18,750	3.70	695	91	3.22	602
1993 ...	35,484	3,625,542	102.2	18,719	3.68	690	90	3.22	602
1994 ...	33,844	3,620,652	107.0	19,129	3.67	705	90	3.23	616
1995 ...	31,628	3,527,187	111.5	19,271	3.67	710	90	3.22	621
1996 ...	29,416	3,486,010	118.5	19,192	3.70	713	90	3.23	620
1997 ...	27,383	3,402,487	124.3	19,815	3.67	731	89	3.23	639
1998 ...	25,738	3,397,396	132.0	20,209	3.68	745	92	3.22	651
1999 ...	24,841	3,449,854	140.9	20,743	3.68	766	93	3.24	673
2000 ...	23,225	3,521,686	151.6	21,092	3.68	781	93	3.15	664
2001 ...	22,095	3,499,214	158.4	21,118	3.66	777	94	3.08	651

[1] The decline in protein production in 2000 reflects a measurement change by the dairy industry from crude to true protein beginning in May 2000. The percentage of milk that is true protein is lower than the percentage that is crude protein by an approximate difference of 0.19 percent.

ARS, Animal Improvement Programs Laboratory, (301) 504–8334, rlaipl@aipl.arsusda.gov or http://aipl.arsusda.gov.

Table 8-10.—Milk and milkfat production: Number of producing cows, production per cow, and total quantity produced, United States, 1992–2001

Year	Number of milk cows [1]	Production of milk and milkfat [2]				
		Per milk cow		Percentage of fat in all milk produced	Total	
		Milk	Milkfat		Milk	Milkfat
	Thousands	*Pounds*	*Pounds*	*Percent*	*Million pounds*	*Million pounds*
1992	9,688	15,570	573	3.68	150,847	5,552
1993	9,581	15,722	576	3.66	150,636	5,517
1994	9,494	16,179	592	3.66	153,602	5,621
1995	9,466	16,405	600	3.66	155,292	5,681
1996	9,372	16,433	606	3.69	154,006	5,679
1997	9,252	16,871	617	3.66	156,091	5,706
1998	9,154	17,189	629	3.66	157,348	5,757
1999	9,156	17,772	652	3.67	162,716	5,975
2000	9,206	18,201	670	3.68	167,559	6,171
2001 [3]	9,115	18,159	666	3.67	165,497	6,072

[1] Average number during year, excluding heifers not yet fresh. [2] Excludes milk sucked by calves. [3] Preliminary.

NASS, Livestock Branch, (202) 720–3570.

Table 8-11.—Milk and milkfat production: Number of milk cows, production per cow, and total quantity produced, by States, 2000

State	Number of milk cows[1]	Production of milk and milkfat[2]				
		Per milk cow		Percentage of fat in all milk produced	Total	
		Milk	Milkfat		Milk	Milkfat
	Thousands	*Pounds*	*Pounds*	*Percent*	*Million pounds*	*Million pounds*
AL	25	13,920	508	3.65	348	12.7
AK	0.9	14,500	494	3.41	13.05	0.4
AZ	139	21,820	788	3.61	3,033	109.5
AR	39	12,436	451	3.63	485	17.6
CA	1,526	21,149	778	3.68	32,273	1,187.6
CO	89	21,618	770	3.56	1,924	68.5
CT	27	17,778	654	3.68	480	17.7
DE	10.0	15,000	564	3.76	150.0	5.6
FL	157	15,688	565	3.60	2,463	88.7
GA	87	16,471	600	3.64	1,433	52.2
HI	8.1	14,358	498	3.47	116.3	4.0
ID	347	20,816	751	3.61	7,223	260.8
IL	120	17,450	654	3.75	2,094	78.5
IN	146	16,568	616	3.72	2,419	90.0
IA	215	18,298	677	3.70	3,934	145.6
KS	91	16,923	623	3.68	1,540	56.7
KY	132	12,841	471	3.67	1,695	62.2
LA	58	12,034	424	3.52	698	24.6
ME	39	17,128	644	3.76	668	25.1
MD	84	16,083	605	3.76	1,351	50.8
MA	22	17,091	634	3.71	376	13.9
MI	300	19,017	696	3.66	5,705	208.8
MN	534	17,777	660	3.71	9,493	352.2
MS	36	15,028	544	3.62	541	19.6
MO	154	14,662	542	3.70	2,258	83.5
MT	19	17,789	637	3.58	338	12.1
NE	76	16,513	611	3.70	1,255	46.4
NV	25	19,040	672	3.53	476	16.8
NH	18	17,333	652	3.76	312	11.7
NJ	16	15,250	557	3.65	244	8.9
NM	250	20,944	746	3.56	5,236	186.4
NY	686	17,378	640	3.68	11,921	438.7
NC	71	16,746	615	3.67	1,189	43.6
ND	48	14,292	526	3.68	686	25.2
OH	262	17,027	639	3.75	4,461	167.3
OK	91	14,231	507	3.56	1,295	46.1
OR	90	18,222	665	3.65	1,640	59.9
PA	617	18,081	673	3.72	11,156	415.0
RI	1.8	15,667	577	3.68	28.2	1.0
SC	23	16,087	592	3.68	370	13.6
SD	102	16,020	593	3.70	1,634	60.5
TN	95	14,789	544	3.68	1,405	51.7
TX	348	16,483	602	3.65	5,736	209.4
UT	96	17,573	638	3.63	1,687	61.2
VT	156	17,199	642	3.73	2,683	100.1
VA	120	15,833	579	3.66	1,900	69.5
WA	247	22,644	827	3.65	5,593	204.1
WV	17	15,588	567	3.64	265	9.6
WI	1,344	17,306	649	3.75	23,259	872.2
WY	5.6	13,571	501	3.69	76.0	2.8
US[3]	9,206	18,201	670	3.68	167,559	6,170.6

[1] Average number during year, excluding heifers not yet fresh. U.S. total may not add due to rounding. [2] Excludes milk sucked by calves. [3] Sum of parts may not equal due to rounding.

NASS, Livestock Branch, (202) 720–3570.

Table 8-12.—Milk and milkfat production: Number of milk cows, production per cow, and total quantity produced, by States, 2001 (preliminary)

State	Number of milk cows[1]	Production of milk and milkfat[2]				
		Per milk cow		Percentage of fat in all milk produced	Total	
		Milk	Milkfat		Milk	Milkfat
	Thousands	*Pounds*	*Pounds*	*Percent*	*Million pounds*	*Million pounds*
AL	21	14,286	523	3.66	300	11.0
AK	1.1	13,055	444	3.40	14.36	0.5
AZ	140	20,679	749	3.62	2,895	104.8
AR	35	12,343	449	3.64	432	15.7
CA	1,590	20,913	772	3.69	33,251	1,227.0
CO	91	21,648	769	3.55	1,970	69.9
CT	25	18,240	671	3.68	456	16.8
DE	9.0	16,778	614	3.66	151.0	5.5
FL	153	15,758	567	3.60	2,411	86.8
GA	86	16,640	606	3.64	1,431	52.1
HI	7.5	14,107	484	3.43	105.8	3.6
ID	366	21,194	763	3.60	7,757	279.3
IL	116	17,414	650	3.73	2,020	75.3
IN	153	16,732	619	3.70	2,560	94.7
IA	210	18,024	669	3.71	3,785	140.4
KS	93	17,312	639	3.69	1,610	59.4
KY	128	12,969	473	3.65	1,660	60.6
LA	54	11,704	412	3.52	632	22.2
ME	38	17,211	633	3.68	654	24.1
MD	82	15,780	578	3.66	1,294	47.4
MA	21	17,048	627	3.68	358	13.2
MI	303	19,323	701	3.63	5,855	212.5
MN	510	17,278	639	3.70	8,812	326.0
MS	35	14,200	517	3.64	497	18.1
MO	145	13,441	496	3.69	1,949	71.9
MT	19	18,211	652	3.58	346	12.4
NE	72	16,056	596	3.71	1,156	42.9
NV	25	19,400	689	3.55	485	17.2
NH	18	17,944	660	3.68	323	11.9
NJ	14	16,643	604	3.63	233	8.5
NM	268	20,750	743	3.58	5,561	199.1
NY	672	17,527	640	3.65	11,778	429.9
NC	67	17,373	632	3.64	1,164	42.4
ND	46	14,000	514	3.67	644	23.6
OH	260	16,612	620	3.73	4,319	161.1
OK	89	14,528	524	3.61	1,293	46.7
OR	95	18,074	662	3.66	1,717	62.8
PA	599	18,112	668	3.69	10,849	400.3
RI	1.4	16,571	610	3.68	23.2	0.9
SC	21	17,476	638	3.65	367	13.4
SD	99	15,960	591	3.70	1,580	58.5
TN	92	14,511	530	3.65	1,335	48.7
TX	325	15,689	574	3.66	5,099	186.6
UT	93	17,581	640	3.64	1,635	59.5
VT	153	17,431	641	3.68	2,667	98.1
VA	118	15,898	576	3.62	1,876	67.9
WA	247	22,324	817	3.66	5,514	201.8
WV	16	15,563	562	3.61	249	9.0
WI	1,292	17,182	641	3.73	22,199	828.0
WY	4.5	14,000	511	3.65	63.0	2.3
US[3]	9,115	18,139	666	3.67	165,336	6,072.3

[1] Average number during year, excluding heifers not yet fresh. U.S. total may not add due to rounding. [2] Excludes milk sucked by calves. [3] Sum of parts may not equal due to rounding.

NASS, Livestock Branch, (202) 720–3570.

Table 8-13.—Milk: Quantities used and marketed by producers, by States, 2001 (preliminary)

State	Milk used where produced			Milk marketed by producers	
	Fed to calves [1]	Used for milk, cream, and butter	Total	Total quantity [2]	Fluid grade [3]
	Million pounds	*Million pounds*	*Million pounds*	*Million pounds*	*Percent*
AL	2	1	3	297	100
AK	0.6	0.2	0.8	13.56	100
AZ	12	1	13	2,882	100
AR	10	3	13	419	100
CA	30	6	36	33,215	99
CO	22	4	26	1,944	100
CT	3.5	0.5	4.0	452.0	100
DE	1.0	0.1	1.1	149.9	100
FL	1	1	2	2,409	100
GA	10	1	11	1,420	100
HI	1.2	0.5	1.7	104.1	100
ID	31	2	33	7,724	99
IL	14	2	16	2,004	98
IN	22	5	27	2,533	96
IA	30	11	41	3,744	97
KS	10	1	11	1,599	99
KY	25	7	32	1,628	99
LA	10	2	12	620	100
ME	4.5	0.5	5.0	649.0	100
MD	7	3	10	1,284	100
MA	3.5	0.5	4.0	354.0	100
MI	50	5	55	5,800	99
MN	95	10	105	8,707	95
MS	1	1	2	495	100
MO	23	5	28	1,921	96
MT	5	4	9	337	100
NE	11	1	12	1,144	97
NV	5	1	6	479	100
NH	2.7	0.3	3.0	320.0	100
NJ	2	1	3	230	100
NM	42	15	57	5,504	100
NY	120	7	127	11,651	100
NC	11	6	17	1,147	100
ND	11	1	12	632	73
OH	20	5	25	4,294	92
OK	13	1	14	1,279	100
OR	7	1	8	1,709	99
PA	50	5	55	10,794	99
RI	0.4	....	0.4	22.8	100
SC	2	1	3	364	100
SD	12	2	14	1,566	93
TN	4	1	5	1,330	99
TX	21	2	23	5,076	100
UT	23	2	25	1,610	96
VT	19	1	20	2,647	100
VA	11	2	13	1,863	100
WA	29	1	30	5,484	100
WV	2	1	3	246	100
WI	240	45	285	21,914	95
WY	1.3	0.2	1.5	61.5	77
US [4]	1,084	180	1,264	164,072	98

[1] Excludes milk sucked by calves. [2] Milk sold to plants and dealers as whole milk and equivalent amounts of milk for cream. Includes milk produced by dealers' own herds and small amounts sold directly to consumers. Also includes milk produced by institutional herds. [3] Percentage of milk sold that is eligible for fluid use (grade A for fluid use in most States). Includes fluid-grade milk used in manufacturing dairy products. [4] May not add due to rounding.

NASS, Livestock Branch, (202) 720–3570.

Table 8-14.—Milk: Cows, yield per cow, and production in specified countries, 1999–2001

Country and continent	Milk cows			Per cow yield			Milk production		
	1999	2000	2001	1999	2000	2001	1999	2000	2001
	1,000 head	*1,000 head*	*1,000 head*	*Kilograms*	*Kilograms*	*Kilograms*	*1,000 metric tons*	*1,000 metric tons*	*1,000 metric tons*
North America:.									
Canada	1,180	1,141	1,091	6,919	7,152	7,430	8,164	8,161	8,106
Mexico	6,700	6,800	6,800	1,325	1,368	1,397	8,877	9,305	9,501
United States	9,156	9,206	9,115	8,061	8,265	8,230	73,805	76,004	74,996
Total	17,096	17,147	17,096	....	....	....	90,846	93,550	91,412
South America:.									
Argentina	2,500	2,450	2,450	4,120	4,000	3,878	10,300	9,800	9,500
Brazil	16,194	16,040	15,900	1,340	1,380	1,403	21,700	22,134	22,300
Chile	613	608	610	3,475	3,388	3,443	2,130	2,060	2,100
Peru	600	610	610	1,750	1,803	1,803	1,050	1,100	1,100
Venezuela	740	735	730	1,773	1,769	1,781	1,312	1,300	1,300
Total	20,647	20,443	20,300	....	....	....	36,492	36,394	36,492
European Union	20,930	20,610	20,480	5,505	5,575	5,623	115,218	114,900	115,165
Eastern Europe:.									
Poland	3,471	3,296	3,130	3,477	3,580	3,800	12,068	11,800	11,894
Romania	1,613	1,580	1,564	3,202	3,228	3,317	5,165	5,100	5,188
Total	5,084	4,876	5,084	....	....	....	17,233	17,020	17,233
Former Soviet Union:.									
Russia	13,500	12,900	12,500	2,370	2,558	2,640	32,000	33,000	33,000
Ukraine	5,841	5,431	4,958	2,250	2,283	2,656	13,140	12,400	13,169
Total	19,341	18,331	19,341	....	....	....	45,140	46,100	45,140
South Asia:.									
India [1]	35,500	33,750	35,900	1,014	1,074	1,014	36,000	36,250	36,400
Total	35,500	33,750	35,500	....	....	....	36,000	36,400	36,000
Asia:.									
China	2,220	2,280	2,220	2,848	3,798	4,619	7,176	8,660	10,255
Japan	1,008	992	971	8,390	8,566	8,548	8,457	8,497	8,300
Total	3,228	3,272	3,228	....	....	....	15,633	16,962	15,633
Oceania:.									
Australia [2]	2,155	2,171	2,281	4,865	5,146	4,761	10,483	11,172	10,860
New Zealand [3]	3,285	3,337	3,557	3,370	3,666	3,700	11,070	12,235	13,162
Total	5,440	5,508	5,440	....	....	....	21,553	24,029	21,553
World total	127,205	125,937	125,495	....	....	....	378,117	382,406	386,296

[1] Year beginning April 1 of the year shown. [2] Year ending June 30 of the year shown. [3] Year ending May 31 of the year shown.

FAS, Dairy, Livestock and Poultry Division, (202) 720–3761. Data from counselor/attaché reports and official statistics.

Table 8-15.—Milk Production: Marketings, income, and value, by States, 2001 (preliminary)

State	Combined marketing of milk and cream: Milk utilized	Average returns[1]: Per cwt milk	Average returns[1]: Per lb. milkfat	Cash receipts from marketings	Used for milk, cream, and butter where produced: Milk utilized	Used for milk, cream, and butter where produced: Value[2]	Gross producer income[3]	Value of milk produced[2 4]
	Million pounds	*Dollars*	*Dollars*	*1,000 dollars*	*Million pounds*	*1,000 dollars*	*1,000 dollars*	*1,000 dollars*
AL	297	16.90	4.62	50,193	1	169	50,362	50,700
AK	13.56	20.50	6.03	2,780	0.2	41	2,821	2,944
AZ	2,882	14.70	4.06	423,654	1	147	423,801	425,565
AR	419	16.10	4.42	67,459	3	483	67,942	69,552
CA	33,215	13.94	3.78	4,630,171	6	836	4,631,007	4,635,189
CO	1,944	14.80	4.17	287,712	4	592	288,304	291,560
CT	452	16.20	4.40	73,224	0.5	81	73,305	73,872
DE	149.9	16.10	4.40	24,134	0.1	16	24,150	24,311
FL	2,409	17.80	4.94	428,802	1	178	428,980	429,158
GA	1,420	15.90	4.37	225,780	1	159	225,939	227,529
HI	104.1	25.50	7.43	26,546	0.5	128	26,674	26,979
ID	7,724	13.50	3.75	1,042,740	2	270	1,043,010	1,047,195
IL	2,004	15.00	4.02	300,600	2	300	300,900	303,000
IN	2,533	16.10	4.35	407,813	5	805	408,618	412,160
IA	3,744	14.70	3.96	550,368	11	1,617	551,985	556,395
KS	1,599	14.50	3.93	231,855	1	145	232,000	233,450
KY	1,628	16.20	4.44	263,736	7	1,134	264,870	268,920
LA	620	16.00	4.55	99,200	2	320	99,520	101,120
ME	649	16.30	4.43	105,787	0.5	82	105,869	106,602
MD	1,284	16.20	4.43	208,008	3	486	208,494	209,628
MA	354	16.20	4.40	57,348	0.5	81	57,429	57,996
MI	5,800	15.20	4.19	881,600	5	760	882,360	889,960
MN	8,707	14.90	4.03	1,297,343	10	1,490	1,298,833	1,312,988
MS	495	16.10	4.42	79,695	1	161	79,856	80,017
MO	1,921	14.90	4.04	286,229	5	745	286,974	290,401
MT	337	15.10	4.22	50,887	4	604	51,491	52,246
NE	1,144	14.60	3.94	167,024	1	146	167,170	168,776
NV	479	13.00	3.66	62,270	1	130	62,400	63,050
NH	320	16.00	4.35	51,200	0.3	48	51,248	51,680
NJ	230	16.10	4.44	37,030	1	161	37,191	37,513
NM	5,504	14.80	4.13	814,592	15	2,220	816,812	823,028
NY	11,651	15.80	4.33	1,840,858	7	1,106	1,841,964	1,860,924
NC	1,147	17.10	4.70	196,137	6	1,026	197,163	199,044
ND	632	14.20	3.87	89,744	1	142	89,886	91,448
OH	4,294	15.20	4.08	652,688	5	760	653,448	656,488
OK	1,279	15.90	4.40	203,361	1	159	203,520	205,587
OR	1,709	15.50	4.23	264,895	1	155	265,050	266,135
PA	10,794	16.60	4.50	1,791,804	5	830	1,792,634	1,800,934
RI	22.8	16.40	4.46	3,739	....	....	3,739	3,805
SC	364	16.50	4.52	60,060	1	165	60,225	60,555
SD	1,566	15.10	4.08	236,466	2	302	236,768	238,580
TN	1,330	16.20	4.44	215,460	1	162	215,622	216,270
TX	5,076	15.80	4.32	802,008	2	316	802,324	805,642
UT	1,610	14.70	4.04	236,670	2	294	236,964	240,345
VT	2,647	15.80	4.29	418,226	1	158	418,384	421,386
VA	1,863	17.00	4.70	316,710	2	340	317,050	318,920
WA	5,484	15.30	4.18	839,052	1	153	839,205	843,642
WV	246	15.80	4.38	38,868	1	158	39,026	39,342
WI	21,914	14.80	3.97	3,243,272	45	6,660	3,249,932	3,285,452
WY	61.5	14.20	3.89	8,733	0.2	28	8,761	8,946
US	164,072	15.05	4.10	24,694,531	180	27,449	24,721,980	24,886,929

[1] Cash receipts divided by milk or milkfat in combined marketings. [2] Value at averaged returns per 100 pounds of milk in combined marketings of milk and cream. [3] Cash receipts from marketings of milk and cream plus value of milk used for home consumption. [4] Includes value of milk fed to calves.

NASS, Livestock Branch, (202) 720–3570.

Table 8-16.—Milk Production: Marketings, income and value, United States, 1992–2001

Year	Combined marketings of milk and cream				Used for milk, cream, and butter on farms where produced		Gross farm income from dairy products[4]	Farm value of all milk produced[3 5]
	Milk utilized	Average returns[2]: Per 100 pounds milk	Average returns[2]: Per pound milkfat	Cash receipts from marketings	Milk utilized	Value[3]		
	Million pounds	*Dollars*	*Dollars*	*1,000 dollars*	*Million pounds*	*1,000 dollars*	*1,000 dollars*	*1,000 dollars*
1992	148,955	13.25	3.60	19,735,601	455	61,470	19,797,071	19,989,087
1993	148,891	12.93	3.53	19,258,442	415	54,699	19,313,144	19,487,104
1994	151,944	13.15	3.59	19,980,583	390	52,378	20,032,961	20,201,639
1995	153,737	12.93	3.53	19,876,353	340	44,522	19,920,875	20,079,217
1996	152,531	14.94	4.05	22,781,435	301	45,304	22,826,739	23,002,715
1997	154,697	13.53	3.70	20,936,726	256	34,854	20,971,580	21,125,886
1998	155,943	15.46	4.22	24,114,036	244	37,842	24,151,878	24,331,981
1999	161,387	14.38	3.92	23,207,099	219	32,038	23,239,137	23,400,050
2000	166,256	12.40	3.37	20,607,889	196	24,573	20,632,462	20,770,589
2001[1]	164,072	15.05	4.10	24,694,531	180	27,449	24,721,981	24,886,929

[1] Preliminary. [2] Cash receipts divided by milk or milkfat represented in combined marketings. [3] Valued at average returns per 100 pounds of milk in combined marketings of milk and cream. [4] Cash receipts from marketings of milk and cream plus value of milk used for home consumption. [5] Includes value of milk fed to calves.

NASS, Livestock Branch, (202) 720–3570.

Table 8-17.—Milk: Quantities used and marketed by farmers, United States, 1992–2001

Year	Milk used on farms where produced			Milk marketed by producers	
	Fed to calves[1]	Consumed as fluid milk or cream	Total	Total[2]	Fluid grade[3]
	Million pounds	*Million pounds*	*Million pounds*	*Million pounds*	*Percent*
1992	1,436	455	1,892	148,993	....
1993	1,330	415	1,745	148,891	95
1994	1,267	390	1,657	151,944	95
1995	1,216	340	1,556	153,737	96
1996	1,175	301	1,476	152,531	96
1997	1,138	256	1,394	154,697	97
1998	1,162	244	1,406	155,943	97
1999	1,109	219	1,328	161,387	98
2000	1,107	196	1,303	166,256	98
2001[4]	1,084	180	1,264	164,072	98

[1] Excludes milk sucked by calves. [2] Milk sold to plants and dealers as whole milk and equivalent amounts of milk for cream. Includes milk produced by dealers' own herds and small amounts sold directly to consumers. Also includes milk produced by institutional herds. [3] Percentage of milk sold that is eligible for fluid use (grade A for fluid use in most States). Includes fluid-grade milk used in manufacturing dairy products. [4] Preliminary.

NASS, Livestock Branch, (202) 720–3570.

Table 8-18.—Federal milk order markets: Measures of growth, 1992–2001 [1]

Year	Number of markets [2]	Population of Federal milk marketing areas	Number of Handlers [2]	Number of Producers [3]	Receipts of producer milk	Producer milk used in Class I	Percentage of producer milk used in Class I
	Number	*Thousands*	*Number*	*Number*	*Million pounds*	*Million pounds*	*Percent*
1992	40	200,530	698	97,803	107,947	44,914	41.6
1993	38	199,604	675	92,934	103,979	44,805	43.1
1994	38	201,561	629	91,397	107,811	44,866	41.6
1995	33	207,548	571	88,717	108,548	45,004	41.5
1996	32	209,599	570	82,947	104,501	45,479	43.5
1997	31	208,379	570	78,422	105,224	44,917	42.7
1998	31	210,484	522	72,402	99,223	44,968	45.3
1999	31	212,118	487	69,008	104,479	45,216	43.3
2000	11	228,899	346	69,590	116,920	45,989	39.3
2001	11	231,487	350	66,423	120,223	45,887	38.2

Year	Prices at 3.5 percent butterfat content per hundredweight [4]		Receipts as percentage of milk sold to plants and dealers		Daily deliveries per producer	Gross value of receipts of producer milk [5]	
	Class I	Blend	Fluid grade	All milk		Per producer	All producers
	Dollars	*Dollars*	*Percent*	*Percent*	*Pounds*	*Dollars*	*1,000 dollars*
1992	14.57	13.12	77	73	3,017	146,452	14,323,698
1993	14.19	12.89	73	69	3,073	145,350	13,507,974
1994	14.75	13.16	75	71	3,232	156,253	14,281,193
1995	14.19	12.79	75	71	3,350	157,754	13,995,454
1996	16.19	14.64	72	69	3,442	187,713	15,570,261
1997	14.36	13.10	71	69	3,676	178,424	13,992,366
1998	16.14	14.92	66	64	3,755	202,770	14,681,340
1999	16.24	14.09	67	65	4,148	216,794	14,960,544
2000	14.24	12.11	72	70	4,590	207,913	14,468,892
2001	16.96	14.90	75	73	4,959	275,642	18,308,968

[1] Over this period, handlers elected periodically not to pool substantial volumes of milk that normally would have been pooled under Federal orders. This decision resulted from intraorder disadvantageous blend/class price relationships and qualification circumstances. This fact should be kept in mind if year-to-year comparisons are made using the various "producer milk" measures of growth. [2] End of year. [3] Average for year. [4] Prices are weighted averages. [5] Based on blend (uniform) price adjusted for butterfat content, and in later years, other milk components of producer milk.

AMS, Dairy Programs, (202) 720–7461.

Table 8-19.—Dairy products: Quantities manufactured, United States, 1997–2001

Product	1997	1998	1999	2000	2001[1]
	1,000 pounds	*1,000 pounds*	*1,000 pounds*	*1,000 pounds*	*1,000 pounds*
Butter	1,151,250	1,167,983	1,277,090	1,256,032	1,236,801
All American cheese	....	....	3,567,924	3,641,624	3,519,162
Cheddar cheese	3,285,558	3,314,652	2,809,534	2,819,023	2,746,477
Swiss cheese	207,583	206,408	220,453	229,322	245,504
Muenster cheese	100,191	94,615	80,273	85,475	81,992
Brick cheese	8,535	7,578	8,114	8,608	8,706
Limburger cheese	746	865	664	637	702
Cream and Neufchatel cheese	614,921	621,335	644,164	687,440	644,692
Hispanic cheese	68,833	75,993	86,370	96,303	102,680
Mozzarella	....	2,366,164	2,527,208	2,634,999	2,665,749
All Italian varieties of cheese	2,881,445	3,004,715	3,151,736	3,288,911	3,328,272
All other varieties of cheese	119,829	165,958	181,550	219,678	197,384
Total of all cheese	7,330,414	7,492,119	7,941,248	8,257,998	8,129,094
Cottage cheese:					
Curd[2]	458,460	465,759	464,785	460,974	454,089
Creamed[2]	359,525	366,761	360,551	371,460	371,525
Lowfat[2]	346,687	361,246	359,315	363,658	370,884
Sweetened condensed milk:					
Bulk goods:					
Skimmed	46,347	37,810	37,510	34,611	32,616
Unskimmed	134,636	79,766	74,104	70,803	70,212
Unsweetened condensed milk:					
Bulk goods:					
Skimmed	1,000,188	1,053,762	1,163,789	1,021,907	931,250
Unskimmed	93,851	91,216	89,294	74,841	70,375
Evaporated and condensed milk:					
Case goods:					
Skimmed	29,836	28,701	27,782	23,488	14,972
Unskimmed	549,129	461,520	470,021	441,986	452,846
Condensed or evaporated buttermilk	30,867	29,771	26,424	19,963	35,063
Dry buttermilk	49,308	50,093	52,107	56,245	51,712
Dry whole milk	122,091	142,523	117,991	111,377	41,307
Nonfat dry milk	1,217,562	1,135,383	1,359,660	1,451,751	1,413,799
Dry skim milk (animal feed)	5,291	4,330	4,817	5,567	5,507
Dry whey	1,136,824	1,178,261	1,147,388	1,187,903	1,045,792
Yogurt plain & fruit flavored	1,574,050	1,638,992	1,717,181	1,836,591	1,998,613
	1,000 gallons	*1,000 gallons*	*1,000 gallons*	*1,000 gallons*	*1,000 gallons*
Ice cream, regular[3]	913,770	935,080	972,210	979,645	981,131
Ice cream, lowfat[4]	385,689	406,679	381,236	373,383	406,746
Ice cream, nonfat	40,640	42,802	39,859	30,735	21,458
Sherbet (does not include water ices)	53,077	54,592	54,479	51,933	52,414
Frozen yogurt	92,167	97,246	90,851	94,478	79,599

[1] Preliminary. [2] Cottage cheese curd includes pot and bakers' cheese. Creamed cottage cheese contains not less than 4 percent milkfat. Lowfat cottage cheese contains less than 4 percent milkfat. [3] Contains minimum milkfat content of 10 percent and not less than 4.5 pounds per gallon. [4] Includes freezer-made milkshake in most States. Contains less than 10 percent milkfat required for ice cream.

NASS, Livestock Branch, (202) 720–3570.

Table 8-20.—Dairy Products: Factory production of specified items, by States, 2000–2001

State	Butter		Total American cheese[2]		Total cheese[3]	
	2000	2001[1]	2000	2001[1]	2000	2001[1]
	1,000 pounds	*1,000 pounds*	*1,000 pounds*	*1,000 pounds*	*1,000 pounds*	*1,000 pounds*
CA	360,013	341,101	641,662	720,311	1,496,653	1,619,391
ID			505,031	470,945	585,228	569,419
IL					99,614	102,690
IA			133,331	131,188	264,175	263,552
MA					1,389	874
MN	57,698		651,560	557,277	693,068	594,085
MO					123,115	99,104
NY	21,819	28,087	95,880	98,476	728,305	702,752
OH			29,074	28,512	139,094	146,184
OR			62,008	70,258		
PA	84,869	74,378			379,671	366,972
SD			73,572	81,182	147,641	146,897
UT			38,047		74,795	62,596
WA	122,012	102,110	142,244	142,776	159,072	146,715
WI	287,271	324,675	906,128	833,232	2,192,171	2,133,156
Other	322,350	366,450	363,087	385,005	1,174,007	1,174,707
US	1,256,032	1,236,801	3,641,624	3,519,162	8,257,998	8,129,094

State	Nonfat dry milk for human food[4]		Total ice cream, regular	
	1999	2000[1]	1999	2000[1]
	Pounds	*Pounds*	*Pounds*	*Pounds*
CA	692,954	694,481	137,534	145,128
FL			23,395	23,413
ID		88,162		
IN			75,409	81,255
KY				5,236
MI			22,781	22,494
MN			44,080	44,609
MO			21,849	22,148
NM	53,141			
NY			38,079	38,226
NC			17,761	21,800
OH			38,071	36,933
OR			11,380	13,478
PA			52,695	52,789
SD			40	52
TN			18,352	19,594
TX			63,542	62,450
UT			16,673	18,865
WA	184,463	187,346	17,878	16,212
Other	521,193	443,810	380,126	356,449
US	1,451,751	1,413,799	979,645	981,131

[1] Preliminary. [2] Includes Colby, washed curd, high and low moisture Jack, and Monterey. [3] Includes full-skim American cheese; excludes cottage cheese. [4] MN and WI data combined and reported under MN.

NASS, Livestock Branch, (202) 720–3570.

Table 8-21.—Milk cows, milk, and fat in cream: Average prices received by farmers, United States, 1992–2001

Year	Milk cows, per head[1]	Milk per 100 pounds[2]					
		Eligible for fluid market[3]		Of manufacturing grade		All milk wholesale	
		Price per 100 lb.	Fat test	Price per 100 lb.	Fat test	Price per 100 lb.	Fat test
	Dollars	*Dollars*	*Percent*	*Dollars*	*Percent*	*Dollars*	*Percent*
1992	1,130.00	13.19	3.68	11.91	3.73	13.15	3.68
1993	1,160.00	12.88	3.66	11.80	3.76	12.84	3.66
1994	1,170.00	13.02	3.65	11.85	3.76	13.01	3.66
1995	1,130.00	12.80	3.65	11.79	3.75	12.78	3.66
1996	1,090.00	14.79	3.69	13.43	3.78	14.75	3.69
1997	1,100.00	13.40	3.65	12.17	3.77	13.36	3.66
1998	1,120.00	15.50	3.65	14.24	3.77	15.46	3.66
1999	1,280.00	14.42	3.67	12.84	3.79	14.38	3.67
2000	1,340.00	12.44	3.68	10.52	3.79	12.40	3.68
2001	1,500.00	15.09	3.67	13.44	3.78	15.05	3.67

[1] Simple average of quarterly prices, by States, weighted by the number of milk cows on farms Jan. 1 of the current year. [2] Average price at average fat test for all milk sold at wholesale to plants and dealers, based on reports from milk-market administrators, cooperative milk-market associations, whole-milk distributors, and milk-products manufacturing plants, f.o.b. plant or receiving station (whichever is the customary place for determining prices) before hauling costs are deducted and including all premiums. [3] Milk eligible for fluid market, including surplus diverted to manufacturing.

NASS, Livestock Branch, (202) 720–3570.

Table 8-22.—Dairy products: Manufacturers' average selling price[1] of specified products, United States, 1992–2001

Year	Dry skim milk for animal feed, per pound, f.o.b. factory	Dry whole milk, per pound, f.o.b. factory
	Cents	*Cents*
1992	56.77	116.11
1993	55.80	113.02
1994	52.54	114.18
1995	50.13	112.70
1996	60.22	128.32
1997	59.49	117.60
1998	51.14	129.47
1999	51.92	125.59
2000	54.32	120.16
2001	....	134.48

[1] Includes milk sold in bulk and in package.

NASS, Livestock Branch, (202) 720–3570.

Table 8-23.—Dairy products: Manufacturers' stocks, end of month, United States, 2000 and 2001

Month	Evaporated and sweetened condensed whole milk (case goods)		Dry whole milk		Nonfat dry milk (human food)	
	2000	2001	2000	2001	2000	2001
	1,000 pounds	*1,000 pounds*	*1,000 pounds*	*1,000 pounds*	*1,000 pounds*	*1,000 pounds*
January	41,342	34,420	6,635	4,032	175,289	146,348
February	44,467	43,272	6,989	2,944	198,129	137,771
March	49,913	53,453	5,332	2,964	190,200	123,481
April	53,063	58,969	8,715	3,534	231,546	127,114
May	57,080	64,606	13,641	3,090	226,017	134,039
June	62,442	75,085	12,692	1,982	203,938	166,665
July	65,252	78,214	11,755	2,163	208,287	147,202
August	65,312	79,532	9,813	2,496	178,983	108,956
September	64,130	82,473	8,470	2,589	154,438	101,889
October	41,623	56,914	5,565	2,921	145,720	102,151
November	42,615	48,278	4,853	3,026	133,340	102,785
December	40,746	40,264	4,390	2,894	146,348	124,458

NASS, Livestock Branch, (202) 720–3570.

Table 8-24.—Milk markets under Federal order program: Whole milk and fat-reduced milk products sold for fluid consumption within defined marketing areas, 2000 [1]

Federal milk order marketing area	Whole milk products [2]		Fat-reduced milk products [3]		Total fluid milk products	
	Quantity	Butterfat content	Quantity	Butterfat content	Quantity	Butterfat content
	Million pounds	*Percent*	*Million pounds*	*Percent*	*Million pounds*	*Percent*
Northeast	4,295	3.18	5,484	1.14	9,779	2.04
Appalachian	1,440	3.26	2,136	1.31	3,577	2.09
Southeast	2,142	3.28	2,804	1.32	4,946	2.17
Florida	1,304	3.32	1,497	1.16	2,801	2.17
Mideast	1,770	3.28	4,857	1.37	6,627	1.88
Upper Midwest	785	3.26	3,602	1.20	4,388	1.57
Central	1,217	3.27	3,427	1.30	4,644	1.82
Southwest	2,065	3.29	2,098	1.35	4,163	2.31
Arizona-Las Vegas	405	3.31	807	1.38	1,212	2.02
Western	153	3.39	737	1.41	890	1.75
Pacific	404	3.43	1,780	1.36	2,184	1.75
Combined areas	15,981	3.26	29,229	1.28	45,210	1.98

[1] In-area sales include total sales in each of the areas by handlers regulated under the respective order, by handlers regulated under other orders, by partially regulated handlers, and by producer-handlers. Sales routes of handlers may extend outside defined marketing areas; therefore, some handlers' in-area sales are partially estimated. [2] Plain, flavored, and miscellaneous whole milk products and eggnog. [3] Plain, fortified and flavored reduced fat milk (2%), low fat milk (1%), and fat-free milk (skim), and miscellaneous fat-reduced milk products, and buttermilk.

AMS, Dairy Programs, (202) 720–7461.

Table 8-25.—Milk markets under Federal order program: Whole milk and fat-reduced milk products sold for fluid consumption within defined marketing areas, 2001 [1]

Federal milk order marketing area	Whole milk products [2]		Fat-reduced milk products [3]		Total fluid milk products	
	Quantity	Butterfat content	Quantity	Butterfat content	Quantity	Butterfat content
	Million pounds	*Percent*	*Million pounds*	*Percent*	*Million pounds*	*Percent*
Northeast	4,171	3.26	5,601	1.13	9,772	2.04
Appalachian	1,349	3.31	2,115	1.33	3,464	2.10
Southeast	2,141	3.29	2,784	1.34	4,925	2.19
Florida	1,302	3.33	1,520	1.17	2,821	2.17
Mideast	1,690	3.29	4,693	1.37	6,383	1.88
Upper Midwest	783	3.28	3,591	1.19	4,374	1.56
Central	1,224	3.29	3,416	1.30	4,640	1.82
Southwest	2,051	3.31	2,118	1.35	4,169	2.32
Arizona-Las Vegas	433	3.29	810	1.38	1,242	2.05
Western	157	3.39	732	1.38	889	1.74
Pacific	416	3.43	1,768	1.36	2,184	1.75
Combined areas	15,715	3.29	29,148	1.28	44,863	1.98

[1] In-area sales include total sales in each of the areas by handlers regulated under the respective order, by handlers regulated under other orders, by partially regulated handlers, and by producer-handlers. Sales routes of handlers may extend outside defined marketing areas; therefore, some handlers' in-area sales are partially estimated. [2] Plain, flavored, and miscellaneous whole milk products and eggnog. [3] Plain, fortified, and flavored reduced fat milk (2%), low fat milk (1%), and fat-free milk (skim), miscellaneous fat-reduced milk products, and buttermilk.

AMS, Dairy Programs, (202) 720–7461.

Table 8-26.—Milk: Supply and utilization, United States, 1992–2001 [1]

Item	1992	1993	1994	1995	1996
	Million pounds	*Million pounds*	*Million pounds*	*Million pounds*	*Million pounds*
Milk production	150,847	150,636	153,602	155,292	154,006
Net imports of ingredients (milk equivalent) [3]	145	127	148	144	185
Net change in storage cream (milk equivalent) [4]	-5	5	-1	1	NA
Total supply	150,987	150,768	153,749	155,437	154,191
Utilization (milk equivalent):					
Used in manufactured products:					
Creamery butter, total milk equivalent	30,478	29,493	29,127	28,388	26,187
Milk equivalent of butter from whey cream	4,150	4,500	4,592	4,735	4,911
Net milk equivalent	26,328	24,993	24,535	23,653	21,276
Cheese:					
American	29,263	29,415	29,651	31,175	32,551
Other	20,195	20,456	21,492	21,412	21,386
Cottage cheese, creamed	592	557	524	494	461
Canned milk:					
Evaporated and sweetened condensed	1,872	1,178	1,184	1,049	1,013
Bulk condensed whole milk:					
Unsweetened	417	244	205	203	242
Sweetened	301	324	277	254	266
Dry whole milk	1,227	1,130	1,227	1,262	83
Ice cream and other frozen dairy products, total milk equivalent	14,191	14,058	15,265	15,094	15,248
Milk equivalent of butter and condensed milk used in ice cream	2,367	1,995	2,083	2,053	2,058
Net milk equivalent	11,824	12,063	13,182	13,041	13,190
Other manufactured products [5]	188	199	216	252	217
Total manufactured products	92,207	90,559	92,493	92,795	91,585
Total available for fluid products [6]	55,294	55,319	54,849	56,408	56,864
Used on farms where produced:					
Fed to calves	1,436	1,330	1,267	1,216	1,175
Used for farm-churned butter	455	415	390	340	301
Total used on farms	1,892	1,745	1,657	1,556	1,476
Residual [7]	1,594	3,145	4,750	4,678	4,266
Total utilization	150,987	150,768	153,749	155,437	154,191

See footnotes at end of table.

Table 8-26.—Milk: Supply and utilization, United States, 1992–2001 [1]—Continued

Item	1997	1998	1999	2000	2001 [2]
	Million pounds	*Million pounds*	*Million pounds*	*Million pounds*	*Million pounds*
Milk production	156,091	157,348	162,716	167,559	165,336
Net imports of ingredients (milk equivalent) [3]	168	656	429	237	688
Net change in storage cream (milk equivalent) [4]	NA	NA	NA	NA	NA
Total supply	156,259	158,004	163,145	167,796	166,024
Utilization (milk equivalent):					
Used in manufactured products:					
Creamery butter, total milk equivalent	25,714	26,211	28,657	28,059	27,675
Milk equivalent of butter from whey cream	4,966	5,094	5,392	5,538	5,525
Net milk equivalent	20,747	21,118	23,265	22,521	22,150
Cheese:					
American	32,760	33,159	35,614	36,267	35,128
Other	22,959	23,668	24,540	25,990	25,960
Cottage cheese, creamed	(8)	(8)	(8)	(8)	(8)
Canned milk:					
Evaporated and sweetened condensed	1,208	1,017	1,037	965	991
Bulk condensed whole milk:					
Unsweetened	227	222	216	180	171
Sweetened	314	186	171	163	163
Dry whole milk	898	1,050	868	815	304
Ice cream and other frozen dairy products, total milk equivalent	15,970	16,452	16,675	16,665	16,984
Milk equivalent of butter and condensed milk used in ice cream	2,112	2,151	2,305	2,218	2,240
Net milk equivalent	13,859	14,301	14,370	14,447	14,744
Other manufactured products [5]	686	697	682	700	701
Total manufactured products	93,657	95,417	100,761	102,047	100,311
Total available for fluid products [6]	56,674	55,267	55,712	55,517	55,097
Used on farms where produced:					
Fed to calves	1,138	1,162	1,109	1,107	1,084
Used for farm-churned butter	256	244	219	196	180
Total used on farms	1,394	1,406	1,328	1,303	1,264
Residual [7]	4,534	5,914	5,344	8,929	9,352
Total utilization	156,259	158,004	163,145	167,796	166,024

[1] Milk production plus the milk equivalent (ME) of net imports of dairy manufacturing ingredients and net out-of-storage movement of cream equals total supply. Data on milk used in manufacturing products are based on total milkfat required to produce the products and include the milk equivalent of any imported dairy ingredients used. [2] Preliminary. [3] Whole milk equivalent (milkfat basis) of net imports of dairy manufacturing ingredients such as frozen cream and butterfat-sugar mixtures. [4] Whole milk equivalent of net out-of-storage movement of fluid and plastic cream during the year. A negative figure represents a net movement of cream into storage. [5] Includes cottage cheese, anhydrous milkfat, butter oil, butterine, and cottage cheese beginning in 1997. [6] Total sales in U.S. (USDA-AMS). [7] Residual, includes minor miscellaneous uses and any inaccuracies in production, utilization estimates, or milk equivalent conversions. [8] Cottage cheese included in other manufactured products beginning 1997. NA-not available.

NASS, Livestock Branch, (202) 720–3570.

Table 8-27.—Milk markets under Federal order program: Blend and Class I milk prices at 3.5 percent fat test, number of producers, producer deliveries, producer deliveries used in Class I, Class I percentage, average daily deliveries per producer, average fat test of milk delivered by markets, 2000

Federal milk order marketing area	Class I price per hundredweight[1]	Blend price per hundredweight[1] [2]	Average number of producers	Total producer deliveries	Producer deliveries used in Cl. I	Class I utilization	Daily deliveries per producer	Average fat test
	Dollars	*Dollars*	*Number*	*Million pounds*	*Million pounds*	*Percent*	*Pounds*	*Percent*
Northeast[3]	14.81	13.02	17,279	23,970	10,513	43.9	3,789	3.71
Appalachian[4]	14.65	13.95	4,213	6,318	4,343	68.8	4,097	3.67
Southeast[5]	14.65	13.59	5,066	7,487	4,867	65.0	4,045	3.64
Florida[6]	15.53	15.22	305	2,867	2,526	88.1	26,128	3.60
Mideast[7] [8]	13.55	12.09	10,030	14,181	6,716	47.4	3,866	3.72
Upper Midwest[8] [9]	13.34	10.55	19,147	23,415	4,092	17.5	3,338	3.73
Central[8] [10]	13.56	11.28	10,709	16,037	4,875	30.4	4,108	3.69
Southwest[8] [11]	14.55	12.57	930	8,712	3,970	45.6	25,793	3.61
Arizona-Las Vegas[12]	13.90	11.81	122	3,110	973	31.3	69,762	3.62
Western[8] [13]	13.45	11.17	743	4,048	1,014	25.1	14,950	3.60
Pacific[8] [14]	13.46	11.70	1,047	6,776	2,100	31.0	17,833	3.65
All markets combined	14.24	12.11	69,590	116,920	45,989	39.3	4,590	3.69

[1] Prices are for milk of 3.5 percent butterfat content and for the principal pricing point of the market. See footnotes 3–14. [2] For those orders that use the component pricing system for paying producers (orders 1, 30, 32, 33, 124, 126, and 135), the figures are the statistical uniform price (the sum of the producer price differential and the Class III price). For those orders that use the skim milk/butterfat pricing system for paying producers (orders 5, 6, 7, and 131), the figures are the uniform price (the sum of the uniform butterfat price times 3.5 and the uniform skim milk price times 0.965). [3] Suffolk Co. (Boston), MA. [4] Mecklenburg Co. (Charlotte), NC. [5] Fulton Co. (Atlanta), GA. [6] Hillsborough Co. (Tampa), Fl. [7] Cuyahoga Co. (Cleveland), OH. [8] Due to disadvantageous intraorder class and uniform price relationships in some months in these markets, handlers elected not to pool milk that normally woulld have been pooled under these orders. [9] Cook Co. (Chicago), IL. [10] Jackson Co. (Kansas City), MO. [11] Dallas Co. (Dallas), TX. [12] Maricopa Co. (Phoenix), Az. [13] Salt Lake Co. (Salt Lake City), UT. [14] King Co. (Seattle), WA.

AMS, Dairy Programs, (202) 720-7461.

Table 8-28.—Milk markets under Federal order program: Blend and Class I milk prices at 3.5 percent fat test, number of producers, producer deliveries, producer deliveries used in Class I, Class I percentage, average daily deliveries per producer, average fat test of milk delivered by markets, 2001

Federal milk order marketing area	Class I price per hundredweight[1]	Blend price per hundredweight[1] [2]	Average number of producers	Total producer deliveries	Producer deliveries used in Cl. I	Class I utilization	Daily deliveries per producer	Average fat test
	Dollars	*Dollars*	*Number*	*Million pounds*	*Million pounds*	*Percent*	*Pounds*	*Percent*
Northeast[3]	17.52	15.68	17,165	24,557	10,642	43.3	3,920	3.68
Appalachian[4]	17.39	16.31	4,232	6,673	4,352	65.2	4,318	3.65
Southeast[5]	17.36	16.07	4,942	7,769	4,805	61.9	4,316	3.64
Florida[6]	18.21	17.76	290	2,772	2,492	89.9	29,845	3.60
Mideast[7] [8]	16.27	14.58	11,395	17,229	6,633	38.5	4,134	3.68
Upper Midwest[8] [9]	16.08	13.70	14,585	20,062	3,969	19.8	3,776	3.72
Central[8] [10]	16.27	14.21	10,912	17,836	4,881	27.4	4,487	3.69
Southwest[8] [11]	17.29	15.48	810	8,604	4,029	46.8	29,184	3.62
Arizona-Las Vegas[12]	16.61	14.43	116	2,956	953	32.2	70,119	3.62
Western[8] [13]	16.19	14.16	791	4,677	1,033	22.1	16,188	3.61
Pacific[8] [14]	16.18	14.32	1,185	7,088	2,098	29.6	16,424	3.66
All markets combined	16.96	14.90	66,423	120,223	45,887	38.2	4,959	3.67

[1] Prices are for milk of 3.5 percent butterfat content and for the principal pricing point of the market. [2] For those orders that use the component pricing system for paying producers (orders 1, 30, 32, 33, 124, 126, and 135), the figures are the statistical uniform price (the sum of the producer price differential and the Class III price). For those orders that use the skim milk/butterfat pricing system for paying producers (orders 5, 6, 7, and 131), the figures are the uniform price (the sum of the uniform butterfat price times 3.5 and the uniform skim milk price times 0.965). [3] Suffolk Co. (Boston), MA. [4] Mecklenburg Co. (Charlotte), NC. [5] Fulton Co. (Atlanta), GA. [6] Hillsborough Co. (Tampa), Fl. [7] Cuyahoga Co. (Cleveland), OH. [8] Due to disadvantageous intraorder class and uniform price relationships in some months in these markets, handlers elected not to pool milk that normally woulld have been pooled under these orders. [9] Cook Co. (Chicago), IL. [10] Jackson Co. (Kansas City), MO. [11] Dallas Co. (Dallas), TX. [12] Maricopa Co. (Phoenix), Az. [13] Salt Lake Co. (Salt Lake City), UT. [14] King Co. (Seattle), WA.

AMS, Dairy Programs, (202) 720-7461.

Table 8-29.—Fluid milk and cream: Total and per capita consumption, United States, 1992–2001 [1]

Year	Consumption	
	Total	Per capita
	Billion pounds	*Pounds*
1992	58.7	229
1993	58.1	224
1994	58.5	223
1995	58.8	221
1996	59.2	220
1997	59.0	216
1998	58.9	213
1999	59.5	213
2000	59.3	210
2001	59.2	208

[1] Sales of beverage, cream, and specialty fluid products plus farm household use.

ERS, Animal Products Branch, (202) 694–5180.

Table 8-30.—Dairy products: Average price per pound for specified products, 1997–2001

Item and market	1997	1998	1999	2000	2001
	Dollars	*Dollars*	*Dollars*	*Dollars*	*Dollars*
Butter, Chicago Mercantile Exchange:					
Grade AA:					
Average [1]	1.1625	1.7685	1.2396	1.1768	1.6630
High [2]	1.9500	2.8100	1.5300	1.8525	2.2250
Low [2]	0.9000	1.1250	0.8750	0.8750	1.1275
Cheese, Cheddar, f.o.b. Midwest Commodity Cheddar: [3]					
Barrels	1.2523	1.5174	1.3644	1.1170	1.4116
40-pound blocks	1.3240	1.5808	1.4227	1.1615	1.4492
Cheese, cheddar, National Cheese Exchange, Inc.: [2 4]					
Barrels:					
High	1.3625	1.8600	1.8850	1.2900	1.6825
Low	1.1225	1.2000	1.0975	0.9900	1.0650
40-pound blocks:					
High	1.4350	1.9000	1.9725	1.3350	1.7800
Low	1.1500	1.1800	1.1000	0.9800	1.0675
Nonfat dry milk, f.o.b.plant, Western Production Area:					
Low/medium heat	1.0715	1.0549	1.0128	1.0056	0.9860
High heat	1.0979	1.0852	1.0450	1.0419	1.0281
Whey Powder, f.o.b. plant, Western Production Area:					
Edible (nonhygroscopic)	0.2260	0.2373	0.1924	0.1852	0.2593

[1] For 1997, Chicago wholesale selling prices. [2] Prices represent high and low for the year. [3] Effective June 1, 1998, this price series was renamed from Wisconsin Assembly Points. [4] Trading on the National Cheese Exchange ended on April 25, 1997. After this date, trading of Cheddar cheese occurred on the Chicago Mercantile Exchange.

AMS, Dairy Programs, (202) 720–7461.

Table 8-31.—Dairy products: Total disappearance, and total and per capita consumption, United States, 1992–2001 [1]

Year	Butter			Cheese [2]			Condensed and evaporated milk [3]		
	Total disappearance	Consumption		Total disappearance	Consumption		Total disappearance	Consumption	
		Total	Per capita		Total	Per capita		Total	Per capita
	Million pounds	*Million pounds*	*Pounds*	*Million pounds*	*Million pounds*	*Pounds*	*Million pounds*	*Million pounds*	*Pounds*
1992	1,464	1,114	4.3	6,720	6,641	25.9	872	820	3.2
1993	1,530	1,209	4.6	6,853	6,774	26.0	843	785	3.0
1994	1,463	1,255	4.8	7,095	6,994	26.5	734	669	2.5
1995	1,329	1,186	4.4	7,279	7,174	26.9	699	608	2.3
1996	1,190	1,148	4.3	7,477	7,364	27.3	696	617	2.3
1997	1,156	1,116	4.1	7,647	7,511	27.5	773	695	2.5
1998	1,217	1,208	4.4	7,797	7,662	27.7	638	553	2.0
1999	1,314	1,307	4.7	8,219	8,086	29.0	648	573	2.1
2000	1,289	1,277	4.5	8,581	8,409	29.8	596	560	2.0
2001	1,280	1,277	4.5	8,611	8,426	29.6	610	561	2.0

Year	Ice cream (product weight)			Dry whole milk			Nonfat dry milk (human food)		
	Total disappearance	Consumption		Total disappearance	Consumption		Total disappearance	Consumption	
		Total	Per capita		Total	Per capita		Total	Per capita
	Million pounds	*Million pounds*	*Pounds*	*Million pounds*	*Million pounds*	*Pounds*	*Million pounds*	*Million pounds*	*Pounds*
1992	4,157	4,157	16.2	167	124	.49	1,008	719	2.8
1993	4,157	4,157	16.0	157	101	.39	946	634	2.4
1994	4,205	4,205	16.0	164	103	.40	1,191	914	3.5
1995	4,139	4,139	15.5	173	106	.40	1,280	910	3.4
1996	4,217	4,217	15.6	137	97	.36	1,081	1,005	3.7
1997	4,386	4,386	16.1	126	102	.38	1,171	906	3.3
1998	4,488	4,488	16.3	149	118	.43	1,120	884	3.2
1999	4,667	4,667	16.7	124	111	.40	1,275	787	2.8
2000	4,702	4,702	16.7	119	80	.28	1,073	741	2.6
2001	4,709	4,709	16.5	50	44	.16	1,156	936	3.3

[1] Total disappearance is based on production, imports, and change in stocks during the year. Production statistics for these commodities appear in other tables in this chapter. The total apparent consumption was obtained by subtracting ending stocks, shipments, and exports, from the total supply. The per capita consumption for each year was obtained by dividing the total apparent consumption by the number of persons. [2] Includes all kinds of cheese except cottage and full-skim American. [3] The evaporated milk is unskimmed, unsweetened, case goods. The condensed milk is unsweetened, unskimmed, bulk goods; and sweetened condensed milk, unskimmed, case and bulk goods.

ERS, Animal Products Branch, (202) 694–5180.

Table 8-32.—Dairy products: Dec. 31 stocks, United States, 1992–2001

Year	Butter [1 2]	Cheese [1 3]	Canned milk [1]	Dry whole milk	Nonfat dry milk for human consumption [1]
	1,000 pounds	*1,000 pounds*	*1,000 pounds*	*1,000 pounds*	*1,000 pounds*
1992	454,211	470,496	44,308	9,144	81,198
1993	243,839	465,762	34,281	6,479	89,568
1994	80,181	437,256	47,014	9,220	131,201
1995	18,628	412,237	31,701	7,318	84,978
1996	13,707	487,174	19,937	6,422	71,414
1997	20,788	480,779	32,466	5,605	124,864
1998	25,910	517,647	36,495	5,161	152,172
1999	25,082	622,197	35,690	5,749	284,542
2000	24,115	708,597	41,228	4,390	662,182
2001	55,915	663,251	40,739	2,894	900,158

[1] Includes Government holdings. [2] Includes butter equivalent of butteroil held by CCC. [3] Excludes cottage and full-skim American cheese. Includes process American cheese held by CCC.

ERS, Animal Products Branch, (202) 694–5180.

Table 8-33.—Butter: Production in specified countries, 1999–2001

Continent and country	1999	2000	2001[1]
	1,000 metric tons	*1,000 metric tons*	*1,000 metric tons*
North America:			
Canada	89	77	82
Mexico	70	70	70
United States	579	570	561
Total	738	717	713
South America:			
Argentina	54	52	47
Brazil	70	72	78
Total	124	124	125
Europe Union	1,731	1,700	1,710
Eastern Europe:			
Poland	168	169	181
Romania	7	6	6
Total	175	175	187
Former USSR:			
Russia	260	265	270
Ukraine	108	135	156
Total	368	400	426
North Africa:			
Egypt	9	10	12
Total	909	1,010	1,212
Southeast Asia:			
India[2]	1,750	1,950	2,250
Total	1,750	1,950	2,250
Asia:			
Japan	86	88	80
Total	86	88	80
Oceania:			
Australia[3]	176	180	160
New Zealand[4]	316	344	352
Total	492	524	512
Grand total	5,474	5,688	6,015

[1] Preliminary. [2] Year ending April 1 of the year shown. [3] Year ending June 30 of the year shown. [4] Year ending May 31 of the year shown.

FAS, Dairy, Livestock, and Poultry Division, (202) 720–3761. Prepared or estimated on the basis of official statistics of foreign governments, other foreign source materials, reports of U.S. Agricultural Counselors, Attachés, and Foreign Service Officers, results of office research, and related information.

Table 8-34.—Cheese: Production in specified countries, 1999–2001

Continent and country	1999	2000	2001[1]
	1,000 metric tons	*1,000 metric tons*	*1,000 metric tons*
North America:			
Canada	329	335	322
Mexico	126	134	140
United States	3,581	3,746	3,687
Total	4,036	4,215	4,149
South America:			
Argentina	446	445	440
Brazil	434	445	460
Total	880	890	900
European Union	5,290	5,437	5,420
Eastern Europe:			
Poland	155	148	161
Romania	95	92	90
Total	250	240	251
Former USSR:			
Russia	185	220	260
Ukraine	53	67	105
Total	238	287	365
North Africa:			
Egypt	382	380	390
Total	382	380	390
Asia:			
Japan	35	34	34
Total	35	34	34
Oceania:			
Australia[2]	320	373	374
New Zealand[3]	245	297	281
Total	565	670	655
World total	11,750	12,230	12,246

[1] Preliminary. [2] Year ending June 30. [3] Year ending May 31.

FAS, Dairy, Livestock and Poultry Division, (202) 720–3761. Prepared or estimated on the basis of official statistics of foreign governments, other foreign source materials, reports of U.S. Agricultural Counselors, Attachés, and Foreign Service Officers, results of office research, and related information.

Table 8-35.—Dairy products: United States imports by country of origin, 1999–2001

Commodity and country of origin	1999	2000	2001 [1]
	Metric tons	*Metric tons*	*Metric tons*
Cheese, all types:			
Canada	7,608	4,372	4,816
Argentina	6,472	7,066	5,620
Austria	12,580	9,048	10,680
Belgium and Luxembourg	1,396	1,245	602
Denmark	14,072	13,796	13,335
Finland	7,011	8,339	7,281
France	13,631	14,678	14,741
Germany, Fed. Rep. of	12,580	9,048	10,680
Greece	13,939	2,075	2,335
Ireland	3,557	4,295	4,655
Italy	29,849	29,617	29,309
Netherlands	10,138	11,783	11,842
Portugal	511	549	539
Spain	708	1,083	1,259
Sweden	877	550	683
United Kingdom	7,487	3,610	6,126
Total EU	105,766	102,517	104,656
Norway	7,276	8,255	7,176
Switzerland	6,813	6,812	5,976
Israel	293	288	260
Czech Republic	577	78	152
Hungary	881	423	795
Poland	2,792	3,562	3,018
Australia	8,259	7,734	8,125
New Zealand	35,696	28,181	39,918
Other countries	15,164	19,419	21,259
Total	197,597	188,707	201,771
Cheese, cheddar: [2]			
Canada	2,272	1,790	1,625
Germany	277	0	20
Ireland	176	192	271
United Kingdo	538	449	600
Australia	3,567	3,442	3,431
New Zealand	23,534	15,622	25,950
Other countries	436	334	318
Total	30,800	21,829	32,215
Cheese, Swiss: [3]			
Canada	183	107	262
Austria	1,109	1,240	984
Denmark	3,417	2,585	2,765
Finland	6,908	8,124	7,153
France	1,039	1,449	2,029
Germany, Fed. Rep. of	6,500	4,665	4,888
Netherlands	429	239	361
Norway	6,886	7,606	6,632
Switzerland	4,671	4,373	3,397
Other countries	2,882	1,854	2,932
Total	34,024	32,242	31,403

[1] Preliminary. [2] Includes American and Colby cheese. [3] Includes Emmenthaler with eye-formation.

FAS, Dairy, Livestock and Poultry Division, (202) 720–3761. Compiled from reports of the U.S. Department of Commerce.

Table 8-36.—Dairy products: United States imports by type of product, 1992–2001

Year	Dried milk [1]	Cheese				Butter [4]	Casein [5]
		Swiss [2]	Cheddar [3]	Other	Total		
	Metric tons	*Metric tons*	*Metric tons*	*Metric tons*	*Metric tons*	*Metric tons*	*Metric tons*
1992	1,238	24,466	8,186	97,013	129,665	947	94,892
1993	1,268	29,265	9,130	107,346	145,741	1,706	80,827
1994	1,593	29,554	7,802	119,349	156,705	1,245	96,008
1995	1,128	28,047	9,472	116,877	154,396	697	93,433
1996	3,968	29,420	12,393	111,457	153,270	4,783	98,547
1997	6,080	25,094	11,566	104,825	141,485	10,956	102,404
1998	8,223	28,865	21,844	117,721	168,430	31,919	111,247
1999	10,557	34,024	30,800	132,771	197,597	18,059	108,271
2000	8,531	32,242	21,829	134,638	188,707	13,689	119,999
2001 [6]	8,093	31,403	32,214	138,151	201,771	34,614	106,787

[1] Includes whole and skimmed milk. [2] Includes Emmenthaler with eye-formation. [3] Includes American and Colby cheese. [4] Includes butter oil. [5] Includes caseinates. [6] Preliminary.

FAS, Dairy, Livestock, and Poultry Division, (202) 720–3761. Compiled from reports of the U.S. Department of Commerce.

Table 8-37.—Dairy products: Exports by principal exporting countries, 1999–2001[1]

Commodity and country	1999	2000	2001[2]
	Metric tons	*Metric tons*	*Metric tons*
Butter:			
United States	2	4	0
EU-15	161	176	174
Poland	3	3	19
Australia[3]	117	139	122
New Zealand[4]	279	338	347
Other	31	52	79
Total	593	712	741
Cheese:[5]			
United States	38	47	52
Canada	25	18	19
EU-15	376	436	447
Poland	12	8	21
Australia[3]	172	223	215
New Zealand[4]	240	253	267
Other	29	44	59
Total	892	1,029	1,080
Milk, dried whole:			
United States	4	15	0
EU-15	577	576	478
Poland	8	7	0
Australia[3]	139	169	183
New Zealand[4]	382	407	511
Other	179	144	207
Total	1,289	1,318	1,379
Milk, nonfat dry milk:			
Canada	41	32	46
United States	217	142	96
EU-15	272	358	137
Poland	83	86	108
Australia[3]	240	253	225
New Zealand[4]	205	175	195
Other	61	156	128
Total	1,119	1,202	935

[1] Within the European Union, exports to other members are not included. [2] Preliminary. [3] Year ending June 30. [4] Year ending May 31. [5] Excludes fresh cheese.

FAS, Dairy, Livestock and Poultry Division, (202) 720–3761. Prepared on the basis of official statistics of foreign governments, other foreign source materials, reports of U.S. Agricultural Counselors, Attachés, and Foreign Service Officers, results of office research, and related information.

Table 8-38.—Dairy products: United States exports by type of product, 1992–2001

Year	Butter	Cheese	Milk and cream			
			Evaporated and condensed	WMP—Whole dried	Nonfat dry milk	Ice cream
	Metric tons	*Metric tons*	*Metric tons*	*Metric tons*	*Metric tons*	*Metric tons*
1992	66,896	15,206	8,800	28,203	74,577	32,583
1993	92,300	16,346	10,734	37,576	75,708	30,865
1994	45,487	22,499	19,595	29,230	49,045	36,320
1995	37,689	29,519	41,378	64,297	59,311	37,827
1996	20,831	32,497	39,582	16,181	18,422	39,765
1997	14,989	37,559	9,347	48,609	62,134	36,767
1998	8,951	36,723	8,021	51,315	72,917	38,206
1999	3,208	38,341	4,821	17,656	141,315	39,701
2000	8,230	47,760	5,215	25,368	84,264	39,366
2001[1]	3,208	52,478	10,672	46,070	96,159	40,058

[1] Preliminary.

FAS, Dairy, Livestock and Poultry Division, (202) 720–3761. Compiled from reports of the U.S. Department of Commerce.

Table 8-39.—Dairy products: United States exports by country of destination, 1999–2001

Commodity and country of destination	1999	2000	2001 [1]
	Metric tons	*Metric tons*	*Metric tons*
Cheese, all types:			
Canada	5,461	9,218	5,659
Mexico	5,090	9,979	16,302
Brazil	474	233	394
Venzuela	1,460	1,432	1,334
United Kingdom	577	1,737	1,334
Saudai Arabia	441	741	529
Philippines	1,016	982	1,322
Korea	1,605	2,649	3,351
Hong Kong	461	544	977
Taiwan	390	387	656
Japan	8,710	7,528	8,445
Other countries	12,656	12,330	12,175
Total	38,341	47,760	52,478
Ice cream:			
Canada	6,851	6,080	4,440
Mexico	6,222	5,046	5,095
United Kin	5,230	6,481	8,414
Russia	73	17	46
Korea	1,084	1,347	2,082
Hong Kong	2,710	3,030	3,643
Japan	10,779	10,335	8,741
Others	6,752	7,030	7,597
Total	39,701	39,366	40,058
Milk, nonfat dry:			
Mexico	60,882	30,650	57,191
Dominican Rep.	8,093	4,893	1,652
Colombia	194	817	529
Algeria	0	1,740	497
Egypt	9,239	3,445	2,446
Thailand	3,230	2,215	1,521
Vietnam	2,869	3,069	784
Indonesia	5,116	3,116	3,860
Philippines	5,979	5,962	5,714
Taiwan	799	540	190
Others	44,914	27,817	21,775
Total	141,315	84,264	96,159
Dry whey:			
Canada	31,472	37,905	35,701
Mexico	12,028	23,330	15,493
Thailand	4,380	12,785	9,119
Philippines	11,355	13,601	13,306
China	12,431	15,819	24,414
Korea	11,106	14,204	5,484
Taiwan	5,668	8,698	9,649
Japan	10,034	10,247	7,107
Other countries	21,698	29,414	22,835
Total	120,172	166,003	143,108

[1] Preliminary.

FAS, Dairy, Livestock and Poultry Division, (202) 720–3761. Compiled from reports of the U.S. Department of Commerce.

Table 8-40.—Dairy products: Price-support operations, United States, 1993–2002

Marketing year[1]	Manufacturing milk: Support level at national average milkfat test, per cwt.	Manufacturing milk: Average price received by farmers per cwt.	Product purchase price per pound[2]: Butter[3]	Product purchase price per pound[2]: Cheddar cheese[4]	Product purchase price per pound[2]: Nonfat milk, spray process[5]
	Dollars	*Dollars*	*Cents*	*Cents*	*Cents*
1992–93	10.10	11.60	76.25	111.75	97.30
	....	....	(6) 65.00	(6) 112.00	(6) 103.40
1993–94	10.10	11.98	65.00	112.00	103.40
1994–95	10.10	11.52	65.00	112.00	103.40
1995–96	10.10	....	65.00	112.00	103.40
	(7) 10.35	(8) 13.50	65.00	(7) 114.50	(7) 106.50
1996–97	10.35	....	65.00	114.50	106.50
	(9) 10.20	11.88	65.00	(9) 113.00	(9) 104.70
1997–98	10.20	....	65.00	113.00	104.70
	(10) 10.05	13.28	65.00	(10) 111.50	(10) 102.80
1998–99	10.05	....	65.00	111.50	102.80
	(11) 9.90	14.04	65.00	(11) 110.00	(11) 101.00
1999–2000	9.90	(12) 11.00	65.00	110.00	101.00
	....	....	(13) 66.80	(13) 112.20	101.00
2000–2001	9.90	12.85	(14) 65.49	(14) 113.14	(14) 100.32
	....	....	(15) 85.48	113.14	(15) 90.00
2001–2002	9.90	11.40	85.48	113.14	90.00

[1] October 1-September 30. [2] Announced purchase prices for products in bulk containers. [3] U.S. Grade A or higher, salted, 25-kg blocks. [4] U.S. Grade A or higher, standard moisture basis 40-pound blocks. [5] U.S. Extra Grade, not more than 3.5 percent moisture content. Prices quoted are for product in 25-kg bags. [6] Effective July 7, 1993. [7] Effective January 1, 1996. [8] Basic Formula Price began May 1995 thru Sept. 1999. [9] Effective January 1, 1997. [10] Effective January 1, 1998. [11] Effective January 1, 1999. [12] Estimated value of milk used in manufactured products. [13] Effective July 31, 2000. [14] Effective January 31, 2001. [15] Effective June 13, 2001.

FSA, Dairy & Sweeteners Analysis, (202) 690–0050

Table 8-41.—Chickens: Number and value, United States, Dec. 1, 1993–2002[1]

Year	Layers 1 year old and older	Layers 20 weeks old but less than 1 year	Pullets: 13 weeks to 20 weeks old	Pullets: Under 13 weeks old	Other chickens	All chickens	Value per head	Total value
	Thousands	*Thousands*	*Thousands*	*Thousands*	*Thousands*	*Thousands*	*Dollars*	*1,000 dollars*
1993	131,688	158,938	33,833	47,941	7,240	379,640	2.37	898,059
1994	135,093	163,432	33,159	46,694	7,369	385,747	2.34	902,815
1995	133,841	165,230	34,004	47,365	7,637	388,077	2.41	934,905
1996	138,048	165,874	33,518	48,054	7,243	392,737	2.65	1,039,071
1997	140,966	171,171	35,578	54,766	7,549	410,030	2.72	1,113,183
1998	151,298	170,420	39,664	55,981	7,682	425,045	2.69	1,143,835
1999	151,914	177,406	38,587	58,775	9,661	436,343	2.65	1,154,898
2000	153,222	179,188	38,395	56,163	8,088	435,056	2.44	1,060,964
2001	153,357	185,271	42,807	51,948	8,126	441,509	2.42	1,068,851
2002[2]	153,032	184,181	39,679	53,706	8,350	438,948	2.38	1,045,222

[1] Does not include commercial broilers. [2] Preliminary.

NASS Livestock Branch, (202) 720-3570.

Table 8-42.—Chickens: Layers, by State and United States, Dec. 1, 2001–2002

State	Layers					
	Layers 1 year old and older		Layers 20 weeks old but less than 1 year		Total	
	2001	2002 [1]	2001	2002 [1]	2001	2002 [1]
	Thousands	*Thousands*	*Thousands*	*Thousands*	*Thousands*	*Thousands*
AL	3,720	3,061	6,670	6,584	10,390	9,645
AR	4,067	3,475	10,668	10,837	14,735	14,312
CA	13,471	13,575	10,292	8,967	23,763	22,542
CO	1,560	1,852	1,997	2,285	3,557	4,137
CT	1,054	1,021	2,054	2,026	3,108	3,047
DE	353	681	885	637	1,238	1,318
FL	6,420	6,060	4,408	4,523	10,828	10,583
GA	6,984	7,811	14,888	12,258	21,872	20,069
HI	373	381	181	142	554	523
ID	472	454	410	374	882	828
IL	2,014	1,679	1,640	1,912	3,654	3,591
IN	11,756	12,291	10,744	10,727	22,500	23,018
IA	16,547	16,588	18,310	21,161	34,857	37,749
KY	1,600	1,375	2,275	2,975	3,875	4,350
LA	945	850	1,111	1,024	2,056	1,874
ME	1,947	1,711	2,069	2,474	4,016	4,185
MD	1,411	1,775	1,955	1,605	3,366	3,380
MA	78	68	215	214	293	282
MI	4,491	5,149	2,243	1,682	6,734	6,831
MN	5,560	6,061	6,655	5,399	12,215	11,460
MS	2,028	1,435	4,722	5,268	6,750	6,703
MO	3,908	3,489	2,955	3,272	6,863	6,761
MT	75	85	265	270	340	355
NE	6,587	6,446	4,907	5,022	11,494	11,468
NH	11	58	132	90	143	148
NJ	1,309	1,039	833	1,030	2,142	2,069
NY	1,326	1,526	2,520	2,170	3,846	3,696
NC	3,616	3,352	7,539	7,224	11,155	10,576
OH	15,157	16,975	15,133	14,020	30,290	30,995
OK	1,468	1,623	2,404	2,571	3,872	4,194
OR	1,565	1,555	1,364	1,228	2,929	2,783
PA	9,403	7,802	14,274	16,378	23,677	24,180
SC	2,775	1,816	3,030	3,480	5,805	5,296
SD	960	948	1,400	1,400	2,360	2,348
TN	210	325	1,155	931	1,365	1,256
TX	7,434	7,247	11,601	11,528	19,035	18,775
UT	1,724	1,781	1,788	1,571	3,512	3,352
VT	84	99	94	95	178	194
VA	1,054	980	2,245	2,182	3,299	3,162
WA	3,120	3,406	1,848	1,756	4,968	5,162
WV	31	86	1,090	1,020	1,121	1,106
WI	2,450	2,500	2,288	1,840	4,738	4,340
WY	8	8	4	4	12	12
Other States [2]	2,231	2,533	2,010	2,025	4,241	4,558
US	153,357	153,032	185,271	184,181	338,628	337,213

[1] Preliminary. [2] AK, AZ, KS, NV, NM, ND, and RI combined to avoid disclosing individual operations.
NASS, Livestock Branch, (202) 720–3570.

Table 8-43.—Chickens: Pullets not of laying age and other chickens, by State and United States, Dec. 1, 2001–2002

State	Pullets 13 weeks old and older but less than 20 weeks		Pullet chicks and pullets under 13 weeks of age		Other chickens	
	2001	2002 [1]	2001	2002 [1]	2001	2002 [1]
	Thousands	*Thousands*	*Thousands*	*Thousands*	*Thousands*	*Thousands*
AL	2,132	2,148	2,281	2,237	1,100	1,226
AR	3,037	2,460	4,321	4,715	1,273	1,300
CA	1,572	1,670	2,629	2,043	78	68
CO	142	156	456	412	55	72
CT	204	241	428	501	8	4
DE	15	59	57	93	22	14
FL	1,024	901	1,202	1,422	69	106
GA	2,750	2,922	5,150	5,022	1,401	1,540
HI	50	44	55	58	0	0
ID	130	185	207	225	6	7
IL	128	122	337	268	17	21
IN	3,582	2,393	2,790	3,625	86	62
IA	3,375	4,987	3,835	3,454	41	50
KY	835	845	690	805	200	195
LA	231	295	327	315	86	76
ME	726	427	859	1,173	6	5
MD	507	468	368	416	73	48
MA	3	2	53	52	0	0
MI	285	606	985	664	1	0
MN	2,090	820	1,439	2,355	59	50
MS	1,594	1,313	1,492	2,134	552	664
MO	779	816	640	793	180	185
MT	72	61	67	63	1	1
NE	539	963	1,671	1,248	0	0
NH	29	14	35	59	8	8
NJ	204	0	0	89	0	0
NY	319	450	550	730	15	16
NC	2,549	2,639	3,179	2,825	1,238	1,002
OH	3,785	3,296	3,900	3,568	34	24
OK	632	380	666	724	225	292
OR	311	195	457	415	7	7
PA	3,479	2,362	2,094	2,623	78	92
SC	311	389	674	745	156	168
SD	206	158	211	171	0	0
TN	264	317	435	453	186	174
TX	2,717	2,452	3,921	3,851	428	428
UT	151	407	350	93	2	1
VT	24	4	3	1	2	2
VA	569	659	443	507	243	245
WA	620	179	782	1,141	2	2
WV	346	310	346	400	140	150
WI	217	330	1,105	891	40	39
WY	2	2	2	2	1	1
Other States [2]	270	232	456	325	7	5
US	42,807	39,679	51,948	53,706	8,126	8,350

[1] Preliminary. [2] AK, AZ, KS, NV, NM, ND, and RI combined to avoid disclosing individual operations.

NASS, Livestock Branch, (202) 720-3570.

Table 8-44.—Chickens: Number, value per head, and total value, by State and United States, Dec. 1, 2001–2002 [1]

State	Number		Value per bird		Total value	
	2001	2002 [2]	2001	2002 [2]	2001	2002 [2]
	1,000 head	*1,000 head*	*Dollars*	*Dollars*	*1,000 dollars*	*1,000 dollars*
AL	15,903	15,256	4.90	3.60	77,925	54,922
AR	23,366	22,787	4.40	4.70	102,810	107,099
CA	28,042	26,323	2.00	1.90	56,084	49,871
CO	4,210	4,777	1.90	2.00	7,999	9,820
CT	3,748	3,793	2.10	2.30	7,871	8,724
DE	1,332	1,484	3.60	3.00	4,795	4,452
FL	13,123	13,012	2.00	1.90	26,246	24,723
GA	31,173	29,553	3.20	3.40	99,754	100,480
HI	659	625	2.10	1.70	1,384	1,063
ID	1,225	1,245	2.10	1.80	2,573	2,241
IL	4,136	4,002	1.10	1.40	4,550	5,603
IN	28,958	29,098	1.30	1.10	37,645	32,008
IA	42,108	46,240	1.20	1.40	50,530	64,736
KY	5,600	6,195	2.80	3.20	15,680	19,824
LA	2,700	2,560	3.00	2.90	8,100	7,424
ME	5,607	5,790	2.10	2.20	11,775	12,738
MD	4,314	4,312	3.10	2.60	13,373	11,211
MA	349	336	2.70	3.10	942	1,042
MI	8,005	8,101	1.70	1.70	13,609	13,772
MN	15,803	14,685	1.60	1.30	25,285	19,091
MS	10,388	10,814	4.90	4.90	50,901	52,989
MO	8,462	8,555	2.30	2.30	19,463	19,677
MT	480	480	3.00	3.10	1,440	1,488
NE	13,704	13,679	1.80	1.80	24,667	24,622
NH	215	229	3.20	3.30	688	756
NJ	2,346	2,158	1.10	1.00	2,581	2,158
NY	4,730	4,892	1.50	1.60	7,095	7,827
NC	18,121	17,042	5.00	4.80	90,605	81,802
OH	38,009	37,883	1.50	1.30	57,014	49,248
OK	5,395	5,590	4.20	5.00	22,659	27,950
OR	3,704	3,400	1.50	2.20	5,556	7,480
PA	29,328	29,257	1.60	1.80	46,925	52,663
SC	6,946	6,598	2.20	2.70	15,281	17,815
SD	2,777	2,677	1.90	1.80	5,276	4,819
TN	2,250	2,200	6.00	6.10	13,500	13,420
TX	26,101	25,506	2.50	2.20	65,253	56,113
UT	4,015	3,853	1.30	1.70	5,220	6,550
VT	207	201	2.50	2.50	518	503
VA	4,554	4,573	4.50	4.80	20,493	21,950
WA	6,372	6,484	2.30	2.40	14,656	15,562
WV	1,953	1,966	4.40	4.70	8,593	9,240
WI	6,100	5,600	2.20	2.00	13,420	11,200
WY	17	17	3.30	3.30	56	56
Oth Sts [3]	4,974	5,120	1.62	1.66	8,061	8,490
US	441,509	438,948	2.42	2.38	1,068,851	1,045,222

[1] Excludes commercial broilers. [2] Preliminary. [3] AK, AZ, KS, NV, NM, ND, and RI combined to avoid disclosure of individual operations.

NASS, Livestock Branch, (202) 720-3570.

Table 8-45.—Poultry Meat: Total imports by specified countries, 1998–2001

Continent and country	1998	1999	2000	2001 [1]
	1,000 tons	*1,000 tons*	*1,000 tons*	*1,000 tons*
Russian Federation	1,020	930	943	1,281
Japan	590	667	721	710
China, Peoples Republic	427	591	608	473
European Union	167	198	299	418
Saudi Arabia	287	364	348	399
Mexico	181	188	219	235
Hong Kong	269	391	239	234
United Arab Emirates	122	117	111	123
Korea, Republic of	13	47	67	83
Canada	55	62	69	73
South Africa, Republic	71	80	72	64
Kuwait	45	54	53	62
Romania	44	19	30	58
Ukraine	44	62	12	35
Malaysia	14	26	28	35
Argentina	63	52	42	24
Hungary	4	2	16	16
Bulgaria	13	14	14	14
Philippines	2	12	18	12
Poland	38	9	8	11
Colombia	14	14	11	11
Taiwan	8	14	12	9
United States	3	2	3	6
Egypt	2	2	2	4
Indonesia	0	4	14	1
Others	59	65	70	0
Grand total	3,555	3,986	4,029	4,391

[1] Estimated.
Prepared or estimated on the basis of official statistics of foreign governments, other foreign source materials, reports of U.S. Agricultural Counselors, Attachés, and Foreign Service Officers, inter-agency analysis, and related information.
FAS, Dairy, Livestock and Poultry Division, (202) 720–8031.

Table 8-46.—Poultry Meat: Total exports by specified countries, 1998–2001

Continent and country	1998	1999	2000	2001 [1]
	1,000 tons	*1,000 tons*	*1,000 tons*	*1,000 tons*
United States	1,978	2,080	2,231	2,521
Brazil	594	750	893	1,241
European Union [2]	788	764	762	718
China, Peoples Republic	323	375	464	489
Thailand	274	288	328	425
Canada	53	47	55	69
Hungary	52	45	35	35
Saudi Arabia	21	16	16	20
United Arab Emirates	23	19	20	20
Australia	13	12	14	19
Poland	24	20	12	13
Others	53	46	38	37
Grand total	4,196	4,462	4,868	5,607

[1] Estimated. [2] EU trade excludes intra-EU trade.
Prepared or estimated on the basis of official statistics of foreign governments, other foreign source materials, reports of U.S. Agricultural Counselors, Attachés, and Foreign Service Officers, interagency analysis, and related information.
FAS, Dairy, Livestock and Poultry Division, (202) 720–8031.

Table 8-47.—Total poultry meat: Production in specified countries, 1998–2001

Continent and country	1998	1999	2000	2001 [1]
	1,000 metric tons	*1,000 metric tons*	*1,000 metric tons*	*1,000 metric tons*
United States	12,525	13,367	13,703	14,033
European Union	6,789	6,614	6,654	6,822
Brazil	4,498	5,526	5,980	6,567
China, Peoples Republic	3,450	4,400	5,050	5,200
Mexico	1,587	1,784	1,936	2,067
India	710	820	1,080	1,250
Thailand	930	980	1,070	1,230
Japan	1,097	1,078	1,091	1,074
Canada	798	847	877	927
Argentina	850	885	870	870
Malaysia	660	684	786	813
South Africa, Republic	667	681	707	730
Taiwan	609	615	631	622
Colombia	589	577	606	606
Philippines	491	500	520	582
Australia	527	550	573	568
Indonesia	285	355	458	522
Egypt	430	500	460	450
Poland	300	355	370	450
Russian Federation	280	350	380	430
Saudi Arabia	435	370	390	424
Korea, Republic of	368	390	394	413
Venezuela	380	375	379	360
Hungary	220	220	205	200
Romania	70	104	121	130
Others	689	485	509	278
Grand total	40,234	43,412	45,800	47,618

[1] Preliminary.

Prepared or estimated on the basis of official statistics of foreign governments, other foreign source materials, reports of U.S. Agricultural Counselors, Attachés, and Foreign Service Officers, inter-agency analysis, and related information.

FAS, Production Estimates and Crop Assessment Division, (202) 720–8031.

Table 8-48.—Mature Chickens: Lost, sold for slaughter, and value of sales, 2001 (preliminary)[1]

State	Number lost[2]	Number sold[3]	Pounds sold[3]	Price per pound[3]	Value of sales[3]
	1,000 head	*1,000 head*	*1,000 pounds*	*Dollars*	*1,000 dollars*
AL	1,863	8,879	64,817	0.072	4,667
AR	2,935	14,310	97,308	0.080	7,785
CA	2,886	10,690	39,553	0.010	396
CO	960	1,457	6,119	0.002	12
CT	286	2,236	8,497	0.005	42
DE	173	1,443	7,215	0.011	79
FL	2,806	2,793	10,613	0.022	233
GA	5,264	19,630	147,225	0.078	11,484
HI	145	182	655	0.169	111
ID	90	375	1,463	0.020	29
IL	391	1,890	6,426	0.010	64
IN	1,970	10,011	34,037	0.010	340
IA	6,867	10,802	43,208	0.001	43
KS	98	641	2,244	0.001	2
KY	537	2,800	16,520	0.060	991
LA	212	1,308	7,456	0.060	447
ME	1,102	2,245	8,756	0.004	35
MD	334	2,373	9,492	0.010	95
MA	22	683	2,664	0.004	11
MI	889	3,657	12,800	0.010	128
MN	1,635	5,792	20,851	0.001	21
MS	1,295	6,371	45,871	0.082	3,761
MO	908	3,375	16,875	0.030	506
MT	35	109	392	0.010	4
NE	2,636	4,830	16,422	0.001	16
NH	11	156	889	0.039	35
NJ	168	590	2,006	0.001	2
NY	482	2,613	10,975	0.001	11
NC	1,326	11,212	86,332	0.080	6,907
OH	4,969	17,038	57,929	0.001	58
OK	733	3,250	19,500	0.050	975
OR	348	1,219	4,388	0.001	4
PA	2,495	14,983	53,939	0.010	539
SC	2,017	2,145	16,517	0.046	760
SD	874	659	2,307	0.003	7
TN	172	1,335	8,811	0.059	520
TX	2,526	14,301	68,645	0.035	2,403
UT	272	1,499	5,247	0.001	5
VT	17	113	531	0.021	11
VA	427	2,943	20,012	0.075	1,501
WA	717	2,617	9,160	0.001	9
WV	210	1,073	8,584	0.090	773
WI	1,432	1,892	7,946	0.024	191
WY	1	12	41	0.020	1
Other States[4]	611	1,568	5,509	0.002	13
Total US	56,147	200,100	1,016,747	0.045	46,027

[1] Estimates cover the 12-month period, Dec. 1, previous year through Nov. 30 and excludes broilers. [2] Includes death, rendered, composted, destroyed, disappeared, and other losses during the 12-month period. [3] Sold for slaughter. [4] AK, AZ, ND, NM, NV, and RI combined to avoid disclosing individual operations.

NASS, Livestock Branch, (202) 720–6351.

Table 8-49.—Broilers: Production and value, United States, 1992–2001[1]

Year	Production		Price per pound[2]	Value of production
	Number	Weight		
	Thousands	*1,000 pounds*	*Cents*	*1,000 dollars*
1992	6,402,490	28,828,872	31.8	9,174,136
1993	6,694,310	30,617,600	34.0	10,416,962
1994	7,017,540	32,528,500	35.0	11,371,723
1995	7,325,670	34,222,000	34.4	11,762,222
1996	7,596,760	36,479,100	38.1	13,903,479
1997	7,764,200	37,540,750	37.7	14,158,926
1998	7,934,280	38,553,600	39.3	15,144,551
1999	8,146,010	40,829,800	37.1	15,128,840
2000	8,284,000	41,623,100	33.6	13,988,434
2001[3]	8,389,100	42,445,900	39.3	16,693,569

[1] Broilers are young chickens of the meat-type strains, raised for the purpose of meat production. These figures are not included in farm production of chickens. Estimates cover the 12-month period, Dec 1 previous year through Nov 30. Excludes States which produced less than 500,000 broilers. [2] Live weight equivalent price. [3] Preliminary.

NASS, Livestock Branch, (202) 720–6351.

Table 8-50.—Mature Chickens: Lost, sold for slaughter, price, and value, United States, 1992–2001 [1]

Year	Number		Pounds (live weight) sold [3]	Price per pound live weight [3]	Value of sales [3]
	Lost [2]	Sold [3]			
	Thousand head	*Thousand head*	*1,000 pounds*	*Dollars*	*1,000 dollars*
1992	35,126	207,689	978,255	0.091	89,105
1993	34,962	198,032	964,356	0.100	96,409
1994	35,621	197,012	992,840	0.074	73,050
1995	61,060	179,503	924,036	0.065	60,153
1996	60,435	174,299	900,652	0.066	59,187
1997	49,256	190,986	925,499	0.077	71,461
1998	53,428	193,768	934,568	0.081	75,542
1999	54,951	210,173	1,034,072	0.071	73,150
2000	50,507	218,212	1,111,717	0.057	63,925
2001	56,147	200,100	1,016,747	0.045	46,027

[1] Estimates cover the 12-month period, Dec. 1, previous year through Nov. 30 and excludes broilers. [2] Includes death, rendered, composted, destroyed, and other losses during the 12-month period. [3] Sold for slaughter.

NASS, Livestock Branch, (202) 720–6146.

Table 8-51.—Chickens: Supply, distribution, and per capita consumption, ready-to-cook basis, United States, 1994–2003

Year	Production			Commercial storage at beginning of year	Exports	Commercial storage at end of year	Consumption	
	Commercial broilers	Other chickens	Total [1]				Total [1] [3]	Per capita
	Million pounds	*Million pounds*	*Million pounds*	*Million pounds*	*Million pounds*	*Million pounds*	*Million pounds*	*Pounds*
1994	23,666	509	24,175	366	2,966	472	21,103	81
1995	24,827	496	25,323	472	3,993	567	21,238	81
1996	26,124	491	26,615	567	4,685	647	21,854	81
1997	27,041	510	27,570	647	4,787	614	22,802	83
1998	27,612	525	28,137	614	4,787	717	23,254	85
1999	29,468	554	30,022	717	4,978	804	24,967	90
2000	30,209	531	31,740	804	5,138	807	25,606	91
2001	30,938	515	31,453	806	5,737	720	25,819	91
2002 [2]	31,955	546	32,501	720	5,002	782	27,416	95
2003 [4]	32,128	520	32,648	772	5,370	706	27,360	94

[1] Totals may not add due to rounding. [2] Preliminary. [3] Shipments to territories now included in total consumption. [4] Forecast.

ERS, Animal Products Branch, (202) 694-5180.

Table 8-52.—Poultry: Feed-price ratios, United States, 1992–2001

Year	Ratios [1]		
	Egg-feed	Broiler-feed	Turkey-feed
	Pounds	*Pounds*	*Pounds*
1992	8.4	5.1	6.4
1993	9.4	5.3	6.5
1994	8.5	5.2	6.6
1995	8.8	5.1	6.3
1996	8.6	4.4	5.3
1997	8.8	4.7	5.7
1998	9.7	6.3	6.7
1999	9.8	7.2	8.6
2000	10.6	6.6	8.7
2001	9.9	7.8	8.2

[1] Number of pounds of poultry feed equivalent in value at local market prices to 1 dozen market eggs, or 1 pound of broiler or 1 pound of turkey live weight. Simple average of monthly feed-price ratios. Egg feed= corn (75 lbs) and soybeans (25 lbs); broiler feed= corn (58 lbs); soybeans (42 lbs); turkey feed= corn (51 lbs), soybeans (28 lbs), and wheat (21 lbs). Monthly equivalent prices of commercial prepared feeds are based on current U.S. prices received for corn, soybeans, and wheat.

NASS, Environmental, Economics, and Demographics Branch, (202) 720–3570.

Table 8-53.—Broilers: Production, price, and value, by States, 2000 and 2001 [1]

State	2000 Production Number	2000 Production Weight	2000 Price per pound [2]	2000 Value of production	2001 [3] Production Number	2001 [3] Production Weight	2001 [3] Price per pound [2]	2001 [3] Value of production
	Thousands	*1,000 pounds*	*Dollars*	*1,000 dollars*	*Thousands*	*1,000 pounds*	*Dollars*	*1,000 dollars*
AL	1,038,700	5,297,400	0.330	1,748,142	1,007,600	5,138,800	0.390	2,004,132
AR	1,191,700	5,839,300	0.330	1,926,969	1,170,900	5,737,400	0.390	2,237,586
DE	247,700	1,461,400	0.340	496,876	257,700	1,494,700	0.400	597,880
FL	119,900	647,500	0.350	226,625	115,300	634,200	0.400	253,680
GA	1,229,700	6,148,500	0.330	2,029,005	1,247,300	6,236,500	0.390	2,432,235
HI	900	3,700	0.545	2,017	900	3,800	0.545	2,071
KY	230,000	1,150,000	0.330	379,500	253,400	1,292,300	0.390	503,997
MD	283,300	1,359,800	0.340	462,332	287,800	1,381,400	0.400	552,560
MN	44,200	221,000	0.330	72,930	43,900	219,500	0.390	85,605
MS	739,900	3,699,500	0.330	1,220,835	765,300	3,826,500	0.390	1,492,335
NE	3,400	19,000	0.330	6,270	3,400	18,000	0.390	7,020
NY	2,100	11,800	0.360	4,248	2,300	12,200	0.410	5,002
NC	698,400	4,050,700	0.350	1,417,745	712,300	4,202,600	0.400	1,681,040
OH	45,700	223,900	0.330	73,887	40,100	212,500	0.390	82,875
OK	223,100	1,093,200	0.330	360,756	226,800	1,111,300	0.390	433,407
PA	133,300	693,200	0.350	242,620	132,300	701,200	0.410	287,492
SC	196,800	1,003,700	0.330	331,221	198,000	1,049,400	0.390	409,266
TN	151,300	696,000	0.330	229,680	198,300	932,000	0.390	363,480
TX	551,000	2,589,700	0.340	880,498	565,500	2,714,400	0.390	1,058,616
VA	264,900	1,298,000	0.340	441,320	271,500	1,330,400	0.390	518,856
WV	91,300	365,200	0.340	124,168	89,800	368,200	0.390	143,598
WI	32,800	147,600	0.340	50,184	31,300	137,700	0.390	53,703
Other States [4]	763,900	3,603,000	0.350	1,260,606	767,400	3,690,900	0.403	1,487,133
Total [5]	8,284,000	41,623,100	0.336	13,988,434	8,389,100	42,445,900	0.393	16,693,569

[1] Broilers are young chickens of the meat-type strains, raised for the purpose of meat production. Estimates cover the 12-month period, Dec. 1, previous year through Nov. 30. [2] Live weight equivalent price. [3] Preliminary. [4] CA, IN, IA, LA, MO, OR, and WA. [5] Excludes States producing less than 500,000 broilers.

NASS, Livestock Branch, (202) 720–3570.

Table 8-54.—Chicks hatched by commercial hatcheries: Number, average price, and value, United States, 1992–2001

Year	Chicks hatched: Broiler-type	Chicks hatched: Egg-type	Chicks hatched: All	Average price of baby chicks per 100: Broiler-type	Average price of baby chicks per 100: Egg-type	Average price of baby chicks per 100: All	Value of chick production [1]
	Thousands	*Thousands*	*Thousands*	*Dollars*	*Dollars*	*Dollars*	*1,000 dollars*
1992	6,892,844	391,319	7,284,163	18.30	54.70	19.30	1,369,893
1993	7,220,768	405,986	7,626,754	18.60	53.80	19.60	1,455,383
1994	7,562,316	381,577	7,943,893	19.30	49.70	20.10	1,557,276
1995	7,932,352	396,501	8,328,853	18.60	49.80	19.40	1,577,782
1996	8,078,159	401,640	8,479,799	18.60	53.80	19.50	1,611,380
1997	8,321,634	424,543	8,746,177	19.70	53.10	20.60	1,756,004
1998	8,491,938	438,273	8,930,211	19.50	53.70	20.30	1,771,713
1999	8,715,423	451,721	9,167,144	20.30	52.60	21.10	1,886,007
2000	8,846,185	430,412	9,276,597	20.50	48.00	21.10	1,913,453
2001	9,006,556	451,813	9,458,369	20.60	51.60	21.40	1,971,783

[1] Excludes egg-type cockerels destroyed.

NASS, Environmental, Economics, and Demographics Branch, (202) 720–6146 and Livestock Branch, (202) 720-3570.

Table 8-55.—Poultry: Slaughtered under Federal inspection, by class, United States, 1999–2001

Class	Number inspected			Pounds inspected (live weight)		
	1999	2000	2001	1999	2000	2001
	Thousands	*Thousands*	*Thousands*	*Thousands*	*Thousands*	*Thousands*
Young chickens	8,111,654	8,261,114	8,406,313	40,444,167	41,293,525	42,355,507
Mature chickens	175,560	165,027	160,069	913,707	876,685	854,308
Total chickens	8,287,215	8,426,141	8,566,382	41,357,874	42,170,210	43,189,815
Young turkeys	263,242	265,642	266,932	6,682,518	6,798,821	6,965,326
Old turkeys	1,951	2,347	2,370	51,815	61,711	63,367
Fryer-roaster turkeys	56	81	....	506	657	....
Total turkeys	265,250	268,069	269,302	6,734,840	6,861,188	7,028,693
Ducks	23,327	24,494	26,260	154,035	161,752	173,983
Other poultry	....	....	....	14,066	14,307	13,498
Total poultry	....	....	....	48,260,815	49,207,458	50,405,989

Class	Pounds certified (ready-to-cook)			Pounds condemned		
				Ante-mortem (live weight)		
	1999	2000	2001	1999	2000	2001
	Thousands	*Thousands*	*Thousands*	*Thousands*	*Thousands*	*Thousands*
Young chickens	29,741,380	30,495,172	31,265,809	170,911	171,243	172,904
Mature chickens	552,342	529,342	515,796	15,792	15,237	14,694
Total chickens	30,293,722	31,024,513	31,781,605	186,703	186,480	187,598
Young turkeys	5,257,442	5,354,433	5,512,988	17,701	19,323	22,243
Old turkeys	38,715	47,260	48,712	512	503	709
Fryer-roaster turkeys	359	495	....	1	17	....
Total turkeys	5,296,516	5,402,188	5,561,700	18,213	19,844	22,952
Ducks	108,147	115,979	124,141	242	279	365
Other poultry	8,183	8,763	8,215	78	60	49
Total poultry	35,706,568	36,551,443	37,475,661	205,236	206,663	210,964

Class	Pounds condemned—Continued		
	Post-mortem (New York dressed weight)		
	1999	2000	2001
	Thousands	*Thousands*	*Thousands*
Young chickens	589,959	540,784	464,846
Mature chickens	56,054	52,346	45,171
Total chickens	646,013	593,130	510,017
Young turkeys	131,965	137,616	138,901
Old turkeys	3,459	3,681	3,831
Fryer-roaster turkeys	5	9	....
Total turkeys	135,428	141,306	142,732
Ducks	3,457	3,866	3,691
Other poultry	154	142	152
Total poultry	785,052	738,443	656,592

NASS, Livestock Branch, (202) 720–3570.

Table 8-56.—Chickens and turkeys: Number classified as "U.S. Pullorum-Typhoid Clean," and number and percentage of reactors, United States, 1992–2001

Year beginning July	Chicken tests				Turkey tests			
	States reporting	Chickens in tested flocks (first test)	Reactors [1]		States reporting	Turkeys in tested flocks (first test)	Reactors	
	Number	*Thousands*	*Number*	*Percent*	*Number*	*Number*	*Number*	*Percent*
1992	47	49,534	11	0.0000	47	5,160,884	0	0.0000
1993	47	49,628	0	0.0000	47	5,339,500	0	0.0000
1994	47	52,668	0	0.0000	47	4,949,240	0	0.0000
1995	48	58,019	0	0.0000	48	4,679,984	0	0.0000
1996	48	58,191	0	0.0000	48	5,905,799	0	0.0000
1997	48	62,402	0	0.0000	48	5,301,183	0	0.0000
1998	48	81,636	0	0.0000	48	5,548,802	0	0.0000
1999	48	79,037	0	0.0000	48	5,516,096	0	0.0000
2000	48	79,407	0	0.0000	48	4,956,140	0	0.0000
2001	48	79,397	0	0.0000	48	5,408,561	0	0.0000

[1] Number of reacting birds and percent of birds tested. Testing year starting July 1, 1989.

APHIS, Veterinary Services, (770) 922–3496.

Table 8-57.—Turkeys: Supply, distribution, and per capita consumption, ready-to-cook basis, United States, 1994–2003

Year	Production	Commercial storage at beginning of year	Exports	Commercial storage at end of year	Consumption	
					Total [1] [2]	Per capita
	Million pounds	*Million pounds*	*Million pounds*	*Million pounds*	*Million pounds*	*Pounds*
1994	4,937	249	280	254	4,652	17.8
1995	5,069	254	348	271	4,706	17.9
1996	5,401	271	438	328	4,906	18.2
1997	5,412	328	606	415	4,720	17.3
1998	5,215	415	446	304	4,880	17.7
1999	5,230	304	378	254	4,902	17.6
2000	5,333	254	445	241	4,902	17.4
2001	5,489	241	487	241	5,004	17.6
2002 [3]	5,642	241	446	334	5,103	17.8
2003 [4]	5,650	334	470	350	5,164	17.8

[1] Totals may not add due to rounding. [2] Shipments to territories now included in consumption. [3] Preliminary. [4] Forecast.

ERS, Animal Products Branch, (202) 694–5180.

Table 8-58.—Turkeys: Production, and value, United States, 1992–2001

Year	Number raised [1]	Pounds (live weight) produced	Price per pound live weight	Value of production
	Thousands	*1,000 pounds*	*Cents*	*1,000 dollars*
1992	289,880	6,355,293	37.7	2,396,364
1993	287,650	6,432,577	39.0	2,509,127
1994	286,585	6,540,295	40.4	2,643,057
1995	292,356	6,761,327	41.0	2,769,397
1996	302,713	7,222,834	43.3	3,124,496
1997	301,251	7,225,059	39.9	2,884,377
1998	285,204	7,050,944	38.0	2,679,301
1999	270,494	6,886,406	40.8	2,809,874
2000	269,969	6,942,809	40.7	2,822,736
2001 [2]	272,059	7,154,781	39.0	2,790,317

[1] Total poults hatched less death loss of poults and young turkeys during the year. [2] Preliminary.
NASS, Livestock Branch, (202) 720–3570.

Table 8-59.—Turkeys: Production and value, by State, 2001 [1]

State	Number raised [2]	Pounds produced	Price per pound [3]	Value of production
	1,000 head	*1,000 pounds*	*Dollars*	*1,000 dollars*
AR	27,000	472,500	0.40	189,000
CA	18,700	450,670	0.47	211,815
CT	5	108	1.13	122
DE	4	109	0.39	43
IL	2,900	82,650	0.38	31,407
IN	14,000	399,000	0.38	151,620
IA	7,100	216,550	0.35	75,793
MD	460	15,042	0.39	5,866
MA	74	1,591	1.44	2,291
MI	4,500	162,000	0.35	56,700
MN	44,000	1,100,000	0.35	385,000
MO	24,000	660,000	0.41	270,600
NH	6	129	1.47	190
NJ	44	893	0.80	714
NY	510	11,067	0.41	4,537
NC	43,000	1,130,900	0.40	452,360
ND	1,600	42,080	0.35	14,728
OH	4,800	181,440	0.35	63,504
PA	9,500	213,750	0.46	98,325
SC	9,200	325,680	0.39	127,015
SD	4,900	166,110	0.35	58,139
VT	45	968	1.40	1,355
VA	24,000	530,400	0.39	206,856
WV	4,100	90,610	0.39	35,338
Other States [4]	27,611	900,534	0.39	346,999
US	272,059	7,154,781	0.39	2,790,317

[1] Preliminary. [2] Based on turkeys placed Sep. 1, 2000, through Aug. 31, 2001. Excludes young turkeys lost. [3] Live weight equivalent price. [4] CO, KS, NE, OK, OR, TX, UT, and WI combined to avoid disclosing individual operations.
NASS, Livestock Branch, (202) 720–3570.

Table 8-60.—Turkeys: Poults placed by commercial hatcheries, United States, 1992–2001

Year	Total all breeds
	Thousands
1992	307,823
1993	308,871
1994	317,468
1995	321,651
1996	327,213
1997	321,487
1998	297,798
1999	296,106
2000	297,299
2001	301,559

NASS, Livestock Branch, (202) 720–3570.

Table 8-61.—Turkeys: Poults placed by commercial hatcheries, U.S. and regions, Monthly, 2000 and 2001

Month	United States			2001				
	2000	2001	2001 as percent of 2000	East North Central	West North Central	North and South Atlantic	South Central	West
	Thousands	*Thousands*	*Percent*	*Thousands*	*Thousands*	*Thousands*	*Thousands*	*Thousands*
All breeds:								
Jan	24,676	25,612	104	3,501	8,775	8,281	2,494	2,561
Feb	24,197	23,775	98	3,559	8,275	7,156	2,323	2,462
Mar	25,741	26,148	102	3,736	9,080	8,195	2,446	2,691
Apr	25,094	25,858	103	3,736	8,759	8,041	2,665	2,657
May	25,816	26,682	103	3,837	9,330	8,552	2,566	2,397
June	26,880	26,121	97	3,542	8,878	8,309	2,693	2,699
July	26,822	27,064	101	3,599	8,956	8,743	2,427	3,339
Aug	24,731	25,032	101	3,605	8,527	8,370	2,429	2,101
Sept	23,016	22,414	97	3,178	7,471	7,508	2,374	1,883
Oct	23,672	24,128	102	3,769	8,300	8,168	2,491	1,400
Nov	23,357	24,226	104	3,839	8,619	7,732	2,518	1,518
Dec	23,297	24,499	105	3,646	8,671	7,649	2,695	1,838
Total	297,299	301,559	101	43,547	103,641	96,704	30,121	27,546

NASS, Livestock Branch, (202) 720–3570.

Table 8-62.—Eggs: Production in specified countries, 1997–2000

Continent and country	1997	1998	1999	2000 [1]
	Million pieces	*Million pieces*	*Million pieces*	*Million pieces*
North America:				
Canada	5,931	5,986	6,151	6,300
Mexico	28,170	29,898	32,428	33,310
United States	77,676	79,896	82,944	84,624
Total	111,777	115,780	121,523	124,234
South America:				
Brazil	12,596	13,636	14,768	15,654
Colombia	7,411	8,014	8,256	8,380
Total	20,007	21,650	23,024	24,034
Eastern Europe:				
Bulgaria	1,538	1,600	1,650	1,650
Poland	7,700	7,700	7,500	7,600
Romania	4,750	4,600	4,500	4,500
Total	13,988	13,900	13,650	13,750
Former Soviet Union:				
Russia	31,900	33,000	33,000	33,200
Ukraine	8,242	8,269	8,740	8,000
Total	40,142	41,269	41,740	41,200
Middle East:				
Turkey	9,100	9,060	7,000	6,250
Total	9,100	9,060	7,000	6,250
Asia:				
China	282,350	307,760	365,300	371,880
Hong Kong	5	1	0	1
India	32,000	34,000	35,000	36,000
Indonesia	11,064	5,300	7,900	9,000
Japan	42,588	42,117	41,975	41,700
Malaysia	5,700	5,860	5,572	5,820
Korea, Republic of	8,790	8,413	8,562	8,600
Taiwan	7,636	7,678	7,800	7,700
Thailand	8,900	8,500	8,200	8,800
Total	399,033	419,629	480,309	489,501
Grand total	594,047	621,288	687,246	698,969

[1] Preliminary.
Prepared or estimated on the basis of official statistics of foreign governments, other foreign source materials, reports of U.S. Agricultural Counselors, Attachés, and Foreign Service Officers, results of office research, and related information.

FAS, Production Estimates and Crop Assessment Division, (202) 720–8031.

Table 8-63.—Eggs: Number of layers and pullets, rate of lay, and production, by State and United States, 2001 and 2002[1]

State	Average number of layers during year		Rate of lay per layer during year[2]		Eggs produced	
	2001	2002	2001	2002	2001	2002
	Thousands	*Thousands*	*Number*	*Number*	*Millions*	*Millions*
AL	10,371	10,173	227	224	2,359	2,281
AR	14,891	14,829	230	224	3,431	3,329
CA	23,437	23,652	256	259	5,998	6,124
CO	3,571	3,736	265	270	946	1,008
CT	3,152	3,106	280	275	884	855
FL	10,756	10,631	254	257	2,737	2,731
GA	20,994	20,452	242	243	5,086	4,961
HI	588	539	220	218	129.4	117.5
IL	3,292	3,431	270	260	888	893
IN	22,958	22,435	262	266	6,025	5,973
IA	32,591	36,980	267	268	8,691	9,910
ME	4,076	4,146	270	260	1,100	1,079
MD	3,182	3,197	273	280	870	894
MI	6,475	6,686	259	265	1,677	1,771
MN	12,142	11,729	256	266	3,112	3,124
MS	6,704	6,795	231	234	1,550	1,588
MO	6,882	6,923	260	265	1,789	1,837
NE	11,650	11,591	258	257	3,001	2,977
NY	4,127	3,887	277	283	1,143	1,100
NC	11,139	11,045	228	228	2,535	2,518
OH	30,160	30,479	262	261	7,900	7,940
OK	4,020	4,173	230	228	926	951
OR	2,958	2,750	277	276	818	760
PA	24,377	23,641	273	276	6,662	6,520
SC	5,586	5,372	254	257	1,416	1,380
SD	2,300	2,335	250	243	575	568
TX	18,744	18,608	253	257	4,734	4,774
VA	3,397	3,134	226	234	766	734
WA	4,987	5,166	269	265	1,339	1,369
WI	4,568	4,272	270	271	1,235	1,158
30 Sts	314,076	315,891	256	257	80,322	81,226
DE	1,399	1,315	264	264	369	347
ID	888	886	283	273	251	242
KY	3,798	3,932	246	234	933	921
LA	1,946	1,982	247	249	480	494
MA	283	269	283	297	80	80
MT	321	341	296	305	95	104
NH	152	159	280	288	43	46
NJ	2,030	1,994	274	267	556	533
TN	1,297	1,354	227	222	294	300
UT	3,282	3,342	264	267	865	894
VT	206	197	298	296	61	59
WV	1,092	1,219	215	214	235	261
WY	12	12	300	300	3.6	3.6
Oth Sts[3]	4,230	4,312	273	276	1,155	1,189
US[4]	335,012	337,202	256	257	85,745	86,698

[1] Annual production estimates for eggs reported on Dec. 1 through Nov. 30 marketing year. [2] Number of eggs producted during the year divided by average of layers hand during the year. [3] AK, AZ, KS, NV, NM, ND, and RI combined to avoid disclosure of individual operations. [4] Sum of States may not add to U.S. total due to rounding.

NASS, Livestock Branch, (202) 720–3570.

Table 8-64.—Eggs: Number, rate of lay, production, and value, United States, 1993–2002 [1]

Year	Layers average number during year	Rate of lay per layer during year [2]	Eggs, total produced	Price per dozen	Value of production
	Thousands	*Number*	*Millions*	*Dollars*	*1,000 dollars*
1993	284,770	253	71,936	0.634	3,800,237
1994	291,035	254	73,903	0.615	3,789,834
1995	294,350	254	74,764	0.625	3,892,912
1996	298,270	256	76,377	0.750	4,776,252
1997	303,604	255	77,532	0.703	4,539,929
1998	312,035	255	79,690	0.668	4,439,446
1999	322,354	257	82,715	0.627	4,322,589
2000	327,985	257	84,386	0.618	4,345,879
2001	335,012	256	85,702	0.622	4,444,864
2002 [3]	337,202	257	86,698	NA	NA

[1] Annual production estimates for eggs reported on Dec. 1 through Nov. 30 marketing year. [2] Number of eggs produced during the year divided by the average of layers on hand during the year. [3] Preliminary. NA-not available.

NASS, Livestock Branch, (202) 720–3570.

Table 8-65.—Eggs: Production and value, by States, 2000–2001 [1]

State	Eggs produced		Price per dozen [2]		Value of production	
	2000	2001	2000	2001	2000	2001
	Millions	*Millions*	*Dollars*	*Dollars*	*1,000 dollars*	*1,000 dollars*
AL	2,371	2,359	1.310	1.350	258,834	265,388
AR	3,559	3,427	1.060	1.060	314,378	302,718
CA	6,293	5,996	0.453	0.472	237,561	235,843
CO	988	946	0.697	0.612	57,386	48,246
CT	863	883	0.555	0.568	39,914	41,810
DE	386	369	0.670	0.676	21,552	20,787
FL	2,716	2,737	0.478	0.536	108,187	122,253
GA	5,114	5,086	0.868	0.868	369,913	367,887
HI	143	129	0.890	0.894	10,598	9,640
ID	249	251	0.611	0.586	12,678	12,257
IL	944	888	0.472	0.486	37,131	35,964
IN	6,098	6,025	0.516	0.515	262,214	258,573
IA	7,554	8,676	0.383	0.390	241,099	281,970
KS	416	434	0.390	0.376	13,523	13,586
KY	940	933	0.904	0.935	70,813	72,696
LA	493	480	0.810	0.793	33,278	31,720
ME	1,135	1,100	0.596	0.618	56,377	56,640
MD	845	870	0.598	0.607	42,109	44,008
MA	93	80	0.627	0.657	4,875	4,380
MI	1,617	1,677	0.419	0.437	56,464	61,063
MN	3,271	3,112	0.428	0.435	116,666	112,810
MS	1,581	1,550	1.180	1.260	155,465	162,750
MO	1,614	1,789	0.520	0.507	69,940	75,585
MT	84	95	0.460	0.410	3,220	3,246
NE	2,999	3,001	0.375	0.382	93,719	95,532
NH	39	42	0.859	0.861	2,813	3,021
NJ	574	556	0.527	0.531	25,208	24,603
NY	1,113	1,143	0.564	0.567	52,311	54,007
NC	2,501	2,535	1.070	1.100	223,006	232,375
OH	8,163	7,900	0.503	0.491	342,166	323,242
OK	931	926	0.838	0.843	65,015	65,052
OR	805	818	0.482	0.600	32,334	40,900
PA	6,309	6,662	0.545	0.547	286,534	303,676
SC	1,245	1,416	0.642	0.674	66,608	79,532
SD	580	575	0.354	0.350	17,110	16,771
TN	278	294	1.240	1.300	28,727	31,850
TX	4,423	4,734	0.697	0.677	256,903	267,077
UT	712	853	0.434	0.440	25,751	31,277
VT	68	62	0.598	0.626	3,389	3,213
VA	824	766	0.963	0.961	66,126	61,344
WA	1,306	1,337	0.549	0.560	59,759	62,412
WV	247	235	1.460	1.410	30,052	27,612
WI	1,225	1,235	0.479	0.500	48,898	51,458
WY	3.6	3.6	0.550	0.560	165	168
Other States [3]	668	719	0.441	0.465	24,613	27,922
US [4]	84,386	85,702	0.618	0.622	4,345,382	4,444,864

[1] Estimates cover the 12-month period Dec. 1, previous year through Nov. 30. [2] Average mid-month price of all eggs sold by producers including hatching eggs. [3] AK, AZ, ND, NM, NV, and RI combined to avoid disclosure of individual operations. [4] Sum of States may not add to U.S. total due to rounding.

NASS, Livestock Branch, (202) 720–3570.

Table 8-66.—Eggs: Supply, distribution, and per capita consumption, United States, 1994–2003 [1]

Year	Total egg production	Storage at beginning of the year [1]	Imports [2]	Exports [2]	Eggs used for hatching	Consumption: Storage at end of the year [2]	Consumption: Total [3]	Consumption: Per capita
	Million dozen	*Million dozen*	*Million dozen*	*Million dozen*	*Million dozen*	*Million dozen*	*Million dozen*	*Number*
1994	6,178	11	4	188	805	15	5,184	239
1995	6,231	15	4	209	847	11	5,182	236
1996	6,378	11	5	253	966	9	5,269	235
1997	6,473	9	7	228	1,008	7	5,359	246
1998	6,658	7	6	219	1,038	8	5,522	240
1999	6,912	8	7	162	1,079	8	5,817	250
2000	7,034	8	8	171	1,081	11	5,927	252
2001	7,155	11	9	190	1,078	10	6,022	254
2002 [4]	7,221	10	15	173	1,073	10	6,094	255
2003 [5]	7,230	10	12	168	1,070	10	6,109	253

[1] Calendar years. [2] Shell eggs and the approximate shell-egg equivalent of egg product. [3] Shipments to territories now included in total consumption. [4] Preliminary. [5] Forecast.

ERS, Animal Products Branch, (202) 694–5180.

Table 8-67.—Eggs, shell: Average price per dozen on consumer Grade A cartoned white eggs to volume buyers, store-door delivery, New York metropolitan area, 1993–2002

Year	Large
	Cents
1993	72.53
1994	67.23
1995	67.26
1996	73.00
1997	81.21
1998	75.80
1999	65.60
2000	68.90
2001	67.14
2002	67.06

AMS, Poultry Division, Market News Branch, (202) 720–6911.

Table 8-68.—Eggs: Broken under Federal inspection, United States, 2001–2002

Item	Quantity: 2001	Quantity: 2002
	1,000 dozen	*1,000 dozen*
Shell eggs broken	1,835,075	1,883,204
	1,000 pounds	1,000 pounds
Edible liquid from shell eggs broken:		
Whole	1,392,298	1,471,215
White	605,635	588,079
Yolk	321,389	320,374
Total	2,319,322	2,379,668
Inedible liquid from shell eggs broken	224,299	230,200

NASS, Livestock Branch, (202) 720–3570.

Table 8-69.—Poultry and poultry products: Cold storage holdings, end of month, United States, 2001 and 2002

Month	Frozen eggs							
	Whites		Yolks		Whole & mixed		Unclassified	
	2001	2002	2001	2002	2001	2002	2001	2002
	1,000 pounds	*1,000 pounds*	*1,000 pounds*	*1,000 pounds*	*1,000 pounds*	*1,000 pounds*	*1,000 pounds*	*1,000 pounds*
January	4,488	2,001	1,042	630	10,282	6,835	1,084	3,669
February	2,985	2,204	1,127	1,036	10,339	7,282	1,017	3,381
March	2,242	1,561	879	906	9,851	6,260	1,629	2,942
April	3,298	1,514	493	979	10,015	4,572	2,078	3,182
May	2,458	1,998	752	840	10,245	4,956	2,346	3,300
June	1,853	1,675	645	819	9,522	6,397	2,370	3,829
July	2,711	1,863	797	1,015	11,045	6,414	2,097	3,651
August	3,370	2,192	685	923	12,343	6,528	1,419	3,535
September	4,251	1,719	777	865	11,274	7,037	1,401	3,577
October	3,024	1,835	744	628	10,298	6,973	1,453	3,649
November	1,815	1,736	580	514	10,132	5,913	1,332	3,036
December	2,467	2,316	480	478	9,463	7,467	1,300	3,264

Month	Frozen eggs, total		Frozen chicken					
			Broilers (Whole)		Hens		Other chicken (parts)	
	2001	2002	2001	2002	2001	2002	2001	2002
	1,000 pounds	*1,000 pounds*	*1,000 pounds*	*1,000 pounds*	*1,000 pounds*	*1,000 pounds*	*1,000 pounds*	*1,000 pounds*
January	16,896	13,135	16,186	24,716	10,982	6,449	730,203	693,414
February	15,468	13,903	14,117	26,981	6,737	4,076	662,452	705,238
March	14,601	11,669	15,585	29,307	6,486	4,924	620,881	774,373
April	15,884	10,247	13,639	30,334	7,343	4,260	633,401	817,836
May	15,801	11,094	15,420	34,544	10,332	6,885	645,492	795,538
June	14,390	12,720	26,457	34,326	8,896	7,081	654,781	814,009
July	16,650	12,943	26,098	36,230	9,096	7,851	607,662	825,145
August	17,817	13,178	20,404	33,700	8,949	6,248	595,056	784,073
September	17,703	13,198	22,308	34,370	8,181	6,293	594,377	794,762
October	15,519	13,085	21,578	33,168	5,972	6,067	607,108	822,049
November	13,859	11,199	22,294	31,590	5,205	5,767	656,518	763,066
December	13,710	13,525	24,087	26,421	7,935	5,420	687,667	736,243

Month	Frozen chicken, total		Frozen turkey					
			Toms		Hens		Total whole	
	2001	2002	2001	2002	2001	2002	2001	2002
	1,000 pounds	*1,000 pounds*	*1,000 pounds*	*1,000 pounds*	*1,000 pounds*	*1,000 pounds*	*1,000 pounds*	*1,000 pounds*
January	757,371	724,579	....	84,826	....	52,298	133,298	137,124
February	683,306	736,295	....	117,250	....	83,929	177,683	201,179
March	642,952	808,604	....	126,315	....	95,572	199,986	221,887
April	654,383	852,430	....	146,752	....	114,196	226,017	260,948
May	671,244	836,967	....	159,553	....	141,343	271,224	300,896
June	690,134	855,416	....	177,789	....	162,639	317,590	340,428
July	642,856	869,226	....	196,773	....	169,506	334,733	366,279
August	624,409	824,021	....	180,685	....	163,528	339,261	344,213
September	624,866	835,425	....	194,953	....	159,630	346,967	354,583
October	634,658	861,284	....	179,724	....	131,549	296,425	311,273
November	684,017	800,423	....	45,638	....	47,060	97,488	92,698
December	719,689	768,084	....	41,324	....	40,552	91,459	81,876

Month	Other		Total		Frozen ducks		Total frozen poultry	
	2001	2002	2001	2002	2001	2002	2001	2002
	1,000 pounds	*1,000 pounds*	*1,000 pounds*	*1,000 pounds*	*1,000 pounds*	*1,000 pounds*	*1,000 pounds*	*1,000 pounds*
January	158,121	189,932	291,419	327,056	3,468	6,462	1,052,258	1,058,097
February	155,838	212,064	333,521	413,243	4,163	6,575	1,020,990	1,156,113
March	155,778	235,678	355,764	457,565	4,883	6,162	1,003,599	1,272,331
April	166,589	254,241	392,606	515,189	6,120	6,436	1,053,109	1,374,055
May	184,806	277,333	456,030	578,229	7,044	7,006	1,134,318	1,422,202
June	189,101	303,672	506,691	644,100	8,670	6,784	1,205,495	1,506,300
July	199,437	339,915	534,170	706,194	8,695	7,173	1,185,721	1,582,593
August	206,050	341,408	545,311	685,621	9,281	7,654	1,179,001	1,517,296
September	195,008	317,799	541,975	672,382	10,133	7,692	1,176,974	1,515,499
October	201,500	313,667	497,925	624,940	9,242	8,875	1,141,825	1,495,099
November	162,541	241,641	260,029	334,339	8,351	7,551	952,397	1,142,313
December	149,065	251,171	240,524	333,047	6,994	6,398	967,207	1,107,529

NASS, Livestock Branch, (202) 720–3570.

Table 8-70.—Dairy products: Cold storage holdings, end of month, United States, 2001 and 2002

Month	Butter		American cheese	
	2001	2002	2001	2002
	1,000 pounds	*1,000 pounds*	*1,000 pounds*	*1,000 pounds*
January	68,405	99,200	507,000	459,707
February	86,078	130,124	505,951	491,147
March	96,208	145,160	500,430	498,290
April	112,318	196,614	502,951	513,576
May	138,022	226,849	497,400	536,080
June	153,486	242,999	521,543	548,675
July	151,123	245,298	527,916	573,368
August	118,049	229,471	499,044	565,419
September	110,857	209,112	487,559	512,887
October	100,791	164,640	463,378	501,401
November	57,916	135,632	438,140	474,055
December	55,915	157,820	449,105	493,262

Month	Swiss cheese		Other natural cheese	
	2001	2002	2001	2002
	1,000 pounds	*1,000 pounds*	*1,000 pounds*	*1,000 pounds*
January	13,968	12,388	188,937	221,541
February	13,027	12,711	204,938	216,463
March	12,680	12,880	198,517	220,552
April	12,340	15,729	196,474	236,327
May	12,404	16,122	202,304	236,787
June	11,990	14,782	205,619	234,099
July	11,479	15,721	213,169	244,489
August	10,668	14,250	211,469	221,829
September	10,439	17,068	210,732	223,979
October	10,671	18,927	198,185	200,064
November	11,335	22,251	181,818	200,757
December	12,051	21,784	198,885	215,015

Month	Total Natural cheese	
	2001	2002
	1,000 pounds	*1,000 pounds*
January	709,905	693,636
February	723,916	720,321
March	711,627	731,722
April	711,765	765,632
May	712,108	788,989
June	739,152	797,556
July	752,564	833,578
August	721,181	801,498
September	708,730	753,934
October	672,234	720,392
November	631,293	697,063
December	660,041	730,061

NASS, Livestock Branch, (202) 720–3570.

CHAPTER IX

FARM RESOURCES, INCOME, AND EXPENSES

The statistics in this chapter deal with farms, farm resources, farm income, and expenses. Many of the series are estimates developed in connection with economic research activities of the Department.

Table 9-1.—Economic trends: Data relating to agriculture, United States, 1992–2001

Year	Prices paid by farmers [1]: Total including interest, taxes, and wage rates	Prices paid by farmers [1]: Production items	Prices received by farmers [1]	Farm income [2]: Gross farm income [6]	Farm income [2]: Production expenses	Farm income [2]: Net farm income
	Index numbers 1990–92=100	*Index numbers 1990–92=100*	*Index numbers 1990–92=100*	*Billion dollars*	*Billion dollars*	*Billion dollars*
1992	101	101	98	200.6	152.8	47.8
1993	103	104	101	205.0	160.4	44.7
1994	106	106	100	216.0	167.2	48.9
1995	109	108	102	210.8	173.8	36.9
1996	115	115	112	235.8	181.0	54.8
1997	118	119	107	238.1	187.6	50.5
1998	115	113	102	232.1	186.5	45.6
1999	115	111	96	234.5	188.3	46.2
2000	120	116	96	241.7	193.7	48.0
2001	124	120	102	246.5	200.8	45.7

Year	National income [3]	Disposable personal income [3]	Industrial production [4]	Consumer prices all items [5]	Producer prices consumer foods [5]
	Billion dollars	*Billion dollars*	*Index numbers 1997=100*	*Index numbers 1982–84=100*	*Index numbers 1982=100*
1992	4,994.9	5,390.4	78.3	140.3	123.3
1993	5,251.9	5,610.0	80.9	144.5	125.7
1994	5,556.8	5,888.0	85.2	148.2	126.8
1995	5,876.7	6,200.9	89.3	152.4	129.0
1996	6,210.4	6,547.4	93.2	156.9	133.6
1997	6,618.4	6,937.0	100.0	160.5	134.5
1998	7,041.1	7,426.0	105.6	163.0	134.3
1999	7,468.7	7,786.5	110.1	166.6	135.1
2000	7,984.4	8,406.6	115.3	172.2	137.2
2001	8,122.0	8,685.3	111.2	177.1	141.3

[1] U.S. Department of Agriculture - NASS. [2] U.S. Department of Agriculture - ERS. [3] U.S. Department of Commerce, Bureau of Economic Analysis. [4] Federal Reserve Board. [5] U.S. Department of Labor, Bureau of Labor Statistics.
[6] Includes cash receipts from farm marketings, government payments, nonmoney income (gross rental value of dwelling and value of home consumption), other income (machine hire custom work and recreational income), and value of change in farm inventories.

ERS, Farm Structure Performance and Well Being Branch, (202) 694–5592. E mail contact is roger@ERS.USDA.gov

Table 9-2.—Farms: Number, land in farms, and average size of farm, U.S., 1993–2002 [1]

Year	Farms [2]	Land in farms	Average size farm
	Number	1,000 acres	Acres
1993	2,201,590	968,845	440
1994	2,197,690	965,935	440
1995	2,196,400	962,515	438
1996	2,190,500	958,675	438
1997	2,190,510	956,010	436
1998	2,191,360	953,500	435
1999	2,192,070	947,440	432
2000	2,172,280	943,090	434
2001	2,155,680	941,310	437
2002 [3]	2,158,090	941,480	436

[1] The farm definition was changed in 1993 to include maple syrup, short rotation woody crops, and places with 5 or more horses. [2] A farm is any establishment from which $1,000 or more of agricultural products were sold or would normally be sold during the year. [3] Preliminary.

NASS, Environmental, Economics, and Demographics Branch, (202) 720–6146.

Table 9-3.—Farms: Percent of farms, land in farms, and average size, by economic sales class, United States, 2001–2002 [1]

Economic sales class	Percent of total				Average size farm	
	Farms		Land		2001	2002 [2]
	2001	2002 [2]	2001	2002 [2]		
	Percent	Percent	Percent	Percent	Acres	Acres
$1,000–$2,499	25.6	26.9	4.3	4.5	73	73
$2,500–$4,999	14.9	14.8	4.0	3.8	117	113
$5,000–$9,999	13.4	12.7	5.0	4.9	163	169
$10,000–$24,999	12.5	12.7	7.5	7.7	264	264
$25,000–$49,999	8.7	8.7	8.4	9.9	424	496
$50,000–$99,999	8.6	8.1	13.2	11.5	675	619
$100,000–$249,999	9.1	8.8	24.1	23.5	1,158	1,161
$250,000–$499,999	4.0	4.1	14.3	16.1	1,564	1,708
$500,000–$999,999	2.0	2.0	10.1	8.6	2,205	1,872
$1,000,000+	1.2	1.2	9.1	9.5	3,311	3,457
Total	100.0	100.0	100.0	100.0	437	436

[1] Economic Sales Classes changed for 2001 to standard government groupings. [2] Preliminary.

NASS, Environmental, Economics, and Demographics Branch, (202) 720–6146.

Table 9-4.—Number of farms: Economic sales class by region and United States, 2000–2002

Region and year	Economic Sales Class						Total
	$1,000-$9,999	$10,000-$99,999	$100,000 & over	$100,000-$249,999	$250,000-$499,999	$500, & over	
	Number	Number	Number	Number			
NE: [1]							
2000	73,900	38,000	22,000	....	....	....	133,900
2001	72,800	38,100	22,200	....	....	....	133,100
2002 [5]	74,000	34,780	23,820	14,140	5,870	3,810	132,600
NC: [2]							
2000	333,800	297,200	183,800	....	....	....	814,800
2001	330,200	289,200	185,900	....	....	....	805,300
2002	332,400	286,000	183,600	109,600	47,600	26,400	802,000
South: [3]							
2000	601,500	227,220	89,280	....	....	....	918,000
2001	598,240	228,580	89,580	....	....	....	916,400
2002	605,410	230,330	86,360	41,900	22,524	21,936	922,100
West: [4]							
2000	164,350	87,130	54,100	....	....	....	305,580
2001	159,750	86,830	54,300	....	....	....	300,880
2002	160,960	86,050	54,380	24,230	13,250	16,900	301,390
US:							
2000	1,173,550	649,550	349,180	....	....	....	2,172,280
2001	1,160,990	642,710	351,980	....	....	....	2,155,680
2002	1,172,770	637,160	348,160	189,870	89,244	69,046	2,158,090

[1] CT, ME, MA, NH, NJ, NY, PA, RI, and VT. [2] IL, IN, IA, KS, MI, MN, MO, NE, ND, OH, SD, WI. [3] AL, AR, DE, FL, GA, KY, LA, MD, MS, NC, OK, SC, TN, TX, VA, WV. [4] AK, AZ, CA, CO, HI, ID, MT, NV, NM, OR, UT, WA, WY. [5] Number of farms estimated for 3 sales classes above $100,00 beginning in 2002.

NASS, Environmental, Economics, and Demographics Branch, (202) 720–6146.

Table 9-5.—Land in farms: Economic sales class by region and United States, 2000–2002

Region and year	Economic Sales Class						Total
	$1,000-$9,999	$10,000-$99,999	$100,000 & over	$100,000-$249,999	$250,000-$499,999	$500,000 & over	
	1,000 Acres	*1,000 Acres*	*1,000 Acres*	*1,000 Acres*			
NE: [1]							
2000	5,470	5,900	8,880	.....	.....	.....	20,250
2001	5,150	5,850	9,130	.....	.....	.....	20,130
2002 [5]	5,340	5,650	9,120	4,235	2,635	2,250	20,110
NC: [2]							
2000	32,500	106,700	214,200	.....	.....	.....	353,400
2001	31,600	103,300	217,900	.....	.....	.....	352,700
2002	31,300	101,600	219,200	104,300	63,600	51,300	352,100
South: [3]							
2000	60,898	96,707	124,875	.....	.....	.....	282,480
2001	60,580	92,085	129,355	.....	.....	.....	282,020
2002	61,280	92,585	129,045	45,615	33,155	50,275	282,910
West: [4]							
2000	28,815	76,885	181,260	.....	.....	.....	286,960
2001	27,615	74,645	184,200	.....	.....	.....	286,460
2002	26,815	74,265	185,280	66,735	52,105	66,440	286,360
US:							
2000	127,683	286,192	529,215	.....	.....	.....	943,090
2001	124,945	275,880	540,485	.....	.....	.....	941,310
2002	124,735	274,100	542,645	220,885	151,495	170,265	941,480

[1] CT, ME, MA, NH, NJ, NY, PA, RI, and VT. [2] IL, IN, IA, KS, MI, MN, MO, NE, ND, OH, SD, WI. [3] AL, AR, DE, FL, GA, KY, LA, MD, MS, NC, OK, SC, TN, TX, VA, WV. [4] AK, AZ, CA, CO, HI, ID, MT, NV, NM, OR, UT, WA, WY. [5] Number of farms estimated for 3 sales classes above $100,000 beginning in 2002.

NASS, Environmental, Economics, and Demographics Branch, (202) 720–6146.

Table 9-6.—Land in farms: Classification by tenure of operator, United States, 1910–2001

Year	Land in farms	Tenure of operator			
		Full owners	Part owners	Managers	All tenants
	Acres	*Percent*	*Percent*	*Percent*	*Percent*
1910	878,798,325	52.9	15.2	6.1	25.8
1920	958,676,612	48.3	18.4	5.7	27.7
1925	924,319,352	45.4	21.3	4.7	28.7
1930 [1]	990,111,984	37.6	24.9	6.4	31.0
1935	1,054,515,111	37.1	25.2	5.8	31.9
1940 [1]	1,065,113,774	35.9	28.2	6.5	29.4
1945	1,141,615,364	36.1	32.5	9.3	22.0
1950 [1]	1,161,419,720	36.1	36.4	9.2	18.3
1954	1,158,191,511	34.2	40.7	8.6	16.5
1959 [1]	1,123,507,574	31.0	44.0	9.8	14.8
1964 [1]	1,110,187,000	28.7	48.0	10.2	13.1
1969 [1]	1,062,892,501	35.3	51.8	.....	13.0
1974 [1]	1,017,030,357	35.3	52.6	.....	12.0
1978 [1]	1,014,777,234	32.7	55.3	.....	12.0
1982 [1]	986,796,579	34.7	53.8	.....	11.5
1987 [1]	964,470,625	32.9	53.9	.....	13.2
1992 [1]	945,531,506	31.3	55.7	.....	13.0
1997 [2]	932,475,414	26.7	62.2	.....	11.2
1998 [2]	900,415,615	28.6	60.2	.....	11.2
1999 [2]	870,720,495	25.6	61.6	.....	12.8
2000 [2]	994,997,682	26.4	62.3	.....	11.4
2001 [2]	959,163,331	24.7	61.2	.....	14.2

[1] Includes Alaska and Hawaii. [2] Excludes Alaska and Hawaii.

ERS, Resource Economics Division, (202) 694–5575. Data for 1910–1992 is from the Census of Agriculture, U.S. Department of Commerce. Data for 1997-2001 is from ERS Agricultural Resource Management Study.

Table 9-7.—Farms: Classification by tenure of operator, United States, 1910–2001

Year	Farms	Tenure of operator			
		Full owners	Part owners	Managers	All tenants
	Number	*Percent*	*Percent*	*Percent*	*Percent*
1910	6,365,822	52.7	9.3	0.9	37.0
1920	6,453,991	52.2	8.7	1.1	38.1
1925	6,371,640	52.0	8.7	0.6	38.6
1930 [1]	6,295,103	46.3	10.4	0.9	42.4
1935	6,812,350	47.1	10.1	0.7	42.1
1940 [1]	6,102,417	50.6	10.1	0.6	38.8
1945	5,859,169	56.4	11.3	0.7	31.7
1950 [1]	5,388,437	57.4	15.3	0.4	26.9
1954	4,783,021	57.4	18.2	0.4	24.0
1959 [1]	3,710,503	57.1	21.9	0.6	20.5
1964 [1]	3,157,857	57.6	24.8	0.6	17.1
1969 [1]	2,730,250	62.5	24.6	....	12.9
1974 [1]	2,314,013	61.5	27.2	....	11.3
1978 [1]	2,257,775	57.5	30.2	....	12.3
1982 [1]	2,240,976	59.2	29.3	....	11.6
1987 [1]	2,087,759	59.3	29.2	....	11.5
1992 [1]	1,925,300	57.7	31.0	....	11.3
1997 [2]	2,049,384	55.3	35.4	....	9.3
1998 [2]	2,054,709	56.5	33.9	....	9.6
1999 [2]	2,186,950	58.3	33.9	....	7.8
2000 [2]	2,166,060	57.7	34.1	....	8.2
2001 [2]	2,149,683	57.2	34.9	....	8.0

[1] Includes Alaska and Hawaii. [2] Excludes Alaska and Hawaii.

ERS, Resource Economics Division, (202) 694–5575. Data for 1910-1992 is from the Census of Agriculture, U.S. Department of Commerce. Data for 1997-2001 is from ERS Agricultural Resource Management Study.

Table 9-8.—Farms: Classification by Tenants and Part Owners, United States, 1900–97

Year	Land in farms	Land rented by operators [1]			Percentage of land rented
		Tenants	Part-owners	Total	
	Million acres	*Million acres*	*Million acres*	*Million acres*	*Percent*
1900	841.8	195.1	[2] 71.1	266.2	31.6
1910	878.8	225.5	[3] 51.3	277.8	31.6
1920	958.7	[4] 265.0	[5] 54.7	319.7	33.3
1925	924.3	264.9	96.3	361.2	39.0
1930	990.1	307.3	125.2	432.5	43.6
1935	1,054.5	336.8	134.3	471.1	44.6
1940	1,165.1	313.2	155.9	469.1	44.0
1945	1,141.6	251.6	178.9	430.5	37.7
1950	1,161.4	212.2	196.2	408.4	35.2
1954	1,158.2	192.6	212.3	404.9	34.9
1959	1,123.0	166.8	234.1	400.9	35.7
1964	1,110.2	144.9	248.1	[6] 393.0	35.4
1969	1,063.3	137.6	241.8	379.4	35.7
1974	1,017.0	122.3	258.4	380.7	37.4
1978	1,029.7	124.1	285.3	406.3	39.4
1982	986.2	113.6	269.9	383.5	38.9
1987	964.5	126.9	275.4	402.3	41.7
1992	945.5	122.7	282.2	404.9	42.8
1997	931.8	108.1	270.0	378.1	40.6

[1] Columns 3,4, and 5 refer only to land rented from others and operated, so subleased land is not included. Numbers of land rented are comparable in the same year, but definitions change over time. Basic sources are 1969 Census of Agriculture, table 5, p.14; 1974 Census of Agriculture, table 3, pp.1-6; 1978 Census of agriculture, vol. 1, part 51, table 5, pp. 124-127; 1982 Census of Agriculture, vol. 1, part 51, table 48, p. 49; 1987 Census of Agriculture vol. 1 part 51, table 48, p.49; 1992 Census of Agriculture vol. 1, part 51, table 46, p.53; 1997 Census of Agriculture, vol. 1, part 51, chapter 1, table 46, p. 57; and earlier census volumes as noted. [2]Sum of part owners and owner/tenant, 1900 Census of Agriculture, table 20, pp.308. [3]Assumes land leased by part-owners is the difference between the average size of full-owner and part-owner farms. Acreage leased by part-owners is this difference times the number of part-owners. 1910 Census of Agriculture, chapter 11, table 1 and 3, pp.97-99. [4]1920 Census of Agriculture, vol. VI, part 1, table 5, p.19. [5]Assumes same proportion of owner and part-owner as in 1910. [6] 1964 Census of Agriculture, vol. II, chapter 8, p.757.

ERS, Resource Economics Division, (202) 694–5572. Data from the Census of Agriculture, National Agricultural Statistics Service and Economic Research Service.

Table 9-9.—Farms: Number and land in farms, by States, 2001 and 2002

State	Farms[1]		Land in farms	
	2001	2002[2]	2001	2002[2]
	Number	*Number*	*1,000 acres*	*1,000 acres*
AL	47,000	47,000	8,900	8,900
AK	580	590	920	920
AZ	7,300	7,300	26,600	26,500
AR	48,000	48,500	14,600	14,600
CA	85,000	84,000	27,700	27,700
CO	30,000	30,000	31,300	31,300
CT	3,900	3,900	360	360
DE	2,500	2,400	570	560
FL	44,000	44,000	10,200	10,200
GA	50,000	50,000	11,000	11,000
HI	5,300	5,300	1,440	1,440
ID	24,000	24,000	11,900	11,900
IL	76,000	76,000	27,700	27,700
IN	63,000	63,000	15,400	15,400
IA	93,500	92,500	32,700	32,600
KS	63,000	63,000	47,400	47,400
KY	88,000	89,000	13,600	13,600
LA	29,000	29,000	8,050	8,050
ME	6,700	6,700	1,260	1,260
MD	12,400	12,200	2,100	2,100
MA	6,000	6,000	560	560
MI	52,000	52,000	10,400	10,400
MN	79,000	79,000	28,500	28,400
MS	42,000	43,000	11,000	11,000
MO	108,000	107,000	29,900	29,800
MT	27,500	28,000	56,700	56,700
NE	53,000	52,000	46,400	46,400
NV	3,000	3,000	6,800	6,800
NH	3,100	3,100	420	410
NJ	9,600	9,600	830	820
NM	15,000	15,000	44,000	44,000
NY	37,500	37,000	7,600	7,600
NC	56,000	56,000	9,100	9,100
ND	30,300	30,000	39,400	39,400
OH	78,000	78,000	14,800	14,700
OK	86,000	87,000	34,000	34,000
OR	40,000	41,000	17,200	17,200
PA	59,000	59,000	7,700	7,700
RI	700	700	60	60
SC	24,000	24,500	4,800	4,800
SD	32,500	32,500	44,000	44,000
TN	91,000	90,000	11,800	11,700
TX	227,000	230,000	130,000	131,000
UT	15,000	15,000	11,600	11,600
VT	6,600	6,600	1,340	1,340
VA	49,000	49,000	8,700	8,700
WA	39,000	39,000	15,700	15,700
WV	20,500	20,500	3,600	3,600
WI	77,000	77,000	16,100	15,900
WY	9,200	9,200	34,600	34,600
US	2,155,680	2,158,090	941,310	941,480

[1] A farm is any establishment from which $1,000 or more of agricultural products were sold or would normally be sold during the year. [2] Preliminary.

NASS, Environmental, Economics, and Demographics Branch, (202) 720–6146.

Table 9-10.—Land: Utilization, by States, 1997

State	Cropland			Grassland pasture [2]	Forest land [3]	Special use areas [4]	Other land [5]	Total land area [6]
	Used for crops [1]	Idle	Used only for pasture					
	1,000 acres	*1,000 acres*	*1,000 acres*	*1,000 acres*	*1,000 acres*	*1,000 acres*	*1,000 acres*	*1,000 acres*
AL	2,298	533	1,639	1,860	21,911	1,423	2,815	32,480
AK	34	26	8	1,226	87,936	143,013	132,796	365,039
AZ	951	187	116	40,509	16,306	10,092	4,571	72,731
AR	7,635	413	2,034	2,006	18,392	1,450	1,398	33,328
CA	8,675	706	1,246	22,343	32,579	20,996	13,277	99,823
CO	8,899	1,780	736	27,867	18,781	5,699	2,623	66,386
CT	131	8	27	30	1,682	299	923	3,101
DE	434	7	11	8	376	102	313	1,251
DC	0	0	0	0	0	0	39	39
FL	2,465	288	896	5,455	14,605	4,676	6,172	34,558
GA	5,00	733	1,395	1,336	23,004	1,854	3,544	37,068
HI	101	150	42	961	1,189	769	898	4,111
ID	4,197	753	816	21,165	17,123	5,266	3,641	52,961
IL	23,140	932	853	1,559	4,058	1,901	3,137	35,580
IN	12,516	516	658	1,158	4,342	1,102	2,666	22,957
IA	24,259	1,578	2,074	1,477	1,944	1,550	2,878	35,760
KS	26,955	2,964	3,789	12,560	1,492	1,620	2,987	52,367
KY	4,889	766	3,205	1,491	12,348	996	1,733	25,429
LA	4,053	563	869	1,582	13,691	1,395	5,729	27,882
ME	324	77	65	37	16,952	520	1,778	19,753
MD	1,344	62	149	208	2,424	731	1,338	6,256
MA	160	12	39	35	2,675	553	1,542	5,016
MI	7,098	705	500	1,606	18,667	2,468	5,313	36,358
MN	20,090	1,707	1,041	1,544	14,820	4,398	7,353	50,954
MS	4,602	677	1,184	1,946	18,589	848	2,179	30,025
MO	12,956	1,645	5,413	6,010	13,411	1,740	2,921	44,095
MT	14,527	2,374	1,672	46,039	19,165	6,414	2,965	93,156
NE	20,314	1,299	1,942	21,828	797	1,423	1,599	49,202
NV	546	56	265	46,278	8,199	5,726	9,204	70,275
NH	84	6	22	40	4,551	317	720	5,740
NJ	526	40	68	29	1,507	728	1,850	4,748
NM	1,313	474	639	52,188	14,084	6,360	2,615	77,673
NY	3,162	317	633	1,314	15,405	3,810	5,581	30,223
NC	4,487	476	927	814	18,638	2,264	3,574	31,180
ND	24,460	2,858	1,500	11,329	441	1,489	2,079	44,156
OH	10,576	556	895	1,376	7,567	1,153	4,087	26,209
OK	9,793	1,184	5,360	17,314	6,233	1,477	2,593	43,954
OR	3,853	566	919	22,395	26,664	3,593	3,450	61,441
PA	4,187	299	695	910	15,852	2,379	4,364	28,685
RI	22	2	6	3	356	61	220	669
SC	1,660	335	538	465	12,418	1,032	2,824	19,271
SD	17,313	1,909	2,542	22,594	1,588	1,575	1,051	48,573
TN	4,307	596	2,587	1,123	13,265	2,203	2,298	26,380
TX	22,613	5,092	12,335	98,059	11,767	5,363	12,396	167,625
UT	1,252	234	558	23,737	13,832	5,058	7,916	52,588
VT	342	10	132	212	4,462	337	425	5,920
VA	2,572	242	1,526	1,533	15,345	1,468	2,657	25,343
WA	6,854	1,018	528	7,406	17,418	6,639	2,749	42,612
WV	664	50	697	481	11,899	699	925	15,415
WI	7,803	799	959	1,844	15,701	2,182	5,472	34,761
WY	2,064	257	759	44,873	5,085	6,332	2,777	62,147
US	348,701	38,839	67,512	580,165	641,536	285,544	300,957	2,263,254

[1] Cropland harvested, crop failure, and cultivated summer fallow. [2] Grassland and other nonforest pasture and range. [3] Excludes reserved and other forest land duplicated in parks and other special uses of land. Includes forested grazing land. [4] Includes rural transportation areas, Federal and State areas used primarily for recreation and wildlife purposes, military areas, farmsteads, and farm roads and lanes. [5] Miscellaneous areas such as marshes, open swamps, bare rock areas, and deserts, including urban and other special uses not inventoried. [6] Approximate land area as established by the Bureau of the Census in conjunction with the 1990 Census of Population.

ERS, Resource Economics Division, (202) 694–5528. Estimates based on reports and records of the U.S. Departments of Agriculture and Commerce, and public land administering and conservation agencies. Estimates developed for years coinciding with a Census of Agriculture.

Table 9-11.—Land in farms: [1] Irrigated land, by States, 1959–97

State	1959	1964	1969	1974	1978 [2]	1982	1987	1992	1997
	1,000 acres	*1,000 acres*	*1,000 acres*	*1,000 acres*	*1,000 acres*	*1,000 acres*	*1,000 acres*	*1,000 acres*	*1,000 acres*
AL	17	12	11	14	59	66	84	82	77
AK	....	(3)	1	1	1	1	2	2	3
AZ	1,152	1,125	1,178	1,153	1,196	1,098	914	956	1,014
AR	712	974	1,010	949	1,683	2,022	2,406	2,702	3,717
CA	7,396	7,599	7,240	7,749	8,506	8,461	7,596	7,571	8,713
CO	2,685	2,690	2,895	2,874	3,431	3,201	3,014	3,170	3,430
CT	5	14	9	7	7	7	7	6	7
DE	16	18	20	20	34	44	61	62	73
FL	414	1,217	1,365	1,559	1,980	1,585	1,623	1,783	1,862
GA	34	64	79	112	463	575	640	725	749
HI	141	144	146	142	159	146	149	134	77
ID	2,577	2,802	2,761	2,859	3,475	3,450	3,219	3,260	3,494
IL	10	14	51	54	130	166	208	328	350
IN	17	17	34	33	75	132	170	241	250
IA	18	22	21	39	101	91	92	116	125
KS	762	1,004	1,522	2,010	2,686	2,675	2,463	2,680	2,707
KY	9	14	20	11	14	23	38	28	58
LA	485	581	702	702	681	694	647	898	943
ME	2	4	6	6	7	6	6	10	22
MD	11	16	22	23	28	39	51	57	69
MA	20	24	19	19	17	17	20	20	25
MI	40	49	77	97	226	286	315	366	393
MN	15	18	36	78	272	315	354	370	380
MS	100	123	150	162	309	431	637	883	1,076
MO	30	59	156	150	320	403	535	709	882
MT	1,875	1,893	1,841	1,759	2,070	2,023	1,997	1,978	1,994
NE	2,078	2,169	2,857	3,967	5,683	6,039	5,682	6,312	6,939
NV	543	825	753	778	881	830	779	556	765
NH	1	3	2	2	2	1	3	2	3
NJ	74	96	72	89	77	83	91	80	93
NM	732	813	823	867	891	807	718	738	805
NY	58	79	55	55	56	52	51	47	69
NC	66	97	59	51	90	81	138	113	156
ND	48	51	63	71	141	163	168	187	180
OH	12	17	22	22	25	28	32	29	34
OK	198	302	524	515	602	492	478	512	506
OR	1,384	1,608	1,519	1,561	1,881	1,808	1,648	1,622	1,949
PA	17	23	19	18	15	18	30	23	36
RI	(3)	1	2	2	3	2	4	3	3
SC	25	19	15	10	32	81	81	76	86
SD	116	130	148	152	335	376	362	371	344
TN	11	11	12	10	13	18	38	37	46
TX	5,656	6,385	6,888	6,594	6,947	5,576	4,271	4,912	5,425
UT	1,062	1,092	1,025	970	1,169	1,082	1,161	1,143	1,212
VT	2	2	(3)	1	1	1	2	2	3
VA	31	51	37	28	42	43	79	62	85
WA	1,007	1,150	1,224	1,309	1,639	1,638	1,519	1,641	1,705
WV	1	2	3	2	1	1	3	3	3
WI	32	62	106	128	235	259	285	331	342
WY	1,470	1,571	1,523	1,460	1,662	1,565	1,518	1,465	1,719
US	33,164	37,057	39,122	41,243	50,350	49,002	46,386	49,404	55,058
PR	76	89	91	70	(4)	(4)	(4)	(4)	(4)
VI	(3)	(3)	(3)	(3)	(4)	(4)	(4)	(4)	(4)
Total	33,240	37,145	39,213	41,313	50,350	49,002	46,386	49,404	55,058

[1] Data may not add because of rounding. [2] Data for 1978 not directly comparable with earlier censuses as it includes estimates from the direct enumeration sample for farms not represented on the mail list. [3] Less than 500 acres. [4] Not available. Note: Data from the Census of Agriculture, U.S. Department of Commerce. Beginning in 1997 Census of Agriculture, U.S. Department of Agriculture.

ERS, Resource Economics Division, (202) 694–5528.

Table 9-12.—Farm real estate: Value of farmland and buildings, by States, 1998–2002[1]

State	Total value of land and buildings				
	Jan. 1, 1998	Jan. 1, 1999	Jan. 1, 2000	Jan. 1, 2001	Jan. 1, 2002
	Million dollars	*Million dollars*	*Million dollars*	*Million dollars*	*Million dollars*
AL	13,680	13,984	15,120	16,020	16,910
AZ[2]	11,741	11,781	11,666	12,725	13,622
AR	16,963	17,873	18,250	18,980	20,002
CA	74,385	77,006	79,230	83,100	85,870
CO	19,900	20,034	21,172	21,754	22,223
CT	2,261	2,331	2,376	2,484	2,628
DE	1,543	1,595	1,624	1,682	1,652
FL	23,744	23,504	24,720	26,520	28,560
GA	17,063	18,256	20,868	23,100	25,300
ID	12,240	12,971	13,923	14,280	14,875
IL	59,214	62,325	65,926	70,081	73,128
IN	32,136	34,410	36,425	38,500	39,886
IA	56,100	58,410	59,696	62,130	64,548
KS	27,408	27,550	28,025	28,914	29,388
KY	20,010	20,808	21,760	24,072	25,160
LA	9,922	9,862	10,125	10,224	10,546
ME	1,523	1,524	1,588	1,638	1,764
MD	6,837	6,930	7,560	7,980	8,400
MA	2,970	3,135	3,363	3,640	4,032
MI	17,368	19,240	22,360	23,920	26,000
MN	33,524	35,424	37,752	38,760	41,180
MS	12,180	12,540	13,098	13,750	14,300
MO	32,207	34,013	37,500	41,262	45,296
MT	16,905	16,872	19,845	21,263	21,773
NE	29,928	31,088	32,248	33,872	35,032
NV[2]	2,680	2,827	2,971	3,114	3,150
NH	945	945	966	1,008	1,066
NJ	5,810	5,810	5,893	6,142	6,560
NM[2]	8,821	8,691	8,539	8,650	8,650
NY	9,984	10,452	10,857	11,400	12,160
NC	19,552	20,925	23,000	25,480	26,390
ND	15,840	15,996	16,351	16,745	17,336
OH	30,396	33,078	34,270	36,704	39,690
OK	20,740	21,250	21,556	22,780	24,140
OR	16,512	17,200	17,544	18,060	18,920
PA	18,403	19,250	20,944	21,868	22,715
RI	423	390	396	414	438
SC	7,252	7,372	7,680	7,920	8,160
SD	15,312	15,840	16,720	17,820	19,360
TN	21,539	23,205	25,155	26,432	27,027
TX	77,980	79,605	81,900	88,400	94,320
UT[2]	6,576	6,978	7,359	7,958	8,557
VT	2,037	2,104	2,211	2,345	2,546
VA	16,896	17,748	19,140	20,445	21,663
WA	18,683	18,683	18,840	18,683	18,683
WV	4,033	3,852	4,140	4,608	4,932
WI	20,336	22,331	27,540	32,200	34,980
WY	7,681	7,612	8,304	8,996	9,861
48 States	900,179	935,610	988,495	1,048,821	1,103,378

[1] Total value of land and buildings is derived by multiplying average value per acre of farm real estate by the land in farms. [2] Value of All land and Buildings adjusted to include reservation land value.

NASS, Environmental, Economics, and Demographics Branch, (202) 720–6146.

Table 9-13.—Land utilization, United States, selected years, 1940–97

Major land uses	1940	1950	1959	1969	1978	1987	1992	1997
	Million acres	*Million acres*	*Million acres*	*Million acres*	*Million acres*	*Million acres*	*Million acres*	*Million acres*
Cropland used for crops [1]	368	377	359	333	369	331	337	349
Idle cropland	31	32	33	51	26	68	56	39
Cropland used only for pasture	68	69	66	88	76	65	67	67
Grassland pasture [2]	650	631	633	604	587	591	591	580
Forest land [3]	608	601	728	723	703	648	648	642
Special uses [4]	....	....	147	174	203	335	340	351
Other land	179	194	305	291	301	227	224	235
Total land area [5]	1,904	1,904	2,271	2,264	2,265	2,265	2,263	2,263

[1] Cropland harvested, crop failure, and cultivated summer fallow. [2] Grassland and other nonforest pasture and range. [3] Excludes reserved and other forest land duplicated in parks and other special uses of land. Includes forested grazing land. [4] Includes urban and transportation areas. Federal and State areas used primarily for recreation and wildlife purposes, military areas, farmsteads and farm roads and lanes. [5] Remeasurement and increases in reserviors account for changes in total land areas except for the major increase in 1959 when data for Alaska and Hawaii were added.

ERS, Resource Economics Division, (202) 694–5528. Estimates based on reports and records of the U.S. Department of Agriculture and Commerce, and public land administering and conservation agencies.

Table 9-14.—Farm real estate: Average value per acre of land and buildings, by State, Mar. 1, 1970, and Jan. 1, 1998–2002

State	Mar. 1, 1970	Jan. 1, 1998	Jan. 1, 1999	Jan. 1, 2000	Jan. 1, 2001	Jan. 1, 2002
	Dollars	*Dollars*	*Dollars*	*Dollars*	*Dollars*	*Dollars*
AL	200	1,440	1,520	1,680	1,800	1,900
AZ [1]	70	987	1,070	1,180	1,360	1,520
AR	260	1,150	1,220	1,250	1,300	1,370
CA	479	2,610	2,770	2,850	3,000	3,100
CO	95	618	630	670	695	710
CT	921	5,950	6,300	6,600	6,900	7,300
DE	499	2,660	2,750	2,800	2,950	2,950
FL	355	2,240	2,260	2,400	2,600	2,800
GA	234	1,510	1,630	1,880	2,100	2,300
ID	177	1,020	1,090	1,170	1,200	1,250
IL	490	2,130	2,250	2,380	2,530	2,640
IN	406	2,060	2,220	2,350	2,500	2,590
IA	392	1,700	1,770	1,820	1,900	1,980
KS	159	577	580	590	*610	620
KY	253	1,450	1,530	1,600	1,770	1,850
LA	321	1,210	1,210	1,250	1,270	1,310
ME	161	1,190	1,200	1,250	1,300	1,400
MD	640	3,180	3,300	3,600	3,800	4,000
MA	565	5,210	5,500	5,900	6,500	7,200
MI	326	1,670	1,850	2,150	2,300	2,500
MN	226	1,160	1,230	1,320	1,360	1,450
MS	234	1,050	1,100	1,180	1,250	1,300
MO	224	1,070	1,130	1,250	1,380	1,520
MT	60	294	296	350	375	384
NE	154	645	670	695	730	755
NV [1]	53	392	420	440	460	460
NH	239	2,250	2,250	2,300	2,400	2,600
NJ	1,092	7,000	7,000	7,100	7,400	8,000
NM [1]	42	217	217	217	220	220
NY	273	1,280	1,340	1,410	1,500	1,600
NC	333	2,080	2,250	2,500	2,800	2,900
ND	94	401	406	415	425	440
OH	399	2,040	2,220	2,300	2,480	2,700
OK	173	610	625	634	670	710
OR	150	960	1,000	1,020	1,050	1,100
PA	373	2,390	2,500	2,720	2,840	2,950
RI	734	6,500	6,500	6,600	6,900	7,300
SC	261	1,480	1,520	1,600	1,650	1,700
SD	84	348	360	380	405	440
TN	268	1,810	1,950	2,150	2,240	2,310
TX	148	593	610	630	680	720
UT [1]	92	807	855	900	975	1,050
VT	224	1,520	1,570	1,650	1,750	1,900
VA	286	1,920	2,040	2,200	2,350	2,490
WA	224	1,190	1,190	1,200	1,190	1,190
WV	136	1,090	1,070	1,150	1,280	1,370
WI	232	1,240	1,370	1,700	2,000	2,200
WY	41	222	220	240	260	285
48 States	196	974	1,020	1,080	1,150	1,210

[1] Excludes Native American Reservation Land.

NASS, Environmental, Economics, and Demographics Branch, (202) 720–6146.

Table 9-15.—Land values, cropland and pasture: By States, 2001–2002

State	2001				2002			
	Cropland [1]	Irrigated cropland	Non-irrigated cropland	Pasture [2]	Cropland [1]	Irrigated cropland	Non-irrigated cropland	Pasture [2]
	Dollars	*Dollars*	*Dollars*	*Dollars*	*Dollars*	*Dollars*	*Dollars*	*Dollars*
AL	1,630	....	....	1,450	1,690	....	....	1,580
AZ	5,400	5,400	....	400	6,000	6,000	....	450
AR	1,140	1,260	1,030	1,000	1,160	1,310	1,030	1,080
CA	6,320	6,800	1,400	1,040	6,510	7,000	1,420	1,060
CO	911	1,750	460	360	929	1,800	460	360
CT	....	....	....	....	....	....	....	....
DE	2,900	....	....	....	2,900	....	....	....
FL	3,540	4,370	1,800	1,800	3,690	4,520	1,950	2,000
GA	1,530	1,700	1,500	2,400	1,660	1,800	1,630	2,600
ID	1,570	2,020	790	810	1,660	2,150	790	820
IL	2,650	....	....	1,180	2,750	....	....	1,220
IN	2,550	....	....	1,700	2,640	....	....	1,730
IA	2,050	....	....	770	2,120	....	....	800
KS	680	1,080	640	400	690	1,100	650	400
KY	2,000	....	....	1,450	2,100	....	....	1,500
LA	1,160	1,070	1,180	1,160	1,200	1,100	1,230	1,200
ME	....	....	....	....	....	....	....	....
MD	3,700	....	....	3,450	3,900	....	....	3,500
MA	....	....	....	....	....	....	....	....
MI	2,100	....	....	1,180	2,300	....	....	1,250
MN	1,340	....	....	500	1,420	....	....	530
MS	1,040	1,120	1,020	1,060	1,080	1,170	1,060	1,100
MO	1,390	1,940	1,350	950	1,470	2,040	1,430	1,040
MT	502	1,500	360	255	510	1,530	365	260
NE	1,170	1,600	950	245	1,210	1,640	980	250
NV	2,000	2,000	....	270	2,000	2,000	....	270
NH	....	....	....	....	....	....	....	....
NJ	8,100	....	....	9,100	8,700	....	....	960
NM	1,450	2,600	270	150	1,450	2,600	270	150
NY	1,240	....	....	620	1,300	....	....	700
NC	2,700	....	....	2,600	2,800	....	....	2,750
ND	430	....	....	165	440	....	....	170
OH	2,600	....	....	1,750	2,750	....	....	1,900
OK	581	780	570	435	620	800	610	460
OR	1,650	2,300	1,120	405	1,680	2,350	1,150	420
PA	3,260	....	....	2,000	3,400	....	....	2,100
RI	....	....	....	....	....	....	....	....
SC	1,300	....	....	1,600	1,350	....	....	1,700
SD	555	780	550	200	605	850	600	210
TN	2,320	....	....	2,320	2,390	....	....	2,390
TX	801	880	780	610	833	920	810	640
UT	2,960	3,500	750	450	3,140	3,700	850	500
VT	....	....	....	....	....	....	....	....
VA	2,300	3,400	680	2,200	2,450	....	....	2,350
WA	1,310	....	....	500	1,290	3,300	690	500
WV	1,700	....	....	960	1,820	....	....	1,020
WI	1,900	1,040	280	900	2,100	....	....	1,050
WY	866	....	....	180	907	1,090	290	200
Other States	4,770	....	....	4,030	5,120	....	....	4,330
US	1,580	....	....	558	1,650	....	....	586

[1] Other cropland States include CT, ME, MA, NH, RI, and VT. [2] Other pasture States include CT, DE, ME, MA, NH, RI, and VT.

NASS, Environmental, Economics, and Demographics Branch, (202) 720–6146.

Table 9-16.—Cash rents, cropland and pasture: By States, 2001–2002

State	2001				2002			
	Cropland	Irrigated cropland	Non-irrigated cropland	Pasture	Cropland	Irrigated cropland	Non-irrigated cropland	Pasture
	Dollars	*Dollars*	*Dollars*	*Dollars*	*Dollars*	*Dollars*	*Dollars*	*Dollars*
AL	36.00			18.00	36.00			18.00
AZ		135.00				135.00		
AR	67.00	78.00	48.00		72.00	81.00	55.00	
CA		310.00		11.00		310.00		11.00
CO	65.00	100.00	22.00	4.80	68.00	95.00	21.00	4.00
CT								
DE	57.50				56.30			
FL			32.00	15.00			32.00	15.00
GA	55.00	103.00	38.00	23.00	55.0	103.00	39.00	23.00
ID	95.00	115.00	48.00		96.0	116.00	50.00	
IL	119.00			33.00	124.0			30.00
IN	100.00				101.0			
IA	117.00			30.00	120.0			29.70
KS	39.00	72.00	36.00	12.60	39.00	70.00	36.00	12.60
KY	72.00				68.00			
LA	62.00	74.00	56.00	16.00	62.00	73.00	57.00	16.00
ME								
MD	55.50				55.50			
MA								
MI	60.00				60.00			
MN	80.50			19.00	80.90			19.00
MS	62.00	79.00	54.00	16.00	60.00	79.00	52.00	18.00
MO			65.00	22.50			65.00	23.60
MT	24.50	46.00	18.00	4.50	24.00	46.50	18.50	5.00
NE	88.00	117.00	65.00	11.30	90.00	121.00	65.00	10.70
NV								
NH								
NJ	47.00				51.00			
NM				1.60				1.60
NY	33.00				35.00			
NC	48.00			22.00	51.00			21.50
ND	36.00			9.80	36.50			9.70
OH	76.50				76.50			
OK			27.00	8.30			27.00	8.50
OR	90.00	110.00			95.00	115.00		
PA	42.00			32.00	40.00			28.00
RI								
SC	27.50				28.50			
SD			40.00	11.00			42.00	10.80
TN	59.50			18.00	60.50			16.50
TX	27.00	53.00	21.00	7.20	27.00	51.00	21.00	7.20
UT		58.00		13.00		62.00		9.00
VT								
VA	36.50			16.50	36.00			16.00
WA		165.00				160.00		
WV	26.00				27.00			
WI	66.00			36.00	66.00			38.00
WY				3.50				3.80
48 Sts	71.00			9.20	71.50			9.20

NASS, Environmental, Economics, and Demographics Branch, (202) 720–6146.

Table 9-17.—Farm assets and claims: Comparative balance sheet of the farming sector, excluding operator households, United States, Dec. 31, 1992–2001 [1]

Item	1992	1993	1994	1995	1996
ASSETS	*Billion dollars*	*Billion dollars*	*Billion dollars*	*Billion dollars*	*Billion dollars*
Physical assets:					
Real estate	640.8	677.6	704.1	740.5	769.5
Non-real estate:					
Livestock [2]	71.0	72.8	67.9	57.8	60.3
Machinery and motor vehicles	84.8	85.4	86.8	87.6	88.8
Crops stored on and off farms [3]	24.2	23.3	23.3	27.4	31.7
Purchased inputs	3.9	3.8	5.0	3.4	4.4
Financial assets: [4]	43.1	46.3	47.6	49.1	49.0
Total [5]	867.8	909.2	934.7	965.7	1003.0
CLAIMS					
Liabilities:					
Real estate debt	75.4	76.0	77.7	79.3	81.7
Non-real estate debt to—					
Reporting institutions [6]	50.4	51.7	53.9	55.3	57.0
Nonreporting creditors [7]	13.2	14.2	15.2	16.2	17.4
Total liabilities [5]	139.1	142.0	146.8	150.8	156.1
Proprietors' equity	728.7	767.2	787.9	815.0	846.9
Total [5]	867.8	909.2	934.7	965.8	1003.0

Item	1997	1998	1999	2000	2001 [8]
ASSETS	*Billion dollars*	*Billion dollars*	*Billion dollars*	*Billion dollars*	*Billion dollars*
Physical assets:					
Real estate	808.2	840.4	889.1	949.4	998.7
Non-real estate:					
Livestock [2]	67.1	63.4	73.2	76.8	73.2
Machinery and motor vehicles	88.7	89.8	90.2	90.3	90.7
Crops stored on and off farms [3]	32.7	29.9	28.3	27.9	252.2
Purchased inputs	4.9	5.0	4.0	4.9	4.2
Financial assets: [4]	49.7	54.8	56.6	57.1	59.0
Total [5]	1051.3	1083.4	1141.1	1206.5	1251.0
CLAIMS					
Liabilities:					
Real estate debt	85.4	89.6	94.2	97.6	103.0
Non-real estate debt to—					
Reporting institutions [6]	61.2	63.6	61.9	65.2	67.4
Nonreporting creditors [7]	18.8	19.6	20.3	21.3	21.7
Total liabilities [5]	165.4	172.9	176.4	184.1	192.0
Proprietors' equity	885.9	910.5	964.7	1022.3	1059.0
Total [5]	1051.3	1083.4	1141.1	1206.4	1251.0

[1] Farms are defined as places with sales greater than $1,000 annually. [2] Horses and mules are excluded. [3] Excludes all crops held on farms including crops under loan to Commodity Credit Corporation, and crops held off farms as security for CCC loans. [4] Includes farm share of currency and demand deposits. [5] Total of rounded data. [6] Loans of all operating banks, the Farm Credit System, and direct loans of the Farm Service Agency. [7] Loans and credits extended by dealers, merchants, finance companies, individuals, and others. [8] Preliminary.

ERS, Farm Sector Performance and Well-Being Branch, (202) 694–5586.

Table 9-18.—Farm labor: Number of workers on farms and average wage rates, United States, 1994–2003 [1]

Year	Total workers	Self-employed and unpaid workers [2]	Ag service workers [3]	Hired workers [3 4]	Hired workers [3 4]
	Number	*Number*	*Number*	*Number*	*Wage rates*
1994.					
Jan	2,480	1,703	140	637	6.54
Apr	2,919	1,876	240	803	6.42
July	3,510	2,129	341	1,040	6.21
Oct	3,165	2,006	279	880	6.46
1995.					
Jan	2,460	1,691	140	629	6.73
Apr	2,857	1,821	214	822	6.49
July	3,672	2,258	348	1,066	6.44
Oct	3,383	2,127	302	954	6.58
Annual average	....	1,967.1	(5)	868.5	6.54
1996.					
Jan	2,462	1,749	115	598	6.89
Apr	2,906	1,919	207	780	6.76
July	(6)	(6)	331	1,015	6.55
Oct	(6)	(6)	291	935	6.97
Annual average	....	2,010	(5)	832	6.78
1997.					
Jan	(6)	(6)	131	624	7.20
Apr	(6)	(6)	207	808	7.03
July	(6)	(6)	340	1,069	6.88
Oct	(6)	(6)	283	1,004	7.31
Annual average	....	1,989.9	(5)	876.5	7.35
1998.					
Jan	(6)	(6)	141	661	7.61
Apr	(6)	(6)	202	803	7.49
July	(6)	(6)	379	1071	7.25
Oct	(6)	(6)	263	983	7.60
Annual average	....	1,946.6	(5)	879.5	7.47
1999.					
Jan	(6)	(6)	157	705	7.94
Apr	(6)	(6)	160	867	7.83
July	(6)	(6)	319	1,155	7.58
Oct	(6)	(6)	290	989	7.83
Annual average	....	2,048.4	(5)	929	7.77
2000.					
Jan	....	(6)	172	685	8.10
Apr	....	(6)	217	840	8.09
July	....	(6)	203	1,084	7.93
Oct	....	(6)	288	952	8.29
Annual average	....	2,062.3	(5)	890.3	8.10
2001.					
Jan	....	(6)	165	691	8.66
Apr	....	(6)	215	804	8.31
July	....	(6)	335	1,039	8.29
Oct	....	(6)	262	991	8.59
Annual average	....	2,049.8	(5)	873.3	8.45
2002.					
Jan	....	(6)	183	707	8.97
Apr	....	(6)	189	890	8.83
July	....	(6)	256	1,006	8.57
Oct	....	(6)	271	940	8.95
Annual average	....	(6)	(5)	884.5	8.80
2003.					
Jan	....	(6)	160	724	9.32

[1] Beginning in July 1984, three surveys a year were conducted in the 48 contiguous States and Hawaii. In 1989, January was added to the survey program. Data are obtained from establishments for the week that includes the 12th. [2] Includes farm operators and partners doing 1 or more hours of farm work and other unpaid workers working 15 hours or more during the survey week without cash wages. [3] Includes all persons doing farm work for pay during the survey week. [4] Excludes agricultural service workers. [5] Annual average not computed. [6] Discontinued.

NASS, Economic, Environmental and Demographics Branch, (202) 720–6146.

Table 9-19.—Farm labor: Number of hired workers on farms and average wage rates, by States and regions, 2002 [1] [2]

State and region [3]	Workers on farms	Farm wage rates			
		Type of worker			
	Hired	Field	Livestock	Field and live-stock	All hired workers [4]
	Thousands	*Dollars per hour*	*Dollars per hour*	*Dollars per hour*	*Dollars per hour*
Jan. 6–12, 2002					
Northeast I	33	9.58	7.80	8.69	9.46
Northeast II	33	8.71	6.88	7.95	8.87
Appalachian I	29	7.73	7.94	7.83	8.72
Appalachian II	25	6.72	7.18	6.93	7.29
Southeast	25	7.56	7.61	7.58	8.08
FL	62	8.15	8.55	8.20	8.97
Lake	38	10.22	8.26	8.94	9.91
Cornbelt I	30	9.03	9.22	9.15	9.75
Cornbelt II	24	8.62	9.41	9.26	9.74
Delta	24	7.90	7.21	7.61	7.95
N. Plains	28	8.34	8.41	8.39	9.00
S. Plains	61	7.82	7.37	7.57	8.05
Mountain I	18	8.02	7.74	7.81	8.34
Mountain II	17	8.54	8.48	8.51	9.15
Mountain III	18	7.57	8.48	7.85	8.42
Pacific	48	8.18	9.64	8.42	9.21
CA	186	8.25	9.69	8.36	9.22
HI	8	9.36	....	9.33	10.90
US (49 States)	707	8.25	8.20	8.23	8.97
Apr. 7–13, 2002					
Northeast I	44	8.35	8.56	8.42	9.14
Northeast II	45	8.20	7.01	7.78	8.83
Appalachian I	36	7.42	8.60	7.86	8.45
Appalachian II	23	6.89	7.87	7.31	7.63
Southeast	44	7.28	8.29	7.42	7.94
FL	52	7.75	7.50	7.72	8.57
Lake	55	9.43	8.44	9.01	9.88
Cornbelt I	43	8.59	8.90	8.70	9.23
Cornbelt II	27	7.84	9.82	8.89	9.34
Delta	30	7.05	7.06	7.05	7.50
N. Plains	37	8.48	8.06	8.31	9.22
S. Plains	60	6.92	7.78	7.31	7.95
Mountain I	28	7.55	7.92	7.80	8.00
Mountain II	25	8.06	8.58	8.26	8.95
Mountain III	20	7.51	8.35	7.71	8.21
Pacific	69	8.18	9.45	8.33	9.10
CA	245	8.33	10.20	8.47	9.13
HI	7	9.40	....	9.43	11.14
US (49 States)	890	8.06	8.43	8.15	8.83

See footnotes at end of table.

Table 9-19.—Farm labor: Number of hired workers on farms and average wage rates, by States and regions [1] [2], 2002—Continued

State and region [3]	Workers on farms	Farm wage rates			
		Type of worker			
	Hired	Field	Livestock	Field and livestock	All hired workers [4]
	Thousands	*Dollars per hour*	*Dollars per hour*	*Dollars per hour*	*Dollars per hour*
July 7–13, 2002					
Northeast I	44	8.41	7.90	8.25	9.12
Northeast II	46	7.87	7.18	7.74	8.35
Appalachian I	60	7.61	7.91	7.68	8.32
Appalachian II	32	7.22	7.64	7.36	7.75
Southeast	34	7.11	8.90	7.50	8.01
FL	43	7.25	7.80	7.38	8.48
Lake	62	7.85	7.85	7.85	8.33
Cornbelt I	54	8.08	9.07	8.36	9.13
Cornbelt II	30	8.42	10.72	8.94	9.20
Delta	37	6.70	7.10	6.80	7.07
No. Plains	39	8.32	8.57	8.41	8.72
So. Plains	74	7.06	7.23	7.14	7.53
Mountain I	30	7.12	7.74	7.42	7.62
Mountain II	19	7.23	7.35	7.27	7.82
Mountain III	17	7.15	8.11	7.43	8.03
Pacific	98	8.32	10.65	8.86	9.36
CA	*280	*8.18	*9.65	*8.32	*8.87
HI	7	9.65	....	9.67	11.29
US (49 States)	*1,006	*7.90	*8.38	8.02	*8.57
October 6–12, 2002					
Northeast I	50	9.16	7.70	8.76	9.36
Northeast II	41	8.61	8.17	8.51	9.29
Appalachian I	42	7.57	8.00	7.70	8.21
Appalachian II	36	7.13	7.46	7.21	7.50
Southeast	33	7.30	8.25	7.56	8.18
FL	57	7.50	8.60	7.72	8.67
Lake	64	9.28	9.11	9.21	9.73
Cornbelt I	42	8.49	8.89	8.54	8.90
Cornbelt II	25	8.69	8.30	8.59	8.81
Delta	38	7.12	7.72	7.27	7.77
No. Plains	34	9.01	9.03	9.02	9.32
So. Plains	58	7.05	7.32	7.18	7.57
Mountain I	30	7.84	7.95	7.89	8.13
Mountain II	16	8.31	8.79	8.50	8.79
Mountain III	19	7.07	8.24	7.42	8.25
Pacific	77	8.91	9.36	8.99	9.47
CA	*270	*8.60	*9.30	*8.68	*9.39
HI	8	9.25	....	9.32	11.02
US (49 States)	*940	*8.34	*8.42	*8.36	*8.95

[1] Excludes agricultural service workers. [2] Includes all persons doing work for pay during the survey week. [3] Regions consist of the following: Northeast I: CT, ME, MA, NH, NY, RI, VT; Northeast II: DE, MD, NJ, PA; Appalachian I: NC, VA; Appalachian II: KY, TN, WV; Southeast: AL, GA, SC; Lake: MI, MN, WI; Cornbelt I: IL, IN, OH; Cornbelt II: IA, MO; Delta: AR, LA, MS; No. Plains: KS, NE, ND, SD; So. Plains: OK, TX; Mountain I: ID, MT, WY; Mountain II: CO, NV, UT; Mountain III: AZ, NM; Pacific: OR, WA. [4] Includes field, livestock, supervisors, and other workers doing work for pay during the survey week. * Revised.

NASS, Economic, Environmental and Demographics Branch, (202) 720–6146.

Table 9-20.—Farm production and output: Index numbers of total output, and production of livestock, crops, and secondary output, by groups, United States, 1990–99

[1992=100]

Year	Farm output	Livestock and products			
		All livestock and products [1]	Meat animals [2]	Dairy products [3]	Poultry and eggs [4]
1990	0.900	0.901	0.920	0.955	0.761
1991	0.906	0.920	0.929	0.955	0.794
1992	0.956	0.946	0.947	0.977	0.833
1993	0.911	0.957	0.952	0.976	0.869
1994	1.018	1.007	1.007	0.996	0.912
1995	0.965	1.022	1.029	1.008	0.947
1996	1.000	1.000	1.000	1.000	1.000
1997	1.044	1.034	1.033	1.014	1.022
1998	1.045	1.044	1.035	1.022	1.041
1999	1.065	1.076	1.066	1.058	1.087

Year	Crops				
	All crops	Cereal crops	Forage crops	Industrial crops [5]	Vegetables and Horticulture crops
1990	0.894	0.940	0.991	0.850	0.823
1991	0.885	0.834	1.019	0.992	0.848
1992	0.964	1.047	0.994	0.939	0.869
1993	0.868	0.791	0.981	0.849	0.909
1994	1.031	1.057	1.011	1.056	1.001
1995	0.920	0.833	1.014	0.932	0.961
1996	1.000	1.000	1.000	1.000	1.000
1997	1.043	1.005	1.040	1.104	1.015
1998	1.031	1.044	1.031	1.044	1.021
1999	1.035	1.003	1.076	1.034	1.061

Year	Crops	
	Fruits and nuts	Secondary output [6]
1990	0.915	0.987
1991	0.907	1.047
1992	0.959	0.981
1993	1.016	1.012
1994	1.071	0.982
1995	0.996	1.050
1996	1.000	1.000
1997	1.184	1.149
1998	1.037	1.207
1999	1.129	1.289

[1] Includes wool, mohair, horses, mules, honey, beeswax, bees, goats, rabbits, aquaculture, and fur animals. These items are not included in the separate groups of livestock and products shown. [2] Cattle and calves, sheep and lambs, and hogs. [3] Butter, butterfat, wholesale milk, retail milk, and milk consumed on farms. [4] Chicken eggs, commercial broilers, chickens, and turkeys. [5] Includes soybeans, peanuts harvested for nuts, sunflower seed, flaxseed, cottonseed, cotton lint, tobacco, sugar crops, forest products, legumes and grass seeds, hops, mint, broomcorn, popcorn, hemp fiber and seed, and flax fiber. [6] These activities are defined as activities closely linked to agriculture for which information on production and input use cannot be separately observed.

ERS, Resources, Technology and Production Branch (202) 694–5601.

Table 9-21.—Hired farmworkers: Number of Workers and Median Weekly Earnings, 2000–2002 [1]

Characteristics	Workers			Median Weekly Earnings [2]		
	2000	2001	2002 [3]	2000	2001	2002 [3]
	Thousands	*Thousands*	*Thousands*	*Dollars*	*Dollars*	*Dollars*
All workers	878	745	....	280	300	....
15–19 years old	123	109	....	140	132	....
20–24 years old	104	87	....	280	280	....
25–34 years old	204	171	....	313	313	....
35–44 years old	208	189	....	300	342	....
45–54 years old	127	108	....	350	345	....
55 years old and older	111	81	....	270	360	....
Male	720	601	....	300	320	....
Female	158	144	....	246	248	....
White [4]	414	366	....	320	320	....
Black and other races [4]	56	38	....	270	314	....
Hispanic	408	341	....	280	300	....
Schooling completed						
Less than 5th grade	117	79	....	260	300	....
5th-8th grade	185	158	....	276	280	....
9th-12th grade (no diploma)	186	174	....	238	250	....
High school diploma	226	203	....	338	346	....
Beyond high school	164	131	....	385	400	....
Full-time (35 or more hours per week)	709	601	....	319	345	....
Part-time (less than 35 hours per week)	169	144	....	120	130	....

[1] Represents average number of persons 15 years old and over in the civilian noninstitutional population who were employed per week as hired farmworkers. Based on the Current Population Survey microdata earnings file. [2] "Median weekly earnings" is the value that divides the earnings into two equal parts, one part having earnings above the median and the other part having earnings below the median. "Earnings" refers to the weekly earnings the farmworker usually earns at a farmwork job, before deductions, and includes any overtime pay or commissions. [3] Preliminary. [4] Excludes persons of Hispanic origin.

ERS, Food Assistance, and Rural Economy Branch, (202) 694–5438.

Table 9-22.—Crops: Area, United States, 1993–2002

Year	Principal crops: Area harvested: Feed grains [1]	Principal crops: Area harvested: Food grains [2]	Principal crops: Area harvested: Total [3]	Principal crops: Area planted total [3]	Commercial vegetables, harvested area [5]	Fruits and nuts, bearing area [6]
	1,000 acres	*1,000 acres*	*1,000 acres*	*1,000 acres*	*1,000 acres*	*1,000 acres*
1993	82,405	65,926	295,503	319,518	3,260.2	3,677.0
1994	92,071	65,493	307,905	323,699	3,484.9	3,748.4
1995	82,694	64,433	301,349	318,289	3,432.3	3,825.6
1996	93,817	65,968	313,202	333,682	3,371.8	3,919.2
1997	90,840	66,259	317,662	332,072	3,270.3	4,004.2
1998	88,931	62,677	311,485	329,983	3,303.6	4,024.9
1999	86,218	57,718	312,222	329,556	3,419.9	4,068.7
2000 [4]	87,708	56,468	307,519	328,325	3,509.6	4,086.0
2001 [4]	83,586	52,202	303,777	324,830	3,371.3	4,069.9
2002 [5]	82,845	49,310	299,855	327,883	3,283.2	3,993.7

[1] Corn for grain, oats, barley, and sorghum for grain. [2] Wheat, rye, and rice. [3] Crops included in area planted and area harvested are corn, sorghum, oats, barley, winter wheat, rye, durum wheat, other spring wheat, rice, soybeans, peanuts, sunflower, cotton, dry edible beans, potatoes, and sugarbeets. Harvested acreage for all hay, tobacco, and sugarcane are used in computing total area planted. [4] For the 2000 crop year many changes occurred to the National Vegetable Estimation Program. Nine new commodities were added to the program. Additionally, States were added or dropped from the seasonal program. Some States were discontinued for the seasonal forecasts but remained in the program on an annual basis. When comparing 2001 and 2000 data to 1999 data, comparable States should be used. [5] For the 2002 crop year, many changes occured to the National Vegetable Estimation Program. Ten fresh market commodities and two processing commodities were removed from the program. States were removed from the program for certain commodities. When comparing 2000 and 2001 data to 2002 data, comparable States should be used. If you need assistance with these comparisons, please contact Biz Wallingsford at (202) 720-2157. For details on the 2002 program changes see the following website: http:/www.usda.gov/nass/events/programchg/vegprogchngs.htm. [6] Includes the following fruits and nuts: Citrus fruits—oranges, tangerines, Temples, grapefruit, lemons, limes, tangelos, and K-Early Citrus (area is for the year of harvest); deciduous fruits—commercial apples, peaches, pears, grapes, cherries, plums, prunes, apricots, bananas, nectarines, figs, kiwifruit, olives, avocados, papayas, dates, berries, guavas, cranberries, pineapples and strawberries; nuts—almonds, hazelnuts, macadamias, pistachios, and walnuts.

NASS, Crops Branch, (202) 720–2127.

Table 9-23.—Crops: Area, yield, production, and value, United States, 2001–2002

Crop	Area harvested		Yield per harvested acre		
	2001	2002 [1]	Unit	2001	2002 [1]
	1,000 acres	*1,000 acres*			
Grains & Hay:					
Barley [2]	4,289.0	4,135.0	Bushel	58.2	54.9
Corn for Grain	68,808.0	69,313.0	Bushel	138.2	130.0
Corn for Silage	6,148.0	7,490.0	Ton	16.6	14.0
Hay, All	63,521.0	64,497.0	Ton	2.47	2.34
Alfalfa	23,822.0	23,135.0	Ton	3.37	3.19
All Other	39,699.0	41,362.0	Ton	1.93	1.86
Oats [2]	1,905.0	2,098.0	Bushel	61.4	56.8
Proso Millet	585.0	220.0	Bushel	33.2	12.5
Rice	3,314.0	3,207.0	Pound	6,496	6,578
Rye [2]	255.0	286.0	Bushel	27.3	24.4
Sorghum for Grain	8,584.0	7,299.0	Bushel	59.9	50.7
Sorghum for Silage	336.0	352.0	Ton	11.1	9.5
Wheat, All [2]	48,633.0	45,817.0	Bushel	40.2	35.3
Winter [2]	31,295.0	29,651.0	Bushel	43.5	38.5
Durum	2,789.0	2,703.0	Bushel	30.0	29.4
Other Spring	14,549.0	13,463.0	Bushel	35.2	29.3
Oilseeds:					
Canola	1,455.0	1,275.0	Pound	1,374	1,218
Cottonseed			Ton		
Flaxseed	578.0	704.0	Bushel	19.8	17.9
Mustard Seed	44.2	175.0	Pound	930	705
Peanuts	1,411.9	1,296.7	Pound	3,029	2,561
Rapeseed	3.1	3.1	Pound	1,306	1,461
Safflower	177.0	196.0	Pound	1,365	1,520
Soybeans for Beans	72,975.0	72,160.0	Bushel	39.6	37.8
Sunflower	2,555.0	2,205.0	Pound	1,338	1,133
Cotton, Tobacco & Sugar Crops:					
Cotton, All	13,827.7	12,413.3	Pound	705	663
Upland	13,559.5	12,171.0	Pound	694	651
Amer-Pima	268.2	242.3	Pound	1,254	1,286
Sugarbeets	1,243.4	1,361.0	Ton	20.7	20.2
Sugarcane	1,027.8	1,026.1	Ton	33.7	35.0
Tobacco	432.3	430.3	Pound	2,293	2,068
Dry Beans, Peas & Lentils:					
Austrian Winter Peas	7.1	11.6	Pound	1,366	1,414
Dry Edible Beans	1,248.5	1,726.9	Pound	1,569	1,736
Dry Edible Peas	192.3	279.7	Pound	1,957	1,517
Lentils	197.0	209.0	Pound	1,471	1,200
Wrinkled Seed Peas			NA		
Potatoes & Misc.:					
Coffee (HI)	6.3	6.2	Pound	1,270	1,370
Ginger Root (HI)	0.4	0.3	Pound	50,000	45,000
Hops	35.9	29.3	Pound	1,861	1,990
Maple syrup			NA		
Mushrooms			NA		
Peppermint Oil	78.5	80.2	Pound	81	85
Potatoes, All	1,222.2	1,276.5	Cwt	358	363
Winter	14.0	15.7	Cwt	294	268
Spring	76.2	86.1	Cwt	286	271
Summer	58.8	59.9	Cwt	310	309
Fall	1,073.2	1,114.8	Cwt	367	374
Spearmint Oil	19.5	18.0	Pound	105	108
Sweet Potatoes	94.4	83.3	Cwt	155	150
Taro (HI)	0.4	0.4	Pound		

See footnotes at end of table.

Table 9-23.—Crops: Area, yield, production, and value, United States, 2001–2002—Continued

Crop	Production			Value of production	
	Unit	2001	2002[1]	2001	2002[1]
		Thousands	*Thousands*	*1,000 dollars*	*1,000 dollars*
Grains & Hay:					
Barley[2]	Bushel	249,420	226,873	536,582	596,759
Corn for Grain	Bushel	9,506,840	9,007,659	18,888,389	21,213,159
Corn for Silage	Ton	102,077	104,979	....	....
Hay, All	Ton	156,764	150,962	12,602,534	12,432,729
Alfalfa	Ton	80,327	73,824	7,524,869	7,172,182
All Other	Ton	76,437	77,138	5,077,665	5,260,547
Oats[2]	Bushel	117,024	119,132	195,711	211,849
Proso Millet	Bushel	19,405	2,755	39,109	20,266
Rice	Cwt	215,270	210,960	925,055	840,727
Rye[2]	Bushel	6,971	6,985	20,422	23,717
Sorghum for Grain	Bushel	514,524	369,758	979,794	883,690
Sorghum for Silage	Ton	3,728	3,360	....	....
Wheat, All[2]	Bushel	1,957,043	1,616,441	5,440,217	5,863,378
Winter[2]	Bushel	1,361,479	1,142,802	3,684,817	3,939,217
Durum	Bushel	83,556	79,450	269,391	331,183
Other Spring	Bushel	512,008	394,189	1,486,009	1,592,978
Oilseeds:					
Canola	Pound	1,998,515	1,552,520	175,351	164,598
Cottonseed	Ton	7,452.2	6,419.3	667,348	638,115
Flaxseed	Bushel	11,455	12,569	49,004	72,944
Mustard Seed	Pound	41,106	123,450	4,976	18,783
Peanuts	Pound	4,276,704	3,320,490	1,000,512	594,426
Rapeseed	Pound	4,050	4,530	437	440
Safflower	Pound	241,665	297,980	26,340	36,723
Soybeans for Beans	Bushel	2,890,682	2,729,709	12,605,717	14,755,470
Sunflower	Pound	3,418,759	2,497,236	325,950	317,244
Cotton, Tobacco & Sugar Crops:					
Cotton, All	Bale	20,302.8	17,145.0	3,121,848	3,593,816
Upland	Bale	19,602.4	16,496.0	2,833,913	3,330,223
Amer-Pima	Bale	700.4	649.0	287,935	263,593
Sugarbeets	Ton	25,764	27,550	1,025,306	....
Sugarcane	Ton	34,587	35,932	1,003,046	....
Tobacco	Pound	991,552	889,632	1,952,120	1,726,013
Dry Beans, Peas & Lentils:					
Austrian Winter Peas	Cwt	97	164	1,067	1,597
Dry Edible Beans	Cwt	19,583	29,974	426,475	519,609
Dry Edible Peas	Cwt	3,763	4,242	20,765	30,263
Lentils	Cwt	2,898	2,508	27,714	30,339
Wrinkled Seed Peas	Cwt	640	457	8,862	6,165
Potatoes & Misc.:					
Coffee (HI)	Pound	8,000	8,500	19,600	19,550
Ginger Root (HI)	Pound	18,000	14,400	8,100	4,320
Hops	Pound	66,832.1	58,336.6	123,843	113,413
Maple syrup[3]	Gallons	....	....	30,116	....
Mushrooms	Pound	860,093	851,211	867,737	911,509
Peppermint Oil	Pound	6,343	6,818	68,964	79,030
Potatoes, All	Cwt	437,888	463,214	3,057,573	3,151,178
Spearmint Oil	Pound	2,052	1,942	18,645	17,696
Sweet Potatoes	Cwt	14,637	12,498	224,795	213,020
Taro (HI)	Pound	6,400	6,100	3,392	3,294

See footnotes at end of table.

Table 9-23.—Crops: Area, yield, production, and value, United States, 2001–2002—Continued

Crop	Area harvested[4]		Yield per harvested acre[5]		
	2001	2002[1]	Unit	2001	2002[1]
	Acres	*Acres*			
Apples, commercial crop	416,550	404,950	Ton	11.30	10.60
Apricots	19,430	18,430	Ton	4.24	4.89
Avocados[6]	64,130	....	Ton	3.45	....
Bananas	1,490	1,250	Ton	9.40	7.60
Blackberries (OR)	6,160	5,980	Ton	3.38	3.53
Blueberries	....	....	Ton	....	....
Cultivated	39,880	40,980	Ton	2.37	2.30
Wild (ME)	....	....	Ton	....	....
Boysenberries	1,610	1,590	Ton	1.87	1.47
Loganberries (OR)	70	60	Ton	0.79	2.17
Raspberries	....	....	Ton	....	....
Black (OR)	1,200	1,100	Ton	1.59	1.24
Red	12,200	12,000	Ton	3.73	3.58
All (CA)	2,300	2,400	Ton	5.40	5.95
Cherries, sweet	68,270	73,220	Ton	3.37	2.46
Cherries, tart	38,540	36,900	Ton	4.80	0.85
Cranberries	34,200	36,400	Ton	7.79	7.75
Dates (CA)	4,500	4,300	Ton	4.38	4.53
Figs (CA)	14,000	12,900	Ton	2.86	3.80
Grapes	930,420	953,850	Ton	7.06	7.49
Guava (HI)	610	....	Ton	12.60	....
Kiwifruit (CA)	4,900	4,500	Ton	5.27	5.58
Nectarines (CA)	36,500	36,500	Ton	7.53	8.19
Olives (CA)	36,000	36,000	Ton	3.72	2.75
Papayas (HI)	1,950	1,700	Ton	14.10	13.40
Peaches	151,650	153,420	Ton	8.00	8.40
Pears	64,030	63,515	Ton	15.60	14.40
Pineapples (HI)	20,100	19,100	Ton	....	....
Plums (CA)	37,000	36,000	Ton	5.68	5.56
Prunes, dried (CA)	86,000	73,000	Ton	4.88	6.49
Prunes and plums, fresh basis (excluding CA)	4,210	4,110	Ton	5.04	3.86
Strawberries	46,000	47,600	Ton	18.00	20.70
Oranges[7]	814,800	796,700	Ton	15.0	15.7
Grapefruit[7]	145,200	138,300	Ton	17.0	17.5
Lemons[7]	64,300	64,300	Ton	15.5	12.9
Limes[7] (FL)	1,200	800	Ton	9.17	8.75
Tangerines[7]	39,900	38,600	Ton	9.35	10.9
Tangelos[7] (FL)	10,800	9,700	Ton	8.80	10.0
Temples[7] (FL)	5,500	4,700	Ton	10.2	14.9
K-Early Citrus[7] (FL)	200	200	Ton	10.0	5.0
Almonds[8] (CA)	525,000	530,000	Ton	1.28	1.62
Hazelnuts (filberts)[8]	28,100	27,800	Ton	1.76	0.65
Macadamia nuts[8] (HI)	17,800	17,800	Ton	1.58	1.46
Pecans[9]	....	....	Ton	....	....
Pistachios[8] (CA)	78,000	83,000	Ton	1.03	1.80
Walnuts[8] (CA)	196,000	200,000	Ton	1.56	1.41

See footnotes at end of table.

Table 9-23.—Crops: Area, yield, production, and value, United States, 2001–2002—Continued

Crop	Total production			Value of production	
	Unit	2001	2002[1]	2001	2002[1]
		Thousands	*Thousands*	*1,000 dollars*	*1,000 dollars*
Apples, commercial crop	Ton	4,714.4	4,296.1	1,448,348	1,622,135
Apricots	Ton	82.5	90.1	26,598	28,326
Avocados[6]	Ton	221.3	....	362,390	....
Bananas[10]	Ton	14.0	9.5	10,640	7,980
Blackberries (OR)	Ton	21.8	21.9	17,267	18,296
Blueberries	Ton	....	....	....	....
Cultivated	Ton	96.6	95.9	163,763	193,787
Wild (ME)	Ton	37.6	31.2	22,945	15,920
Boysenberries	Ton	3.0	2.4	4,133	3,142
Loganberries (OR)	Ton	0.1	0.1	50	145
Raspberries	Ton	....	....	....	....
Black (OR)	Ton	2.0	1.4	1,724	1,119
Red	Ton	46.0	42.9	45,940	45,359
All (CA)	Ton	12.4	14.3	41,168	40,005
Cherries, sweet	Ton	230.4	180.2	270,914	273,694
Cherries, tart	Ton	185.0	31.3	57,150	27,879
Cranberries	Ton	266.5	282.1	110,342	154,255
Dates (CA)	Ton	19.7	19.5	26,004	30,810
Figs (CA)	Ton	40.1	49.0	14,849	16,907
Grapes	Ton	6,569.6	7,144.0	2,929,038	2,912,742
Guava[10] (HI)	Ton	7.7	....	2,157	....
Kiwifruit (CA)	Ton	25.8	25.1	15,340	....
Nectarines (CA)	Ton	275.0	299.0	127,642	114,517
Olives (CA)	Ton	134.0	99.0	90,096	58,562
Papayas[10] (HI)	Ton	27.5	22.8	14,598	11,778
Peaches	Ton	1,216.6	1,287.7	493,298	507,089
Pears	Ton	1,001.8	911.5	271,788	297,410
Pineapples[10] (HI)	Ton	323.0	320.0	96,337	100,616
Plums (CA)	Ton	210.0	200.0	64,362	77,200
Prunes, dried (CA)	Ton	420.0	474.0	97,605	120,690
Prunes and plums, fresh basis (excluding CA)	Ton	21.2	15.9	5,459	4,271
Strawberries[10]	Ton	826.2	985.1	1,070,052	1,220,504
Oranges[7 11]	Ton	12,221	12,543	1,682,790	1,834,028
Grapefruit[7 11]	Ton	2,462	2,427	285,065	285,546
Lemons[7 11]	Ton	996	828	237,362	341,174
Limes[7 11] (FL)	Ton	11	7	4,249	1,732
Tangerines[7 11]	Ton	373	419	96,789	124,789
Tangelos[7 11] (FL)	Ton	95	97	8,193	10,967
Temples[7 11] (FL)	Ton	56	70	5,282	7,054
K-Early Citrus[7 11] (FL)	Ton	2	1	187	115
Almonds[10] (CA)	Ton	671.5	857.6	740,012	1,049,188
Hazelnuts (filberts)[10]	Ton	49.5	18.0	34,700	18,009
Macadamia nuts[10] (HI)	Ton	28.0	26.0	33,040	28,600
Pecans[10]	Ton	169.3	89.2	201,101	169,299
Pistachios[10] (CA)	Ton	80.5	150.0	162,610	333,000
Walnuts[10] (CA)	Ton	305.0	282.0	341,600	....

See footnotes at end of table.

Table 9-23.—Crops: Area, yield, production, and value, United States, 2001–2002—Continued

Crop	Area harvested		Yield per harvested acre		
	2001	2002[1]	Unit	2001	2002[1]
	Acres	*Acres*			
Commercial Vegetables:					
Fresh Market					
Artichokes[12]	8,000	8,200	Cwt	125	115
Asparagus[12]	70,150	66,000	Cwt	30	28
Beans, lima[13]	7,090	....	Cwt	28	....
Beans, snap	95,900	100,000	Cwt	64	60
Broccoli[12]	133,000	126,000	Cwt	145	139
Brussel sprouts[11 13]	2,200	....	Cwt	180	....
Cabbage	79,240	77,080	Cwt	329	317
Cantaloups	96,130	93,900	Cwt	236	245
Carrots	101,760	97,400	Cwt	309	298
Cauliflower[12]	44,050	42,400	Cwt	159	152
Celery[12]	27,800	27,100	Cwt	678	691
Corn, sweet	251,600	248,800	Cwt	109	106
Cucumbers	54,400	55,500	Cwt	198	205
Eggplant[13]	5,300	....	Cwt	255	....
Escarole/endive[13]	5,170	....	Cwt	180	....
Garlic[12]	35,200	32,800	Cwt	167	172
Honeydew melons	24,200	24,600	Cwt	189	206
Lettuce, head	184,300	184,400	Cwt	382	367
Lettuce, leaf	52,500	54,800	Cwt	225	226
Lettuce, Romaine	54,600	54,500	Cwt	284	290
Onions[12]	161,590	160,220	Cwt	419	420
Peppers, bell[12]	56,020	55,800	Cwt	287	290
Spinach	28,650	30,900	Cwt	129	156
Tomatoes	124,250	124,900	Cwt	286	299
Watermelons	160,820	149,900	Cwt	253	262
Greens:[12 13]					
Collards	14,100	....	Cwt	119	....
Kale	4,110	....	Cwt	185	....
Mustard	9,360	....	Cwt	128	....
Turnip	10,500	....	Cwt	107	....
Okra[12 13]	3,020	....	Cwt	60	....
Peppers, Chile[12]	30,550	28,300	Cwt	93	103
Pumpkins[12]	35,600	37,600	Cwt	228	210
Radishes[13]	14,300	....	Cwt	87	....
Squash[12]	52,200	53,100	Cwt	150	163
Processing:					
Beans, lima	51,200	48,500	Ton	1.31	1.30
Beans, snap	195,480	210,100	Ton	3.56	3.96
Beets[13]	6,650	....	Ton	16.72	....
Cabbage for sauerkraut[13]	6,950	....	Ton	25.09	....
Carrots	18,680	15,400	Ton	24.21	25.69
Corn, sweet	446,450	419,900	Ton	7.04	7.35
Cucumbers for pickles	108,260	118,900	Ton	5.37	5.19
Peas, green	211,140	211,300	Ton	1.85	1.64
Spinach	13,940	13,120	Ton	9.12	9.24
Tomatoes	274,860	311,800	Ton	33.65	37.39

See footnotes at end of table.

Table 9-23.—Crops: Area, yield, production, and value, United States, 2001–2002—Continued

Crop	Production			Value of production	
	Unit	2001	2002[1]	2001	2002[1]
		Thousands	*Thousands*	*1,000 dollars*	*1,000 dollars*
Commercial Vegetables:					
Fresh Market					
Artichokes[12]	Cwt	1,000	943	58,559	66,764
Asparagus[12]	Cwt	2,078	1,868	228,925	172,876
Beans, lima[13]	Cwt	202	....	5,861	....
Beans, snap	Cwt	6,155	5,958	277,611	282,163
Broccoli[12]	Cwt	19,280	17,475	500,987	551,280
Brussel sprouts[12 13]	Cwt	396	....	14,471	....
Cabbage	Cwt	26,069	24,415	340,241	301,482
Cantaloups	Cwt	22,718	23,046	431,381	404,685
Carrots	Cwt	31,464	29,027	541,859	551,433
Cauliflower[12]	Cwt	7,018	6,428	199,044	207,629
Celery[12]	Cwt	18,856	18,737	272,391	241,571
Corn, sweet	Cwt	27,383	26,430	534,586	531,159
Cucumbers	Cwt	10,756	11,379	210,609	214,318
Eggplant[13]	Cwt	1,354	....	36,599	....
Escarole/endive[13]	Cwt	933	....	27,538	....
Garlic[12]	Cwt	5,877	5,650	173,020	153,763
Honeydew melons	Cwt	4,576	5,059	96,663	90,546
Lettuce, head	Cwt	70,350	67,726	1,258,418	1,456,472
Lettuce, leaf	Cwt	11,824	12,362	324,328	408,943
Lettuce, Romaine	Cwt	15,487	15,828	298,284	395,770
Onions[12]	Cwt	67,653	67,365	697,950	715,979
Peppers, bell[12]	Cwt	16,062	16,168	458,750	498,650
Spinach	Cwt	3,709	4,810	120,371	165,764
Tomatoes	Cwt	35,527	37,302	1,080,166	1,170,922
Watermelons	Cwt	40,648	39,204	275,035	328,550
Greens:[12 13]					
Collards	Cwt	1,683	....	36,392	....
Kale	Cwt	762	....	20,048	....
Mustard	Cwt	1,196	....	35,190	....
Turnip	Cwt	1,124	....	19,758	....
Okra[12 13]	Cwt	181	....	8,339	....
Peppers, Chile[12]	Cwt	2,839	2,909	89,384	84,153
Pumpkins[12]	Cwt	8,108	7,907	70,912	82,744
Radishes[13]	Cwt	1,244	....	50,034	....
Squash[12]	Cwt	7,805	8,631	173,753	204,288
Processing:					
Beans, lima	Ton	67,160	63,220	30,854	27,167
Beans, snap	Ton	695,190	831,260	112,137	126,761
Beets[13]	Ton	111,180	....	7,317	....
Cabbage for sauerkraut[13]	Ton	174,360	....	8,499	....
Carrots	Ton	452,240	395,700	33,685	27,717
Corn, sweet	Ton	3,142,840	3,084,290	229,204	210,335
Cucumbers for pickles	Ton	581,540	617,300	168,958	171,016
Peas, green	Ton	390,220	347,090	103,207	87,494
Spinach	Ton	127,100	121,250	14,698	13,319
Tomatoes	Ton	9,248,720	11,656,820	547,473	683,083

[1] Preliminary. [2] Includes area seeded in preceding fall. [3] Value for 2002 is not available. [4] Bearing acreage for citrus and noncitrus fruits and nuts. [5] Yield per bearing acreage for citrus and noncitrus fruits and nuts. [6] Year of bloom. [7] Crop year begins with bloom in one year and ends with completion of harvest the following year. Citrus production is for the year of harvest. [8] Yield derived from the in-shell basis. [9] Bearing acreage and yield not calculated. [10] Utilized production. [11] Equivalent packinghouse–door returns. [12] Includes processing total for dual usage crops. [13] Estimates discontinued for 2002.

NASS, Crops Branch, (202) 720–2127.

Table 9-24.—Total farm input: Index numbers of farm input, by major subgroups, United States, 1990–99

[1992=100]

Year	Total input	Farm labor	Capital	Land	Energy	Agricultural chemicals [1]	Feed, seed, and livestock purchases [2]	Other purchased inputs [3]
1990	1.016	1.046	1.132	1.028	0.940	0.949	1.014	0.894
1991	1.022	1.078	1.121	1.021	0.944	0.963	1.006	0.915
1992	1.001	1.018	1.092	1.016	0.936	0.978	1.009	0.872
1993	1.004	0.978	1.071	1.011	0.939	0.944	1.026	0.981
1994	1.013	0.990	1.038	1.007	0.967	1.008	1.033	1.010
1995	1.035	1.027	1.029	1.003	1.015	0.923	1.086	1.071
1996	1.000	1.000	1.000	1.000	1.000	1.000	1.000	1.000
1997	1.029	1.006	0.996	0.997	1.022	1.085	1.053	1.061
1998	1.034	0.978	0.988	0.994	1.037	1.047	1.114	1.077
1999	1.059	1.008	0.984	0.990	1.054	1.038	1.173	1.106

[1] Includes fertilizer, lime, and pesticide. [2] Includes broilers- and egg-type chicks and turkey poults and imports of livestock for purposes other than immediate slaughter. [3] Includes purchased services and miscellaneous inputs.

ERS, Resources, Technology and Productivity Branch (202) 694–5601.

Table 9-25.—Livestock and livestock products: Production and value, United States, 1999–2001

Product	Production [1]			Value of production		
	1999	2000	2001 [2]	1999	2000	2001 [2]
	1,000 pounds	*1,000 pounds*	*1,000 pounds*	*1,000 dollars*	*1,000 dollars*	*1,000 dollars*
Cattle and calves	42,468,398	42,841,607	42,365,499	26,051,477	28,391,942	29,267,264
Sheep and lambs	533,610	508,935	498,763	348,742	361,231	299,998
Hogs	25,790,623	25,730,208	25,936,846	7,766,241	10,817,518	11,442,171
Broilers [3]	40,829,800	41,623,100	42,445,900	15,128,840	13,988,434	16,693,569
Mature chickens	1,034,072	1,111,717	1,016,747	73,150	63,925	46,027
Turkeys	6,886,406	6,942,809	7,154,781	2,809,874	2,822,736	2,790,317
Milk	162,716,000	167,559,000	165,336,000	23,400,050	20,770,589	24,886,929
Wool (shorn)	46,592	46,446	43,016	17,860	15,377	15,311
Mohair [4]	2,896	2,628	1,970	10,016	10,609	4,200
Catfish [5]	....	....	....	489,291	501,400	443,681
Trout [6]	....	....	....	76,506	75,373	76,241
Honey	205,250	220,339	185,461	126,075	132,742	132,225
	Millions	*Millions*	*Millions*			
Eggs	82,715	84,386	85,702	4,287,164	4,345,382	4,444,864

[1] For cattle, sheep, and hogs, the quantity of net production is the live weight actually produced during the year, adjustments having been made for animals shipped in and changes in inventory. Estimates for broilers and eggs cover the 12-month period Dec. 1, previous year through Nov. 30. [2] Preliminary, except for wool shorn and mohair. [3] Young chickens of meat–type strains raised for meat production. [4] AZ, NM, and TX. [5] Value of fish sold. [6] Value of fish and eggs sold.

NASS, Livestock Branch, (202) 720–3570.

Table 9-26.—Agricultural productivity: Index numbers (1996=100) of farm output per unit of input, United States, 1990–99

Year	Productivity [1]
1990	0.886
1991	0.887
1992	0.955
1993	0.908
1994	1.005
1995	0.932
1996	1.000
1997	1.015
1998	1.011
1999	1.006

[1] Productivity is the output-input ratio. The ratio is obtained by dividing the index of farm output in table 9–25 by the index of total input in table 9–26.

ERS, Resources, Technology and Productivity Branch (202) 694–5601.

Table 9-27.—U.S. farm foods: Marketing bill, farm value, and consumer expenditures, 1992–2001 [1]

Year	Total marketing bill	Farm value	Expenditures for farm foods
	Billion dollars	*Billion dollars*	*Billion dollars*
1992	369.4	105.1	474.5
1993	379.6	109.6	489.2
1994	402.6	109.6	512.2
1995	415.7	113.8	529.5
1996	424.5	122.2	546.7
1997	444.6	121.9	566.5
1998	465.4	119.6	585.0
1999	503.1	122.2	625.3
2000	537.8	123.3	661.1
2001 [2]	557.5	130.0	687.5

[1] The total marketing bill is the difference between total expenditures for domestic farm-originated food products and the farm value or payment farmers received for the equivalent farm products. It relates only to food purchased by consumers that is not imported or exported. [2] Preliminary.

ERS, Food Markets Branch, (202) 694–5375.

Table 9-28.—Farm food products: Marketing costs, United States, 1992–2001

Year	Labor [1]	Packaging materials	Intercity transportation, rail and truck	Fuels and electricity	Corporate profits before taxes	Other [2]	Total marketing bill [3]
	Billion dollars	*Billion dollars*	*Billion dollars*	*Billion dollars*	*Billion dollars*	*Billion dollars*	*Billion dollars*
1992	168.4	40.1	20.6	16.7	15.7	107.9	369.4
1993	178.0	40.9	21.2	17.2	18.1	104.2	379.6
1994	186.1	43.3	21.8	17.9	20.9	112.6	402.6
1995	196.6	48.2	22.3	18.6	19.5	110.5	415.7
1996	204.6	47.7	22.9	19.6	20.7	109.0	424.5
1997	216.9	48.7	23.6	20.2	22.3	112.9	444.6
1998	229.9	50.4	24.4	20.7	25.5	114.5	465.4
1999	241.5	50.9	25.2	22.0	29.2	134.3	503.1
2000	252.9	53.5	26.4	23.1	31.1	150.8	537.8
2001 [4]	263.8	55.0	27.5	24.1	32.0	155.1	557.5

[1] Includes employee wages or salaries, and their health and welfare benefits. Also includes imputed earnings of proprietors, partners, and family workers not receiving stated remuneration. [2] Includes depreciation, rent, advertising and promotion, interest, taxes, licenses, insurance, professional services, local for-hire transportation, food service in schools, colleges, hospitals, and other institutions, and miscellaneous items. [3] The marketing bill is the difference between the farm value or payments to farmers for foodstuffs and consumer expenditures for these foods both at foodstores and away from home eating places. Thus, it covers processing, wholesaling, transportation, and retailing costs and profits. [4] Preliminary.

ERS, Food Markets Branch, (202) 694–5375.

Table 9-29.—Price components: Market basket of farm-originated food products by food group, United States, 1992–2001 [1]

Year	Market basket of food products				Bakery and cereal products			
	Retail cost [2]	Farm value [3]	Farm to retail spread [4]	Farm value share of retail cost	Retail cost	Farm value	Farm to retail spread	Farm value share of retail cost
	Index 1982–84=100	*Index 1982–84=100*	*Index 1982–84=100*	*Percent*	*Index 1982–84=100*	*Index 1982–84=100*	*Index 1982–84=100*	*Percent*
1992	138	103	157	26	152	94	160	8
1993	142	105	162	26	157	92	166	7
1994	145	101	169	24	164	103	171	8
1995	149	103	175	24	168	110	176	8
1996	156	111	180	25	174	126	181	9
1997	160	106	189	23	178	108	187	7
1998	163	103	195	22	181	94	193	6
1999	167	98	205	21	185	83	199	6
2000	171	97	210	20	188	75	204	5
2001 [5]	177	106	215	21	194	79	210	5

Year	Meat products				Fruits and vegetables, fresh			
	Index 1982–84=100	*Index 1982–84=100*	*Index 1982–84=100*	*Percent*	*Index 1982–84=100*	*Index 1982–84=100*	*Index 1982–84=100*	*Percent*
1992	131	105	158	41	174	122	199	23
1993	135	107	163	40	182	131	207	24
1994	135	96	176	36	191	119	225	21
1995	136	94	178	35	210	133	248	21
1996	140	100	181	36	216	133	257	20
1997	144	101	189	36	220	128	265	20
1998	142	85	200	30	237	133	288	19
1999	142	82	205	29	252	136	308	18
2000	150	88	214	30	252	131	310	17
2001 [5]	159	97	223	31	261	138	321	17

Year	Dairy products				Fats and oils			
	Index 1982–84=100	*Index 1982–84=100*	*Index 1982–84=100*	*Percent*	*Index 1982–84=100*	*Index 1982–84=100*	*Index 1982–84=100*	*Percent*
1992	129	96	159	36	130	93	143	19
1993	129	93	163	35	130	108	138	22
1994	132	94	166	34	134	126	137	25
1995	133	92	170	33	137	121	143	24
1996	142	107	174	36	141	112	151	22
1997	146	98	189	32	142	109	154	21
1998	151	113	186	36	147	119	157	22
1999	160	108	207	32	148	89	170	16
2000	161	99	218	30	147	81	172	15
2001 [5]	167	119	212	34	156	77	185	13

Year	Poultry				Fruits and vegetables, processed			
	Index 1982–84=100	*Index 1982–84=100*	*Index 1982–84=100*	*Percent*	*Index 1982–84=100*	*Index 1982–84=100*	*Index 1982–84=100*	*Percent*
1992	131	104	163	42	134	129	135	23
1993	137	112	166	44	132	107	139	19
1994	142	115	173	43	135	113	141	20
1995	144	114	178	42	138	121	143	21
1996	152	126	183	44	144	122	152	20
1997	157	121	198	41	148	116	158	19
1998	157	126	193	43	151	115	162	18
1999	158	119	203	40	155	114	168	17
2000	160	117	209	39	154	106	168	17
2001 [5]	165	126	209	41	159	108	175	16

[1] The market basket consists of foods that mainly originate on U.S. farms bought in foodstores in a base period, currently 1982–84. [2] Indexes of retail cost are components of the Consumer Price Index published by the Bureau of Labor Statistics. [3] Gross return or payment to farmers for the farm products equivalent to foods in the market basket. [4] The spread between the retail cost and farm value is an estimate of the gross margin received by marketing firms for assembling, processing, transporting, and distributing the products. [5] Preliminary.

ERS, Food Markets Branch (202) 694–5375.

Table 9-30.—Farm product prices: Marketing year average prices received by farmers; Parity prices for January, United States, 2000 and 2001

Commodity and unit		Marketing year average price [1]		Parity price [3]	
		2000 [2]	2001 [2]	2000	2001
		Dollars	*Dollars*	*Dollars*	*Dollars*
Basic commodities:					
Cotton:					
American Upland	pound	0.498	0.324	1.76	1.66
Extra long staple	pound	1.003	0.932	2.25	2.32
Wheat	bushel	2.62	2.80	9.59	9.61
Rice	cwt	5.61	4.25	28.30	26.70
Corn	bushel	1.85	2.00	6.61	6.56
Peanuts	pound	0.274	0.234	0.671	0.675
Tobacco:					
Flue-cured, types 11–14	pound	1.839	1.868	3.98	4.18
Va., fire-cured, type 21	pound	1.637	1.759	4.00	4.19
Ky.-Tenn., fire-cured, types 22–23	pound	2.163	2.49	5.01	5.27
Burley, type 31	pound	1.963	1.973	4.25	4.45
Maryland, type 32 [4]	pound	1.690	1.680	3.60	3.67
Dark air-cured, types 35–36	pound	1.965	1.829	4.28	4.45
Sun-cured, type 37	pound	1.800	1.686	3.64	3.78
Pa., seedleaf, type 41	pound	....	1.650	3.02	3.10
Cigar binder type 51	pound	4.916	5.589	9.63	10.60
Puerto Rican filler, type 46	pound	....	....	2.72	2.82
Cigar filler types 54–55	pound	1.550	1.650	3.34	3.56
Designated nonbasic commodities:					
All milk, sold to plants	cwt	12.40	15.05	31.30	32.00
Fluid market	cwt	12.44	15.09	....	....
Manufacturing grade	cwt	10.52	13.44	....	....
Honey, all	pound	0.597	0.669	1.42	1.50
Wool and mohair:					
Wool [5]	pound	0.330	0.360	1.79	1.65
Mohair [6]	pound	4.04	2.13	3.78	4.39
Other nonbasic commodities:					
Field crops and miscellaneous:					
Barley	bushel	2.11	2.25	5.90	6.41
Beans, dry edible	cwt	15.50	19.40	49.00	47.60
Cottonseed	ton	105.00	89.50	246.00	253.00
Crude pine gum	barrel	....	....	219.00	....
Flaxseed	bushel	3.30	4.25	11.60	11.30
Hay, all, baled	ton	85.00	97.30	195.00	201.00
Hops	pound	1.87	1.91	3.86	4.09
Oats	bushel	1.10	1.50	3.35	3.75
Peas, dry edible	cwt	5.31	5.81	28.50	....
Peppermint oil	pounds	10.80	10.80	30.10	3.30
Popcorn, shelled basis	cwt	....	....	....	....
Potatoes	cwt	5.08	6.60	13.40	13.60
Rye	bushel	2.60	2.90	5.16	5.46
Sorghum grain	cwt	3.37	3.50	10.40	11.20
Soybeans	bushel	4.54	4.30	13.80	....
Spearmint oil	pound	9.06	9.10	28.30	27.90
Sweetpotatoes	cwt	15.30	15.40	32.60	34.10
Tobacco:					
Cigar wrapper, type 61	pound	25.30	....	43.10	43.40
Fruits:					
Citrus (equiv. on-tree): [7]					
Grapefruit	box	3.89	2.24	8.14	8.04
Lemons	box	9.49	5.00	19.20	19.60
Limes, Florida	box	11.21	12.00	20.50	20.70
Oranges	box	3.58	3.56	11.50	11.30
Tangelos, Florida	box	2.52	1.06	....	....
Tangerines	box	7.15	8.37	28.30	27.40
Temples, Florida	box	2.55	1.91	9.99	9.64
Deciduous and other:					
Apples:					
For all sales	pound	....	....	....	....
For fresh consumption [8]	pound	0.178	0.229	0.461	0.486
For processing [9]	ton	101.00	105.00	305.00	320.00
Apricots:					
For all sales	ton	....	....	....	....
For fresh consumption [10]	ton	369.00	353.00	1,840.00	1,790.00
Dried, California (dried basis) [9]	ton	288.00	260.00	4,580.00	4,810.00
For processing (except dried) [9]	ton	283.00	257.00	648.00	672.00
Avocados [10]	ton	1,380.00	1,400.00	3,260.00	3,430.00

See footnotes at end of table.

Table 9-30.—Farm product prices: Marketing year average prices received by farmers; Parity prices for January, United States, 2000 and 2001—Continued

Commodity and unit		Marketing year average price [1]		Parity price [3]	
		2000 [2]	2001 [2]	2000	2001
		Dollars	Dollars	Dollars	Dollars
Deciduous and other—Con.					
Berries for processing:					
Blackberries (Oregon)	pound	0.477	0.415	0.976	1.010
Boysenberries (California & Oregon)	pound	0.581	0.689	1.380	1.430
Gooseberries	pound	....	....	0.587	0.608
Loganberries (Oregon)	pound	0.643	0.455	0.949	0.903
Raspberries, black (Oregon)	pound	1.480	0.452	1.67	1.73
Raspberries, red (Oregon & Washington)	pound	0.392	0.505	1.330	1.380
Cherries:					
Sweet	ton	1,340.00	1,230.00	2,570.00	2,770.00
Tart	pound	0.187	0.186	0.406	0.440
Cranberries [11]	barrel	16.80	20.80	117.00	115.00
Dates, California [10]	ton	1,230.00	1,410.00	2,250.00	2,450.00
Figs, California	ton	272.00	365.00	....	....
Grapes:					
For all sales	ton	403.00	446.00	....	....
Raisin varieties dried, California (dried basis) [9]	ton	510.00	500.00	2,380.00	2,480.00
Nectarines (California):					
For all sales	ton	398.00	464.00	....	....
For fresh consumption [19]	ton	....	....	1,000.00	1,020.00
For processing [19]	ton	....	....	64.40	65.90
Olives (California): [12]					
For all sales	ton	656.00	672.00	....	....
Crushed for oil	ton	11.00	6.00	....	25.80
For all sales (except crushed)	ton	....	....	1,260.00	1,350.00
For canning	ton	769.00	770.00	1,450.00	1,570.00
Peaches:					
For all sales	pound	....	....	....	....
For fresh consumption [8]	pound	0.281	0.310	0.611	0.623
Dried, California (dried basis) [9]	ton	740.00	514.00	1,810.00	1,840.00
For processing California (except dried):					
Clingstone [12]	ton	250.00	244.00	501.00	530.00
Freestone [9]	ton	187.00	200.00	431.00	449.00
Pears:					
For all sales	ton	264.00	282.00	....	....
For fresh consumption [8]	ton	318.00	365.00	829.00	848.00
Dried, California (dried basis) [9]	ton	909.00	1,210.00	2,390.00	2,530.00
For processing (except dried) [9]	ton	190.00	175.00	466.00	480.00
Plums (California):					
For all sales [10]	ton	442.00	306.00	....	....
For fresh consumption [19]	ton	....	....	844.00	889.00
For processing [19]	ton	....	....	33.90	41.40
Prunes, dried (California) [9]	ton	770.00	723.00	2,140.00	2,270.00
Prunes and plums (excl. California):					
For fresh consumption [13]	ton	321.00	345.00	839.00	863.00
For processing (except dried) [9]	ton	178.00	185.00	371.00	398.00
Strawberries:					
For fresh consumption [14]	pound	0.649	0.758	1.410	1.490
For processing [9]	pound	0.245	0.292	0.639	0.651
Sugar crops:					
Maple sirup	gallon	27.10	....	....	....
Sugarbeets	ton	34.20	39.40	92.10	94.30
Sugarcane for sugar	ton	26.10	28.40	66.00	67.70
Tree nuts: [15]					
Almonds	pound	0.97	0.90	3.46	3.67
Hazelnuts	ton	891.00	701.00	1,860.00	1,960.00
Pecans, all	pound	1.140	0.594	4,500.00	4,630.00
Improved	pound	....	....	....	....
Seedling	pound	....	....	....	....
Walnuts	ton	1,240.00	1,120.00	2,860.00	2,940.00

See footnotes at end of table.

Table 9-30.—Farm product prices: Marketing year average prices received by farmers; Parity prices for January, United States, 2000 and 2001—Continued

Commodity and unit		Marketing year average price [1]		Parity price [3]	
		2000 [2]	2001 [2]	2000	2001
		Dollars	*Dollars*	*Dollars*	*Dollars*
Vegetables for fresh market: [14]					
Artichokes, California	cwt....	60.30	58.20	71.60	74.20
Asparagus	cwt....	117.00	140.00	231.00	251.00
Broccoli	cwt....	31.00	25.10	59.50	63.90
Cabbage	cwt....	12.80	14.20	21.70	22.50
Cantaloups	cwt....	17.50	18.50	33.50	34.70
Carrots [16]	cwt....	13.10	17.40	29.40	31.50
Cauliflower [16]	cwt....	33.10	27.30	68.00	72.90
Celery [16]	cwt....	18.50	14.70	29.00	32.00
Cucumbers	cwt....	19.90	19.50	....	37.20
Eggplant	cwt....	28.60	31.70	37.90	39.30
Escarole/Endive	cwt....	30.90	29.20	48.80	50.60
Garlic, California	cwt....	27.80	30.90	46.10	47.80
Green peppers [16]	cwt....	31.30	28.40	56.70	58.70
Honeydew melons	cwt....	19.20	20.60	42.00	44.00
Lettuce	cwt....	17.40	17.60	34.50	37.20
Onions [16]	cwt....	11.30	11.40	27.10	27.90
Snap beans	cwt....	42.60	45.40	71.00	73.60
Spinach	cwt....	36.40	40.80	62.20	64.40
Sweet corn	cwt....	18.20	19.60	38.60	40.90
Tomatoes	cwt....	30.80	30.20	68.90	72.40
Watermelons	cwt....	6.41	6.86	13.30	13.80
Vegetables for processing: [9]					
Asparagus	ton....	1,180.00	1,040.00	2,500.00	2,640.00
Beets	ton....	61.20	65.80	123.00	127.00
Cabbage	ton....	47.40	48.70	94.60	98.00
Cucumbers	ton....	269.00	291.00	....	....
Green peas	ton....	248.00	264.00	609.00	631.00
Lima beans	ton....	432.00	459.00	1,050.00	1,090.00
Snap beans	ton....	171.00	161.00	401.00	413.00
Spinach	ton....	121.00	116.00	228.00	236.00
Sweet corn	ton....	73.40	72.90	165.00	173.00
Tomatoes	ton....	59.80	59.20	149.00	153.00
Livestock and livestock products:					
All beef cattle	cwt....	68.60	71.30	153.00	157.00
Cows	cwt....	37.90	40.90	....	....
Steers and heifers	cwt....	72.30	75.10	....	....
Calves	cwt....	104.00	106.00	197.00	207.00
Beeswax	pound...	....	....	4.58	4.75
Chickens:					
Excluding broilers, live	pound...	0.057	0.045	....	....
Broilers, live [20]	pound...	0.336	0.393	....	....
All Eggs	dozen...	0.618	0.622	1.50	1.55
Hogs	cwt....	42.30	44.30	101.00	103.00
Lambs	cwt....	79.80	66.90	162.00	173.00
Milk cows [17]	head....	1,340.00	....	....	....
Milk, retail [18]	quart...	....	....	....	....
Sheep	cwt....	34.30	34.60	67.10	72.10
Turkeys, live	pound...	0.407	0.390	0.913	0.954

[1] Marketing year average prices for crops; weighted calendar year average for livestock and livestock products, except chickens, eggs, and hogs, which are on a Nov.-Dec. marketing year basis. Unless otherwise noted, these are averages for marketing season or calendar year computed by weighing State prices by quantities sold, or by production for those commodities for which virtually all the production is sold. [2] Preliminary. [3] Parity prices are for January of the year shown as published in the January issue of Agricultural Prices. [4] Previous year. [5] Average local market price for wool sold excluding incentive payment. [6] Average local market price for mohair sold excluding incentive payment. Texas only prior to 1988. [7] Crop year begins with bloom in one year and ends with completion of harvest the following year. Prices refer to the year harvest begins. Thus the prices shown for 1996 relate to the citrus crop designated as 1996–97 in the production reports. [8] Equivalent packinghouse-door returns for California, Oregon (pears only), Washington, and New York (apples only), and prices as sold for other States. [9] Equivalent returns at processing plant-door. [10] Equivalent returns at packinghouse-door. [11] Weighted average of co-op and independent sales. Co-op prices represent pool proceeds excluding returns from non-cranberry products and before deductions for capital stock and other retains. [12] Equivalent per unit returns for bulk fruit at first delivery point. [13] Average price as sold. [14] FOB shipping point when available. Weighted average of prices at points of first sale when FOB shipping point price not available. [15] Prices are in-shell basis except almonds which are shelled basis. [16] Includes some processing. [17] Simple average of States weighted by estimated Jan. 1 head for U.S. average. [18] Sold by farmers directly to consumers. [19] Prices for fresh and processing breakdown no longer published to avoid disclosure of individual operations. [20] Live weight equivalent price.

NASS, Livestock and Economic Branch, (202) 720–3570.

Table 9-31.—Producer prices: Index numbers, by groups of commodities, United States, 1993–2002

[1982=100]

Year	Total finished goods	Consumer foods	Total consumer goods	Total intermediate materials	Total crude materials
1993	124.7	125.7	123.0	116.2	102.4
1994	125.5	126.8	123.3	118.5	101.8
1995	127.9	129.0	125.6	124.9	102.7
1996	131.3	133.6	129.5	125.7	113.8
1997	131.8	134.5	130.2	125.6	111.1
1998	130.7	134.3	128.9	123.0	96.8
1999	133.0	135.1	132.0	123.2	98.2
2000	138.0	137.2	138.2	129.2	120.6
2001	140.7	141.3	141.5	129.7	121.0
2002 [1]	138.8	140.0	139.3	127.8	108.1

[1] Preliminary.

ERS, Food Marketing Branch, (202) 694–5349. Compiled from reports of the U.S. Department of Labor.

Table 9-32.—Prices received by farmers: Index numbers by groups of commodities and parity ratio, United States, 1993–2002 [1]

[1910–14=100]

Year	Food grains	Feed grains and hay	Cotton	Tobacco	Oil-bearing crops	Fruit & nuts [2]	Commercial vegetables	Other crops
1993	332	354	455	1,521	590	646	785	510
1994	377	381	559	1,538	600	632	731	517
1995	426	400	653	1,559	568	680	806	525
1996	497	521	626	1,592	700	824	740	532
1997	406	418	573	1,570	715	770	792	532
1998	328	356	546	1,572	588	788	818	532
1999	287	307	436	1,536	452	804	736	532
2000	272	308	421	1,614	467	689	807	541
2001	290	325	328	1,614	436	761	844	554
2002 [4]	330	356	280	1,639	480	780	944	558

Year	Potatoes, and dry edible beans	All crops	Meat animals	Dairy products	Poultry and eggs	Livestock and livestock products	All farm products	Parity ratio [3]
1993	539	505	1,011	787	295	768	643	47
1994	556	520	910	797	298	727	634	45
1995	541	553	868	783	300	707	646	45
1996	576	624	882	914	337	761	712	47
1997	457	568	933	820	319	755	678	43
1998	500	527	804	953	328	740	645	42
1999	507	476	840	882	310	731	607	40
2000	472	473	955	760	299	744	612	38
2001	497	487	989	922	325	814	649	40
2002 [4]	675	525	884	745	273	697	627	38

[1] These indexes are computed using the price estimates of averages for all classes and grades for individual commodities being sold in local farm markets. In computing the group indexes, prices of individual commodities have been compared with 1990–92 weighted average prices. The resulting ratios are seasonally weighted by average quantities sold for the most recent previous 5–year period. For example, 1994 indexes use quantities sold for the period 1988-92. Then, the 1990–92 indexes are adjusted to a 1910–14 reference. [2] Fresh market for noncitrus, and fresh market and processing for citrus. [3] Ratio of Index of Prices Received to the Index of Prices Paid by Farmers for Commodities and Services, Interest, Taxes, and Farm Wage Rates. [4] Preliminary.

NASS, Environmental, Economics, and Demographics Branch, (202) 720–6146.

Table 9-33.—Prices received by farmers: Index numbers by groups of commodities and ratio, United States, 1993–2002[1]

(1990–92=100)

Year	Food grains	Feed grains and hay	Cotton	Tobacco	Oil-bearing crops	Fruit & Nuts[2]	Commercial vegetables	Other Crops
1993	105	99	89	101	108	93	117	103
1994	119	106	109	102	110	90	109	105
1995	134	112	127	103	104	97	121	106
1996	157	146	122	105	128	118	111	108
1997	128	117	112	104	131	110	118	108
1998	103	100	107	104	107	113	123	108
1999	91	86	85	102	83	115	110	108
2000	85	86	82	107	85	99	121	110
2001	91	91	64	107	80	109	126	112
2002[4]	104	100	55	108	88	112	141	113

Year	Potatoes and dry edible beans	All crops	Meat animals	Dairy products	Poultry and eggs	Livestock and livestock products	All farm products	Ratio[3]
1993	107	102	100	98	105	100	101	97
1994	110	105	90	99	106	95	100	94
1995	107	112	85	98	107	92	102	93
1996	114	127	87	114	120	99	112	98
1997	90	115	92	102	113	98	107	90
1998	99	107	79	119	117	97	102	89
1999	100	96	83	110	110	95	96	83
2000	93	96	94	95	106	97	96	80
2001	98	99	97	115	116	106	102	83
2002[4]	134	106	87	93	97	91	99	80

[1] These indexes are computed using the price estimates of averages for all classes and grades for individual commodities being sold in local farm markets. In computing the group indexes, prices of individual commodities have been compared with 1990–92 weighted average prices. The resulting ratios are seasonally weighted by average quantities sold for the most recent previous 5–year period. For example, 1994 indexes use quantities sold for the period 1988–92. [2] Fresh market for noncitrus, and fresh market and processing for citrus. [3] Ratio of Index of Prices Received (1990–92=100) to Index of Prices Paid by Farmers for Commodities & Services, Interest, Taxes, and Wage Rates (1990–92=100). [4] Preliminary.

NASS, Environmental, Economics, and Demographics Branch, (202) 720–6146.

Table 9-34.—Prices paid by farmers: Index numbers, by groups of commodities, United States, 1993–2002

(1990–92=100)

Year	Production indexes								
	Production (all commodities)	Feed	Livestock & Poultry	Seeds	Fertilizer	Agricultural chemicals	Fuels	Supplies and Repairs	Autos and trucks
1993	104	102	104	101	96	109	93	107	107
1994	106	106	94	108	105	112	89	109	111
1995	108	103	82	110	121	116	89	112	115
1996	115	129	75	115	125	119	102	115	118
1997	119	125	94	119	121	121	106	118	119
1998	113	110	88	122	112	122	84	119	119
1999	111	100	95	121	105	121	93	121	119
2000	116	102	110	124	110	120	134	124	119
2001	120	109	111	132	123	120	119	128	118
2002[4]	119	113	102	142	109	118	112	131	116

Year	Production indexes - continued				Interest	Taxes	Wage rates[2]	Production, interest, taxes, and wage rates	Family living	Commodities, interest, taxes, and wage rates[3]
	Farm machinery	Building Materials	Farm services[1]	Rent						
1993	107	106	110	100	87	108	108	103	107	104
1994	113	109	115	108	94	106	111	106	110	106
1995	120	114	116	117	102	109	114	108	113	109
1996	125	115	116	128	106	112	117	115	116	115
1997	128	118	116	136	105	115	123	118	119	118
1998	132	118	115	120	104	119	129	114	121	115
1999	135	120	116	113	106	120	135	113	124	115
2000	139	121	119	110	113	123	140	118	128	120
2001	144	121	121	117	114	124	146	122	131	124
2002[4]	145	122	120	119	109	126	152	122	133	124

[1] The Farm Service and Rent indexes were combined prior to 1992. [2] Simple average of seasonally adjusted quarterly indexes. [3] Family Living component included. [4] Preliminary.

NASS, Environmental, Economics, and Demographics Branch, (202) 720–6146.

Table 9-35.—Prices paid by farmers: Index numbers, by groups of commodities, United States, 1993–2002[1]

[1910–14=100]

Year	Family living	Production indexes							
		Production (all commodities)	Feed	Livestock and poultry	Seed	Fertilizer	Agricultural chemicals	Fuels	Supplies and repairs
1993	1,369	1,006	497	1,327	1,002	352	672	717	757
1994	1,404	1,030	518	1,207	1,071	387	695	689	776
1995	1,447	1,051	502	1,047	1,088	444	717	690	798
1996	1,490	1,118	631	962	1,142	458	736	789	816
1997	1,525	1,151	612	1,200	1,179	443	745	816	835
1998	1,548	1,092	539	1,123	1,208	412	756	646	846
1999	1,582	1,078	486	1,217	1,201	385	746	720	861
2000	1,636	1,124	496	1,402	1,227	404	741	1,033	881
2001	1,682	1,161	530	1,418	1,306	450	745	915	906
2002[3]	1,708	1,157	553	1,304	1,402	399	733	868	927

Year	Production indexes—Continued				Interest	Taxes	Wage rates	Production, interest, taxes, and wage rates	Commodities, interest, taxes, and wage rates[2]
	Autos and trucks	Farm machinery	Building materials	Farm services and rent					
1993	2,833	2,689	1,436	1,262	2,191	2,886	4,055	1,384	1,381
1994	2,958	2,843	1,488	1,304	2,548	2,828	4,140	1,419	1,416
1995	3,053	3,009	1,553	1,384	2,548	2,915	4,278	1,456	1,454
1996	3,126	3,128	1,569	1,442	2,652	3,001	4,389	1,540	1,531
1997	3,161	3,216	1,602	1,477	2,621	3,093	4,591	1,585	1,574
1998	3,152	3,313	1,605	1,394	2,617	3,185	4,838	1,528	1,532
1999	3,166	3,393	1,628	1,364	2,663	3,214	5,037	1,520	1,531
2000	3,160	3,490	1,647	1,374	2,825	3,281	5,235	1,585	1,594
2001	3,141	3,601	1,646	1,422	2,845	3,330	5,468	1,636	1,644
2002[3]	3,082	3,678	1,654	1,426	2,728	3,378	5,702	1,636	1,649

[1] Based on Consumer Price Index-Urban of Bureau of Labor Statistics. [2] The index known as the Parity Index is the Index of Prices Paid by Farmers for Commodities and Services, Interest, Taxes, and Wage Rates expressed on the 1910–14=100 base. [3] Preliminary.

NASS, Environmental, Economics, and Demographics Branch, (202) 720–6146.

Table 9-36.—Prices paid by farmers: April prices, by commodities, United States, 2000–2002[1]

Commodity	Unit	2000	2001	2002
		Dollars	*Dollars*	*Dollars*
Fuels and energy:				
Diesel fuel[2 3]	Gal	1.080	1.080	0.964
Gasoline, service station, unleaded[4]	Gal	1.47	1.43	1.36
Gasoline, service station, bulk delivery[4]	Gal	1.48	1.47	1.37
L. P. gas, bulk delivery[2]	Gal	0.963	1.160	0.925
Feeds:				
Alfalfa Meal	Cwt	12.50	13.40	14.40
Alfalfa Pellets	Cwt	12.40	13.40	14.40
Bran	Cwt	13.20	13.40	13.60
Beef Cattle Concentrate.				
32-36% Protein	Ton	261	279	277
Corn Meal	Cwt	7.67	7.59	8.06
Cottonseed Meal, 41%	Cwt	14.90	15.70	15.80
Dairy Feed				
14% Protein	Ton	170	176	175
16% Protein	Ton	175	184	190
18% Protein	Ton	179	185	194
20% Protein	Ton	183	184	191
32% Protein Conc.	Ton	287	291	298
Hog Feed				
14-18% Protein	Ton	203	215	216
38-42% Protein Conc.	Ton	297	302	307
Molasses, Liquid	Cwt	12.00	12.40	12.90
Poultry Feed:.				
Broiler Grower	Ton	226	232	259
Chick Starter	Ton	249	239	240
Laying Feed	Ton	206	212	225
Turkey Grower	Ton	255	259	264
Soybean Meal, 44%	Cwt	13.00	13.40	13.50
Stock Salt	50 Lb	4.01	4.05	4.12
Trace Mineral Blocks	50 Lb	5.02	5.14	5.24

See footnotes at end of table.

Table 9-36.—Prices paid by farmers: April prices, by commodities, United States, 2000–2002 [1]—Continued

Commodity	Unit	2000	2001	2002
		Dollars	*Dollars*	*Dollars*
Fertilizer: [5]				
0-15-40	Ton	183	189	178
0-18-36	Ton	182	184	175
0-20-20	Ton	189	192	193
3-10-30	Ton	168	173	172
5-10-10	Ton	149	150	146
5-10-15	Ton	164	170	165
5-10-30	Ton	177	184	182
5-20-20	Ton	182	188	182
6- 6- 6	Ton	195	215	224
6- 6-18	Ton	214	213	207
6-12-12	Ton	159	173	156
6-24-24	Ton	221	221	214
8- 8- 8	Ton	160	173	166
8-20- 5	Ton	237	232	224
8-32-16	Ton	225	228	229
9-23-30	Ton	202	208	195
10- 3- 3	Ton	190	207	190
10- 6- 4	Ton	160	157	166
10-10-10	Ton	171	179	174
10-20-10	Ton	191	199	194
10-20-20	Ton	207	217	209
10-34- 0	Ton	243	260	238
11-52- 0	Ton	259	260	246
13-13-13	Ton	194	210	201
15-15-15	Ton	216	230	223
16- 0-13	Ton	176	185	161
16- 4- 8	Ton	216	237	235
16- 6-12	Ton	173	203	186
16-16-16	Ton	268	284	292
16-20- 0	Ton	240	246	243
17-17-17	Ton	212	230	212
18-46- 0 (DAP)	Ton	240	244	227
19-19-19	Ton	213	234	207
24- 8- 0	Ton	176	205	167
Ammonium Nitrate	Ton	194	260	195
Anhydrous Ammonia	Ton	227	399	250
Aqua Ammonia	Ton	98.30	119	104
Limestone, Spread on field	Ton	18.90	19.10	19.00
Muriate of Potash, 60–62% K2O	Ton	165	170	164
Nitrate of Soda	Ton	260	274	282
Nitrogen Solutions.				
28% N	Ton	125	204	135
30% N	Ton	131	189	127
32% N	Ton	137	221	148
Sulfate of Ammonia	Ton	167	192	187
Superphosphate, 44-46% P2O5	Ton	233	236	221
Urea, 44-46% Nitrogen	Ton	200	280	191
Farm Machinery:				
Baler, Pick-Up, Automatic Tie, P.T.O.				
Square Conventional, Under 200 Lb Bales	Each	16,100	16,300	16,800
Round, 1200-1500 Lb Bale	Each	17,300	17,600	17,900
Round, 1900-2200 Lb Bale	Each	23,800	23,000	25,200
Chisel Plow, Maxiumum 1 Foot Depth of Tillage, Chisel or Sweep Type, Drawn or. Mounted, 16-20 Foot	Each	13,400	13,000	13,400
Combine, Self Propelled with Grain head				
Extra-large capacity	Each	176,000	181,000	187,000
Large capacity	Each	146,000	152,000	156,000
Corn Head for Combine				
6 Row	Each	23,400	24,500	25,700
8 Row	Each	30,400	32,100	33,200
Cotton Picker, Self Propelled, with sprindle, 4-Row	Each	206,000	222,000	225,000
Cultivator, Row Crop				
6-Row	Each	5,860	6,040	5,990
12-Row, Flexible	Each	13,400	13,800	13,900
Disk Harrow, Tandem, Drawn [7]				
15-17 Foot	Each	13,500	13,700	14,300
18-20 foot	Each	17,400	18,000	18,500

See footnotes at end of table.

Table 9-36.—Prices paid by farmers: April prices, by commodities, United States, 2000–2002 [1]—Continued

Commodity	Unit	2000	2001	2002
		Dollars	*Dollars*	*Dollars*
Elevator, Portable, Without Power Unit,				
Auger Type, 8 Inch Diameter, 60 Foot	Each	3,820	3,820	4,020
Feed Grinder-Mixer, Trailer Mtd., P.T.O.	Each	13,600	14,600	14,900
Field Cultivator, Mounted or Drawn				
17-19 Foot	Each	10,300	10,700	11,000
20-25 Foot, Flexible	Each	14,400	14,300	15,900
Forage Harvester, P.T.O., Shear Bar,				
With Pick-Up Attachment	Each	28,200	28,900	30,100
With Row Crop Unit, 2-Row	Each	30,400	31,400	32,800
Forage Harvester, Self-propelled, Shear Bar,				
With 4–6 row	Each	216,000	228,000	237,000
Front-End Loader, Hydraulic, Tractor Mounted				
1800-2500 Lb. Capacity, 60 Inch Bucket	Each	4,880	4,910	4,990
Grain Drill, Most Common Spacing				
Plain, 15-17 Openers	Each	11,200	12,100	14,000
Press, 23-25 Openers	Each	17,500	18,500	23,100
With Fertilizer Attachment, 20-24 Openers	Each	15,200	15,700	18,600
Min/No-Till W/Fert. Attach., 15 Foot	Each	26,300	27,100	28,100
Hayrake, Side-Delivery, or Wheel Rake,				
Traction Drive, 8-12 Foot Working Width	Each	4,670	4,980	5,020
Hay Tedder, 15-18 Foot	Each	4,480	4,560	4,830
Manure Spreader, Conveyor Type, P.T.O.,				
2-Wheel, with Tires.				
141-190 Bushel Capacity	Each	6,720	6,630	6,760
225-300 Bushel Capacity	Each	9,640	9,980	9,950
Mower-Conditioner, P.T.O., Pull Type, with				
8-10 Foot, Sickle (Cutter) Bar or Disc	Each	13,200	13,700	13,700
14-16 Foot, Sickle (Cutter) Bar or Disc	Each	21,000	21,500	21,800
Mower, Mounted or Drawn,				
7-8 ft Sickle (Cutter) Bar	Each	4,360	4,730	4,840
13-14 Foot, Sickle (Cutter) Bar or Disc	Each	13,700	14,200	16,100
Planter, Row Crop				
With Fertilizer Attachment, 4-Row	Each	15,300	15,500	15,100
With Fertilizer Attachment, 8-Row	Each	26,900	28,800	29,000
With Fertilizer Attachment, 24-Row	Each	92,600	95,000	97,600
12-Row Conservation (No-Till Cond), w/Fert	Each	46,400	47,400	50,400
Rotary Hoe, 20-25 Foot	Each	6,280	6,920	6,490
Rotary Cutter, 7-8 Foot	Each	2,970	3,100	3,240
Sprayer, Field Crop, Power, Boom Type				
(Excl. Self-Propelled and Orchard).				
Tractor Mounted, w/ 300 Gal. Spray Tank	Each	5,260	5,500	5,460
Trailer Type, w/ 500-700 Gal. Spray Tank	Each	11,100	11,300	12,000
Tractor, 2-Wheel Drive				
30-39 P.T.O. horsepower	Each	16,400	16,600	16,400
50-59 P.T.O. horsepower	Each	21,800	22,000	21,900
70-89 P.T.O. horsepower	Each	33,700	34,200	34,500
110 - 129 P.T.O. horsepower	Each	62,400	63,000	63,700
140 - 159 P.T.O. horsepower	Each	81,700	82,300	83,200
190 - 220 P.T.O. horsepower	Each	113,000	114,000	118,000
Tractor, 4-Wheel Drive				
200 - 280 P.T.O. horsepower	Each	120,000	127,000	132,000
Wagon, Gravity Unload, W/Box and Running				
Gear, and Tires,				
200-400 Bushel Capacity				
Without Side Extensions	Each	4,040	4,160	4,320
Wagon, Running Gear, W/O Box				
8-10 Ton Capacity	Each	1,660	1,800	1,730
Windrower, Self-Propelled,				
14-16 Foot	Each	61,600	63,200	62,900
Agricultural Chemicals: [8]				
Fungicides:				
Basic Copper Sulfate, 53% WP	Lb	1.17	1.15	1.16
Benomyl (Benlate), 50% WP	Lb	18.60	18.40	18.40
Calcium Polysulfide (Lime Sulfur) Liq.Conc	Gal	8.32	8.42	8.36
Captan 50% WP	Lb	3.45	3.61	3.76
Chlorothalonil (Bravo), 6#/Gal EC	Gal	56.20	56.10	49.70
Copper Hydroxide (Kocide), 77% WP	Lb	2.58	2.60	2.62
Dodine (Cyprex Syllit), 65% WP	Lb	11.50	11.50	11.50
Fenarimol (Rubigan), 1#/Gal EC	Gal	310	321	320
Ferbam (Carbamate), 76% WP	Lb	4.29	4.17	4.25
Fosethyl-AL (Aliette), 80% WP	Lb	12.00	12.30	12.60
Iprodione (Rovral), 50% WP	Lb	23.40	23.70	24.10
Mancozeb (Dithane 80% WP,Manzate 75% DF)	Lb	3.05	3.03	2.95
Maneb, 80% WP, 75% DF	Lb	3.19	3.16	3.20
Metalaxyl (Ridomil), 2#/Gal EC	Gal	178	190	183
Myclobutanil (Systhane, Nova, Rally), 40% WP	Lb	73.00	69.70	67.90
Oxytetraycline (Mycoshield), 17% WP	Lb	20.90	22.00	22.00
Sulfur, 95% WP	Lb	0.313	0.320	0.330
Triforine (Funginex), 1.6#/Gal EC	Gal	93.30	93.30	98.50
Triadimefon (Bayleton), 50% WP	Lb	67.30	69.20	70.00
Ziram, 76% WP	Lb	2.72	2.82	2.82

See footnotes at end of table.

Table 9-36.—Prices paid by farmers: April prices, by commodities, United States, 2000–2001 [1]—Continued

Commodity	Unit	2000	2001	2002
		Dollars	*Dollars*	*Dollars*
Fumigants:				
Methyl Bromide (Terr-o-gas 98)	Lb	3.51	4.87	5.31
Herbicides:				
2,4-D, 4#/Gal EC	Gal	14.70	14.90	14.90
Acetochlor (Harness, Surpass), 6.4–7#/Gal EC	Gal	68.40	68.90	68.10
Alachlor (Lasso), 4#/Gal EC	Gal	24.50	24.80	24.30
Atrazine(AAtrex), 4#/Gal L	Gal	13.60	12.50	12.20
Bentazon (Basagran), 4#/Gal EC	Gal	79.40	83.00	83.20
Butylate (Sutan), 6.7#/Gal EC	Gal	21.20	22.50	23.60
Chlorimuron-ethyl (Classic), 25% DF	Oz	11.90	12.40	12.40
Chlorsulfuron (Glean), 75%	Oz	19.00	20.20	18.70
Cyanazine (Bladex), 4#/Gal EC	Gal	33.10	33.10	33.00
DCPA (Dacthal), 75% WP	Lb	13.20	13.50	14.00
Dicamba (Banvel), 4#/Gal EC	Gal	95.10	94.20	92.10
Diuron (Karmex, Diurex), 80% WP	Lb	5.21	4.98	4.90
EPTC (Eptan), 7E-(Eradicane),6.7#/Gal EC	Gal	33.30	33.40	33.70
Glyphosate (Roundup), 4#/Gal EC	Gal	43.30	44.50	43.50
Linuron (Lorox, Linex), 50% DF	Lb	11.90	11.60	11.90
MCPA, 4#/Gal, EC	Gal	16.60	16.90	17.10
Metolachlor (Dual), 8#/Gal EC	Gal	82.60	94.50	99.00
Metribuzin (Lexone or Sencor), 75% DF	Lb	20.90	20.60	20.10
Paraquat (Gramoxone Extra), 2.5#/Gal EC	Gal	23.50	21.80	21.60
MSMA (Super Arsonade), 4-6# Gal EC	Gal	9.61	9.85	9.40
Napropamide (Devrinol), 50% WP	Lb	34.30	35.20	37.60
Pendimethalin (Prowl),3.3#/Gal EC	Gal	23.60	23.10	22.10
Sethoxydim (Poast), 1.5#/Gal EC	Gal	78.70	77.90	74.60
Simazine (Princep), 4#/Gal EC	Gal	18.20	18.00	17.60
Terbacil (Sinbar), 80% WP	Lb	30.10	30.90	32.70
Trifluralin (Treflan), 4#/Gal EC	Gal	25.50	25.30	24.40
Insecticides:				
Acephate (Orthene), 75% SP	Lb	12.70	12.80	12.60
Aldicarb (Temik), 15% G	Lb	3.73	3.76	3.70
Azinphos-methyl (Guthion), 50% WP	Lb	9.68	9.87	10.60
Bt (Dipel 2X), WP	Lb	12.40	12.60	12.20
Carbaryl, (Sevin), 80% S, SP or WP	Lb	5.44	5.75	5.41
Carbofuran (Furadan), 4F	Gal	73.50	74.80	77.80
Chlorpyrifos (Lorsban), 4#/Gal EC	Gal	43.60	42.40	41.60
Cyfluthrin (Baythroid) 2#/Gal EC	Gal	511	493	397
Cypermethrin,(Ammo 2.5-Cymbush 3#G)EC	Gal	262	255	196
Diazinon, 4#/Gal EC	Gal	36.20	35.90	36.10
Dicofol (Kelthane), 35% WP	Lb	11.90	12.20	13.10
Dicrotophos (Bidrin), 8#/Gal EC	Gal	88.50	89.70	91.00
Dimethoate (Cygon), 2.67#/Gal EC	Gal	36.40	35.50	36.60
Disulfoton (Di-Syston), 8#/Gal EC	Gal	82.10	84.80	87.60
Endosulfon (Thiodan, Phaser), 3#/Gal EC	Gal	35.90	35.90	33.70
Esfenvalerate (Asana XL),0.66#/Gal EC	Gal	125	123	107
Ethion 4#/Gal EC	Gal	38.50	35.10	39.30
Fonofos (Dyfonate II), 20% G	Lb	2.16	2.15	1.29
Imidacloprid (Admire, Provado),. 1.6–2#/Gal EC	Gal	565	565	575
Malathion, 5#/Gal EC	Gal	26.20	27.40	28.40
Methidathion (Supracide), 25% WP	Lb	6.97	7.51	7.20
Methomyl (Lannate) L), 1.81 #/Gal Liq.	Gal	49.30	49.50	51.90
Methyl Parathion, 4#/Gal EC	Gal	31.10	33.40	32.00
Oil, Superior Oil, Supreme, Volck	Gal	5.22	5.38	5.82
Oxamyl (Vydate-L), 2# L	Gal	65.50	67.90	68.80
Oxydemeton-Methyl (Metasystox-R). 2#/Gal EC	Gal	72.10	77.90	76.00
Oxythioquinox (Morestan), 25% WP	Lb	17.60	18.60	20.30
Phorate (Thimet), 20% G	Lb	2.49	2.41	2.45
Phosmet (Imidan, Prolate), 50% WP	Lb	6.59	6.98	7.30
Propargite (Comite, Omite), 30% WP	Lb	6.86	6.05	6.26
Synthetic Pyrethroids,. (Pounce 2.0, Ambush 3.2 #/Gal) EC	Gal	141	137	136
Terbufos (Counter), 15% G	Lb	2.71	2.67	2.72
Zeta–Cyermethrin (Fury), 1.5#/Gal EC	Gal	301	282	195
Other:				
Gibberellic Acid,(Ry3Up,Pro-Gibb)4.0% L	Gal	183	170	164
Nad Napthalene Acetamide, 8.4 WP	Lb	59.50	63.50	63.70

[1] Prices paid by famers are collected, for the most part, from retail establishments located in smaller cities and towns in rural areas. Prior to 1995, recorded prices reflected a modified annual average based on frequency item was surveyed during the year. Recorded item values, 1995-99, are the U.S. April average price. [2] Excludes Federal, State, and local per gallon taxes where applicable. [3] Excludes Federal excise tax. [4] Includes Federal, State, and local per gallon taxes. [5] Excludes cost of application, except for limestone. [6] Discontinued in 2000. [7] With hydraulic lift, transport wheels, and tires. [8] Formulation abbreviations: EC–Emulsifiable Concentrate, DF–Dry Flowable, DG–Dry Granular, G–Granular, L–Liquid, S–Solution, SP–Soluble Powder, and WP–Wettable Powder.

NASS, Environmental, Economics, and Demographics Branch, (202) 720–6146.

Table 9-37.—Agricultural commodities: Support prices per unit, United States, 1993–2002 [1 2]

Commodity and unit		1993	1994	1995	1996	1997
		Dollars	*Dollars*	*Dollars*	*Dollars*	*Dollars*
Basic commodities:						
Corn:						
Target price	Bushel	2.75	2.75	2.75	(9)	(9)
Loan rate	do	1.72	1.89	1.89	1.89	1.89
Cotton:						
American upland: [3]						
Target price	Pound	0.7290	0.7290	0.7290	(9)	(9)
Loan rate	do	0.5235	0.5000	0.5192	0.5192	0.5192
Extra-long staple:						
Target price	do	1.0570	1.0200	0.9560	(9)	(9)
Loan rate	do	0.8812	0.8503	0.7965	0.7965	0.7965
Peanuts: [4]	do	0.3375	0.3392	0.3392	0.3050	0.3050
Rice:						
Target price	Cwt.	10.71	10.71	10.71	(9)	(9)
Loan rate	do	6.50	6.50	6.50	6.50	6.50
Wheat:						
Target price	Bushel	4.00	4.00	4.00	(9)	(9)
Loan rate	do	2.45	2.58	2.58	2.58	2.58
Tobacco:						
Flue-cured, types 11-14	Pound	1.577	1.583	1.597	1.601	1.621
Fire-cured, type 21	do	1.395	1.407	1.430	1.455	1.498
Fire-cured, types 22-23	do	1.464	1.483	1.518	1.557	1.623
Burley, type 31	do	1.683	1.714	1.725	1.737	1.760
Dark air-cured, types 35-36	do	1.255	1.273	1.304	1.339	1.398
Virginia sun-cured, type 37	do	1.233	1.245	1.265	1.288	1.326
Cigar filler, Puerto Rican, type 46	do	0.834	0.844	0.861	(8)	(8)
Ohio filler and Wisconsin binder,. types 42-44 and 53-55	do	1.074	1.084	1.101	1.120	1.169
Barley: [6]						
Target price	Bushel	2.36	2.36	2.36	(9)	(9)
Loan rate	do	1.40	1.54	1.54	1.55	1.57
Sorghum grain: [6]						
Target price	Cwt.	4.66	4.66	4.66	(9)	(9)
Loan rate	do	2.91	3.21	3.21	3.23	3.14
Oats: [6]						
Target price	Bushel	1.45	1.45	1.45	(9)	(9)
Loan rate	do	0.88	0.97	0.97	1.03	1.11
Rye: [6]	do	1.46	1.61	1.61	(8)	(8)
Nonbasic commodities:						
Beans, dry edible	Cwt.	(8)	(8)	(8)	(8)	(8)
Cottonseed	Ton	(8)	(8)	(8)	(8)	(8)
Minor oilseeds [7]	Pound	0.089	0.087	0.087	0.0891	0.0930
Soybeans	Bushel	5.02	4.92	4.92	4.99	5.26
Sugar, raw	Pound	0.1800	0.1800	0.1800	0.1800	0.1800
Milk for manufacturing	Cwt.	10.10	10.10	10.10	(10) 10.35	(13) 10.20
Honey, extracted	Pound	0.538	0.500	0.500	(11)	(11)
Mohair	do	4.738	4.739	4.657	(12)	(12)
Wool	Pound	2.04	2.09	2.12	(12)	(12)

See footnotes at end of table.

Table 9-37.—Agricultural commodities: Support prices per unit, United States, 1993–2002 [1] [2]—Continued

Commodity	Unit	1998	1999	2000	2001	2002
		Dollars	*Dollars*	*Dollars*	*Dollars*	*Dollars*
Basic commodities:						
Corn:						
Target price	Bushel	(9)	(9)	(9)	(9)	(16) 2.60
Loan rate	do	1.89	1.89	1.89	1.89	1.98
Cotton:						
American upland: [3]						
Target price	Pound	(9)	(9)	(9)	(9)	(16) 0.742
Loan rate	do	0.5192	0.5192	0.5192	0.5192	0.5200
Extra-long staple:						
Target price	do	(9)	(9)	(9)	(9)	....
Loan rate	do	0.7965	0.7965	0.7965	0.7965	0.7977
Peanuts: [4]						
Target price	do	....	....	....	....	0.2475
Loan rate	do	0.3050	0.3050	0.3050	0.3050	0.1775
Rice:						
Target price	Cwt.	(9)	(9)	(9)	(9)	(16) 10.50
Loan rate	do	6.50	6.50	6.50	6.50	6.50
Wheat:						
Target price	Bushel	(9)	(9)	(9)	(9)	(16) 3.86
Loan rate	do	2.58	2.58	2.58	2.58	2.80
Tobacco:						
Flue-cured, types 11-14	Pound	1.628	1.632	1.640	1.660	1.656
Fire-cured, type 21	do	1.536	1.559	1.559	1.572	1.603
Fire-cured, types 22-23	do	1.681	1.716	1.716	1.736	1.767
Burley, type 31	do	1.778	1.789	1.805	1.826	1.835
Dark air-cured, types 35-36	do	1.450	1.481	1.481	1.499	1.526
Virginia sun-cured, type 37	do	1.360	1.380	1.380	1.392	1.429
Cigar filler, Puerto Rican, type 46	do	(8)	(8)	(8)	(8)	(8)
Ohio filler and Wisconsin binder,. types 42-44 and 53-55	do	1.212	1.238	1.238	1.252	1.286
Barley: [6]						
Target price	Bushel	(9)	(9)	(9)	(9)	(16) 2.21
Loan rate	do	1.56	1.59	1.62	1.65	1.88
Sorghum grain: [6]						
Target price	Cwt.	(9)	(9)	(9)	(9)	(16) 4.54
Loan rate	do	3.11	3.11	3.05	3.05	3.54
Oats: [6]						
Target price	Bushel	(9)	(9)	(9)	(9)	(16) 1.40
Loan rate	do	1.11	1.13	1.16	1.21	1.35
Rye: [6]	do	(8)	(8)	(8)	(8)	(8)
Nonbasic commodities:						
Beans, dry edible	Cwt.	(8)	(8)	(8)	(8)	(8)
Cottonseed	Ton	(8)	(8)	(8)	(8)	(8)
Other oilseeds [7]:						
Target price	do	....	....	....	....	(16) 9.80
Loan rate	do	9.30	9.30	9.30	9.30	9.60
Soybeans:						
Target price	Bushel	....	....	....	....	(16) 5.80
Loan rate	do	5.26	5.26	5.26	5.26	5.00
Dry peas	Cwt.	....	....	....	....	(16) 6.33
Sugar, raw	Pound	0.1800	0.1800	0.1800	0.1800	0.1800
Milk for manufacturing	Cwt.	(14) 10.05	(15) 9.90	9.90	9.90	9.90
Honey, extracted	Pound	(11)	0.59	0.59	(18) 0.65	(18) 0.60
Mohair	do	(12)	2.00	2.00	....	(17) 4.20
Wool	Pound	(12)	(12)	(12)	(12)	(17) 0.40/1

[1] National averages during the marketing years for the individual crops, beginning in the years shown. [2] The target price is known in the statute as the "established price". [3] 1 1/16 strict low middling, micronaire 3.5 through 4.9. [4] For quota portion of crop (1993 through 2001). Enactment of the Farm Security and Rural Investment Act of 2002 (2002 Act) repealed the peanut quota marketing program; and established payment rates for the 2002/2003 and subsequent crops according to the provisions of the Direct Payment Program. [5] Grade No. 2 or better except for oats which is Grade No. 3. [6] The rye price support program was terminated by the Federal Agriculture Improvement and Reform Act of 1996. Rye was not reestablished with the 2002 Act. [7] Includes flaxseed, sunflower seed (oil and other), safflower, rapeseed (industrial), canola and mustard seed. [8] No support program. [9] The Federal Agriculture Improvement and Reform Act of 1996 replaced the deficiency payment/production adjustment programs for the program crops with a Production Flexibility Contract program, making target prices no longer applicable beginning with the 1996/97 marketing year. [10] As of January 1, 1996. [11] The honey price support program was terminated by the Federal Agriculture Improvement and Reform Act of 1996. [12] The wool and mohair support programs terminated as of December 31, 1995, as required by Public Law 103-130. [13] As of January 1, 1997. [14] As of January 1, 1998. [15] As of January 1, 1999. [16] The Farm Security and Rural Investment Act of 2002 (2002 Act) reestablished target prices, now including soybeans and other oilseeds. The 2002 Act also established, for the first time, loan rates for dry peas, small chickpeas and lentils under the marketing loan program. [17] Wool and mohair programs were reestablished following enactment of the Farm Security and Rural Investment Act of 2002 (2002 Act). First wool number is for ungraded/second is graded. [18] There was no support for honey produced in 2001.

FSA, Economic Policy and Analysis Staff, (202) 720–2711.

Table 9-38.—Farm income: Cash receipts by commodity groups and selected commodities, United States, 1994–2001 [1]

Commodity	1994	1995	1996	1997
	1,000 dollars	*1,000 dollars*	*1,000 dollars*	*1,000 dollars*
All commodities	181,273,883	188,048,906	199,296,420	207,687,971
Livestock and products	88,293,071	87,217,115	92,948,988	96,473,145
Cattle and calves	36,252,219	34,044,036	30,976,868	35,999,622
Hogs	9,898,434	10,254,866	12,565,187	13,053,680
Sheep and lambs	510,536	566,100	612,170	632,602
Dairy products	19,983,347	19,879,611	22,785,017	20,940,261
Broilers	11,371,723	11,762,222	13,904,023	14,158,926
Farm chickens	73,071	60,218	59,043	71,219
Chicken eggs	3,789,877	3,892,912	4,776,255	4,539,929
Turkeys	2,643,057	2,769,397	3,045,718	2,814,997
Miscellaneous livestock	3,770,807	3,987,753	4,224,707	4,261,909

Commodity	1998	1999	2000	2001
All commodities	195,982,024	187,480,551	193,695,233	202,849,408
Livestock and products	94,112,123	95,610,896	99,559,392	106,431,172
Cattle and calves	33,415,407	36,529,975	40,711,033	40,439,877
Hogs	9,444,082	8,621,759	11,803,046	12,455,792
Sheep and lambs	479,760	462,297	466,474	393,823
Dairy products	24,114,036	23,207,093	20,607,889	24,694,531
Broilers	15,144,551	15,128,118	13,988,946	16,688,339
Farm chickens	75,262	72,712	64,376	46,016
Chicken eggs	4,439,446	4,287,164	4,347,994	4,444,864
Turkeys	2,616,280	2,754,104	2,765,347	2,729,457
Miscellaneous livestock	4,383,299	4,547,674	4,804,287	4,538,473

Commodity	1994	1995	1996	1997
Crops	92,980,812	100,831,791	106,347,432	111,214,826
Food grains	9,548,491	10,356,077	10,795,279	10,410,552
Feed crops	20,310,488	24,516,001	27,250,921	27,141,212
Cotton	6,738,770	6,852,715	6,983,125	6,345,803
Tobacco	2,656,352	2,548,399	2,794,668	2,873,917
Oil crops	14,652,170	15,492,965	16,344,700	19,758,302
Vegetables	14,058,378	14,984,318	14,457,168	14,668,839
Fruits/nuts	10,325,680	11,075,008	11,903,599	13,144,089
All other crops	14,690,483	15,006,308	15,817,972	16,872,112

Commodity	1998	1999	2000	2001
Crops	101,869,901	91,869,655	94,135,841	96,418,236
Food grains	8,808,347	6,969,197	6,758,338	6,595,289
Feed crops	22,633,962	19,554,900	20,775,440	23,245,354
Cotton	6,072,960	4,630,256	3,839,868	4,954,043
Tobacco	2,802,755	2,272,785	2,314,528	1,880,300
Oil crops	17,371,709	13,355,112	13,825,588	14,317,312
Vegetables	15,160,371	15,127,492	15,599,877	15,511,608
Fruits/nuts	11,840,208	11,953,032	12,625,868	11,742,483
All other crops	17,179,563	18,006,882	18,396,333	18,171,847

[1] USDA estimates and publishes individual cash receipt values only for major commodities and major producing States. The U.S. receipts for individual commodities, computed as the sum of the reported States, may understate the value of sales for some commodities, with the balance included in the appropriate category labeled "other" or "miscellaneous." The degree of underestimation in some of the minor commodities can be substantial.

ERS, Farm Structure Performance and Well-Being Branch, (202) 694–5592. E-mail contact is rogers@ERS.USDA.gov.

Table 9-39.—Farm income, United States, 1994–2001 [1]

Item	1994	1995	1996	1997
	Thousand dollars	*Thousand dollars*	*Thousand dollars*	*Thousand dollars*
Total gross farm income	216.0	210.8	235.8	238.1
Value of Production [2]	208.2	203.5	228.4	230.6
Crops	100.3	95.7	115.5	112.4
Livestock and products	89.8	87.8	92.1	96.5
Services and forestry	18.1	19.9	20.8	21.8
Direct government payments	7.9	7.3	7.3	7.5
Total production expenses	167.2	173.8	181.0	187.6
Net farm income	48.9	36.9	54.8	50.5
Gross cash income	198.2	205.9	217.7	227.3
Cash expenses	147.5	153.3	159.9	166.9
Net cash income	50.7	52.5	57.7	60.4

Item	1998	1999	2000	2001
Total gross farm income	232.1	234.5	241.7	246.5
Value of production [2]	219.7	212.9	218.8	225.8
Crops	101.7	92.4	95.0	93.9
Livestock and product	94.2	95.3	99.3	106.3
Services and forestry	23.8	25.2	24.4	25.6
Direct government payments	12.4	21.5	22.9	20.7
Total production expenses	186.5	188.3	193.7	200.8
Net farm income	45.6	46.2	48.0	45.7
Gross cash income	222.3	224.0	230.4	238.5
Cash expenses	165.5	166.9	172.0	178.8
Net cash income	56.8	57.1	58.4	59.7

[1] Component values and additional details may be found in the value-added and cash income tables on the internet at http://www.ers.usda.gov/data/farmincome/finfidmu.htm.

[2] Includes cash receipts, value of change in inventories, and home consumption.

ERS, Farm Structure Performance and Well-Being Branch, (202) 694–5592. E-mail contact is rogers@ERS.USDA.gov

Table 9-40.—Expenses: Farm production expenses, United States, 1994–2001

Item	1994	1995	1996	1997
	Thousand dollars	*Thousand dollars*	*Thousand dollars*	*Thousand dollars*
Total production expenses	167,190,751	173,818,816	180,970,473	187,636,607
Feed purchased	22,635,018	23,831,792	25,236,518	26,334,299
Livestock and poultry purchased	13,304,604	12,505,851	11,293,747	13,820,317
Seed purchased	5,372,789	5,462,110	6,211,914	6,711,329
Fertilizer and lime	9,177,120	10,029,109	10,928,918	10,927,288
Pesticides	7,219,623	7,718,667	8,518,367	9,017,509
Fuel and oil	5,312,504	5,426,405	5,978,952	6,242,613
Electricity	2,682,739	2,969,776	3,163,860	3,043,921
Other [1]	37,773,985	40,213,495	40,171,057	41,649,074
Interest	11,555,708	12,576,470	12,961,117	13,132,569
Contract and hired labor expenses	15,310,370	16,293,790	17,331,338	18,410,052
Net rent to nonoperator landlords [2]	11,765,291	10,914,217	12,971,368	11,869,985
Capital consumption	18,605,561	19,165,709	19,387,396	19,538,617
Property taxes	6,475,439	6,711,426	6,815,921	6,939,033

Item	1998	1999	2000	2001
Total production expenses	186,502,131	188,273,666	193,659,988	200,770,202
Feed purchased	25,031,106	24,527,489	24,526,178	25,225,280
Livestock and poultry purchased	12,576,095	13,840,120	16,058,038	15,685,326
Seed purchased	7,211,523	7,211,963	7,512,200	8,312,167
Fertilizer and lime	10,624,176	9,921,918	10,021,273	10,323,092
Pesticides	9,017,479	8,618,108	8,517,483	8,616,897
Fuel and oil	5,599,819	5,588,728	7,208,151	7,208,229
Electricity	2,908,384	2,986,750	2,999,827	3,531,487
Other [1]	42,711,831	43,929,252	42,857,452	45,933,067
Interest	13,365,025	13,584,609	14,277,531	14,144,430
Contract and hired labor expenses	19,120,408	19,807,826	20,636,566	22,189,470
Net rent to nonoperator landlords [2]	11,443,054	11,308,009	11,831,521	12,029,308
Capital consumption	19,880,510	20,141,414	20,306,158	20,563,736
Property taxes	7,012,721	6,807,480	6,907,609	7,007,714

[1] Includes repair and maintenance, machine hire and customwork, marketing, storage and transportation, insurance premiums, and miscellaneous other expenses. [2] Includes landlord capital consumption.

ERS, Farm Structure Performance and Well-Being Branch, (202) 694–5592. E-mail contact is rogers@ERS.USDA.gov

Table 9-41.—Farm marketings: Cash receipts, government payments, and principal commodities, by States, 2001

State	Cash receipts			Government payments	Rank
	Total	Crops	Livestock and roducts		
	1,000 dollars	*1,000 dollars*	*1,000 dollars*	*1,000 dollars*	
AL	3,519,731	705,216	2,814,515	230,734	23-Broilers, cattle/calves, chicken eggs, greenhouse (81%).
AK	51,865	23,853	28,012	2,173	49-Greenhouse, potatoes, hay, dairy (45%).
AZ	2,574,698	1,409,090	1,165,608	99,254	29-Cattle /calves, lettuce, dairy, cotton (71%).
AR	5,131,964	1,624,569	3,507,395	832,545	13-Broilers, cotton, Cattle/calves, soybeans (70%).
CA	25,892,319	18,545,880	7,346,439	586,761	1-Dairy, grapes, greenhouse, lettuce (44%).
CO	4,728,954	1,354,465	3,374,489	319,599	16-Cattle/calves, dairy, corn, hogs (72%).
CT	476,150	298,829	177,321	7,540	44-Greenhouse, dairy, chicken eggs, aquaculture (66%).
DE	847,718	185,719	661,999	25,004	39-Broilers, corn, soybeans, greenhouse (83%).
FL	6,415,882	4,957,896	1,457,986	108,011	9-Greenhouse, oranges, tomatoes, sugar cane (51%).
GA	5,514,952	1,975,220	3,539,732	427,261	11-Broilers, cotton, peanuts, chicken eggs (68%).
HI	510,507	419,298	91,209	3,860	42-Pineapples, greenhouse, sugar cane, macadamia nuts (53%).
ID	3,847,926	1,787,513	2,060,413	207,664	21-Dairy, cattle/calves, potatoes, hay (73%).
IL	7,547,087	5,704,242	1,842,845	1,849,769	8-Corn, soybeans, hogs, cattle/calves (87%).
IN	5,105,437	3,235,048	1,870,389	925,278	14Corn, soybeans, hogs, dairy (76%).
IA	11,550,109	5,614,520	5,935,589	1,971,677	3-Hogs, corn, soybeans, cattle/calves (89%).
KS	8,121,044	2,585,380	5,535,664	1,068,706	5-Cattle/calves, wheat, corn, sorghum grain (85%).
KY	3,548,328	1,280,795	2,267,533	293,379	22-Horses/mules, tobacco, broilers, cattle/calves (67%).
LA	1,817,088	1,115,957	701,131	434,612	33-Sugar cane, cotton, cattle/calves, dairy (49%).
ME	485,064	210,774	274,290	7,819	43-Dairy, potatoes, aquaculture, chicken eggs (68%).
MD	1,596,085	646,712	949,373	86,626	36-Broilers, greenhouse, dairy, corn (71%).
MA	366,611	272,925	93,686	10,138	47-Greenhouse, dairy, cranberries, sweet corn (65%).
MI	3,469,122	1,979,799	1,489,323	352,766	24-Dairy, greenhouse, corn, soybeans (58%).
MN	8,101,875	3,813,440	4,288,435	1,242,141	6-Corn, hogs, dairy, soybeans (67%).
MS	3,146,582	871,056	2,275,526	517,007	26-Broilers, cotton, aquaculture, cattle/calves (74%).
MO	4,824,141	2,144,809	2,679,332	817,044	15-Cattle/calves, soybeans, corn, hogs (61%).
MT	1,785,002	657,248	1,127,754	476,158	34-Cattle/calves, wheat, hay, barley (85%).
NE	9,488,580	3,402,349	6,086,231	1,297,623	4-Cattle/calves, corn, soybeans, hogs (91%).
NV	424,596	153,300	271,296	5,864	45-Cattle/calves, hay, dairy, greenhouse (86%).
NH	155,478	89,644	65,834	2,815	48-Greenhouse, dairy, apples, cattle/calves (77%).
NJ	821,070	617,316	203,754	16,403	40-Greenhouse, horses/mules, dairy, blueberries (61%).
NM	2,215,122	545,019	1,670,103	93,729	31-Dairy, cattle/calves, hay, greenhouse (83%).
NY	3,419,790	1,199,163	2,220,627	114,039	25-Dairy, greenhouse, cattle/calves, apples (71%).
NC	7,730,633	3,086,554	4,644,079	330,730	7-Hogs, broilers, greenhouse, tobacco (67%).
ND	2,978,548	2,258,615	719,933	944,591	28-Wheat, cattle/calves, soybeans, corn (61%).
OH	4,682,011	2,818,473	1,863,538	681,651	17-Soybeans, corn, dairy, greenhouse (62%).
OK	4,026,680	873,802	3,152,878	392,822	20-Cattle/calves, hogs, broilers, wheat (79%).
OR	3,122,641	2,297,688	824,953	104,946	27-Greenhouse, cattle/calves, dairy, hay (56%).
PA	4,454,979	1,308,750	3,146,229	103,462	18-Dairy, cattle/calves, agaricus mushrooms, greenhouse (65%).
RI	47,438	39,735	7,703	292	50-Greenhouse, dairy, sweet corn, cattle/calves (78%).
SC	1,646,020	763,677	882,343	130,287	35-Broilers, greenhouse, tobacco, cattle/calves (57%).
SD	4,107,879	1,852,454	2,255,425	715,264	19-Cattle/calves, corn, soybeans, hogs (78%).
TN	2,160,707	1,033,948	1,126,759	247,485	32-Cattle/calves, broilers, dairy, greenhouse (55%).
TX	13,795,618	4,456,153	9,339,465	1,703,168	2-Cattle/calves, greenhouse, broilers, cotton (73%).
UT	1,116,343	263,082	853,261	39,754	37-Cattle/calves, dairy, hay, hogs (76%).
VT	556,779	66,719	490,060	7,877	41-Dairy, cattle/calves, greenhouse, hay (91%).
VA	2,443,987	770,785	1,673,202	117,158	30-Broilers, cattle/calves, dairy, turkeys (57%).
WA	5,191,920	3,464,259	1,727,661	298,784	12-Dairy, apples, cattle/calves, potatoes (53%).
WV	407,570	59,315	348,255	9,842	46-Broilers, cattle/calves, dairy, turkeys (75%).
WI	5,896,293	1,432,106	4,464,187	415,110	10-Dairy, cattle/calves, corn, greenhouse (79%).
WY	982,545	145,086	837,459	50,272	38-Cattle/calves, hay, hogs, sugar beets (88%).
US	202,849,408	96,418,236	106,431,172	20,727,496	Cattle/calves, dairy, corn, broilers (49%).

ERS, Farm Sector Performance and Well-Being Branch, (202) 694–5592. E-mail contact is rogers@ERS.USDA.gov

Table 9-42.—Average Income to Farm Operator Households, United States, 1999–2002 [1] [2]

Item	1999 [3]	2000	2001	2002
	Dollars per farm			
Net cash farm business income [4]	13,194	11,175	14,311	10,970
Less depreciation [5]	7,027	7,357	7,609	NA
Less wages paid to operator [6]	499	608	932	NA
Less farmland rental income [7]	802	757	477	NA
Less adjusted farm business income due to other household(s) [8]	1,262	801	1,083	NA
	Dollars per farm operator household			
Equals adjusted farm business income	3,603	1,652	4,211	NA
Plus wages paid to operator	499	608	932	NA
Plus net income from farmland rental [9]	1,312	NA	NA	NA
Equals farm self-employment income *	5,415	2,260	5,143	NA
Plus other farm-related earnings [10]	944	339	396	NA
Equals earnings of the operator household from farming activities	6,359	2,598	5,539	2,439
Plus earnings of the operator household from off-farm sources [11]	57,988	59,349	58,578	60,077
Equals average farm operator household income comparable to U.S. average household income, as measured by the CPS	64,347	61,947	64,117	62,515
	Dollars per U.S. household			
U.S. average household income [12]	54,842	57,045	58,208	NA
	Percent			
Average farm operator household income as percent of U.S. average household income	117.3	108.6	110.2	NA
Average operator household earnings from farming activities as percent of average operator household income	9.9	4.2	8.6	3.9

NA = Not available. * The relative standard error exceeds 25 percent, but is no more than 50 percent. [1] This table derives farm operator household income estimates from the Agricultural Resource Management Study (ARMS) that are consistent with Current Population Survey (CPS) methodology. The CPS, conducted by the Census Bureau, is the source of official U.S. household income statistics. The CPS defines income to include any income received as cash. The CPS definition departs from a strictly cash concept by including depreciation as an expense that farm operators and other self-employed people subtract from gross receipts when reporting net cash income. [2] Preliminary. [3] Prior to 2000, net cash income from operating another farm and net cash income from farm land rental were included in earnings from farming activities. However, because of a change in the ARMS survey design, net cash income from a farm other than the one being surveyed and net income from farm land rentals are not separable from total off-farm income. Although there is no effect upon estimates of farm operator household income in 2000, estimates of farm self-employment, other farm related earnings, earnings of the household from farming activities, and earnings of the farm from off-farm sources are not strictly comparable to those from previous years. [4] A component of farm sector income. Excludes income of contractors and landlords as well as the income of farms organized as non-family corporations or cooperatives and farms run by a hired manager. Includes the income of farms organized as proprietorships, partnerships, and family corporations. [5] Consistent with the CPS definition of self-employment income, reported depreciation expenses are subtracted from net cash income. The ARMS collects farm business depreciation used for tax purposes. [6] Wages paid to the operator are subtracted here because they are not shared among other households that have claims on farm business income. These wages are added to the operator household's adjusted farm business income to obtain farm self-employment income. [7] Gross rental income is subtracted here because net rental income from the farm operation is added below to income received by the household. [8] More than one household may have a claim on the income of a farm business. On average,1.1 households share the income of a farm business. [9] Includes net rental income from the business. Also includes net rental income from farmland held by household members that is not part of the farm business. Beginning in 2000, net income from farmland rental is considered as part of off-farm income. (See footnote 2.) [10] Wages paid to other operator household members by the farm business and net cash income from a farm business other than the one being surveyed. In 2000, however, net cash income from a farm businesses other than the one being surveyed is included in off-farm earnings. (See footnote 2.) Beginning in 1996, also includes the value of commodities provided to household members for farm work. [11] Wages, salaries, net income from nonfarm businesses, interest, dividends, transfer payments, etc. Beginning in 2000, also includes net cash income from other farm and net cash income from farm rental. (See footnote 2.) [12] From the CPS. Sources: U.S. Dept. of Agriculture, Economic Research Service, 1999, 2000, and 2001 Agricultural Resource Management Study (ARMS) for farm operator household data. U.S. Dept. of Commerce, Bureau of the Census, Current Population Survey (CPS), for U.S. average household income. For information on household income contact: Bob Green (202) 694-5568. Email rgreen@ers.usda.gov or Bob Hoppe (202) 694-5572. Email rhoppe@ers.usda.gov.

ERS, Farm Structure and Performance Branch, (202) 694-5568. Sources: U.S. Dept. of Agriculture, Economic Research Service, 1999-2003 Agricultural Resource Management Study (ARMS) for farm operator household data. U.S. Dept. of Commerce, Bureau of the Census, Current Population Survey (CPS), for U.S. average household income.

Table 9-43.—Grazing fees: Rates for cattle by selected States and regions, 2001–2002

State	Monthly lease rates for private non-irrigated grazing land [1]					
	Animal unit [2]		Cow-calf		Per head	
	2001	2002	2001	2002	2001	2002
	Dollars per month	*Dollars per month*	*Dollars per month*	*Dollars per month*	*Dollars per month*	*Dollars per month*
AZ	7.00	7.30	9.75	(7)	7.20	7.80
CA	13.00	12.80	16.00	16.50	13.50	14.00
CO	11.90	12.60	14.20	14.90	12.30	12.80
ID	11.50	11.70	13.40	13.70	12.00	12.20
KS	12.50	13.00	15.50	16.00	12.50	13.00
MT	14.90	15.10	16.70	17.30	16.00	16.30
NE	20.60	20.90	24.60	25.00	21.50	22.00
NV	10.00	10.50	13.00	11.50	10.70	11.50
NM	10.20	8.80	12.50	10.70	10.50	9.90
ND	10.30	12.50	12.20	13.50	12.50	12.00
OK	8.00	7.50	9.00	8.50	7.50	7.50
OR	12.10	11.80	12.80	12.60	9.50	10.20
SD	15.70	16.90	18.30	19.10	17.20	18.00
TX	8.50	9.00	9.50	10.00	9.00	9.20
UT	11.00	11.60	14.00	13.70	11.50	12.10
WA	9.10	9.60	11.80	11.00	10.20	9.60
WY	12.90	13.50	15.00	15.60	13.10	14.00
17-State [3]	11.90	12.30	13.90	14.20	12.40	12.70
16-State [4]	13.10	13.50	15.50	15.80	13.70	14.00
11-State [5]	12.30	12.50	14.50	14.60	12.60	13.00
9-State [6]	11.70	12.20	13.70	14.10	12.30	12.60

[1] The average rates are estimates (rates over $10.00 are rounded to the nearest dime) based on survey indications of monthly lease rates for private, non-irrigated grazing land from the January Agricultural Survey. [2] Includes animal unit plus cow-calf rates. Cow-calf rate converted to animal unit (AUM) using (1 aum=cow-calf *0.833). [3] Seventeen Western States: All States listed. [4] Sixteen Western States: All States, except Texas. [5] Eleven Western States: AZ, CA, CO, ID, MT, NV, NM, OR, UT, WA, WY. [6] Nine Great Plains States: CO, KS, NE, NM, ND, OK, SD, TX, WY. [7] Insufficient data.

NASS, Environmental, Economics, and Demographics Branch, (202) 720–6146.

CHAPTER X

INSURANCE, CREDIT, AND COOPERATIVES

The statistics in this chapter deal with taxes, insurance, agricultural credit, and farm cooperatives. Some of the series were developed in connection with research activities of the Department, while others, such as data from agricultural credit agencies, are primarily records of operations.

Table 10-1.—Crop losses: Average percentage of indemnities attributed to specific hazards, by crops, 1948–2000

Crop	Year	Drought heat (excess)	Hail	Precip. (excess poor drain-age)	Frost freeze, (other cold dam-age)	Flood	Cy-clone, tor-nado, wind, hot wind	In-sects	Dis-ease	All others
		Per-cent	*Per-cent*	*Per-cent*	*Per-cent*	*Per-cent*	*Per-cent*	*Per-cent*	*Per-cent*	*Per-cent*
Adjusted gross revenue	2001-2001	1	15	2	0	0	0	0	0	83
All other citrus tree	2000-2001	0	0	0	23	0	0	0	0	77
All other grapefruit	2001-2001	0	0	0	0	0	100	0	0	0
Almonds	1981-2002	1	4	60	29	0	6	0	0	0
Apples	1963-2002	6	26	4	52	0	4	0	0	8
Avocado trees	1996-2001	0	0	82	0	0	18	0	0	0
Avocados	1998-2001	3	0	13	20	0	33	20	0	10
Barley	1956-2002	40	16	29	4	1	2	2	3	3
Blueberries	1995-2002	21	12	0	60	1	1	0	0	5
Burley tobacco	1997-2002	23	8	34	4	7	2	0	21	1
Cabbage	1999-2002	4	1	9	3	1	20	60	1	1
Canning peaches	1986-1987	4	0	92	0	0	0	0	0	4
Canola	1995-2002	15	11	52	16	0	2	3	2	0
Carambola trees	2001-2001	0	0	100	0	0	0	0	0	0
Cherries	1963-2002	0	11	34	47	0	7	0	0	0
Chile peppers	2000-2001	2	27	5	21	0	11	17	1	16
Cigar binder tobacco ...	1997-2002	6	5	37	0	0	1	0	52	0
Cigar filler tobacco	1998-1999	84	0	0	0	0	0	2	14	0
Cigar wrapper tobacco	1997-2001	0	0	64	0	0	0	0	36	0
Citrus	1952-1997	8	2	0	88	0	2	0	0	0
Citrus I	1998-2002	0	1	0	82	0	17	0	0	0
Citrus II	2000-2001	0	0	0	100	0	0	0	0	0
Citrus III	2001-2001	0	0	0	100	0	0	0	0	0
Citrus IV	1998-2002	0	24	0	64	0	12	0	0	0
Citrus V	1999-2002	0	7	0	75	0	18	0	0	0
Citrus VII	1998-2002	0	51	0	35	0	14	0	0	0
Citrus tree	1984-1997	0	0	0	100	0	0	0	0	0
Clams	2001-2002	0	0	0	1	0	4	0	0	95
Combined crop	1981-1987	0	0	0	0	0	0	0	0	100
Corn	1948-2002	25	12	12	11	10	12	2	5	10
Cotton	1948-2002	25	11	13	13	6	19	5	2	7
Cotton ex long staple ..	1984-2002	16	16	20	12	0	12	16	1	7
Crambe	1999-2001	7	15	32	12	0	26	0	7	0
Cranberries	1984-2002	12	15	7	54	0	1	4	1	5
Cultivated wild rice	1999-2002	7	33	3	3	0	19	2	0	32
Dark air tobacco	1997-2001	54	2	16	5	0	4	0	18	0
Dry beans	1948-2002	17	31	20	23	1	3	0	4	1
Dry peas	1963-2002	54	18	13	11	0	1	0	3	0
Early & Mid oranges	1998-2002	0	0	0	68	0	31	0	0	1
Figs	1988-2001	9	0	44	35	0	3	0	0	9
Fire cured tobacco	1997-2002	30	10	13	18	1	4	0	8	16
Flax	1948-2002	36	8	49	2	0	2	2	1	1
Flue cured tobacco	1997-2002	33	14	6	2	2	17	0	25	1
Forage production	1979-2002	42	2	23	28	0	1	1	0	3
Forage seeding	1978-2002	28	0	21	50	0	1	0	0	0
Fresh apricots	1997-2002	0	58	17	23	0	2	0	0	0
Fresh freestone peach-es	1997-2002	3	30	45	22	0	0	0	0	0
Fresh market beans	2000-2002	0	0	74	16	0	4	0	5	0
Fresh market sweet corn	1985-2002	23	2	25	41	0	8	0	0	1
Fresh market tomatoes	1984-2002	4	20	39	15	0	10	3	9	0
Fresh nectarines	1997-2002	1	60	19	20	0	0	0	0	0
Fresh plums	1990-1997	0	59	8	32	0	1	0	0	0
Grain sorghum	1959-2002	38	7	25	14	2	10	3	0	1
Grapefruit	1997-2002	20	6	1	51	0	23	0	0	0
Grapefruit trees	2000-2002	0	0	0	3	0	0	0	0	97
Grapes	1967-2002	15	4	22	57	0	1	0	0	1
Green peas	1962-2002	38	4	51	3	0	1	0	2	1
Hybrid corn seed	1983-2002	55	3	32	3	0	3	0	3	0
Hybr sorghum seed	1988-2001	13	12	1	59	0	13	0	0	1
Income protection corn	1996-1996	3	0	93	3	0	0	0	0	0

See end of table.

Table 10-1.—Crop losses: Average percentage of indemnities attributed to specific hazards, by crops, 1948–2002—Continued

Crop	Year	Drought heat (excess)	Hail	Precip. (excess poor drain-age)	Frost freeze, (other cold dam-age)	Flood	Cy-clone, tor-nado, wind, hot wind	In-sects	Dis-ease	All others
		Per-cent	*Per-cent*	*Per-cent*	*Per-cent*	*Per-cent*	*Per-cent*	*Per-cent*	*Per-cent*	*Per-cent*
Income protection cotton	1996-1996	96	0	4	0	0	0	0	0	0
Income protection wheat	1996-1996	9	0	90	0	1	0	0	0	0
Late oranges	1998-1999	0	0	0	19	0	41	0	0	41
Lemons	1997-2002	2	0	0	98	0	0	0	0	0
Lime trees	1998-2002	0	0	0	1	0	0	0	0	99
Macadamia nuts	1996-2003	84	0	0	0	0	0	13	0	3
Macadamia trees	2000-2000	0	0	100	0	0	0	0	0	0
Mandarins	1997-2002	38	0	13	48	0	1	0	0	0
Mango trees	1997-1997	0	0	0	100	0	0	0	0	0
Maryland tobacco	1997-2001	63	6	9	5	0	4	0	14	0
Millet	1996-2002	68	19	11	0	0	1	0	0	0
Minneola tangelos	1998-2002	8	0	10	78	0	4	0	0	0
Mint	2000-2002	24	0	20	55	0	0	0	0	0
Mustard	1999-2002	18	59	6	13	0	3	0	0	0
Navel oranges	1998-2002	50	4	4	40	0	1	0	0	0
Nursery	1987-1999	24	0	7	11	7	42	3	5	0
Nursery (F,G&C)	2001-2002	0	4	27	21	44	3	0	1	0
Oats	1956-2002	44	12	36	4	0	1	1	1	1
Onions	1988-2002	11	18	40	6	0	2	0	22	1
Orange trees	1996-2002	0	0	2	1	0	0	0	0	97
Oranges	1958-1980	0	0	0	100	0	0	0	0	0
Oranges	1997-1997	30	0	0	37	0	21	0	0	11
Orlando tangelos	1998-2001	0	0	0	100	0	0	0	0	0
Peaches	1957-2002	3	20	1	73	0	0	0	0	3
Peanuts	1962-2002	43	0	19	7	1	3	0	24	3
Pears	1989-2002	0	22	1	74	0	0	0	0	2
Pecans	1998-2001	68	1	1	5	0	21	0	1	4
Peppers	1984-2002	0	15	44	27	0	11	0	1	1
Plums	1998-2002	1	49	29	21	0	0	0	0	0
Popcorn	1984-2002	57	10	21	5	1	2	1	4	0
Potatoes	1962-2002	23	6	23	24	0	1	0	20	1
Prevented planting	1982-1985	0	0	97	0	3	0	0	0	0
Prev planting endorse	1986-1994	25	0	14	0	57	0	0	0	4
Processing apricots	1997-2002	2	9	20	34	0	35	0	0	0
Processing beans	1988-2002	46	2	48	2	0	1	0	1	0
Procesing cling pea	1997-2002	24	3	33	38	0	1	0	0	0
Processing cucumbers	2000-2002	83	0	11	4	0	1	0	1	0
Processing freestone	1998-2002	8	1	10	82	0	0	0	0	0
Prunes	1986-2002	24	1	18	53	0	1	0	0	2
Raisins	1961-2001	0	0	100	0	0	0	0	0	0
Rangeland	1999-2001	0	0	0	0	0	0	0	0	100
Revenue coverage corn	1996-1996	20	44	3	5	26	0	0	1	0
Revenue coverage soybeans	1996-1996	1	24	55	5	13	0	0	3	0
Rice	1960-2002	19	0	44	11	7	2	0	6	11
Rio red & star ruby	1998-2002	0	0	0	69	0	14	3	0	13
Ruby red grapefruit	1998-2000	0	0	0	98	0	2	0	0	0
Rye	1980-2002	46	9	31	10	0	2	0	0	0
Safflower	1964-2002	37	4	19	18	0	20	0	1	0
Soybeans	1955-2002	29	13	23	13	9	3	1	2	5
Special citrus	1992-1994	6	12	0	82	0	0	0	0	0
Stonefruit	1988-1996	1	28	44	19	0	2	0	0	7
Strawberries	2000-2002	60	0	1	16	0	0	0	22	0
Sugar beets	1965-2002	14	11	29	16	2	11	2	15	1
Sugarcane	1967-2002	16	0	7	20	0	4	1	19	33
Sunflowers	1976-2002	26	16	26	11	0	4	6	7	3
Sweet corn	1978-2002	47	0	32	17	0	3	0	0	1
Sweet oranges	1998-2002	8	0	5	85	0	0	1	0	0
Sweetpotatoes	1998-2002	34	0	48	0	0	14	3	0	0
Table grapes	1984-2002	27	12	30	29	0	1	0	0	1
Tangelos	1997-1997	3	0	0	97	0	0	0	0	0
Tobacco	1948-1996	24	18	21	1	2	15	0	17	2
Tomatoes	1963-2002	20	1	68	6	1	1	0	2	1
Tung nuts	1965-1970	0	0	0	100	0	0	0	0	0
Valencia oranges	1998-2002	8	5	2	84	0	1	0	0	0
Walnuts	1984-2002	27	2	60	10	0	2	0	0	0
Watermelons	1999-1999	8	7	38	1	0	14	0	29	2
Wheat	1948-2002	21	10	12	26	4	10	3	8	6
Winter squash	1999-2002	36	5	56	1	0	0	0	2	1

GRP crops do not have any specific cause of loss.

RMA, Program Automation Branch, (816) 926–7910.

Table 10-2.—Crop insurance programs: Coverage, amount of premiums and indemnities, by crops, United States, 2000–2003[1]

Commodity and year	Coverage: County programs	Coverage: Insured units[2]	Coverage: Area insured[3]	Coverage: Maximum insured production	Amount of premium	Indemnities: Number	Indemnities: Area indemnified[3]	Indemnities: Amount
	Number	*Number*	*1,000 acres*	*1,000 dollars*	*1,000 dollars*		*1,000 acres*	*1,000 dollars*
Adjusted gross revenue:								
2000	88	75	0	9,331	388	23	...	398
2001	214	553	0	195,143	7,053	87	...	11,087
2002	214	753	0	254,658	9,682	0	...	0
Alfalfa seed:								
2002	10	115	8	4,468	309	0	0	0
All other citrus trees:								
2000	28	992	0	43,745	852	1	0	7
2001	28	1,013	0	45,115	1,114	3	7	56
2002	28	1,017	0	56,186	1,377	0	...	0
All other grapefruit:								
2001	3	2	0	1	0	1	0	1
2002	3	3	0	2	0	0	...	0
2003	3	2	0	1	0	0	...	0
Almonds:								
2000	16	4,336	312	326,703	21,535	841	50	20,146
2001	16	4,217	320	281,141	18,861	442	38	12,215
2002	16	4,193	338	284,054	18,844	28	2	441
Apples:								
2000	332	4,628	258	291,849	17,515	726	24	19,344
2001	332	5,079	263	352,500	23,922	950	27	28,987
2002	332	5,069	253	320,160	22,307	259	5	5,550
Avocado trees:								
2000	...	207	0	3,078	93	3	0	4
2001	1	231	0	5,892	172	3	0	13
2002	1	208	0	5,504	160	0	...	0
Avocados:								
2000	1	46	1	642	44	0	...	0
2001	6	1,111	27	33,916	4,965	25	0	278
2002	6	1,258	31	39,124	5,640	0	...	0
2003	6	1,078	30	39,092	5,707	0	...	0
Barley:								
2000	1,539	45,440	3,933	195,711	18,423	10,230	1,238	26,539
2001	1,539	42,402	3,765	203,807	21,103	14,426	1,726	42,970
2002	1,539	43,113	4,004	244,702	26,175	11,082	1,105	35,207
Blueberries:								
2000	32	467	27	15,749	1,364	32	1	729
2001	32	539	32	20,386	1,647	83	2	1,045
2002	32	529	32	22,358	1,840	38	1	808
Burley tobacco:								
2000	277	33,465	95	198,871	16,068	6,081	25	29,726
2001	277	27,762	96	183,551	14,926	4,185	16	18,880
2002	277	24,384	80	172,977	13,939	197	1	1,617
Cabbage:								
2000	27	247	18	17,244	1,244	52	3	2,453
2001	27	351	15	11,376	727	49	1	516
2002	27	447	16	13,432	891	50	1	657
Canola:								
2000	206	17,266	1,347	107,048	12,381	3,612	344	11,504
2001	225	18,941	1,517	115,987	15,251	5,144	465	17,254
2002	225	17,743	1,450	125,776	16,188	5,372	522	26,196
Carambola trees:								
2000	1	20	0	380	9	0	...	0
2001	1	17	0	242	6	1	0	4
2002	1	15	0	244	6	0	...	0
Cherries:								
2000	21	1,464	31	47,929	4,332	252	6	4,746
2001	21	1,748	32	50,296	4,682	285	5	4,855
2002	21	1,837	35	52,815	4,973	283	5	3,926
Chili peppers:								
2000	3	71	7	4,084	279	7	0	157
2001	3	103	8	4,836	355	31	2	551
2002	3	86	7	4,513	336	0	...	0
Cigar binder tobacco:								
2000	16	835	3	16,752	991	297	2	11,612
2001	16	778	3	16,327	1,083	121	0	1,719
2002	16	766	4	16,512	1,143	50	0	956
Cigar filler tobacco:								
2000	3	4	0	41	1	0	...	0
2001	3	5	0	46	1	0	...	0
2002	3	38	0	506	20	0	...	0
Cigar wrapper tobacco:								
2000	5	37	1	21,557	1,185	10	0	4,126
2001	5	50	1	24,074	1,497	1	0	104
2002	5	45	1	22,698	1,370	0	...	0

See footnotes at end of table.

Table 10-2.—Crop insurance programs: Coverage, amount of premiums and indemnities, by crops, United States, 2000–2003 [1]—Continued

Commodity and year	Coverage				Amount of premium	Indemnities		
	County programs	Insured units [2]	Area insured [3]	Maximum insured production		Number	Area indemnified [3]	Amount
	Number	*Number*	*1,000 acres*	*1,000 dollars*	*1,000 dollars*		*1,000 acres*	*1,000 dollars*
Citrus I:								
2001	29	2,493	233	139,142	3,099	53	1	490
2002	29	2,643	234	140,204	3,069	1	0	6
2003	29	2,550	231	115,799	2,526	0	....	0
Citrus II:								
2001	29	1,908	230	172,336	4,355	45	2	1,113
2002	29	1,947	229	175,860	4,454	0	....	0
2003	29	1,953	233	142,468	3,631	0	....	0
Citrus III:								
2001	29	104	4	1,673	42	3	0	11
2002	29	116	6	2,646	65	0	....	0
2003	29	110	8	3,095	70	0	....	0
Citrus IV:								
2001	29	1,158	36	23,523	681	93	2	853
2002	29	1,614	34	23,383	686	19	0	163
2003	29	1,486	31	20,224	598	0	....	0
Citrus trees I:								
2000	3	617	7	18,048	785	0	....	0
2001	3	608	7	18,123	785	0	....	0
2002	3	584	6	17,752	764	0	....	0
Citrus trees II:								
2000	3	110	1	1,952	91	0	....	0
2001	3	109	1	2,228	103	0	....	0
2002	3	108	1	2,560	119	0	....	0
Citrus trees III:								
2000	3	6	0	92	5	0	....	0
2001	3	7	0	106	6	0	....	0
2002	3	7	0	109	6	0	....	0
Citrus trees IV:								
2000	3	859	14	35,628	1,980	0	....	0
2001	3	873	14	37,018	2,047	0	....	0
2002	3	851	14	37,412	2,060	0	....	0
Citrus trees V:								
2000	3	228	3	7,528	492	0	....	0
2001	3	214	3	7,077	462	0	....	0
2002	3	190	2	6,636	430	0	....	0
Citrus V:								
2001	29	447	12	20,668	717	96	3	3,562
2002	29	497	12	20,726	699	14	0	383
2003	29	464	13	18,776	639	0	....	0
Citrus VI:								
2001	5	11	1	1,746	35	0	....	0
2002	5	6	1	1,205	24	0	....	0
2003	5	8	1	915	19	0	....	0
Citrus VII:								
2001	29	1,565	132	70,939	2,134	86	6	1,380
2002	29	1,579	129	74,521	2,184	8	1	194
2003	29	1,472	123	65,332	1,943	0	....	0
Clams:								
2001	13	422	0	41,215	1,401	96	....	2,143
2002	13	481	0	59,241	2,167	79	....	2,094
Corn:								
2001	7,806	896,618	56,863	10,183,315	740,383	123,990	9,242	403,393
2002	8,510	884,341	55,846	10,702,495	865,585	157,864	11,329	565,659
2003	8,524	905,000	58,792	11,435,285	911,076	96,366	6,650	443,894
Cotton ELS:								
2000	31	456	150	42,755	1,965	29	4	593
2001	31	777	271	109,040	6,104	115	29	11,292
2002	31	808	255	87,430	5,316	58	10	4,235
Cotton:								
2000	1,491	176,302	14,599	2,862,825	372,555	74,633	9,043	584,482
2001	1,491	181,900	14,685	3,289,450	444,182	66,589	8,533	653,646
2002	1,491	167,988	12,977	2,254,634	319,291	17,276	2,354	139,044
Crambe:								
2000	7	224	18	1,453	179	32	3	87
2001	7	138	11	950	106	16	1	34
2002	7	12	1	55	6	0	....	0
Cranberries:								
2000	30	654	28	74,118	2,684	42	1	780
2001	30	629	28	27,622	1,037	62	2	545
2002	30	594	30	37,038	1,285	4	0	24
Cultivated wild rice:								
2000	10	62	20	5,311	328	18	4	426
2001	10	71	22	5,631	345	27	11	870
2002	10	67	22	5,575	343	17	2	119

See footnotes at end of table.

Table 10-2.—Crop insurance programs: Coverage, amount of premiums and indemnities, by crops, United States, 2000–2002 [1]—Continued

Commodity and year	Coverage				Amount of premium	Indemnities		
	County programs	Insured units [2]	Area insured [3]	Maximum insured production		Number	Area indemnified [3]	Amount
	Number	*Number*	*1,000 acres*	*1,000 dollars*	*1,000 dollars*		*1,000 acres*	*1,000 dollars*
Dark air tobacco:								
2000	37	1,233	2	4,719	201	42	0	180
2001	37	1,076	2	4,350	204	64	0	100
2002	37	921	1	3,867	188	0	....	0
Dry beans:								
2000	292	22,549	1,383	238,364	29,918	6,039	466	40,501
2001	292	22,093	1,325	201,202	28,805	9,400	679	57,805
2002	286	27,467	1,663	279,153	38,342	2,986	189	14,200
Dry Peas:								
2000	97	2,927	224	12,908	1,292	350	36	762
2001	97	3,493	260	16,159	1,675	390	38	762
2002	97	4,662	357	26,653	2,921	618	56	1,770
Early and Midseason oranges:								
2000	3	362	6	2,650	142	1	0	1
2001	3	342	6	2,821	157	2	0	5
2002	3	322	6	2,767	150	0	....	0
Figs:								
2000	4	103	9	4,694	307	9	2	324
2001	4	111	9	4,094	292	15	2	349
2002	4	134	8	5,210	365	0	....	0
Fired cured tobacco:								
2000	43	2,734	9	25,843	1,127	230	1	1,554
2001	43	2,421	8	25,531	1,173	149	1	1,133
2002	43	2,099	6	20,709	997	23	0	148
Flax:								
2000	107	6,302	421	20,131	2,533	992	68	1,477
2001	107	7,595	492	24,676	3,302	1,414	96	2,120
2002	107	9,611	670	36,718	4,723	1,460	116	3,063
Flue cured tobacco:								
2000	173	31,788	224	598,688	22,266	1,967	22	28,852
2001	173	29,444	214	599,526	22,786	1,898	19	21,504
2002	173	28,402	224	653,889	25,612	3,202	33	42,361
Forage production:								
2000	713	13,872	2,005	254,152	11,348	2,603	631	25,618
2001	570	26,030	2,517	244,964	16,910	7,303	842	28,284
2002	570	29,477	2,867	239,079	18,117	10,875	1,270	37,082
Forage seeding:								
2000	263	3,251	102	8,044	843	414	15	758
2001	263	2,860	91	8,871	1,106	518	20	1,287
2002	263	3,768	149	15,475	1,953	235	13	1,033
Fresh apricots:								
2000	13	167	3	5,077	500	61	2	1,470
2001	13	158	3	5,317	567	51	2	1,148
2002	13	167	3	4,700	484	26	1	337
Fresh freestone peaches:								
2000	7	595	17	13,382	750	17	0	113
2001	7	558	18	15,911	907	46	1	321
2002	7	579	19	15,977	932	8	0	49
Fresh market beans:								
2000	5	412	20	16,240	1,487	141	7	4,889
2001	5	466	21	16,773	1,839	160	8	3,793
2002	5	673	28	25,448	3,231	215	10	8,795
Fresh market sweet corn:								
2000	227	1,287	70	28,485	2,620	140	6	2,273
2001	227	1,444	74	32,624	3,306	227	9	3,948
2002	227	1,431	71	27,165	2,608	107	2	834
Fresh market tomatoes:								
2000	49	785	64	115,940	10,907	235	8	14,262
2001	49	826	66	126,048	13,340	177	5	9,376
2002	49	802	58	118,803	10,572	94	3	4,359
Fresh nectarines:								
2000	7	642	22	19,424	1,254	30	0	191
2001	7	614	21	21,801	1,422	104	2	1,178
2002	7	677	24	23,005	1,567	9	0	69
Grain sorghum:								
1998	2,909	128,095	6,389	406,012	49,944	43,972	4,437	87,119
1998	2,909	147,919	7,775	531,994	75,901	52,128	5,918	112,796
2000	2,909	139,341	7,337	548,041	83,686	24,438	2,847	86,640
Grapefruit trees:								
1998	28	1,149	0	102,094	2,346	7	16	1,306
1999	28	1,180	0	110,567	3,236	8	29	301
2000	28	1,158	0	129,965	3,823	2	20	75

See footnotes at end of table.

Table 10-2.—Crop insurance programs: Coverage, amount of premiums and indemnities, by crops, United States, 1999–2002 [1]—Continued

Commodity and year	Coverage				Amount of premium	Indemnities		
	County programs	Insured units [2]	Area insured [3]	Maximum insured production		Number	Area indemnified [3]	Amount
	Number	*Number*	*1,000 acres*	*1,000 dollars*	*1,000 dollars*		*1,000 acres*	*1,000 dollars*
Grapefruit:								
1998	8	143	6	5,853	324	4	1	66
1999	8	128	7	7,523	412	1	0	5
2001	8	118	5	5,426	287	0	....	0
Grapes:								
1998	83	12,015	567	513,277	28,774	476	14	6,695
1999	83	12,286	566	564,481	33,141	1,537	52	19,345
2000	83	12,414	578	535,041	32,780	283	11	4,690
Green peas:								
1998	145	3,709	192	33,522	3,384	700	35	2,580
1999	145	2,898	148	27,029	2,739	727	41	2,233
2000	145	3,120	163	31,072	3,489	1,032	64	3,978
Hybrid corn seed:								
1998	384	7,101	417	104,563	10,460	822	61	3,078
1999	384	7,121	408	114,551	11,531	613	42	2,667
2000	384	5,930	346	107,307	11,121	73	5	648
Hybrid sorghum seed:								
2000	21	553	29	4,501	727	137	12	734
2001	21	739	46	8,359	1,360	136	12	1,092
2002	21	910	57	11,485	1,929	0	....	0
Late oranges:								
2001	3	81	1	382	51	0	....	0
2002	3	77	1	473	68	0	....	0
2003	3	84	1	488	66	0	....	0
Lemon trees:								
2000	5	2	0	929	16	0	....	0
2001	4	2	0	1,689	29	0	....	0
2002	4	2	0	1,957	39	0	....	0
Lemons:								
2001	15	701	41	56,894	3,289	20	1	574
2002	15	688	38	64,648	3,594	30	2	2,252
2003	15	693	39	70,195	3,921	0	....	0
Lime trees:								
2000	3	87	0	6,852	263	44	6	3,721
2001	3	49	0	4,137	157	9	1	679
2002	3	34	0	3,068	100	3	0	1,207
Macadamia nuts:								
2001	3	61	12	13,103	168	11	2	1,110
2002	3	95	12	18,774	298	7	2	728
2003	3	119	13	17,897	281	1	1	302
Macadamia trees:								
2000	3	131	14	80,429	860	4	1	304
2001	3	123	14	78,514	839	0	....	0
2002	3	124	14	64,287	764	0	....	0
Mandarins:								
2001	7	64	1	1,340	121	4	0	39
2002	7	68	2	1,448	124	9	0	147
2003	7	88	2	1,790	151	0	....	0
Mango trees:								
2000	....	43	0	446	12	0	....	0
2001	1	37	0	429	12	0	....	0
2002	1	29	0	429	11	0	....	0
Maryland tobacco:								
2000	6	128	1	706	19	4	0	25
2001	6	50	0	287	10	4	0	15
2002	6	20	0	194	6	0	....	0
Millet:								
2000	5	1,101	67	2,362	257	661	59	803
2001	5	1,813	111	5,439	708	307	29	510
2002	5	1,501	85	4,365	590	440	40	851
Minneola tangelos:								
2001	8	163	5	4,560	362	20	1	441
2002	8	165	6	7,891	561	6	0	71
2003	8	160	4	5,198	427	0	....	0
Mint:								
2000	9	203	16	5,193	301	51	2	186
2001	9	238	14	6,124	311	46	2	393
2002	9	238	14	6,916	358	49	1	90
Mustard:								
2000	19	184	16	735	105	32	3	76
2001	19	212	17	918	136	15	2	48
2002	19	922	90	6,693	1,028	229	27	995
Naval oranges:								
2001	17	2,762	109	109,033	6,324	106	4	1,005
2002	17	2,712	108	131,858	7,629	232	8	3,258
2003	16	2,815	108	139,118	7,454	0	....	0
Nursery:								
2001	3,087	3,761	0	2,597,874	47,147	244	....	40,682
2002	3,087	4,184	0	3,026,281	54,324	77	....	3,837

See footnotes at end of table.

Table 10-2.—Crop insurance programs: Coverage, amount of premiums and indemnities, by crops, United States, 1999–2002 [1]—Continued

Commodity and year	Coverage				Amount of premium	Indemnities		
	County programs	Insured units [2]	Area insured [3]	Maximum insured production		Number	Area indemnified [3]	Amount
	Number	*Number*	*1,000 acres*	*1,000 dollars*	*1,000 dollars*		*1,000 acres*	*1,000 dollars*
Oats:								
2000	1,648	19,535	886	24,845	3,349	3,176	178	3,032
2001	1,648	18,317	824	25,402	3,413	3,563	169	2,969
2002	1,648	23,896	1,135	44,045	6,224	9,626	641	16,984
Onions:								
2000	97	1,460	78	98,771	10,476	347	14	13,160
2001	97	1,751	79	100,191	10,971	377	13	11,256
2002	95	1,769	77	99,253	10,900	346	18	14,497
Orange trees:								
2000	28	3,489	0	543,311	10,389	6	281	1,381
2001	28	3,690	0	670,064	15,233	14	926	2,624
2002	28	4,137	0	894,082	21,402	6	482	1,273
Orlando tangelos:								
2001	5	17	0	235	15	2	0	7
2002	5	10	0	127	8	0	....	0
2003	5	7	0	83	5	0	....	0
Peaches:								
2000	220	1,393	41	38,433	6,145	404	12	7,382
2001	220	1,428	42	44,467	7,298	414	10	7,905
2002	220	1,417	41	49,302	8,377	397	10	6,922
Peanuts:								
2000	327	30,728	1,394	562,727	47,202	11,468	658	127,655
2001	327	31,793	1,413	600,997	54,713	5,073	401	62,543
2002	326	28,323	1,244	337,564	29,920	942	44	4,814
Pears:								
2000	26	1,125	36	37,065	1,324	46	1	380
2001	26	1,873	37	45,102	1,676	72	2	1,217
2002	26	1,909	36	48,209	1,775	33	0	199
Pecans:								
2000	7	195	43	24,228	1,740	18	2	414
2001	7	200	39	24,443	1,690	25	3	743
2002	7	215	43	29,216	2,253	0	....	0
Peppers:								
2000	13	165	10	26,591	3,916	50	3	4,950
2001	13	234	12	38,066	6,177	83	3	3,913
2002	13	263	13	40,007	6,255	103	5	9,954
Plums:								
2000	7	1,084	22	20,862	1,566	224	2	1,333
2001	7	1,355	22	22,175	1,724	237	2	1,594
2002	7	1,399	21	21,880	1,791	52	1	333
Popcorn:								
2000	338	2,156	135	28,906	2,322	236	18	1,168
2001	338	2,665	181	37,379	3,213	256	21	1,126
2002	319	2,757	192	44,501	3,740	242	20	1,424
Potatoes:								
2000	331	6,645	955	704,735	49,991	1,338	114	51,833
2001	331	7,671	935	718,223	58,596	1,808	162	60,698
2002	325	8,363	1,018	813,883	67,434	368	29	11,359
Processing apricots:								
2000	13	181	9	7,864	773	33	2	1,103
2001	13	154	8	7,596	796	41	2	1,056
2002	13	128	6	4,640	483	19	1	251
Processing beans:								
2000	132	1,129	78	16,642	1,651	209	14	2,027
2001	132	1,141	78	16,417	1,659	326	16	1,764
2002	124	1,263	89	19,993	2,039	119	7	1,034
Processing cling peaches:								
2000	10	875	21	23,399	1,263	75	1	485
2001	10	1,263	19	23,060	1,261	127	1	931
2002	10	1,296	20	24,664	1,389	23	0	115
Processing cucumbers:								
2000	11	265	13	4,916	373	73	6	984
2001	11	295	15	6,113	468	103	6	1,300
2002	11	373	19	8,202	666	207	12	1,774
Processing freestone:								
2000	7	116	3	3,426	205	3	0	27
2001	7	108	3	3,390	208	12	0	115
2002	7	112	3	3,221	193	1	0	1
Prunes:								
2000	75	59,157	....	5,387	192	9	....	3,383
2001	75	64,308	....	6,002	665	36	....	16,898
2002	68	56,408	....	5,358	45	2	....	751
Raisins:								
2000	7	3,830	371	246,056	21,215	317	30	6,438
2001	7	2,815	213	90,118	7,728	3	0	5
Rangeland:								
2000	12	355	2,797	13,009	407	116	698	384
2001	12	568	5,484	27,561	1,112	367	3,695	6,536
2002	12	900	8,123	46,724	2,223	0	....	0

See footnotes at end of table.

Table 10-2.—Crop insurance programs: Coverage, amount of premiums and indemnities, by crops, United States, 1999–2002 [1]—Continued

Commodity and year	Coverage				Amount of premium	Indemnities		
	County programs	Insured units [2]	Area insured [3]	Maximum insured production		Number	Area indemnified [3]	Amount
	Number	*Number*	*1,000 acres*	*1,000 dollars*	*1,000 dollars*		*1,000 acres*	*1,000 dollars*
Raspberry and blackberry:								
2002	7	52	2	1,077	74	0	....	0
Rice:								
2000	276	20,405	2,318	388,228	20,355	1,083	151	11,905
2001	276	21,439	2,518	424,731	19,829	1,749	269	13,890
2002	276	20,428	2,436	391,007	19,288	564	71	4,960
Rio Red & Star Ruby:								
2001	3	520	12	8,097	1,177	3	0	31
2002	3	510	13	8,525	1,240	2	0	12
2003	3	474	12	9,263	1,372	0	....	0
Ruby red grapefruit:								
2001	3	134	2	1,217	152	0	....	0
2002	3	127	2	1,221	162	0	....	0
2003	3	109	2	1,150	146	0	....	0
Rye:								
2000	48	437	31	940	99	53	4	70
2001	48	430	31	947	111	90	7	85
2002	47	399	31	1,024	113	119	14	232
Safflower:								
2000	71	1,133	120	6,843	670	237	35	1,164
2001	71	886	103	6,117	585	239	35	1,098
2002	71	1,058	125	7,584	809	170	29	951
Soybeans:								
2000	5,790	860,265	55,049	6,709,931	456,172	171,004	13,215	397,043
2001	6,415	896,526	56,225	6,992,286	509,679	153,049	10,841	322,915
2002	6,422	883,158	56,103	6,926,110	496,210	45,237	2,671	78,566
Strawberries:								
2000	21	217	11	36,750	1,470	30	1	1,190
2001	21	232	11	50,609	2,022	56	1	2,950
2002	21	303	14	63,575	2,723	62	1	2,340
Sugarbeets:								
2000	181	18,062	1,227	623,753	31,977	3,429	232	41,840
2001	181	17,128	1,161	549,992	31,038	3,235	232	45,110
2002	176	17,419	1,192	546,434	31,013	3,540	252	17,607
Sugarcane:								
2000	31	5,755	830	247,698	8,380	229	16	2,377
2001	31	5,806	890	249,613	8,896	108	8	994
2002	31	6,059	858	209,694	6,873	12	0	164
Sunflowers:								
2000	329	32,715	2,466	170,206	21,553	12,516	1,028	34,563
2001	329	36,822	2,703	199,288	28,820	15,868	1,408	52,177
2002	329	30,078	2,371	201,309	26,475	6,314	542	27,078
Sweet corn:								
2000	178	3,698	244	43,985	2,609	378	24	1,916
2001	178	3,730	264	53,280	3,274	389	23	1,423
2002	170	3,721	265	58,774	3,773	259	14	1,573
Sweet oranges:								
2001	7	45	0	289	21	1	0	2
2002	7	45	0	334	23	3	0	9
2003	6	52	0	402	29	0	....	0
Sweet potatoes:								
2000	8	895	30	23,271	2,007	580	16	8,303
2001	8	1,113	31	27,430	2,601	763	17	10,746
2002	8	986	28	23,891	2,491	135	2	1,172
Table grapes:								
2000	12	1,066	90	140,065	5,790	65	1	1,779
2001	12	1,127	92	158,875	6,651	108	3	3,136
2002	12	1,186	88	156,013	7,027	29	1	1,384
Tomatoes:								
2000	92	2,770	275	223,324	9,599	194	14	4,925
2001	92	2,719	251	222,884	10,431	263	26	5,530
2002	86	2,944	280	266,015	12,309	65	4	1,181
Valencia oranges:								
2001	13	1,819	54	57,537	3,988	143	4	1,519
2002	13	1,759	53	70,600	4,919	59	2	874
2003	13	1,782	51	66,255	4,717	0	....	0
Walnuts:								
2000	26	1,421	81	53,326	2,101	113	5	1,924
2001	26	1,396	84	52,842	2,131	79	3	1,153
2002	26	1,367	84	55,206	2,299	4	0	116
Wheat:								
2000	5,822	574,998	45,796	3,130,552	332,788	121,513	15,768	437,114
2001	6,686	607,345	45,872	3,298,942	409,663	166,772	20,332	601,379
2002	6,702	596,480	45,481	3,431,465	423,125	207,397	26,017	754,130
Winter squash:								
2000	18	104	3	777	74	28	1	117
2001	18	144	3	1,021	93	19	0	68
2002	18	147	3	1,272	124	3	0	16

[1] Data for 2000 are preliminary. [2] Number of farms on which the insured crop was planted including duplication where both the landlord and tenant are insured. Insured farms on which no insured crop was planted are not included. [3] The insured's share of the planted area on the farm.

RMA, Program Automation Branch, (816) 926–7910.

Table 10-3.—Farm real estate debt: Amount outstanding by farming regions, December 31, 1992–2001 [1]

Year	Northeast	Lake States	Corn Belt	Northern Plains	Appalachian	Southeast
	1,000 dollars	*1,000 dollars*	*1,000 dollars*	*1,000 dollars*	*1,000 dollars*	*1,000 dollars*
1992	4,112,518	7,955,498	18,571,881	8,735,914	6,160,201	5,750,907
1993	4,225,679	8,054,166	18,925,500	9,012,984	6,098,602	5,774,239
1994	4,279,584	8,326,818	19,647,100	9,209,160	6,225,866	5,986,644
1995	4,209,999	8,590,997	20,173,969	9,475,739	6,318,832	5,963,890
1996	4,361,893	8,806,742	20,992,414	9,698,135	6,406,038	6,094,847
1997	4,501,110	9,354,068	21,972,845	9,961,567	6,661,529	6,400,437
1998	4,331,507	9,863,472	23,325,630	10,465,276	6,955,985	6,629,569
1999	4,758,063	10,104,094	24,393,299	10,967,135	7,360,575	7,108,485
2000	4,937,759	10,246,259	24,841,240	11,106,788	7,800,900	7,596,411
2001 [2]	5,232,962	10,821,794	26,175,314	11,715,092	8,275,929	7,992,414

Year	Delta States	Southern Plains	Mountain	Pacific	United States [3]
	1,000 dollars	*1,000 dollars*	*1,000 dollars*	*1,000 dollars*	*1,000 dollars*
1992	3,904,423	6,508,897	6,623,770	11,220,642	79,738,417
1993	4,081,189	6,477,892	6,620,553	11,280,603	80,738,585
1994	4,236,316	6,610,909	6,755,581	11,509,349	82,971,225
1995	4,407,071	6,867,236	6,858,828	11,576,106	84,560,910
1996	4,511,602	7,169,360	7,010,136	11,877,465	87,079,331
1997	4,748,915	7,590,562	7,155,300	12,524,045	91,031,125
1998	5,010,012	7,824,385	7,307,074	13,673,234	95,559,725
1999	5,405,321	8,271,031	7,641,196	14,204,770	100,381,473
2000	5,634,340	8,466,555	7,772,905	15,376,732	103,946,407
2001 [2]	5,947,764	8,975,572	8,195,792	16,150,339	109,658,518

[1] Includes operator households. Data for 1987–1992 include Commodity Credit Corporation storage and drying facility loans. States included in regions are as follows: Northeast—Maine, New Hampshire, Vermont, Massachusetts, Rhode Island, Connecticut, New York, New Jersey, Pennsylvania, Delaware, and Maryland; Lake States—Michigan, Wisconsin, and Minnesota; Corn Belt—Ohio, Indiana, Illinois, Iowa, and Missouri; Northern Plains—North Dakota, South Dakota, Nebraska, and Kansas; Appalachian—Virginia, West Virginia, North Carolina, Kentucky, and Tennessee; Southeast—South Carolina, Georgia, Florida, and Alabama; Delta States—Mississippi, Arkansas, and Louisiana; Southern Plains—Oklahoma and Texas; Mountain—Montana, Idaho, Wyoming, Colorado, New Mexico, Arizona, Utah, and Nevada; Pacific—Washington, Oregon, and California. [2] Preliminary. [3] Includes Alaska and Hawaii.

ERS, Farm Sector Performance and Well-Being Branch, (202) 694–5586.

Table 10-4.—Farm real estate debt: Amount outstanding, by States, Dec. 31, 1996–2001 [1]

State	1996	1997	1998	1999	2000	2001 [2]
	1,000 dollars	*1,000 dollars*	*1,000 dollars*	*1,000 dollars*	*1,000 dollars*	*1,000 dollars*
AL	879,233	932,469	1,009,369	1,137,319	1,206,551	1,286,247
AK	8,883	10,103	10,833	12,052	14,035	14,698
AZ	491,501	534,882	554,595	602,136	592,456	621,394
AR	2,192,741	2,410,180	2,512,826	2,671,067	2,758,805	2,918,634
CA	8,774,191	9,267,141	10,165,531	10,640,871	11,793,494	12,406,327
CO	1,811,867	1,797,355	1,823,382	1,915,820	1,963,844	2,091,475
CT	87,321	125,038	117,815	144,224	168,498	180,147
DE	163,195	194,957	193,133	194,672	186,071	197,302
FL	2,961,251	3,121,134	3,233,279	3,462,963	3,681,780	3,837,533
GA	1,841,792	1,911,317	1,939,948	2,039,264	2,217,107	2,344,181
HI	141,818	150,644	162,747	155,452	152,483	160,847
ID	1,347,508	1,379,239	1,379,086	1,418,831	1,422,449	1,499,102
IL	5,072,178	5,296,091	5,650,360	5,854,421	5,894,665	6,220,178
IN	3,405,800	3,583,698	3,749,248	3,949,761	4,060,338	4,274,361
IA	6,740,976	6,986,449	7,359,183	7,732,959	7,932,470	8,330,671
KS	2,874,099	2,946,857	3,047,772	3,206,112	3,270,360	3,467,272
KY	2,061,015	2,150,103	2,256,230	2,404,634	2,520,099	2,656,847
LA	843,255	832,820	875,907	938,216	955,657	1,011,122
ME	132,683	133,985	134,910	143,046	135,434	138,816
MD [3]	684,282	699,354	702,044	720,738	720,214	769,610
MA	133,264	153,995	147,696	209,637	230,194	246,733
MI	1,633,925	1,729,891	1,735,177	1,789,040	1,777,331	1,886,983
MN	4,261,585	4,587,238	4,966,921	4,987,247	5,052,604	5,334,899
MS	1,475,607	1,505,916	1,621,279	1,796,039	1,919,879	2,018,007
MO	3,425,862	3,612,044	3,941,689	4,079,740	4,092,525	4,311,199
MT	1,661,463	1,704,825	1,715,012	1,767,583	1,742,045	1,814,630
NE	3,483,338	3,581,096	3,867,709	4,102,475	4,224,782	4,443,713
NV	162,679	164,952	167,949	178,348	184,254	195,862
NH	42,142	41,089	38,156	46,929	51,502	55,168
NJ	408,294	409,591	278,247	323,170	313,366	333,973
NM	706,136	704,905	754,877	809,663	833,593	885,639
NY	930,135	916,444	907,235	1,068,961	1,118,228	1,184,210
NC	1,494,808	1,505,228	1,569,398	1,618,503	1,758,626	1,862,278
ND	1,702,010	1,765,246	1,772,287	1,812,826	1,796,077	1,899,213
OH	2,347,598	2,494,563	2,625,148	2,776,418	2,861,242	3,038,905
OK	1,996,282	2,006,917	2,073,866	2,130,719	2,121,121	2,224,304
OR	1,454,679	1,550,798	1,677,402	1,709,033	1,693,411	1,766,620
PA	1,563,501	1,613,670	1,609,824	1,672,201	1,773,629	1,872,507
RI	18,544	18,634	16,707	21,419	23,266	25,144
SC	412,572	435,516	446,973	468,940	490,972	524,453
SD	1,638,689	1,668,368	1,777,508	1,845,722	1,815,568	1,904,895
TN	1,316,148	1,426,654	1,560,550	1,686,915	1,801,593	1,921,225
TX	5,173,078	5,583,646	5,750,519	6,140,312	6,345,434	6,751,268
UT	377,418	400,845	404,075	403,652	491,172	518,989
VT	198,532	194,355	185,739	213,067	217,357	229,352
VA	1,241,964	1,287,883	1,276,304	1,345,300	1,414,127	1,515,591
WA	1,648,595	1,706,107	1,830,301	1,854,866	1,889,827	1,977,393
WV	292,104	291,661	293,502	305,223	306,455	319,988
WI	2,911,231	3,036,939	3,161,373	3,327,807	3,416,324	3,599,912
WY	451,566	468,296	508,098	545,162	543,092	568,701
US	87,079,331	91,031,125	95,559,725	100,381,473	103,946,407	109,658,518

[1] Includes operator households. Includes Commodity Credit Corporation storage and drying facility loans. [2] Preliminary. [3] Includes District of Columbia.

ERS, Farm Sector and Performance and Well-Being Branch, (202) 694–5586.

Table 10-5.—Farm real estate debt: Amount outstanding by lender, by States, Dec. 31, 2000 and 2001 [1]

State	Federal Credit System [2]		Farm Service Agency [3]		Life insurance companies [4]		All operating banks		Individuals and others [5]	
	2000	2001	2000	2001	2000	2001	2000	2001	2000	2001
	1,000 dollars	*1,000 dollars*	*1,000 dollars*	*1,000 dollars*	*1,000 dollars*	*1,000 dollars*	*1,000 dollars*	*1,000 dollars*	*1,000 dollars*	*1,000 dollars*
AL	496,695	550,189	36,627	36,198	21,452	21,771	515,860	539,409	135,917	138,680
AK	3,871	4,288	1,626	1,616	0	0	3,286	3,436	5,252	5,359
AZ	196,827	218,025	26,046	25,316	144,005	146,143	68,910	72,056	156,668	159,853
AR	933,558	1,034,102	111,912	109,230	338,207	343,227	1,144,918	1,197,184	230,211	234,891
CA	4,142,968	4,589,165	134,662	129,024	4,272,193	4,355,616	1,694,560	1,771,917	1,549,111	1,580,605
CO	920,710	1,019,870	54,493	53,651	159,014	161,374	398,343	416,527	431,284	440,052
CT	96,200	106,561	5,942	5,862	38,105	38,671	8,993	9,404	19,258	19,650
DE	72,868	80,716	3,854	3,806	29,089	29,521	53,996	56,461	26,264	26,798
FL	753,738	834,916	54,548	52,268	1,446,608	1,468,084	1,041,548	1,089,095	385,337	393,171
GA	629,437	697,228	56,867	55,860	152,140	154,399	1,184,978	1,239,073	193,684	197,622
HI	55,348	61,309	21,606	21,242	12,825	13,016	51,431	53,779	11,273	11,502
ID	598,153	662,574	98,188	95,448	274,552	278,628	67,801	70,896	383,755	391,556
IL	1,657,807	1,836,353	156,575	154,603	369,860	375,351	2,687,974	2,810,681	1,022,405	1,043,190
IN	1,129,894	1,251,584	106,063	104,307	377,634	383,240	1,530,034	1,599,881	916,712	935,349
IA	1,974,276	2,186,905	218,448	215,649	481,179	488,322	2,935,168	3,069,159	2,323,400	2,370,635
KS	1,170,017	1,296,027	140,045	138,354	143,997	146,134	1,324,196	1,384,646	492,106	502,110
KY	656,255	726,934	134,968	132,899	70,639	71,687	1,318,233	1,378,411	340,004	346,916
LA	333,916	369,879	37,536	36,764	114,116	115,810	356,411	372,681	113,678	115,989
ME	20,325	22,514	31,278	30,775	52,918	53,704	11,162	11,672	19,751	20,153
MD	355,965	394,303	16,399	16,073	26,613	27,008	176,118	184,158	145,118	148,068
MA	143,617	159,084	19,429	19,095	36,852	37,400	9,573	10,010	20,723	21,144
MI	770,215	853,167	86,299	84,251	52,587	53,368	407,477	426,048	460,781	470,149
MN	1,641,497	1,818,286	154,697	152,327	290,928	295,247	1,708,822	1,786,830	1,256,660	1,282,208
MS	527,776	584,617	98,097	94,793	317,333	322,044	790,876	826,980	185,796	189,573
MO	1,047,875	1,160,731	161,265	158,963	173,260	175,832	1,992,520	2,083,480	717,605	732,194
MT	385,779	427,327	100,719	98,191	240,356	243,924	369,528	386,397	645,664	658,790
NE	1,150,441	1,274,343	182,223	178,291	467,714	474,657	1,687,486	1,764,520	736,919	751,901
NV	94,302	104,458	8,443	8,118	12,159	12,340	7,391	7,728	61,959	63,218
NH	29,686	32,883	3,762	3,670	0	0	7,623	7,971	10,432	10,644
NJ	157,352	174,298	20,158	19,783	0	0	50,277	52,573	85,579	87,319
NM	369,532	409,331	32,624	32,102	89,738	91,071	177,322	185,417	164,377	167,719
NY	467,912	518,305	96,766	94,737	14,178	14,388	254,325	265,935	285,049	290,844
NC	592,228	656,010	81,676	79,504	79,661	80,843	806,717	843,544	198,345	202,377
ND	670,973	743,237	170,406	167,725	36,236	36,774	566,466	592,325	351,997	359,153
OH	1,085,326	1,202,215	73,530	72,459	104,770	106,326	1,098,274	1,148,410	499,343	509,495
OK	548,708	607,804	163,712	160,831	222,519	225,823	772,076	807,321	414,106	422,525
OR	455,553	504,615	58,286	56,874	381,431	378,093	144,927	151,543	653,215	666,495
PA	469,614	520,191	69,214	68,263	356	341	953,785	997,326	280,679	286,385
RI	16,460	18,232	2,819	2,791	0	019,773	2,033	2,125	1,955	1,995
SC	253,429	280,723	38,475	37,913	19,484	106,283	110,937	116,001	68,648	70,044
SD	555,824	615,686	210,970	206,472	104,728	34,164	521,914	545,739	422,132	430,714
TN	790,136	875,233	88,909	87,203	33,665	523,176	697,856	729,713	191,028	194,912
TX	2,835,071	3,140,407	211,964	208,349	515,522	19,044	1,575,111	1,647,015	1,207,766	1,232,320
UT	199,700	221,207	35,837	35,235	18,765	1,882	71,780	75,057	165,090	168,446
VT	78,090	86,500	27,071	26,541	1,855	53,568	72,838	76,163	37,503	38,265
VA	735,186	814,365	40,177	39,493	52,785	486,156	405,625	424,142	180,355	184,022
WA	521,610	577,788	66,519	64,575	479,044	62,741	375,055	392,176	447,598	456,698
WV	62,105	68,794	28,038	27,653	61,823	84,385	125,221	130,938	29,268	29,863
WI	905,318	1,002,820	104,895	103,245	83,151	99,651	1,511,135	1,621,945	771,825	787,516
WY	147,217	163,073	22,416	21,891	98,193	127,349	127,349	133,163	147,916	150,924
US	33,907,358	37,559,173	3,907,122	3,830,284	12,514,220	12,700,000	34,016,207	35,569,061	19,601,500	20,000,000

[1] Includes operator households. [2] Includes mortgages in process of foreclosure. [3] Includes farm ownership loans, soil and water loans to individuals, rural and labor housing loans on farms and association loans for grazing, Indian tribe land acquisition loans, and one-half of economic emergency loans. [4] Includes U.S. legal reserve companies only. Includes regular mortgages and purchase-money mortgages. [5] Estimated by ERS, USDA.

ERS, Farm Sector and Performance and Well-Being Branch, (202) 694–5586.

Table 10-6.—All operating banks: Amount of agricultural loans outstanding, by type, and by States, specified dates, 1998–2001 [1]

State and Territory	Farm real estate loans				Nonreal estate farm loans			
	1998	1999	2000	2001 [2]	1998	1999	2000	2001 [2]
AL	440,228	482,853	515,860	539,409	357,767	350,674	372,218	370,805
AK	2,804	3,076	3,286	3,436	3,526	3,456	3,668	3,654
AZ	58,807	64,501	68,910	72,056	443,123	434,337	461,021	459,270
AR	977,059	1,071,661	1,144,918	1,197,184	871,290	854,016	906,483	903,041
CA	1,446,116	1,586,134	1,694,560	1,771,917	4,035,124	3,955,124	4,198,108	4,182,166
CO	339,941	372,855	398,343	416,527	930,823	912,369	968,420	964,743
CT	7,675	8,418	8,993	9,404	4,491	4,402	4,672	4,655
DE	46,079	50,541	53,996	56,461	44,399	43,519	46,193	46,017
FL	888,844	974,905	1,041,548	1,089,095	477,095	467,637	496,366	494,481
GA	1,011,246	1,109,158	1,184,978	1,239,073	555,341	544,331	577,772	575,578
HI	43,891	48,140	51,431	53,779	39,098	38,323	40,678	40,523
ID	57,860	63,462	67,801	70,896	1,033,280	1,012,794	1,075,015	1,070,933
IL	2,293,883	2,515,986	2,687,974	2,810,681	2,681,268	2,628,109	2,789,568	2,778,975
IN	1,305,712	1,432,136	1,530,034	1,599,881	1,050,479	1,029,652	1,092,909	1,088,759
IA	2,504,835	2,747,362	2,935,168	3,069,159	4,015,616	3,936,002	4,177,811	4,161,946
KS	1,130,052	1,239,468	1,324,196	1,384,646	2,650,180	2,597,638	2,757,224	2,746,754
KY	1,124,963	1,233,887	1,318,233	1,378,411	572,757	561,402	595,892	593,629
LA	304,156	333,606	356,411	372,681	416,789	408,526	433,623	431,977
ME	9,526	10,448	11,162	11,672	52,935	51,886	55,073	54,864
MD [3]	150,297	164,849	176,118	184,158	59,246	58,072	61,639	61,405
MA	8,170	8,961	9,573	10,010	58,000	56,850	60,343	60,114
MI	347,710	381,377	407,447	426,048	452,448	443,477	470,723	468,935
MN	1,458,287	1,599,484	1,708,822	1,786,830	2,533,697	2,483,464	2,636,036	2,626,026
MS	674,924	740,273	790,876	826,980	564,433	553,242	587,231	585,001
MO	1,700,392	1,865,030	1,992,520	2,083,480	1,561,156	1,530,204	1,624,213	1,618,045
MT	315,350	345,884	369,528	386,397	741,587	726,884	771,540	768,610
NE	1,440,079	1,579,513	1,687,486	1,764,520	3,682,500	3,609,491	3,831,241	3,816,692
NV	6,307	6,918	7,391	7,728	17,162	16,822	17,855	17,787
NH	6,505	7,135	7,623	7,971	484	474	504	502
NJ	42,906	47,060	50,277	52,573	28,253	27,693	29,395	29,283
NM	151,324	165,976	177,322	185,417	281,456	275,876	292,824	291,712
NY	217,037	238,052	254,325	265,935	437,817	429,137	455,501	453,772
NC	688,442	755,099	806,717	843,544	508,128	498,054	528,652	526,644
ND	483,415	530,221	566,466	592,325	1,274,986	1,249,708	1,326,484	1,321,447
OH	937,253	1,028,001	1,098,274	1,148,410	652,991	640,045	679,366	676,787
OK	658,880	722,675	772,076	807,321	1,762,560	1,727,615	1,833,751	1,826,788
OR	123,679	135,654	144,927	151,543	586,964	575,326	610,672	608,353
PA	813,948	892,758	953,785	997,326	306,430	300,355	318,807	317,597
RI	1,735	1,902	2,033	2,125	239	235	249	248
SC	94,672	103,838	110,937	116,001	82,937	81,293	86,287	85,959
SD	445,394	488,519	521,914	545,739	1,738,660	1,704,189	1,808,886	1,802,017
TN	595,541	653,204	697,856	729,713	399,400	391,481	415,532	413,954
TX	1,344,180	1,474,328	1,575,111	1,647,015	3,615,694	3,544,010	3,761,736	3,747,452
UT	61,256	67,188	71,780	75,057	202,585	198,569	210,768	209,968
VT	62,159	68,178	72,838	76,163	19,888	19,493	20,691	20,612
VA	346,155	379,671	405,625	424,142	193,581	189,743	201,400	200,635
WA	320,067	351,057	357,055	392,176	1,261,879	1,236,860	1,312,847	1,307,862
WV	106,862	117,209	125,221	130,938	30,084	29,487	31,299	31,180
WI	1,323,719	1,451,886	1,551,135	1,621,945	1,409,606	1,381,658	1,466,541	1,460,972
WY	108,678	119,201	127,349	133,163	396,945	389,075	412,978	411,410
US	29,029,001	31,839,697	34,016,207	35,569,061	45,097,177	44,203,079	46,918,704	46,740,538

[1] Includes operator households. Includes loans of national and commercial, mutual savings, stock savings, and private banks. Loans are classified according to location of bank and, therefore, are not strictly comparable with data for other lenders which are classified according to location of borrower. [2] Preliminary. [3] Includes District of Columbia.

ERS, Farm Sector and Performance and Well-Being Branch, (202) 694–5586.

Table 10-7.—Farm real estate debt: Average interest rates on loans outstanding, by farming regions, [1] Dec. 31, 1992–2001 [2]

Year	Northeast	Lake States	Corn Belt	Northern Plains	Appalachian	Southeast	Delta States	Southern Plains	Mountain	Pacific	United States
	Percent	*Percent*	*Percent*	*Percent*	*Percent*	*Percent*	*Percent*	*Percent*	*Percent*	*Percent*	*Percent*
1992	8.3	8.4	8.5	8.4	8.5	8.7	8.6	8.4	8.3	8.6	8.5
1993	7.7	7.7	7.8	7.7	7.8	8.0	7.9	7.7	7.6	7.9	7.8
1994	7.9	7.9	8.0	7.9	8.0	8.2	8.1	7.9	7.8	8.1	8.0
1995	7.9	8.0	8.1	8.0	8.1	8.2	8.2	7.9	7.8	8.0	8.0
1996	8.1	8.1	8.2	8.1	8.3	8.5	8.3	7.9	7.9	8.2	8.1
1997	7.9	8.0	8.1	8.0	8.1	8.2	8.2	7.9	7.8	8.0	8.0
1998	7.9	8.0	8.0	7.9	8.1	8.1	8.1	7.9	7.8	7.9	8.0
1999	7.6	7.6	7.7	7.6	7.7	7.7	7.7	7.6	7.4	7.5	7.6
2000	7.8	7.8	7.9	7.8	7.9	8.0	7.9	7.8	7.6	7.7	7.8
2001 [3]	7.4	7.4	7.5	7.4	7.5	7.5	7.5	7.4	7.2	7.3	7.4

[1] Includes operator households. For States included in regions, see footnote 1, table 10-5. [2] Contract rates. Excludes Alaska and Hawaii. [3] Preliminary.

ERS, Farm Sector Performance and Well- Being Branch, (202) 694–5586.

Table 10-8.—Farm real estate debt: Interest charges on debt outstanding, by farming regions [1], 1992–2001 [2]

Year	Northeast	Lake States	Corn Belt	Northern Plains	Appalachian	Southeast
	1,000 dollars	*1,000 dollars*	*1,000 dollars*	*1,000 dollars*	*1,000 dollars*	*1,000 dollars*
1992	336,369	664,675	1,560,895	724,934	528,778	500,659
1993	319,704	617,365	1,465,078	684,447	479,959	463,735
1994	334,572	646,787	1,541,089	720,994	495,254	485,115
1995	333,765	675,051	1,611,737	744,051	507,155	491,280
1996	347,696	703,277	1,679,010	775,107	528,323	511,785
1997	350,540	726,944	1,741,748	785,712	529,043	515,126
1998	348,553	764,419	1,822,966	811,740	548,519	529,763
1999	343,985	759,983	1,833,463	815,285	552,230	531,987
2000	377,270	797,200	1,946,281	863.893	601,009	584,616
2001 [3]	375,377	783,153	1,912,101	846,978	603,228	585,136

Year	Delta States	Southern Plains	Mountain	Pacific	United States
	1,000 dollars	*1,000 dollars*	*1,000 dollars*	*1,000 dollars*	*1,000 dollars*
1992	334,927	552,996	550,603	961,067	6,733,192
1993	316,449	500,736	503,406	889,014	6,254,851
1994	338,366	517,086	520,084	921,092	6,535,257
1995	353,230	533,017	527,831	919,063	6,708,371
1996	367,974	555,265	549,862	957,375	6,988,375
1997	378,392	585,733	554,381	978,942	7,158,756
1998	394,658	607,429	561,115	1,032,822	7,434,976
1999	402,257	607,761	554,858	1,048,096	7,462,658
2000	438,061	649,943	588,395	1,143,326	8,002,859
2001 [3]	434,588	642,149	578,554	1,151,863	7,925,623

[1] Includes operator households. For States included in regions, see footnote 1, table 10-5. [2] Interest charges during calendar year. [3] Preliminary.

ERS, Farm Sector Performance and Well-Being Branch, (202) 694–5586.

Table 10-9.—Farm real estate debt: Amount outstanding by lender, United States, Dec. 31, 1992–2001 [1]

Year	Farm Credit System	Farm Service Agency [2]	Life insurance companies [3]	All operating banks [4]	Individuals and others [5]	CCC storage and drying facility	Total farm mortgage debt
	1,000 dollars	*1,000 dollars*	*1,000 dollars*	*1,000 dollars*	*1,000 dollars*	*1,000 dollars*	*1,000 dollars*
1992	26,886,261	6,779,546	9,208,000	19,862,622	17,000,000	1,984	79,738,413
1993	26,460,450	6,216,178	9,469,174	20,847,783	17,450,000	0	80,738,585
1994	26,300,421	5,852,920	9,562,841	22,555,042	18,700,000	0	82,971,224
1995	26,529,840	5,403,307	9,622,280	23,805,146	19,200,000	0	84,560,906
1996	27,462,253	5,025,262	10,021,976	24,870,048	19,700,000	0	87,079,339
1997	28,922,818	4,663,567	10,267,452	26,968,325	20,200,000	0	91,022,162
1998	30,824,704	4,352,326	11,353,694	29,029,001	20,000,000	0	95,559,725
1999	32,339,403	4,136,820	12,165,553	31,839,697	19,900,000	0	100,381,473
2000	33,907,358	3,907,122	12,514,220	34,016,207	19,543,500	58,000	103,946,407
2001 [6]	37,559,173	3,830,284	12,700,000	35,569,061	19,847,000	153,000	109,658,518

[1] Includes operator households. Includes regular mortgages, purchase-money mortgages, and sales contracts. [2] Includes farm ownership loans, soil and water loans to individuals, rural and labor housing loans, association loans for grazing, Indian tribe land acquisition loans, and one-half of economic emergency loans. [3] Compiled by American Council of Life Insurance. [4] Includes all operating commercial, savings, and private banks. [5] Estimated by ERS. [6] Preliminary.

ERS, Farm Sector Performance and Well-Being Branch, (202) 694–5586.

Table 10-10.—Nonreal estate farm debt: Amount outstanding, by lender, United States, Dec. 31, 1992–2001 [1]

Year	Debt owed to reporting institutions (excluding CCC)				Debts owed to individuals and others	Total excluding CCC loans	Price-support loans made or guaranteed by CCC [2]	Total including CCC loans
	All operating banks	Farm Credit System [2]	Farm Service Agency	Total				
	Million dollars	*Million dollars*	*Million dollars*	*Million dollars*	*Million dollars*	*Million dollars*	*Million dollars*	*Million dollars*
1992	34,644	10,777	8,118	53,539	13,500	67,038	4,771	71,809
1993	36,778	10,979	7,090	54,846	14,500	69,346	3,170	72,517
1994	38,663	11,646	6,841	57,150	15,500	72,650	6,237	78,887
1995	39,735	12,992	5,786	58,513	16,500	75,013	2,979	77,992
1996	40,362	14,599	5,243	60,204	17,800	78,004	3,508	81,512
1997	43,908	15,878	4,899	64,685	19,200	83,885	1,982	85,867
1998	45,097	17,314	4,538	66,949	20,000	86,949	5,230	92,179
1999	44,203	16,579	4,557	65,340	20,700	86,040	5,681	91,721
2000	46,919	17,455	4,402	68,776	21,735	90,511	4,253	94,764
2001 [3]	46,741	19,962	4,309	71,012	22,100	93,112	5,464	98,576

[1] Includes operator households. [2] Although price-support loans of the Commodity Credit Corporation (CCC) are nonrecourse loans, they are treated as income in the year received. They are not considered farm debt even though borrowers must either pay them or deliver the commodities on which they are based. [3] Preliminary.

ERS, Farm Sector Performance and Well-Being Branch, (202) 694–5586.

Table 10-11.—Farm Service Agency: Loans made to individuals and associations for farming purposes, and amount outstanding, United States and Territories, 1993–2002 [1]

Year	Loans to individuals						
	Farm ownership			Soil and water			Recreation
	New borrowers	Loans made	Outstanding Jan. 1	New borrowers	Loans made	Outstanding Jan. 1	Outstanding Jan. 1
	Number	*1,000 dollars*	*1,000 dollars*	*Number*	*1,000 dollars*	*1,000 dollars*	*1,000 dollars*
1993	3,767	609,312	5,399,659	122	2,705	159,155	5,474
1994	4,418	681,217	5,012,489	98	3,146	138,686	4,521
1995	3,727	616,870	4,715,766	0	0	124,733	4,278
1996	3,630	624,316	6,816,032	0	0	118,484	....
1997	3,482	613,877	7,008,911	0	0	98,774	3,874
1998	2,899	508,466	6,831,520	0	0	81,067	2,949
1999	4,308	944,694	6,675,272	0	0	76,042	2,694
2000	4,552	1,106,492	6,755,110	0	0	66,602	2,221
2001	3,704	1,015,634	7,287,728	0	0	52,883	1,784
2002	4,107	1,279,027	7,495,449	0	0	46,284	1,447

Year	Loans to individuals					
	Operating			Emergency		
	New borrowers	Loans made	Outstanding Jan. 1	New borrowers	Loans made	Outstanding Jan. 1
	Number	*1,000 dollars*	*1,000 dollars*	*Number*	*1,000 dollars*	*1,000 dollars*
1993	10,582	1,626,821	3,112,516	578	58,571	4,244,237
1994	13,137	2,014,715	2,805,783	2,469	90,013	3,649,004
1995	10,517	1,816,177	2,711,011	1,146	68,823	3,273,315
1996	10,377	1,882,431	5,945,331	2,163	176,500	3,046,279
1997	9,065	1,560,559	6,100,452	1,760	144,880	2,423,475
1998	9,433	1,568,071	6,040,488	1,045	97,569	2,039,657
1999	14,525	2,564,767	5,961,862	2,846	329,848	1,940,961
2000	12,979	2,464,802	6,570,523	1,557	150,852	1,915,780
2001	10,732	2,152,814	6,823,828	962	90,026	1,712,807
2002	10,476	2,217,735	6,639,837	501	57,608	1,523,438

Year	Loans to associations					Economic opportunity individual loans	Economic emergency loans
	Indian tribe land acquisition			Grazing association	Irrigation, drainage, and soil conservation		
	New borrowers	Loans made	Outstanding Jan. 1	Outstanding Jan. 1	Outstanding Jan. 1	Outstanding Jan. 1	Outstanding Jan. 1
	Number	*1,000 dollars*	*1,000 dolllars*	*1,000 dollars*	*1,000 dollars*	*1,000 dollars*	*1,000 dollars*
1993	1	1,367	78,385	34,427	11,492	37	1,634,898
1994	1	137	75,575	31,267	10,428	28	1,375,611
1995	1	550	74,171	28,929	10,183	25	1,185,261
1996	0	641	73,479	28,613	6,657	14	1,082,954
1997	0	224	62,603	23,878	6,229	14	874,601
1998	0	500	59,856	19,654	5,956	12	685,147
1999	0	0	58,461	17,855	5,666	11	653,953
2000	1	673	57,117	15,660	5,449	10	545,423
2001	1	590	62,738	12,785	5,177	8	427,176
2002	1	74	60,777	10,849	3,729	8	364,377

[1] Includes loans made directly by FmHA and those guaranteed by the Agency. Amounts of loans made represent obligations and include loans to new borrowers and subsequent loans to borrowers who received an initial loan in a prior year. Amounts outstanding are loan advances less principal repayments for loans made directly by the Agency.

FSA, Loan Making Division, (202) 690–4006.

Table 10-12.—Farmers' marketing, farm supply, and related service cooperatives: Number, memberships, and business, United States, 1992–2001

Year[1]	Cooperatives[2]				Estimated memberships[4]				Estimated service receipts[5]
	Marketing	Farm supply	Related service[3]	Total	Marketing	Farm supply	Related service[3]	Total	
	Number	*Number*	*Number*	*Number*	*1,000 members*	*1,000 members*	*1,000 members*	*1,000 members*	*Million dollars*
1992	2,218	1,618	479	4,315	1,839	2,020	212	4,072	2,575
1993	2,214	1,547	483	4,244	1,830	1,977	216	4,023	2,724
1994	2,173	1,496	505	4,174	1,805	1,936	245	3,986	2,986
1995	2,074	1,458	474	4,006	1,712	1,846	210	3,767	3,284
1996	2,012	1,403	469	3,884	1,682	1,795	187	3,664	3,100
1997[6]	1,941	1,386	464	3,791	1,498	1,706	183	3,387	3,647
1998	1,863	1,347	441	3,651	1,398	1,774	181	3,353	3,473
1999	1,749	1,313	404	3,466	1,283	1,731	159	3,173	3,905
2000	1,672	1,277	397	3,346	1,243	1,718	124	3,085	3,510
2001[7]	1,606	1,234	389	3,229	1,160	1,746	128	3,034	3,471

Year[1]	Marketing volume		Farm supply volume		Total marketing and farm supply volume and service receipts	
	Estimated gross business[8]	Estimated net business[9]	Estimated gross business[8]	Estimated net business[9]	Estimated gross business[8]	Estimated net business[9]
	Million dollars	*Million dollars*	*Million dollars*	*Million dollars*	*Million dollars*	*Million dollars*
1992	63,790	58,196	27,046	18,513	93,411	79,284
1993	66,839	60,930	28,180	19,218	97,744	82,872
1994	72,148	65,545	30,405	20,779	105,539	89,309
1995	77,946	69,321	30,965	21,213	112,195	93,818
1996	90,270	79,429	34,728	23,653	128,098	106,182
1997[6]	85,949	77,843	37,076	25,181	126,673	106,670
1998	84,524	76,642	32,964	24,551	120,961	104,667
1999	80,506	71,982	30,879	23,177	115,291	99,064
2000	80,400	72,065	36,809	24,085	120,719	99,659
2001[7]	83,954	75,042	36,141	24,756	123,566	103,269

[1] Reports of cooperatives are included for the calendar year. [2] Includes independent local cooperatives, centralized cooperatives, federations of cooperatives and cooperatives with mixed organizational structures. Cooperatives are classified according to their major activity. If, for example, more than 50 percent of a cooperative's business is derived from marketing activities, it is included as a marketing cooperative. [3] Includes cooperatives whose major activity is providing services related to marketing and farm supply activities. [4] Includes members (those entitled to vote for directors) but does not include nonvoting patrons. (Some duplication exists because some farmers belong to more than one cooperative.) [5] Receipts for services related to marketing and purchasing activities, but not included in the volumes reported for these activities. [6] Revised. [7] Preliminary. [8] Estimated gross business includes all business reported between cooperatives, such as the wholesale business of farm supply cooperatives with other cooperatives or terminal market sales for local cooperatives. [9] Estimated net business represents the value at the first level at which cooperatives transact business for farmers. Figures are adjusted for duplication resulting from intercooperative business.

Rural Business-Cooperative Service (RBS), Statistics, (202) 720–2480 or celestine.adams@usda.gov. Based on records from cooperatives reporting to the Service.

Table 10-13.—Farmers' cooperatives: Business volume of marketing, farm supply, and related service cooperatives, United States, 2000 and 2001 (preliminary)

Item	Gross business		Net business [1]	
	2000	2001	2000	2001
	1,000 dollars	*1,000 dollars*	*1,000 dollars*	*1,000 dollars*
Products marketed:				
Beans and peas (dry edible)	147,021	154,120	146,709	151,427
Cotton and cotton products	2,790,550	2,457,339	2,730,540	2,400,371
Dairy products	25,776,488	29,408,877	22,720,634	26,186,721
Fruits and vegetables	10,064,884	9,685,688	9,569,963	8,822,247
Grain and oilseeds excluding cottonseeds	22,851,883	22,619,166	18,369,559	18,055,321
Livestock and livestock products	8,042,891	9,558,957	8,042,690	9,558,957
Nuts	720,617	893,281	720,617	893,281
Poultry products	2,245,769	2,216,694	2,133,369	2,216,694
Rice	814,793	756,262	814,793	756,262
Sugar products	2,680,755	2,738,229	2,680,755	2,647,675
Tobacco	563,077	440,759	563,077	440,759
Wool and mohair	6,337	7,781	4,375	7,781
Miscellaneous [2]	3,694,502	3,016,585	3,567,569	2,904,315
Total farm products	80,399,567	83,953,737	72,064,650	75,041,810
Supplies purchased:				
Crop protectants	4,025,484	3,643,793	3,027,592	2,956,777
Feed	6,622,171	5,485,950	4,691,123	3,998,286
Fertilizer	7,320,431	7,402,223	4,574,367	4,966,445
Petroleum	13,333,374	14,084,842	7,456,720	8,445,751
Seed	1,381,600	1,520,107	915,667	1,051,064
Other supplies [3]	4,126,228	4,004,391	3,419,061	3,337,902
Total farm supplies	36,809,288	36,141,305	24,084,530	24,756,224
Receipts for services: [4]				
Trucking, cotton ginning, storage, grinding, locker plants, miscellaneous	3,510,216	3,470,717	3,510,216	3,470,717
Total business	120,719,071	123,565,760	99,659,396	103,268,751

[1] Represents value at the first level at which cooperatives transact business for farmers. [2] Includes coffee, fish, forest products, hay, hops, seed marketed for growers, nursery stock, other farm products not separately classified, and sales of farm products not received directly from member-patrons. Also includes manufactured food products and resale items marketed by cooperatives. [3] Includes automotive supplies, building materials, chicks, containers, farm machinery and equipment, hardware, meats and groceries, and other supplies not separately classified. [4] Charges for services related to marketing or purchasing but not included in the volume reported for those activities, plus other income.

RBS, Statistics, (202) 720–2480 or celestine.adams@usda.gov. Based on records from cooperatives reporting to the Service.

Table 10-14.—Farmers' cooperatives: Types, numbers, and memberships, United States, 2001

Type	Year or date of data	Associations	Estimated memberships or participants
Marketing and farm supply, and related service: [1]			
Marketing	2001	1,606	1,160,000
Farm supply	2001	1,234	1,745,771
Related services [2]	2001	389	128,136
Service:			
Production credit associations [3]	Sep. 31, 2001	15	NA
Rural credit unions [4]	Dec. 31, 2001	542	3,798
Rural electric cooperatives [5]	Dec. 31, 2001	601	10,894
Production:			
Dairy herd improvement associations [6]	Dec. 31, 2001	NA	29,466

[1] Rural Business - Cooperative Service, U.S. Department of Agriculture. [2] Includes trucking, storage, grinding, locker plant, and other services. [3] Farm Credit Administration. [4] Credit Union National Association, Inc. [5] Rural Utilities Service, U.S. Department of Agriculture. [6] Agriculture Research Service, U.S. Department of Agriculture. NA-not available.

RBS, Statistics, (202) 720–2480 or celestine.adams@usda.gov.

Table 10-15.—Farmers' cooperatives: Number of cooperatives, memberships, and business volume of marketing, farm supply, and related service cooperatives, by States, 1999 to 2001 (preliminary)

State	Cooperatives headquartered in State		Memberships in State [1]		Net business [1]	
	2000	2001	1999	2001	1999	2001
	Number	*Number*	*Number*	*Number*	*1,000 dollars*	*1,000 dollars*
AL	63	62	53,886	42,222	1,169,594	949,788
AK	9	9	1,815	4,812	76,496	43,561
AZ	9	9	3,376	2,710	927,983	607,896
AR	52	52	56,038	55,431	1,643,198	1,670,856
CA	178	171	53,604	49,553	7,823,548	7,572,686
CO	46	47	32,613	29,122	928,817	1,118,632
CT	4	5	2,728	1,899	175,897	214,144
DE	3	3	43,422	43,430	89,741	103,958
FL	40	40	26,050	29,753	2,419,178	2,413,535
GA	16	17	26,673	21,748	2,006,532	1,676,955
HI	26	20	2,768	904	103,807	28,829
ID	38	37	18,583	20,259	1,331,705	1,422,387
IL	179	175	195,258	175,871	5,316,036	5,470,384
IN	52	46	75,409	66,818	1,931,213	2,070,633
IA	179	166	180,168	162,249	7,932,170	9,232,231
KS	133	133	133,269	129,576	4,658,873	4,794,447
KY	42	42	231,862	242,873	773,265	707,833
LA	48	47	12,294	13,032	564,193	735,703
ME	27	26	8,348	7,922	203,284	225,165
MD	18	17	71,953	74,404	352,277	411,482
MA	13	12	5,551	4,573	627,730	637,056
MI	62	62	29,788	28,660	1,925,074	2,198,121
MN	302	293	186,902	176,594	9,306,888	9,895,732
MS	73	70	105,239	111,043	949,425	1,055,738
MO	66	66	142,771	109,134	4,860,441	4,611,124
MT	68	65	28,720	33,066	685,725	719,154
NE	103	95	92,353	84,834	4,543,391	4,720,598
NV	(2)	(2)	(2)	(2)	(2)	(2)
NH	(2)	(2)	(2)	(2)	(2)	(2)
NJ	16	16	7,347	7,052	312,984	388,148
NM	9	10	2,974	2,705	554,906	507,304
NY	95	92	17,770	37,431	3,235,290	2,825,792
NC	24	22	98,404	96,351	825,510	975,355
ND	251	239	118,435	120,303	3,041,338	3,292,172
OH	78	73	55,970	53,407	2,277,303	2,792,014
OK	89	81	68,611	65,640	1,386,496	1,562,099
OR	33	33	24,638	27,043	1,716,532	1,703,893
PA	55	54	45,008	33,442	1,353,040	1,440,730
RI	(2)	(2)	(2)	(2)	(2)	(2)
SC	5	5	3,458	2,639	246,516	185,466
SD	133	126	107,046	98,395	2,451,052	2,699,501
TN	79	79	139,124	137,805	729,982	808,937
TX	244	238	118,064	109,926	3,400,868	3,876,461
UT	17	17	10,013	9,354	461,461	512,095
VT	5	5	5,981	4,557	471,561	523,119
VA	59	57	174,308	193,470	996,687	846,781
WA	82	78	33,903	31,664	3,325,162	3,331,110
WV	26	26	75,083	83,051	88,703	99,280
WI	181	175	205,690	186,490	8,018,466	8,738,431
WY	12	12	5,145	5,485	258,813	241,042
US	3,346	3,229	3,168,534	3,031,142	98,574,157	102,769,087
Foreign [3]	....	....	4,789	2,765	490,163	499,664
Total	3,346	3,229	3,173,323	3,033,907	99,064,320	103,268,751

[1] Represents value at the first level at which cooperatives transact business for farmers. These statistics for 2000 are presented on a national basis only. Totals may not add due to rounding. [2] Dollar volume or membership is not shown to avoid disclosing operations of individual cooperatives. [3] Sales outside the United States, sales to domestic military installations, and sales of certain products not received directly from member-patrons.

RBS, Statistics, (202) 720–2480 or celestine.adams@usda.gov.

Table 10-16.—Rural Utilities Service: Long-term electric financing approved by purpose, by States as of December 31, 2001 [1]

State	Borrowers	Total financing approved						Loan estimates	
		RUS loans	Non-RUS financing		Financing approved by purpose			Miles of line	Consumers
			With RUS guarantee	Without RUS guarantee [2]	Distribution	Generation and transmission [2]	Consumer facilities		
	Number	*1,000 dollars*	*1,000 dollars*	*1,000 dollars*	*1,000 dollars*	*1,000 dollars*	*1,000 dollars*	*Number*	*Number*
AL	27	708,815	919,913	280,177	736,420	1,171,127	1,359	62,272	553,058
AK	17	820,292	295,240	132,206	551,229	695,420	1,089	10,133	179,300
AZ	15	292,778	391,345	110,887	316,264	478,257	490	19,443	175,812
AR	20	915,421	1,039,813	455,399	1,029,645	1,376,516	4,471	70,149	493,456
CA	10	83,836	....	6,803	73,302	17,280	56	6,543	76,837
CO	25	1,122,283	2,070,869	680,265	984,126	2,889,216	74	69,954	400,266
CT	0	....	....	....	....	....	....	0	0
DE	1	56,030	....	18,666	73,832	861	3	4,161	47,939
FL	18	978,900	1,216,234	609,514	1,335,386	1,466,743	2,519	65,221	816,024
GA	50	1,581,099	5,129,666	1,840,527	2,590,524	5,957,809	2,959	152,474	1,575,090
HI	0	....	....	....	....	....	....	0	0
ID	10	155,838	20,134	36,612	184,439	26,949	1,197	12,325	64,241
IL	29	560,496	1,169,444	179,645	555,345	1,354,028	212	55,014	248,749
IN	46	449,144	1,474,183	519,176	494,718	1,947,068	717	55,183	432,018
IA	47	745,363	386,434	140,040	586,314	685,125	397	66,007	204,384
KS	32	610,914	699,885	102,528	570,384	842,528	415	71,763	216,084
KY	27	1,360,743	2,115,822	624,700	1,491,024	2,609,083	1,159	85,422	812,564
LA	20	583,798	2,789,955	370,523	737,715	3,006,383	177	48,970	429,138
ME	4	32,043	3,871	17,998	34,989	18,879	44	2,062	19,193
MD	2	263,343	18,355	137,356	310,841	108,213	....	13,526	155,563
MA	0	....	....	....	....	....	....	0	0
MI	10	511,238	827,416	92,692	481,427	949,446	473	34,249	273,626
MN	48	1,206,477	2,031,449	436,528	1,269,563	2,400,545	4,346	115,124	610,696
MS	29	762,922	791,296	283,796	901,950	935,369	694	83,365	660,476
MO	48	1,445,562	995,781	567,370	1,337,848	1,670,044	821	118,003	688,001
MT	26	362,663	34,888	49,161	376,404	70,079	229	45,417	125,799
NE	35	468,920	15,936	41,577	412,423	113,449	562	75,465	173,041
NV	8	75,008	1,241	10,441	65,315	21,127	248	6,265	24,889
NH	1	100,398	143,839	8,696	103,050	149,850	32	4,616	64,601
NJ	2	18,173	....	5,377	22,250	1,295	4	1,000	13,334
NM	17	455,738	24,088	68,799	490,249	55,367	3,010	42,230	211,989
NY	6	34,332	13,112	9,464	54,138	2,686	85	5,186	26,775
NC	34	1,249,831	1,463,711	450,585	1,517,434	1,642,057	4,635	89,902	883,182
ND	24	1,014,431	2,247,020	868,851	648,817	3,479,500	1,986	69,056	134,098
OH	27	475,791	448,749	468,530	585,987	806,864	218	46,750	342,852
OK	29	1,003,309	721,996	278,755	1,041,849	959,535	2,676	98,079	486,220
OR	18	268,473	69,128	83,272	298,803	121,834	237	23,372	130,147
PA	13	346,533	598,833	126,794	447,130	624,791	239	26,814	216,483
RI	1	....	3,420	....	340	3,080	....	4	160
SC	28	1,009,967	857,285	379,965	1,464,773	780,213	2,231	69,342	670,747
SD	30	624,747	81,715	56,256	598,910	162,927	881	65,521	149,159
TN	33	629,139	83,772	190,919	887,106	16,500	223	84,187	957,458
TX	99	2,226,166	1,520,777	771,500	2,500,631	2,015,289	2,524	263,460	1,417,350
UT	6	72,455	1,031,811	216,171	62,920	1,257,393	124	5,860	25,033
VT	3	63,917	50,726	7,310	53,749	67,300	903	2,924	24,048
VA	18	693,719	309,794	239,828	947,326	295,526	488	43,302	401,705
WA	23	232,546	7,234	40,181	260,843	18,866	252	20,904	106,850
WV	1	13,736	....	1,059	14,790	....	5	820	7,024
WI	27	507,855	453,946	156,639	403,875	712,756	1,809	45,538	221,932
WY	13	274,025	14,015	23,297	246,249	64,987	101	28,832	84,705
MH	1	....	12,500	....	325	12,175	....	0	2,624
PR	1	300,981	....	31,424	292,851	39,554	....	16,633	624,343
VI	1	430	....	....	234	197	....	85	912
US [3]	1,060	27,770,617	34,596,640	12,228,258	30,446,055	44,102,087	47,374	2,402,927	16,659,975

[1] Includes $628,346,164 discounted principal from 219 prepaid borrowers. [2] Includes loans obtained by RUS borrowers' affiliates specifically organized to facillitate non-RUS finanacing. [3] Includes figures not shown elsewhere in this table for two borrowers whose loans have been foreclosed. The total amount of these loans was $37,237.

RD, Planning and Policy Branch, (202)692-0341

Table 10-17.—Rural Utilities Service: Composite revenues and patronage capital, average number of consumers and megawatt-hour sales reported by rus electric borrowers operating distribution systems - calendar years 1999–2001

Item	1999		2000		2001	
	Amount	Percent of total	Amount	Percent of total	Amount	Percent of total
Number of borrowers reporting	650	....	633	....	621	....
Average number of consumers served:						
Residential service (farm & non-farm)	9,585,118	89.6	9,695,066	89.5	9,712,110	89.4
Commercial & industrial, small	956,253	8.9	974,722	9.0	990,256	9.1
Commercial & industrial, large	7,076	0.1	7,596	0.1	6,773	0.1
Irrigation	97,775	0.9	99,585	0.9	98,645	0.9
Other electric service	50,567	0.5	52,878	0.5	55,276	0.5
To others for resale	191	*	190	*	201	*
Total	10,696,896	100.0	10,829,973	100.0	10,863,261	100.0
Megawatt-hour sales:						
Residential service (farm & non-farm)	122,031,252	57.3	127,208,955	57.0	128,738,978	57.2
Commercial & industrial, small	37,738,823	17.7	39,272,310	17.6	39,933,548	17.7
Commercial & industrial, large	45,857,319	21.5	48,560,956	21.8	49,328,266	21.9
Irrigation	3,440,908	1.6	4,274,009	1.9	3,784,221	1.7
Other electric service	1,818,007	0.9	1,965,798	0.9	1,980,123	0.9
To others for resale	2,119,032	1.0	1,784,234	0.8	1,467,556	0.7
Total	213,005,320	100.0	223,066,273	100.0	225,232,664	100.0
Revenue and patronage capital:						
Residential service (farm & non-farm)	9,196,884	64.1	9,647,246	63.9	9,993,895	64.0
Commercial & industrial, small	2,650,940	18.5	2,764,267	18.3	2,869,885	18.4
Commercial & industrial, large	1,800,675	12.6	1,915,687	12.7	2,010,623	12.9
Irrigation	225,101	1.6	273,626	1.8	257,340	1.6
Other electric service	136,707	1.0	145,657	1.0	152,966	1.0
To others for resale	80,940	0.6	85,421	0.6	69,015	0.4
Total from sales of electric energy	14,091,247	98.3	14,831,904	98.2	15,353,724	98.4
Other operating revenue	250,491	1.7	269,311	1.8	254,173	1.6
Total operating revenue	14,341,738	100.0	15,101,214	100.0	15,607,897	100.0

*Less than 0.05 percent.

Rural Development, Planning and Policy Branch, (202) 692-0341

Table 10-18.—Rural Utilities Service: Advances to, and operating statistics of, electric borrowers, United States and Territories, 1993–2002

Year	Total advances as of Jan. 1 [1] [2]	Statistics for borrowers in operation								
		As of Jan. 1			During year				Average monthly kw.-hr. consumption per consumer	
		Borrowers	Miles energized [3]	Consumers served [3]	Kw.-hr. generated	Kw.-hr. purchased [4]	Kw.-hr. sales [4]	Total revenue [4]	All consumers	Residential consumers [5]
	1,000 dollars	*Numbers*	*Miles*	*Number*	*1,000 kw.-hr.*	*1,000 kw.-hr.*	*1,000 kw.-hr.*	*1,000 dollars*	Kw. hr.	Kw. hr.
1993	21,410,549	1,052	2,230,138	12,905,821	157,375,158	144,159,840	277,205,315	17,168,816	1,530	1,021
1994	21,839,184	1,049	2,247,943	13,170,387	159,463,827	138,222,711	274,658,217	16,827,484	1,528	1,001
1995	22,535,357	1,050	2,235,156	12,835,552	152,549,534	147,000,468	275,749,181	16,802,281	1,570	1,036
1996	23,235,820	1,047	2,265,042	13,172,812	155,359,522	153,407,637	287,682,820	16,781,276	1,632	1,071
1997	23,917,858	1,043	2,302,221	13,569,867	147,746,172	156,904,182	284,072,253	16,427,652	1,612	1,037
1998	24,503,187	1,033	2,303,030	13,899,900	157,896,742	153,038,893	292,649,616	16,772,806	1,670	1,073
1999	25,150,466	1,027	2,329,844	14,183,217	153,450,950	156,465,574	288,039,703	16,644,924	1,668	1,067
2000	25,725,433	1,021	2,352,646	14,469,337	163,594,498	168,685,303	306,028,142	16,233,044	1,716	1,093
2001	26,317,572	1,016	2,390,313	14,985,239	160,777,217	173,957,934	311,458,938	18,742,342	1,728	1,105
2002	26,838,962	1,014	2,378,754	15,062,084	NA	NA	NA	NA	NA	NA

[1] Cumulative from organization, May 11, 1935. [2] Actual funds advanced out of RUS loans approved, as shown in table 10-16. [3] Includes data at time of repayment of loan for borrowers whose loans have been repaid in full. As of Jan. 1, 2002, there were 294 such borrowers with 614,452 miles energized and 4,090,487 consumers served. [4] Excludes energy sales and revenues of power sold by one RUS borrower to another. Includes patronage capital. [5] Includes rural nonfarm and farm consumers. NA-not available.

RD, Planning and Policy Branch, (202) 692–0341.

Table 10-19.—Rural Utilities Service: Annual revenues and expenses reported by electric borrowers, United States, 1992–2001

Year	Operating revenue	Operating expense	Interest expense	Depreciation and amortization expense	Net margins	Total utility plant
	1,000 dollars	*1,000 dollars*	*1,000 dollars*	*1,000 dollars*	*1,000 dollars*	*1,000 dollars*
1992	23,325,191	18,643,390	2,778,213	1,709,293	710,890	60,669,964
1993	24,872,684	19,907,666	2,667,343	1,767,525	1,034,849	62,654,291
1994	24,861,855	20,138,000	2,501,585	1,824,860	807,741	64,391,203
1995	24,609,188	21,741,162	2,171,170	1,779,568	1,236,050	61,867,838
1996	24,438,558	19,777,504	2,054,058	1,788,164	1,328,237	61,443,968
1997	23,321,068	18,987,613	1,908,451	1,726,982	1,018,702	60,770,878
1998	23,987,773	19,491,238	1,890,050	1,732,056	1,274,324	61,720,967
1999	23,823,791	19,536,422	1,832,553	1,746,681	1,112,665	62,684,354
2000	25,628,917	21,161,991	1,905,043	1,819,616	1,164,076	66,353,227
2001	26,402,496	21,825,642	1,907,432	1,888,516	1,212,914	69,407,710

RD, Planning and Policy Branch, (202) 692–0341.

Table 10-20.—Loans to farmers' cooperative organizations: Outstanding amounts held by the banks for cooperatives, and agricultural credit banks classified by type of loan, United States, Jan. 1, 1993–2002 [1]

Year	Operating capital loans	Facility loans
	1,000 dollars	*1,000 dollars*
1993	7,478,199	6,474,582
1994	7,393,147	6,143,719
1995	7,003,776	8,200,439
1996	9,119,835	9,128,364
1997	7,332,313	10,657,055
1998	6,358,665	11,684,548
1999	6,227,194	10,911,590
2000	6,196,401	11,602,316
2001	7,293,142	11,348,179
2002	7,660,584	11,311,516

[1] Includes Puerto Rico.

FCA, Office of Policy and Analysis, (703) 883–4073.

CHAPTER XI

STABILIZATION AND PRICE-SUPPORT PROGRAMS

The statistics in this chapter relate to activities of the Commodity Credit Corporation (CCC), cropland diversion and production adjustment programs, and marketing agreement and order programs for fruits and vegetables. Statistics for Federal Milk Marketing Order programs are contained in chapter VIII.

Table 11-1.—Commodity Credit Corporation: Price-supported commodities owned as of Dec. 31, 1992–2001 [1] (Inventory quantity)

Year	Barley	Butter and butter oil	Cheese	Corn	Cotton upland	Sorghum grain	Nonfat dry milk	Oils and oilseeds
	Million bushels	*Million pounds*	*Million pounds*	*Million bushels*	*1,000 bales*	*Million bushels*	*Million pounds*	
1992	5	628	15	125	(6) 7	8	(7) 62	(2)
1993	5	564	(3)	54	(6) 146	3	25	(2)
1994	5	111	0	44	(6) 90	1	44	(2)
1995	5	(3)	0	42	(8)	1	25	(2)
1996	(3)	0	0	30	(8)	0	(3)	0
1997	(3)	0	(3)	2	(2)	(2)	30	(2)
1998	1	0	0	15	(8)	1	111	0
1999	1	0	0	26	(8)	1	161	(3)
2000	(3)	0	0	36	(2)	(3)	602	(2)
2001	(3)	0	5	24	(2)	(2)	844	(2)

Year	Oats	Rice [4]	Rye	Soybeans	Honey	Wheat	Value of all commodities owned [5]
	Million bushels	*Million cwt.*	*Million bushels*	*Million bushels*	*Million pounds*	*Million bushels*	*Million dollars*
1992	(3)	1	(2)	0	(3)	165	1,719
1993	(2)	(2)	0	(3)	4	168	833
1994	0	(2)	0	(3)	13	144	715
1995	0	(2)	0	(2)	1	141	654
1996	0	0	0	0	0	96	435
1997	0	0	0	0	0	93	364
1998	(2)	0	0	3	0	107	363
1999	(3)	(3)	0	7	0	104	347
2000	(2)	(2)	0	10	0	109	790
2001	(2)	(3)	0	4	0	118	844

[1] Commodities which were owned by CCC in some years but not shown in this table are as follows: blended foods, cottonseed and products, naval stores, wheat products, corn products, oat products, rice products, vegetable oil, mixed feed, linseed oil, evaporated milk, meat, tallow, egg mix-dry, foundation seeds, peanut products, and sugar. [2] Less than 50,000 units. [3] Less than 500,000 units. [4] Rough basis; includes milled rice in rough equivalent. [5] The total value of all commodities owned by CCC, including price-supported commodities not shown and commodities acquired under programs other than price-support programs, less reserve for losses on inventory. [6] Includes extra long staple, cotton. [7] Includes infant formula. [8] Less than 500 units.

FSA, Financial Management Division, (703) 305–1277.

Table 11-2.—Commodity Credit Corporation: Loans made, United States and Territories, by crop years, 1995–2000 [1]

Commodity	Unit	1995 Quantity pledged	1995 Face amount	1996 Quantity pledged	1996 Face amount
			1,000 dollars		*1,000 dollars*
Barley	1,000 bushels	17,953	26,697	28,836	41,110
Corn	1,000 bushels	677,115	1,232,669	970,590	1,764,291
Cotton	1,000 bales	3,495	880,572	3,381	869,911
Seed cotton upland	1,000 pounds	5,137	2,662	131,356	4,765
Sugar Cane and Beet	1,000 pounds	4,805,154	965,070	1,837,547	383,460
Flaxseed	1,000 Cwt	29	256	21	190
Honey	1,000 pounds	54,427	27,211	....	....
Oats	1,000 bushels	704	673	1,526	1,484
Peanuts	1,000 Pounds	819,032	255,629	322,668	34,856
Rice	1,000 cwt	100,992	649,003	68,920	447,790
Rye	1,000 bushels	102	159	....	....
Sorghum grain	1,000 bushels	7,248	12,856	20,383	36,106
Soybeans	1,000 bushels	182,027	889,582	195,997	957,948
Tobacco	1,000 pounds	12,255	24,601	1,719	3,813
Wheat	1,000 bushels	114,182	294,851	194,795	505,268
Sunflower Seed	1,000 cwt	1,130	9,717	2,534	22,380
Canola Seed	1,000 Cwt	95	880	96	902
Safflower Seed	1,000 Cwt	....	....	5	39
Mustard Seed	1,000 Cwt	....	....	....	....
Sunflower Seed (non-oil)	1,000 Cwt	....	....	....	....

Commodity	Unit	1997 Quantity pledged	1997 Face amount	1998 Quantity pledged	1998 Face amount
			1,000 dollars		*1,000 dollars*
Barley	1,000 bushels	32,894	49,108	32,894	49,108
Corn	1,000 bushels	1,129,915	2,062,308	1,129,915	2,062,308
Cotton [3]	1,000 bales	4,315	1,109,629	4,315	1,109,629
Seed cotton	1,000 pounds	2,420	1,245	2,420	1,245
Sugar Cane and Beet	1,000 pounds	1,593,076	344,775	1,593,076	344,775
Flaxseed	1,000 Cwt	59	550	59	550
Oats	1,000bushels	2,380	2,533	2,380	2,533
Peanuts	1,000 Pounds	418,969	41,182	418,969	41,182
Rice	1,000 cwt	67,614	427,602	67,614	427,602
Sorghum grain	1,000 bushels	17,487	30,346	17,487	30,346
Soybeans	1,000 bushels	265,224	1,378,812	265,224	1,378,812
Tobacco	1,000 pounds	263,998	631,219	263,998	631,219
Wheat	1,000 bushels	262,852	677,866	262,852	677,866
Sunflower Seed	1,000 cwt	2,801	25,730	2,801	25,730
Canola Seed	1,000 Cwt	334	3,214	334	3,214
Safflower Seed	1,000 Cwt	10	69	10	69
Mustard Seed	1,000 Cwt	19	178	19	178
Sunflower Seed (non-oil)	1,000 Cwt	231	2,147	231	2,147

Commodity	Unit	1999 Quantity pledged	1999 Face amount	2000 [2] Quantity pledged	2000 [2] Face amount
			1,000 dollars		*1,000 dollars*
Barley	1,000 bushels	13,556	22,712	16,024	25,648
Corn	1,000 bushels	1,420,878	2,590,443	1,393,947	2,562,172
Cotton [3]	1,000 bales	9,109	2,339,291	8,959	2,216,075
Seed cotton	1,000 pounds	1,553	1,033	2,519	1,350
Sugar Cane and Beet	1,000 pounds	(531,880)	830,856	3,472,531	735,558
Flaxseed	1,000 Cwt	242	2,249	197	1,820
Honey	1,000 pounds	(86,491)	12,386	53,972	35,068
Oats	1,000bushels	1,652	1,803	1,696	1,892
Peanuts	1,000 Pounds	471,028	128,283	477,285	47,173
Rice	1,000 cwt	110,806	711,875	97,430	625,715
Sorghum grain	1,000 bushels	17,193	29,794	15,303	25,781
Soybeans	1,000 bushels	286,823	1,488,423	312,916	1,627,897
Tobacco	1,000 pounds	364,012	741,506	51,801	109,246
Wheat	1,000 bushels	154,276	398,367	181,133	468,418
Sunflower Seed	1,000 cwt	1,994	18,326	2,385	21,836
Canola Seed	1,000 Cwt	206	1,998	748	7,144
Safflower Seed	1,000 Cwt	35	247	36	251
Mustard Seed	1,000 Cwt	28	259	23	208
Sunflower Seed (non-oil)	1,000 Cwt	404	3,724	294	2,736
Crambe Oilseed	1,000 Cwt	142	1,232	236	2,042
Mohair	1,000 Pounds	879	2,466	4,128	7,465

[1] Includes loans made directly by Commodity Credit Corporation. [2] Loans through Sept. 30, 1996. [3] Includes extra long staple cotton and upland cotton.

FSA, Financial Management Division, (703) 305–1277.

Table 11-3.—Commodity Credit Corporation: Loan transactions for fiscal year 2000, by commodities [1]

Commodity	Unit	Loans outstanding Oct. 1, 2000 [2]	New loans made	Repayments	Collateral acquired in settlement	Loans written off [3]	Loans outstanding Sept. 30, 2001	
							Value [2]	Quantity collateral remaining pledged
		1,000 dollars	*1,000 dollars*	*1,000 dollars*	*1,000 dollars*	*1,000 dollars*	*1,000 dollars*	*1,000 units*
Basic commodities:								
Corn	Bushel	345,172	2,450,851	2,234,188	46,457	222,533	292,845	158,869
Cotton	Bale	129,172	2,141,812	1,992,420	12,939	73,380	192,245	774
Seed cotton	Pound	0	1,350	1,350	0	0	0	0
Peanuts	Pound	2,202	63,234	45,163	0	4	20,269	86,421
Rice	Cwt	172,054	648,077	473,441	355	150,418	195,917	29,671
Tobacco [4]	Pound	1,544,142	114,692	263,053	609,014	76,000	710,767	176,816
Wheat	Bushel	331,769	424,141	350,721	50,577	31,147	323,465	125,990
Total	....	2,524,511	5,844,157	5,360,336	719,342	553,482	1,735,508	XXXXX
Designated nonbasic commodities:								
Barley	Bushel	17,682	16,925	21,599	1,851	1,199	9,958	6,118
Sorghum	Bushel	8,625	19,813	19,905	1,032	732	6,769	3,925
Honey	Pound	9,675	30,869	31,510	0	5,677	3,357	5,188
Oats	Bushel	1,559	1,665	1,506	59	295	1,364	1,163
Sugar, beet	Pound	231,409	536,739	515,125	237,855	....	15,168	66,571
Sugar, cane	Pound	125,519	198,819	198,981	88,231	....	37,126	207,969
Sunflower seed	Cwt	2,912	21,519	16,723	1,536	5,536	636	70
Flaxseed	Cwt	721	966	848	328	296	215	23
Canola seed	Cwt	1,252	3,816	2,119	147	361	2,441	257
Safflower seed	Cwt	92	205	272	0	....	25	4
Rapeseed	Cwt	0	0	0	0	....	0	....
Mustard seed	Cwt	97	197	214	3	1	76	9
Crambe Oilseed	Cwt	31	1,290	1,284	0	13	24	3
Sunflower seed, non oil.	Cwt	953	2,736	2,697	39	636	317	35
Total	....	400,527	835,559	812,783	331,081	14,746	77,476	XXXXX
Other nonbasic commodities:								
Soybeans	Bushel	137,182	1,586,594	1,346,997	34,965	259,489	82,325	15,822
Mohair	Pound	8,351	905	8,039	0	430	787	487
Total	....	145,533	1,587,499	1,355,036	34,965	259,919	83,112	XXXXX
Other loans:								
Farm Storage facility [5]	....	10,039	86,059	2,138	0	....	93,960	....
Bollweevil	....	....	10,000	....	....	....	10,000	....
Total	....	10,039	96,059	2,138	0	0	103,960	0
Grand total [6]	....	3,080,610	8,267,215	7,528,155	1,085,388	828,147	1,896,096	XXXXX

[1] Loans made directly by Commodity Credit Corporation. [2] Book value of outstanding loans; includes face amounts and any charges paid. [3] Includes transfers to accounts receivable. [4] Charge offs represents pre-No Net Cost Tobacco loans - 1981 and prior crop loans. [5] This program was closed out during FY 1993. [6] Table may not add due to rounding.

FSA, Financial Management Division, (703) 305–1277

Table 11-4.—Commodity Credit Corporation: Selected inventory transactions, programs and commodity, fiscal year 2001

Program and commodity	Quantity					
	Unit	Inventory Oct. 1, 2000	Purchases	Collateral acquired from loans	Sales[1]	Inventory Sept. 30, 2001
		Thousands	*Thousands*	*Thousands*	*Thousands*	*Thousands*
Feed grains:.						
Barley	Bushel	407	487	1,045	1,659	280
Corn	Bushel	30,758	77,713	26,130	112,234	22,368
Corn products	Pound	2,391	257,401	0	257,766	2,026
Grain sorghum	Bushel	317	1,814	566	2,604	93
Sorghum grits	Pound	0	21,395	0	21,395	0
Oats	Bushel	29	2	56	67	20
Oats, rolled	Pound	0	0	0	0	0
Rye	Bushel	0	0	0	0	0
Total feed grains		xxx	xxx	xxx	xxx	xxx
Wheat (A)	Bushel	107,220	103,074	20,434	112,613	118,115
Wheat flour	Pound	2,706	725,350	0	728,030	27
Wheat products, other	Pound	0	270,299	0	269,700	600
Rice, milled	Cwt	0	4,312	0	4,313	0
Rice, rough	Cwt	9	51,616	53	51,647	32
Rice, brown	Pound	0	0	0	0	0
Rice, cereal	Pound	0	0	0	0	0
Cotton, extra long staple	Bale	4	0	11	4	10
Upland Cotton	Bale	2	7,327	35	7,343	21
Dairy products:						
Butter	Pound	0	0	0	0	0
Butter oil	Pound	0	0	0	0	0
Cheese	Pound	1,754	13,056	0	9,385	5,426
Milk, dried	Pound	0	0	0	0	0
Milk, UHT	Pound	0	0	0	0	0
Dry Whole Milk	Pound	0	987	0	987	0
Non fat dry milk	Pound	553,864	418,407	0	115,219	857,052
Total dairy products		xxx	xxx	xxx	xxx	xxx
Oils and oilseeds:						
Canola seed	Cwt.	1	97	15	113	0
Sunflower seed	Cwt.	5	954	164	1,122	1
Sunflower seed, non-oil	Cwt.	0	62	3	65	0
Sunflower seed oil, processed	Cwt.	0	14,320	0	14,320	0
Peanuts, farmers' stock	Pound	0	0	0	0	0
Peanut products	Pound	0	0	0	0	0
Peanut butter	Pound	0	0	0	0	0
Soybeans	Bushel	9,237	60,500	6,749	73,501	2,985
Soybean meal	Pound	0	556,502	0	556,502	0
Flaxseed	Cwt.	3	37	36	76	0
Totals oils and oilseeds		xxx	xxx	xxx	xxx	xxx
Blended foods	Pound	11,228	524,879	0	518,984	17,123
Grains and seeds:						
Feed for Government facilities	Cwt	0	0	0	0	0
Foundation seeds	Pound	0	0	0	0	0
Total grains and seeds		xxx	xxx	xxx	xxx	xxx
Peas, dry whole	Pound	1,438	175,266	0	174,231	2,473
Honey	Pound	0	5,884	0	5,884	0
Denatured alcohol	Gallon	0	0	0	0	0
Sugar, cane and beet	Pound	556,046	0	1,532,629	583,782	1,504,893
Vegetable oil products	Pound	1,279	757,093	0	751,825	6,548
Egg mix, dry	Pound	0	0	0	0	0
Pork Bellies	Pound	0	405	0	405	0
Plants & Seeds	Pound	0	1	0	1	0
Tallow	Pound	0	5,511	0	5,511	0
Other (B)		89	698	0	787	0
Total inventory operations		xxx	xxx	xxx	xxx	xxx
Additional Adjustment for lag activity	0	0	0	0	0	

See footnotes at end of table.

Table 11-4.—Commodity Credit Corporation: Selected inventory transactions, programs and commodity, fiscal year 2000—Continued

Program and commodity	Unit	Value				
		Inventory Oct. 1, 2000	Purchases	Collateral acquired from loans	Sales [1]	Inventory Sept. 30, 2001
		1,000 dollars	*1,000 dollars*	*1,000 dollars*	*1,000 dollars*	*1,000 dollars*
Feed grains:.						
Barley	Bushel	802	820	1,740	2,872	490
Corn	Bushel	69,834	157,034	47,850	230,192	44,526
Corn products	Pound	251	29,595	0	29,563	283
Grain sorghum	Bushel	622	4,612	1,085	6,139	180
Sorghum grits	Pound	0	2,580	0	2,580	0
Oats	Bushel	41	2	53	72	23
Oats, rolled	Pound	0	0	0	0	0
Rye	Bushel	0	0	0	0	0
Total feed grains		71,550	194,644	50,728	271,418	45,503
Wheat (A)	Bushel	399,400	352,056	50,111	397,927	403,640
Wheat flour	Pound	221	78,018	0	78,236	3
Wheat products, other	Pound	0	25,555	0	25,497	58
Rice, milled	Cwt	4	47,708	0	47,712	0
Rice, rough	Cwt	65	335,613	300	335,785	193
Rice, brown	Pound	0	0	0	0	0
Rice, cereal	Pound	0	0	0	0	0
Cotton, extra long staple	Bale	1,552	0	4,332	1,592	4,292
Upland Cotton	Bale	713	1,786,360	9,842	1,790,765	6,150
Dairy products:						
Butter	Pound	0	0	0	0	0
Butter oil	Pound	0	0	0	0	0
Cheese	Pound	1,984	15,532	0	11,053	6,463
Milk, dried	Pound	0	0	0	0	0
Milk, UHT	Pound	0	0	0	0	0
Dry Whole Milk	Pound	0	1,306	0	1,306	0
Non fat dry milk	Pound	560,172	415,760	0	115,561	860,371
Total dairy products		562,156	432,598	0	127,920	866,834
Oils and oilseeds:						
Canola seed	Cwt.	7	928	146	1,079	3
Mustard seed	Cwt.	0	0	3	0	3
Sunflower seed	Cwt.	51	8,736	1,476	10,254	10
Sunflower seed, non-oil	Cwt.	0	572	26	597	(0)
Sunflower seed oil, processed	Cwt.	0	4,735	0	4,735	0
Peanuts, farmers' stock	Pound	0	0	0	0	0
Peanut products	Pound	0	0	0	0	0
Peanut butter	Pound	0	0	0	0	0
Soybeans	Bushel	48,380	176,287	35,426	244,399	15,694
Soybean meal	Pound	0	50,370	0	50,370	0
Flaxseed	Cwt.	35	344	336	711	4
Totals oils and oilseeds		48,474	241,973	37,413	312,145	15,714
Blended foods	Pound	1,231	68,978	0	67,959	2,250
Grains and seeds:						
Feed for Government facilities	Cwt	0	0	0	0	0
Foundation seeds	Pound	0	0	0	0	0
Total grains and seeds		0	0	0	0	0
Peas, dry whole	Pound	178	23,531	0	23,357	352
Honey	Pound	0	3,824	0	3,824	0
Denatured alcohol	Gallon	0	0	0	0	0
Sugar, cane and beet	Pound	115,306	18	326,086	112,671	328,739
Vegetable oil products	Pound	587	173,578	0	171,999	2,166
Egg mix, dry	Pound	0	0	0	0	0
Pork Bellies	Pound	0	309	0	309	0
Plants & Seeds	Pound	0	1	0	1	0
Tallow	Pound	0	594	0	594	0
Other (B)		2,263	12,773	0	15,036	0
Total inventory operations		1,203,699	3,778,129	478,812	3,784,746	1,675,893
Additional Adjustment for lag activity						
		1,203,699	3,778,129	478,812	3,784,746	1,675,893

[1] Includes sales, commodity donations, transfers to other government agencies and inventory adjustment.
(A) Excludes wheat set aside for Food Security Wheat Reserve (FSWR).
(B) Includes beans, dry edible, and fish, canned salmon.
Table may not add due to rounding.

FSA, Financial Management Division, (703) 305–1277.

Table 11-5.—Commodity Credit Corporation: Cost value of export and domestic commodity dispositions, by type of disposition, fiscal year 2001 [1]

(In Thousands)

Commodity	Domestic				
	Dollar sales	Transfers to other Government agencies	Donations [1]	Inventory adjustments and other recoveries (domestic)	Total domestic
	1,000 dollars	*1,000 dollars*	*1,000 dollars*	*1,000 dollars*	*1,000 dollars*
Feed grains:					
Barley	2,872,119	0	0	0	2,872,119
Corn	192,852,255	210	15,055	0	192,867,520
Corn products	0	0	0	1,127	1,127
Grain sorghum	2,246,646	0	0	(1,012)	2,245,634
Sorghum grits	0	0	0	0	0
Oats	72,017	0	0	0	72,017
Oats, rolled	0	0	0	0	0
Wheat	79,107,108	0	0	12,298	79,119,406
Wheat flour	0	0	0	3,995	3,995
Wheat products, other	0	0	0	66,686	66,686
Rice, milled	0	0	0	0	0
Rice, rough	335,784,510	0	0	0	335,784,510
Rice, brown and textured soy	0	0	0	0	0
Cotton, extra long staple and upland	1,792,356,494	0	0	0	1,792,356,494
Rye	0	0	0	0	0
Dairy products:					
Butter oil	0	0	0	0	0
Butter	0	0	0	0	0
Cheese	749,831	2,035,322	8,048,903	219,307	11,053,361
Nonfat dry milk	4,498,005	0	48,110,074	1,850,693	54,458,772
Milk, dried	0	0	0	0	0
Oils and oilseeds:.					
Peanut products	0	0	0	0	0
Peanut butter	0	0	0	0	0
Peanuts, farmer's stock	0	0	0	0	0
Soya flour	0	0	0	0	0
Flaxseed	711,490	0	0	0	711,490
Sunflower Seed (oil & non-oil)	10,851,154	0	0	0	10,851,154
Soybeans	242,347,775	0	0	0	242,347,776
Emergency food ration bars	0	0	0	0	0
Blended foods	0	0	0	(161)	(161)
Dry Vegetable Burger	0	0	0	0	0
Grains and seeds:.					
Feed for Government facilities	2,331,501	0	0	0	2,331,501
Foundation seeds	0	0	0	0	0
Field Seeds	0	0	0	119,983	119,983
Vegetable Seeds	0	0	0	0	0
Peas, dried whole	0	0	0	3,416	3,416
Dry edible beans	0	0	0	132	132
Honey	3,824,313	0	0	0	3,824,313
Sugar	109,720,122	0	1,979,986	0	111,700,108
Vegetable oil products	0	0	0	5,846	5,846
Meat	0	0	0	0	0
Poultry, Frozen Chicken	0	0	0	0	0
Other	0	0	0	0	0
Total [2]	2,780,325,341	2,035,531	58,154,017	2,282,309	2,842,797,199

See footnotes at end of table.

Table 11-5.—Commodity Credit Corporation: Cost value of export and domestic commodity dispositions, by type of disposition, fiscal year 2001[1]—Continued

(In Thousands)

Commodity	Export			Total export and domestic
	Public law 480 Title II/III	Donations[1]	Total export[3]	
	1,000 dollars	*1,000 dollars*	*1,000 dollars*	*1,000 dollars*
Feed Grains:.				
Barley	0	0	0	2,872,119
Corn	13,118,704	25,291,444	38,410,149	231,277,669
Corn products	21,708,383	8,569,995	30,278,378	30,279,505
Grain sorghum	3,939,129	0	3,939,129	6,184,763
Sorghum grits	2,580,322	0	2,580,322	2,580,322
Oats	0	0	0	72,017
Oats, rolled	0	0	0	0
Wheat	125,887,482	141,477,210	267,364,692	346,484,098
Wheat flour	35,813,083	99,311,275	135,124,358	135,128,354
Wheat products, other	34,989,509	(1,032,416	) 33,957,093	34,023,779
Rice, milled	0	0	0	0
Rice, rough	0	0	0	335,784,510
Rice, brown and textured soy	0	0	0	0
Cotton, extra long staple and upland	0	0	0	1,792,356,494
Rye	0	0	0	0
Dairy products:				
Butter oil	0	0	0	0
Butter	0	0	0	0
Cheese	0	0	0	11,053,361
Nonfat dry milk	13,144,260	47,169,892	60,314,152	114,772,924
Milk, dried	0	1,305,928	1,305,928	1,305,928
Oils and oilseeds:.				
Peanut products	0	0	0	0
Peanut butter	0	0	0	0
Peanuts, farmer's stock	0	0	0	0
Soya flour	0	0	0	0
Flaxseed	0	0	0	711,490
Sunflower Seed (oil & non-oil)	1,468,540	3,266,702	4,735,241	15,586,395
Soybeans	10,033,991	42,355,055	52,389,046	294,736,822
Emergency food ration bars	(12)	0	(12)	(12)
Blended foods	62,851,332	28,550,504	91,401,836	91,401,674
Dry Vegetable Burger	0	0	0	0
Grains and seeds:.				
Feed for Government facilities	0	0	0	2,331,501
Foundation seeds	20,218,270	14,018,208	34,236,478	34,236,478
Field Seeds	(20,217,555)	0	(20,217,555)	(20,097,572)
Vegetable Seeds	0	0	0	0
Peas, dried whole	22,748,231	605,004	23,353,235	23,356,650
Dry edible beans	10,225,314	4,810,526	15,035,840	15,035,972
Honey	0	0	0	3,824,313
Sugar	0	0	0	111,700,108
Vegetable oil products	92,624,601	75,493,342	168,117,943	168,123,790
Meat	309,469	593,644	903,113	903,113
Poultry, Frozen Chicken	0	0	0	0
Other	20,085,107	26,994,678	47,079,785	47,079,785
Total[2]	471,528,162	518,780,989	990,309,150	3,833,106,349

[1] Includes donations under section 202,407,416, Section 210, P.L. 85-540 and miscellaneous donations under various other authorizations. [2] Totals may not aαd due to rounding. [3] Includes Sales, PL 480 Titles I, II and III, Transfers to Other Agencies and Donations.

FSA, Financial Management Division, (703) 305–1277.

Table 11-6.—Commodity Credit Corporation: Investment in price-support operations, by quarters, 1992–2001 [1]

Date	Inventory after revaluation	Loans held CCC	Total loans outstanding	Total investment
	Million dollars	*Million dollars*	*Million dollars*	*Million dollars*
1992:				
March	2,167.0	5,045.0	5,045.0	7,212.0
June	2,047.0	2,944.2	2,944.2	4,991.2
1993:				
March	1,688.2	6,104.7	6,104.7	7,792.9
June	1,606.1	4,564.0	4,564.0	6,170.1
1994:				
March	1,281.3	4,263.2	4,263.2	5,544.5
June	1,211.6	3,048.1	3,048.1	4,259.6
1995:				
March	774.2	7,624.1	7,624.1	8,398.3
June	705.1	4,969.6	4,969.6	5,674.7
1996:				
March	649.6	3,715.1	3,715.1	4,364.7
June	574.6	2,177.9	2,177.9	2,752.5
1997:				
March	436.0	3,753.0	3,753.0	4,189.0
June	406.0	2,154.0	2,154.0	2,560.0
1998:				
March	386.1	5,058.9	5,058.9	5,445.0
June	436.1	3,376.7	3,376.7	3,812.8
1999:				
March	384.3	6,231.0	6,366.5	6,615.3
June	480.7	5,022.0	5,157.5	5,502.7
2000:				
March	500.8	5,160.5	5,160.5	5,661.3
June	650.7	4,545.3	4,739.6	5,196.0
2001:				
March	1,635.1	5,627.9	5,627.9	7,263.0
June	2,299.0	3,663.3	3,663.3	5,962.3

[1] Reflects total CCC loans and inventories.

FSA, Financial Management Division, (703) 305–1277.

Table 11-7.—Farm Service Agency programs: Payments to producers, by program and commodity, United States, calendar years 1997–2001

Program and commodity	1997	1998	1999	2000	2001
	1,000 dollars	*1,000 dollars*	*1,000 dollars*	*1,000 dollars*	*1,000 dollars*
Production flexibility	6,119,814	6,000,580	5,046,071	5,048,750	4,040,639
Agricultural Management Assist	......	......	......	......	1,376
Quality Losses	......	......	......	......	52,478
Supl Oilseed Payment Program	......	......	......	......	422,392
WAMLAP III - Apportioned	......	......	......	......	16,442
Peanut Marketing Asst Pgm III	......	......	......	......	53,911
Supplemental Tobacco Loss	......	......	......	......	128,259
Wamlap II - Apportioned	......	......	......	......	18,671
AMLAP - Apportioned	......	......	......	......	95,079
Bioenergy	......	......	......	......	5,294
Citrus Losses in California	......	......	......	......	2,154
Poultry Enteritis Sydnrome	......	......	......	......	1,788
Acreage Grazing Payments	......	......	......	......	6,245
Nursery Losses - Florida	......	......	......	......	6,581
AILFP -- Apportioned	......	......	......	......	6,020
Crop Disaster Program	......	......	......	......	1,841,870
Cottonseed Payment Program	......	......	......	77,627	81,385
Sugar PIK Diversion	......	......	......	105,649	44,193
Wool and mohair	14	90	(4)	7,424	2,838
Peanut Marketing Assistance	......	......	......	94,089	24,904
Oilseed Program	......	......	......	459,985	498,413
Loan deficiency [1]	(574,637)	1,777,895	5,893,142	6,480,751	5,703,964
Emergency feed/livestock assistance	17,638	(41)	269,510	189,028	427,071
Lamb Meat Adjustment assistance	......	......	......	12,244	11,207
Tobacco loss/disaster assistance	......	......	......	345,202	1,018
Conservation reserve [2]	1,610,742	1,429,561	1,452,520	1,551,429	1,769,997
Pasture Recovery Program	......	......	......	27,007	26,365
Agricultural conservation	44,453	17,527	9,009	2,846	1,393
Emergency conservation	33,726	24,547	40,394	60,627	31,183
Environ. quality incentives program	44,662	76,621	93,274	95,516	97,079
American Indian livestock assist. pro	......	......	6,235	5,334	885
Options pilot program	(3)	......	......	43	(172)
Potato diversion program	5,220	......	......	......	11,327
Market gains	(34)	171,127	884,649	1,127,283	707,909
Noninsured assistance program (NAP) [3]	43,382	22,661	53,283	39,574	55,571
Karnal bunt fungus	8,614	8,760	1,200	1,373	2,727
Marketing Loss Assistance	......	2,809,144	5,464,805	5,462,333	4,644,030
Dairy market loss assistance	......	......	200,085	670,788	123,668
Other [4]	103,319	19,080	2,115,066	1,222,545	24,687
Grand Total	7,456,910	12,357,552	21,529,239	23,087,447	20,990,842

[1] Includes Crop Special Grade Rice LDP, Rice Deficiency, Cotton Deficiency, Feed Grain Deficiency, Wheat Deficiency, Loan Deficiency, and LDP - Non-Contract. [2] Includes CRP Cost-Shares, CRP Incentives, and CRP Annual Rent. [3] Includes Nap-Supplemental and Noninsured Assistance Program. [4] Includes Wetlands Reserve Program, Soil and Water Conservation Program, Settlement Payments, Dairy Indemnity Program, Disaster and Disaster Reserve Refunds, Dairy Disaster Refunds, Small Hog Operation Program, Crop Loss Disaster Payments, Interest Payments, LIP Contract Growers, Flood Compensation, National Wool Act, Finanlity Rule, and Payment limitation refund.

FSA Budget/Corporate Programs Branch, (202) 720–5148.

Table 11-8.—Farm Service Agency programs: Payments received, by States, 1999–2001

State	Payments		
	1999	2000	2001
AL	178,825	170,399	230,126
AK	1,756	1,669	2,173
AZ	108,080	107,169	99,254
AR	808,137	899,901	832,135
CA	663,873	671,494	586,699
CO	373,653	357,587	319,271
CT	8,708	10,994	7,540
DE	19,795	25,908	24,963
FL	76,390	58,844	107,311
GA	361,746	389,170	426,534
HI	824	939	3,860
ID	211,179	282,633	207,636
IL	1,800,700	1,944,781	1,849,734
IN	852,891	938,907	925,249
IA	2,061,918	2,302,163	1,971,615
KS	1,402,069	1,309,667	1,068,601
KY	232,028	448,329	293,367
LA	418,987	451,117	434,012
ME	11,656	13,829	7,794
MD	68,275	88,415	86,543
MA	10,163	10,974	10,129
MI	401,856	392,311	352,730
MN	1,411,338	1,537,407	1,242,073
MS	440,187	462,509	516,314
MO	717,167	869,595	817,027
MT	492,065	492,296	475,972
NE	1,412,514	1,410,502	1,297,564
NV	2,676	3,823	5,860
NH	3,921	4,744	2,774
NJ	10,254	22,491	16,399
NM	92,209	79,403	93,560
NY	120,206	159,977	114,009
NC	287,833	446,588	330,312
MD	975,752	1,183,748	944,546
OH	650,898	678,514	681,519
OK	531,995	439,780	391,712
OR	105,161	139,406	104,725
PA	95,670	147,944	103,435
RI	877	1,218	292
SC	127,478	143,970	129,742
SD	791,450	790,036	714,936
TN	226,550	298,917	247,454
TX	1,926,582	1,646,642	1,702,477
UT	30,379	36,184	39,689
VT	12,239	26,150	7,863
VA	100,712	152,096	116,801
WA	270,363	353,732	298,547
WV	11,218	23,487	9,807
WI	503,168	603,736	414,981
WY	39,831	37,217	50,171
PR	64,741	17,996	10,160
VI	73	89	97
GU	211	6	0
MI	8	44	3
AS	4	0	0
Undistributed			262,745
Total [1]	21,529,239	23,087,447	20,990,842

[1] Total may not add due to rounding.

FSA, Budget, Corporate Programs Branch, (202) 720–5148.

Table 11-9.—Commodity Credit Corporation: Loans made in fiscal year 2001 for crop year 2000, by States and Territories [1]

State or Territory	Barley	Corn	Cotton	Flaxseed	Honey	Oats	Oilseeds	Peanuts
	1,000 dollars	*1,000 dollars*	*1,000 dollars*	*1,000 dollars*	*1,000 dollars*	*1,000 dollars*	*1,000 dollars*	*1,000 dollars*
Alabama	0	1,232	37,439	0	120	4	0	215
Alaska	0	0	0	0	0	0	0	0
Arizona	396	1,214	9,860	0	370	0	0	17
Arkansas	0	3,722	76,829	0	1,353	0	0	0
California	105	458	379,191	0	3,628	0	0	0
Colorado	1,008	19,512	0	0	276	15	439	0
Connecticut	0	0	0	0	0	0	0	0
Delaware	111	3,583	0	0	0	0	0	0
Florida	0	138	1,470	0	764	0	0	739
Georgia	0	3,909	18,971	0	30	43	0	639
Hawaii	0	0	0	0	0	0	0	0
Idaho	4,308	175	0	22	1,034	5	16	0
Illinois	3	276,889	0	0	0	3	0	0
Indiana	13	181,345	0	0	0	0	0	0
Iowa	5	598,677	0	0	886	49	0	0
Kansas	0	33,758	0	0	160	0	281	0
KCMO Commodity OPS	0	0	434	0	0	0	0	0
Kentucky	56	23,796	0	0	0	0	0	0
Louisiana	0	1,169	75,629	0	139	0	0	0
Maine	17	0	0	0	0	68	0	0
Maryland	155	9,009	0	0	0	0	0	0
Massachusetts	0	63	0	0	0	0	0	0
Michigan	0	58,877	0	0	671	40	11	0
Minnesota	1,410	464,547	0	37	752	161	2,509	0
Mississippi	0	2,460	792,834	0	129	0	0	0
Missouri	9	56,162	18,264	0	41	0	0	0
Montana	1,687	0	0	21	1,346	9	261	0
Nebraska	26	253,034	0	0	1,004	44	194	0
Nevada	0	0	0	0	0	0	0	0
New Jersey	0	1,929	0	0	76	26	0	0
New Mexico	0	974	630	0	15	0	0	370
New York	47	15,161	0	0	514	133	0	0
North Carolina	72	12,139	73,046	0	30	11	0	3
North Dakota	4,773	34,809	0	664	4,509	218	11,442	0
Ohio	0	90,129	0	0	0	31	18	0
Oklahoma	10	2,873	671	0	2	0	20	0
Oregon	244	417	0	0	595	0	0	0
Peanut Associations	0	0	0	0	0	0	0	59,561
Pennsylvania	38	13,089	0	0	34	71	14	0
South Carolina	4	5,820	1,659	0	9	19	0	0
South Dakota	333	147,622	0	213	8,307	633	14,494	0
Tennessee	103	8,254	148,511	0	0	0	0	0
Texas	0	32,456	505,565	0	1,845	42	0	0
Utah	75	157	0	0	236	1	0	0
Vermont	0	29	0	0	9	0	0	0
Virginia	331	10,405	807	0	16	0	0	1,691
Washington	1,568	913	0	9	712	29	59	0
West Virginia	0	927	0	0	0	0	0	0
Wisconsin	18	78,586	0	0	1,257	8	0	0
Wyoming	0	429	0	0	0	2	3	0
Total	16,925	2,450,847	2,141,810	966	30,869	1,665	29,761	63,235

See footnotes at end of table.

Table 11-9.—Commodity Credit Corporation: Loans made in fiscal year 2001 for crop year 2000, by States and Territories[1]—Continued

State or Territory	Rice	Seed Cotton	Sorghum	Soybeans	Sugar	Tobacco	Wheat	Mohair	Total
	1,000 dollars	*1,000 dollars*	*1,000 dollars*	*1,000 dollars*	*1,000 dollars*	*1,000 dollars*	*1,000 dollars*	*1,000 dollars*	*1,000 dollars*
Alabama	0	0	0	319	0	0	150	0	39,479
Alaska	0	0	0	0	0	0	0	0	0
Arizona	0	0	0	0	0	0	243	0	12,100
Arkansas	406,359	0	14	45,241	0	0	9,393	0	136,552
California	176,526	0	0	0	0	0	3,774	0	387,156
Colorado	0	0	345	93	33,812	0	13,892	0	69,392
Connecticut	0	0	0	0	0	0	0	0	0
Delaware	0	0	0	3,561	0	0	233	0	7,488
Florida	0	0	0	0	111,834	0	0	0	114,945
Georgia	0	269	9	356	0	0	368	0	24,594
Hawaii	0	0	0	0	0	0	0	0	0
Idaho	0	0	0	0	0	0	29,550	0	35,110
Illinois	0	0	837	186,776	0	0	3,349	0	467,857
Indiana	0	0	15	152,001	0	0	2,285	0	335,659
Iowa	0	0	0	355,773	0	0	48	8	955,446
Kansas	0	0	6,922	11,766	0	0	72,093	0	124,980
KCMO Commodity OPS	0	0	0	0	0	(2,803,230)	0	0	0
Kentucky	13	0	14	15,685	0	1,108,858	2,895	0	1,151,304
Louisiana	13,509	0	63	2,549	71,979	0	12	0	151,540
Maine	0	0	0	0	0	0	0	0	85
Maryland	0	0	0	8,025	0	0	680	0	17,869
Massachusetts	0	0	0	0	0	0	0	0	63
Michigan	0	0	0	33,158	116,531	0	2,440	0	211,728
Minnesota	0	0	0	270,479	84,360	0	24,930	0	849,185
Mississippi	31,884	0	0	4,978	0	0	96	0	800,497
Missouri	11,009	0	1,962	92,129	0	0	3,449	10	172,026
Montana	0	0	0	0	0	0	17,648	0	20,972
Nebraska	0	0	3,550	84,149	0	0	9,893	1	351,895
Nevada	0	0	0	0	0	0	0	0	0
New Jersey	0	0	0	1,324	0	0	59	0	3,414
New Mexico	0	0	47	5	0	0	2,710	0	4,751
New York	0	0	0	4,162	0	0	3,936	0	23,953
North Carolina	0	847	2	19,298	0	1,478,457	1,160	0	1,585,065
North Dakota	0	0	29	29,360	11,628	0	71,013	0	168,445
Ohio	0	0	31	92,375	0	0	2,019	0	184,603
Oklahoma	0	0	166	197	0	0	35,277	0	39,216
Oregon	0	0	0	0	0	0	6,393	0	7,649
Peanut Associations	0	0	0	0	0	0	0	0	0
Pennsylvania	0	0	7	6,614	0	0	525	0	20,392
South Carolina	0	0	60	6,515	0	0	2,942	0	17,028
South Dakota	0	0	538	116,757	0	0	32,133	19	321,049
Tennessee	53	0	54	7,957	0	327,693	1,463	0	494,035
Texas	8,724	0	5,141	511	15,005	0	20,619	867	582,051
Utah	0	0	0	0	258,446	0	1,820	0	260,735
Vermont	0	0	0	0	0	0	0	0	38
Virginia	0	234	7	10,984	0	989	1,437	0	26,901
Washington	0	0	0	0	31,962	0	42,183	0	77,435
West Virginia	0	0	0	666	0	0	167	0	1,760
Wisconsin	0	0	0	22,830	0	1,924	474	0	105,097
Wyoming	0	0	0	0	0	0	387	0	821
Total	648,077	1,350	19,813	1,586,593	735,557	114,691	424,138	905	7,619,125

[1] Loans made directly by Commodity Credit Corporation. As far as possible, loans have been distributed according to the location of producers receiving the loans. Direct loans to cooperative associations for the benefit of members have been distributed according to the location of the association.

FSA, Financial Management Division, (703) 305–1277.

Table 11-10.—Fruit, vegetable, and tree nut marketing agreement and order programs, 2001–2002

Program	Estimated number of commercial producers	Farm value
	Number	*1,000 dollars*
Citrus fruits (2001-02 season):		
Florida oranges, grapefruit, tangerines, and tangelos	11,000	269,808
Florida limes	54	1,688
Texas oranges and grapefruit	410	29,026
Deciduous fruits (2001 season):		
California fresh pears and peaches [1]	1,800	143,942
California nectarines	1,800	127,642
California olives	1,200	90,096
California desert grapes	80	135,371
California kiwifruit	360	15,340
Florida avocados	203	15,548
Washington apricots	200	4,072
Washington sweet cherries	1,500	132,720
Washington, Oregon, and California winter pears	1,700	128,702
Tart cherries (7 States) [2]	900	56,950
Washington and Oregon Bartlett pears	1,600	33,805
Washington and Oregon fresh prunes [3]	100	2,425
Cranberries (10 States) [4]	1,200	99,455
Hawaiian papayas	400	14,508
Dried fruits (2001 season):		
California dates	100	27,777
California dried prunes	1,250	101,250
California raisins	4,500	228,003
Vegetables (2001-02 season):		
Florida tomatoes	100	427,531
Idaho and Eastern Oregon onions	280	85,300
South Texas melons	33	83,328
South Texas onions	78	85,378
Georgia onions (Vidalia)	144	49,393
Walla Walla onions	130	5,047
Potatoes (2001-02 season):		
Colorado	261	116,705
Idaho and Eastern Oregon	950	221,553
Oregon and Northern California	238	18,577
Southeastern States (Virginia-North Carolina)	90	8,466
Washington	380	74,233
Nuts (2001 season):		
California almonds	6,000	731,880
California walnuts	5,000	341,600
Oregon and Washington hazelnuts	800	34,700
Peanuts [5]	25,000	1,003,099
Spearmint oil (2001 season) [6]	179	15,194
(Total 36 programs) [7]	....	4,960,112

[1] Value of non-Bartlett pears is not available. [2] The tart cherry order covers the States of Michigan, New York, Pennsylvania, Oregon, Utah, Washington, and Wisconsin. [3] Farm value is available only for fresh and processed combined. [4] Massachusetts, Rhode Island, Connecticut, New Jersey, Wisconsin, Michigan, Minnesota, Oregon, Washington, and Long Island in New York. (Only top 5 are reported). [5] There is only a marketing agreement (No. 146) for peanuts. It covers the States of Alabama, Florida, Georgia, Mississippi, South Carolina, Arizona, Arkansas, California, Louisiana, New Mexico, Oklahoma, Texas, Missouri, North Carolina, Tennessee, and Virginia. The marketing agreement covers about 95 percent of all peanuts; figures here are for all U.S. peanuts. Non-signers to the Agreement must meet the same quality and inspection requirements that apply under the Agreement 146. [6] The marketing order regulates the handling of spearmint oil produced in the States of Washington, Idaho, Montana, Nevada, Utah, Oregon, and California. The farm value is the sum of values for Idaho, Oregon, and Washington, the only significant producing States in the marketing order area. [7] Total number of producers cannot be determined from totals for individual commodities; some producers produce more than one commodity.

AMS, Fruit and Vegetable Programs, (202) 720–2615.

CHAPTER XII

AGRICULTURAL CONSERVATION AND FORESTRY STATISTICS

Statistics in this chapter concern conservation of various natural resources, particularly soil, water, timber, wetlands, wildlife, and improvement of water quality. Forestry statistics include area of private and public-owned forest land, timber production, imports and exports, pulpwood consumption and paper and board production, area burned over by forest fires, livestock grazing, and recreational use of national forest lands.

Conservation Practices on Active Contracts

Practice code	Practice	Acres
CP01	Introduced grasses and legumes	4,229,575
CP02	Native grasses	6,225,279
CP03	Tree planting	1,154,746
CP04	Wildlife habitat with woody veg.	2,282,315
CP05 *	Field windbreaks	50,417
CP06	Diversions	887
CP07	Erosion control structures	763
CP08 *	Grass waterways	80,822
CP09 *	Shallow water areas for wildlife	44,044
CP10 [1]	Existing grasses and legumes	14,941,153
CP11	Existing trees	1,036,530
CP12	Wildlife food plots	68,880
CP13	Vegetative filter strips	31,472
CP15 *	Contour grass strips	63,114
CP16 *	Shelterbelts	22,556
CP17 *	Living snow fences	2,860
CP18 *	Salinity reducing vegetation	278,590
CP19	Alley cropping	52
CP20	Alternative perennials	15
CP21 *	Filter strips (grass)	847,555
CP22 *	Riparian buffers (trees)	485,171
CP23	Wetland restoration	1,660,078
CP24 *	Cross wind trap strips	521
CP25	Rare and declining habitat	388,280
CP27	Farmable wetland pilot (wetland)	21,781
CP28	Farmable wetland (upland)	53,730
....	Total	33,971,186

[1] Includes both introduced grasses and legumes and native grasses. * Continuous non-CREP signup acreage includes practices and acreage in designated wellhead protection areas (CP1, CP2, CP3, CP4, or CP10). CREP acreage may include general or continuous signup practice acreage.

CRP enrollment: By sign up and initial contract year [1], as of January 2003

Sign up	Before 1993	1993	1996	1997	1998	1999	2000	2001	2002	2003	Total
1-11	841,539	0	0	0	0	0	0	0	0	0	841,539
12	0	861,360	0	0	0	0	0	0	0	0	861,360
13	0	0	435,113	159,333	0	0	0	0	0	0	594,446
14	0	0	0	99,105	460,885	0	0	0	0	0	559,989
15	0	0	0	0	16,166,758	355,172	0	0	0	0	16,521,930
16	0	0	0	0	1,771,181	4,070,173	0	0	0	0	5,841,354
17	0	0	0	0	112,874	103,943	0	0	0	0	216,817
18	0	0	0	0	0	0	4,742,246	0	0	0	4,742,246
19	0	0	0	0	0	135,244	131,014	0	0	0	266,258
20	0	0	0	0	0	0	0	2,247,986	0	0	2,247,986
21	0	0	0	0	0	0	106,260	12,730	0	0	118,990
22	0	0	0	0	0	0	33,424	171,917	0	0	205,341
23	0	0	0	0	0	0	0	221,068	248,834	0	469,902
24	0	0	0	0	0	0	0	0	289,079	134,546	423,625
25	0	0	0	0	0	0	0	0	0	59,292	59,292
All	841,539	861,360	435,113	258,438	18,511,697	4,664,532	5,012,943	2,653,702	537,912	193,838	33,971,075

[1] For CRP, contract year is the same as fiscal year, which begins October 1. Note: There are no contracts with initial contract years 1994 or 1995. Note: General Signup Numbers: 1-13, 15, 16, 18, 20. Continuous Signup Numbers: 14, 17, 19, 21-25.
FSA, Conservation and Environmental Protection Division, (202) 720–6675.

Table 12-1.—Conservation Reserve Program (CRP): Enrollment by practice, under contract, January 2003

(CP 1 and CP 2)

State	CP 1 Establishment of permanent introduced grasses and legumes			CP 2 Establishment of permanent native grasses		
	Total acres treated	Total cost share	Cost share per acre treated[1]	Total acres treated	Total cost share	Cost share per acre treated[1]
Alabama	15,684.2	844,193	58.61	3,625.5	243,546	82.20
Alaska	5,746.4	438,254	76.27	0.0	0	....
Arizona	0.0	0	....	32.8	0	....
Arkansas	5,574.6	241,476	44.98	1,322.1	131,067	108.95
California	4,918.4	137,030	32.68	417.2	28,128	111.53
Colorado	56,591.1	2,172,756	44.36	550,314.3	30,206,818	58.06
Connecticut	70.3	7,091	178.61	34.3	3,630	105.83
Delaware	27.8	2,780	100.00	0.0	0	....
Florida	1,010.9	46,101	46.04	150.5	335	67.00
Georgia	2,613.1	166,296	63.85	368.0	31,528	85.67
Hawaii	0.0	0	....	0.0	0	....
Idaho	137,184.4	4,435,951	33.41	8,225.9	351,428	65.69
Illinois	243,385.0	9,384,276	43.91	32,237.5	2,102,019	83.01
Indiana	79,264.8	4,626,858	61.07	23,332.1	1,465,903	74.92
Iowa	398,457.5	19,301,840	65.00	137,599.5	7,777,907	68.39
Kansas	27,426.1	764,870	41.06	755,907.0	19,078,354	34.95
Kentucky	89,850.5	5,970,326	69.72	28,496.9	2,582,587	98.33
Louisiana	1,395.8	55,785	38.06	1,916.0	81,741	43.96
Maine	1,806.8	244,444	139.65	112.4	13,681	123.92
Maryland	11,146.7	1,244,579	153.59	3,107.9	606,167	221.69
Massachusetts	0.0	0	....	0.0	0	....
Michigan	89,355.0	3,183,364	37.04	12,650.0	1,145,219	109.67
Minnesota	281,203.1	12,258,336	44.86	134,962.1	9,467,512	73.49
Mississippi	15,924.7	641,838	40.87	371.9	29,362	85.01
Missouri	456,596.6	17,727,160	44.09	188,037.9	11,183,675	61.80
Montana	750,667.4	15,481,674	21.14	822,763.9	28,004,691	35.30
Nebraska	68,322.3	1,433,286	22.84	335,487.7	11,267,742	40.15
Nevada	0.0	0	....	0.0	0	....
New Hampshire	10.0	1,200	120.00	0.0	0	....
New Jersey	1,150.4	188,643	164.40	393.3	76,132	193.57
New Mexico	970.1	28,604	29.49	182,845.2	7,611,149	43.00
New York	6,731.3	648,388	102.33	937.1	91,677	105.56
North Carolina	4,973.7	361,102	76.61	1,462.5	112,028	89.86
North Dakota	428,973.4	8,190,022	19.86	65,032.7	3,658,739	59.57
Ohio	93,368.8	5,015,479	51.31	24,110.5	1,708,243	81.08
Oklahoma	34,186.9	1,237,112	39.26	381,277.4	17,637,497	46.47
Oregon	105,759.7	2,770,533	29.65	31,131.4	1,341,976	50.70
Pennsylvania	41,765.0	5,336,280	130.23	15,610.3	2,493,905	164.36
Puerto Rico	108.0	17,550	162.50	0.0	0	....
South Carolina	1,133.2	78,043	73.99	107.2	9,804	108.09
South Dakota	182,462.5	5,261,963	28.88	221,640.7	9,661,549	44.28
Tennessee	48,655.9	2,789,730	59.01	17,457.0	1,313,342	80.44
Texas	179,783.1	6,572,449	38.82	1,680,856.3	78,874,584	50.23
Utah	59,421.7	1,925,466	35.26	14,058.0	624,431	45.94
Vermont	6.0	750	125.00	0.0	0	....
Virginia	6,089.0	335,548	56.10	2,419.5	174,719	80.89
Washington	162,143.5	7,589,031	50.77	481,106.0	45,220,237	96.06
West Virginia	10.2	500	49.02	22.4	1,053	47.01
Wisconsin	70,562.9	3,347,366	50.91	55,717.3	4,265,006	80.32
Wyoming	57,086.3	1,835,844	33.11	7,620.7	258,896	33.99
United States, total	4,229,575.1	154,342,167	39.45	6,225,278.9	300,938,007	52.87

[1] Not including acres which receive no cost share. Note: Total acres treated may not add due to rounding.

FSA, Conservation and Environmental Protection Division, (202) 720–6675.

Table 12-2.—Conservation Reserve Program (CRP): Enrollment by practice, under contract, January 2003

(CP 3 and CP 4)

State	CP 3 Tree planting			CP 4 Permanent wildlife habitat		
	Total acres treated	Total cost share	Cost share per acre treated [1]	Total acres treated	Total cost share	Cost share per acre treated [1]
Alabama	146,961.2	10,598,447	79.27	10,693.4	42,231	35.95
Alaska	0.0	0	....	11.3	0	....
Arizona	0.0	0	....	0.0	0	....
Arkansas	40,732.6	2,629,234	67.05	3,201.9	137,456	53.06
California	67.0	2,350	335.71	896.5	97,789	507.20
Colorado	142.1	125,309	886.20	219,792.1	14,173,262	69.97
Connecticut	0.0	0	....	0.0	0	....
Delaware	2,582.3	874,363	345.68	2,093.3	449,702	230.83
Florida	27,756.6	1,140,268	44.47	3,398.1	88,296	57.98
Georgia	167,411.1	13,292,985	80.34	7,497.3	236,936	61.19
Hawaii	0.0	0	....	0.0	0	....
Idaho	3,525.6	437,497	130.55	128,962.2	3,698,032	29.26
Illinois	46,919.0	3,949,124	96.30	119,273.0	6,489,608	83.54
Indiana	19,349.9	2,936,115	160.36	11,301.9	952,559	111.49
Iowa	16,976.5	3,367,283	211.71	316,294.4	13,758,399	77.70
Kansas	900.5	103,383	120.62	16,102.5	632,415	49.17
Kentucky	5,510.6	540,658	102.61	726.7	57,025	115.20
Louisiana	114,809.2	8,897,640	78.59	1,713.6	110,853	77.10
Maine	239.7	29,470	131.21	923.7	35,463	100.18
Maryland	1,273.9	227,041	224.24	1,979.8	352,961	235.17
Massachusetts	0.0	0	....	0.0	0	....
Michigan	12,314.3	1,256,347	103.73	25,316.4	1,188,658	64.63
Minnesota	37,427.9	4,438,622	120.41	349,775.1	23,236,200	69.68
Mississippi	302,630.4	14,067,366	47.93	9,183.3	267,039	68.55
Missouri	24,175.8	2,725,342	126.18	4,792.7	293,424	79.95
Montana	223.7	51,253	234.35	32,355.8	1,225,777	44.35
Nebraska	2,368.2	200,482	115.10	46,078.6	1,523,748	42.30
Nevada	0.0	0	....	151.3	4,100	27.10
New Hampshire	0.0	0	....	0.0	0	....
New Jersey	88.5	16,156	182.55	22.7	5,199	229.03
New Mexico	80.0	2,120	26.50	0.0	0	....
New York	1,689.3	243,174	146.98	489.8	43,781	102.43
North Carolina	18,027.5	849,470	63.01	2,953.1	260,327	105.37
North Dakota	593.5	141,378	212.35	559,570.7	14,240,555	29.45
Ohio	7,977.5	975,990	131.87	21,262.9	846,219	66.60
Oklahoma	626.8	39,570	60.94	3,316.9	192,991	62.15
Oregon	2,074.8	182,306	145.17	13,270.7	737,892	64.14
Pennsylvania	1,141.7	607,741	551.33	869.2	362,359	503.70
Puerto Rico	91.0	13,741	151.00	0.0	0	....
South Carolina	56,439.2	3,110,461	58.42	10,883.1	288,354	63.27
South Dakota	911.9	223,539	272.64	89,676.0	4,704,885	60.82
Tennessee	14,024.8	822,214	64.57	8,865.1	581,770	73.77
Texas	4,456.5	189,126	49.87	34,966.7	1,628,996	53.38
Utah	0.0	0	....	774.4	2,851	3.68
Vermont	0.0	0	....	0.0	0	....
Virginia	4,123.6	187,876	50.21	955.8	63,836	70.86
Washington	1,623.8	350,114	217.09	184,730.6	18,490,499	101.06
West Virginia	134.5	6,980	51.90	0.0	0	....
Wisconsin	66,331.4	9,731,826	148.18	10,362.0	870,619	106.56
Wyoming	11.8	14,755	1,261.11	26,830.0	496,508	38.62
United States, total	1,154,746.2	89,599,116	81.48	2,282,314.6	112,869,574	59.87

[1] Not including acres which receive no cost share. Note: Total acres treated may not add due to rounding.

FSA, Conservation and Environmental Protection Division, (202) 720–6675.

Table 12-3.—Conservation Reserve Program (CRP): Enrollment by practice, under contract, January 2003
(CP 5, CP 6 and CP 7)

State	CP 5 Establishment of field windbreaks			CP 6 Diversions I			CP 7 Erosion control structures		
	Total acres reated	Total cost share	Cost share per acre treated [1]	Total acres treated	Total cost share	Cost share per acre treated [1]	Total acres treated	Total cost share	Cost share per acre treated [1]
AL	0.0	0	....	3.6	4,407	1,224.17	5.0	6,467	1,293.40
AK	0.0	0	....	0.0	0	....	0.0	0	....
AZ	0.0	0	....	0.0	0	....	0.0	0	....
AR	0.0	0	....	0.0	0	....	2.0	729	364.50
CA	0.0	0	....	0.0	0	....	0.0	0	....
CO	880.5	696,623	802.56	0.0	0	....	225.9	2,731	151.72
CT	0.0	0	....	0.0	0	....	0.0	0	....
DE	0.0	0	....	0.0	0	....	0.0	0	....
FL	0.0	0	....	0.0	0	....	0.0	0	....
GA	0.0	0	....	0.0	0	....	7.7	0	....
HI	1.0	800	800.00	0.0	0	....	0.0	0	....
ID	298.9	688,146	2,302.26	0.0	0	....	4.2	7,000	1,666.67
IL	2,020.2	442,368	228.72	21.1	12,589	1,430.57	57.3	81,005	1,579.04
IN	1,815.4	342,496	190.24	0.4	764	1,910.00	10.8	24,116	2,460.82
IA	4,683.3	1,193,847	266.77	10.0	1,500	150.00	19.8	17,430	1,613.89
KS	1,035.9	421,281	447.66	33.7	12,997	397.46	139.4	16,057	729.86
KY	0.5	1,010	2,020.00	0.5	3,000	6,000.00	4.9	10,472	2,137.14
LA	0.0	0	....	5.0	476	95.20	2.0	200	100.00
ME	0.0	0	....	0.0	0	....	0.0	0	....
MD	0.4	1,375	3,437.50	0.1	276	2,760.00	7.9	12,338	1,561.77
MA	0.0	0	....	0.0	0	....	0.0	0	....
MI	1,402.7	375,728	286.51	5.0	6,450	1,290.00	11.2	25,350	2,263.39
MN	6,409.1	1,705,460	277.36	0.0	0	....	0.5	2,500	5,000.00
MS	0.0	0	....	3.0	3,500	1,166.67	1.1	2,175	21,750.00
MO	89.0	13,883	182.43	563.8	59,928	106.29	208.7	86,417	677.78
MT	325.2	140,713	431.64	0.0	0	....	0.0	0	....
NE	15,928.8	6,021,035	494.56	0.0	0	....	9.9	0	....
NV	0.0	0	....	0.0	0	....	0.0	0	....
NH	0.0	0	....	0.0	0	....	0.3	700	2,333.33
NJ	7.7	12,305	1,598.05	4.5	750	166.67	0.0	0	....
NM	0.0	0	....	0.0	0	....	0.0	0	....
NY	10.9	12,455	1,142.66	0.0	0	....	1.0	3,500	3,500.00
NC	12.3	1,882	153.01	0.0	0	....	0.0	0	....
ND	3,267.5	1,379,323	428.06	0.6	143	238.33	0.0	0	....
OH	1,453.8	401,007	294.58	0.0	0	....	1.0	750	750.00
OK	46.9	7,764	213.88	58.7	15,840	276.92	20.0	1,741	87.05
OR	32.4	1,398	43.15	0.0	0	....	0.0	0	....
PA	4.3	430	100.00	7.6	6,428	845.79	1.0	919	919.00
PR	0.0	0	....	0.0	0	....	0.0	0	....
SC	51.4	3,488	77.17	0.0	0	....	0.0	0	....
SD	10,217.6	6,859,078	709.92	0.0	0	....	0.0	0	....
TN	0.0	0	....	1.1	1,250	1,136.36	14.6	25,180	1,724.66
TX	14.6	6,585	451.03	21.0	1,155	55.00	0.0	0	....
UT	3.7	7,974	2,155.14	0.0	0	....	0.0	0	....
VT	5.0	1,010	202.00	0.0	0	....	0.0	0	....
VA	3.0	117	39.00	0.0	0	....	0.0	0	....
WA	11.8	6,703	788.59	0.0	0	....	4.0	2,035	508.75
WV	0.0	0	....	0.0	0	....	0.0	0	....
WI	221.8	39,973	193.57	0.7	1,390	1,985.71	2.3	18,408	8,003.48
WY	161.8	162,357	1,057.01	146.2	0	....	0.0	0	....
US	50,417.4	20,948,614	464.24	886.6	132,843	183.08	762.5	348,220	1,083.45

[1] Not including acres which receive no cost share. Note: Total acres treated may not add due to rounding.

FSA, Conservation and Environmental Protection Division, (202) 720–6675.

Table 12-4.—Conservation Reserve Program (CRP): Enrollment by practice, under contract, January 2003

(CP 8, CP 9 and CP 10)

State	CP 8 Grass waterways			CP 9 Shallow water areas for wildlife			CP 10 Vegetative-cover-grass-already established		
	Total acres treated	Total cost share	Cost share per acre treated [1]	Total acres treated	Total cost share	Cost share per acre treated [1]	Total acres treated	Total cost share	Cost share per acre treated [1]
AL	44.3	11,553	394.30	119.6	114,829	1,003.62	113,818.2	0	....
AK	0.5	14,022	28,044.00	4.6	56,864	12,361.74	23,543.3	0	....
AZ	3.6	2,880	800.00	0.0	0	....	0.0	0	....
AR	9.8	1,505	396.05	723.3	158,300	343.31	27,833.3	0	....
CA	0.0	0	....	1,274.6	60,757	47.67	127,540.2	0	....
CO	876.3	220,799	240.52	49.2	12,853	261.24	1,367,147.8	0	....
CT	0.0	0	....	0.0	0	....	130.5	0	....
DE	2.9	4,728	1,891.20	359.3	1,009,209	2,848.46	30.2	0	....
FL	0.0	0	....	0.0	0	....	1,780.0	0	....
GA	74.5	38,103	548.24	27.8	29,577	1,063.92	7,507.9	0	....
HI	0.0	0	....	0.0	0	....	0.0	0	....
ID	13.3	15,578	1,876.87	89.0	87,093	978.57	499,627.0	0	....
IL	21,725.5	26,463,771	1,289.54	4,535.3	1,991,601	471.51	209,881.3	0	....
IN	11,301.6	30,802,242	2,871.67	1,315.2	770,138	669.57	82,256.4	0	....
IA	21,284.8	19,314,414	1,005.93	14,945.7	3,174,681	243.76	570,835.2	0	....
KS	5,935.1	2,048,893	378.87	585.6	143,115	272.75	1,706,160.3	0	....
KY	2,282.8	2,554,841	1,169.48	2,285.6	1,077,862	509.92	144,087.9	0	....
LA	52.6	54,800	1,091.63	449.9	80,894	254.22	18,427.1	0	....
ME	24.9	182,524	7,636.99	0.0	0	....	19,965.2	0	....
MD	165.2	461,826	3,514.66	1,085.7	1,307,462	1,416.84	4,236.7	0	....
MA	1.0	5	5.00	0.2	0	....	52.8	0	....
MI	597.5	1,354,438	2,439.49	1,517.9	869,725	617.28	106,804.2	0	....
MN	3,453.3	3,638,049	1,122.93	988.5	134,953	161.74	300,018.6	0	....
MS	61.1	5,711	514.50	685.5	79,946	267.20	129,934.3	0	....
MO	1,145.8	1,011,176	986.41	2,447.8	646,839	293.42	771,150.8	0	....
MT	99.4	7,042	89.59	11.2	10,238	914.11	1,509,392.9	0	....
NE	1,295.4	406,902	338.76	165.1	48,455	305.52	572,962.4	0	....
NV	0.0	0	....	0.0	0	....	0.0	0	....
NH	0.0	0	....	0.0	0	....	0.0	0	....
NJ	20.0	140,214	7,010.70	2.8	7,181	2,564.64	407.3	0	....
NM	0.0	0	....	0.0	0	....	406,050.0	0	....
NY	52.3	105,378	2,582.79	84.5	15,661	219.96	39,568.6	0	....
NC	104.1	135,432	1,926.49	2,815.9	1,565,254	665.44	17,453.5	0	....
ND	97.7	41,187	434.00	35.1	730	34.60	1,377,824.7	0	....
OH	4,983.4	13,281,574	2,877.90	858.0	550,089	668.88	92,623.8	0	....
OK	282.7	68,144	248.34	84.1	12,888	377.95	589,358.6	0	....
OR	73.2	41,821	729.86	16.7	8,267	495.03	289,302.7	0	....
PA	350.7	851,969	2,493.88	66.8	119,539	1,822.24	54,893.1	0	....
PR	0.0	0	....	0.0	0	....	351.0	0	....
SC	64.5	118,338	1,872.81	1,910.1	2,813,030	1,486.88	11,770.2	0	....
SD	953.4	390,211	517.83	294.1	221,711	854.05	493,012.2	0	....
TN	94.4	57,965	658.69	83.8	39,129	466.93	129,644.6	0	....
TX	1,657.3	1,004,831	648.15	96.8	39,084	425.75	2,095,095.2	0	....
UT	6.3	252	40.00	0.0	0	....	126,797.8	0	....
VT	1.0	1,595	1,595.00	0.0	0	....	116.2	0	....
VA	27.9	23,920	889.22	75.0	27,232	389.03	13,478.8	0	....
WA	296.8	106,756	418.65	61.4	43,262	710.38	355,534.8	0	....
WV	0.0	0	....	0.0	0	....	675.1	0	....
WI	1,292.8	2,143,622	1,720.95	3,892.4	1,284,114	396.55	348,569.4	0	....
WY	12.6	5,611	445.32	0.0	0	....	183,500.8	0	....
US	80,822.3	107,134,622	1,431.13	44,044.1	18,612,562	481.74	14,941,152.9	0	....

[1] Not including acres which receive no cost share. Note: Total acres treated may not add due to rounding.

FSA, Conservation and Environmental Protection Division, (202) 720–6675.

Table 12-5.—Conservation Reserve Program (CRP): Enrollment by practice, under contract, January 2003
(CP 11, CP 12 and CP 13)

State	CP 11 Vegetative-cover-trees-already established			CP 12 Wildlife food plots			CP 13 Filter strips		
	Total acres treated	Total cost share	Cost share per acre treated [1]	Total acres treated	Total cost share	Cost share per acre treated [1]	Total acres treated	Total cost share	Cost share per acre treated [1]
AL	171,371.9	7,562	23.76	1,603.8	0	....	112.3	8,478	91.06
AK	0.0	0	....	20.4	0	....	0.0	0	....
AZ	0.0	0	....	0.0	0	....	0.0	0	....
AR	53,370.2	0	....	366.2	0	....	122.7	7,416	60.89
CA	49.5	0	....	279.7	0	....	0.0	0	....
CO	230.2	0	....	1,010.7	0	....	96.0	12,729	132.59
CT	0.0	0	....	0.0	0	....	0.0	0	....
DE	56.4	0	....	30.8	0	....	0.0	0	....
FL	52,396.5	1,955	29.05	115.1	0	....	4.8	293	61.04
GA	118,550.7	348	5.01	1,593.7	0	....	881.3	41,782	45.69
HI	0.0	0	....	0.0	0	....	0.0	0	....
ID	2,311.5	0	....	1,056.5	0	....	5.9	2,427	411.36
IL	12,407.3	4,337	91.69	3,753.9	0	....	5,287.2	300,193	54.78
IN	6,190.7	0	....	598.0	0	....	1,717.1	131,493	78.26
IA	5,556.1	125	250.00	5,229.4	0	....	3,814.0	188,648	49.78
KS	1,318.2	1,000	29.94	4,701.5	0	....	1,193.5	56,963	47.54
KY	1,725.9	0	....	942.5	0	....	625.5	31,992	53.58
LA	37,725.5	2,912	4.62	1,215.6	0	....	47.1	1,675	35.56
ME	727.6	0	....	1.6	0	....	0.0	0	....
MD	520.9	0	....	177.9	0	....	957.5	214,432	223.95
MA	0.0	0	....	0.0	0	....	46.9	0	....
MI	5,936.8	3,885	54.34	1,948.8	0	....	886.9	37,240	49.55
MN	18,649.1	14,467	64.90	4,543.0	0	....	8,483.6	309,366	38.55
MS	326,627.2	3,063	10.30	4,951.5	0	....	480.8	20,548	43.36
MO	4,547.3	0	....	2,588.8	0	....	532.4	34,261	64.75
MT	1,077.0	0	....	2,540.5	0	....	26.0	2,566	73.31
NE	2,927.3	0	....	2,328.8	0	....	313.5	8,630	33.44
NV	0.0	0	....	0.0	0	....	0.0	0	....
NH	0.0	0	....	0.0	0	....	0.2	175	875.00
NJ	5.0	0	....	2.0	0	....	9.0	2,385	265.00
NM	79.7	0	....	38.0	0	....	0.0	0	....
NY	1,138.5	0	....	66.6	0	....	141.6	3,731	64.77
NC	39,114.0	489	31.55	52.9	0	....	37.3	6,295	168.77
ND	1,464.2	0	....	3,961.1	0	....	424.9	7,879	19.61
OH	5,012.4	0	....	767.0	0	....	1,025.8	58,201	61.09
OK	442.2	63	6.12	1,285.4	0	....	169.1	7,135	81.73
OR	1,527.6	695	24.30	209.5	0	....	0.0	0	....
PA	539.8	0	....	394.2	0	....	6.5	270	50.00
PR	121.0	0	....	0.0	0	....	0.0	0	....
SC	101,261.1	0	....	1,066.5	0	....	1,739.3	78,481	45.09
SD	1,167.8	0	....	8,338.4	0	....	612.3	35,733	58.47
TN	16,084.1	0	....	193.7	0	....	350.5	16,721	51.24
TX	5,901.3	0	....	6,441.0	0	....	102.4	4,593	55.01
UT	0.0	0	....	32.7	0	....	0.0	0	....
VT	0.0	0	....	0.0	0	....	0.0	0	....
VA	13,854.4	0	....	112.1	0	....	51.5	7,150	138.83
WA	917.9	536	85.08	843.2	0	....	711.5	26,439	45.61
WV	9.0	0	....	0.3	0	....	0.0	0	....
WI	23,543.4	0	....	3,337.5	0	....	455.1	30,348	72.46
WY	72.6	0	....	138.9	0	....	0.0	0	....
US	1,036,529.8	41,437	22.79	68,879.7	0	....	31,472.0	1,696,668	55.81

[1] Not including acres which receive no cost share. Note: Total acres treated may not add due to rounding.
FSA, Conservation and Environmental Protection Division, (202) 720–6675.

Table 12-6.—Conservation Reserve Program (CRP): Enrollment by practice, under contract, January 2003
(CP 14, CP 15 and CP 16)

State	CP 14 Wetland trees			CP 15 Contour grass strips			CP 16 Shelter belts		
	Total acres treated	Total cost share	Cost share per acre treated [1]	Total acres treated	Total cost share	Cost share per acre treated [1]	Total acres treated	Total cost share	Cost share per acre treated [1]
AL	…	…	…	188.3	3,442	80.23	0.0	0	…
AK	…	…	…	0.0	0	…	0.0	0	…
AZ	…	…	…	0.0	0	…	4.3	1,262	293.49
AR	…	…	…	0.0	0	…	0.0	0	…
CA	…	…	…	0.0	0	…	0.0	0	…
CO	…	…	…	444.4	438	55.19	3,012.7	2,759,113	1,073.08
CT	…	…	…	0.0	0	…	0.0	0	…
DE	…	…	…	0.0	0	…	0.0	0	…
FL	…	…	…	0.0	0	…	0.0	0	…
GA	…	…	…	39.1	3,023	125.81	0.0	0	…
HI	…	…	…	0.0	0	…	0.0	0	…
ID	…	…	…	65.9	7,262	126.79	218.3	468,750	2,147.27
IL	…	…	…	1,586.2	74,801	55.57	81.6	20,628	270.35
IN	…	…	…	181.0	12,490	87.33	17.1	3,014	176.26
IA	…	…	…	28,162.6	888,706	51.60	1,543.3	1,022,928	712.69
KS	…	…	…	4,603.6	162,996	53.34	422.9	205,106	527.94
KY	…	…	…	60.6	4,322	98.60	0.0	0	…
LA	…	…	…	0.0	0	…	0.0	0	…
ME	…	…	…	0.0	0	…	0.0	0	…
MD	…	…	…	0.0	0	…	0.0	0	…
MA	…	…	…	0.0	0	…	0.0	0	…
MI	…	…	…	14.9	1,788	166.63	57.7	9,302	220.43
MN	…	…	…	989.6	65,074	72.73	2,550.3	771,679	319.73
MS	…	…	…	36.6	0	…	0.0	0	…
MO	…	…	…	2,009.7	61,356	56.28	34.0	2,428	90.60
MT	…	…	…	0.0	0	…	224.4	108,233	497.17
NE	…	…	…	360.7	13,102	55.32	1,780.6	844,080	496.31
NV	…	…	…	…	…	…	0.0	0	…
NH	…	…	…	0.0	0	…	0.0	0	…
NJ	…	…	…	4.4	1,045	237.50	0.3	175	583.33
NM	…	…	…	0.0	0	…	0.0	0	…
NY	…	…	…	4.0	495	123.75	0.0	0	…
NC	…	…	…	0.0	0	…	13.4	644	67.08
ND	…	…	…	0.0	0	…	2,796.8	1,335,681	486.00
OH	…	…	…	9.9	642	156.45	65.5	13,480	213.63
OK	…	…	…	1.8	0	…	21.1	4,783	226.68
OR	…	…	…	26.8	0	…	2.3	710	887.50
PA	…	…	…	115.0	17,376	174.58	0.0	0	…
PR	…	…	…	…	…	…	0.0	0	…
SC	…	…	…	0.3	37	123.33	0.0	0	…
SD	…	…	…	126.6	7,946	214.51	9,627.7	6,701,834	721.88
TN	…	…	…	26.4	2,340	99.15	0.0	0	…
TX	…	…	…	215.5	11,443	55.58	29.0	10,612	384.49
UT	…	…	…	0.0	0	…	0.0	0	…
VT	…	…	…	0.0	0	…	0.0	0	…
VA	…	…	…	0.0	0	…	0.0	0	…
WA	…	…	…	22,959.1	1,453,540	76.94	8.8	26,447	4,722.68
WV	…	…	…	0.0	0	…	0.0	0	…
WI	…	…	…	879.9	58,106	87.08	19.6	4,946	252.35
WY	…	…	…	0.8	166	…	24.7	26,176	1,059.76
US	…	…	…	63,113.7	2,851,936	67.14	22,556.4	14,342,011	673.23

[1] Not including acres which receive no cost share. Note: Total acres treated may not add due to rounding.
FSA, Conservation and Environmental Protection Division, (202) 720–6675.

Table 12-7.—Conservation Reserve Program (CRP): Enrollment by practice, under contract, January 2003

(CP 17, CP 18 and CP 19)

State	CP 17 Living snow fences			CP 18 Salt tolerant grasses			CP 19 Alley cropping		
	Total acres treated	Total cost share	Cost share per acre treated [1]	Total acres treated	Total cost share	Cost share per acre treated [1]	Total acres treated	Total cost share	Cost share per acre treated [1]
AL	0.0	0	......	0.0	0	......	0.0	0	......
AK	0.0	0	......	0.0	0	......	0.0	0	......
AZ	0.0	0	......	0.0	0	......	0.0	0	......
AR	0.0	0	......	0.0	0	......	0.0	0	......
CA	0.0	0	......	0.0	0	......	0.0	0	......
CO	40.0	43,550	1,088.75	125.6	9,869	78.57	0.0	0	......
CT	0.0	0	......	0.0	0	......	0.0	0	......
DE	0.0	0	......	0.0	0	......	0.0	0	......
FL	0.0	0	......	0.0	0	......	0.0	0	......
GA	0.0	0	......	0.0	0	......	0.0	0	......
HI	0.0	0	......	0.0	0	......	0.0	0	......
ID	65.4	54,900	839.45	0.0	0	......	0.0	0	......
IL	24.4	10,758	440.90	5.7	1,150	294.87	0.0	0	......
IN	1.8	135	75.00	0.5	85	170.00	0.0	0	......
IA	89.8	23,714	273.83	0.0	0	......	0.0	0	......
KS	47.9	21,964	515.59	2,409.7	53,389	22.46	0.0	0	......
KY	0.0	0	......	0.0	0	......	0.0	0	......
LA	0.0	0	......	0.0	0	......	0.0	0	......
ME	0.0	0	......	0.0	0	......	0.0	0	......
MD	0.0	0	......	0.0	0	......	0.0	0	......
MA	0.0	0	......	0.0	0	......	0.0	0	......
MI	2.5	900	360.00	0.0	0	......	0.0	0	......
MN	2,013.1	316,565	161.43	5,784.3	335,266	59.95	0.0	0	......
MS	0.0	0	......	0.0	0	......	0.0	0	......
MO	0.0	0	......	0.0	0	......	0.0	0	......
MT	17.8	11,922	669.78	148,042.7	1,558,523	12.57	0.0	0	......
NE	126.8	43,782	436.08	279.9	7,532	26.91	0.0	0	......
NV	0.0	0	......	0.0	0	......	0.0	0	......
NH	0.0	0	......	0.0	0	......	0.0	0	......
NJ	0.0	0	......	0.0	0	......	0.0	0	......
NM	0.0	0	......	0.0	0	......	0.0	0	......
NY	0.0	0	......	0.0	0	......	0.0	0	......
NC	0.0	0	......	0.0	0	......	0.0	0	......
ND	228.9	102,991	466.23	105,699.7	2,368,106	27.20	0.0	0	......
OH	2.8	400	142.86	2.1	79	37.62	0.0	0	......
OK	3.7	0	......	9,873.6	230,301	24.06	0.0	0	......
OR	0.0	0	......	0.0	0	......	0.0	0	......
PA	0.0	0	......	0.0	0	......	0.0	0	......
PR	0.0	0	......	0.0	0	......	0.0	0	......
SC	0.0	0	......	0.0	0	......	0.0	0	......
SD	167.7	83,506	506.10	4,862.8	215,885	46.34	0.0	0	......
TN	0.0	0	......	0.0	0	......	0.0	0	......
TX	0.0	0	......	1,080.7	54,808	53.60	0.0	0	......
UT	0.0	0	......	0.0	0	......	0.0	0	......
VT	0.0	0	......	0.0	0	......	0.0	0	......
VA	2.5	243	97.20	0.0	0	......	0.0	0	......
WA	2.6	2,268	872.31	402.7	16,977	52.95	0.0	0	......
WV	0.0	0	......	0.0	0	......	0.0	0	......
WI	20.8	4,232	203.46	0.0	0	......	52.1	4,557	87.47
WY	1.0	2,367	2,367.00	19.8	744	37.58	0.0	0	......
US	2,859.5	724,197	262.68	278,589.8	4,852,714	20.66	52.1	4,557	87.47

[1] Not including acres which receive no cost share. Note: Total acres treated may not add due to rounding.

FSA, Conservation and Environmental Protection Division, (202) 720–6675.

Table 12-8.—Conservation Reserve Program (CRP): Enrollment by practice, under contract, January 2003

(CP 20, CP 21 and CP 22)

State	CP 20 Alternative perennials			CP 21 Filter strips			CP 22 Riparian buffer		
	Total acres treated	Total cost share	Cost share per acre treated [1]	Total acres treated	Total cost share	Cost share per acre treated [1]	Total acres treated	Total cost share	Cost share per acre treated [1]
AL	0.0	0	....	601.7	48,019	126.97	18,237.3	2,583,766	146.73
AK	0.0	0	....	7.8	780	100.00	197.9	36,696	185.43
AZ	0.0	0	....	0.0	0	....	0.0	0	....
AR	0.0	0	....	3,878.0	238,338	71.44	14,825.9	1,446,429	107.23
CA	0.0	0	....	0.0	0	....	2,909.8	666,533	233.62
CO	0.0	0	....	321.8	6,687	61.24	709.7	751,797	1,124.43
CT	0.0	0	....	33.9	3,336	98.41	63.1	30,292	480.06
DE	0.0	0	....	1,279.2	320,694	264.41	157.6	57,891	367.33
FL	0.0	0	....	0.0	0	....	67.5	1,505	22.30
GA	0.0	0	....	516.1	17,786	39.18	833.2	239,488	391.58
HI	0.0	0	....	0.0	0	....	19.6	19,936	1,017.14
ID	0.0	0	....	1,053.4	72,784	81.29	5,361.1	2,156,276	402.22
IL	10.0	0	....	127,936.1	6,527,962	53.88	90,018.4	13,542,626	160.11
IN	0.0	0	....	46,370.1	4,717,437	110.54	3,765.0	833,418	244.68
IA	0.0	0	....	213,056.5	11,459,186	61.29	53,281.5	13,484,267	268.65
KS	4.6	0	....	19,409.7	1,021,582	59.98	3,930.7	169,979	47.44
KY	0.0	0	....	27,758.9	3,075,604	131.01	8,580.7	4,763,287	592.45
LA	0.0	0	....	536.9	20,238	43.41	2,079.9	196,915	100.96
ME	0.0	0	....	126.2	29,873	461.72	187.6	411,518	2,193.59
MD	0.0	0	....	28,830.6	4,294,225	169.65	14,973.7	4,781,369	388.22
MA	0.0	0	....	14.6	1,072	73.42	5.0	750	150.00
MI	0.0	0	....	35,909.0	4,251,532	127.10	2,651.1	624,033	252.34
MN	0.0	0	....	135,040.6	8,636,033	67.98	36,368.7	5,710,781	167.87
MS	0.0	0	....	6,276.9	341,069	67.89	67,774.8	3,924,226	64.16
MO	0.0	0	....	37,563.9	2,022,323	63.91	17,974.1	2,980,895	199.39
MT	0.0	0	....	95.1	2,104	29.18	1,924.5	305,500	164.49
NE	0.0	0	....	16,838.8	946,125	61.31	3,000.3	550,457	193.09
NV	0.0	0	....	0.0	0	....	0.0	0	....
NH	0.0	0	....	162.1	2,554	16.53	22.2	23,828	1,309.23
NJ	0.0	0	....	137.7	36,012	261.53	17.8	6,923	388.93
NM	0.0	0	....	0.0	0	....	3,201.3	951,912	297.35
NY	0.0	0	....	359.0	90,175	256.69	8,042.3	4,776,670	659.39
NC	0.0	0	....	6,625.1	1,239,054	190.70	18,366.9	1,497,034	86.72
ND	0.0	0	....	7,083.3	230,699	40.03	486.8	110,024	250.51
OH	0.0	0	....	37,460.3	1,846,811	54.90	3,512.7	710,239	228.02
OK	0.0	0	....	798.7	36,450	52.64	1,028.6	156,934	160.38
OR	0.0	0	....	2,015.4	108,334	62.87	9,794.5	2,894,885	430.18
PA	0.0	0	....	1,270.7	285,325	248.11	5,203.9	7,984,516	1,567.92
PR	0.0	0	....	0.0	0	....	0.0	0	....
SC	0.0	0	....	4,349.3	68,725	36.49	26,785.4	1,276,872	65.31
SD	0.0	0	....	5,320.9	220,909	50.02	2,254.0	1,234,183	569.06
TN	0.0	0	....	7,119.2	512,576	89.54	2,940.5	466,937	165.93
TX	0.0	0	....	1,607.3	290,447	203.95	10,148.2	906,189	100.59
UT	0.0	0	....	12.2	1,162	95.25	75.1	29,947	398.76
VT	0.0	0	....	101.3	20,196	199.37	920.3	637,588	756.96
VA	0.0	0	....	2,208.7	254,759	126.80	12,855.7	13,682,545	1,143.46
WA	0.0	0	....	48,625.7	2,712,676	60.06	14,559.9	15,775,346	1,132.99
WV	0.0	0	....	12.3	1,150	93.50	786.4	195,065	259.12
WI	0.0	0	....	18,820.3	1,523,926	102.57	12,124.2	3,036,973	253.92
WY	0.0	0	....	9.4	1,382	147.02	2,146.0	499,596	233.07
US	14.6	0	....	847,554.7	57,538,111	75.55	485,171.4	117,124,836	265.34

[1] Not including acres which receive no cost share. Note: Total acres treated may not add due to rounding.

FSA, Conservation and Environmental Protection Division, (202) 720–6675.

Table 12-9.—Conservation Reserve Program (CRP): Enrollment by practice, under contract, January 2003
(CP 23, CP 24 and CP 25)

State	CP 23 Wetland restoration			CP 24 Cross wind trap strips			CP 25 Rare and declining habitat		
	Total acres treated	Total cost share	Cost share per acre treated [1]	Total acres treated	Total cost share	Cost share per acre treated [1]	Total acres treated	Total cost share	Cost share per acre treated [1]
AL	72.6	4,279	77.52	0.0	0		509.8	1,754	55.51
AK	0.0	0		0.0	0		0.0	0	
AZ	0.0	0		0.0	0		0.0	0	
AR	13,998.6	583,120	69.50	0.0	0		0.0	0	
CA	5,109.4	125,262	24.56	0.0	0		0.0	0	
CO	982.4	107,531	109.68	109.1	28,009	256.73	0.0	0	
CT	0.0	0		0.0	0		0.0	0	
DE	237.0	196,232	936.67	0.0	0		0.0	0	
FL	0.0	0		0.0	0		0.0	0	
GA	326.7	12,920	88.01	0.0	0		0.0	0	
HI	0.0	0		0.0	0		0.0	0	
ID	1,401.8	46,875	41.55	0.0	0		0.0	0	
IL	42,763.1	4,761,860	141.82	0.0	0		1,632.2	153,606	94.96
IN	6,341.6	474,821	148.61	0.0	0		4.7	0	
IA	16,392.8	1,305,772	142.46	41.3	2,245	102.98	17,867.9	2,479,598	171.80
KS	4,167.4	111,704	51.77	140.8	5,223	45.42	100,027.0	3,948,814	44.63
KY	34.7	1,749	145.75	0.0	0		0.0	0	
LA	23,523.5	512,292	53.69	0.0	0		0.0	0	
ME	0.0	0		0.0	0		0.0	0	
MD	1,779.7	1,135,769	751.82	0.0	0		0.0	0	
MA	0.0	0		0.0	0		0.0	0	
MI	10,584.2	2,387,243	241.06	0.0	0		0.0	0	
MN	305,062.9	19,501,211	69.50	7.2	838	116.39	51,972.6	5,673,330	111.32
MS	11,825.2	373,298	64.85	0.0	0		0.0	0	
MO	3,890.8	163,315	101.80	0.0	0		25,647.6	2,079,509	88.01
MT	5,020.3	298,955	72.02	26.5	110	12.09	131,645.9	5,497,587	42.29
NE	14,903.5	273,150	27.61	42.5	66	36.67	49,023.0	2,861,855	67.17
NV	0.0	0		0.0	0		0.0	0	
NH	0.0	0		0.0	0		0.0	0	
NJ	1.0	1,500	1,500.00	0.0	0		0.0	0	
NM	0.0	0		0.0	0		0.0	0	
NY	50.5	10,925	225.26	0.0	0		0.0	0	
NC	1,257.4	149,171	130.78	0.0	0		0.0	0	
ND	769,641.7	11,561,821	23.60	0.0	0		21.7	1,769	81.52
OH	3,488.6	800,745	311.40	3.5	1,656	473.14	0.0	0	
OK	1,301.6	25,510	37.24	0.0	0		0.0	0	
OR	469.8	66,772	287.19	0.0	0		0.0	0	
PA	228.3	390,425	1,752.36	0.0	0		0.0	0	
PR	0.0	0		0.0	0		0.0	0	
SC	283.6	4,391	18.86	0.0	0		0.0	0	
SD	389,212.7	11,061,447	36.22	14.8	1,116	75.89	1,013.1	74,786	81.93
TN	861.8	12,147	50.03	0.0	0		0.0	0	
TX	8,768.1	300,193	56.61	105.6	1,670	51.23	0.0	0	
UT	0.0	0		0.0	0		0.0	0	
VT	0.0	0		0.0	0		0.0	0	
VA	189.9	171,353	1,235.42	0.0	0		0.0	0	
WA	3,748.6	326,224	90.89	13.6	1,220	89.71	0.0	0	
WV	0.0	0		0.0	0		0.0	0	
WI	12,155.9	944,738	134.98	0.0	0		8,914.9	2,103,531	237.74
WY	0.0	0		16.5	1,764	106.91	0.0	0	
US	1,660,077.7	58,204,720	48.51	521.4	43,917	127.35	388,280.4	24,876,139	69.02

[1] Not including acres which receive no cost share. Note: Total acres treated may not add due to rounding.
FSA, Conservation and Environmental Protection Division, (202) 720–6675.

Table 12-10.—Emergency Conservation Program: Assistance, by State and Caribbean area, fiscal years 1993–2002 [1]

Year	Emergency Conservation Program
	1,000 dollars
1993	23,709
1994	35,825
1995	27,348
1996	26,867
1997	30,847
1998	20,533
1999	40,226
2000	97,970
2001	55,246
2002	32,601

[1] Totals are from unrounded data.

FSA, Conservation and Environmental Protection Division, (202) 720–6675.

Table 12-11.—Forestry Incentives Program: Practices performed, by States and Caribbean area, 2001 and 2002 fiscal year and cumulative 1975–2002 [1]

State	2001			2002			Cumulative 1975–2002		
	Planting trees	Improving a stand of forest trees	Site prep for natural regeneration	Planting trees	Improving a stand of forest trees	Site prep for natural regeneration	Planting trees	Improving a stand of forest trees	Site prep for natural regeneration
	Acres	*Acres*	*Acres*	*Acres*	*Acres*	*Acres*	*Acres*	*Acres*	*Acres*
AL	10,337	1,679	0	6,951	1,325	50	381,319	37,881	1,499
AK	25	0	0	142	0	0	335	44	8,707
AZ	0	0	0	0	0	0	13	1,807	0
AR	8,835	440	0	6,940	258	0	264,882	113,049	607
CA	102	275	94	155	32	0	10,731	12,377	326
CO	10	0	0	10	0	0	407	131	10
CT	0	0	0	0	10	0	12	7,864	0
DE	807	0	0	730	0	0	12,854	4,296	1,297
FL	10,153	0	0	7,455	0	0	359,243	360	233
GA	8,801	0	0	10,019	0	0	407,493	27,584	1,481
HI	0	0	0	2	0	0	274	0	0
ID	190	84	0	260	33	0	3,469	4,164	32
IL	0	763	0	0	341	0	133	42,072	397
IN	10	888	0	96	1,098	0	4,351	90,638	960
IA	0	354	0	0	429	0	3,461	10,846	31
KS	0	175	0	4	77	0	461	5,430	0
KY	306	299	20	20	550	5	5,567	59,455	1,986
LA	3,791	308	0	5,277	406	0	244,274	48,287	6,440
ME	0	199	0	0	105	0	6,878	21,742	10
MD	1,527	178	0	1,773	0	0	38,125	18,779	121
MA	0	282	0	0	480	24	703	38,774	244
MI	299	185	0	283	65	0	31,437	38,638	25
MN	298	124	110	278	0	100	18,827	14,041	1,881
MS	9,884	0	0	13,697	0	0	434,407	31,732	245
MO	10	492	0	30	333	6	12,275	86,576	398
MT	0	87	0	0	94	0	341	5,730	117
NE	0	59	0	7	54	0	576	242	0
NV	0	0	0	0	0	0	526	273	0
NH	4	397	185	39	389	5	315	32,351	1,557
NJ	160	42	0	135	16	0	340	12,348	20
NM	0	0	0	0	0	0	65	7,341	0
NY	10	396	22	21	390	0	112	68,797	142
NC	6,234	4,860	0	3,731	3,540	0	390,797	36,843	1,308
ND	0	0	0	0	0	0	207	141	0
OH	502	1,203	0	96	40	0	16,027	90,330	2,125
OK	1,005	160	56	924	260	0	27,235	37,212	618
OR	788	2,042	0	940	1,141	0	65,412	39,621	0
PA	100	579	0	167	370	0	6,197	41,458	351
PR	10	0	0	44	0	0	1,572	10	0
RI	0	0	0	0	0	3	1,153	2,653	13
SC	4,000	1,200	400	10,473	1,352	0	360,787	31,182	10,353
SD	0	105	18	0	116	0	29	5,948	28
TN	286	416	0	0	432	0	35,973	17,644	425
TX	12,645	1,160	0	5,722	0	0	256,683	59,360	1,243
UT	0	0	0	0	0	0	0	12	0
VT	10	360	30	10	343	0	491	23,389	280
VA	10,378	1,959	0	8,740	1,330	0	450,278	67,027	582
WA	906	1,778	0	660	1,517	0	51,736	24,641	20
WV	38	729	0	141	749	0	7,294	101,892	20
WI	869	127	37	925	290	72	38,920	36,240	5,133
WY	0	844	0	0	407	0	4	22,016	0
Total	93,330	25,228	972	47,063	13,228	186	3,955,001	1,481,268	51,265

[1] In 1974, the program was part of the Rural Environmental Conservation Program (now called the Agricultural Conservation Program). Data for the program year were published in the 1974 Rural Environmental Conservation Program Summary. In 1975 it became a separately funded program.

NRCS, Conservation Operations Division (202) 720–1845.

Table 12-12.—Forestry Incentives Program: Participation and assistance, by States and Caribbean area, 2001 and 2002 fiscal years

State	2001			2002		
	Number of participants	Area served	Cost share paid	Number of participants	Area served	Cost share paid
	Number	*Acres*	*Dollars*	*Number*	*Acres*	*Dollars*
AL	250	12,016	627,300	312	8,326	521,449
AK	2	25	3,478	12	142	53,202
AZ	0	0	0	0	0	0
AR	246	9,275	392,860	131	7,198	407,607
CA	23	471	71,927	16	187	46,675
CO	1	10	3,737	1	10	4,300
CT	0	0	0	1	10	750
DE	23	807	31,655	14	730	34,800
FL	209	10,153	580,176	161	7,455	453,225
GA	335	8,801	697,751	442	10,019	611,745
HI	0	0	0	1	2	1,354
ID	12	274	31,629	21	293	37,070
IL	20	763	29,100	12	341	13,039
IN	20	898	23,730	21	1,194	51,500
IA	12	354	23,031	12	429	17,795
KS	8	175	10,671	6	81	5,754
KY	31	625	29,820	13	575	16,936
LA	87	4,099	180,497	153	5,683	351,975
ME	21	199	14,082	31	105	26,390
MD	57	1,705	65,470	59	1,773	72,332
MA	20	282	11,309	11	504	14,492
MI	22	484	35,373	21	348	29,112
MN	18	532	68,958	17	378	41,187
MS	345	9,884	379,760	453	13,697	282,257
MO	19	502	11,194	18	369	18,027
MT	3	87	4,100	4	94	9,119
NE	6	59	5,479	8	61	5,828
NV	0	0	0	0	0	0
NH	25	586	35,489	29	433	39,288
NJ	7	202	3,877	5	151	8,105
NM	0	0	0	0	0	0
NY	34	428	30,708	39	411	35,933
NC	279	11,094	508,807	184	7,271	335,813
ND	0	0	0	0	0	0
OH	89	1,705	126,383	11	136	13,460
OK	24	1,221	57,390	18	1,184	57,617
OR	99	2,830	259,699	82	2,081	225,455
PA	47	679	29,405	46	537	34,926
PR	1	10	1,920	3	44	16,560
RI	0	0	0	2	3	975
SC	400	5,600	400,000	352	11,825	758,454
SD	9	123	13,737	7	116	9,860
TN	10	702	38,714	5	432	17,389
TX	286	13,805	581,897	264	5,722	557,593
UT	0	0	0	0	0	0
VT	30	400	26,300	25	353	22,735
VA	367	12,337	640,468	222	10,070	244,747
WA	135	2,684	240,416	122	2,177	263,442
WV	58	767	31,813	43	890	33,666
WI	136	1,033	162,779	76	1,287	135,794
WY	15	844	42,500	9	407	27,905
Total	3,841	119,530	6,565,389	3,495	105,534	5,967,637

NRCS,Conservation Operations Division (202)720–1845.

Table 12-13.—Soil conservation districts: Funds appropriated by State and local governments for conservation programs, fiscal year ending Sept. 30, 1998

State or Territory	Actual appropriations
	1,000 dollars
AK	466
AL	8,248
AR	13,840
AZ	2,995
CA	82,565
CO	7,960
CT	4,655
DE	14,095
FL	9,809
GA	5,910
HI	2,458
IA	17,254
ID	10,307
IL	15,148
IN	12,542
KS	6,687
KY	20,908
LA	3,916
MA	9,338
MD	43,313
ME	2,888
MI	15,029
MN	48,119
MO	35,374
MS	5,680
MT	2,231
NC	24,464
ND	5,759
NE	60,929
NH	736
NJ	83,430
NM	3,358
NV	8,793
NY	34,630
OH	27,291
OK	16,888
OR	4,466
PA	70,530
Pacific Basin [1]	353
Caribbean [2]	7,561
RI	2,951
SC	3,845
SD	2,449
TN	6,819
TX	17,195
UT	85,704
VA	11,409
VT	7,971
WA	31,418
WI	40,023
WV	10,867
WY	3,782
Total	975,359

[1] Pacific Basin includes Guam and Northern Marianas. [2] Caribbean includes Puerto Rico and Virgin Islands.

NRCS, Conservation Operations Division, (202) 720-1845.

Table 12-14.—Soil conservation districts: Number organized each year, approximate total area and land in farms, United States and Territories, 1986–95

Year [1]	Number districts organized [2]		Total area in districts	
	Net increase for the year	Cumulative total	New districts and additions	Cumulative total
	Number	*Number*	*1,000 acres*	*1,000 acres*
1986	7	2,947	—24,839	2,194,315
1987	—6	2,941	3,219	2,197,534
1988	15	2,956	—13,837	2,183,697
1989	....	2,956	....	2,183,697
1990	—1	2,955	—3,227	2,180,470
1991	0	2,955	29,692	2,210,162
1992	0	2,955	—64,101	2,146,061
1993	—2	2,953	37,627	2,183,688
1994	0	2,953	0	2,183,688
1995	0	2,953	0	2,183,688

[1] Fiscal year Oct. 1 through Sept. 30. [2] Totals at the end of each year are net, including corrections and adjustments made during the year.

NRCS, Resource Conservation and Community Development Division, (202) 720–4072.

Table 12-15.—Soil conservation districts: Number of districts organized, approximate total area and number of farms, cumulative through Sept. 30, 1995

State or Territory	Districts organized[1]	Approximate area and number of farms within organized districts	
		Total area	Farms (and ranches)
	Number	*1,000 acres*	*Thousands*
Alabama	67	32,454	116
Alaska	10	358,267	1
Arizona	38	73,090	6
Arkansas	75	33,074	52
California	117	64,351	55
Colorado	79	58,774	31
Connecticut	8	398	4
Delaware	3	1,309	8
District of Columbia	1	44	0
Florida	63	32,686	60
Georgia	40	37,261	78
Hawaii	16	4,035	6
Idaho	51	52,416	32
Illinois	98	28,502	89
Indiana	92	23,165	76
Iowa	100	36,016	107
Kansas	105	52,286	134
Kentucky	121	25,396	170
Louisiana	43	26,414	28
Maine	16	18,735	17
Maryland	24	6,636	33
Massachusetts	16	5,065	6
Michigan	83	36,034	59
Minnesota	90	50,762	124
Mississippi	82	29,816	126
Missouri	113	41,446	146
Montana	59	91,141	30
Nebraska	23	47,795	60
Nevada	29	70,759	3
New Hampshire	10	5,955	3
New Jersey	16	4,983	9
New Mexico	47	74,453	27
New York	57	30,488	49
North Carolina	100	33,709	154
North Dakota	60	42,958	63
Ohio	88	26,260	101
Oklahoma	89	44,000	128
Oregon	45	58,730	31
Pennsylvania	66	28,906	52
Rhode Island	3	675	1
South Carolina	46	19,912	26
South Dakota	69	49,353	42
Tennessee	95	25,353	163
Texas	212	165,607	285
Utah	39	52,410	15
Vermont	14	6,186	7
Virginia	45	25,346	52
Washington	47	38,239	42
West Virginia	14	14,439	17
Wisconsin	72	35,938	83
Wyoming	32	59,326	10
US	2,928	2,181,353	3,017
Caribbean	18	2,311	31
Pacific Basin[2]	7	23	6
Totals	2,953	2,183,687	3,054

[1] For specific procedure on organization of soil conservation districts, reference should be made to each of the respective State soil conservation district's laws. [2] Guam, Northern Mariana Island, American Samoa, and Federated States of Micronesia.

NRCS, Resource Conservation and Community Development Division, (202) 720–4072.

Table 12-16.—Soil surveys: Total area mapped for resource planning and other uses under all programs, by States and Territories, year ending Sept. 30, 1998, and cumulative through Sept. 30, 1998

State or Territory	Progressive Soil Survey Program [1]		
	Work during year ending Sept. 30, 1998		Total to date
	Acres mapped	*Acres updated*	*Acres*
AK	1,170,754	165,000	44,549,753
AL	206,693	0	30,039,863
AR	94,101	70,261	33,643,815
AZ	648,640	0	49,082,581
CA	569,437	15,000	76,402,501
CO	371,768	62,560	64,328,486
CT	0	48,000	3,211,700
DE	0	18,500	1,308,500
FL	78,244	0	34,733,118
GA	122,053	193,321	35,196,871
HI	0	140,938	4,042,820
IA	0	338,036	36,016,200
ID	173,866	117,500	36,089,913
IL	0	1,978,902	36,060,840
IN	0	383,071	23,158,600
KS	0	77,454	52,657,500
KY	0	139,092	25,815,272
LA	0	64,267	30,561,000
MA	0	55,105	5,301,800
MD	0	268,698	6,695,000
ME	126,340	54,139	17,090,862
MI	454,927	7,920	35,003,991
MN	428,599	360,381	45,387,084
MO	914,672	1,156	41,735,508
MS	121,359	0	28,554,970
MT	1,264,206	82,626	81,909,846
NC	268,465	87,212	31,126,800
ND	188,922	229,259	44,966,592
NE	59,235	328,325	50,004,936
NH	45,135	37,558	5,413,458
NJ	4,275	189,100	4,865,375
NM	132,570	491,707	69,638,366
NV	193,631	0	60,023,901
NY	100,751	35,720	29,128,330
OH	0	109,349	26,414,944
OK	0	207,828	44,771,700
OR	217,872	243,620	38,632,727
PA	0	305,000	28,997,200
PB [2]	0	0	545,032
PR [3]	0	11,325	2,333,800
RI	0	0	775,900
SC	0	98,757	19,912,100
SD	0	147,990	49,354,542
TN	263,723	238,625	26,317,003
TX	495,448	1,775,225	163,616,542
UT	406,000	77,182	47,816,853
VA	352,363	0	23,751,921
VT	57,060	12,995	5,184,845
WA	240,600	81,900	38,721,119
WI	185,124	429,040	32,915,071
WV	109,701	261,804	14,697,657
WY	415,042	4,744	45,875,357
Total	10,481,576	10,046,192	1,784,380,465

[1] The National Cooperative Soil Survey of the Natural Resources Conservation Service and cooperating agencies is designed to locate, describe, and show on maps the different kinds of soil and make multiple-use interpretations of each kind of soil for town-and-country planning and for other purposes. [2] Pacific Basin (PB) is a new addition. It includes the American Trust Territories of American Samoa, Islands of Aunuu, Ofu, Olosega, Tau, and Tutuila; Federated States of Micronesia Islands of Kosrae, Ponape, Truk and Yap; Territory of Guam; Islands of Airik, Arno, Majuro, Mili, and Taroa, Republic of the Marshall Islands; Islands of Aguijan, Rota, Saipan and Tinian, Commonwealth of the Northern Mariana Islands; and Islands of Palau, Republic of Palau. [3] Puerto Rico (PR) includes acreages for Virgin Islands. This was previously designated as Caribbean.

NRCS, Soil Survey Division, (202) 720–1820.

Table 12-17.—Land protected from erosion by soil conservation practices: Applied through farmer-district cooperation and cooperation with other agencies, with Natural Resources Conservation Service (NRCS) assistance, by States and Territories, year ending Sept. 30, 1995

State or Territory	Cropland	Pasture and hayland	Range native pasture	Woodland	Wildlife	Recreation	Other land	Total land protected
	Acres	*Acres*	*Acres*	*Acres*	*Acres*	*Acres*	*Acres*	*Acres*
AL	16,813	11,368	11,418	97	119	2	0	39,817
AK	5,674	2,175	0	21	240	0	27,687	35,797
AZ	1,268	157	1,157,896	0	0	0	0	1,159,321
AR	171,086	183,690	4,838	36,221	19,282	7	300	415,424
CA	350,443	5,790	242,707	7,018	29,055	217	4,572	639,802
CO	617,111	58,664	460,378	937	12,405	5	5,461	1,154,961
CT	2,735	5	0	0	0	0	0	2,740
DE	12,688	438	0	25	0	0	1	13,152
FL	317,341	337,925	274,838	26,287	4,219	482	1,953	963,045
GA	134,345	17,901	0	18,721	369	0	3,486	174,822
HI	62,436	5,208	58,984	41	0	0	33	126,702
ID	284,170	11,273	17,848	560	30	72	0	313,953
IL	1,622,002	38,204	0	1,901	3,586	5	1,893	1,667,591
IN	405,628	1,197	2,494	957	299	2	5,297	415,874
IA	1,133,191	12,754	117	1,869	15,296	71	4,537	1,167,835
KS	809,085	33,396	199,418	58	1,164	0	524	1,043,645
KY	0	0	0	0	0	0	0	0
LA	275,325	73,096	14,112	97,380	132,344	287	1,915	594,459
ME	770	2,134	0	47	0	0	2	2,953
MD	33,066	894	0	0	0	0	5	33,965
MA	6,073	134	33	0	16	0	2	6,258
MI	170,678	17,095	269	9,536	6,162	326	43,726	247,792
MN	316,831	6,091	611	1,629	7,613	111	3,798	336,684
MS	325,793	125,133	0	79,118	33,731	261	3,305	567,341
MO	214,859	40,405	35	10,556	9,594	1	11,741	287,191
MT	844,064	106,881	899,675	1,005	23,461	370	195	1,875,651
NE	201,450	17,666	406,764	147	1,206	0	738	627,971
NV	491	51,881	397,963	0	166,398	2	44,435	661,170
NH	5,450	6,440	0	4,621	312	499	514	17,836
NJ	13,592	458	0	0	37	0	121	14,208
NM	188,165	13,786	2,574,056	8,973	194,874	45	151	2,980,050
NY	127,621	6,284	33	0	0	0	142	134,080
NC	311,862	12,588	4	1,867	128	13	1,668	328,130
ND	420,526	52,730	238,702	948	22,668	15	14,011	749,600
OH	118,346	8,608	28	3,073	289	0	635	130,979
OK	126,928	78,662	47,065	4,037	4,348	7	0	261,047
OR	424,395	31,200	176,443	2,147	7,801	0	236	642,222
PA	0	0	0	0	0	0	0	0
RI	119	27	0	0	0	4	0	150
SC	82,929	32,786	0	38,685	18,718	119	4,145	177,382
SD	844,468	125,808	1,105,167	263	21,745	200	9,940	2,107,591
TN	285,600	14,149	0	564	0	0	201	300,514
TX	4,343,415	2,050,347	8,088,141	2,816,961	537,553	31,358	62,393	17,930,168
UT	18,415	11,615	194,402	0	9,731	3	176	234,342
VT	6,207	39	0	0	0	0	3	6,249
VA	124,559	84,893	24	3,184	1,538	0	153	214,351
WA	3,991	16,787	0	3,652	378	23	2,706	27,537
WV	648	5,928	0	74	700	17	52	7,419
WI	371,277	4,251	45	2,519	2,761	0	0	380,853
WY	40,824	27,274	359,287	328	2,779	10	76	430,578
PR[1]	16,813	11,368	11,418	97	119	2	0	39,817
Total	16,211,566	3,757,583	16,945,213	3,186,124	1,293,068	34,536	262,929	41,691,019

[1] Puerto Rico and Virgin Islands.

NRCS, Operations Management and Oversight Division, (202) 720–3906.

Table 12-18.—Land protected from erosion by soil conservation practices: Newly applied in soil conservation districts with Natural Resources Conservation Service (NRCS) assistance, fiscal years 1991–95

Land protected from erosion	Unit	1991	1992	1993	1994	1995
Cropland	Acres	23,301,183	27,221,042	29,011,585	22,707,109	16,211,566
Pasture and hayland	......do	4,387,931	4,547,619	4,321,933	5,027,169	3,757,583
Range and native pasture	......do	28,413,329	24,831,891	25,472,427	29,943,787	16,945,213
Woodland	......do	1,481,606	1,387,574	2,564,233	4,919,308	3,186,124
Wildlife	......do	2,142,766	1,858,827	1,922,912	2,288,598	1,293,068
Recreation	......do	81,489	127,621	49,530	63,612	34,536
All others	......do	304,115	283,633	246,933	303,440	262,929
Total land protected	Acres	60,112,419	60,258,207	63,589,553	65,253,023	41,691,019

NRCS, Operations Management and Oversight Division, (202) 720–3906.

Table 12-19.—Conservation plans: Active district cooperators and conservation plans prepared under all programs and area covered in soil conservation districts with Natural Resources Conservation Service assistance, by States and Territories, year ending Sept. 30, 1995

State or Territory	Progress, year ending Sept. 30, 1995		
	New SCD cooperators		New conservation plans[1]
	Number	*Acres*	*Acres*
Alabama	136	6,030	16,368
Alaska	6	1,685	21,842
Arizona	74	356,016	652,834
Arkansas	444	82,519	375,299
California	359	158,942	108,749
Colorado	379	294,077	546,157
Connecticut	64	10,109	1,131
Delaware	7,154	1,106,007	373,240
Florida	328	164,471	217,744
Georgia	2	1,445	56,816
Hawaii	221	128,614	34,505
Idaho	122	27,332	206,980
Illinois	501	91,295	463,250
Indiana	71,745	4,735,275	307,759
Iowa	2,873	417,687	230,577
Kansas	157	21,823	125,463
Kentucky	595	60,476	140,788
Louisana	583	127,008	241,713
Maine	47	5,204	20,790
Maryland	87	8,990	52,279
Massachusetts	20	1,369	10,785
Michigan	39	4,126	167,356
Minnesota	2,965	83,320	1,676,000
Mississippi	640	56,286	177,414
Missouri	1,113	275,978	145,533
Montana	7	10,328	419,498
Nebraska	204	245,352	147,087
Nevada	172	62,523	41,659
New Hampshire	28	4,424	9,122
New Jersey	65	11,381	21,583
New Mexico	118	167,568	1,978,963
New York	18	973	8,041
North Carolina	877	60,847	152,688
North Dakota	19	17,626	562,714
Ohio	256	35,132	91,943
Oklahoma	1,511	1,314,339	2,638,733
Oregon	101	52,530	370,144
Pennsylvania	164	14,494	52,738
Rhode Island	8	1,116	1,157
South Carolina	877	150,082	130,763
South Dakota	319	42,075	191,338
Tennessee	1,507	135,189	2,490
Texas	2,486	1,172,265	1,336,056
Utah	123	179,604	472,126
Vermont	12	2,281	10,517
Virginia	278	15,158	137,572
Washington	9	934	50,312
West Virginia	73	17,162	32,488
Wisconsin	45	5,755	116,870
Wyoming	203	11,796	505,028
Puerto Rico[2]	136	6,030	16,368
Total	100,270	11,963,048	15,869,370

[1] Includes conservation plans prepared on farms and ranches of SCD cooperators who signed agreements in prior years. [2] Puerto Rico and Virgin Islands.

NRCS, Operations Management and Oversight Division, (202) 720–3906.

Table 12-20.—Interim EQIP/GPCP [1]: Status of cost-share contracts, by States, year ending Sept. 30, 2001 [2]

State	Designated counties	Active land treatment contracts in operation	
		2001	2002
	Number	*Number*	*1,000 acres*
CO	38	92	419.9
KS	62	34	26.0
MT	46	23	167.1
NE	65	10	27.5
NM	27	25	634.8
ND	48	72	148.0
OK	44	71	44.3
SD	51	33	127.0
TX	156	119	365.1
WY	19	29	234.4
Total	556	508	2,194.1

[1] EQIP -- Environmental Quality Incentives Program; GPCP -- Great Plains Conservation Program. [2] As authorized by the Act of Congress April 4, 1996, (Public Law 127, 104th Congress).

NRCS, Conservation Operations Division, (202) 720–1845.

Table 12-21.—Great Plains Conservation Program: Status of cost-share contracts, by States, year ending Sept. 30, 2001 [1]

State	Designated counties	Active land treatment contracts in operation	
		2001	2001
	Number	*Number*	*1,000 acres*
CO	38	69	257.4
KS	62	22	15.0
MT	46	25	132.6
NE	65	120	319.7
NM	27	27	324.0
ND	48	99	246.8
OK	44	75	52.0
SD	51	92	348.0
TX	156	76	113.6
WY	19	53	634.0
Total	556	658	2,443.1

[1] As authorized by the Act of Congress August 7, 1956 (Public Law 1021, 84th Congress).

NRCS, Conservation Operations Division, (202) 720–1845.

Table 12-22.—Watershed protection and flood prevention projects: Works of improvement installed with the assistance of the Natural Resources Conservation Service with watershed protection and flood prevention funds for year ending Sept. 30, 1995 [1] [2]

Item	Unit	Watershed and flood prevention projects
Flood prevention subwatersheds in operation during year	Number	318
Watershed projects in operation during year	......do	1,591
Land protected from erosion:		
Cropland	Acres	291,118
Pasture and hayland	......do	158,665
Range and native pasture	......do	340,303
Woodland	......do	15,597
Wildlife	......do	30,226
Recreation	......do	3,973
Other	......do	13,843
Total land [3]	......do	855,634

[1] As authorized by the Soil Conservation Act of 1935 (Public Law 46, 74th Congress) and Watershed Protection and Flood Prevention Act P.L. 566, as amended. No update available. [2] All of the 11 watersheds were approved Dec. 22, 1944; however, flood control funds did not become available for works of improvement until 1946. [3] Land protected data included in Table 12-22 "Land Protected From Erosion by Soil Conservation Practices."

NRCS, Watershed and Wetlands Division, (202) 720–3527.

Table 12-23.—Small watershed protection and flood prevention projects: Accomplishments for years ending Sept. 30, 1993–97

Item	Unit of measure	1993	1994	1995	1996	1997
Small watershed protection: [1]						
Land treatment: [2]						
Forest land	Acres	38,322	16,806	1,905	2,193	8,402
Cropland	......do	501	626	0	1,160	741
Pastureland	......do	170	28	7,284	45	88
Total land treatment	......do	38,993	17,460	9,189	3,398	9,233
Land owners assisted	Number	3,534	1,483	1,465	1,348	1,186
Flood prevention: [3]						
Land treatment: [2]						
Forest land	Acres	2,196	6,335	63,028	8,682	6,541
Cropland	......do	....	....	575	1,668	20
Pastureland	......do	....	40	83	92	78
Total land treatment	......do	2,196	6,375	63,686	10,442	6,639
Land owners assisted	Number	1,452	1,528	2,461	2,265	1,183

[1] As authorized by the Watershed Protection and Flood Prevention Act of 1954 (Public Law 83–566), as amended. Accomplishments are limited to activities accomplished solely by small watershed protection program funds. [2] Reported in land use categories consistent with those reported by the National Resources Conservation Service. [3] As authorized by the Navigation and Flood Control Act of 1944 (Public Law 78–534), as amended. Accomplishments are limited to activities accomplished solely by small watershed protection program funds.

FS, Timber Demand and Technology Assessment, RWU-4851, (608) 231–9376.

Table 12-24.—Tree planting: Acres seeded and acres of tree planting, in States and Territories, fiscal year 2001

State or other area	Total	Federal lands: Total	Federal lands: National Forest System	Federal lands: Other[3]	Non-federal public[1] lands	Private[2] lands
	Acres	*Acres*	*Acres*	*Acres*	*Acres*	*Acres*
AL	141,382	1,691	986	705	2,000	136,000
AK	3,069	333	329	4	502	1,901
AZ	512	56	56	0	0	400
AR	19,699	1,919	1,919	0	0	15,861
CA	44,108	15,667	15,649	18	317	12,457
CO	4,307	774	773	1	0	2,759
CT	52	4	0	4	1	43
DE	2,894	0	0	0	31	2,863
FL	179,883	7,895	4,374	3,521	97,911	66,182
GA	272,155	2,371	266	2,105	382	267,031
HI	1,500	0	0	0	13	1,487
ID	28,311	11,464	11,464	0	3,620	1,763
IL	41,788	1,525	1,525	0	0	38,738
IN	9,249	97	42	55	174	8,881
IA	8,247	0	0	0	175	8,072
KS	1,730	7	0	7	0	1,716
KY	3,893	39	36	3	0	3,815
LA	107,337	953	908	45	0	105,431
ME	305	0	0	0	26	279
MD	5,456	3	0	3	500	4,950
MA	39	0	0	0	0	39
MI	12,105	3,844	3,812	32	2,937	1,480
MN	32,466	3,472	3,472	0	9,802	15,720
MS	228,969	3,179	3,036	143	12,939	209,672
MO	13,738	267	231	36	250	12,954
MT	18,705	8,651	8,651	0	840	563
NE	4,088	0	0	0	5	4,083
NV	430	40	40	0	0	350
NH	259	0	0	0	44	215
NJ	116	1	0	1	39	75
NM	423	135	135	0	0	153
NY	1,175	0	0	0	0	1,175
NC	102,045	1,293	467	826	42	99,417
ND	14,539	13	0	13	0	14,513
OH	3,363	138	138	0	280	2,807
OK	6,320	25	0	25	0	6,270
OR	56,740	13,914	13,914	0	4,320	24,592
PA	2,687	153	153	0	1,738	643
RI	0	0	0	0	0	0
SC	83,809	1,116	83	1,033	1,742	79,835
SD	8,479	0	0	0	0	8,479
TN	9,255	543	444	99	106	8,063
TX	56,406	493	183	310	100	55,320
UT	4,284	1,871	1,277	594	0	542
VT	86	43	40	3	0	0
VA	80,211	193	54	139	166	79,659
WA	58,728	7,974	7,637	337	16,000	26,780
WV	1,418	0	0	0	11	1,407
WI	22,067	1,435	1,410	25	764	18,433
WY	1,666	457	457	0	0	752
State totals	1,700,493	94,048	83,961	10,087	157,777	1,354,620
PR	187,755	0	0	0	106	187,649
Other[4]	344	0	0	0	2	342
Total	1,888,592	94,048	83,961	10,087	157,885	1,542,611

[1] State forest, other State, and other public agencies lands. [2] Forest industry, other industry, and nonindustrial lands. [3] U.S. Department of Interior and Indian Reservations, and other federal lands. [4] Guam and the Trust Territories of the Pacific Islands.

FS, Timber Demand and Technology Assessment, RWU-4851, (608) 231-9376.

Table 12-25.—Forest land: Total forest land and area and ownership of timberland, by regions, Jan. 1, 1996 [1]

Region	Total forest land [2]	Timberland [3]							
		All ownerships	Federal			State, county, and municipal	Private		
			Total	National forest	Other		Total	Forest industry	Farmer and other private [4]
	1,000 acres	*1,000 acres*	*1,000 acres*	*1,000 acres*	*1,000 acres*	*1,000 acres*	*1,000 acres*	*1,000 acres*	*1,000 acres*
Northeast	85,484	86,035	9,603	2,029	7,574	7,112	69,320	10,996	58,324
North Central	84,795	94,220	22,555	7,881	14,674	13,746	57,919	3,796	54,123
North	170,279	180,255	32,158	9,910	22,248	20,858	127,239	14,792	112,447
Southeast	88,662	84,803	6,851	4,594	2,257	2,522	75,430	14,508	60,922
South Central	125,448	116,306	9,000	6,604	2,396	2,562	104,744	22,526	82,218
South	214,110	201,109	15,851	11,198	4,653	5,084	180,174	37,034	143,140
Great Plains	4,787	4,317	1,075	999	76	180	3,062	0	3,062
Intermountain	138,447	63,455	42,755	42,580	175	2,637	18,063	2,926	15,137
Rocky Mountains	143,234	67,772	43,830	43,579	251	2,817	21,125	2,926	18,199
Alaska	127,380	11,988	3,899	3,780	119	4,299	3,790	0	3,790
Pacific Northwest	51,581	38,454	18,173	18,035	138	2,614	17,667	9,193	8,474
Pacific Southwest [5]	40,214	18,447	10,097	10,086	11	505	7,845	2,932	4,913
Pacific Coast	219,175	68,889	32,169	31,901	268	7,418	29,302	12,125	17,177
All regions	746,798	518,025	124,008	96,588	27,420	36,177	357,840	66,877	290,963

[1] Data may not add to totals because of rounding. [2] Forest land is land at least 10 percent stocked by forest trees of any size, including land that formerly had such tree cover and that will be naturally or artificially regenerated. Forest land includes transition zones, such as areas between heavily forested and nonforested lands that are at least 10 percent stocked with forest trees, and forest areas adjacent to urban and built-up lands. Also included are pinyon-juniper and chaparral areas in the West and afforested areas. The minimum area for classification of forest land is 1 acre. Roadside, streamside, and shelterbelt strips of timber must have a crown width at least 120 feet wide to qualify as forest land. Unimproved roads and trails, streams, and clearings in forest areas are classified as forest if less than 120 feet in width. [3] Timberland is forest land that is producing or is capable of producing crops of industrial wood and that is not withdrawn from timber utilization by statute or administrative regulation. Areas qualifying as timberland have the capability of producing more than 20 cubic feet per acre per year of industrial wood in natural stands. Currently inaccessible and inoperable areas are included. [4] Includes Indian lands. [5] Includes Hawaii.

FS, Timber Demand and Technology Assessment, RWU-4851, (608) 231–9376.

Table 12-26.—Timber volume: Net volume of growing stock and sawtimber on timberland, by softwoods and hardwoods, and regions, Jan. 1, 1996 [1]

Region	Growing stock [2]			Sawtimber [3]		
	All species	Softwoods	Hardwoods	All species	Softwoods	Hardwoods
	Million cubic feet	*Million cubic feet*	*Million cubic feet*	*Million board feet*	*Million board feet*	*Million board feet*
Northeast	121,179	30,945	90,234	321,066	91,564	229,502
North Central	91,393	18,355	73,038	253,427	54,676	198,751
North	212,572	49,300	163,272	574,493	146,240	428,253
Southeast	122,985	51,861	71,124	391,863	174,866	216,997
South Central	133,377	52,985	80,392	465,944	218,316	247,628
South	256,362	104,846	151,516	857,807	393,182	464,625
Great Plains	5,539	1,687	3,852	13,915	5,642	8,273
Intermountain	134,989	124,598	10,391	468,261	452,160	16,101
Rocky Mountains	140,528	126,285	14,243	482,176	457,802	24,374
Alaska	35,059	31,707	3,352	146,787	140,563	6,224
Pacific Northwest	155,758	141,845	13,913	844,138	796,907	47,231
Pacific Southwest [4]	59,366	49,859	9,507	321,615	297,121	24,494
Pacific Coast	250,183	223,411	26,772	1,312,540	1,234,591	77,949
All regions	859,645	503,842	355,803	3,227,016	2,231,815	995,201

[1] Data may not add to totals because of rounding. [2] Live trees of commercial species meeting specified standards of quality or vigor. Cull trees are excluded. Includes only trees 5.0-inches diameter or larger at 4½ feet above ground. [3] Live trees of commercial species containing at least one 12-foot sawlog or two noncontiguous 8-foot logs, and meeting regional specifications for freedom from defect. Softwood trees must be at least 9.0-inches diameter and hardwood trees must be at least 11.0-inches diameter at 4½ feet above ground. [4] Includes Hawaii.

FS, Timber Demand and Technology Assessment, RWU-4851, (608) 231–9376.

Table 12-27.—Timber removals: Roundwood product output, logging residues and other removals from growing stock and other sources, by softwoods and hardwoods, 1996 [1]

Roundwood products, logging residues, and other removals	All sources			Growing stock [2]			Other sources [3]		
	All species	Soft-woods	Hard-woods	All species	Soft-woods	Hard-woods	All species	Soft-woods	Hard-woods
	Million cubic feet	*Million cubic feet*	*Million cubic feet*	*Million cubic feet*	*Million cubic feet*	*Million cubic feet*	*Million cubic feet*	*Million cubic feet*	*Million cubic feet*
Roundwood products:									
Sawlogs	7,120	5,143	1,977	6,712	4,928	1,784	408	215	193
Pulpwood	1,269	1,111	158	1,241	1,090	151	28	21	7
Veneer logs	5,208	2,858	2,350	4,615	2,560	2,055	593	298	295
Other products [4]	504	399	105	469	384	85	35	15	20
Fuelwood [5]	2,306	493	1,813	818	187	631	1,488	306	1,182
Total	16,407	10,004	6,403	13,855	9,149	4,706	2,552	855	1,697
Logging residues [6]	3,374	1,349	2,025	1,292	615	677	2,082	734	1,348
Other removals [7]	1,404	455	949	936	355	581	468	100	368
Total	4,778	1,804	2,974	2,228	970	1,258	2,550	834	1,716

[1] Data may not add to totals because of rounding. [2] Includes live trees of commercial species meeting specified standards of quality or vigor. Cull trees are excluded. Includes only trees 5.0-inches diameter or larger at 4½ feet above ground. [3] Includes salvable dead trees, rough and rotten trees, trees of noncommercial species, trees less than 5.0-inches diameter at 4½ feet above ground, tops, and roundwood harvested from nonforest land (for example, fence rows). [4] Includes such items as cooperage, pilings, poles, posts, shakes, shingles, board mills, charcoal and export logs. [5] Downed and dead wood volume left on the ground after trees have been cut on timberland. [6] Net of wet rot or advanced dry rot, and excludes old punky logs; consists of material sound enough to chip; excludes stumps and limbs. [7] Unutilized wood volume from cut or otherwise killed growing stock, from nongrowing stock sources on timberland (for example, precommercial thinnings), or from timberland clearing. Does not include volume removed from inventory through reclassification of timberland to reserved timberland.

FS, Timber Demand and Technology Assessment, RWU-4851, (608) 231–9376.

Table 12-28.—Timber growth, removals and mortality: Net annual growth, removals, and mortality of growing stock on timberland by softwoods and hardwoods and regions, 1997 [1]

Region	Growth [2]			Removals [3]			Mortality [4]		
	All species	Soft-woods	Hard-woods	All species	Soft-woods	Hard-woods	All species	Soft-woods	Hard-woods
	Million cubic feet	*Million cubic feet*	*Million cubic feet*	*Million cubic feet*	*Million cubic feet*	*Million cubic feet*	*Million cubic feet*	*Million cubic feet*	*Million cubic feet*
Northeast	2,869	646	2,223	1,275	414	861	788	274	514
North Central	2,482	521	1,961	1,498	255	1,243	817	181	636
North	5,351	1,167	4,184	2,773	669	2,104	1,605	455	1,150
Southeast	4,731	2,779	1,952	4,459	2,947	1,512	1,234	630	604
South Central	5,981	3,110	2,871	5,726	3,531	2,195	1,003	406	597
South	10,712	5,889	4,823	10,185	6,478	3,707	2,237	1,036	1,201
Great Plains	98	52	46	35	20	15	48	10	38
Intermountain	2,390	1,956	434	497	481	16	993	890	103
Rocky Mountains	2,488	2,008	480	532	501	31	1,041	900	141
Alaska	223	137	86	182	177	5	205	195	10
Pacific Northwest	3,582	3,156	426	1,720	1,621	99	779	697	82
Pacific Southwest [5]	1,288	1,155	133	628	618	10	315	263	52
Pacific Coast	5,093	4,448	645	2,530	2,416	114	1,299	1,155	144
All regions	23,644	13,512	10,132	16,020	10,064	5,956	6,182	3,546	2,636

[1] Data may not add to totals because of rounding. [2] The net increase in the volume of trees during a specified year. Components include the increment in net volume of trees at the beginning of the specific year surviving to its end, plus the net volume of trees reaching the minimum size class during the year, minus the volume of trees that died during the year, and minus the net volume of trees that became cull trees during the year. [3] The net volume of trees removed from the inventory during a specified year by harvesting, cultural operations such as timber stand improvement, or land clearing. [4] The volume of sound wood in trees that died from natural causes during a specified year. [5] Includes Hawaii.

FS, Timber Demand and Technology Assessment, RWU-4851, (608) 231–9376.

Table 12-29.—Timber volume: Net volume of sawtimber on timberland in the West, by regions and species, Jan. 1, 1997[1]

Species	Total West	Inter-mountain	Alaska	Pacific Northwest	Pacific South-west[2]	Great Plains
	Million board feet	*Million board feet*	*Million board feet*	*Million board feet*	*Million board feet*	*Million board feet*
Softwoods:						
Douglas-fir	640,832	127,211	0	429,850	83,771	0
Ponderosa and Jeffrey pines	203,733	76,732	0	63,637	59,647	3,717
True fir	238,850	65,806	9	93,369	79,666	0
Western hemlock	171,001	4,692	51,911	114,240	158	0
Sugar pine	31,257	3,824	0	7,558	19,875	0
Western white pine	7,300	3,118	0	2,355	1,827	0
Redwood	27,663	0	0	193	27,470	0
Sitka spruce	47,279	0	44,944	2,335	0	0
Engelmann and other spruces	103,028	68,997	17,669	16,011	216	135
Western larch	29,228	17,537	0	11,691	0	0
Incense cedar	19,567	14	0	3,971	15,582	0
Lodgepole pine	82,647	64,408	131	12,638	5,470	0
Western Red Cedar	43,065	9,473	4,945	28,641	6	0
Other	47,112	14,383	13,828	13,678	3,433	1,790
Total	1,692,562	456,195	133,437	800,167	297,121	5,642
Hardwoods:						
Cottonwood and aspen	23,744	15,558	4,276	3,746	158	6
Red alder	29,013	0	91	28,336	586	0
Oak	11,833	0	0	809	11,024	0
Other	37,734	542	1,921	14,276	12,727	8,268
Total	102,324	16,100	6,288	47,167	24,495	8,274
All species	1,794,886	472,295	139,725	847,334	321,616	13,916

[1] International ¼-inch rule. Data may not add to totals because of rounding. [2] Includes Hawaii.

FS, Timber Demand and Technology Assessment, RWU-4851, (608) 231–9376.

Table 12-30.—Timber volume: Net volume of sawtimber on timberland in the East, by regions and species, Jan. 1, 1997[1]

Species	Total East	North			South		
		Total	Northeast	North Central	Total	Southeast	South Central
	Million board feet	*Million board feet*	*Million board feet*	*Million board feet*	*Million board feet*	*Million board feet*	*Million board feet*
Softwoods:							
Longleaf and slash pines	54,280	0	0	0	54,280	34,512	19,768
Loblolly and shortleaf pines	275,854	4,445	1,952	2,493	271,409	96,186	175,223
Other yellow pines	33,095	4,938	4,092	846	28,157	19,115	9,042
White and red pines	67,091	57,840	36,641	21,199	9,251	7,979	1,272
Jack pine	4,195	4,195	35	4,160	0	0	0
Spruce and balsam fir	30,102	30,022	20,434	9,588	80	80	0
Eastern hemlock	29,806	27,205	22,095	5,110	2,601	1,779	822
Cypress	24,656	152	15	137	24,504	14,016	10,488
Other	20,344	17,442	6,299	11,143	2,902	1,200	1,702
Total	539,423	146,239	91,563	54,676	393,184	174,867	218,317
Hardwoods:							
Select white oaks	90,351	38,343	12,799	25,544	52,008	23,157	28,851
Select red oaks	78,236	48,582	27,463	21,119	29,654	11,737	17,917
Other white oaks	53,997	14,424	10,528	3,896	39,573	18,716	20,857
Other red oaks	128,701	33,441	14,599	18,842	95,260	39,930	55,330
Hickory	47,734	15,756	6,649	9,107	31,978	10,327	21,651
Yellow birch	9,467	9,220	6,955	2,265	247	235	12
Hard maple	54,904	50,015	30,336	19,679	4,889	1,310	3,579
Soft maple	68,782	51,678	34,551	17,127	17,104	12,657	4,447
Beech	27,790	18,594	14,426	4,168	9,196	3,065	6,131
Sweetgum	64,140	3,148	1,354	1,794	60,992	17,031	43,961
Tupelo and black gum	57,861	3,427	1,204	2,223	54,434	21,329	33,105
Ash	113,916	35,575	11,740	23,835	78,341	34,848	43,493
Basswood	65,143	20,039	4,698	15,341	45,104	21,801	23,303
Yellow-poplar	231,176	50,740	17,497	33,243	180,436	73,778	106,658
Cottonwood and aspen	159,429	44,764	7,440	37,324	114,665	55,752	58,913
Black walnut	15,171	4,093	743	3,350	11,078	4,959	6,119
Black cherry	79,946	33,755	13,843	19,912	46,191	22,573	23,618
Other	169,342	44,105	12,679	31,426	125,237	51,028	74,209
Total	1,516,086	519,699	229,504	290,195	996,387	424,233	572,154
All species	2,055,509	665,938	321,067	344,871	1,389,571	599,100	790,471

[1] International ¼-inch rule. Data may not add to totals because of rounding.

FS, Timber Demand and Technology Assessment, RWU-4851, (608) 231–9376.

Table 12-31.—National Forest System: National Forest System lands and other lands in States and Territories, Sept. 30, 2001

State or other area	Gross acreage	National Forest System acreage [1]	Other acreage [2]
	1,000 acres	*1,000 acres*	*1,000 acres*
AL	1,288	666	622
AK	24,355	21,987	2,368
AZ	11,891	11,262	629
AR	3,540	2,587	953
CA	24,430	20,709	3,721
CO	16,015	14,481	1,533
CT	0	0	0
FL	1,434	1,153	281
GA	1,856	865	991
HI	0	0	0
ID	21,653	20,463	1,190
IL	857	293	564
IN	644	199	446
KS	116	108	8
KY	2,210	805	1,405
LA	1,025	604	420
ME	93	53	40
MI	4,894	2,864	2,030
MN	5,467	2,839	2,628
MS	2,320	1,169	1,151
MO	3,060	1,494	1,566
MT	19,108	16,903	2,205
NE	442	352	90
NV	6,275	5,835	440
NH	828	728	100
NM	10,455	9,417	1,039
NY	16	16	0
NC	3,167	1,247	1,919
ND	1,106	1,106	0
OH	834	233	601
OK	772	398	375
OR	17,501	15,662	1,839
PA	743	513	230
SC	1,376	617	759
SD	2,369	2,013	356
TN	1,276	700	576
TX	1,994	755	1,239
UT	9,209	8,189	1,019
VT	817	386	431
VA	3,224	1,661	1,563
WA	10,110	9,252	858
WV	1,869	1,034	836
WI	2,023	1,523	500
WY	9,703	9,238	466
PR	56	28	28
VI	0	0	0
Total	232,423	192,407	40,016

[1] *National Forest System acreage.*—A nationally significant system of Federally owned units of forest, range, and related land consisting of national forests, purchase units, national grasslands, land utilization project areas, experimental forest areas, experimental range areas, designated experimental areas, other land areas; water areas, and interests in lands that are administered by USDA Forest Service or designated for administration through the Forest Service.

National forests.—Units formally established and permanently set aside and reserved for national forest purposes.

Purchase units.—Units designated by the Secretary of Agriculture or previously approved by the National Forest Reservation Commission for purposes of Weeks Law Acquisition.

National grasslands.—Units designated by the Secretary of Agriculture and permanently held by the Department of Agriculture under Title III of the Bankhead-Jones Farm Tenant Act.

Land utilization projects.—Units designated by the Secretary of Agriculture for conservation and utilization under Title III of the Bankhead-Jones Farm Tenant Act.

Research and experimental areas.—Units reserved and dedicated by the Secretary of Agriculture for forest or range research and experimentation.

Other areas.—Units administered by the Forest Service that are not included in the above groups. [2] *Other acreage.*—Lands within the unit boundaries in private, State, county, and municipal ownership and Federal lands over which the Forest Service has no jurisdiction. Areas of such lands which have been offered to the United States and have been approved for acquisition and subsequent Forest Service administration, but to which title had not yet been accepted by the United States.

FS, Timber, Demand and Technology Assessment, RWU-4851, (608) 231–9376.

Table 12-32.—Forest products cut on National Forest System lands: Volume and value of timber cut and value of all products, United States, fiscal years 1992–2001

Year[1]	Timber cut[2]		Value of miscellaneous forest products[4]	Total value including free-use timber
	Volume	Value[3]		
	Million bd. ft.	*1,000 dollars*	*1,000 dollars*	*1,000 dollars*
1992	7,290	934,504	2,655	937,991
1993	5,917	914,646	2,791	918,225
1994	4,815	783,038	3,138	796,926
1995	3,866	616,117	2,935	619,732
1996	3,725	544,349	3,262	547,428
1997	3,285	497,957	3,262	500,896
1998	3,298	445,774	3,262	448,752
1999	2,939	339,471	3,262	342,599
2000	2,542	302,934	3,262	305,921
2001	2,306	177,634	3,262	180,708

[1] Fiscal years Oct. 1–Sept. 30. [2] Commercial and cost sales and land exchanges. [3] Includes collections for forest restoration or improvement under the Knutson-Vandenberg Act, 1930. [4] Includes materials not measurable in board feet, such as Christmas trees, tanbark, turpentine, seedlings, Spanish moss, etc.

FS, Timber Demand and Technology Assessment, RWU-4851, (608) 231–9376.

Table 12-33.—National Forest System lands: Receipts, United States and Puerto Rico, fiscal years 1992–2001

Year[1]	From the use of timber[2]	From the use of grazing	From special land uses, water power, etc.	Total[2]
	1,000 dollars	*1,000 dollars*	*1,000 dollars*	*1,000 dollars*
1992	520,003	10,780	83,505	614,288
1993	425,105	10,518	67,955	503,578
1994	431,615	11,056	72,196	514,867
1995	303,046	8,756	74,943	386,745
1996	195,000	7,352	71,183	273,535
1997	197,194	6,972	80,588	284,754
1998	207,938	6,992	78,869	293,799
1999	NA	NA	NA	NA
2000	NA	NA	NA	NA
2001	NA	NA	NA	NA

[1] Fiscal years Oct. 1–Sept. 30. [2] Includes receipts from Oregon and California Railroad Grant Lands.

FS, Timber Demand and Technology Assessment, RUW-4851, (608) 231–9376.

Table 12-34.—National forests: Payments to States and Puerto Rico from receipts from timber sales, grazing fees, and miscellaneous uses, fiscal years 1997–99 [1] [2]

State or other areas	1997	1998	1999
	1,000 dollars	*1,000 dollars*	*1,000 dollars*
AL	2,050	964	1,133
AK	5,906	1,187	1,820
AZ	1,632	2,215	2,113
AR	6,648	5,954	6,584
CA	36,158	33,963	30,533
CO	5,956	4,388	5,045
FL	1,066	1,000	1,435
GA	908	697	328
ID	17,458	14,267	12,468
IL	28	17	394
IN	7	26	138
KY	494	433	255
LA	2,736	2,949	2,361
ME	35	30	37
MI	2,384	2,759	2,996
MN	3,179	2,922	3,412
MS	8,276	4,919	5,399
MO	1,232	1,149	1,237
MT	9,383	8,558	10,367
NE	31	32	33
NV	299	388	330
NH	510	440	549
NM	653	923	854
NY	6	6	2
NC	692	654	594
ND	3	3	3
OH	11	16	2
OK	883	1,017	1,034
OR	95,239	92,243	85,505
PA	6,207	5,998	5,800
SC	960	1,292	557
SD	2,350	3,814	3,663
TN	319	440	327
TX	4,337	2,264	5,621
UT	1,831	1,599	1,512
VT	257	226	436
VA	822	790	767
WA	29,429	28,425	27,073
WV	1,861	1,624	1,944
WI	1,621	1,861	2,166
WY	1,844	1,849	2,184
PR	21	23	24
Total	255,719	234,321	229,032

[1] Fiscal years Oct. 1–Sept. 30. [2] Payments under the acts of May 23, 1908 (as amended), July 24, 1956, and Oct. 22, 1976, are 25 percent of total receipts remaining after deducting (a) payments to Arizona and New Mexico on account school section lands administered by Forest Service, (b) appropriations of receipts under laws authorizing such appropriations for acquisition of lands in specified national forests or portions thereof, and (c) receipts from an area of the Superior National Forest, Minnesota, on account of which the State (for the counties) is paid 0.75 percent of the appraised valuation in lieu of 25 percent of the receipts. Payments made in the following year.

FS, Timber Demand and Technology Assessment, RWU-4851, (608) 231–9376.

Table 12-35.—Livestock on National Forest System lands: Number grazed and grazing receipts, United States, 1989–1998

Year	Number grazed [1]		Receipts from grazing [2]
	Cattle, horses, and burros	Sheep and goats	
	Thousands	*Thousands*	*1,000 dollars*
1989	1,526	972	10,950
1990	1,236	958	10,419
1991	1,265	1,029	11,457
1992	1,408	1,183	10,780
1993	1,318	1,111	10,518
1994	1,229	941	11,056
1995	1,227	940	8,756
1996	1,174	868	7,352
1997	1,225	932	6,972
1998	1,208	909	6,992

[1] Calendar year data for number actually grazed. [2] Fiscal years Oct. 1–Sept. 30.

FS, Timber Demand and Technology Assessment, RWU-4851, (608) 231–9376.

Table 12-36.—Livestock on National Forest System lands: Number grazed and grazing receipts for fiscal year 1999, by States

State or other area	Head months [1]: Cattle, horses, and burros	Head months [1]: Sheep and goats	Receipts from grazing, 1992 [2]
	Number	*Number*	*Dollars*
AL	303	0	619
AZ	723,978	47,385	1,647,142
AR	4,160	0	39,045
CA	293,376	155,700	592,633
CO	593,907	445,914	1,310,957
FL	600	0	18,885
GA	3,091	0	8,080
ID	442,272	532,074	977,514
IL	101	0	167
KS	34	0	0
KY	31,905	0	25,701
LA	15,916	0	25,719
MI	1,752	0	1,557
MN	41	0	136
MS	2,705	0	3,426
MO	26,850	0	27,154
MT	420,212	49,335	820,929
NE	105,479	0	158,367
NV	155,020	166,744	402,127
NM	749,223	79,974	1,235,864
NY	7,751	0	57
ND	391,513	0	443,840
OH	709	0	961
OK	19,523	0	28,620
OR [3]	335,644	64,966	618,786
SD	389,493	19,321	547,595
TX	32,469	0	97,788
UT	335,358	513,261	836,937
VT	108	0	0
VA	6,656	0	10,694
WA	65,643	35,456	169,044
WV	5,688	321	18,585
WY	433,442	369,419	709,486
National forests	5,594,922	2,479,870	9,464,412
National grasslands and land utilization project land			1,314,323
Total	5,594,922	2,479,870	10,778,735

[1] A head month is the billing unit for permitted grazing and is equal to 1 month's occupancy. [2] 1992 is the most recent year for which grazing receipts by state are available. [3] Figure does not include $2,285.44 receipts from Oregon & CA Railroad Grant.

FS, Timber Demand and Technology Assessment, RWU-4851, (608) 231–9376.

Table 12-37.—Timber prices: Average stumpage prices for sawtimber sold from national forests, by selected species, 1992–2001

Year	Douglas-fir [1]	Southern pine [2]	Ponderosa pine [3]	Western hemlock [4]	All eastern hardwoods [5]	Oak, white, red, and black [5]	Maple, sugar [6]
	Dollars per 1,000 bd. ft.	*Dollars per 1,000 bd. ft.*	*Dollars per 1,000 bd. ft.*	*Dollars per 1,000 bd. ft.*	*Dollars per 1,000 bd. ft.*	*Dollars per 1,000 bd. ft.*	*Dollars per 1,000 bd. ft.*
1992	477.20	198.40	292.30	164.60	166.60	211.20	144.60
1993	317.75	217.21	535.17	363.66	264.06	194.57	219.52
1994	652.36	265.85	291.35	334.80	352.07	317.40	313.37
1995	453.54	248.49	149.94	297.09	313.92	296.59	285.57
1996	453.04	251.05	269.97	289.30	312.57	264.44	213.20
1997	331.40	307.30	270.20	211.30	286.88	264.50	357.12
1998	254.20	287.80	204.90	161.40	240.90	270.20	394.80
1999	314.70	268.50	181.00	95.70	195.10	317.40	448.10
2000	433.40	141.80	154.60	46.12	220.43	191.90	314.10
2001	381.07	91.55	71.63	33.98	199.68	223.89	322.62

[1] Western Washington and western Oregon. [2] Southern region. [3] Pacific Southwest region. Includes Jeffrey pine. [4] Pacific Northwest region. [5] Eastern and Southern regions. [6] Eastern region.

Forest Service National Forest prices in this table are for timber sold on a Scribner Decimal C log rule basis, except in the Northeastern States where International 1/4-inch log rule is used. Prices include KV payments; exclude timber sold by land exchanges and from land utilization project lands. Data for 1983 are statistical high bid prices; beginning in 1984, data are high bid prices which include specified road costs.

FS, Timber Demand and Technology Assessment, RWU-4851, (608) 231–9376.

Table 12-38.—National Forest System lands: Number of visitor-days, 1986–96, by activity, States, and Puerto Rico, 1996 [1]

Year and activity	Visitor-days	Percent	1997			
			State or other area	Visitor-days	State or other area	Visitor-days
	Thousands			*Thousands*		*Thousands*
			AL	689	NH	3,354
1986	226,533	...	AK	6,962	NM	9,326
1987	238,458	...	AZ	35,000	NY	39
1988	242,316	...	AR	2,210	NC	6,979
1989	252,495	...	CA	71,165	ND	133
1990	263,051	...	CO	30,971	OH	524
1991	278,849	...	FL	2,960	OK	393
1992	287,690	...	GA	2,925	OR	37,029
1993	295,473	...	ID	15,365	PA	3,268
1994	330,348	...	IL	1,188	SC	1,011
1995	345,083	...	IN	684	SD	3,571
1996	341,200	...	KS	86	TN	3,309
1996 activity:			KY	2,326	TX	2,302
Mechanized travel and viewing scenery	122,141	35.8	LA	599	UT	19,378
Camping, picnicking and swimming	87,082	25.5	ME	158	VT	1,395
Hiking, horseback riding and water travel	33,099	9.7	MI	4,866	VA	4,927
Winter sports	19,708	5.8	MN	5,982	WA	24,797
Hunting	19,384	5.7	MS	1,828	WV	1,499
Resorts, cabins and organization camps	17,702	5.2	MO	2,518	WI	2,527
Fishing	18,160	5.3	MT	13,495	WY	9,114
Nature studies	3,299	1.0	NE	320	PR	171
Other [2]	20,627	6.0	NV	3,857	...	
Total [3]	341,200	100.0			Total [3]	341,200

[1] Recreational use of national forest land and water that aggregates 12 visitor-hours. This may entail one person for 12 hours, 12 persons for 1 hour, or any equivalent combination of individual or group use, either continuous or intermittent. Fiscal years Oct. 1–Sept. 30. [2] Includes team sports, gathering forest products, attending talks and programs, and other uses. [3] Data may not add to total because of rounding.

FS, Timber Demand and Technology Assessment, RWU -4851 (608) 231–9376.

Table 12-39.—Timber products: Production, imports, exports, and consumption, United States, 1992–2001[1]

Year	Industrial roundwood used for—											
	Lumber				Plywood and veneer				Pulp products			
	Production	Imports	Exports	Consumption	Production	Imports	Exports	Consumption	Production	Imports[2]	Exports[2]	Consumption
	Million cu. ft.[3]	*Million cu. ft.*[3]	*Million cu. ft.*[3]	*Million cu. ft.*[3]	*Million cu. ft.*[3]	*Million cu. ft.*[3]	*Million cu. ft.*[3]	*Million cu. ft.*[3]	*Million cu. ft.*[3]	*Million cu. ft.*[3]	*Million cu. ft.*[3]	*Million cu. ft.*[3]
1992	6,983	1,960	561	8,383	1,294	100	106	1,288	5,516	992	801	5,706
1993	6,887	2,240	532	8,595	1,293	100	100	1,293	5,853	1,151	786	6,218
1994	7,052	2,395	512	8,935	1,320	94	86	1,328	6,005	1,188	821	6,372
1995	6,815	2,522	460	8,877	1,303	107	89	1,321	6,086	1,250	906	6,429
1996	6,886	2,616	449	9,053	1,281	97	87	1,291	5,908	1,144	891	6,161
1997	7,103	2,619	452	9,270	1,213	114	103	1,224	6,097	1,249	929	6,417
1998	7,093	2,690	350	9,432	1,201	131	55	1,276	6,114	1,269	818	6,565
1999	7,371	2,810	404	9,777	1,208	160	46	1,322	5,813	1,352	768	6,397
2000	7,201	2,832	421	9,612	1,186	155	42	1,295	5,881	1,462	822	6,521
2001	6,870	2,896	347	9,418	1,143	173	32	1,283	5,668	1,193	794	6,068

Year	Industrial roundwood used for—Continued									Fuelwood production and consumption	Production, all products	Consumption, all products
	Other industrial products,[4] production and consumption	Logs		Pulpwood chip imports	Pulpwood chip exports	Total						
		Imports	Exports			Production	Imports	Exports	Consumption			
	Million cu. ft.[3]	*Million cu. ft.*[3]	*Million cu. ft.*[3]	*Million cu. ft.*[3]	*Million cu. ft.*[3]	*Million cu. ft.*[3]	*Million cu. ft.*[3]	*Million cu. ft.*[3]	*Million cu. ft.*[3]	*Million cu. ft.*[3]	*Million cu. ft.*[3]	*Million cu. ft.*[3]
1992	610	7	524	1	351	15,280	3,060	2,343	16,026	3,044	18,324	19,070
1993	621	15	460	5	326	15,011	3,511	2,204	16,334	3,084	18,095	19,418
1994	576	18	429	10	354	15,306	3,705	2,202	16,800	3,134	18,440	19,934
1995	513	13	451	16	377	15,430	3,908	2,283	17,082	2,937	18,367	20,019
1996	500	18	422	10	416	15,258	3,885	2,265	16,918	2,739	17,997	19,658
1997	485	20	384	5	422	14,790	4,007	2,290	16,519	2,542	17,332	19,061
1998	485	30	316	7	412	14,899	4,127	1,951	17,065	2,523	17,422	19,588
1999	485	47	326	2	422	15,032	4,371	1,966	17,438	2,542	17,574	19,980
2000	485	68	331	2	353	14,798	4,519	1,969	17,289	2,561	17,530	19,850
2001	485	70	307	1	265	14,166	4,333	1,745	17,254	2,571	16,737	19,825

[1] Data may not add to totals because of rounding. [2] Includes both pulpwood and the pulpwood equivalent of woodpulp, paper, and board. [3] Roundwood equivalent. [4] Includes cooperage logs, poles and piling, fence posts, hewn ties, round mine timbers, box bolts, excelsior bolts, chemical wood, shingle bolts, and miscellaneous items.

FS, Timber Demand and Technology Assessment, RWU-4851, (608) 231–9376.

Table 12-40.—Lumber: Production, United States, 1992–2001

Year	Total	Softwoods	Hardwoods
	Million bd. ft.	*Million bd. ft.*	*Million bd. ft.*
1992	45,949	34,526	11,423
1993	45,166	32,947	12,219
1994	46,492	34,107	12,385
1995	44,877	32,233	12,644
1996	45,754	33,266	12,488
1997	47,340	34,667	12,673
1998	47,407	34,677	12,730
1999	49,501	36,674	12,827
2000	48,543	35,965	12,578
2001	46,170	34,596	11,574

FS, Timber Demand and Technology Assessment, RWU-4851, (608) 231–9376. From data published by the American Forest and Paper Association.

Table 12-41.—Timber products: Pulpwood consumption, woodpulp production, and paper and board production and consumption, United States, 1992–2001 [1]

Year	Pulpwood consumption [2]	Woodpulp production [3]	Paper and board [4]		
			Production	Consumption or new supply [5]	Per capita consumption
	1,000 cords [6]	*1,000 tons*	*1,000 tons*	*1,000 tons*	*Pounds*
1992	96,146	93,642	82,868	87,577	686
1993	96,089	97,628	84,857	91,013	705
1994	98,142	98,009	89,080	95,195	730
1995	94,125	94,125	89,509	96,180	731
1996	88,246	88,246	90,381	94,287	710
1997	92,312	92,312	95,029	99,186	740
1998	90,591	90,591	94,510	99,945	740
1999	87,878	87,878	97,038	104,478	765
2000	87,453	87,453	94,613	102,161	742
2001	83,384	83,384	89,007	91,261	638

[1] Revised to match data from American Forest and Paper Association and American Pulpwood Association. [2] Includes changes in stocks. [3] Excludes defibrated and exploded woodpulp used for hard pressed board. [4] Excludes hardboard. [5] Production plus imports and minus exports (excludes products); changes in inventories not taken into account. [6] One cord equals 128 cubic feet.

FS, Timber Demand and Technology Assessment, RWU-4851, (608) 231–9376. Compiled from U.S. Department of Commerce and American Forest and Paper Association.

Table 12-42.—Timber products: Producer price indexes, selected products, United States, 1992–2001

[1982=100]

Year	Lumber	Softwood plywood	Woodpulp	Paper	Paperboard
1992	80.4	84.5	97.6	82.4	86.1
1993	101.9	97.5	83.6	82.9	83.3
1994	104.9	101.7	92.4	84.2	89.8
1995	96.4	108.2	140.2	106.4	117.3
1996	100.0	100.0	100.0	100.0	100.0
1997	108.2	100.8	96.1	96.2	92.8
1998	98.4	100.7	91.6	97.7	97.3
1999	103.1	118.9	93.3	94.8	98.2
2000	97.9	98.7	108.6	100.6	112.9
2001	93.9	95.8	107.8	100.6	110.0

FS, Timber Demand and Technology Assessment, RWU-4851, (608) 231–9376. Compiled from reports of the U.S. Department of Labor, Bureau of Labor Statistics.

Table 12-43.—Timber products: Structual panels, LVL, and lumber production, United States, 1992–2001

Year	Laminated veneer lumber [1]	Oriented strand board	Plywood	Medium-density fiberboard	Lumber	
					Hardwood	Softwood [2]
	Million cubic meters	*Million cubic meters*	*Million cubic meters*	*Million cubic meters*	*Million cubic meters*	*Million cubic meters*
1992	0.48	5.89	17.11	1.89	27.00	58.70
1993	0.59	6.20	17.09	2.05	28.80	56.00
1994	0.65	6.63	17.38	2.21	29.20	58.00
1995	0.79	6.99	17.14	1.96	29.80	54.80
1996	0.91	8.24	16.98	2.21	29.50	56.60
1997	1.08	9.32	15.90	2.45	29.90	58.90
1998	1.16	9.94	15.73	2.48	29.97	59.00
1999	1.36	10.28	15.77	2.50	30.44	62.39
2000	1.47	10.54	15.47	2.60	29.74	61.20
2001	1.67	10.71	14.58	2.51	27.14	58.82

[1] Prior to 1994, data are estimates from various articles and reports. [2] Revised due to softwood conversion factor of 1.7 (2.36 was previously used).

FS, Timber Demand and Technology Assessment, RWU-4851, (608) 231–9376.

CHAPTER XIII

CONSUMPTION AND FAMILY LIVING

The statistics in this chapter deal with the consumption of food by both rural and urban people, retail price levels, and other aspects of family living of farm people. Data presented here on quantities of food available for consumption are based on material presented in the earlier commodity chapters, but they are shown here at the retail level, a form that is more useful for an analysis of the demand situation faced by the producer. Data on quantities of farm-produced food consumed directly by farm households are presented in the commodity chapters. Its value and the rental value of the farm home are given in the section on farm income.

Table 13-1.—Population: Number of people eating from civilian food supplies, United States, Jan. 1 and July 1, 1993–2002

Year	Jan. 1	July 1
	Millions	*Millions*
1993	255.0	256.3
1994	257.7	258.9
1995	260.2	261.4
1996	262.8	263.9
1997	265.3	266.5
1998	267.9	269.0
1999	270.4	271.5
2000	272.8	281.0
2001	282.6	284.1
2002	285.0	286.2

ERS,Rural Economy Branch (202) 694–5436. Compiled from reports of the U.S. Department of Commerce, Census Bureau.

Table 13-2.—Macronutrients: Quantities available for consumption per capita per day, United States, 1970–2000 [1]

Year	Food energy	Protein	Fat				Choles-terol	Carbo-hydrate	Dietary fiber
			Total fat	Monounsaturated	Saturated	Polyunsaturated			
	Kilocalories	*Grams*	*Grams*	*Grams*	*Grams*	*Grams*	*Milligrams*	*Grams*	*Grams*
1970	3,300	96	151	61	53	26	470	395	19
1971	3,300	97	152	61	53	26	470	396	19
1972	3,300	96	152	61	53	27	460	390	19
1973	3,200	94	147	59	50	27	430	395	19
1974	3,200	95	149	60	51	27	440	391	19
1975	3,200	93	144	57	49	27	420	391	19
1976	3,300	97	150	60	51	29	430	404	20
1977	3,300	96	147	59	50	28	430	402	19
1978	3,200	95	148	60	50	29	430	394	19
1979	3,300	95	150	61	50	30	430	400	19
1980	3,300	95	151	61	51	30	420	399	19
1981	3,300	95	151	61	51	30	420	398	19
1982	3,300	95	151	61	50	30	420	399	20
1983	3,300	96	155	62	52	31	420	403	20
1984	3,400	97	157	64	52	31	430	408	20
1985	3,500	100	162	66	54	32	420	423	21
1986	3,500	102	161	66	54	32	420	428	21
1987	3,500	103	159	65	53	32	420	442	22
1988	3,600	104	160	65	52	33	420	448	22
1989	3,500	103	155	64	51	32	410	446	22
1990	3,600	105	157	65	51	33	400	462	23
1991	3,600	106	155	65	50	33	400	467	23
1992	3,700	108	159	67	51	33	400	480	24
1993	3,800	108	162	69	52	34	400	487	24
1994	3,800	110	159	68	51	33	410	494	24
1995	3,700	108	156	66	50	33	400	492	24
1996	3,800	109	154	66	49	32	400	500	24
1997	3,800	109	154	65	49	33	400	502	24
1998	3,800	110	159	68	51	33	410	503	25
1999	3,900	111	162	69	51	34	420	508	25
2000	3,900	112	171	72	54	37	430	505	25

See footnote at end of table.

Table 13-2.—Vitamins: Quantities available for consumption per capita per day, United States, 1970–2000 [1]—Continued

Year	Vitamins										
	Vitamin A	Carotenes	Vitamin E	Vitamin C	Thiamin	Riboflavin	Niacin	Vitamin B_6	Total Folate	Folate DFE	Vitamin B_{12}
	Micrograms retinol equivalent	*Micrograms retinol equivalent*	*Milligrams alpha-tocopherol*	*Milligrams*	*Milligrams*	*Milligrams*	*Milligrams*	*Milligrams*	*Micrograms*	*Micrograms*	*Micrograms*
1970	1,260	480	13.3	104	1.9	2.3	22	2.0	291	286	9.5
1971	1,260	490	13.0	105	2.0	2.3	22	2.0	293	287	9.5
1972	1,250	520	13.4	106	2.0	2.3	22	2.0	290	284	9.4
1973	1,210	550	13.9	104	2.0	2.2	22	1.9	296	290	8.9
1974	1,230	580	13.8	111	2.4	2.6	26	2.1	324	281	9.1
1975	1,210	590	14.0	114	2.4	2.6	26	2.0	336	297	8.5
1976	1,240	590	14.4	116	2.4	2.7	27	2.1	341	301	8.9
1977	1,190	550	14.0	115	2.4	2.7	27	2.1	340	300	8.8
1978	1,180	550	14.3	111	2.4	2.6	27	2.0	328	290	8.5
1979	1,190	580	14.4	112	2.5	2.7	27	2.1	341	301	8.2
1980	1,170	570	14.3	115	2.5	2.7	27	2.1	335	295	8.2
1981	1,170	580	14.5	113	2.5	2.6	27	2.1	334	295	8.3
1982	1,160	600	14.8	114	2.5	2.6	27	2.1	340	300	7.9
1983	1,160	570	15.1	118	2.5	2.7	27	2.1	345	305	8.2
1984	1,170	600	15.5	116	2.5	2.7	28	2.2	339	298	8.3
1985	1,160	600	16.0	117	2.6	2.7	29	2.2	354	311	8.3
1986	1,160	580	16.2	121	2.6	2.8	29	2.3	358	313	8.2
1987	1,170	610	16.3	117	2.7	2.9	30	2.3	349	302	8.2
1988	1,120	570	16.9	119	2.8	2.9	30	2.3	363	316	8.0
1989	1,150	610	16.5	119	2.8	2.9	30	2.3	358	311	8.1
1990	1,160	630	16.9	115	2.9	2.9	31	2.3	366	317	8.0
1991	1,130	600	17.2	118	2.9	2.9	31	2.4	376	326	7.9
1992	1,160	640	17.5	126	3.0	3.0	32	2.5	393	341	7.9
1993	1,190	710	17.9	126	3.0	2.9	32	2.5	388	334	7.8
1994	1,230	780	17.4	128	3.0	3.0	32	2.5	390	335	8.0
1995	1,200	720	16.9	126	3.0	2.9	31	2.4	380	326	8.0
1996	1,220	780	16.7	133	3.0	2.9	32	2.4	382	326	8.0
1997	1,240	820	16.9	128	3.0	2.9	32	2.4	376	321	7.8
1998	1,250	780	17.4	130	3.0	2.9	32	2.4	703	924	8.0
1999	1,210	730	17.6	129	3.0	2.9	32	2.5	712	935	8.0
2000	1,200	720	19.2	126	3.0	2.9	33	2.5	711	936	8.3

Year	Minerals								
	Calcium	Phosphorus	Magnesium	Iron	Zinc	Copper	Selenium	Potassium	Sodium [2]
	Milligrams	*Milligrams*	*Milligrams*	*Milligrams*	*Milligrams*	*Milligrams*	*Micrograms*	*Milligrams*	*Milligrams*
1970	930	1,520	330	15.7	12.3	1.5	128	3,490	1,370
1971	940	1,530	330	15.9	12.4	1.5	128	3,490	1,400
1972	930	1,510	330	15.7	12.2	1.6	128	3,460	1,370
1973	930	1,490	330	15.9	11.9	1.6	123	3,450	1,340
1974	900	1,490	330	16.5	13.4	1.6	121	3,410	1,350
1975	880	1,440	330	16.6	13.1	1.6	139	3,390	1,320
1976	920	1,500	340	17.1	13.7	1.7	143	3,500	1,380
1977	910	1,490	330	16.9	13.6	1.6	135	3,420	1,360
1978	900	1,470	330	16.5	13.3	1.6	137	3,350	1,360
1979	910	1,490	330	17.0	13.4	1.6	136	3,430	1,370
1980	890	1,470	330	16.8	13.3	1.6	135	3,380	1,330
1981	880	1,470	330	17.0	13.4	1.6	135	3,350	1,320
1982	890	1,470	330	17.2	13.4	1.6	137	3,370	1,310
1983	910	1,500	340	19.6	13.7	1.7	140	3,430	1,330
1984	910	1,510	340	19.8	13.8	1.7	140	3,450	1,360
1985	940	1,560	350	20.6	14.1	1.7	143	3,540	1,370
1986	950	1,580	360	20.9	14.4	1.8	145	3,590	1,380
1987	950	1,600	360	21.2	14.3	1.8	146	3,550	1,370
1988	940	1,610	370	21.8	14.5	1.8	148	3,580	1,340
1989	940	1,600	370	21.7	14.5	1.8	148	3,570	1,350
1990	960	1,650	380	22.6	15.0	1.8	151	3,610	1,370
1991	960	1,650	390	23.0	15.1	1.9	159	3,660	1,360
1992	970	1,690	400	23.6	15.6	1.9	163	3,740	1,380
1993	960	1,680	390	23.5	15.3	1.9	163	3,730	1,380
1994	990	1,700	390	23.4	15.3	1.9	164	3,770	1,380
1995	960	1,670	380	23.0	15.0	1.9	161	3,690	1,360
1996	970	1,680	390	23.2	14.9	1.9	164	3,760	1,340
1997	970	1,670	380	23.0	14.6	1.9	165	3,720	1,340
1998	970	1,680	390	23.2	14.9	1.9	178	3,760	1,360
1999	970	1,700	390	23.7	15.2	1.9	179	3,790	1,330
2000	970	1,700	390	23.7	15.2	1.9	181	3,770	1,340

[1] Computed by Center for Nutrition Policy and Promotion (CNPP), USDA. Based on Economic Research Service estimates of per capita quantities of food available for consumption (retail weight) and on CNPP estimates of quantities of produce from home gardens and certain other foods. No deduction is made in food supply estimates for loss of food or nutrients in further processing, in marketing, or in the home. Data include iron, thiamin, riboflavin, niacin, vitamin A, vitamin B_6, vitamin B_{12}, ascorbic acid, and zinc added by enrichment and fortification. [2] Sodium levels do not reflect sodium from most processed foods and therefore underestimate total sodium available in the U.S. food supply.

Center for Nutrition Policy and Promotion (CNPP), (703) 305–2563

Table 13-3.—Food nutrients: Percentage of total contributed by major food groups, 1970[1]

Nutrient	Meat, poultry, fish	Dairy[2] products	Eggs	Fats,[3] oils	Fruits		
					Citrus	Non- citrus	Total[5]
	Percent	*Percent*	*Percent*	*Percent*	*Percent*	*Percent*	*Percent*
Food energy	19.5	10.9	2.0	17.7	0.9	2.0	2.9
Carbohydrate	0.1	6.6	0.1	0	1.8	4.1	5.9
Protein	39.8	22.0	5.6	0.2	0.5	0.7	1.2
Total fat	34.7	12.8	2.9	43.3	0.1	0.3	0.4
Saturated fat	37.8	22.7	2.5	33.3	0	0.2	0.2
Monounsaturated fat	38.5	9.1	2.7	44.2	0	0.4	0.4
Polyunsaturated fat	19.1	2.5	2.2	64.4	0.1	0.4	0.5
Cholesterol	39.2	15.6	39.5	5.7	0	0	0
Dietary fiber	0	0.4	0	0	3.0	9.9	12.9
Vitamin A (retinol equivalents)	34.7	21.1	6.5	11.2	0.3	1.5	1.8
Carotene (retinol equivalents)	0	3.3	0	4.8	1.6	8.0	9.6
Vitamin E	5.4	3.9	3.4	65.7	0.9	3.2	4.1
Vitamin C	2.4	4.2	0	0	25.5	14.5	40.0
Thiamin	25.0	8.8	1.3	0.1	2.3	1.9	4.2
Riboflavin	21.6	38.4	9.5	0.3	0.6	1.7	2.3
Niacin	43.8	2.2	0.1	0	0.7	2.0	2.7
Vitamin B[6]	38.1	12.1	2.9	0.1	1.6	7.1	8.7
Folate	9.8	9.5	7.0	0.1	6.6	2.7	9.3
Folate DFE	8.0	9.3	7.1	0.1	6.7	2.8	9.5
Vitamin B[12]	73.3	20.3	4.5	0.2	0	0	0
Calcium	2.8	75.5	2.3	0.6	1.2	1.1	2.3
Phosphorus	25.5	36.6	5.1	0.3	0.6	1.0	1.6
Magnesium	12.7	20.7	1.3	0.1	2.[illegible]	3.9	5.8
Iron	22.6	2.4	4.0	0.1	0.[illegible]	2.4	3.1
Zinc	46.7	19.4	3.9	0.1	0	0.9	1.3
Copper	19.7	3.5	0.3	0	1	5.0	6.7
Selenium	18.4	16.1	10.4	0.1	0	0.4	0.6
Potassium	16.5	23.7	1.5	0.2	3	6.0	9.3
Sodium	24.5	25.8	4.0	13.2		1.4	1.4

Nutrient	Vegetables				Legumes, nuts, soy	Grain products	Sugar sweeteners	Miscellaneous[4]	Total[5]
	White potatoes	Dark green, deep yellow	Other	Total[5]					
	Percent	*Percent*	*Percent*	*Percent*	*Percent*	*Percent*	*Percent*	*Percent*	*Percent*
Food energy	2.7	0.4	1.4	5.1	2.9	19.7	18.5	0.8	100
Carbohydrate	5.2	0.7	2.6	9.7	2.1	34.3	40	1.1	99.9
Protein	2.4	0.4	2.2	5.6	5.3	18.8	0	1.5	100
Total fat	0.1	0	0.2	0.4	3.5	1.4	0	0.7	100.1
Saturated fat	0.1	0	0.1	0.2	1.9	0.7		0.7	100
Monounsaturated fat	0	0	0.1	0.1	3.8	0.5		0.6	99.9
Polyunsaturated fat	0.2	0.1	0.6	1.1	6.2	3.3		0.6	99.9
Cholesterol	0	0	0	0	0	0		0	100
Dietary fiber	11.3	3.4	13.7	32.6	14.1	31.2		8.8	100
Vitamin A (retinol equivalents)	0	15.8	2.3	19.5	0	0.1		5	99.9
Carotene (retinol equivalents)	0	62.4	11.7	78.2	0.1	0.6		3.3	99.9
Vitamin E	0.3	1.1	2.7	8.1	6.3	2.8		0.4	100.1
Vitamin C	18.5	6.4	14.7	48.9	0	0		4.5	100
Thiamin	5.5	0.8	3.9	11.6	5.3	42.9	0.[illegible]	0.6	99.9
Riboflavin	1.2	0.9	2.7	5.9	1.5	18.6	0.7	1.2	100
Niacin	6.1	0.8	3.1	12.3	4.9	29.4	0	4.6	100
Vitamin B[6]	14	2.2	4.8	24.1	3.4	9.4	0.3	1.1	100.2
Folate	5.5	2.8	15.9	27.2	19.4	15.5	0	2.3	100.1
Folate DFE	5.6	2.9	16.2	27.8	19.8	16.2	0	2.3	100.1
Vitamin B[12]	0	0	0	0	0	1.6	0	0	99.9
Calcium	1.2	0.9	3.5	6.5	3.6	3.5	0.7	2.2	100
Phosphorus	3.6	0.6	3.0	8.1	5.1	14.5	0.4	2.9	100.1
Magnesium	6.6	1.3	6.3	16.5	12	17.2	0.8	12.9	100
Iron	4.9	1.2	5.4	13.7	9.2	36.6	1.3	7.0	100
Zinc	2.6	0.5	3.0	6.8	5.9	12.3	0.5	3.2	100.1
Copper	6.3	1.5	5.3	17.4	17.1	17.9	4.4	13.0	100
Selenium	1.7	0.2	0.7	2.8	9.6	39.6	0.9	1.5	100
Potassium	12.2	1.8	6.8	25.3	7.8	6.6	0.5	8.6	100
Sodium	2.8	1.0	15.3	27.3	0.2	0.6	2.7	0.3	100

[1] Percentages of food groups are based on aggregate data. [2] Excludes buter. [3] Includes butter. [4] Coffee, tea, spices, chocolate liquor equivalent of cocoa beans, and fortification not assigned to a specific group. [5] Components may not add to total due to rounding.

Center for Nutrition Policy and Promotion, (703) 305–2563.

Table 13-4.—Food nutrients: Percentage of total contributed by major food groups, 2000[1]

Nutrient	Meat, poultry, fish	Dairy products[2]	Eggs	Fats, oils[3]	Fruits		
					Citrus	Non-citrus	Total[5]
	Percent	*Percent*	*Percent*	*Percent*	*Percent*	*Percent*	*Percent*
Food energy	13.8	8.9	1.3	21.4	0.9	2.1	3.0
Carbohydrate	0.1	4.4	0.1	0	1.7	4.0	5.7
Protein	39.1	19.1	3.9	0.1	0.5	0.7	1.2
Total fat	22.9	11.7	2.1	55.6	0	0.4	0.5
Saturated fat	25.6	23.4	2.0	43.7	0	0.3	0.3
Monounsaturated fat	25.1	7.9	1.8	58.9	0	0.4	0.5
Polyunsaturated fat	13.6	1.8	1.3	72.0	0	0.4	0.5
Cholesterol	43.5	16.0	35.2	5.3	0	0	0
Dietary fiber	0	0.3	0	0	2.2	8.8	11.0
Vitamin A (retinol equivalents)	28.1	23.0	5.6	9.5	0.3	1.9	2.2
Carotene (retinol equivalents)	0	2.1	0	2.6	1.1	6.4	7.5
Vitamin E	4.0	2.4	1.9	71.4	0.8	2.4	3.2
Vitamin C	2.0	2.5	0	0	25.8	16.1	41.9
Thiamin	16.9	4.6	0.7	0	1.8	1.6	3.4
Riboflavin	16.5	26.0	6.0	0.2	0.4	1.8	2.3
Niacin	35.1	1.2	0.1	0	0.5	1.5	2.0
Vitamin B^6	34.4	8.6	1.9	0	1.5	8.3	9.7
Folate	3.5	3.3	2.3	0	4.0	1.6	5.7
Folate DFE	2.6	2.5	1.8	0	3.1	1.2	4.3
Vitamin B^{12}	75.2	20.3	4.3	0.2	0	0	0
Calcium	3.2	71.9	1.8	0.4	1.2	1.3	2.5
Phosphorus	24.3	32.0	3.7	0.2	0.7	1.1	1.8
Magnesium	12.2	15.4	0.9	0.1	2.0	4.2	6.2
Iron	15.2	1.9	2.1	0.1	0.4	1.9	2.3
Zinc	37.2	16.5	2.5	0.1	0.3	0.9	1.2
Copper	14.1	2.6	0.2	0	1.6	4.7	6.3
Selenium	27.4	10.7	6.0	0	0.1	0.4	0.5
Potassium	16.6	17.9	1.1	0.1	3.8	7.4	11.2
Sodium	19.3	32.6	3.3	11.1	0.1	1.3	1.4

Nutrient	Vegetables				Legumes, nuts, soy	Grain products	Sugars, sweeteners	Miscellaneous[4]	Total[5]
	White potatoes	Dark-green, deep-yellow	Other	Total[5]					
	Percent	*Percent*	*Percent*	*Percent*	*Percent*	*Percent*	*Percent*	*Percent*	*Percent*
Food energy	2.3	0.4	1.1	4.5	3.0	24.8	18.6	0.8	100.1
Carbohydrate	4.1	0.8	2.0	7.9	2.0	40.3	38.3	1.2	100
Protein	2.2	0.6	1.7	5.1	6.2	23.4	0	1.8	99.9
Total fat	0.1	0.1	0.2	0.4	3.5	2.5	0	0.9	100.1
Saturated fat	0.1	0	0.1	0.2	2.2	1.6	0	1.1	100.1
Monounsaturated fat	0	0	0.1	0.1	3.7	1.3	0	0.7	100
Polyunsaturated fat	0.1	0.1	0.4	0.8	5.0	4.3	0	0.7	100
Cholesterol	0	0	0	0	0	0	0	0	100
Dietary fiber	8.2	4.2	9.5	25.7	14.4	37.7	0	10.9	100
Vitamin A (retinol equivalents)	0	21.6	1.8	25.0	0	0.4	0	6.1	99.9
Carotene (retinol equivalents)	0	72.8	5.9	82.1	0.1	0.6	0	5.0	100
Vitamin E	0.2	1.8	1.5	6.7	5.3	4.6	0	0.5	100
Vitamin C	15	12.7	9.8	45.1	0.1	4.3	0	4.2	100.1
Thiamin	4.2	0.9	2.3	8.5	4.4	60.8	0.1	0.6	100
Riboflavin	1.0	1.2	2.2	5.4	1.6	40.0	0.8	1.4	100.2
Niacin	4.6	0.9	2.0	9.1	3.7	45.8	0	3.0	100
Vitamin B^6	11.2	2.8	4.6	21.3	3.8	18.5	0.2	1.6	100
Folate	2.2	2.3	5.7	11.4	9.7	62.8	0	1.3	100
Folate DFE	1.7	1.7	4.3	8.7	7.3	71.8	0	1.0	100
Vitamin B^{12}	0	0	0	0	0	0.1	0	0	100.1
Calcium	1.1	1.5	3.4	6.9	4.5	5.2	0.7	3.0	100.1
Phosphorus	3.0	1.0	2.7	7.6	6.1	20.3	0.3	3.7	100
Magnesium	5.2	1.8	4.4	13.6	13.1	24.6	0.7	13.2	100
Iron	3.5	1.3	3.3	9.6	7.7	53.4	0.9	6.9	100.1
Zinc	2.1	0.7	2.3	5.8	5.6	26.8	0.5	3.8	100
Copper	4.9	1.7	4.2	14.6	20.4	23.8	3.7	14.3	100
Selenium	1.3	0.2	0.7	2.3	6.0	45.0	0.9	1.2	100
Potassium	11.2	3.2	6.1	25.0	9.5	9.9	0.5	8.1	99.9
Sodium	3.0	1.0	11.8	27.0	0.3	0.9	3.7	0.4	100

[1] Percentages of food groups are based on aggregate nutrient data [2] Excludes butter. [3] Includes butter. [4] Coffee, tea, spices, chocolate liquor equivalent of cocoa beans, and fortification not assigned to a specific food group. [5] Components may not add to total due to rounding.

Center for Nutrition Policy and Promotion, (703) 305–2563.

Table 13-5.—Consumption: Per capita consumption of major food commodities, United States, 1993–2000 [1]

Commodity	1993	1994	1995	1996	1997	1998	1999	2000 [2]
	Pounds	*Pounds*	*Pounds*	*Pounds*	*Poun ds*	*Pounds*	*Pounds*	*Pounds*
Red meats [3 4]	111.2	113.5	113.6	111.0	109.0	113.2	115.1	113.7
Beef	61.0	62.9	63.5	64.0	62.6	63.6	64.3	64.6
Veal	0.8	0.8	0.8	1.0	0.8	0.7	0.6	0.5
Lamb and mutton	1.0	0.9	0.9	0.8	0.8	0.9	0.8	0.8
Pork	48.5	49.0	48.4	45.2	44.7	48.2	49.3	47.8
Fish [3]	14.8	15.0	14.8	14.5	14.3	14.5	14.8	15.2
Canned	4.4	4.4	4.6	4.4	4.3	4.3	4.5	4.7
Fresh and frozen	10.1	10.2	9.8	9.8	9.7	9.9	10.1	10.2
Cured	0.3	0.3	0.3	0.3	0.3	0.3	0.3	0.3
Poultry [3 4]	62.0	62.6	62.1	63.1	63.1	63.7	66.7	66.9
Chicken	48.1	48.7	48.2	48.8	49.4	49.7	52.8	53.2
Turkey	13.9	13.9	13.9	14.3	13.6	13.9	13.8	13.7
Eggs	30.1	30.3	29.8	29.9	30.2	30.8	32.1	32.2
Dairy products: [5]								
Total dairy products	569.4	579.7	576.2	566.2	567.2	572.4	584.6	593.4
Fluid milk and cream	223.0	221.9	219.6	218.7	215.8	212.5	212.0	209.2
Plain and flavored whole milk	79.1	77.2	74.0	73.0	71.0	69.5	70.1	69.3
Plain reduced fat and light milk (2%, 1%, and 0.5%)	95.5	94.0	90.9	89.1	87.0	85.0	84.5	83.8
Plain fat free milk (skim)	26.5	28.5	31.5	33.2	33.7	33.7	32.5	30.0
Flavored lower fat fat free milk	6.8	7.0	7.2	7.7	7.9	8.2	8.4	8.7
Buttermilk	3.0	2.9	2.8	2.6	2.5	2.5	2.4	2.2
Eggnog	0.4	0.4	0.4	0.4	0.4	0.4	0.4	0.3
Yogurt (excl. frozen)	4.2	4.6	5.0	4.8	5.1	5.0	4.9	5.4
Heavy cream, light cream and half and half	4.9	4.9	5.0	5.3	5.6	5.6	6.1	6.3
Sour cream and dip	2.7	2.7	2.9	2.8	2.9	3.0	3.0	3.3
Cheese (excluding cottage) [6]	26.0	26.5	26.9	27.3	27.5	27.8	29.0	29.8
American	11.3	11.4	11.7	11.8	11.8	11.9	12.6	12.7
Cheddar	9.1	9.0	9.0	9.0	9.4	9.4	9.8	9.7
Italian	9.7	10.2	10.3	10.6	10.8	11.1	11.5	12.1
Mozzarella	7.5	7.9	8.0	8.3	8.2	8.6	9.0	9.3
Cottage cheese	2.9	2.8	2.7	2.6	2.6	2.7	2.6	2.6
Condensed and evaporated milk	8.2	8.4	6.8	6.3	6.5	6.1	6.5	5.8
Ice cream	16.0	16.0	15.5	15.6	16.1	16.3	16.7	16.5
Fats and oils [7]	69.1	67.3	65.4	64.2	63.7	64.3	67.0	74.5
Butter	4.6	4.8	4.4	4.3	4.1	4.4	4.7	4.6
Margarine	11.0	9.8	9.1	9.0	8.4	8.2	7.9	8.2
Shortening	24.9	23.9	22.2	21.9	20.5	20.5	21.1	23.1
Lard (direct use)	1.1	1.8	1.6	1.7	1.9	2.0	2.0	1.9
Edible tallow (direct use)	2.2	2.4	2.7	2.9	2.1	3.1	3.6	4.0
Salad and cooking oils	26.6	25.9	26.5	25.7	28.0	27.3	28.8	33.7
Fruits and vegetables [4 8]	689.6	701.8	690.0	697.9	707.6	698.8	705.1	708.6
Fruits	280.6	287.5	281.8	279.1	289.5	283.9	289.8	280.0
Fresh	123.5	124.9	122.5	126.2	129.4	128.8	129.4	126.9
Citrus	25.8	24.7	23.8	24.6	26.5	26.6	20.3	23.4
Noncitrus	97.7	100.2	98.7	101.6	102.9	102.3	109.1	103.4
Processing	157.1	162.6	159.3	153.0	160.1	155.1	160.3	153.2
Citrus	86.9	93.2	93.4	87.0	94.9	90.5	93.4	89.0
Noncitrus	70.2	69.5	65.9	65.9	65.2	64.6	67.0	64.1
Vegetables	409.0	414.3	408.2	418.8	418.1	414.9	415.4	428.5
Fresh	180.7	186.6	180.8	185.9	190.1	186.3	191.8	201.8
Processing	228.3	227.7	227.4	232.9	228.0	228.6	223.6	226.7
Flour and cereal products [4]	189.1	191.8	190.2	196.2	197.3	196.0	196.8	200.0
Wheat flour [9]	142.1	142.9	140.0	146.4	146.8	144.8	144.0	146.3
Rice (milled basis)	16.6	18.0	18.6	17.6	18.1	18.3	19.5	19.7
Corn products	23.1	24.0	24.9	25.9	26.5	27.2	27.8	28.4
Oat products	6.0	5.7	5.4	5.0	4.6	4.4	4.4	4.3

See footnotes at end of table.

Table 13-5.—Consumption: Per capita consumption of major food commodities, United States, 1993–2000 [1]—Continued

Commodity	1993	1994	1995	1996	1997	1998	1999	2000 [2]
	Pounds	*Pounds*	*Pounds*	*Pounds*	*Pounds*	*Pounds*	*Pounds*	*Pounds*
Barley and rye products	1.3	1.3	1.3	1.3	1.2	1.3	1.2	1.2
Caloric sweeteners (dry weight basis) [4]	139.1	141.5	143.8	145.0	148.1	149.1	151.3	148.9
Sugar (refined)	63.8	64.3	64.7	65.5	65.3	65.1	66.3	65.6
Corn sweeteners [10]	73.9	75.9	77.9	78.2	81.5	82.7	83.5	81.9
Honey and edible syrups	1.4	1.4	1.3	1.4	1.3	1.3	1.5	1.5
Other:								
Coffee (green bean equivalent)	9.0	8.1	7.9	8.7	9.1	9.3	9.8	10.3
Cocoa (chocolate liquor equivalent) [11]	4.3	3.8	3.6	4.2	4.0	4.3	4.5	4.7
Tea (dry leaf equivalent)	0.9	0.9	0.8	0.8	0.8	0.9	0.9	0.8
Peanuts (shelled)	6.0	5.7	5.6	5.6	5.7	5.8	6.0	5.8
Tree nuts (shelled)	2.3	2.3	1.9	1.9	2.1	2.2	2.5	2.5

[1] Quantity in pounds, retail weight unless otherwise shown. [2] Preliminary. [3] Boneless, trimmed weight equivalent. [4] Total may not add due to rounding. [5] Total dairy products reported on a milk-equivalent, milkfat basis. All other dairy categories reported on a product weight basis. [6] Natural equivalent of cheese and cheese products. [7] Total fats and oils reported on a fat content basis. All other fats and oils categories reported on a product weight basis. [8] Farm weight. [9] White, whole wheat, semolina, and durum flour. [10] High fructose, glucose, and dextrose. [11] Chocolate liquor is what remains after cocoa beans have been roasted and hulled; it is sometimes called ground or bitter chocolate.--= Not available

ERS, Food and Rural Economics Division, (202) 694-5400. Historical consumption and supply-utilization data for food may be found at,www.ers.USDA.gov/DATA/Food Consumption/, ERS, USDA, 2002.

Table 13-6.—Food plans: Food cost at home, at four cost levels, for families and individuals in the United States, for week and month, June 2002 [1]

Age-gender groups	Weekly cost				Monthy cost			
	Thrifty plan	Low-cost plan	Moderate-cost plan	Liberal plan	Thrifty plan	Low-cost plan	Moderate-cost plan	Liberal plan
	Dollars	*Dollars*	*Dollars*	*Dollars*	*Dollars*	*Dollars*	*Dollars*	*Dollars*
Individuals: [2].								
Child:.								
1 year	16.70	20.40	23.90	29.00	72.40	88.40	103.60	125.70
2 year	16.70	20.40	23.90	29.00	72.40	88.40	103.60	125.70
3-5 years	18.10	22.40	27.70	33.10	78.40	97.10	120.00	143.40
6-8 years	22.60	29.80	37.10	43.10	97.90	129.10	160.80	186.80
9-11 years	26.70	33.80	43.30	50.00	115.70	146.50	187.60	216.70
Male:.								
12-14 years	27.70	38.20	47.30	55.70	120.00	165.50	205.00	241.30
15-19 years	28.70	39.50	49.20	56.70	124.40	171.20	213.20	245.70
20-50 years	30.60	39.30	48.90	59.20	132.60	170.30	211.90	256.50
51 years and over	27.60	37.40	45.90	55.10	119.60	162.10	198.90	238.70
Female:.								
12-19 years	27.60	32.90	40.00	48.20	119.60	142.60	173.30	208.90
20-50 years	27.60	34.30	41.70	53.50	119.60	148.60	180.70	231.80
51 years and over	26.90	33.40	41.30	49.30	116.60	144.70	179.00	213.60
Families:.								
Family of 2: [3].								
20-50 years	64.00	81.00	99.70	124.00	277.40	350.80	431.90	537.10
51 years and over	60.00	77.90	95.90	114.80	259.80	337.50	415.70	497.50
Family of 4:.								
Couple, 20-50 years and children.								
2 and 3-5 years	93.00	116.40	142.20	174.80	403.00	504.40	616.20	757.40
6-8 and 9-11 years	107.50	137.20	171.00	205.80	465.80	594.50	741.00	891.80

[1] Basis is that all meals and snacks are purchased at stores and prepared at home. For specific foods and quantities of foods in the Low-Cost, Moderate-Cost, and Liberal Plans, see *Family Economics Review,* No. 2 (1983); for specific foods and quantities of foods in the Thrifty Food Plan, see *Thrifty Food Plan, 1999, Executive Summary,* CNPP-7A. The Thrifty Food Plan is based on 1989-91 data, and the other three food plans are based on 1977-78 data updated to current dollars using the Consumer Price Index for specific food items. [2] The costs given are for individuals in 4-person families. For individuals in other size families, the following adjustments are suggested: 1-person--add 20 percent; 2-person--add 10 percent; 3-person--add 5 percent; 5- or 6-person--subtract 5 percent; 7- (or more) person--subtract 10 percent. [3] Ten percent added for family size adjustment.

Center for Nutrition Policy and Promotion, (703) 605–4266.

Table 13-7.—Food Stamp Program: Participation and Federal costs, fiscal years 1992–2001

Fiscal year[1]	Average monthly participation[2]		Recipient benefits	Total cost[3]	Average monthly benefit	
	Persons	Housholds			Per person	Per household
	1,000	*1,000*	*1,000 dollars*	*1,000 dollars*	*Dollars*	*Dollars*
1992	25,406	10,060	20,905,665	22,462,321	68.57	173.18
1993	26,982	10,788	22,006,031	23,652,972	67.96	169.99
1994	27,468	11,089	22,748,559	24,492,689	69.02	170.96
1995	26,619	10,879	22,764,067	24,619,600	71.27	174.37
1996	25,542	10,549	22,441,461	24,327,016	73.22	177.28
1997	22,858	9,455	19,550,178	21,486,911	71.27	172.31
1998	19,788	8,249	16,889,069	18,892,566	71.12	170.62
1999	18,183	7,668	15,755,368	17,698,259	72.21	171.22
2000	17,158	7,334	14,984,803	17,058,127	72.78	170.28
2001[4]	17,313	7,447	15,546,941	17,796,683	74.83	173.98

[1] October 1 to September 30. [2] Participation data are 12-month averages. [3] Total cost includes matching funds for state administrative expenses (e.g., certification of households, quality control, anti-fraud activities; employment and training); other Federal costs (e.g., benefit redemption processing; computer support; electronic benefit transfer systems; retailer redemption and monitoring; certification of SSI recipients; nutrition education and program information). [4] Preliminary.

FNS, Budget Division/Program Reports, Analysis and Monitoring Branch, (703) 305–2163

Table 13-8.—Food and Nutrition Service Programs: Federal costs of the National School Lunch, School Breakfast, Child Care Food, Summer Food Service, WIC, Special Milk, and Food Distribution Programs, fiscal years 1992–2001[1]

Fiscal year[2]	Child Nutrition					WIC[6]	Special Milk	Food Distribution Programs[7]
	Cash payments[3]				Cost of food distributed[5]			
	School Lunch	School Breakfast	Child & Adult Care[4]	Summer Food				
	1,000 dollars	*1,000 dollars*	*1,000 dollars*	*1,000 dollars*	*1,000 dollars*	*1,000 dollars*	*1,000 dollars*	*1,000 dollars*
1992	3,856,480	786,653	1,063,696	201,288	734,399	2,596,570	19,537	707,723
1993	4,081,343	868,794	1,179,904	217,459	703,945	2,828,914	18,738	693,934
1994	4,290,808	959,044	1,303,271	227,742	764,414	3,169,503	17,751	696,805
1995	4,466,446	1,048,266	1,411,382	235,451	732,886	3,436,121	16,984	512,837
1996	4,661,542	1,118,738	1,479,107	248,555	733,718	3,683,572	16,755	406,746
1997	4,934,039	1,214,288	1,514,531	242,499	661,288	3,844,142	17,436	514,874
1998	5,101,623	1,272,211	1,489,658	261,586	774,276	3,889,874	16,943	551,605
1999	5,314,245	1,344,758	1,556,849	266,879	753,936	3,939,717	16,487	594,966
2000	5,493,323	1,393,377	1,619,644	265,674	704,192	3,981,028	15,443	531,208
2001[8]	5,611,942	1,449,887	1,667,195	269,627	917,060	4,149,521	15,532	707,654

[1] See table 13-7 for Food Stamp Program costs. [2] October 1–September 30. [3] Includes sponsor administrative costs for the Child and Adult Care Food Program (CACFP) and the Summer Food Service Programs (SFS), and State administrative and health clinic expenses for SFS. Excludes CACFP audit and startup costs and School Breakfast startup costs. [4] The Adult Care component was initiated in fiscal year 1989. [5] Includes entitlement commodities, bonus commodities, and cash-in-lieu for the National School Lunch, School Breakfast, Child and Adult Care Food, and Summer Food Service Programs. [6] Includes food costs, administrative costs, program evaluation funds, special grants, and Farmers Market projects for the Special Supplemental Food Program for Women, Infants and Children. [7] Includes entitlement and bonus commodities, cash-in-lieu of commodities, and administrative costs of the following programs: Indian Reservations (Needy Family), Nutrition for the Elderly, Commodity Supplemental Food, Charitable Institutions, Summer Camps, Emergency Food Assistance Program (TEFAP), Soup Kitchens/Food Banks, and Disaster Feeding. [8] Preliminary.

FNS, Budget Division/Program Reports, Analysis and Monitoring Branch, (703) 305–2163

Table 13-9.—Food and Nutrition Service program benefits: Cash payments made under the National School Lunch, School Breakfast, Child and Adult Care, Summer Food and Special Milk Programs and the value of food benefits provided under the Food Stamp, WIC, Commodity Distribution and the Emergency Feeding Food Programs, fiscal year 2001 [1]

State/Territory	Child Nutrition Program (cash payments only) [2]					Special Supplemental Food (WIC) [3]	Commodity distribution [4]	Food Stamp Program [5]	Emergency food assistance (TEFAP)	Total [5]
	Child and Adult Care Food	Summer Food	Special Milk	National School Lunch	Breakfast					
	1,000 dollars	*1,000 dollars*	*1,000 dollars*	*1,000 dollars*	*1,000 dollars*	*1,000 dollars*	*1,000 dollars*	*1,000 dollars*	*1,000 dollars*	*1,000 dollars*
Alabama	29,003	3,907	54	111,753	27,554	49,140	19,555	364,736	5,134	610,837
Alaska	5,356	150	6	16,764	2,783	12,180	2,161	46,477	1,199	87,077
Am. Samoa [5]	0	0	0	0	0	3,842	0	0	0	3,842
Arizona	34,316	3,922	156	106,496	26,401	63,448	28,405	280,403	6,693	550,239
Arkansas	16,312	1,449	16	65,462	20,733	30,635	13,065	223,351	1,615	372,639
California	186,588	18,004	787	774,886	198,240	555,466	123,778	1,583,143	48,546	3,489,438
Colorado	19,022	1,134	150	53,193	9,863	27,104	16,843	130,771	3,903	261,984
Connecticut	8,301	2,643	428	45,421	9,879	25,027	10,709	135,683	2,386	240,477
Delaware	6,489	930	50	11,905	2,997	6,291	3,098	31,927	1,210	64,897
District of Col.	2,799	1,975	7	16,351	4,326	6,615	3,363	70,341	980	106,757
Florida	73,150	18,708	125	319,202	84,958	140,976	50,586	770,726	17,611	1,476,042
Georgia	55,365	9,687	38	213,110	67,515	86,855	39,268	515,153	5,255	992,246
Guam	81	0	0	3,585	1,199	3,950	281	37,172	276	46,545
Hawaii	4,246	575	3	25,351	5,761	17,651	4,323	149,759	1,147	208,816
Idaho	3,690	1,048	200	25,088	4,628	11,044	5,362	47,225	1,816	100,099
Illinois	71,043	9,622	2,590	225,130	36,360	112,373	40,850	810,104	17,206	1,325,280
Indiana	23,443	2,551	321	91,118	19,648	50,159	24,653	317,041	5,860	534,793
Iowa	12,649	584	133	46,814	9,465	22,031	16,183	107,295	2,125	217,279
Kansas	22,685	979	131	46,048	11,751	19,569	11,545	91,868	3,453	208,029
Kentucky	21,789	3,559	84	94,743	31,672	45,326	20,052	350,490	5,279	572,994
Louisiana	37,592	5,496	54	139,073	43,824	55,081	37,105	482,657	5,639	806,521
Maine	7,936	684	124	17,446	4,046	6,623	4,362	86,333	1,950	129,504
Maryland	26,473	2,780	441	75,736	20,227	35,784	15,470	190,821	2,173	369,904
Massachusetts	35,748	4,533	491	82,581	20,840	41,371	23,223	172,989	3,292	385,069
Michigan	39,250	3,806	841	143,699	34,598	88,668	50,830	504,150	10,614	876,454
Minnesota	47,271	2,931	959	70,977	15,447	36,127	23,474	172,485	3,277	372,948
Mississippi	21,110	3,208	9	100,853	34,678	39,907	14,460	254,371	4,650	473,247
Missouri	28,487	5,026	409	99,947	27,615	47,864	24,840	394,611	7,628	636,427
Montana	7,879	478	37	14,413	3,059	8,774	6,765	53,847	933	96,185
Nebraska	18,612	598	123	30,036	5,792	13,807	11,893	62,711	1,247	144,820
Nevada	2,631	1,819	174	28,113	7,275	15,354	5,872	65,330	1,097	127,666
New Hampshire	2,332	376	199	11,154	2,147	5,504	5,968	28,051	1,393	57,125
New Jersey	31,387	6,053	865	116,577	17,366	51,548	24,221	292,111	7,333	547,461
New Mexico	27,681	5,479	9	50,357	15,416	23,248	14,375	136,264	3,195	276,024
New York	109,324	38,832	1,006	403,293	92,417	200,811	85,476	1,365,353	21,722	2,318,235
North Carolina	56,838	4,385	160	162,998	49,696	73,930	37,676	424,835	8,944	819,461
North Dakota	7,227	270	74	10,592	1,945	6,438	6,120	27,126	827	60,617
Ohio	44,490	4,922	847	159,581	36,166	91,654	35,808	572,860	16,119	962,445
Oklahoma	32,630	1,880	67	77,683	25,346	38,393	30,837	236,136	6,870	449,842
Oregon	20,599	2,012	149	51,416	17,295	38,073	11,279	239,554	5,081	385,457
Pennsylvania	35,656	13,326	754	166,594	35,822	93,664	42,140	639,175	13,584	1,040,716
Puerto Rico [5]	16,416	4,695	0	112,504	26,552	131,873	14,033	0	9,145	315,216
Rhode Island	4,852	961	99	16,656	3,626	8,721	3,218	59,064	990	98,187
South Carolina	18,726	6,486	8	101,030	32,092	43,751	16,859	269,270	4,787	493,010
South Dakota	5,051	835	41	15,791	3,353	7,875	9,847	39,349	1,349	83,490

See footnotes at end of table.

Table 13-9.—Food and Nutrition Service program benefits: Cash payments made under the National School Lunch, School Breakfast, Child and Adult Care, Summer Food and Special Milk Programs and the value of food benefits provided under the Food Stamp, WIC, Commodity Distribution and the Emergency Feeding Food Programs, fiscal year 2001 [1]—Continued

State/Territory	Child Nutrition Program (cash payments only) [2]					Special Supplemental Food (WIC) [3]	Commodity distribution [4]	Food Stamp Program [5]	Emergency Food Assistance (TEFAP)	Total [5]
	Child and Adult Care Food	Summer Food	Special Milk	National School Lunch	Breakfast					
	1,000 dollars	*1,000 dollars*	*1,000 dollars*	*1,000 dollars*	*1,000 dollars*	*1,000 dollars*	*1,000 dollars*	*1,000 dollars*	*1,000 dollars*	*1,000 dollars*
Tennessee	30,542	5,364	28	114,129	31,390	66,348	23,804	454,424	6,170	732,198
Texas	119,789	19,421	99	574,710	183,237	242,320	88,293	1,269,894	25,484	2,523,246
Utah	19,459	1,751	83	40,034	6,389	21,577	19,850	66,827	2,107	178,078
Vermont	3,119	333	91	7,537	2,155	6,848	2,905	31,103	737	54,827
Virginia	643	523	2	3,854	441	3,613	550	17,631	256	27,513
Virgin Islands	21,523	3,554	322	106,383	27,199	57,654	22,982	262,923	5,246	507,787
Washington	29,843	2,995	292	88,326	21,897	66,201	19,832	260,878	7,877	498,141
West Virginia	11,479	1,437	31	38,744	13,628	19,485	7,191	178,432	2,181	272,608
Wisconsin	25,897	2,101	1,341	72,533	9,560	39,510	22,601	152,453	6,208	332,205
Wyoming	3,691	65	23	7,725	1,568	4,049	2,836	19,259	659	39,873
Outlying Areas [6]	0	0	0	6,448	50	0	1,233	0	0	7,731
Dpt. of Defense	0	0	0	0	0	0	59	0	0	59
United States	1,548,510	240,845	15,532	5,611,942	1,449,887	3,022,127	1,202,372	15,546,941	332,455	28,970,609

[1] Preliminary. Excludes all administrative and program evaluation costs. [2] Excludes $1.7 million for Food Safety Education, $13.7 million for the School Meals Initiative for Healthy Children (Team Nutrition), and $8.2 million for the Free School Breakfast Pilot Project. [3] Includes $14.2 million for WIC Farmers' Market Nutrition Program benefits. [4] Includes distribution of bonus and entitlement commodities to the National School Lunch, Child and Adult Care, Summer Food Service, Charitable Institutions, Summer Camps, Food Distribution on Indian Reservations, Nutrition for the Elderly, Commodity Supplemental Food, and Disaster Feeding programs. Also includes cash-in-lieu of commodities for the National School Lunch, Child and Adult Care, and Nutrition for the Elderly programs. [5] Excludes Nutrition Assistance grants of $1,296 million for Puerto Rico, $6.1 million for the Northern Marianas, $5.3 million for American Samoa, and $0.6 million for nuclear affected areas of the Marshall Islands. [6] Outlying Areas include the Northern Marianas and the Marshall Islands.

FNS, Budget Division/Program Reports, Analysis and Monitoring Branch (703) 305–2163.

Table 13-10.—Food and Nutrition Service Programs: Persons participating, fiscal years 1992–2001

Fiscal year	National School Lunch Program [1]	School Breakfast Program [1]	Child and Adult Care Program [2]	Summer Food Service [3]	WIC Program [4]
	Thousands	*Thousands*	*Thousands*	*Thousands*	*Thousands*
1992	24,606	4,918	1,872	1,922	5,403
1993	24,855	5,358	1,977	2,119	5,921
1994	25,281	5,835	2,187	2,207	6,477
1995	25,685	6,318	2,354	2,107	6,894
1996	25,942	6,583	2,415	2,213	7,191
1997	26,341	6,922	2,472	2,176	7,407
1998	26,598	7,142	2,601	2,201	7,367
1999	26,946	7,371	2,670	2,172	7,311
2000	27,239	7,554	2,707	2,104	7,192
2001 [5]	27,504	7,792	2,725	2,115	7,306

[1] Average monthly participation (excluding summer months). [2] Average daily attendance (data reported quarterly). [3] Average daily attendance for peak month (July). [4] Average monthly participation. WIC is an abbreviation for the Special Supplemental Food Program for Women, Infants and Children. [5] Preliminary.

FNS, Budget Division/Program Reports, Analysis and Monitoring Branch (703) 305–2163.

Table 13-11.—Consumers' prices: Index number of prices paid for goods and services, United States, 1993–2002 [1]

[1982–84=100]

Year	Food	Nonfood items					All items
		Apparel and upkeep	Housing		Transportation	Medical care	
			Total	Rent			
1993	140.9	133.7	141.2	162.0	130.4	201.4	144.5
1994	144.3	133.4	144.8	167.0	134.3	211.0	148.2
1995	148.4	132.0	148.5	172.4	139.1	220.5	152.4
1996	153.3	131.7	152.8	178.0	143.0	228.2	156.9
1997	157.3	132.9	156.8	183.4	144.3	234.6	160.5
1998	160.7	133.0	160.4	189.6	141.6	242.1	163.0
1999	164.1	131.3	163.9	195.0	144.4	250.6	166.6
2000	167.8	129.6	169.6	201.3	153.3	260.8	172.2
2001	173.1	127.3	176.4	208.9	154.3	272.8	177.1
2002	176.2	124.0	180.3	216.7	152.9	285.6	179.9

[1] Reflects retail prices of goods and services usually bought by average families in urban areas of the United States. This index is the official index released monthly by the U.S. Department of Labor. Beginning 1978 data are for all urban consumers; earlier data are for urban wage earners and clerical workers.

ERS, Food Markets Branch, (202) 694–5349. Compiled from data of the U.S. Department of Labor.

Table 13-12.—Food service: Sales by industry segment, 1995–2001 [1]

Industry segment	1995 [2]	1996 [2]	1997 [2]	1998 [2]	1999 [2]	2000 [2]	2001 [2]
	Million dollars						
Commercial:							
Separate eating places.							
Full Service restaurants	88,774	93,386	102,461	107,016	111,357	120,094	122,901
Limited service restaurants	108,216	110,196	113,302	121,139	126,710	136,176	143,126
Other places	3,988	4,421	5,090	5,434	5,929	6,401	6,896
Separate drinking places	1,564	1,561	1,546	1,623	1,698	1,895	1,926
Total	202,542	209,564	222,399	235,212	245,694	264,566	274,849
Lodging places	12,526	12,807	13,599	13,907	14,650	15,353	15,740
Retail hosts	11,194	11,404	11,628	12,363	13,370	13,357	13,604
Recreation and entertainment	9,938	10,669	8,043	9,422	10,058	10,799	11,398
Commercial feeding total	236,200	244,444	255,669	270,904	283,772	304,075	315,591
Noncommercial:							
Education.							
Elementary and secondary	10,400	10,960	11,452	11,866	12,536	12,845	13,265
Colleges and universities	10,992	11,416	11,943	12,544	12,910	13,152	13,561
Total	21,392	22,376	23,395	24,410	25,446	25,997	26,826
Military services—							
Troop feeding	1,130	1,102	1,070	1,054	1,040	1,000	1,000
Clubs and exchanges	1,700	1,704	1,716	1,762	1,800	1,846	2,022
Total	2,830	2,806	2,786	2,816	2,840	2,846	3,022
Plants and office buildings	5,757	5,594	5,645	5,896	6,121	6,346	6,472
Hospitals	3,570	3,602	3,576	3,514	3,508	3,526	3,618
Nursing homes	5,758	6,158	6,234	6,268	6,332	6,482	6,716
Vending	3,133	3,079	2,800	2,874	3,084	3,282	3,256
Transportation	4,162	4,266	4,710	4,794	5,890	5,342	5,336
Associations	969	784	1,125	1,222	1,309	1,404	1,486
Correctional facilities	3,050	3,072	5,578	6,068	6,080	6,300	6,602
Child daycare centers	3,350	3,490	3,800	4,042	4,452	4,866	5,300
Elderly feeding programs	344	338	348	346	354	370	408
Other group	4,128	4,332	4,762	4,640	4,856	4,972	4,590
Total	34,221	34,715	38,578	39,664	41,986	42,890	43,784
Noncommercial feeding total	58,443	59,897	64,759	66,890	70,272	71,733	73,632
Total	294,643	304,341	320,428	337,794	354,044	375,808	389,223

[1] Revised using new NAICS data. [2] Includes sales tax, excludes tips.
ERS, Food Markets Branch, 202-694-5384.

CHAPTER XIV

STATISTICS OF FERTILIZERS AND PESTICIDES

This chapter contains statistics on percentages of crop acres treated by various types of fertilizers and pesticides. Nitrogen, phosphate, and potash are the most common fertilizers; herbicides, insecticides, fungicides, and other chemicals are the main categories of pesticides. Other chemicals include soil fumigants, vine killers, and dessicants. The tables show data for field crops for 1998–2001, fruits for 1999, and vegetables for 2001. NASS collects data for field crops on an annual basis and data for fruits and vegetables on a bi-yearly alternating basis. The surveyed States are generally the major producing States for each crop shown in the tables and represent 65–95 percent of the U.S. planted acres, depending on the selected crop. Quantities and rates of active chemical ingredients applied to each crop at State levels are available in the NASS series of ''Agricultural Chemical Usage'' reports.

Table 14-1.—Field crops: Fertilizer, total acreage, and percent of area receiving applications, all States surveyed, 1998–2001 [1]

Crop	Nitrogen	Phosphate	Potash
	Percent	*Percent*	*Percent*
1998:			
Corn	98	83	66
Cotton, Upland	84	66	53
Potatoes, Fall	100	97	96
Soybeans	17	24	27
Wheat, Durum	94	76	5
Wheat, Other Spring	87	77	25
Wheat, Winter	89	63	22
1999:			
Corn	98	82	67
Cotton, Upland	86	59	52
Peanuts	64	70	88
Potatoes, Fall	100	98	88
Soybeans	18	26	28
Sunflower, All	95	33	....
Wheat, Durum [2]	....	....	....
Wheat, Other Spring [2]	....	....	....
Wheat, Winter	97	91	90
2000:			
Corn	98	84	66
Cotton, Upland	83	63	53
Potatoes, Fall [2]	....	....	....
Rice	100	59	47
Soybeans	18	24	27
Sugarbeets	98	92	50
Wheat, Durum	86	66	5
Wheat, Other Spring	95	84	27
Wheat, Winter	87	54	17
2001:			
Corn	96	79	65
Cotton, Upland	76	48	41
Potatoes, Fall	98	95	86
Soybeans	11	17	20

[1] Refers to acres receiving one or more applications of a specific fertilizer ingredient. See tables 14-2 through 14-11 for surveyed States. Note: Acreage estimates are on page I–24 for corn, page II–1 for cotton, page III–13 for soybeans, and page I–1 for wheat. [2] Data not available for all states for all years.

NASS, Environmental, Economics, and Demographics Branch, (202) 720–6146.

Table 14-2.—Corn: Pesticide usage, 1997–2001 [1]

State and Year	Percent treated and amount applied			
	Herbicide [2]		Insecticide [3]	
	Area applied	Pounds applied	Area applied	Pounds applied
	Percent	*Thousands*	*Percent*	*Thousands*
CO:.				
1998	90	1,595	29	530
1999	93	1,763	45	479
2000	97	1,501	59	505
2001	92	1,506	51	431
GA:.				
2001	95	398	34	57
IL:.				
1997	98	32,733	44	4,266
1998	94	31,723	31	1,996
1999	98	28,467	38	1,833
2000	100	28,190	43	3,131
2001	100	31,868	42	1,787
IN:.				
1997	94	18,127	31	1,023
1998	99	18,373	45	1,595
1999	99	14,819	36	1,156
2000	99	15,460	30	797
2001	99	16,007	47	1,103
IA:.				
1997	98	36,144	19	2,323
1998	98	31,911	18	1,534
1999	99	27,966	25	2,462
2000	100	24,158	16	635
2001	99	20,627	7	864
KS:.				
1998	95	5,357	49	400
1999	98	6,619	32	385
2000	93	7,765	31	287
2001	95	9,958	24	657
KY:.				
1998	99	4,174	*	*
1999	94	3,487	50	22
2000	95	2,600	26	65
2001	97	2,834	18	43
MI:.				
1997	98	6,912	11	200
1998	97	5,740	17	299
1999	99	6,128	22	214
2000	99	5,658	10	131
2001	88	4,944	22	288
MN:.				
1997	91	13,956	10	291
1998	97	14,248	10	353
1999	98	11,126	11	280
2000	99	10,597	8	369
2001	99	13,446	*	*
MO:.				
1997	97	8,203	35	475
1998	95	7,718	44	291
1999	98	7,988	38	218
2000	87	5,988	20	114
2001	97	7,232	37	167
NE:.				
1997	98	19,970	62	3,531
1998	93	19,459	44	1,667
1999	99	19,747	39	1,295
2000	97	16,862	55	1,470
2001	99	15,159	48	1,104
NY:.				
2000	92	2,312	31	204
2001	96	2,610	19	69
NC:.				
1998	96	2,150	32	283
1999	82	1,340	35	222
2000	93	1,732	46	363
2001	96	1,558	37	181
ND:.				
2000	71	1,284	*	*
2001	90	745	*	*
OH:.				
1997	100	12,971	18	711
1998	99	9,722	41	1,094
1999	99	10,136	7	98
2000	99	10,339	24	603
2001	99	9,986	26	647
PA:.				
1998	97	4,436	44	262
2000	100	4,419	57	302
2001	99	4,484	60	550

See footnotes at end of table.

Table 14-2.—Corn: Pesticide usage, 1997–2001 [1]—Continued

State and Year	Herbicide [2] Area applied	Herbicide [2] Pounds applied	Insecticide [3] Area applied	Insecticide [3] Pounds applied
	Percent	*Thousands*	*Percent*	*Thousands*
SD:.				
1997	93	6,346	10	317
1998	95	9,947	*	*
1999	95	5,862	18	520
2000	100	5,790	15	44
2001	96	5,622	8	87
TX:.				
1998	94	2,520	68	1,191
1999	93	3,190	54	458
2000	81	2,039	55	426
2001	90	1,990	76	664
WI:.				
1998	98	8,689	19	433
1998	97	7,939	24	593
1999	96	5,421	31	473
2000	95	6,410	20	365
2001	98	6,265	16	155

[1] Data not available for all States for all years. [2] Insufficient number of reports to publish data for fungicides and other chemicals. [3] Amount applied excludes Bt (bacillus thuringiensis). Note: Planted acres are on page I-24. * Insufficient number of reports to publish data.

NASS, Environmental, Economics, and Demographics Branch, (202) 720–6146.

Table 14-3.—Corn: Fertilizer usage, 1997–2001 [1]

State and Year	Nitrogen Area applied	Nitrogen Pounds applied	Phosphate Area applied	Phosphate Pounds applied	Potash Area applied	Potash Pounds applied
	Percent	*Millions*	*Percent*	*Millions*	*Percent*	*Millions*
CO:.						
1998	93	110.3	78	18.0	49	4.9
1999	98	165.6	65	30.3	16	3.4
2000	95	182.0	78	42.2	17	7.4
2001	93	141.5	65	32.1	24	10.8
GA:.						
2001	97	28.6	91	12.6	87	20.8
IL:.						
1997	99	1,689.5	87	747.9	87	1,046.8
1998	99	1,636.8	74	567.8	70	785.9
1999	98	1,639.8	80	603.2	81	1,003.0
2000	99	1,797.7	83	739.3	82	1,028.5
2001	99	1,682.8	81	720.6	85	1,092.2
IN:.						
1997	100	876.7	96	410.9	82	525.9
1998	100	846.3	97	341.0	90	619.4
1999	99	881.8	92	299.1	88	593.3
2000	99	864.8	90	366.1	85	625.9
2001	98	837.4	85	331.7	86	660.0
IA:.						
1997	99	1,464.3	75	575.4	75	668.0
1998	96	1,529.0	81	613.8	81	803.0
1999	98	1,502.8	75	604.9	75	734.7
2000	95	1,533.0	74	503.2	74	630.9
2001	87	1,272.8	62	415.8	60	482.4
KS:.						
1998	100	514.3	83	101.4	21	21.6
1999	99	443.3	70	86.2	22	20.5
2000	100	506.0	78	97.3	39	37.1
2001	97	444.4	71	93.5	19	24.8
KY:.						
1998	100	227.3	94	103.7	95	140.4
1999	100	234.9	81	66.6	50	64.5
2000	99	198.7	81	88.3	80	92.0
2001	91	173.4	87	92.5	82	99.9
MI:.						
1997	100	309.2	91	117.9	94	263.8
1998	95	228.9	91	90.7	87	179.2
1999	100	277.9	92	91.9	91	174.4
2000	99	240.1	96	96.9	83	154.3
2001	91	251.3	78	85.9	78	175.2
MN:.						
1997	97	750.9	79	270.4	81	309.6
1998	96	851.2	91	352.3	87	447.9
1999	92	702.9	90	299.6	86	312.9
2000	97	786.4	91	404.2	76	377.9
2001	97	750.2	90	283.4	81	340.5

See footnotes at end of table.

Table 14-3.—Corn: Fertilizer usage, 1997–2001 [1]—Continued

State and Year	Percent treated and amount applied					
	Nitrogen		Phosphate		Potash	
	Area applied	Pounds applied	Area applied	Pounds applied	Area applied	Pounds applied
	Percent	*Millions*	*Percent*	*Millions*	*Percenta*	*Millions*
MO:.						
1997	100	447.1	84	131.3	84	176.4
1998	99	466.7	92	138.2	93	184.6
1999	100	422.3	84	136.1	84	169.4
2000	100	422.7	82	136.3	82	169.1
2001	99	411.6	82	129.6	83	161.2
NE:.						
1997	100	1,313.1	80	205.2	26	33.3
1998	99	1,106.1	69	215.1	21	33.1
1999	99	1,115.2	75	232.8	18	22.1
2000	99	1,260.7	82	243.2	22	21.5
2001	100	1,067.0	77	219.4	25	42.8
NY:.						
2000	99	71.2	89	45.6	78	41.8
2001	100	76.8	98	49.4	90	45.6
NC:.						
1998	98	105.1	92	42.2	91	76.1
1999	99	83.2	82	36.3	88	66.3
2000	96	86.0	88	37.5	86	52.7
2001	98	81.8	70	41.6	96	56.6
ND:.						
2000	98	103.0	80	38.8	29	8.7
2001	94	89.9	83	33.8	38	10.1
OH:.						
1997	99	567.5	89	234.6	89	313.6
1998	100	587.5	96	243.0	74	310.3
1999	100	527.0	97	236.1	94	324.2
2000	100	572.8	92	224.2	83	287.0
2001	100	572.1	92	210.8	89	338.9
PA:.						
1998	88	128.5	71	54.4	69	41.4
2000	95	103.8	87	59.9	67	35.9
2001	98	130.2	79	55.8	76	43.4
SD:.						
1997	96	303.1	80	113.9	31	25.5
1998	94	305.9	78	117.4	25	21.5
1999	98	334.6	88	136.2	49	42.5
2000	99	418.9	92	153.6	39	36.1
2001	95	393.8	69	119.4	32	38.9
TX:.						
1998	99	319.4	87	89.3	21	15.4
1999	100	304.5	80	74.5	40	22.4
2000	98	304.0	85	80.3	27	15.9
2001	100	245.6	83	66.3	40	18.4
WI:.						
1997	98	285.2	97	154.0	93	244.1
1998	97	326.8	96	148.2	96	188.0
1999	98	305.1	82	104.2	91	177.8
2000	97	300.7	89	120.6	90	161.0
2001	98	355.3	95	120.9	89	169.5

[1] Data not available for all States for all years. Note: Planted acres are on page I-24.

NASS, Environmental, Economics, and Demographics Branch, (202) 720–6146.

Table 14-4.—Upland Cotton: Pesticide usage, 1997–2001 [1]

State and Year	Percent treated and amount applied							
	Herbicide		Insecticide [2]		Fungicide		Other Chemicals	
	Area applied	Pounds applied	Area applied	Pounds applied	Area applied	Pounds applied	Area applied	Pounds applied
	Percent	*Thousands*	*Percent*	*Thousands*	*Percent*	*Thousands*	*Percent*	*Thousands*
AL:.								
1997	100	1,667	85	469	17	22	69	482
1998	99	1,300	91	422	16	52	85	454
1999	99	1,154	87	436	30	130	78	617
2000	97	1,435	67	270	16	84	58	398
AZ:.								
1997	87	534	85	705	*	*	86	770
1998	95	426	91	677	4	6	97	947
1999	90	519	60	360	**	**	95	1,361
2000	94	497	66	455	10	31	79	670
AR:.								
1997	89	2,882	77	678	10	83	84	1,335
1998	93	2,119	98	886	19	71	93	1,490
2000	95	1,993	82	1,610	17	57	89	1,459
2001	96	2,312	53	2,038	8	9	78	1,395
CA:.								
1997	93	1,227	92	2,242	*	*	98	3,471
1998	99	879	98	800	*	*	99	1,611
1999	98	1,006	94	861	1	7	100	2,406
2000	99	1,475	90	1,051	1	9	99	2,714
2001	*	*	*	*	*	*	*	*
GA:.								
1997	100	4,623	90	895	**	**	85	4,397
1998	99	3,629	84	869	*	*	72	2,322
1999	98	4,249	92	816	*	3	78	2,992
2000	98	3,526	81	725	**	**	78	3,258
2001	93	2,958	59	366	**	**	65	1,902
LA:.								
1997	90	2,331	85	1,789	19	85	66	469
1998	96	1,655	98	2,385	22	76	83	499
1999	98	1,763	98	4,206	9	40	88	707
2000	96	1,825	98	4,795	23	229	88	749
2001	95	2,552	93	2,217	16	70	88	931
MS:.								
1997	100	3,124	100	3,972	30	447	97	1,556
1998	100	2,588	98	4,757	16	115	92	1,103
1999	100	3,821	98	6,580	17	180	99	1,980
2000	98	3,557	99	6,112	15	131	99	1,986
2001	99	3,913	92	3,306	5	22	95	2,461
MO:.								
1997	100	839	71	210	*	*	99	573
2000	94	677	90	360	*	*	97	695
NC:.								
1997	97	1,832	92	339	*	*	96	1,093
1998	95	1,494	92	363	9	30	89	909
1999	96	2,079	91	533	6	42	57	996
2000	99	2,375	94	510	4	19	91	1,921
2001	*	*	*	*	*	*	*	*
SC:.								
1997	100	875	98	241	18	5	96	467
TN:.								
1997	98	1,275	85	417	29	123	79	551
1998	100	1,127	97	1,297	37	61	93	547
1999	96	1,385	95	1,222	27	132	89	585
2000	99	1,347	100	4,333	20	77	93	691
TX:.								
1997	97	6,401	62	6,327	**	**	53	2,398
1998	93	6,989	47	2,833	*	*	45	2,113
1999	97	7,081	76	23,417	1	49	32	1,840
2000	92	7,847	69	20,639	*	*	29	1,593
2001	85	5,921	58	14,587	1	19	20	1,330

[1] Data not available for all States for all years. [2] Amount applied excludes Bt (bacillus thuringiensis). * Insufficient number of reports to publish data. ** No reports received for this pesticide class. Note: Planted acres are on page II-1.

NASS, Environmental, Economics, and Demographics Branch, (202) 720–6146.

Table 14-5.—Upland Cotton: Fertilizer usage, 1997–2001 [1]

State and Year	Percent treated and amount applied					
	Nitrogen		Phosphate		Potash	
	Area applied [2]	Pounds applied	Area applied [2]	Pounds applied	Area applied [2]	Pounds applied
	Percent	*Millions*	*Percent*	*Millions*	*Percent*	*Millions*
AL:.						
1997	100	47.8	93	26.8	95	40.1
1998	99	38.3	94	27.3	94	35.5
1999	97	46.5	94	36.3	95	45.3
2000	100	60.5	95	35.2	91	46.7
AZ:.						
1997	99	41.8	29	5.2	4	0.6
1998	98	34.5	38	6.1	13	0.3
1999	99	39.6	22	5.0	15	0.7
2000	98	35.6	30	4.7	8	0.9
AR:.						
1997	92	67.2	83	42.5	91	57.6
1998	98	82.4	88	33.8	88	61.6
1999	97	88.0	82	31.8	85	63.5
2000	100	84.2	78	30.5	84	66.1
2001	93	80.3	63	24.6	68	54.0
CA:.						
1997	96	122.7	25	13.4	26	16.7
1998	98	81.8	23	11.2	13	6.9
1999	99	92.6	51	19.1	19	11.1
2000	98	105.4	29	12.6	12	5.3
2001	*	*	*	*	*	*
GA:.						
1997	95	126.3	94	81.4	99	132.0
1998	96	119.6	90	71.3	92	123.4
1999	100	127.2	98	81.3	100	160.3
2000	96	124.9	94	77.6	93	117.7
2001	99	116.2	92	71.9	93	119.3
LA:.						
1997	98	48.5	71	22.8	76	33.8
1998	99	47.9	69	15.5	73	29.8
1999	100	52.4	43	14.7	45	18.9
2000	100	60.7	64	20.1	66	33.0
2001	95	70.8	50	18.4	52	35.1
MS:.						
1997	100	107.3	46	22.6	77	72.1
1998	97	98.5	51	27.6	67	62.9
1999	100	133.3	36	21.2	65	85.8
2000	100	147.7	44	29.5	68	86.1
2001	99	179.9	31	25.8	46	72.5
MO:.						
1997	100	42.6	72	10.0	95	28.1
2000	100	40.4	86	11.7	95	33.5
NC:.						
1997	92	38.4	64	16.6	85	56.8
1998	98	60.2	90	35.0	93	71.6
1999	96	66.3	89	37.0	96	90.3
2000	96	76.0	80	34.9	91	98.5
2001	*	*	*	*	*	*
SC:.						
1997	100	26.8	100	15.8	100	34.2
TN:.						
1997	100	44.8	99	27.4	99	42.9
1998	99	42.6	100	28.4	100	41.0
1999	100	51.2	99	30.2	100	50.9
2000	99	47.5	93	29.8	98	50.4
TX:.						
1997	82	280.9	62	126.3	29	25.8
1998	68	237.7	56	122.0	27	28.5
1999	71	281.8	45	112.8	23	26.6
2000	63	263.4	54	136.9	26	31.1
2001	52	195.9	37	85.2	14	16.4

[1] Data not available for all States for all years. [2] Planted acres are on page II-1.

NASS, Environmental, Economics, and Demographics Branch, (202) 720–6146.

Table 14-6.—Fall potatoes: Pesticide usage, 1997–1999 [1]

State and Year	Percent treated and amount applied							
	Herbicide		Insecticide [2]		Fungicide		Other Chemicals	
	Area applied [3]	Pounds applied	Area applied [3]	Pounds applied	Area applied [3]	Pounds applied	Area applied	Pounds applied
	Percent	*Thousands*	*Percent*	*Thousands*	*Percent*	*Thousands*	*Percent*	*Thousands*
CO:.								
1999	86	175	76	39	98	387	57	14,056
ID:.								
1997	92	962	92	1,057	100	2,233	59	40,356
1999	92	953	92	1,066	92	1,502	56	53,358
2001	75	714	93	853	70	691	59	46,698
ME:.								
1997	96	39	97	68	99	641	96	1,609
1999	100	25	97	29	100	553	24	89
2001	92	28	88	13	98	530	97	405
MI:.								
1999	100	101	100	52	99	609	56	137
MN:.								
1997	28	35	99	84	98	816	82	113
1999	86	82	91	54	93	577	16	2,103
2001	78	53	95	18	97	431	56	456
NY:.								
1995	86	52	93	68	91	211	65	16
ND:.								
1997	63	134	77	161	99	1,232	36	22
1999	83	94	95	121	99	966	5	1,315
2001	*	*	*	*	*	*	*	*
OR:.								
1997	94	142	85	178	93	346	69	8,306
1999	100	129	89	183	97	314	65	7,489
2001	*	*	*	*	*	*	*	*
PA:.								
1998	90	36	99	32	99	152	69	5
1999	94	35	99	47	95	125	3	4
WA:.								
1997	85	264	99	644	95	1,084	71	9,658
1998	98	360	99	810	97	1,206	75	19,377
2001	92	290	95	647	91	1,108	78	14,470
WI:.								
1997	98	70	95	95	100	1,103	87	3,601
1998	96	85	97	119	99	1,065	91	2,538
1999	98	84	100	193	98	921	16	1,104
2001	88	73	100	110	97	1,193	86	2,644

[1] Data not available for all States for all years. [2] Amount applied excludes Bt (bacillus thuringiensis). [3] Planted acres are on page IV-21.

NASS, Environmental, Economics, and Demographics Branch, (202) 720–6146.

Table 14-7.—Fall potatoes: Fertilizer usage, 1997–2000 [1]

State and Year	Percent treated and amount applied					
	Nitrogen		Phosphate		Potash	
	Area applied	Pounds applied	Area applied	Pounds applied	Area applied	Pounds applied
	Percent	*Millions*	*Percent*	*Millions*	*Percent*	*Millions*
CO:.						
1999	98	14.6	95	13.3	74	5.6
ID:.						
1997	100	103.6	97	72.3	88	41.7
1999	100	91.0	99	78.5	82	42.7
2001	99	79.6	97	63.2	77	35.1
IN:.						
1999	100	0.6	100	0.5	100	0.5
ME:.						
1997	100	12.9	100	13.3	100	13.5
1999	100	11.5	100	12.3	100	12.4
2001	98	11.0	98	11.4	98	11.8
MI:.						
1999	100	10.1	98	6.6	100	10.0
MN:.						
1997	96	11.9	99	6.1	97	6.6
1999	99	8.0	91	5.3	91	9.6
2001	93	6.4	89	4.5	89	7.6
ND:.						
1997	100	16.7	96	11.7	80	7.7
1999	99	15.4	98	10.9	83	9.2
2001	*	*	*	*	*	*
OR:.						
1997	100	15.1	100	10.8	87	11.1
1999	100	13.5	100	8.2	91	7.5
2001	*	*	*	*	*	*
PA:.						
1998	100	2.1	97	1.6	96	2.1
1999	97	2.2	97	1.8	97	2.0
WA:.						
1997	100	47.9	99	42.6	98	31.6
1999	100	55.5	99	40.7	97	43.7
2001	97	37.6	92	33.0	92	37.4
WI:.						
1997	100	15.0	100	9.5	100	22.2
1999	100	20.8	100	12.0	99	20.4
2001	100	22.0	98	13.7	100	24.3

[1] Data not available for all states for all years. Note: Planted acres are on page IV-21.

NASS, Environmental, Economics, and Demographics Branch, (202) 720–6146.

Table 14-8.—Soybeans: Pesticide usage, 1997–2001 [1]

State and Year	Percent treated and amount applied [2]			
	Herbicide		Insecticide [3]	
	Area applied	Pounds applied	Area applied	Pounds applied
	Percent	*Thousands*	*Percent*	*Thousands*
AR:.				
1997	97	5,019	*	*
1998	75	3,058	4	37
1999	94	3,670	9	17
2000	86	2,918	3	4
2001	80	2,440	*	*
DE:.				
1997	78	314	*	*
IL:.				
1997	98	11,136	*	*
1998	95	11,354	*	*
1999	96	10,290	*	20
2000	98	10,582	1	3
2001	96	10,102	**	**
IN:.				
1997	99	7,062	**	**
1998	98	5,798	**	**
1999	89	5,750	**	**
2000	99	5,414	*	*
2001	98	5,612	**	**
IA:.				
1997	99	13,691	**	**
1998	100	11,866	**	**
1999	99	11,995	**	**
2000	98	13,053	*	*
2001	95	11,704	*	*
KS:.				
1997	94	2,947	*	*
1998	95	2,156	*	*
1999	97	3,273	*	1
2000	94	2,953	*	*

See footnotes at end of table.

Table 14-8.—Soybeans: Pesticide usage, 1997–2001 [1]—Continued

State and Year	Percent treated and amount applied [2]			
	Herbicide		Insecticide [3]	
	Area applied	Pounds applied	Area applied	Pounds applied
	Percent	*Thousands*	*Percent*	*Thousands*
KY:.				
1997	91	1,460	*	*
1998	98	1,239	*	*
1999	94	1,037	*	*
2000	88	1,151	1	6
LA:.				
1997	90	1,843	29	331
1998	89	1,442	32	217
1999	94	1,123	53	229
2000	96	1091	56	173
MI:.				
1997	98	2,452	**	**
1998	98	2,620	*	*
1999	97	2,342	**	**
2000	98	2,094	*	*
MN:.				
1997	96	6,902	**	**
1998	97	6,071	*	*
1999	97	6,203	**	**
2000	95	7,151	**	**
2001	99	6,363	**	**
MS:.				
1997	98	2,453	*	*
1998	100	2,948	6	33
1999	99	2,967	9	78
2000	99	2,096	5	23
MO:.				
1997	94	5,521	**	*
1998	92	6,152	**	**
1999	97	5,556	**	**
2000	98	5,867	**	**
2001	95	4,691	*	*
NE:.				
1997	99	4,093	*	*
1998	88	4,226	*	*
1999	96	4,758	1	10
2000	98	5,795	*	*
2001	96	5,336	*	*
NC:.				
1997	98	1,625	35	130
1998	84	1,440	3	20
1999	88	1,283	3	3
2000	92	1,016	7	15
ND:.				
2000	99	2,046	**	**
OH:.				
1997	99	5,307	*	*
1998	99	5,435	*	*
1999	99	4,705	*	3
2000	98	4,586	1	2
2001	96	4,216	*	*
PA:.				
1997	86	661	**	**
1999	99	429	11	20
SD:.				
1997	90	3,059	*	*
1998	96	3,706	*	*
2000	98	4,863	**	**
TN:.				
1997	100	1,664	*	*
1998	98	1,926	*	*
1999	98	1,405	2	19
2000	95	1,319	1	8
WI:.				
1997	100	998	**	**
2000	85	1,169	**	**

[1] Data not available for all States for all years. [2] Insufficient number of reports to publish data for fungicides and other chemicals. [3] Amount applied excludes Bt (bacillus thuringiensis). * Insufficient number of reports to publish data. ** No reports received for this pesticide class. Note: Planted acres are on page III-13.

NASS, Environmental, Economics, and Demographics Branch, (202) 720–6146.

Table 14-9.—Soybeans: Fertilizer usage, 1997–2001 [1]

State and Year	Percent treated and amount applied					
	Nitrogen		Phosphate		Potash	
	Area applied	Pounds applied	Area applied	Pounds applied	Area applied	Pounds applied
	Percent	*Millions*	*Percent*	*Millions*	*Percent*	*Millions*
AR:.						
1997	6	9.3	29	60.8	30	71.4
1998	5	8.6	29	65.3	29	75.4
1999	17	17.3	43	78.0	40	90.0
2000	10	21.0	30	43.4	31	73.0
2001	3	3.4	30	42.8	24	54.9
DE:.						
1997	37	1.5	38	3.8	29	5.7
IL:.						
1997	11	12.6	23	160.3	34	352.5
1998	7	17.2	12	78.7	24	321.4
1999	7	16.2	14	64.1	28	304.0
2000	11	16.8	16	77.5	29	286.0
2001	10	42.8	12	95.8	22	250.5
IN:.						
1997	16	40.8	22	65.2	36	213.0
1998	15	25.0	26	70.4	51	255.3
1999	28	33.6	36	105.3	36	219.8
2000	7	11.0	15	53.9	33	207.8
2001	12	11.4	20	58.1	36	222.4
IA:.						
1997	16	30.4	23	129.3	25	205.4
1998	10	20.4	13	62.1	14	79.0
1999	7	23.5	17	103.5	22	173.7
2000	15	81.0	22	110.1	22	138.0
2001	5	9.9	9	47.9	10	71.3
KS:.						
1997	20	12.1	18	14.8	15	18.9
1998	16	7.5	21	16.6	11	8.3
1999	22	14.9	22	19.4	15	7.6
2000	18	10.3	16	16.9	*	*
KY:.						
1997	32	22.7	42	36.9	41	59.4
1998	35	17.0	58	58.9	63	73.3
1999	17	4.8	25	18.3	26	24.2
2000	13	7.7	40	31.7	39	37.7
LA:.						
1997	13	5.8	23	13.8	23	21.3
1998	3	0.4	25	12.0	26	19.4
1999	5	1.4	14	7.2	11	6.8
2000	6	1.5	20	7.3	26	15.6
MI:.						
1997	63	21.3	49	49.9	71	100.9
1998	72	24.3	73	54.6	75	99.5
1999	31	9.5	45	27.7	65	109.5
2000	37	11.1	40	44.8	72	131.2
MN:.						
1997	16	15.2	20	55.6	22	141.5
1998	18	27.5	17	38.1	9	33.1
1999	13	18.7	13	29.5	13	54.5
2000	8	10.2	9	24.1	24	118.6
2001	13	15.3	13	32.3	12	41.5
MS:.						
1997	16	5.4	23	25.5	26	48.4
1998	6	2.1	10	10.0	16	23.2
1999	10	4.2	15	14.1	22	23.9
2000	9	3.4	19	14.3	20	23.5
MO:.						
1997	15	17.2	28	60.4	35	136.2
1998	24	25.9	47	119.8	53	198.2
1999	15	11.7	23	54.8	23	87.3
2000	20	27.5	28	98.1	27	94.2
2001	6	5.4	24	52.2	22	61.7

See footnotes at end of table.

Table 14-9.—Soybeans: Fertilizer usage, 1997–2001 [1]—Continued

State and Year	Percent treated and amount applied					
	Nitrogen		Phosphate		Potash	
	Area applied	Pounds applied	Area applied	Pounds applied	Area applied	Pounds applied
	Percent	*Millions*	*Percent*	*Millions*	*Percent*	*Millions*
NE:.						
1997	31	19.5	31	45.9	16	11.3
1998	22	12.1	19	27.0	8	7.3
1999	25	17.8	25	31.7	16	17.0
2000	30	19.8	20	36.7	15	6.2
2001	22	23.4	21	38.3	10	13.2
NC:.						
1997	52	46.7	67	36.8	77	103.3
1998	36	12.4	34	19.4	39	47.3
1999	54	15.8	71	53.9	71	85.0
2000	38	12.6	62	64.7	47	47.7
ND:.						
2000	46	27.8	41	25.3	*	*
OH:.						
1997	16	11.9	26	56.8	60	308.4
1998	19	16.5	29	71.9	42	179.3
1999	21	14.4	35	81.6	47	205.6
2000	25	21.7	32	70.2	47	192.8
2001	17	19.1	30	63.9	41	164.7
PA:.						
1997	53	3.4	55	8.7	59	19.5
1999	37	2.8	41	7.5	43	10.0
SD:.						
1997	35	43.3	34	42.2	18	14.5
1998	32	29.7	32	38.1	11	2.9
1999	47	41.3	47	88.3	19	21.3
2000	38	24.3	43	66.0	12	12.2
TN:.						
1997	29	7.4	48	33.1	52	52.6
1998	19	4.5	36	20.7	39	29.4
1999	34	7.1	46	25.9	48	38.4
2000	18	3.0	29	14.3	31	22.2
WI:.						
1997	53	8.2	54	11.7	69	56.0
2000	24	6.5	30	16.6	40	46.2

[1] Data not available for all States for all years. Note: Planted acres are on page III-13. * Insufficient number of reports to publish data.

NASS, Environmental, Economics, and Demographics Branch, (202) 720–6146.

Table 14-10.—Wheat: Pesticide usage, 1997–2001 [1]

State and Year	Percent treated and amount applied [2]			
	Herbicide		Insecticide [3]	
	Area applied	Pounds applied	Area applied	Pounds applied
	Percent	*Thousands*	*Percent*	*Thousands*
Winter.				
AR:.				
2000	41	239	**	**
CA:.				
1998	47	146	*	*
CO:.				
1997	64	803	13	321
1998	61	610	*	*
2000	23	281	*	*
GA:.				
1998	38	80	*	*
ID:.				
1997	98	631	**	**
1998	88	495	*	*
2000	89	411	4	15
IL:.				
1997	40	16	**	**
1998	47	17	*	*
2000	44	21	**	**
IN:.				
1999	39	28	*	*
KS:.				
1997	31	819	**	**
1998	35	1,620	**	**
2000	31	478	8	395
KY:.				
2000	51	57	8	15
LA:.				
1998	*	*	*	*
MS:.				
1998	55	78	*	*
MO:.				
1997	33	67	**	**
1998	28	12	**	**
2000	51	47	*	*
MT:.				
1997	88	1,089	*	*
1998	89	889	**	**
2000	91	745	*	*
NE:.				
1997	53	189	**	**
1998	52	320	*	*
2000	26	248	**	**
NC:.				
1998	60	92	13	11
2000	65	206	19	3
OH:.				
1997	20	56	**	**
1998	13	75	**	**
2000	18	53	**	**
OK:.				
1997	38	435	13	234
1998	42	827	6	89
2000	25	94	*	*
OR:.				
1997	100	516	*	*
1998	100	415	*	*
2000	99	550	**	**
PA:.				
1997	21	8	*	*
SD:.				
1997	89	383	**	**
1998	88	589	*	*
2000	56	415	**	**
TX:.				
1997	24	181	18	351
1998	27	435	7	177
2000	12	441	1	26
WA:.				
1997	98	1,584	*	*
1998	97	1,718	*	*
1999	95	847	**	**

See footnotes at end of table.

Table 14-10.—Wheat: Pesticide usage, 1997–2001 [1]—Continued

State and Year	Percent treated and amount applied [2]			
	Herbicide		Insecticide [3]	
	Area applied	Pounds applied	Area applied	Pounds applied
	Percent	*Thousands*	*Percent*	*Thousands*
Durum.				
ND:.				
1997	93	2,221	2	12
1998	98	2,631	*	*
2000	97	2,807	*	*
Other Spring.				
ID:.				
1998	95	392	*	*
MN:.				
1997	94	1,434	*	*
1998	97	1,396	11	65
2000	92	1,845	*	*
MT:.				
1997	94	3,254	*	*
1998	81	1,816	*	*
2000	92	2,955	**	**
ND:.				
1997	88	4,583	*	*
1998	98	4,053	7	176
2000	97	4,205	*	*
OR:.				
1998	98	87	**	**
SD:.				
1997	86	886	**	**
1998	73	698	*	*
2000	93	619	**	**
WA:.				
1998	100	552	*	*

[1] Data not available for all States for all years. [2] Insufficient number of reports to publish data for fungicides and other chemicals. [3] Amount applied excludes Bt (bacillus thuringiensis). * Insufficient number of reports to publish data. ** No reports received for this pesticide class. Note: Planted acres are on page I-2.

NASS, Environmental, Economics, and Demographics Branch, (202) 720–6146.

Table 14-11.—Wheat: Fertilizer usage, 1997–2001 [1]

State and Year	Percent treated and amount applied					
	Nitrogen		Phosphate		Potash	
	Area applied	Pounds applied	Area applied	Pounds applied	Area applied	Pounds applied
	Percent	*Millions*	*Percent*	*Millions*	*Percent*	*Millions*
Winter.						
AR:.						
2000	92	110.1	28	12.3	28	16.0
CA:.						
1998	*	*	*	*	*	*
CO:.						
1997	68	69.0	39	18.7	*	*
1998	78	108.5	33	22.2	4	0.7
2000	87	85.2	14	5.6	*	*
GA:.						
1998	98	27.7	90	12.3	86	17.2
ID:.						
1997	98	105.4	62	19.4	11	4.3
1998	97	93.6	67	20.6	23	7.0
2000	90	75.5	54	12.1	13	2.7
IL:.						
1997	91	103.8	69	62.1	77	86.9
1998	98	119.9	82	78.5	70	94.7
2000	98	80.1	82	55.5	78	65.7
IN:.						
1999	97	46.3	91	31.6	90	39.0
KS:.						
1997	78	509.0	56	186.8	8	19.5
1998	92	596.7	74	248.3	13	50.7
2000	94	522.9	65	178.7	6	11.2
KY:.						
2000	80	52.0	62	25.9	60	29.2
LA:.						
1998	92	9.2	32	1.6	30	1.8
MS:.						
1998	100	22.2	14	1.4	14	1.7
MO:.						
1997	92	98.2	81	45.4	70	49.4
1998	98	138.4	86	51.1	86	74.9
2000	96	86.8	76	39.9	84	59.1
MT:.						
1997	95	64.5	78	31.4	23	4.3
1998	90	67.9	88	30.7	31	5.1
2000	82	74.2	77	34.0	43	8.2
NE:.						
1997	92	94.6	74	47.8	*	*
1998	85	69.9	59	25.3	12	1.0
2000	90	76.5	68	31.5	*	*
NC:.						
1998	91	63.9	76	24.1	84	53.8
2000	88	78.3	48	15.8	56	30.9
OH:.						
1997	100	93.1	92	63.7	98	87.6
1998	100	106.9	93	66.8	94	80.2
2000	94	107.0	81	64.1	82	74.0
OK:.						
1997	90	317.5	47	83.6	7	6.4
1998	95	381.0	64	130.8	15	10.7
2000	97	393.3	62	148.4	5	8.3
OR:.						
1997	100	75.3	15	4.7	*	*
1998	100	57.8	9	1.7	1	3
2000	99	46.1	11	1.8	7	1.4
PA:.						
1997	81	7.3	60	5.2	59	5.2
SD:.						
1997	78	38.7	58	15.3	*	*
1998	94	79.7	92	36.6	*	*
2000	91	60.8	61	26.6	12	1.3
TX:.						
1997	78	267.2	36	49.1	16	10.8
1998	75	337.2	50	111.7	22	24.6
2000	55	280.2	35	79.7	14	32.0
WA:.						
1997	98	145.4	34	16.5	12	7.6
1998	100	155.8	30	14.7	10	3.8
2000	100	111.7	30	10.2	6	1.3
Durum.						
ND:.						
1997	95	170.9	77	48.7	8	4.1
1998	98	175.0	79	49.0	3	1.7
2000	86	173.8	66	47.6	5	2.1

See footnotes at end of table.

Table 14-11.—Wheat: Fertilizer usage, 1997–2001 [1]—Continued

State and Year	Nitrogen: Area applied	Nitrogen: Pounds applied	Phosphate: Area applied	Phosphate: Pounds applied	Potash: Area applied	Potash: Pounds applied
	Percent	*Millions*	*Percent*	*Millions*	*Percent*	*Millions*
Other.						
Spring:.						
ID:.						
1998	96	59.4	83	17.9	33	2.9
MN:.						
1997	98	209.1	91	77.4	73	73.9
1998	100	166.5	97	64.5	64	10.3
2000	94	169.8	85	51.8	73	29.3
MT:.						
1997	79	153.5	66	68.2	15	10.2
1998	61	129.6	55	64.5	22	10.3
2000	90	167.6	84	75.5	36	15.6
ND:.						
1997	99	621.8	92	248.3	24	43.8
1998	97	472.8	87	166.8	20	9.0
2000	97	501.8	83	170.1	12	13.3
SD:.						
1997	90	140.2	70	61.6	*	*
1998	85	92.2	66	45.0	11	5.7
2000	95	98.1	83	36.7	12	2.8

[1] Data not available for all States for all years. * Insufficient number of reports to publish data. Note: Planted acres are on page I-2.

NASS, Environmental, Economics, and Demographics Branch, (202) 720–6146.

Table 14-12.—Fruits: Percent of acres receiving applications, for surveyed States, 2001 [1]

Crop	Herbicide	Insecticide	Fungicide
	Percent		
Apples	52	95	85
Apricots	24	79	74
Avocados	24	44	*
Blackberries	80	71	80
Blueberries	65	89	87
Cherries, Sweet	41	79	79
Cherries, Tart	62	99	99
Dates	*	18	*
Figs	50	11	*
Grapefruit	90	90	83
Grapes, All	65	60	84
Grapes, Non bearing	2	17	19
Grapes, Raisin	59	59	78
Grapes, Table	75	80	96
Grapes, Wine	65	58	87
Kiwifruit	28	19	*
Lemons	51	55	25
Nectarines	70	87	89
Olives	38	31	6
Oranges - except Temples	85	82	48
Oranges - Temples	93	98	66
Peaches	59	91	92
Pears	49	91	85
Plums	60	85	66
Prunes	34	58	36
Raspberries	89	81	94
Tangelos	79	86	71
Tangerines	80	78	62

[1] Refers to acres receiving one or more applications of a specific agricultural chemical. * Insufficient number of reports to publish data.

NASS, Environmental, Economics, and Demographics Branch, (202) 720–6146.

Table 14-13.—Vegetables: Percent of acres receiving applications, for surveyed States, 2000[1]

Crop	Herbicide	Insecticide	Fungicide
	Percent		
Artichokes	58	73	*
Asparagus	84	84	42
Beans, Lima, Fresh	94	38	2
Beans, Lima, Processing	53	71	7
Beans, Snap, Fresh	53	89	77
Beans, Snap, Processing	96	81	57
Beets	98	*	*
Broccoli	51	92	15
Brussels Sprouts	*	74	75
Cabbage, Fresh	64	96	60
Cabbage, Kraut	85	99	6
Carrots, Fresh	58	13	49
Carrots, Processing	77	54	65
Cauliflower	48	94	9
Celery	75	97	87
Collards	55	87	35
Corn, Sweet, Fresh	79	84	38
Corn, Sweet, Processing	90	73	22
Cucumbers, Fresh	31	75	79
Cucumbers, Pickles	85	40	45
Eggplant	41	89	81
Garlic	45	29	61
Greens, Mustard	40	86	52
Greens, Turnip	55	72	44
Kale	54	88	45
Lettuce, Head	57	92	54
Lettuce, Other	74	93	66
Melons, Cantaloupe	43	74	60
Melons, Honeydew	12	80	26
Melons, Watermelon	52	47	79
Okra	28	70	27
Onions, Dry	83	72	80
Peas, Green, Processing	94	53	5
Peppers, Bell	54	93	74
Pumpkins	57	60	59
Radishes	2	66	17
Spinach, Fresh	75	85	78
Spinach, Processing	38	86	44
Squash	38	72	60
Strawberries	39	81	84
Tomatoes, Fresh	63	87	86
Tomatoes, Processing	78	64	73

[1] Refers to acres receiving one or more applications of a specific agricultural chemical. * Insufficient number of reports to publish data.

NASS, Environmental, Economics, and Demographics Branch, (202) 720–6146.

CHAPTER XV

MISCELLANEOUS AGRICULTURAL STATISTICS

This chapter contains miscellaneous data which do not fit into the preceding chapters. Included here are summary tables on foreign trade in agricultural products; statistics on fishery products; tables on refrigerated warehouses; and statistics on crops in Alaska.

Foreign Agricultural Trade Statistics

Agricultural products, sometimes referred to as food and fiber products, cover a broad range of goods from unprocessed bulk commodities like soybeans, feed corn and wheat to highly-processed, high-value foods and beverages like sausages, bakery goods, ice cream, or beer sold in retail stores and restaurants. All of the products found in Chapters 1-24 (except for fishery products in Chapter 3) of the U.S. Harmonized Tariff Schedule are considered agricultural products. These products generally fall into the following categories: grains, animal feeds, and grain products (like bread and pasta); oilseeds and oilseed products (like canola oil); livestock, poultry and dairy products including live animals, meats, eggs, and feathers; horticultural products including all fresh and processed fruits, vegetables, tree nuts, as well as nursery products and beer and wine; unmanufactured tobacco; and tropical products like sugar, cocoa, and coffee. Certain other products are considered ''agricultural,'' the most significant of which are essential oils (Chapter 33), raw rubber (Chapter 40), raw animal hides and skins (Chapter 41), and wool and cotton (Chapters 51-52). Manufactured products derived from plants or animals, but which are not considered ''agricultural'' are cotton yarn, textiles and clothing; leather and leather articles of apparel; and cigarettes and spirits.

U.S. foreign agricultural trade statistics are based on documents filed by exporters and importers and compiled by the Bureau of the Census. Puerto Rico is a Customs district within the U.S. Customs territory, and its trade with foreign countries is included in U.S. export and import statistics. U.S. export and import statistics include merchandise trade between the U.S. Virgin Islands and foreign countries even though the Virgin Islands of the United States are not officially a part of the U.S. Customs territory.

Data on trade of other U.S. outlying possessions with foreign countries is not compiled by the United States. Export statistics are fully compiled on shipments to all countries, except Canada, where the value of commodities classified under each individual Schedule B number is over $2,500. Value data for such commodities valued under $2,501 are estimated for individual countries using factors based on the ratios of low-valued shipments to individual country totals for past periods. The estimates for low-valued shipments are shown under a single Schedule B number and are omitted from the statistics for the detailed commodity classifications. Shipments valued under $2,501 to all counties, except Canada, represent slightly less that 2.5 percent of the monthly value of U.S. exports to those countries. As a result of the data exchange between the United States and Canada, the United States has adopted the Canadian import exemption level for its export statistics on shipments to Canada. The Canadian import exemption level is based on total value per shipment rather than value per commodity classification line item.

The export value, the value at the port of exportation, is based on the selling price and includes inland freight, insurance, and other charges to the port. The country of destination is the country of ultimate destination or where the commodities are consumed or further processed. When the shipper does not know the ultimate destination, the shipments are credited to the last country, as known at the time of shipment from the United States.

Agricultural products, like manufactured goods, are often transhipped from the one country to another. Shippers are asked to identify the ultimate destination of a shipment. However, transhipment points are often recorded as the ultimate destination even though the actual point of consumption may be in a neighboring state. Thus, exports to countries which act as transhipment points are generally overstated, while exports to neighboring countries are often understated. Major world transhipment points include the Netherlands, Hong Kong, and Singapore. In such cases, exports are over reported for the Netherlands, but under reported for Germany, Belgium and the United Kingdom. They are overstated to Hong Kong, but under reported to China, and they overstated to Singapore, but understated to Malaysia and Indonesia. After the collapse of communism in Eastern Europe and Russia, Germany and the Baltic countries became important transhipment points to those countries further east.

Imports for consumption are a combination of entries for immediate consumption and withdrawals from warehouses for consumption. The import value, defined generally as the market value in the foreign country, excludes import duties, ocean freight, and marine insurance. The country of origin is defined as the country where the commodities were grown or processed. Where the country of origin is not known, the imports are credited to the country of shipment.

Import statistics are fully compiled on shipments valued over $1,250. Value data for shipments valued under $1,251 are not required to be reported on formal entries. They are estimated for individual countries using factors based on the ratios of low-valued shipments to individual country totals for past periods. The estimates for low-valued shipments are shown under a single HTS number. The total value excluded represents slightly less than 1 percent of the monthly import value.

Table 15-1.—Foreign trade: Value of total agricultural exports and imports, United States, fiscal years 1993–2002

Year ending Sept. 30[1]	Total domestic exports	Agricultural exports[2]		Total imports for consumption	Agricultural imports (for consumption)[3]			Competitive		Excess of domestic agricultural exports over competitive imports
		Domestic	Percentage (domestic) of total exports		Total	Competitive	Noncompetitive	Percentage of total agricultural exports	Percentage of total agricultural imports	
	Million dollars	*Million dollars*	*Percent*	*Million dollars*	*Million dollars*	*Million dollars*	*Million dollars*	*Percent*	*Percent*	*Million dollars*
1993	433,373	42,832	10	562,038	24,610	19,066	5,544	45	77	23,766
1994	469,272	43,911	9	631,162	26,562	20,351	6,211	46	77	23,560
1995	534,248	54,644	10	728,886	29,788	21,325	8,463	39	72	33,319
1996	574,646	59,752	10	795,289	32,444	24,269	8,175	41	75	35,483
1997	629,317	57,269	9	865,346	35,654	26,547	9,107	46	74	30,722
1998	639,556	53,653	8	895,900	36,837	27,608	9,229	51	75	26,045
1999	635,754	49,043	8	976,258	37,293	29,220	8,074	60	78	19,823
2000	701,651	50,744	7	1,167,768	38,857	30,833	8,024	61	79	19,911
2001	690,634	52,699	8	1,152,642	39,027	32,409	6,618	61	83	20,290
2002[4]	628,583	53,294	8	1,121,960	40,979	34,166	6,813	64	83	19,128

[1] Fiscal years Oct. 1–Sept. 30. [2] Includes food exported for relief or charity by individuals and private agencies. [3] Competitive agricultural imports consist of all imports similar to agricultural commodities produced commercially in the United States, together with all other agricultural imports interchangeable to any significant extent with such U.S. commodities. Noncompetitive agricultural imports include all others, about 90 percent of which consist of rubber, coffee, raw silk, cacao beans, wool for carpets, bananas, tea, and vegetable fibers. [4] Preliminary.

ERS, Market and Trade Economics Division, (202) 694–5272.

Table 15-2.—Agricultural exports: Quantity Indexes, United States, fiscal years 1993–2002

[Calendar year 1987=100]

Year ending Sept. 30[1]	Total agricultural exports[2]	Animals and animal products	Meat and meat products[3]	Animal fats[3]	Dairy products[3]	Poultry and poultry products[3]	Cotton and linters	Tobacco unmanufactured	Grains and feeds	Vegetable oils and oilseeds	Fruits and vegetables
1993	122	151	217	112	81	251	87	118	108	101	178
1994	117	173	246	110	75	345	122	101	92	85	186
1995	148	214	305	144	69	475	154	101	122	121	197
1996	145	235	349	115	63	581	127	112	113	106	198
1997	145	234	340	86	46	633	123	122	96	120	209
1998	146	254	386	112	56	667	119	107	90	129	208
1999	143	238	390	114	58	511	68	105	107	122	206
2000	155	287	456	99	70	693	114	92	105	124	214
2001	157	294	456	86	67	699	126	90	99	126	223
2002[4]	163	294	484	110	72	644	164	83	98	139	220

[1] Fiscal years, Oct. 1–Sept. 30. [2] Excludes quantities not in metric tons; accounts for 86 percent of total value of U.S. agricultural exports in 1987. [3] Included in "Animals and animal products." [4] Preliminary.

ERS, Market and Trade Economics Division, (202) 694–5272.

Table 15-3.—Agricultural imports: Quantity indexes, United States, fiscal years 1993–2002

[Calendar year 1987=100]

Year ending Sept. 30 [1]	Total agricultural imports [2]	Non-competitive imports [3]	Competitive imports [3]	Animals and animal products	Dairy products [4]	Hides and skins [4]	Meat and meat products [4]	Wool excluding free in bond [4]	Grains and feeds	Vegetable oils and oilseeds	Sugar cane or beet	Tobacco unmanufactured
1993	114	115	113	98	115	62	90	93	165	135	132	164
1994	117	110	121	98	131	62	93	87	253	168	137	135
1995	113	105	117	96	124	62	84	85	225	159	132	95
1996	121	117	124	88	124	62	82	71	220	157	230	115
1997	132	122	139	92	124	62	91	53	264	188	248	151
1998	136	127	141	99	129	62	98	67	270	203	183	107
1999	142	133	147	109	164	62	112	34	285	179	144	96
2000	151	144	154	118	162	62	124	30	293	197	115	98
2001	148	128	160	122	159	62	127	20	317	205	116	94
2002 [5] ...	157	134	170	122	184	62	127	11	347	189	113	121

[1] Fiscal years Oct. 1–Sept. 30. [2] Excludes quantities not in metric tons; accounts for 86 percent of total value of U.S. agricultural imports in 1987. [3] Competitive agricultural imports consist of all imports similar to agricultural commodities produced commercially in the United States, together with all other agricultural imports interchangeable to any significant extent with such U.S. commodities. Non-competitive agricultural imports include all others, about 90 percent of which consist of rubber, coffee, tea, cacao beans, raw silk, wool for carpets, bananas, spices, and vegetable fibers. [4] Included in "animals and animal products." [5] Preliminary.

ERS, Market and Trade Economics Divison, (202) 694–5272.

Table 15-4.—Foreign trade in agricultural products: Value of exports by principal groups, United States, fiscal years 1998–2002 [1]

Product	1998	1999	2000	2001	2002 [2]
	1,000 dollars	*1,000 dollars*	*1,000 dollars*	*1,000 dollars*	*1,000 dollars*
All commodities	639,555,800	635,753,900	701,650,500	690,634,100	628,582,800
Nonagricultural commodities	585,902,862	586,710,600	650,906,659	637,935,582	575,289,174
Agricultural commodities	53,652,938	49,043,300	50,743,841	52,698,518	53,293,626
Animals and animal products [3]	10,957,574	9,886,839	11,490,433	12,186,033	11,637,841
Animals, live excluding poultry	505,306	476,147	608,522	727,346	696,177
Cattle	162,957	132,139	237,516	271,792	226,154
Other	342,349	344,008	371,006	455,554	470,023
Dairy products	925,058	914,890	998,500	1,120,921	1,030,729
Butter and anhydrous milkfat	13,532	3,115	3,274	4,033	4,632
Cheese and curd	121,318	126,748	138,343	157,403	164,896
Milk, condensed, or evaporated	7,814	3,674	4,508	5,936	11,383
Milk, nonfat dry	152,854	151,078	193,000	196,839	166,751
Other	629,540	630,275	659,375	756,710	683,067
Fats, oils and greases	635,815	543,852	421,463	319,994	454,361
Lard and other rendered pig fat	36,229	37,091	44,856	29,856	24,793
Tallow, inedible	461,011	350,106	249,262	176,953	254,441
Other	138,575	156,655	127,345	113,185	175,127
Meats and meat preparations	4,507,334	4,479,304	5,428,603	5,192,599	5,113,422
Beef and veal, except offals	2,380,903	2,536,307	3,091,104	2,644,357	2,578,337
Pork, except offals	1,066,814	977,463	1,268,001	1,320,582	1,383,885
Offals, edible, variety meats	671,771	667,709	743,194	919,576	831,429
Others	387,846	297,825	326,304	308,084	319,771
Poultry and poultry products	2,707,351	2,059,801	2,233,143	2,515,748	2,277,963
Poultry, live:					
Breeding chicks	98,824	95,932	83,920	85,064	92,051
Baby chicks, excluding breeding	1,207	2,076	2,212	3,434	5,371
Other	16,958	14,432	13,976	11,028	10,372
Poultry meat; fresh, frozen:					
Chickens	1,790,232	1,320,758	1,443,312	1,580,480	1,431,066
Turkeys	226,539	165,868	194,156	224,008	178,090
Other	11,332	10,292	10,876	9,221	7,470
Eggs	224,042	179,843	171,570	179,822	173,086
Other	338,217	270,600	313,121	422,691	380,457
Hides and skins, including furskins	1,358,335	1,110,124	1,428,429	1,932,948	1,776,050
Cattle hides, whole	969,325	846,137	1,116,545	1,437,387	1,120,937
Furskins	161,065	123,733	138,216	159,041	161,449
Other	227,945	140,254	173,668	336,520	493,664
Other animal products	318,375	302,721	371,773	376,477	289,139
Hair, animal, excluding wool or fine hair	2,848	4,193	3,272	2,622	2,531
Sausage casings	61,184	66,613	65,657	63,750	57,870
Wool, unmanufactured, including fine hair	11,910	20,272	21,105	11,878	22,394
Other	242,433	211,643	281,739	298,227	206,344
Non-animal products [3]	42,695,364	39,156,461	39,253,408	40,512,485	41,655,785
Cotton, unmanufactured	2,537,336	1,305,858	1,829,242	2,092,502	2,052,053
Cotton, raw	2,517,288	1,293,154	1,809,059	2,078,664	2,036,608
Linters	20,048	12,704	20,183	13,838	15,445
Fruits and preparations	3,217,857	3,285,531	3,380,170	3,501,670	3,433,533
Canned	143,542	146,126	142,128	144,095	139,740
Dried	390,018	388,950	334,306	358,399	344,037
Fresh	1,857,835	1,843,663	2,009,065	2,158,242	2,096,830
Fruit juices	652,983	735,872	715,289	680,488	694,417
Other	173,479	170,920	179,382	160,446	158,509
Nuts and preparations	1,411,661	1,261,118	1,212,130	1,280,055	1,466,309
Grains and feeds	13,960,820	14,266,363	13,789,244	13,817,741	14,068,898
Feed grains and products	5,187,463	5,824,746	5,483,422	5,469,963	5,660,049
Feed grain	4,990,045	5,611,346	5,283,397	5,238,695	5,270,841
Barley	141,515	67,411	100,070	136,007	78,332
Corn	4,261,466	5,047,840	4,601,162	4,514,076	4,580,776
Grain sorghum	582,774	492,725	578,743	583,795	606,586
Oats	4,134	3,148	2,487	4,073	4,789
Rye	157	222	934	744	357
Other	197,418	213,400	200,025	231,268	389,208
Rice	1,131,632	1,006,864	904,908	754,014	730,183
Wheat and products	3,953,271	3,913,248	3,610,991	3,446,090	3,711,172
Wheat	3,758,840	3,648,568	3,384,325	3,247,918	3,490,539
Wheat flour	116,782	176,410	133,955	107,273	116,482
Other wheat products	77,649	88,270	92,711	90,899	104,151
Blended food products	77,581	76,819	78,336	62,763	86,758
Other	1,189,669	1,192,976	1,228,949	1,316,520	1,324,833
Feeds and fodders, excluding oilcake and meal	2,421,204	2,251,710	2,482,638	2,768,391	2,555,903
Corn byproducts	691,437	565,993	594,287	594,583	571,256
Alfalfa meal and cubes	43,717	39,902	34,073	24,331	22,612
Other	1,686,050	1,645,815	1,854,278	2,149,477	1,962,035

See footnotes at end of table.

Table 15-4.—Foreign trade in agricultural products: Value of exports by principal groups, United States, fiscal years 1998–2002 [1]—Continued

Product	1998	1999	2000	2001	2002 [2]
	1,000 dollars	*1,000 dollars*	*1,000 dollars*	*1,000 dollars*	*1,000 dollars*
Non-animal products--Continued					
Oilseeds and products	11,016,559	8,610,138	8,391,157	8,698,601	9,676,084
Oilcake and meal	1,974,755	1,100,918	1,258,027	1,427,251	1,334,617
Soybean oilseed cake and meal	1,914,405	1,028,563	1,198,925	1,367,215	1,275,685
Other	60,350	72,355	59,102	60,036	58,932
Oilseeds	6,817,706	5,692,241	5,783,926	6,096,768	6,705,428
Soybeans	6,116,765	4,758,444	5,071,500	5,089,138	5,468,217
Other	311,287	341,752	290,507	317,264	365,759
Protein substances	389,654	592,045	421,919	690,366	871,452
Vegetable oils and waxes	2,224,098	1,816,979	1,349,203	1,174,582	1,636,039
Soybean oil	881,449	585,297	278,396	239,574	454,202
Other	1,342,649	1,231,682	1,070,807	935,008	1,181,837
Tobacco, unmanufactured	1,447,912	1,376,389	1,226,797	1,181,298	1,147,403
Flue-cured	757,698	715,941	608,081	573,540	606,431
Other	690,214	660,448	618,716	607,758	540,972
Vegetable and preparations	4,167,912	4,253,347	4,440,373	4,511,168	4,551,008
Canned	350,964	332,948	346,214	328,172	328,598
Pulses	318,994	268,729	240,062	251,225	244,732
Fresh	1,050,101	1,044,318	1,166,679	1,207,720	1,225,560
Hops, including extract	89,133	76,504	90,706	121,192	97,557
Other	2,358,720	2,530,848	2,596,712	2,602,859	2,654,561
Essential oils	533,004	507,687	591,583	674,581	764,031
Seeds, except oilseeds	827,255	794,160	771,828	727,121	839,179
Other vegetable products	3,575,048	3,495,870	3,620,884	4,027,748	3,657,287

[1] Fiscal years, Oct. 1–Sept. 30. [2] Preliminary. [3] Totals may not add due to rounding.

ERS, Market and Trade Economics Division, (202) 694–5272. Compiled from reports of the U.S. Department of Commerce.

Table 15-5.—Foreign trade in agricultural products: Value of imports by principal groups, United States, fiscal years 1998–2002 [1]

Product	1998	1999	2000	2001	2002 [2]
	1,000 dollars	*1,000 dollars*	*1,000 dollars*	*1,000 dollars*	*1,000 dollars*
All commodities	895,900,100	976,258,200	1,167,768,000	1,152,642,400	1,121,960,300
Nonagricultural commodities	859,063,507	938,964,957	1,128,910,922	1,113,615,317	1,080,980,960
Agricultural commodities	36,836,593	37,293,243	38,857,078	39,027,083	40,979,340
Competitive [3]	27,607,814	29,219,502	30,833,096	32,409,320	34,166,271
Animals and animal products	6,776,868	7,029,248	8,149,902	9,030,126	9,093,899
Animals, live, excluding poultry	1,649,873	1,414,222	1,735,191	2,197,823	2,021,755
Cattle, live	1,144,070	1,005,208	1,089,231	1,443,573	1,376,488
Meats and meat products	2,717,895	3,108,044	3,722,739	4,091,262	4,186,999
Meats, P.L. 88-482	1,562,421	1,808,452	2,218,177	2,471,300	2,551,805
Meats, fresh, frozen	2,079,121	2,381,678	2,977,663	3,347,497	3,360,794
Beef and veal	1,550,139	1,798,508	2,203,297	2,451,983	2,527,344
Pork	446,960	479,726	655,981	764,561	711,168
Mutton, goat, and lamb	159,829	180,241	205,238	234,890	274,875
Meats, prepared, preserved	450,252	490,604	503,144	467,261	501,948
Beef and veal	210,976	249,017	201,246	192,765	221,568
Pork	239,276	241,588	301,898	274,496	280,380
Mutton, goat, and lamb	0	0	0	0	0
Poultry and products	206,912	201,455	287,487	258,312	316,547
Dairy products	1,368,213	1,572,543	1,652,514	1,727,726	1,840,799
Milk and cream, fresh or dried	31,140	46,867	35,823	40,783	43,822
Cheese	588,801	718,987	697,133	703,518	807,808
Casein and mixtures	453,810	410,529	469,936	526,474	461,938
Fats, oils and greases	50,664	55,740	69,412	62,437	63,027
Hides and skins	184,293	146,184	160,065	162,127	135,998
Sheep and lamb skins	7,970	4,646	4,083	5,353	3,811
Furskins	69,833	49,062	54,155	59,731	55,710
Wool, unmanufactured excluding free	118,344	48,810	41,410	28,135	16,226
Sausage casings	76,405	70,676	60,848	59,443	70,305
Other animal products	404,269	411,574	420,236	442,861	442,243
Non-animal products	30,059,725	30,263,995	30,707,176	29,996,957	31,885,441
Grains and feeds	2,919,119	2,945,658	3,037,963	3,189,465	3,593,944
Wheat, excluding seed, edible	302,330	295,314	204,543	259,222	312,943
Barley	100,590	72,458	83,818	83,013	70,944
Biscuits, wafers	716,591	825,690	935,616	999,975	1,152,460
Feeds and fodders	506,283	464,466	477,335	488,550	502,874

See footnotes at end of table.

Table 15-5.—Foreign trade in agricultural products: Value of imports by principal groups, United States, fiscal years 1998–2002 [1]—Continued

Product	1998	1999	2000	2001	2002 [2]
Competitive [3]—Continued	*1,000 dollars*	*1,000 dollars*	*1,000 dollars*	*1,000 dollars*	*1,000 dollars*
Non-animal products—Continued					
Fruits and preparations	2,794,216	3,470,779	3,408,893	3,447,704	3,742,804
Fresh or frozen	1,525,090	1,998,902	1,953,050	2,131,664	2,300,608
Strawberries	87,282	100,300	83,060	73,257	107,710
Prepared or preserved	600,284	699,947	672,942	667,446	789,486
Pineapples	183,139	242,781	198,644	162,227	189,753
Fruit juices	668,841	771,930	782,901	648,595	652,709
Nuts and preparations	642,967	709,030	790,720	655,628	648,341
Vegetables and preparation	4,248,748	4,528,661	4,660,455	5,181,907	5,442,609
Fresh or frozen	2,552,127	2,578,208	2,701,440	3,100,462	3,121,233
Tomatoes	735,180	713,121	608,428	755,074	740,244
Prepared or preserved	1,696,621	1,950,453	1,959,015	2,084,339	2,324,762
Olives	177,029	196,611	183,204	198,644	209,704
Mushrooms, canned	105,703	99,581	128,517	113,750	100,222
Hops and extract	30,192	29,884	28,762	22,475	23,365
Sugar and related products	1,674,685	1,585,898	1,514,662	1,618,326	1,710,879
Sugar, cane and beet	758,459	614,670	484,415	524,251	519,138
Molasses, inedible	80,360	56,294	54,671	96,120	98,769
Confectionery products	601,674	683,983	742,937	769,217	844,775
Spices	52,907	63,646	67,802	68,936	87,073
Tobacco, unmanufactured	821,970	741,828	651,192	648,100	734,158
Tobacco, filler	758,895	695,762	606,925	597,074	676,322
Tobacco, scrap	19,614	16,440	16,220	14,642	22,845
Beverages, excluding fruit juices	3,705,464	4,259,986	4,701,324	4,990,694	5,570,932
Wine	1,824,900	2,132,267	2,218,399	2,225,061	2,520,668
Malt beverages	1,677,002	1,865,038	2,126,018	2,296,189	2,526,251
Oilseeds and products	2,100,605	1,873,024	1,859,844	1,679,534	1,678,786
Oilseeds and oilnuts	361,930	315,773	298,126	266,491	216,770
Flaxseed	62,171	40,860	30,237	12,952	16,382
Sesame seed	48,524	52,375	56,437	47,684	41,778
Oils and waxes--vegetables	1,551,141	1,410,351	1,409,838	1,261,042	1,313,889
Castor oil	40,039	35,842	42,355	33,836	21,394
Coconut oil	361,177	235,861	251,103	182,730	168,917
Olive oil	355,521	349,619	416,151	386,836	420,741
Palm oil	59,440	62,513	50,024	54,828	63,285
Palm kernel oil	110,778	131,162	118,407	74,344	54,595
Cotton, excluding linters	3,494	141,960	22,613	9,767	15,105
Cotton linters	7,552	8,090	5,160	17,353	7,928
Seeds, field and garden	431,294	468,311	503,101	442,989	416,901
Nursery stock	1,081,966	1,075,996	1,165,142	1,156,023	1,134,840
Other vegetable products	345,959	317,387	294,322	271,321	286,378
Noncompetitive [3]	9,228,779	8,073,740	8,023,982	6,617,762	6,813,069
Bananas, fresh	1,110,175	1,083,000	1,020,220	1,047,856	1,077,784
Plaintains, fresh	74,327	94,733	75,430	75,694	76,723
Coffee, green	3,237,116	2,592,263	2,546,284	1,446,207	1,281,646
Coffee, roasted	169,464	188,529	176,322	157,210	159,428
Coffee soluble, extracts	180,042	179,751	182,282	157,332	168,115
Cocoa bean	662,585	582,299	459,842	323,814	519,114
Cocoa butter	306,071	224,968	216,974	178,631	134,339
Cocoa, cake, powder, etc	127,838	122,854	149,200	161,065	223,887
Chocolate preparation	604,005	601,227	637,959	725,875	836,856
Tea, crude or prepared	256,398	241,228	252,106	264,523	278,940
Spices	459,270	503,986	526,411	495,362	483,085
Pepper, unground, black	154,594	217,007	277,845	131,843	82,752
Vanilla beans	36,435	31,262	30,640	130,136	150,739
Other	268,241	255,717	217,926	233,383	249,594
Rubber, crude, natural:					
Rubber-latex	89,456	73,412	65,032	61,835	70,311
Rubber-dry form	936,585	664,383	766,530	598,257	584,207
Silk, raw	417	527	155	177	164
Wool, unmanufactured, duty-free	33,008	26,345	24,337	25,027	14,461
Fibers, excluding cotton, jute	44,950	41,677	39,476	32,629	23,631
Essential oils	366,197	329,566	322,393	307,624	339,206
Drugs, crude natural	539,704	487,393	518,666	516,745	506,167
Other	31,171	35,599	44,363	41,899	35,005

[1] Fiscal years, Oct. 1–Sept. 30. [2] Preliminary. [3] Competitive agricultural import products consists of all products similar to agricultural commodities produced commercially in the United States, together with all other agricultural products interchangeable to any significant extent with such U.S. commodities. Noncompetitive agricultural products include all others, about 90 percent of which consist of rubber, coffee, raw silk, cacao beans, wool for carpets, bananas, tea, spices, and vegetable fibers.

ERS, Market and Trade Economics Division, (202) 694–5272. Compiled from reports of the U.S. Department of Commerce.

Table 15-6.—Agricultural exports: Value of U.S. exports to the European Union, by commodity, fiscal years 1999/2000–2001/2002 [1]

Commodity	Value		
	1999/2000	2000/2001	2001/2002 [2]
	1,000 dollars	*1,000 dollars*	*1,000 dollars*
Total agricultural exports	6,192,666	6,249,303	6,273,219
Variable-levy commodities:.			
Beef and veal	20,577	14,419	9,750
Pork	10,287	13,422	11,244
Other meats—fresh or prepared	2,765	2,009	2,327
Poultry—live	3,958	3,353	5,010
Baby chicks	3,442	3,115	4,723
Poultry meats	30,069	24,725	18,818
Chickens—fresh or frozen	28,039	22,970	16,362
Turkeys—fresh or frozen	952	1,467	1,926
Other poultry—fresh or frozen	95	19	26
Poultry meats—prepared or preserved	983	268	503
Eggs	21,116	23,365	23,226
Dairy products	39,324	45,617	41,927
Lard	149	22	21
Wheat, unmilled	213,855	204,618	222,084
Wheat, flour	4,952	346	17,814
Other wheat products	1,816	1,007	1,054
Rice-paddy, milled	110,495	85,385	66,062
Feed grains and products	45,957	20,362	25,412
Feed grains	31,923	12,218	14,006
Barley	7,159	9,950	10,305
Corn	7,936	1,929	2,784
Grain sorghums	16,598	0	850
Oats	215	139	66
Rye	15	200	0
Blended food products	1,328	86	4,361
Other grain products	61,597	57,488	53,467
Sugar	2,506	1,375	3,991
Related sugar products	56,309	40,007	53,021
Chocolate and prep	5,764	9,324	7,544
Nonvariable-levy commodities.			
Animals live—excluding poultry	163,872	199,936	184,150
Horses, mules, burros—live	163,651	199,714	184,106
Horsemeat—fr. chill, froz	22,460	27,468	18,203
Variety meats, edible offals	13,174	13,521	6,469
Feathers and downs, crude	0	0	0
Tallow—inedible	17,758	22,752	8,346
Other animal fats and oils	10,588	4,537	5,821
Hides and skins, incl. furs	118,198	141,730	120,256
Cattle hides, whole	74,462	80,818	59,027
Furskins	15,050	21,757	16,181
Wool and mohair	10,826	4,751	5,303
Sausage casings	1,946	2,508	2,075
Bull semen	19,733	17,856	15,138
Misc. animal products—other	47,315	40,865	31,440
Corn by-products	392,495	361,177	365,770
Alfalfa meal and cubes	92	15	6
Beet pulp	21,574	22,946	18,439
Citrus pulp pellets	45,328	27,243	30,029
Other feeds and fodders	203,092	205,791	191,017
Fruits and prep, excluding juice	317,464	323,806	301,505
Fruits, fresh	146,491	127,392	128,765
Fruits, fresh, citrus	54,215	53,406	53,595
Grapefruit, fresh	52,845	52,625	52,284
Lemons and limes, fresh	125	163	332
Oranges and tangerines, fresh	1,089	584	975
Other citrus, fresh	156	35	4
Fruits, fresh, noncitrus	92,276	73,985	75,170
Apples, fresh	18,543	21,709	20,860
Grapes, fresh	26,806	21,602	24,825
Other noncitrus, fresh	46,927	30,674	29,485
Fruits, dried	127,919	157,457	132,632
Raisins, dried	51,581	63,718	55,473
Prunes, dried	63,148	85,373	66,750
Other dried fruits	13,190	8,365	10,408
Fruits—canned excl. juice	22,537	25,901	24,444
Fruits—froz. excl. juice	4,443	1,385	2,251
Other fruits—prep. or pres	16,073	11,671	13,413
Fruit juices incl. frozen	149,700	141,157	155,584
Orange juice	96,046	78,253	92,080
Other fruit juices	53,654	62,904	63,504
Wine	274,605	322,433	289,135

See footnotes at end of table.

Table 15-6.—Agricultural exports: Value of U.S. exports to the European Union, by commodity, fiscal years 1999/2000–2001/2002 [1]—Continued

Commodity	Value		
	1999/2000	2000/2001	2001/2002 [2]
	1,000 dollars	*1,000 dollars*	*1,000 dollars*
Nuts and preparations	562,493	564,843	703,597
Almonds (shelled basis)	297,090	349,903	385,355
Peanuts, shelled or prepared	104,741	44,333	116,839
Walnuts, shelled/unshelled	73,023	83,062	93,100
Vegetables and preparations	364,114	378,762	393,314
Vegetables—fresh	28,311	22,320	23,080
Vegetables—frozen	15,951	22,745	8,935
Vegetables—canned	30,751	24,748	27,254
Pulses	57,736	70,595	52,906
Dried beans	40,522	55,058	34,940
Dried peas	7,158	6,564	6,595
Dried lentils	10,057	8,974	11,371
Hops, including hop extract	31,083	40,029	29,624
Other veg.—prep. or pres	200,282	198,325	197,514
Oilseeds and products	1,460,365	1,482,979	1,591,336
Oilcake and meal	68,174	126,845	76,419
Soybean meal	31,286	104,256	47,732
Other oilcake and meal	36,888	22,589	28,687
Oilseeds	1,260,652	1,238,924	1,388,794
Soybeans	1,107,303	1,069,503	1,211,802
Sunflowerseeds	60,147	58,670	72,018
Peanuts and oilstock	13,124	9,390	11,160
Other oilseeds	4,621	11,392	11,246
Protein substances	75,457	89,969	82,568
Vegetable oils	131,539	117,209	126,123
Soybean oil	11,343	5,549	10,723
Cottonseed oil	494	973	544
Sunflower oil	1,408	23,632	12,398
Other vegetable oils and waxes	118,294	87,055	102,458
Tobacco—unmfg	551,589	598,826	542,963
Cotton, excl. linters	63,383	59,862	61,518
Cotton linters	1,159	774	4,452
Essential oils	173,345	214,302	178,801
Seeds—field and garden	205,461	161,207	181,616
Coffee	6,863	24,799	12,308
Cocoa	1,359	3,699	1,444
Tea and mate	17,567	26,469	18,436
Spices	19,998	11,550	12,106
Flavoring sirups and extract	0	0	0
Rubber—crude natural	5,288	3,970	4,764
Fibers excl. cotton	291	294	1,192
Nursery and greenhouse products	61,680	69,276	65,679
Beverages excl. juices	80,304	33,253	37,260
Other miscellaneous vegetable products	135,795	168,994	186,941

[1] Fiscal years Oct. 1–Sept. 30. [2] Preliminary.

ERS, Market and Trade Economics Division, (202) 694–5272.

Table 15-7.—Agricultural imports for consumption: Value by country of origin, United States, fiscal years 2000–2002 [2]

Country	2000			2001			2002		
	Total	Non-competitive [2]	Competitive [2]	Total	Non-competitive [2]	Competitive [2]	Total	Non-competitive [2]	Competitive [2]
	Million dollars	*Million dollars*	*Million dollars*	*Million dollars*	*Million dollars*	*Million dollars*	*Million dollars*	*Million dollars*	*Million dollars*
Canada	8,519.8	613.5	7,906.3	9,520.5	641.5	8,879.0	10,188.2	720.1	9,468.1
Mexico	4,988.6	589.6	4,399.0	5,310.9	314.8	4,996.1	5,288.4	295.1	4,993.3
Australia	1,517.2	39.0	1,478.2	1,798.4	41.5	1,757.0	1,894.4	40.4	1,854.0
Italy	1,561.5	65.2	1,499.6	1,531.4	65.7	1,465.7	1,723.4	71.6	1,651.8
Netherlands	1,526.7	131.0	1,395.7	1,605.0	145.8	1,459.2	1,715.5	166.0	1,549.5
France	1,491.0	117.8	1,372.7	1,360.8	131.9	1,228.8	1,420.8	112.6	1,308.2
New Zealand	1,105.7	18.3	1,087.4	1,228.1	19.7	1,208.5	1,230.4	10.4	1,220.0
Chile	974.0	15.2	958.8	1,022.9	11.3	1,011.4	1,123.9	12.6	1,111.3
Brazil	1,249.9	484.5	765.5	984.6	296.4	688.1	1,080.6	299.2	781.4
China	810.6	167.0	643.6	787.0	137.8	649.1	973.9	145.2	828.7
Indonesia	1,000.9	786.2	214.8	858.5	713.4	145.0	910.0	768.2	141.8
Colombia	1,162.5	710.4	452.1	962.4	540.9	421.6	903.2	514.2	389.0
Costa Rica	797.6	480.3	317.3	816.8	432.0	384.8	814.5	409.8	404.6
Germany	810.0	187.4	622.6	753.4	174.5	578.9	761.7	174.0	587.7
Spain	685.7	41.7	644.0	682.0	40.0	642.0	735.4	42.2	693.2
Thailand	756.2	275.9	480.4	725.5	217.6	508.0	708.4	192.7	515.6
India	866.1	262.9	603.2	688.2	196.9	491.4	669.6	181.9	487.7
Guatemala	734.2	503.9	230.2	625.0	423.1	201.8	644.9	412.6	232.3
Argentina	684.2	67.8	616.3	606.8	71.9	534.9	580.8	75.3	505.5
United Kingdom	611.9	116.3	495.6	572.5	112.7	459.8	567.7	117.1	450.6
Ecuador	479.0	335.6	143.4	469.7	302.7	166.9	503.6	343.4	160.2
Denmark	470.4	53.8	416.6	477.8	66.2	411.6	501.2	80.4	420.9
Philippines	463.7	13.7	450.0	429.6	9.7	419.9	436.7	7.8	428.9
Japan	376.4	70.0	306.4	355.6	71.1	284.5	374.0	72.6	301.4
Ireland	314.0	22.9	291.1	277.2	16.8	260.3	333.5	20.5	313.0
Turkey	310.3	43.9	266.5	293.6	37.1	256.5	321.0	42.8	278.2
Malaysia	370.2	166.7	203.5	288.3	117.1	171.2	267.9	106.1	161.9
Dominican Republic	234.8	39.3	195.5	260.2	45.4	214.9	264.9	54.0	211.0
Ivory Coast	290.7	284.1	6.6	163.4	157.5	5.9	241.0	237.5	3.5
Honduras	230.5	162.3	68.2	235.4	145.8	89.6	226.2	144.4	81.8
Peru	219.1	108.6	110.5	188.4	63.5	124.9	220.7	70.0	150.7
Belgium/Luxembourg	171.5	59.6	111.9	172.7	57.9	114.8	192.0	60.8	131.2
Vietnam	200.9	145.1	55.8	157.2	99.8	57.4	175.7	82.6	93.1
Taiwan	188.1	8.0	180.0	182.6	7.6	175.0	172.2	8.2	164.0
Greece	141.9	2.5	139.5	141.3	2.6	138.7	165.6	2.0	163.6
Switzerland	144.4	56.0	88.4	152.5	50.7	101.8	161.8	56.2	105.6
Poland	95.7	5.7	90.0	108.2	5.9	102.3	145.2	8.1	137.1
Korea, Rep. of	112.0	7.7	104.3	122.8	5.7	117.1	143.9	5.3	138.6
Israel	120.4	23.4	97.0	131.2	26.0	105.2	136.2	20.2	116.0
Rep. S. Africa	119.2	6.6	112.6	113.6	10.2	103.4	120.7	10.8	109.9
Austria	51.9	2.1	49.8	115.7	2.5	113.2	111.9	8.9	102.9
Madagascar (Malagasy)	31.0	27.1	3.9	95.3	91.6	3.7	110.3	106.9	3.4
Nicaragua	108.1	59.5	48.6	96.2	45.2	51.0	94.2	36.6	57.6
Hong Kong	83.2	11.3	71.9	80.1	11.5	68.6	90.3	10.4	79.9
El Salvador	164.7	133.4	31.3	78.8	33.9	34.0	87.7	29.6	58.2
Sweden	82.8	29.6	53.1	85.0	27.2	57.8	85.5	28.8	56.7
Finland	64.4	13.6	50.8	73.2	10.0	63.3	81.5	13.5	68.1
Portugal	72.8	0.6	72.2	64.7	0.4	64.2	72.1	0.5	71.5
Malawi	57.5	12.0	45.4	52.0	12.1	39.9	62.5	12.0	50.5
Jamaica	48.1	6.7	41.4	50.7	7.3	43.4	56.3	7.3	49.0
Other	1,185.1	438.7	743.6	1,073.4	347.4	737.2	1,087.3	363.7	723.6
Total [4]	38,857.1	8,024.0	30,833.1	39,027.1	6,617.8	32,409.3	40,979.3	6,813.1	34,166.3

[1] Fiscal years Oct. 1–Sept. 30. [2] Competitive agricultural import products consist of all products similar to agricultural commodities produced commercially in the United States, together with all agricultural products interchangeable to any significant extent with such U.S. commodities. Noncompetitive agricultural products include all others, about 90 percent of which consist of rubber, coffee, raw silk, cacao beans, wool for carpets, bananas, tea, spices, and vegetable fibers. [3] Beginning January 1, 1984 includes Canary Islands. [4] Totals may not add due to rounding.

ERS, Market and Trade Economics Division, (202) 694–5272. Compiled from reports of the U.S. Department of Commerce.

Table 15-8.—European Union: Value of agricultural imports by origin, 1992–2001 [1]

Year [2]	United States	EU countries	Other countries	Total
	Million dollars	*Million dollars*	*Million dollars*	*Million dollars*
1992	8,810	119,124	47,092	175,026
1993	8,125	99,265	42,655	150,045
1994	8,405	109,913	51,108	169,426
1995	8,567	122,003	54,104	184,674
1996	9,026	127,148	54,939	191,112
1997	9,105	129,520	52,439	191,064
1998	7,961	133,739	52,482	194,182
1999	6,603	132,666	49,032	188,301
2000	6,312	117,228	48,673	172,213
2001	6,429	117,910	48,004	172,343

[1] EU-15. Based on bilateral import data from the United Nations. [2] Data on calendar year basis.

ERS, Market and Trade Economics Division, (202) 694–5273.

Table 15-9.—Fisheries: Landings and value of principal species: 1994–2001 [1]

[Preliminary]

Species	Landings							
	1994	1995	1996	1997	1998	1999	2000	2001
	Mil. lbs.	*Mil. lbs.*	*Mil. lbs.*	*Mil. lbs.*	*Mil. lbs.*	*Mil. lbs.*	*Mil. l bs.*	*Mil. lbs.*
Fish:								
Cod, Atlantic	39	30	31	29	25	21	25	33
Flounder	427	423	460	566	391	331	413	352
Haddock	1	1	1	3	6	7	9	13
Halibut	58	45	49	70	73	80	75	78
Herring, sea	214	265	318	348	272	267	235	300
Jack mackerel	6	4	5	3	3	2	3	8
Menhaden	2,324	1,847	1,755	2,028	1,706	1,989	1,760	1,741
Ocean perch, Atlantic	1	1	1	1	1	1	1	1
Pollock	3,133	2,853	2,630	2,522	2,729	2,336	2,616	3,188
Salmon, Pacific	901	1,137	877	568	644	815	629	723
Tuna	72	14	85	83	85	58	51	52
Whiting	36	34	35	34	33	31	27	28
Shellfish:								
Clams (meats)	131	134	123	114	108	112	118	123
Crabs	447	364	392	430	553	458	299	272
Lobsters, American	66	66	71	84	80	87	83	74
Oysters (meats)	38	40	38	40	34	27	41	33
Scallops (meats)	25	20	18	15	13	27	33	47
Shrimp	293	283	307	317	290	278	304	324
	Value							
	Mil. dol.	*Mil. dol.*	*Mil. dol.*	*Mil. dol.*	*Mil. dol.*	*Mil. dol.*	*Mil. dol.*	*Mil. dol.*
Fish:								
Cod, Atlantic	36	28	27	24	25	24	26	32
Flounder	127	150	154	131	97	90	110	105
Haddock	1	1	1	4	8	9	12	15
Halibut	85	67	83	117	104	125	144	115
Herring, sea	31	59	81	23	22	26	22	26
Jack mackerel	(2)	(2)	(2)	(2)	(2)	(2)	(2)	(2)
Menhaden	128	99	94	112	104	113	112	103
Ocean perch, Atlantic	1	5	(2)	(2)	(2)	(2)	(2)	(2)
Pollock	383	266	243	248	198	171	168	237
Salmon, Pacific	456	527	369	270	257	360	270	209
Tuna	108	103	110	110	94	86	95	93
Whiting	14	15	13	15	13	14	11	13
Shellfish:								
Clams (meats)	122	140	128	130	135	135	154	162
Crabs	533	512	427	430	473	521	405	382
Lobsters, American	196	215	242	267	254	323	301	254
Oysters (meats)	133	102	115	117	89	73	91	81
Scallops (meats)	124	95	102	94	80	129	165	175
Shrimp	564	570	509	544	516	561	690	569

[1] Data exclude landings by U.S. flag vessels at Puerto Rico and other ports outside the 50 States, and production of artificially cultivated fish and shellfish. [2] Less than $500.000.

U.S. Department of Commerce, NOAA, NMFS, Fisheries Statistics and Economics Division. (301) 713–2328.

Table 15-10.—Fresh and frozen fishery products: Production and value, 1994–2001 [1]

[2000 is preliminary]

Product	1994	1995	1996	1997	1998	1999	2000	2001
	Production							
	Mil. lb.	*Mil. lb.*	*Mil. lb.*	*Mil. lb.*	*Mil. lb.*	*Mil. lb.*	*Mil. lb.*	*Mil. lb.*
Fish fillets and steaks [2]	425	385	423	410	422	362	368	477
Cod	58	65	67	79	67	61	56	40
Flounder	35	35	29	27	24	23	27	31
Haddock	3	3	4	7	6	5	6	6
Ocean perch, Atlantic	(3)	(3)	2	1	1	1	1	(3)
Rockfish	23	25	20	17	16	11	11	9
Pollock, Atlantic	7	4	2	1	4	2	2	2
Pollock, Alaska	141	135	136	112	161	144	160	271
Other	158	118	163	166	143	115	105	118
	Value							
	Mil. dol.	*Mil. dol.*	*Mil. dol.*	*Mil. dol.*	*Mil. dol.*	*Mil. dol*	*Mil. dol.*	*Mil. dol.*
Fish fillets and steaks [2]	836	841	904	961	961	807	823	915
Cod	135	152	163	179	161	108	167	123
Flounder	89	86	79	79	70	67	71	76
Haddock	12	11	14	24	22	20	24	27
Ocean perch, Atlantic	1	1	4	2	2	2	1	1
Rockfish	33	38	42	33	33	23	25	21
Pollock, Atlantic	16	10	5	2	7	4	4	8
Pollock, Alaska	141	184	159	129	190	169	178	296
Other	406	359	438	513	476	414	353	363

[1] Excludes Alaska and Hawaii, except frozen products includes Alaska and Hawaii. [2] Fresh and frozen. [3] Less than 500,000 lb.

U.S. Department of Commerce, NOAA, NMFS, Fisheries Statistics and Economics Division (301) 713–2328.

Table 15-11.—Canned fishery products: Production and value, 1994–2001 [1]

[1999 is preliminary]

Product	1994	1995	1996	1997	1998	1999	2000	2001
	Production							
	Mil. lb.	*Mil. lb.*	*Mil. lb.*	*Mil. lb.*	*Mil. lb.*	*Mil. lb.*	*Mil. lb.*	*Mil. lb.*
Total [2]	1,768	1,927	1,877	1,565	1,533	1,897	1,747	1,664
Tuna	610	667	676	627	681	694	671	507
Salmon	207	249	197	162	159	234	171	212
Clam products	122	129	129	127	113	123	127	126
Sardines, Maine	16	14	18	16	12	12	(3)	(3)
Shrimp	(5)	(5)	(5)	1	2	2	2	2
Crabs	(5)	(5)	(5)	(5)	(5)	(5)	(5)	(5)
Oysters [4]	(5)	(5)	(5)	(5)	(5)	(5)	(5)	1
	Value							
	Mil. dol.	*Mil. dol.*	*Mil. dol.*	*Mil. dol.*	*Mil. dol.*	*Mil. dol.*	*Mil. dol.*	*Mil. dol.*
Total [2]	1,796	1,887	1,800	1,593	1,765	1,861	1,626	1,402
Tuna	963	939	957	919	983	946	856	658
Salmon	329	419	284	253	274	393	288	259
Clam products	106	111	117	115	105	110	120	125
Sardines, Maine	28	24	30	29	19	20	(3)	(3)
Shrimp	4	7	6	5	11	10	11	10
Crabs	(5)	(5)	(5)	(5)	(5)	(5)	(5)	(5)
Oysters [4]	(5)	(5)	(5)	(5)	(5)	(5)	1	1

[1] Natural pack only. [2] Includes other products not shown separately. [3] Confidential data. [4] Includes oyster specialties. [5] Less than 500,000 pounds or $500,000.

U.S. Dept. of Commerce, NOAA, NMFS, Fisheries Statistics and Economics Division (301) 713–2328.

Table 15-12.—Fisheries: Fishermen and craft, 1977, and catch, 1996–2001 by area

[1995–2000 are preliminary]

Area	1977[1]			1996		1997	
	Fishermen	Fishing vessels	Fishing boats[2]	Total catch	Value	Total catch	Value
	1,000	*Number*	*1,000*	*Mil. lb.*	*Mil. dol.*	*Mil. lb.*	*Mil. dol.*
United States	182.1	17,545	89.2	9,565	3,487	9,843	3,448
New England States	31.7	929	15.4	642	564	641	556
Middle Atlantic States	17.3	573	11.3	242	182	237	200
Chesapeake Bay States	27.9	2,086	19.0	729	159	688	169
South Atlantic States	11.6	1,463	6.7	269	210	299	213
Gulf States	29.3	5,328	11.0	1,497	680	1,790	759
Pacific Coast States	54.0	7,643	15.4	6,129	1,611	6,126	1,464
Great Lakes States	1.2	217	0.5	25	17	26	17
Hawaii	2.7	101	1.3	32	64	37	69

Area	1998		1999		2000		2001	
	Total catch	Value	Total catch	Value	Total catch	Value	Total catch	Value
	Mil. lb.	*Mil. dol.*	*Mil. lb.*	*Mil. dol.*	*Mil. lb.*	*Mil. dol.*	*Mil. lb.*	*Mil. dol.*
United States	9,194	3,128	9,339	3,467	9,069	3,549	9,492	3,228
New England States	596	537	584	655	571	681	635	646
Middle Atlantic States	262	181	225	181	220	173	217	173
Chesapeake Bay States	653	180	527	172	492	172	617	175
South Atlantic States	240	197	230	198	221	204	200	176
Gulf States	1,537	719	1,945	758	1,760	911	1,606	798
Pacific Coast States	5,843	1,236	5,766	1,472	5,750	1,321	6,174	1,187
Great Lakes States	27	16	24	16	22	19	19	18
Hawaii	36	62	37	65	33	68	24	55

[1] Exclusive of duplication among regions. Computation of area amounts will not equal U.S. total. Mississippi River data included with total. [2] Refers to craft having capacity of less than 5 net tons. Note: Table may not add due to rounding.

U.S. Department of Commerce, NOAA, NMFS, Fisheries Statistics and Economics Division (301) 713–2328.

Table 15-13.—Fisheries: Quantity and value of domestic catch, 1992–2001

[1991–2000 are preliminary]

Year	Quantity[1]			Ex-vessel value	Average price per lb.
	Total	For human food	For industrial products[2]		
	Mil. lb.	*Mil. lb.*	*Mil. lb.*	*Mil. dol.*	*Cents*
1992	9,637	7,618	2,019	3,678	38.2
1993	10,467	8,214	2,253	3,471	33.2
1994	10,461	7,936	2,525	3,809	36.4
1995	9,788	7,667	2,121	3,770	38.5
1996	9,565	7,474	2,091	3,487	36.5
1997	9,842	7,244	2,598	3,448	35.0
1998	9,194	7,173	2,021	3,128	34.0
1999	9,339	6,832	2,507	3,467	37.1
2000	9,069	6,912	2,157	3,550	39.1
2001	9,492	7,314	2,178	3,228	34.0

[1] Live weight. [2] Meals, oil, fish solubles, homogenized condensed fish, shell products, bait, and animal food.

U.S. Department of Commerce, NOAA, NMFS Fisheries Statistics and Economics Division (301) 723–2328.

Table 15-14.—Fishery products: Supply, 1992–2001 [1]

[1993–2000 are preliminary]

Item	1992	1993	1994	1995	1996
	Mil. lbs	*Mil. lbs*	*Mil. lbs*	*Mil. lbs*	*Mil. lbs*
Total	16,106	20,334	19,309	16,484	16,474
For human food	13,242	13,821	13,714	13,584	13,626
Finfish	10,297	10,796	10,719	10,692	10,699
Shellfish [2]	2,945	3,025	2,995	2,891	2,927
For industrial use	2,864	6,513	5,595	2,900	2,848
Domestic catch	9,637	10,467	10,461	9,788	9,565
Percent of total	59.8	51.5	54.2	59.4	58.1
For human food	7,618	8,214	7,936	7,667	7,476
Finfish	6,182	6,770	6,612	6,414	6,205
Shellfish [2]	1,436	1,444	1,324	1,252	1,271
For industrial use	2,019	2,253	2,525	2,121	2,089
Imports [3]	6,469	9,867	8,848	6,696	6,909
Percent of total	40.2	48.5	45.8	40.6	41.9
For human food	5,624	5,607	5,778	5,917	6,150
Finfish	4,115	4,026	4,107	4,278	4,494
Shellfish [2]	1,509	1,581	1,671	1,639	1,656
For industrial use [4]	845	4,260	3,070	779	759

Item	1997	1998	1999	2000	2001
	Mil. lbs	*Mil. lbs*	*Mil. lbs*	*Mil. lbs*	*Mil. lbs*
Total	17,133	16,898	17,378	17,339	18,119
For human food	13,740	14,175	14,462	14,740	15,306
Finfish	10,580	10,837	10,831	11,006	11,330
Shellfish [2]	3,160	3,338	3,630	3,734	3,977
For industrial use	3,393	2,723	2,916	2,599	2,812
Domestic catch	9,843	9,194	9,339	9,068	9,492
Percent of total	57.4	54.4	53.7	52.3	52.4
For human food	7,245	7,174	6,832	6,912	7,314
Finfish	5,969	5,935	5,490	5,637	6,162
Shellfish [2]	1,277	1,238	1,341	1,275	1,152
For industrial use	2,598	2,021	2,507	2,157	2,178
Imports [3]	7,290	7,704	8,039	8,271	8,627
Percent of total	42.5	45.6	46.3	47.7	47.6
For human food	6,495	7,001	7,630	7,828	7,992
Finfish	4,612	4,901	5,341	5,369	5,168
Shellfish [2]	1,883	2,100	2,289	2,459	2,828
For industrial use [4]	795	702	409	442	634

[1] Live weight, except percent. May not add due to rounding. [2] For univalve and bivalves mollusks (conchs, clams, oysters, scallops, etc.), the weight of meats, excluding the shell is reported. [3] Excluding imports of edible fishery products consumed in Puerto Rico; includes landings of tuna caught by foreign vessels in American Samoa. [4] Fish meal and sea herring.

U.S. Department of Commerce, NOAA, NMFS Fisheries Statistics and Economics Division (301) 713–2328.

Table 15-15.—Fisheries: Disposition of domestic catch, 1992–2001 [1]

[1990–99 are preliminary]

Disposition	1992	1993	1994	1995	1996	1997	1998	1999	2000	2001
	Mil. lbs.	*Mil. lbs.*	*Mil. lbs.*	*Mil. lbs.*	*Mil. lbs.*	*Mil. lbs.*	*Mil. lbs.*	*Mil. lbs.*	*Mil. lbs.*	*Mil. lbs.*
Fresh and frozen	7,288	7,744	7,475	7,099	7,054	6,873	6,870	6,416	6,657	7,085
Canned	543	649	622	769	678	648	516	712	530	536
Cured	110	115	95	90	93	108	129	133	119	123
Reduced to meal, oil, etc	1,696	1,959	2,269	1,830	1,740	2,213	1,679	2,078	1,763	1,748
Total	9,637	10,467	10,461	9,788	9,565	9,842	9,194	9,339	9,069	9,492

[1] Live weight catch. In addition to whole fish, a large portion of waste (400–500 mil. lb.) derived from canning, filleting, and dressing fish and shellfish is utilized in production of fish meal and oil in each year shown.

U.S. Department of Commerce, NOAA, NMFS Fisheries Statistics and Economics Division (301) 713–2328.

Table 15-16.—Processed fishery products: Production and value, 1994–2001 [1]

Item	Production							
	1994	1995	1996	1997	1998	1999	2000 [3]	2001 [3]
	Mil. lb.	*Mil. lb.*	*Mil. lb.*	*Mil. lb.*	*Mil. lb.*	*Mil. lb.*	*Mil. lb.*	*Mil. lb.*
Fresh and frozen	NA	NA	NA	NA	NA	NA	NA	NA
Fillets	370	355	372	355	391	337	336	447
Steaks	55	30	51	55	31	25	32	30
Fish sticks	58	74	65	69	69	65	40	43
Fish portions	196	251	214	196	185	203	183	189
Breaded shrimp	113	100	108	117	109	119	121	152
Canned products [2]	1,768	1,927	1,877	1,565	1,533	1,897	1,747	1,664
Fish and shellfish	986	1,085	1,053	953	989	1,100	1,008	885
Animal feed	782	842	824	612	544	797	739	779
Industrial products	NA	NA	NA	NA	NA	NA	NA	NA
Meal and scrap	808	667	643	725	613	672	627	635
Oil (body and liver)	291	242	248	283	223	286	192	279
Other	NA	NA	NA	NA	NA	NA	NA	NA

Item	Value							
	1994	1995	1996	1997	1998	1999	2000 [3]	2001 [3]
	Mil. lb.	*Mil. lb.*	*Mil. lb.*	*Mil. lb.*	*Mil. lb.*	*Mil. lb.*	*Mil. lb.*	*Mil. lb.*
Fresh and frozen:.								
Fillets	747	767	807	845	887	739	741	845
Steaks	89	73	97	116	74	68	82	70
Fish sticks	51	73	56	64	63	63	43	42
Fish portions	268	357	307	285	211	269	233	235
Breaded shrimp	305	299	342	335	333	352	375	540
Canned products [2]	1,796	1,887	1,800	1,593	1,775	1,861	1,626	1,402
Fish and shellfish	1,471	1,544	1,429	1,361	1,425	1,522	1,334	1,111
Animal feed	325	343	371	232	350	340	292	291
Industrial products	248	247	274	347	233	268	511	520
Meal and scrap	134	122	138	174	117	147	115	126
Oil (body and liver)	37	42	44	55	56	42	21	48
Other	62	74	86	118	60	79	83	56

[1] Includes cured fish. [2] Includes salmon eggs for baits. [3] Preliminary. NA-not available.

U.S. Department of Commerce, NOAA, NMFS, Fisheries Statistics and Economics Division (301) 713–2328.

Table 15-17.—Selected fishery products: Imports and exports, 1994–2001 [1]

Product	Quantity							
	1994	1995	1996	1997	1998	1999	2000	2001
	Mil. lb.	*Mil. lb.*	*Mil. lb.*	*Mil. lb.*	*Mil. lb.*	*Mil. lb.*	*Mil. lb.*	*Mil. lb.*
Imports								
Edible	3,035	3,066	3,170	3,339	3,647	3,888	3,978	4,102
Fresh or frozen	2,528	2,597	2,727	2,861	3,119	3,227	3,310	3,449
Salmon [2]	117	126	142	163	152	156	151	159
Tuna	458	496	529	438	571	491	445	405
Groundfish fillets, blocks [3]	347	372	385	384	376	410	393	310
Other fillets and steaks	292	316	326	339	392	429	510	601
Scallops (meats)	57	48	59	60	52	44	54	40
Lobster, American and spiny	63	66	63	65	64	81	95	92
Shrimp and prawn	622	591	579	645	692	728	757	878
Canned	419	378	354	387	428	546	556	539
Sardines, in oil	14	14	13	13	15	16	26	19
Sardines and herring, not in oil	30	28	30	31	33	38	46	42
Tuna	249	215	193	212	240	335	313	292
Oysters	12	11	10	10	14	13	43	12
Pickled or salted	42	39	37	37	42	38	42	43
Cod, haddock, hake, pollock, cusk	7	5	5	5	7	7	9	8
Nonedible scrap and metal	548	139	136	142	125	73	79	113
Exports								
Canned salmon	91	98	96	82	77	114	81	110
Fish oil, nonedible	243	260	187	215	197	233	142	249

Product	Value							
	1994	1995	1996	1997	1998	1999	2000	2001
	Mil. lb.	*Mil. lb.*	*Mil. lb.*	*Mil. lb.*	*Mil. lb.*	*Mil. lb.*	*Mil. lb.*	*Mil. lb.*
Imports								
Edible	6,645	6,792	6,730	7,754	8,173	9,014	10,054	9,864
Fresh or frozen	5,917	6,072	6,044	7,022	7,356	8,043	9,120	8,832
Salmon [2]	253	281	305	344	319	345	333	323
Tuna	426	458	510	494	556	550	520	515
Groundfish fillets, blocks [3]	487	511	498	534	579	674	589	479
Other fillets and steaks	551	632	645	727	837	982	1,233	1,263
Scallops (meats)	216	174	198	237	218	193	212	128
Lobster, American and spiny	448	492	446	481	476	628	712	727
Shrimp and prawn	2,654	2,565	2,448	2,943	3,102	3,131	3,749	3,617
Canned	542	503	474	525	588	682	670	774
Sardines, in oil	25	26	23	25	28	28	39	30
Sardines and herring, not in oil	24	23	26	26	29	36	44	39
Tuna	286	234	216	250	289	336	258	314
Oysters	31	31	29	25	26	27	26	24
Pickled or salted	47	47	47	47	57	59	60	61
Cod, haddock, hake, pollock, cusk	14	10	9	8	13	15	19	16
Nonedible scrap and metal	77	25	34	36	34	17	18	27
Exports								
Canned salmon	164	175	154	135	143	198	146	168
Fish oil, nonedible	30	44	36	54	60	36	24	42

[1] Includes Puerto Rico. [2] Excludes fillets. [3] Includes cod, cusk, haddock, hake, pollock, ocean perch, and whiting.

U.S. Dept. of Commerce, NOAA, NMFS, Fisheries Statistics and Economics Division (301) 713–2328.

Table 15-18.—Fishery products: Imports and exports, 1992–2001 [1]

Year	Imports [2]				Exports			
	Total value	Edible products		Non-edible, value	Total value	Edible products		Non-edible, value
		Quantity	Value			Quantity	Value	
	Mil. dol.	*Mil. dol.*	*Mil. dol.*	*Mil. dol.*	*Mil. dol.*	*Mil. dol.*	*Mil. dol.*	*Mil. dol.*
1992	9,871	2,894	5,706	4,165	7,120	2,088	3,466	3,654
1993	10,622	2,917	5,849	4,774	6,925	1,986	3,077	3,848
1994	11,987	3,035	6,645	5,342	7,381	1,979	3,126	4,255
1995	12,452	3,066	6,792	5,660	8,268	2,047	3,262	5,006
1996	13,060	3,170	6,730	6,331	8,653	2,112	3,032	5,621
1997	14,528	3,339	7,754	6,774	9,354	2,019	2,713	6,640
1998	15,633	3,647	8,173	7,459	8,697	1,664	2,260	6,437
1999	17,040	3,888	9,014	8,026	10,007	1,961	2,849	7,158
2000	19,013	3,978	10,054	8,959	10,782	2,165	2,952	7,830
2001	18,547	4,102	9,864	8,683	11,834	2,565	3,195	8,639

[1] Includes Puerto Rico. [2] Includes landings of tuna by foreign vessels in American Samoa.

U.S. Department of Commerce, NMFS, Fisheries Statistics and Economics Division (301) 713–2328.

Table 15-19.—Fish trips: Estimated number of fishing trips taken by marine recreational fishermen by subregion and year, Atlantic and Gulf and Pacific Coasts, 1998–2001

Subregion	1998	1999	2000	2001
	Thousands	*Thousands*	*Thousands*	*Thousands*
Atlantic and Gulf: [1]				
North Atlantic	6,796	6,478	8,765	9,035
Mid-Atlantic	14,453	14,105	19,451	21,206
South Atlantic [2]	16,837	14,435	20,075	21,596
Gulf [2]	16,703	15,894	21,018	22,890
Total	54,789	50,912	69,309	74,727

Subregion	1998	1999	2000	2001
	Thousands	*Thousands*	*Thousands*	*Thousands*
Pacific: [3]				
Southern California	2,973	2,437	3,787	4,052
Northern California	1,932	1,713	2,158	2,208
Oregon	649	554	930	1,170
Washington	1,463	1,256	1,643	2,191
Total	7,017	5,960	8,518	9,621

[1] 1998–2000 were revised. Data do not include recreational catch in Texas. [2] Does not include catch from headboats (party boats) in the South Atlantic and Gulf of Mexico. [3] Data do not include recreational catch in Hawaii or Alaska. Pacific state estimates do not include salmon data collected by recreational surveys.

U.S. Department of Commerce, NOAA, NMFS, Fisheries Statistics and Economics Division (301) 713–2328.

Table 15-20.—Fish harvested: Estimated number of fish harvested by marine recreational anglers by subregion and year, Atlantic, Gulf Coasts, and Pacific Coasts, 1998–2001

Subregion	1998	1999	2000	2001
	Thousands	*Thousands*	*Thousands*	*Thousands*
Atlantic and Gulf: [1]				
North Atlantic	6,783	8,841	17,185	12,153
Mid-Atlantic	29,447	24,756	50,652	34,704
South Atlantic [2]	24,704	33,914	40,414	43,824
Gulf [2]	60,561	55,525	67,365	76,571
Total	121,495	123,036	175,616	167,252
Subregion	1998	1999	2000	2001
	Thousands	*Thousands*	*Thousands*	*Thousands*
Pacific: [3]				
Southern California	5,827	5,098	7,494	7,726
Northern California	6,123	3,909	3,787	4,799
Oregon	1,712	1,153	1,848	2,123
Washington	5,220	2,486	3,238	4,798
Total	18,882	12,646	16,367	19,446

[1] 1998–2000 were revised. Data do not include recreational catch in Texas. [2] Does not include catch for headboats (party boats) in the South Atlantic and Gulf of Mexico. [3] Data do not include recreational catch in Hawaii or Alaska. Pacific estimates do not include salmon data collected by State recreational surveys. Note: "Harvested" includes dead discards and fish used for bait but does not include fish released alive.

U.S. Department of Commerce, NOAA, NMFS, Fisheries Statistics and Economics Division (301) 713–2328.

Table 15-21.—Fish harvested: Estimated number of fish harvested by marine recreational anglers by mode and year, Atlantic, Gulf Coasts, and Pacific Coasts, 1998–2001

Mode	1998	1999	2000	2001
	Thousands	*Thousands*	*Thousands*	*Thousands*
Atlantic and Gulf: [1]				
Shore	32,132	33,908	60,706	53,092
Party/charter [2]	10,191	10,920	12,434	11,628
Private/rental	79,172	78,208	102,475	102,532
Total	121,495	123,036	175,615	167,252
Mode	1998	1999	2000	2001
	Thousands	*Thousands*	*Thousands*	*Thousands*
Pacific: [3]				
Shore	9,123	3,803	4,785	8,040
Party/charter	3,489	4,484	6,408	5,176
Private/rental	6,269	4,358	5,172	6,230
Total	18,881	12,645	16,365	19,446

[1] 1998–2000 were revised. Data do not include recreational catch in Texas. [2] Does not include catch for headboats (party boats) in the South Atlantic or Gulf of Mexico. [3] Data do not include recreational catch in Hawaii or Alaska. Pacific estimates do not include salmon data collected by State recreational surveys. Note: "Harvested" includes dead discards and fish used for bait but does not include fish released alive.

U.S. Department of Commerce, NOAA, NMFS, Fisheries Statistics and Economics Division (301) 713–2328.

Table 15-22.—Fish harvested: Estimated number of fish harvested by marine recreational anglers by species group and year, Atlantic and Gulf coasts, 1998–2001 [1]

Species group	1998	1999	2000	2001
	Thousands	*Thousands*	*Thousands*	*Thousands*
Barracudas	150	139	135	163
Bluefish	4,430	3,856	5,048	7,016
Dogfish sharks	50	37	50	49
Other sharks	259	185	316	343
Skates/rays	53	70	118	57
Freshwater catfishes	448	249	239	118
Saltwater catfishes	525	460	683	629
Atlantic cod	466	375	812	1,118
Other cods/hakes	79	66	151	168
Pollock	185	218	437	356
Red hake	115	118	101	58
Dolphins	1,875	2,064	2,455	2,088
Other croaker	0	0	0	0
Atlantic croaker	10,010	10,090	12,494	14,681
Black drum	715	723	1,320	990
Kingfishes	3,030	4,118	5,474	7,456
Other drum	213	765	448	278
Red drum	2,152	2,448	3,672	3,475
Sand seatrout	2,808	4,998	4,508	3,308
Silver perch	511	373	654	404
Spot	6,992	3,734	5,010	7,308
Spotted seatrout	7,694	10,603	12,946	10,200
Weakfish	2,390	1,652	2,089	1,527
Eels	14	14	56	54
Gulf flounder	174	162	169	212
Other flounders	98	99	87	88
Southern flounder	801	958	1,023	1,128
Summer flounder	7,003	4,123	7,820	5,307
Winter flounder	648	724	1,459	964
Other grunts	694	577	608	463
Pigfish	933	933	1,128	1,552
White grunt	1,809	1,749	2,112	2,772
Herrings	25,638	22,613	29,327	33,473
Blue runner	1,869	860	2,057	3,160
Crevalle Jack	511	612	601	812
Florida pompano	522	407	650	614
Greater amberjack	91	140	92	135
Other jacks	514	1,313	1,826	2,965
Mullets	3,240	5,710	7,074	7,435
Other fishes	4,791	5,971	20,176	4,769
Other porgies	149	170	326	228
Pinfishes	7,593	5,112	6,892	9,469
Red porgy	81	81	59	75
Scup	1,212	3,251	7,244	5,099
Sheepshead	1,742	1,905	2,113	2,267
Puffers	148	175	219	346
Sculpins	3	1	([2])	([2])
Black sea bass	1,697	2,000	4,546	3,932
Epinephelus groupers	154	213	355	248
Mycteroperca groupers	641	652	765	540
Other sea basses	308	274	276	337
Searobins	202	122	170	143
Gray snapper	1,072	1,038	1,265	1,199
Lane snapper	205	195	212	392
Other snappers	82	78	95	79
Red snapper	1,336	1,266	841	900
Vermilion snapper	294	540	374	613
Yellowtail snapper	352	268	245	189
Other temperate basses	6	0	0	1
Striped bass	1,395	1,368	1,993	2,039
White perch	1,834	1,113	1,758	664
Toadfishes	10	9	4	7
Triggerfishes/filefishes	390	390	263	308
Atlantic mackerel	1,874	3,236	4,194	4,127
King mackerel	965	812	930	691
Little tunny/Atlantic bonito	378	382	348	260
Other tunas/mackerels	567	656	655	737
Spanish mackerel	1,787	2,540	3,075	3,747
Cunner	94	103	77	56
Other wrasses	64	92	46	79
Tautog	358	688	853	792
Total [3]	121,494	123,036	175,616	167,252

[1] 1998–2000 were revised. Data does not include recreational catch in Texas or headboats (party boats) in the South Atlantic and the Gulf of Mexico. [2] Less than one thousand. [3] Totals may not add due to rounding. Note: "Harvested" includes dead discards and fish used for bait but does not include fish released alive.

U.S. Department of Commerce, NOAA, NMFS, Fisheries Statistics and Economics Division (301) 713–2328.

Table 15-23.—Fish harvested: Estimated number of fish harvested by marine recreational anglers by species group and year, Pacific coast[1], 1998–2001

Species group	1998	1999	2000	2001
	Thousands	*Thousands*	*Thousands*	*Thousands*
Northern anchovy	3	104	419	579
Other anchovies	0	0	8	0
California scorpionfish	160	280	230	293
Dogfish sharks	116	23	25	20
Other sharks	27	24	21	36
Skates/rays	17	11	19	29
Other cods/hakes	0	0	0	0
Pacific cod	1	0	0	1
Pacific hake	2	3	0	0
Pacific tomcod	1	1	6	2
California corbina	15	17	7	14
Other croakers	119	93	94	172
Queenfish	42	122	121	76
White croaker	434	301	372	389
Dolphins	14	1	84	0
Other drum	5	13	3	4
California halibut	106	124	168	202
Other flounders	87	69	171	80
Rock sole	14	5	19	12
Sanddabs	259	225	840	451
Starry flounder	17	11	8	14
Kelp greenling	98	89	153	153
Lingcod	160	172	139	113
Other greenlings	12	8	31	28
Herrings	1,289	666	245	799
Other jacks	20	46	8	21
Yellowtail	673	56	164	87
Mullets	1	0	16	5
Other fishes	1,139	788	1,433	2,305
Pacific barracuda	450	423	354	311
Black rockfish	1,178	956	1,423	1,119
Blue rockfish	841	673	415	464
Bocaccio	40	92	203	199
Brown rockfish	81	121	107	185
Canary rockfish	107	163	147	78
Chilipepper rockfish	9	44	73	77
Copper rockfish	157	136	106	78
Greenspotted rockfish	40	133	123	83
Olive rockfish	77	68	68	159
Other rockfishes	1,108	1,914	1,450	994
Quillback rockfish	53	42	34	26
Gopher rockfish	102	154	166	272
Widow rockfish	67	44	82	19
Yellowtail rockfish	274	403	319	162
Sablefishes	1	0	1	1
Cabezon	91	60	57	69
Sculpins	35	41	23	49
Barred sand bass	408	456	1,260	1,119
Kelp bass	326	236	648	633
Other sea basses	9	8	11	24
Spotted sand bass	17	30	287	361
Halfmoon	21	41	76	132
Opaleye	87	39	49	59
Jacksmelt	366	272	283	614
Other silversides	29	44	30	46
Other smelts	0	4	0	0
Surf smelt	4,825	1,217	1,965	3,661
Sturgeons	32	19	21	17
Barred surfperch	646	210	105	147
Black perch	94	38	54	54
Other surfperches	135	65	93	108
Pile perch	60	18	19	32
Redtail surfperch	92	60	126	123
Shiner perch	85	91	292	183
Silver surfperch	43	86	15	16
Striped seaperch	199	44	47	96
Walleye surfperch	59	68	51	163
White seaperch	22	21	8	32
Striped bass	89	83	97	44
Other tunas/mackerels	1,565	703	880	1,755
Pacific bonito	71	2	85	31
California sheephead	49	63	113	75
Other wrasses	7	5	4	5
Total[2]	18,882	12,644	16,366	19,446

[1] 1998–2000 data were revised. Data do not include recreational catch in Hawaii or Alaska. Pacific estimates do not include salmon data collected by State recreational surveys. [2] Totals may not add due to rounding. Note:"Harvested" includes dead discards and fish used for bait but does not include fish released alive.

U.S. Department of Commerce, NOAA, NMFS, Fisheries Statistics and Economics Division. (301) 713–2328.

Table 15-24.—Fish harvested: Estimated number of fish harvested by marine recreational anglers, by area of fishing and year, Atlantic and Gulf and Pacific Coast, 1998–2001

Area	1998	1999	2000	2001
	Thousands	*Thousands*	*Thousands*	*Thousands*
Atlantic and Gulf: [1]				
Inland	76,748	71,918	112,431	96,875
State Territorial Sea [2]	30,062	33,318	42,821	49,450
Federal Exclusive Ecomomic Zone [3]	14,684	17,800	20,364	20,928
Total	121,494	123,036	175,616	167,253

Area	1998	1999	2000	2001
	Thousands	*Thousands*	*Thousands*	*Thousands*
Pacific: [4]				
Inland	6,647	3,804	4,758	7,528
State Territorial Sea [2]	8,585	6,257	7,430	9,182
Federal Exclusive Ecomomic Zone [3]	3,649	2,585	4,178	2,736
Total	18,881	12,646	16,366	19,446

[1] 1998–2000 data were revised. Data do not include recreational catch in Texas or headboats (party boats) in the South Atlantic and the Gulf of Mexico. [2] Open Ocean extending 0 to 3 miles from shore, except West Florida (10 miles). [3] Open ocean extending to 200 miles offshore from the outer edge of the State Territorial Sea. [4] Data do not include recreational catch in Hawaii or Alaska. Pacific state estimates do not include salmon data collected by recreational surveys. Note: "Harvested" includes dead discards and fish used for bait but does not include fish released alive.

U.S. Department of Commerce, NOAA, NMFS, Fisheries Statistics and Economics Division. (301) 713–2328.

Table 15-25.—Farm-raised catfish: Processed, sales, inventory, and imports, 1993–2002

Year	Round [1] weight processed	Prices paid to producer	Fresh sales	Frozen sales	Total sales	Inventory end of year	Imports [2]
	(000) pounds	*Cents per pounds*	*(000) pounds*	*(000) pounds*	*(000) pounds*	*(000) pounds*	*(000) pounds*
1993	459,013	70.9	93,758	139,717	233,475	9,502	4,110
1994	439,269	78.4	87,211	129,266	216,477	11,623	3,486
1995	446,886	78.6	89,735	137,237	226,972	10,909	2,426
1996	472,123	77.3	96,722	140,458	237,180	11,894	2,482
1997	524,949	71.2	106,512	155,248	261,760	11,911	942
1998	564,355	74.3	113,092	168,306	281,398	10,807	1,386
1999	596,628	73.7	116,697	175,968	292,665	12,551	3,451
2000	593,603	75.1	116,734	180,422	297,156	13,598	8,236
2001	597,108	64.7	120,775	175,592	296,367	14,997	18,079
2002	603,601	56.8	NA	NA	NA	NA	NA

[1] Price for fish delivered to processing plant door. [2] Data furnished by U.S. Bureau of Census. NA-not available.

NASS, Livestock Branch, (202) 720–6146.

Table 15-26.—Farm-raised catfish: Prices received by processors, 1993–2002

Year	Fresh			Frozen		
	Whole fish [1]	Fillets [2]	Other [3]	Whole fish [1]	Fillets [2]	Other [3]
	Dollars per/lb	*Dollars per/lb*	*Dollars per/lb*	*Dollars per/lb*	*Dollars per/lb*	*Dollars per/lb*
1993	1.50	2.62	1.89	1.73	2.60	1.92
1994	1.70	2.88	1.92	1.98	2.86	1.97
1995	1.68	2.90	1.88	2.00	2.85	2.01
1996	1.68	2.87	1.80	1.99	2.78	1.88
1997	1.55	2.75	1.67	1.93	2.63	1.76
1998	1.59	2.80	1.72	1.94	2.69	1.73
1999	1.59	2.81	1.64	1.99	2.76	1.69
2000	1.66	2.86	1.68	2.03	2.83	1.65
2001	1.57	2.74	1.60	1.98	2.61	1.63
2002	NA	NA	NA	NA	NA	NA

[1] Dressed weight, (head, visera, and skin removed). [2] Includes regular, shank, and strip fillets; excludes any breaded product. [3] Includes nuggets, steaks, and all other products not already reported, includes weight of breading and added ingredients. NA-not available.

NASS, Livestock Branch, (202) 720–6146.

Table 15-27.—Catfish: Number of operations and water surface acres used for production, 2002–03, and total sales, 2001–02 by State and United States

State	Number of operations on Jan. 1		Water surface accres used for production during Jan 1 - Jun 30		Total sales	
	2002 [1]	2003	2002 [1]	2003	2001 [1]	2002
	Number	*Number*	*Acres*	*Acres*	*1,000 dollars*	*1,000 dollars*
AL	240	231	25,900	25,500	69,965	76,045
AR	195	155	38,000	34,000	65,306	56,380
CA	42	38	2,200	2,400	8,365	7,875
FL	45	34	880	660	1,327	756
GA	50	43	1,220	960	2,577	1,411
IL	15	12	260	130	309	226
KY	50	60	400	570	1,097	1,180
LA	70	57	12,100	9,900	26,207	15,812
MS	395	405	111,500	109,000	260,852	243,226
MO	40	31	1,800	1,400	2,840	1,070
NC	45	46	1,400	1,700	3,151	3,143
SC	15	13	520	130	923	617
TX	34	30	580	395	762	2,087
US	1,236	1,155	196,760	186,745	443,681	409,828

[1] Revised.

NASS, Livestock Branch, (202) 720-0585.

Table 15-28.—Catfish production: Water surface acre usage by State and United States, 2002–03

State	Acres intended for utilization during Jan 1-Jun 30 for:					Acres taken out of production during Jul 1-Dec 31 prev. year
	Foodsize	Fingerlings	Broodfish	Currently under or scheduled for: Renovation	Currently under or scheduled for: New construction	
2002 [1]						
AL	22,800	1,800	720	200	820	430
AR	31,500	5,100	700	750	*	1,100
CA	1,670	330	90	130	40	60
FL	700	125	51	*	25	*
GA	930	190	80	5	15	70
IL	170	55	25	30		
KY	200	70	5	15	150	5
LA	10,400	1,350	220	590	130	1,000
MS	87,800	18,100	3,000	3,700	1,400	1,200
MO	1,160	560	45	45	15	25
NC	1,200	145	40	20	400	
SC	390	85	30	*	*	*
TX	220	80	40	30		
Oth Sts				20	100	55
US	159,140	27,990	5,046	5,535	3,095	3,945
2003						
AL	22,900	1,500	630	460	180	830
AR	28,500	4,200	650	800	*	1,000
CA	1,810	360	90	110	90	100
FL	590	45	15	*	*	*
GA	700	115	60	15		65
IL	65	45	10	*	*	*
KY	460	95	15	*	30	25
LA	8,600	1,050	170	530	35	1,350
MS	86,000	16,800	3,000	3,300	650	4,400
MO	690	590	55	90	65	55
NC	1,480	140	60	30	40	30
SC	70	25	20	*	*	*
TX	175	105	55	50	5	10
Oth Sts				58	130	85
US	152,040	25,070	4,830	5,443	1,225	7,950

[1] Revised. * Included in other States to avoid disclosure of individual operations.

NASS, Livestock Branch, (202) 720–0585.

Table 15-29.—Catfish: Sales by size category, by State and United States, 2001–02

Size category and State	Number of fish		Live weight		Sales			
					Total		Average per pound	
	2001 [1]	2002	2001 [1]	2002	2001 [1]	2002	2001 [1]	2002
	1,000	*1,000*	*1,000 pounds*	*1,000 pounds*	*1,000 dollars*	*1,000 dollars*	*Dollars*	*Dollars*
Foodsize:								
AL	60,000	75,000	115,000	142,000	67,850	73,840	0.59	0.52
AR	51,600	59,400	94,000	97,000	59,220	53,350	0.63	0.55
CA	1,850	1,860	4,160	3,900	8,154	7,644	1.96	1.96
FL	1,400	900	1,700	1,300	1,071	637	0.63	0.49
GA	1,650	1,240	2,000	1,670	1,900	1,202	0.95	0.72
IL	120	180	209	200	257	178	1.23	0.89
KY	460	730	910	1,470	1,001	1,088	1.10	0.74
LA	29,000	18,200	40,000	26,600	24,800	15,428	0.62	0.58
MS	255,600	244,000	381,000	392,000	240,030	219,520	0.63	0.56
MO	1,010	350	2,180	765	1,984	635	0.91	0.83
NC	2,540	2,900	4,430	4,900	2,968	2,891	0.67	0.59
SC	1,380	485	1,440	988	893	613	0.62	0.62
TX	290	550	430	950	559	1,435	1.30	1.51
US	406,900	405,795	647,459	673,743	410,687	378,461	0.63	0.56
Broodfish:								
AL	*	*	*	*	*	*	*	*
AR	45	215	165	600	111	390	0.67	0.65
CA	*	*	*	*	*	*	*	*
FL	*	*	*	*	*	*	*	*
GA	*	....	*	....	*	....	*	....
IL	....	*	....	*	....	*	....	*
KY	*	....	*	....	*	....	*	....
LA	*	*	*	*	*	*	*	*
MS	100	300	410	1,420	443	951	1.08	0.67
MO	*	*	*	*	*	*	*	*
NC	*	*	*	*	*	*	*	*
SC	*	*	*	*	*	*	*	*
TX	....	*	....	*	....	*	....	*
Oth Sts	33	14	128	66	205	94	1.60	1.42
US	178	529	703	2,086	759	1,435	1.08	0.69

See footnotes at end of table.

Table 15-29.—Catfish: Sales by size category, by State and United States, 2001–02—Continued

Size category and State	Number of fish		Live weight		Sales			
					Total		Average per pound	
	2001 [1]	2002	2001 [1]	2002	2001 [1]	2002	2001 [1]	2002
	1,000	*1,000*	*1,000 pounds*	*1,000 pounds*	*1,000 dollars*	*1,000 dollars*	*Dollars*	*Dollars*
Stockers:								
AL	*	*	*	*	*	*	*	*
AR	18,300	12,000	2,900	1,500	2,726	1,200	0.94	0.80
CA	*	*	*	*	*	*	*	*
FL	*	*	*	*	*	*	*	*
GA	*	....	*	....	*	....	*	....
IL	132	*	23	*	52	*	2.27	*
KY	*	*	*	*	*	*	*	*
LA	*	*	*	*	*	*	*	*
MS	42,400	78,600	3,530	9,680	4,059	9,196	1.15	0.95
MO	*	*	*	*	*	*	*	*
NC	*	*	*	*	*	*	*	*
SC	*	*	*	*	*	*	*	*
TX	350	*	90	*	153	*	1.70	*
Oth Sts	4,291	6,494	2,046	1,918	1,428	2,197	0.70	1.15
US	65,473	97,094	8,589	13,098	8,418	12,593	0.98	0.96
Fingerlings and fry:								
AL	36,200	18,200	1,600	370	1,920	670	1.20	1.81
AR	123,800	64,000	1,900	900	3,249	1,440	1.71	1.60
CA	2,160	1,950	50	43	147	155	2.93	3.60
FL	3,500	1,750	110	73	220	110	2.00	1.50
GA	3,000	3,340	70	95	203	209	2.90	2.20
IL	....	....	....	....	....	....	....	....
KY	*	*	*	*	*	*	*	*
LA	*	*	*	*	*	*	*	*
MS	805,700	417,600	10,200	10,350	16,320	13,559	1.60	1.31
MO	*	2,790	*	46	*	138	*	3.00
NC	2,330	2,400	96	135	147	212	1.53	1.57
SC	*	*	*	*	*	*	*	*
TX	930	2,500	10	105	50	515	5.00	4.90
Oth Sts	29,725	3,502	570	111	1,561	331	2.74	2.98
US	1,007,345	518,032	14,606	12,228	23,817	17,339	1.63	1.42

[1] Revised. * Included in other States to avoid disclosure of individual operations.

NASS, Livestock Branch, (202) 720–0585.

Table 15-30—Trout: Number of operations by State and United States, 2002–03

State	Total		Selling trout		Distributing trout[1]	
	2002[2]	2003	2002[2]	2003	2002[2]	2003
	Number	*Number*	*Number*	*Number*	*Number*	*Number*
AR	5	5	....	....	5	5
CA	30	27	20	15	14	14
CO	34	34	15	15	19	19
CT	6	6	3	3	4	3
GA	9	9	7	7	2	2
ID	53	52	30	28	23	24
ME	14	14	7	7	8	8
MA	14	11	9	7	5	5
MI	33	22	31	20	2	5
MO	10	12	7	7	3	5
NY	38	39	27	30	11	13
NC	54	54	50	50	5	5
OR	48	51	14	21	37	30
PA	60	61	49	51	17	16
TN	16	15	9	8	8	8
UT	26	23	15	12	11	11
VA	27	22	22	18	9	5
WA	59	59	14	15	45	48
WV	29	25	19	17	12	11
WI	72	65	55	47	21	21
US	637	606	403	378	261	258

[1] Trout distributed for restoration, conservation or recreational purposes. [2] Revised.

NASS, Livestock Branch, (202) 720-0585.

Table 15-31.—Trout: Value of fish sold and distributed, by State (excluding eggs), and United States (including and excluding eggs), 2001–02

State	Total value of fish sold		Total value of distributed fish	
	2001[1]	2002	2001[1]	2002
	1,000 dollars	*1,000 dollars*	*1,000 dollars*	*1,000 dollars*
AR	....	....	*	*
CA	6,020	5,091	7,347	8,405
CO	2,794	2,503	7,643	5,477
CT	222	275	*	*
GA	535	579	*	*
ID	34,823	30,456	2,794	3,949
ME	182	130	*	*
MA	375	369	*	*
MI	823	663	*	*
MO	1,893	1,858	1,770	1,837
NY	707	567	*	*
NC	6,475	6,829	623	883
OR	2,205	801	6,480	4,861
PA	4,894	4,562	10,763	5,881
TN	362	400	*	*
UT	1,324	1,081	*	*
VA	1,318	1,259	1,462	1,513
WA	3,516	5,385	5,277	5,081
WV	827	583	1,407	1,256
WI	1,833	1,760	1,827	2,011
Oth Sts	....	....	17,432	17,402
US[2]	71,128	65,151	64,825	58,556
US[3]	76,241	69,597	66,634	60,332

[1] Revised. [2] Excludes value of eggs. [3] Includes value of eggs. *Included in other States to avoid disclosure of individual operations.

NASS, Livestock Branch, (202) 720-0585.

Table 15-32.—Trout: Sales by size category, by State and United States, 2001–02

Size category and State	Number of fish		Live weight		Sales			
					Total		Average per pound	
	2001 [1]	2002	2001 [1]	2002	2001 [1]	2002	2001 [1]	2002
	1,000 dollars	*1,000 dollars*	*1,000 dollars*	*1,000 dollars*	*1,000 dollars*	*1,000 dollars*	*Dollars*	*Dollars*
12 inch or longer:								
CA	1,890	1,800	2,500	2,300	5,125	4,416	2.05	1.92
CO	660	420	665	475	1,995	1,330	3.00	2.80
CT	*	*	*	*	*	*	*	*
GA	265	160	265	215	527	529	1.99	2.46
ID	40,000	38,000	39,500	37,400	34,365	29,920	0.87	0.80
ME	*	*	*	*	*	*	*	*
MA	67	45	83	40	282	190	3.40	4.75
MI	275	180	330	215	660	553	2.00	2.57
MO	*	*	*	*	*	*	*	*
NY	80	90	83	90	354	327	4.27	3.63
NC	4,550	4,830	4,550	4,600	5,915	6,348	1.30	1.38
OR	1,063	240	1,477	395	1,905	691	1.29	1.75
PA	1,740	1,400	1,640	1,410	4,100	3,722	2.50	2.64
TN	98	130	113	148	293	297	2.59	2.01
UT	720	470	705	496	1,114	893	1.58	1.80
VA	676	670	630	578	1,215	1,168	1.93	2.02
WA	480	650	2,590	4,550	2,590	4,505	1.00	0.99
WV	340	300	467	349	785	559	1.68	1.60
WI	740	470	618	554	1,644	1,490	2.66	2.69
Oth Sts	877	376	662	636	1,527	1,396	2.31	2.19
US	54,521	50,231	56,878	54,451	64,396	58,334	1.13	1.07
6 inch-12 inch:								
CA	580	*	290	*	766	*	2.64	*
CO	550	1,060	225	450	720	1,125	3.20	2.50
CT	*	*	*	*	*	*	*	*
GA	*	*	*	*	*	*	*	*
ID	*	*	*	*	*	*	*	*
ME	*	*	*	*	*	*	*	*
MA	*	*	*	*	*	*	*	*
MI	110	90	42	30	116	83	2.75	2.77
MO	*	*	*	*	*	*	*	*
NY	145	120	66	48	312	198	4.72	4.13
NC	660	330	170	90	214	158	1.26	1.75
OR	325	160	113	39	277	99	2.45	2.53
PA	540	550	232	221	742	774	3.20	3.50
TN	*	*	*	*	*	*	*	*
UT	170	260	85	74	178	181	2.09	2.44
VA	*	*	*	*	*	*	*	*
WA	1,370	670	275	200	729	492	2.65	2.46
WV	*	*	*	*	*	*	*	*
WI	165	220	54	80	155	212	2.87	2.65
Oth Sts	3,137	3,883	711	935	1,233	2,098	1.73	2.24
US	7,752	7,343	2,263	2,167	5,442	5,420	2.40	2.50

See footnotes at end of table.

Table 15-32.—Trout: Sales by size category, by State and United States, 2001–02—Continued

Size category and State	Number of fish/eggs		Live weight		Sales			
					Total		Average per 1,000 fish eggs	
	2001 [1]	2002	2001 [1]	2002	2001 [1]	2002	2001 [1]	2002
	1,000 dollars	*1,000 dollars*	*1,000 dollars*	*1,000 dollars*	*1,000 dollars*	*1,000 dollars*	*Dollars*	*Dollars*
1 inch-6 inch:								
CA	560	*	8	*	129	*	231.00	*
CO	235	190	10	5	79	48	335.00	250.00
CT	....	....	....	....	....	....	....	....
GA	*	*	*	*	*	*	*	*
ID	*	*	*	*	*	*	*	*
ME	*	*	*	*	*	*	*	*
MA	*	*	*	*	*	*	*	*
MI	170	100	4	3	47	27	275.00	266.00
MO	*	*	*	*	*	*	*	*
NY	160	180	3	6	41	42	256.00	236.00
NC	3,840	3,590	43	55	346	323	90.00	90.00
OR	80	30	1	1	23	11	286.00	359.00
PA	220	220	6	7	52	66	237.00	298.00
TN	*	*	*	*	*	*	*	*
UT	210	36	10	1	32	7	151.00	196.00
VA	*	*	*	*	*	*	*	*
WA	940	1,630	16	39	197	388	210.00	238.00
WV	*	*	*	*	*	*	*	*
WI	280	340	3	3	34	58	122.00	172.00
Oth Sts	2,328	2,647	101	43	310	427	133.00	161.00
US	9,023	8,963	205	163	1,290	1,397	143.00	156.00
Trout eggs	....	....	....	....	....	....	....	....
Region [2] [3]	....	....	....	....	....	....	....	....
North East	1,100	1,210	....	....	25	27	22.70	22.30
South and Central	595	715	....	....	11	15	18.50	21.00
West	336,530	296,300	....	....	5,077	4,404	15.10	14.90
WA	317,000	287,000	....	....	4,755	4,305	15.00	15.00
US	338,225	298,225	....	....	5,113	4,446	15.10	14.90

[1] Revised. [2] Data published at the regional level to avoid disclosure of individual operations. [3] Regions are defined as follows - North East: CT, MA, ME, NY, PA, and WV; South: AR, GA, NC, TN, and VA; Central: MI, MO, and WI; West: CA, CO, ID, OR, UT, and WA. *Included in other States to avoid disclosure of individual operations.

NASS, Livestock Branch, (202) 720-0585.

Table 15-33.—Refrigerated warehouses: Gross refrigerated space by type of plant, United States, biennially, October 1983–2001 [1] [2]

Type	1983	1985	1987	1988	1991
	1,000 Cubic Feet				
General:					
Public	1,028,300	1,110,394	1,285,860	1,391,901	1,572,879
Private and Semiprivate	573,463	589,132	676,369	603,402	624,005
Total	1,601,763	1,699,526	1,962,229	1,995,303	2,196,884
Apple:					
Public	18,871	19,383	19,750	21,945	27,227
Private and Semiprivate	445,996	466,768	494,404	554,150	584,296
Total	464,867	486,151	514,154	576,095	611,523
Total, all	2,066,631	2,185,677	2,476,384	2,571,397	2,808,407

Type	1993	1995	1997	1999	2001
	1,000 Cubic Feet				
General:					
Public	1,678,461	1,741,585	2,043,908	2,146,643	2,251,943
Private and Semiprivate	658,893	674,649	683,372	756,505	788,853
Total	2,337,354	2,416,234	2,727,280	2,903,152	3,040,796
Apple:					
Public	21,645	23,419	23,907	21,690	14,183
Private and Semiprivate	613,093	647,993	675,838	680,736	712,412
Total	634,737	671,412	699,745	702,426	726,595
Total, all	2,972,092	3,087,646	3,427,025	3,605,578	3,767,394

[1] Warehouse space is defined as all space artificially cooled to temperatures of 50 degrees F. or less, in which food commodities are normally held for 30 days or longer. [2] Totals may not add due to rounding.
NASS, Livestock Branch, (202) 720–8784.

Table 15-34.—Apple and pear storages: Number of refrigerated warehouses, gross and usable refrigerated space, regular and CA capacity, by State and United States, October 1, 2001 [1] [2]

State	Number of warehouses	Refrigerated space		Apple & pear storage capacity		
		Gross	Usable	Regular	Controlled atmosphere	Total
		1,000 Cubic feet	*1,000 Cubic feet*	*1,000 Bushels*	*1,000 Bushels*	*1,000 Bushels*
CA	256	28,873	21,664	6,140	1,538	7,678
CO	24	2,584	2,115	646	163	809
CT	36	1,698	1,346	351	187	538
ID	29	5,278	4,573	1,479	994	2,473
IL	53	1,234	1,033	356	80	436
IN	49	2,280	1,789	375	206	581
KY	10	141	110	36	....	36
ME	40	3,640	3,124	718	715	1,433
MD	18	1,390	1,115	173	354	527
MA	93	4,153	3,409	797	514	1,311
MI	228	33,880	29,742	5,104	7,537	12,641
MN	60	761	656	261	33	294
NH	29	2,021	1,747	330	476	806
NJ	55	2,155	1,768	512	165	677
NY	237	36,446	31,552	5,292	7,590	12,882
NC	38	6,194	5,069	1,072	480	1,552
OH	78	3,425	2,791	779	332	1,111
OR	103	50,944	39,809	7,935	3,774	11,709
PA	158	27,685	21,659	5,368	2,344	7,712
RI	10	139	126	34	8	42
SC	18	383	301	124	....	124
UT	39	2,678	2,358	507	494	1,001
VT	13	2,221	1,884	317	507	824
VA	63	11,767	10,234	2,059	2,026	4,085
WA	344	483,786	389,550	62,135	132,668	194,803
WV	19	7,135	5,859	1,870	418	2,288
WI	143	740	625	243	65	308
Oth Sts	498	2,964	2,331	430	236	666
US	2,741	726,595	588,339	105,443	163,904	269,347

[1] Totals may not add due to rounding. [2] Firms in this table store only apples or pears.
NASS, Livestock Branch, (202) 720–8784.

Table 15-35.—General storages: Gross and usable cooler and freezer space, by State and United States, October 1, 2001 [1] [2]

State	Cooler		Freezer		Total	
	Gross	Usable	Gross	Usable	Gross	Usable
	1,000 Cubic Feet					
AL	5,040	4,354	24,069	19,928	29,110	24,283
AK	136	130	3,735	3,125	3,871	3,254
AZ	*	*	*	*	*	*
AR	3,201	2,431	79,435	65,148	82,637	67,579
CA	163,395	129,243	269,244	209,864	432,639	339,108
CO	5,178	3,758	12,570	9,041	17,748	12,799
CT	1,026	717	4,235	3,513	5,261	4,230
DE	*	*	*	*	26,161	18,602
FL	75,674	61,767	155,802	127,797	231,477	189,563
GA	39,499	31,444	92,102	74,671	131,601	106,115
HI	*	*	*	*	3,898	1,867
ID	2,212	1,755	58,526	51,305	60,737	53,060
IL	22,527	16,870	92,287	70,025	114,814	86,894
IN	4,534	3,547	38,930	30,423	43,463	33,970
IA	14,700	10,873	58,451	49,020	73,150	59,893
KS	*	*	*	*	83,464	64,366
KY	*	*	*	*	22,337	19,487
LA	1,796	1,490	*	7,140	10,345	8,630
ME	286	245	11,930	7,784	12,216	8,029
MD	1,215	942	26,269	19,704	27,483	20,646
MA	5,132	4,153	49,526	40,336	54,658	44,489
MI	16,835	14,023	77,098	59,869	93,933	73,892
MN	15,915	11,411	48,294	34,866	64,211	46,277
MS	*	*	*	*	22,895	16,996
MO	31,685	26,318	69,132	55,686	100,817	82,004
MT	*	*	*	*	*	*
NE	9,467	7,517	31,592	25,059	41,059	32,576
NV	1,491	1,340	5,879	5,197	7,370	6,536
NH	*	*	*	*	*	*
NJ	23,725	18,370	46,479	33,703	70,204	52,073
NM	*	*	*	*	1,156	1,026
NY	24,882	18,394	61,321	49,501	86,203	67,895
NC	4,256	3,599	54,897	44,924	59,154	48,522
ND	*	*	*	*	8,126	6,222
OH	5,896	4,473	64,752	49,461	70,647	53,934
OK	3,669	2,998	7,819	5,296	11,488	8,292
OR	4,762	3,617	99,604	80,421	104,364	84,037
PA	19,439	14,082	114,830	92,983	134,269	107,064
RI	*	*	*	*	*	*
SC	636	429	24,337	19,767	24,972	20,197
SD	6,676	3,106	5,647	3,443	12,323	6,549
TN	2,224	1,411	38,173	28,960	40,398	30,370
TX	37,698	27,034	118,953	81,933	156,650	108,968
UT	1,522	1,156	25,020	20,775	26,542	21,932
VT	*	*	*	*	*	*
VA	18,800	15,913	42,062	32,391	60,862	48,304
WA	17,160	12,963	171,980	139,242	189,140	152,205
WV	*	*	*	*	2,419	1,096
WI	65,954	48,140	95,629	80,570	161,583	128,710
WY	*	*	*	*	*	*
Oth Sts	42,667	30,557	159,282	117,993	22,941	18,888
US	700,910	540,570	2,339,891	1,850,864	3,040,796	2,391,429

[1] Totals may not add due to rounding. [2] Excludes storages used exclusively for storing apples and pears. Includes frozen juice tank storage capacity. * Not published to avoid disclosure of individual operations. Included in "Other States" and U.S. totals.

NASS, Livestock Branch, (202) 720–8784.

Table 15-36.—Alaska crops: Acreage harvested, volume harvested, and value of production, 1993–2002

Year	Oats for grain	Barley for grain	All hay	Grain silage	Grass silage	Potatoes	Other vegetables [1]
	Acreage harvested						
	Acres	*Acres*	*Acres*	*Acres*	*Acres*	*Acres*	*Acres*
1993	900	4,200	19,900	400	700	680	290
1994	1,200	6,400	18,300	400	900	780	271
1995	1,200	7,300	19,200	(3)	(3)	1,040	300
1996	700	6,900	20,200	(3)	(3)	630	343
1997	1,500	7,000	22,500	(3)	(3)	820	337
1998	1,500	6,500	22,000	(3)	(3)	820	340
1999	1,500	4,600	20,300	(3)	(3)	850	357
2000	300	3,300	18,000	(3)	(3)	840	370
2001	1,200	5,100	23,000	(3)	(3)	910	361
2002 [2]	1,200	3,600	22,000	(3)	(3)	850	NA

Year	Oats for grain	Barley for grain	All hay	Grain silage	Grass silage	Potatoes	Other vegetables [1]
	Volume harvested						
	Bushels	*Bushels*	*Tons*	*Tons*	*Tons*	*Cwt.*	*Cwt.*
1993	46,000	164,000	18,000	1,200	2,600	157,000	40,222
1994	48,000	230,000	20,900	1,200	2,900	134,000	37,115
1995	70,200	372,000	22,800	(3)	(3)	220,000	41,645
1996	31,500	283,000	14,400	(3)	(3)	126,000	43,232
1997	65,300	164,500	26,000	(3)	(3)	168,000	46,723
1998	45,000	122,900	23,760	(3)	(3)	150,000	41,846
1999	62,100	154,800	23,200	(3)	(3)	185,000	53,745
2000	7,000	102,500	17,000	(3)	(3)	129,000	58,042
2001	61,000	208,000	30,000	(3)	(3)	230,000	49,989
2002 [2]	48,000	158,000	26,000	(3)	(3)	154,000	NA

Year	Oats for grain	Barley for grain	All hay	All silage	Potatoes	Other vegetables [1]
	Value of production					
	Dollars	*Dollars*	*Dollars*	*Dollars*	*Dollars*	*Dollars*
1993	115,000	549,000	3,330,000	211,000	2,967,000	1,115,000
1994	120,000	771,000	3,762,000	228,000	2,600,000	1,083,000
1995	179,000	1,228,000	4,104,000	(3)	4,092,000	1,242,000
1996	79,000	891,000	2,736,000	(3)	2,494,000	1,443,000
1997	163,000	526,000	4,940,000	(3)	3,360,000	1,620,000
1998	117,000	442,000	4,752,000	(3)	3,135,000	1,397,000
1999	152,000	581,000	4,524,000	(3)	3,830,000	1,897,000
2000	22,000	369,000	3,740,000	(3)	2,670,000	2,080,000
2001	153,000	707,000	6,300,000	(3)	4,669,000	2,169,000
2002 [2]	125,000	561,000	5,590,000	(3)	3,065,000	NA

[1] Excludes greenhouse-grown vegetables. [2] Preliminary. [3] Data no longer estimated. NA-not available.

NASS, Crops Branch, (202) 720–2127.

Table 15-37.—Crop ranking: Major field crops, rank by production, major States, 2002

Rank	State	Corn, grain	State	Soybeans	State	All wheat
		1,000 Bushels		*1,000 Bushels*		*1,000 Bushels*
1	IA	1,963,500	IA	494,880	KS	267,300
2	IL	1,496,000	IL	449,780	ND	216,610
3	MN	1,051,900	MN	308,850	WA	129,695
4	NE	940,800	IN	235,750	MT	109,895
5	IN	631,620	NE	176,330	OK	98,000
6	WI	391,500	MO	170,000	ID	87,660
7	SD	304,000	OH	141,300	TX	78,300
8	KS	290,000	SD	126,790	MN	62,240
9	MO	283,500	AR	96,480	OH	50,220
10	OH	252,560	ND	86,790	NE	48,640
	US	9,007,659	US	2,729,709	US	1,616,441

Rank	State	Winter wheat	State	Durum wheat	State	Other spring wheat
		1,000 Bushels		*1,000 Bushels*		*1,000 Bushels*
1	KS	267,300	ND	48,750	ND	165,200
2	WA	103,250	MT	12,995	MT	75,900
3	OK	98,000	CA	9,000	MN	61,200
4	TX	78,300	AZ	8,455	ID	33,150
5	ID	54,510	MN	140	WA	26,445
6	OH	50,220	SD	110	SD	24,000
7	NE	48,640		....	OR	4,900
8	AR	38,640		....	CO	2,400
9	CO	36,300		....	UT	517
10	MO	34,200		....	WI	231
	US	1,142,802	US	79,450	US	394,189

Rank	State	Sorghum, grain	State	Barley	State	Oats
		1,000 bushels		*1,000 bushels*		*1,000 bushels*
1	KS	135,000	ND	57,040	MN	15,960
2	TX	130,050	ID	53,960	WI	15,000
3	AR	17,710	MT	39,900	IA	13,300
4	MO	15,725	WA	18,360	ND	12,760
5	NE	15,000	CO	7,200	TX	7,040
6	OK	14,850	MN	6,435	PA	7,015
7	LA	13,365	CA	5,100	SD	4,500
8	IL	6,391	WY	4,900	MI	4,160
9	MS	6,237	PA	4,440	OH	3,720
10	SD	3,060	AZ	4,400	NY	3,630
	US	369,758	US	226,873	US	119,132

Rank	State	All cotton	State	Peanuts	State	Rice
		1,000 bales		*1,000 pounds*		*1,000 cwt.*
1	TX	5,039.0	GA	1,313,000	AR	96,752
2	CA	2,010.0	TX	868,000	CA	42,989
3	MS	1,980.0	AL	379,250	LA	29,400
4	AR	1,650.0	NC	210,000	MS	16,192
5	GA	1,650.0	FL	197,800	TX	14,616
6	TN	813.0	OK	159,600	MO	11,011
7	NC	790.0	VA	119,700		....
8	LA	750.0	NM	54,000		....
9	MO	610.0	SC	19,140		....
10	AZ	576.0		....		....
	US	17,145.0	US	3,320,490	US	210,960

Rank	State	All hay, baled	State	Alfalfa hay, baled	State	Other hay, baled
		1,000 tons		*1,000 tons*		*1,000 tons*
1	TX	13,850	CA	8,094	TX	13,200
2	CA	9,594	MN	5,280	MO	6,460
3	MO	7,840	ID	5,000	KY	4,620
4	KS	6,965	IA	4,875	TN	4,400
5	MN	6,610	WI	4,620	OK	3,840
6	NE	5,950	NE	4,050	AR	3,510
7	IA	5,645	KS	3,515	KS	3,450
8	ID	5,608	SD	3,360	NY	2,415
9	KY	5,520	MI	3,150	ND	2,035
10	WI	5,340	MT	2,940	OH	1,980
	US	150,962	US	73,824	US	77,138

Rank	State	All tobacco	State	Dry edible beans	State	Potatoes
		1,000 pounds		*1,000 cwt.*		*1,000 cwt.*
1	NC	357,350	ND	10,626	ID	133,385
2	KY	226,430	MI	4,903	WA	95,200
3	TN	72,540	NE	3,465	WI	31,125
4	VA	66,180	MN	2,475	CO	30,189
5	SC	59,475	ID	1,907	OR	24,936
6	GA	55,650	CA	1,762	ND	23,460
7	FL	11,960	CO	1,519	MN	18,700
8	OH	9,460	WA	820	CA	17,695
9	IN	8,000	WY	624	ME	16,960
10	PA	6,815	MT	361	MI	13,878
	US	889,632	US	29,974	US	463,214

NASS, Crops Branch, (202) 720–2127.

Table 15-38.—U.S. crop progress: 2002 crop and 5-year average

[In percent]

Week-ending date	Winter wheat							
	Planted		Emerged		Headed		Harvested	
	2002	Avg	2002	Avg	2002	Avg	2002	Avg
2001:[1]								
Sep 9	8	8						
Sep 16	19	16						
Sep 23	32	27	14	10				
Sep 30	52	40	25	19				
Oct 7	70	56	40	30				
Oct 14	80	71	55	43				
Oct 21	85	81	67	58				
Oct 28	90	87	74	69				
Nov 4	94	91	81	78				
Nov 11	96	93	86	84				
Nov 18			90	88				
Nov 25			93	91				
2002:								
Mar 31					1	2		
Apr 7					2	3		
Apr 14					4	7		
Apr 21					9	12		
Apr 28					22	21		
May 5					37	33		
May 12					53	49		
May 19					64	65		
May 26					74	77		
Jun 2					82	85	5	3
Jun 9					88	91	9	9
Jun 16					92	95	17	19
Jun 23					96	98	34	32
Jun 30							61	47
Jul 7							74	64
Jul 14							81	76
Jul 21							86	82
Jul 28							88	87
Aug 4							91	91
Aug 11							94	94
Aug 18							96	97

Week-ending date	Spring wheat							
	Planted		Emerged		Headed		Harvested	
	2002	Avg	2002	Avg	2002	Avg	2002	Avg
2002:								
Mar 31	1	2						
Apr 7	2	5						
Apr 14	5	8						
Apr 21	12	15						
Apr 28	21	30	4	9				
May 5	38	49	10	21				
May 12	47	63	16	36				
May 19	71	78	33	53				
May 26	89	89	52	69				
Jun 2	97	95	77	83				
Jun 9			93	92				
Jun 16					4	10		
Jun 23					14	21		
Jun 30					32	41		
Jul 7					52	63		
Jul 14					80	81		
Jul 21					92	91		
Jul 28					97	97		
Aug 4							15	12
Aug 11							28	29
Aug 18							41	47
Aug 25							54	65
Sep 2							89	78
Sep 9							96	87

See footnote at end of table.

Table 15-38.—U.S. crop progress: 2002 crop and 5-year average—Continued

[In percent]

Week-ending date	Rice								Sorghum									
	Planted		Emerged		Headed		Harvested		Planted		Headed		Coloring		Mature		Harvested	
	2002	Avg	2002	Avg	2002	Avg	2002	Avg	2002	Avg	2002	Avg	2002	Avg	2002	Avg	2002	Avg
2002:																		
Mar 31	9	7							12	11								
Apr 7	14	13							14	13								
Apr 14	22	21	12	10					16	15								
Apr 21	41	36	22	15					18	16								
Apr 28	62	53	37	25					20	19								
May 5	73	67	54	39					25	24								
May 12	83	79	67	54					30	29								
May 19	89	90	76	70					36	38								
May 26	95	95	86	82					45	52								
Jun 2			92	91					58	63								
Jun 9			96	95					74	75								
Jun 16					4	3			81	86								
Jun 23					8	6			88	92	19	15						
Jun 30					14	11			95	96	24	18						
Jul 7					17	14					25	22	14	14				
Jul 14					24	21					30	28	16	16				
Jul 21					35	29					35	36	18	18				
Jul 28					48	41					49	47	22	22				
Aug 4					62	56	8	6			58	61	25	27				
Aug 11					77	72	12	9			68	76	32	33				
Aug 18					91	84	16	13			76	86	38	42				
Aug 25					96	92	18	17			87	93	52	54	25	27		
Sep 1							24	23			93	96	66	68	33	34		
Sep 8							33	32					74	80	44	44	25	28
Sep 15							45	45					81	88	52	53	30	34
Sep 22							57	60					87	94	60	65	35	39
Sep 29							71	72					91	97	71	75	43	45
Oct 6							83	83							79	85	50	52
Oct 13							88	91							85	92	53	61
Oct 20							93	94							89	96	61	72
Oct 27							96	98							93	98	67	82
Nov 3																	70	87
Nov 10																	76	91
Nov 17																	84	94
Nov 24																	90	96

Week-ending date	Corn													
	Planted		Emerged		Silked		Dough		Dent		Mature		Harvested	
	2002	Avg	2002	Avg	2002	Avg	2002	Avg	2002	Avg	2002	Avg	2002	Avg
2002:														
Apr 7	2	2												
Apr 14	4	5												
Apr 21	13	10												
Apr 28	26	25	7	NA										
May 5	42	51	12	NA										
May 12	62	72	21	NA										
May 19	70	87	34	NA										
May 26	83	94	53	NA										
Jun 2	92	98	75	NA										
Jun 9	98	99	90	NA										
Jun 16			97	NA										
Jun 23														
Jun 30					5	5								
Jul 7					11	13								
Jul 14					24	28	3	4						
Jul 21					51	53	7	8						
Jul 28					77	78	15	16						
Aug 4					91	91	31	29	8	6				
Aug 11					96	97	48	48	15	15				
Aug 18							67	67	28	27				
Aug 25							82	82	44	44	9	8		
Sep 1							93	91	63	65	16	17		
Sep 8							97	96	78	79	29	30		
Sep 15									90	90	45	48	9	8
Sep 22									96	95	65	66	13	14
Sep 29											83	82	20	22
Oct 6											92	92	28	32
Oct 13											96	96	37	45
Oct 20													51	60
Oct 27													66	74
Nov 3													77	84
Nov 10													86	92
Nov 17													93	96
Nov 24													97	98

See footnote at end of table.

Table 15-38.—U.S. crop progress: 2002 crop and 5-year average—Continued

[In percent]

Week-ending date	Soybeans											
	Planted		Emerged		Blooming		Pods set		Leaf drop		Harvested	
	2002	Avg	2002	Avg	2002	Avg	2002	Avg	2002	Avg	2002	Avg
2002:												
Apr 28	3	4	...	...	...	...	...	...	...	...	...	
May 5	7	13	...	...	...	...	...	...	...	...	...	
May 12 ...	17	29	3	...	...	...	...	...	...	...	...	
May 19 ...	30	50	6	...	...	...	...	...	...	...	...	
May 26 ...	51	69	15	...	...	...	...	...	...	...	...	
Jun 2	70	81	39	...	...	...	...	...	...	...	...	
Jun 9	85	88	66	...	...	...	...	...	...	...	...	
Jun 16	92	93	83	...	...	...	...	...	...	...	...	
Jun 23	97	95	91	...	...	...	...	...	...	...	...	
Jun 30	...	...	96	6	9	...	...	...	...	...	...	
Jul 7	...	...	...	19	21	...	...	...	...	...	...	
Jul 14	...	...	...	39	39	6	7	...	...	...	...	
Jul 21	...	...	...	59	60	16	18	...	...	...	...	
Jul 28	...	...	...	76	77	34	34	...	...	...	...	
Aug 4	...	...	...	88	88	53	54	...	...	...	...	
Aug 11	...	...	...	94	93	71	72	...	...	...	...	
Aug 18	...	...	...	97	97	85	85	...	...	...	...	
Aug 25	...	...	...	...	...	93	93	3	3	...	...	
Sep 1	...	...	...	...	...	97	97	7	9	...	...	
Sep 8	...	...	...	...	...	...	...	19	20	...	...	
Sep 15	...	...	...	...	...	...	...	42	41	...	...	
Sep 22	...	...	...	...	...	...	...	65	65	6	9	
Sep 29	...	...	...	...	...	...	...	83	82	17	22	
Oct 6	...	...	...	...	...	...	...	91	91	31	44	
Oct 13	...	...	...	...	...	...	...	96	96	53	63	
Oct 20	...	...	...	...	...	...	...	...	...	74	77	
Oct 27	...	...	...	...	...	...	...	...	...	84	87	
Nov 3	...	...	...	...	...	...	...	...	...	87	92	
Nov 10	...	...	...	...	...	...	...	...	...	91	96	
Nov 17	...	...	...	...	...	...	...	...	...	94	97	
Nov 24	...	...	...	...	...	...	...	...	...	97	98	

Week-ending date	Cotton									
	Planted		Squaring		Bolls set		Bolls open		Harvested	
	2002	Avg	2002	Avg	2002	Avg	2002	Avg	2002	Avg
2002:										
Mar 31	4	4	...	...	...	...	...	...	...	...
Apr 7	6	5	...	...	...	...	...	...	...	...
Apr 14	9	8	...	...	...	...	...	...	...	...
Apr 21	15	12	...	...	...	...	...	...	...	...
Apr 28	26	19	...	...	...	...	...	...	...	...
May 5	41	31	...	...	...	...	...	...	...	...
May 12 ...	55	49	...	...	...	...	...	...	...	...
May 19 ...	69	67	...	...	...	...	...	...	...	...
May 26 ...	82	79	6	5	...	...	...	...	...	...
Jun 2	88	87	10	9	...	...	...	...	...	...
Jun 9	94	93	18	16	...	...	...	...	...	...
Jun 16	...	...	31	30	6	5	...	...	...	...
Jun 23	...	...	47	46	10	9	...	...	...	...
Jun 30	...	...	63	62	18	15	...	...	...	...
Jul 7	...	...	74	75	27	24	...	...	...	...
Jul 14	...	...	85	85	41	40	...	...	...	...
Jul 21	...	...	91	91	59	57	...	...	...	...
Jul 28	...	...	94	95	73	74	...	...	...	...
Aug 4	...	...	...	...	85	85	8	7	...	...
Aug 11	...	...	...	...	91	91	12	11	...	...
Aug 18	...	...	...	...	94	96	20	18	...	...
Aug 25	...	...	...	...	...	...	29	27	...	...
Sep 1	...	...	...	...	...	...	40	39	...	...
Sep 8	...	...	...	...	...	...	52	51	...	...
Sep 15	...	...	...	...	...	...	64	64	9	10
Sep 22	...	...	...	...	...	...	75	75	13	15
Sep 29	...	...	...	...	...	...	85	83	17	22
Oct 6	...	...	...	...	...	...	90	89	24	32
Oct 13	...	...	...	...	...	...	94	93	30	41
Oct 20	...	...	...	...	...	...	...	...	40	52
Oct 27	...	...	...	...	...	...	...	...	48	61
Nov 3	...	...	...	...	...	...	...	...	53	69
Nov 10	...	...	...	...	...	...	...	...	59	77
Nov 17	...	...	...	...	...	...	...	...	67	83
Nov 24	...	...	...	...	...	...	...	...	77	87

See footnote at end of table.

Table 15-38.—U.S. crop progress: 2002 crop and 5-year average—Continued

[In percent]

Week-ending date	Oats								Barley							
	Planted		Emerged		Headed		Harvested		Planted		Emerged		Headed		Harvested	
	2002	Avg	2002	Avg	2002	Avg	2002	Avg	2002	Avg	2002	Avg	2002	Avg	2002	Avg
2002:																
Mar 31	4	5	...	...	...	...	...	...	2	4	...	...	...	...	...	...
Apr 07	9	12	...	...	...	...	...	...	5	7	...	...	...	...	...	...
Apr 14	17	19	2	4	...	...	...	...	11	12	...	...	...	...	...	...
Apr 21	32	28	9	9	...	...	...	...	16	20	...	...	...	...	...	...
Apr 28	44	45	19	18	...	...	...	...	23	33	7	12	...	...	...	...
May 05	58	63	30	33	...	...	...	...	39	51	14	24	...	...	...	...
May 12	68	76	43	52	...	...	...	...	48	65	21	37	...	...	...	...
May 19	83	87	56	68	...	...	...	...	74	78	34	53	...	...	...	...
May 26	94	94	72	82	...	...	...	...	92	90	55	70	...	...	...	...
Jun 02	98	97	85	91	...	...	...	...	98	96	80	83	...	...	...	...
Jun 09	...	...	96	96	...	...	...	...	...	...	95	93	...	...	...	...
Jun 16	...	...	...	...	21	22	...	...	...	...	...	...	6	13	...	...
Jun 23	...	...	...	...	40	41	...	...	...	...	...	...	15	24	...	...
Jun 30	...	...	...	...	60	63	...	...	...	...	...	...	28	39	...	...
Jul 07	...	...	...	...	78	79	...	...	...	...	...	...	51	59	...	...
Jul 14	...	...	...	...	91	89	7	4	...	...	...	...	78	78	...	...
Jul 21	...	...	...	...	95	95	21	13	...	...	...	...	91	91	...	...
Jul 28	...	...	...	...	...	...	36	27	...	...	...	...	96	97	...	...
Aug 04	...	...	...	...	...	...	50	47	...	...	...	...	...	...	7	12
Aug 11	...	...	...	...	...	...	70	65	...	...	...	...	...	...	17	27
Aug 18	...	...	...	...	...	...	79	79	...	...	...	...	...	...	32	45
Aug 25	...	...	...	...	...	...	88	88	...	...	...	...	...	...	56	65
Sep 02	...	...	...	...	...	...	98	94	...	...	...	...	...	...	89	80
Sep 09	...	...	...	...	...	...	...	...	...	...	...	...	...	...	96	89

Week-ending date	Peanuts					
	Planted		Pegging		Harvested	
	2002	Avg	2002	Avg	2002	Avg
2002:						
Apr 28	6	6	...	...	...	...
May 5	17	18	...	...	...	...
May 12	37	39	...	...	...	...
May 19	61	62	...	...	...	...
May 26	80	78	...	...	...	...
Jun 2	91	87	...	...	...	...
Jun 9	96	92	...	...	...	...
Jun 16	...	...	8	10	...	...
Jun 23	...	...	18	18	...	...
Jun 30	...	...	36	32	...	...
Jul 7	...	...	52	47	...	...
Jul 14	...	...	71	62	...	...
Jul 21	...	...	80	75	...	...
Jul 28	...	...	89	85	...	...
Aug 4	...	...	94	91	...	...
Aug 11	...	...	...	...	...	...
Aug 18	...	...	...	...	...	...
Aug 25	...	...	...	...	...	...
Sep 1	...	...	...	...	...	...
Sep 8	...	...	...	...	...	...
Sep 15	...	...	...	...	7	9
Sep 22	...	...	...	...	13	18
Sep 29	...	...	...	...	20	29
Oct 6	...	...	...	...	32	41
Oct 13	...	...	...	...	47	55
Oct 20	...	...	...	...	62	67
Oct 27	...	...	...	...	71	78
Nov 3	...	...	...	...	75	86
Nov 10	...	...	...	...	80	90
Nov 17	...	...	...	...	87	93
Nov 24	...	...	...	...	95	96

[1] Planted the preceding fall. NA-not available.

NASS, Crops Branch, (202) 720–2127.

Appendix I

Telephone Contact List

Appreciation is expressed to the following agencies for their help in this publication. The information offices are listed to provide help to those users who require additional information about specific tables in this publication.

Agricultural Marketing Service:
USDA/AMS
Room 3510 South Bldg.
Washington, DC 20250
202–720–8998

Agricultural Research Service:
USDA/ARS
5601 Sunnyside Ave
Bldg 1, Rm 2250
Beltsville, MD 20705–5128
301–504–1638

Animal and Plant Health Inspection Service:
USDA/APHIS
USDA Center at Riverside
4B09 - Unit 51
Riverdale, MD 20737
301–734–7280

Center for Nutrition Policy and Promotion:
USDA/CNPP
3101 Park Center Drive
Alexandria, VA 22302
703–605–4266

Economic Research Service:
USDA/ERS
1800 M St, NW
Washington, DC
202–694–5050

Farm Credit Administration:
FCA
1501 Farm Credit Dr.
McLean, VA 22102
703–883–4000

Farm Service Agency:
USDA/FSA
Room 3624 South Bldg.
Washington, DC 20250
202–720–7809

Food and Consumer Service:
USDA/FCS
Office of Public Information
3101 Park Center Drive, Room 819
Alexandria, VA 22302
703–305–2286

Foreign Agricultural Service:
USDA/FAS
Room 5074 South Bldg.
Washington, DC 20250
202–720–7115

Forest Service:
USDA/FS
2nd Floor Central Wing, Aud. Bldg.
Washington, DC 20250
202–205–1273

National Agricultural Statistics Service:
USDA/NASS
Room 5829 South Bldg.
Washington, DC 20250
202–720–3878

National Marine Fisheries Service:
USDC/NOAA/NMFS
1315 East/West Highway,
SSMC III - Room 12340
Silver Spring, MD 20910–3282
301–713–2328

Natural Resources Conservation Service:
USDA/NRCS
Room 6121 South Bldg.
Washington, DC 20250
202–720–3210

Rural Business-Cooperatives Service:
USDA/RECD/RBS
Room 5039 South Bldg.
Washington, DC 20250
202–720–1019

Rural Utilities Service:
USDA/RD/RUS
Room 4043 South Bldg.
Washington, DC 20250
202–720–1255

INDEX

INDEX

INDEX

INDEX

INDEX

INDEX

INDEX

INDEX

INDEX

ISBN 0-16-051387-1

90000

9 780160 513879